THE CRUSTACEA

REVISED AND UPDATED, AS WELL AS EXTENDED FROM THE

TRAITÉ DE ZOOLOGIE

VOLUME 4

PART B

Cover: Oil lamp decorated with a crab, presumably *Pachygrapsus marmoratus*; early 2nd century AD. Potter mark on the back side. Inscription: LHOSCRI. Terracotta, beige paste, brown varnish; 10.9 × 7.9 × 3.1 cm. Musée de l'Arles Antique, France, inv. FAN.91.00.2044, © M. Lacanaud. [Reproduced with permission.] [See fig. 27B.2 on pp. 158 and 399 herein.]

Treatise on Zoology – Anatomy, Taxonomy, Biology

THE CRUSTACEA

Revised and updated, as well as extended from the

TRAITÉ DE ZOOLOGIE

[*Founded by* P.-P. GRASSÉ (†)]

Edited by
J. C. von VAUPEL KLEIN, M. CHARMANTIER-DAURES
and F. R. SCHRAM

Volume 4

Part B

With contributions by
A. P. Ariani, H.-M. Cauchie, G. Charmantier, J.-P. Lagardère, L. Laubier (†),
Th. Monod (†), P. Noël, K. J. Wittmann

English translations by
J. C. von Vaupel Klein and F. R. Schram

BRILL
LEIDEN · BOSTON
2014

Original edition published as:
Traité de Zoologie – Anatomie, Systématique, Biologie. [Series editor P.-P. Grassé.] Vol. VII, Crustacés (fasc. II [pro parte]), edited by J. Forest. ISBN 2-225-84973-0. Masson Publishers, Paris, 1996.

Traité de Zoologie – Anatomie, Systématique, Biologie. [Series editor P.-P. Grassé.] Vol. VII, Crustacés (fasc. III (A) [pro parte]), edited by J. Forest. Mémoires de l'Institut Océanographique, Monaco, n° **19**. ISBN 2-7260-0202-1. Musée Océanographique, Monaco, 1999.

English edition by Brill Academic Publishers, 2014, updated and enhanced by the original authors and/or additional authors and by the editors. Edited by J. C. von Vaupel Klein (former affiliation: Division of Systematic Zoology, Leiden University, c/o Naturalis Biodiversity Center, P.O. Box 9517, NL-2300 RA Leiden, Netherlands); Mrs. M. Charmantier-Daures (former affiliation: Equipe Adaptation Ecophysiologique et Ontogenèse, UMR 5119 Ecosym, Université Montpellier 2, Cc 092, Place Eugène Bataillon, F-34095 Montpellier Cedex 05, France); and F. R. Schram (current affiliation: Burke Museum of Natural History and Cultures, University of Washington, c/o Post Box 1567, Langley, WA 98260, U.S.A.).

Translated from the French by J. C. von Vaupel Klein (Bilthoven, Netherlands).

Despite our best efforts, we have not been able to trace all rights holders to some copyrighted material. The publisher welcomes communications from copyrights holders, so that the appropriate acknowledgements can be made in future editions, and to settle other permission matters.

This book is printed on acid-free paper.
Library of Congress Cataloging-in-Publication Data
The Library of Congress Cataloging-in-Publication Data is available from the Publisher.

ISBN-13: 978 90 04 26492 2
E-ISBN: 978 90 04 26493 9

©*Copyright 2014 by Koninklijke Brill NV, Leiden, The Netherlands*
Koninklijke Brill NV incorporates the imprints Brill, Brill Nijhoff, Global Oriental and Hotei Publishing.

All rights reserved. No part of this publication may be reproduced, translated, stored in a retrieval system, or transmitted in any form or by any means, electronic, mechanical, photocopying, recording or otherwise, without written permission of the publisher.
Authorization to photocopy items for internal or personal use is granted by Brill provided that the appropriate fees are paid directly to Copyright Clearance Center, 222 Rosewood Drive, Suite 910, Danvers, MA 01923, USA. Fees are subject to change.

Printed by Printforce, the Netherlands

CONTENTS

Preface	1
PIERRE NOËL, THÉODORE MONOD (†) & LUCIEN LAUBIER (†), Crustacea in the biosphere	3
HENRY-MICHEL CAUCHIE, THÉODORE MONOD (†) & LUCIEN LAUBIER (†), Crustaceans and mankind	117
GUY CHARMANTIER, Crustaceans in art	139
KARL J. WITTMANN, ANTONIO P. ARIANI & JEAN-PAUL LAGARDÈRE, Orders Lophogastrida Boas, 1883, Stygiomysida Tchindonova, 1981, and Mysida Boas, 1883 (also known collectively as Mysidacea)	189
Colour figures of vol. 4B	397
List of contributors	407
Taxonomic index	411
Subject index	439
Errata TOZ-C vol. 4A	463

This work has been published with the help of the French Ministère de la Culture – Centre National du Livre.

PREFACE

This fascicle comprises part B of volume 4, part A having been published in 2013. The present volume thus sequentially comprises the fifth tome in the series The Crustacea, i.e., the revised and updated texts from the Traité de Zoologie – Crustacea. The chapters in this book emanated from those in the French edition volumes 7(II) and 7(III)(A), with the exception of chapter 27B, which has been newly conceived and was never published in French.

Overall, this constitutes the seventh volume published in this English series, viz., preceded by volumes 1 (2004), 2 (2006), 9A (2010), 9B (2012), 3 (2012), and 4A (2013). As already stated in the Preface of vol. 4A, we have had to abandon publishing the chapters in the serial sequence as originally conceived, because these various contributions become available in a more or less random order. Since this necessarily results in a non-serial sequence of the chapters in vols. 4-8, we shall publish an overview of the total series after completion, in which an enumeration of the volumes and the chapters included will be given, so as to facilitate quick reference throughout the series as a whole.

We are confident that also this fascicle will make a useful compendium for carcinologists all over the globe, both as a guide in their studies on Crustacea, and in carefully interpreting their results.

May, 2014

Montpellier, MIREILLE CHARMANTIER-DAURES
Bilthoven, J. CAREL VON VAUPEL KLEIN
Langley (WA), FREDERICK R. SCHRAM

CHAPTER 26

CRUSTACEA IN THE BIOSPHERE[1])

BY

PIERRE NOËL, THÉODORE MONOD (†) AND LUCIEN LAUBIER (†)

Contents. – **Introduction: the presence of Crustacea in the biosphere. The extremes: sizes and environmental conditions** – Size – Temperature – Altitude – Deep sea – Salinity. **Habitats of crustaceans. The oceanic environment** – Coastal zone: littoral and neritic bottoms – Deep benthos – The pelagic realm. **Fresh waters** – Continental waters – Thermal waters – Subterranean waters – Temporary waters – Hypersaline water bodies – Phytothelmes – Muscicolous crustaceans. **Terrestrial habitats** – "Flying" Crustacea. **Commensal and parasitic associations of Crustacea. Epizoonts, commensals and parasites of crustaceans. Appendices. Bibliography.**

INTRODUCTION: THE PRESENCE OF CRUSTACEA IN THE BIOSPHERE

P. J. van Beneden wrote in 1861: « *Aucune classe du règne animal ne nous semble présenter une diversité plus remarquable. Tous les genres d'habitation possibles semblent avoir été épuisés, et, depuis l'état de libre vagabondage en pleine mer jusqu'à la dépendance la plus complète de l'hôte qui les héberge ou du gîte qui les abrite, toutes les nuances intermédiaires imaginables semblent avoir été épuisées.* » ["*Not a single class of the animal kingdom appears to present a more remarkable diversity [than Crustacea]. All types of possible lodging seem to have been exhausted, from the state of vagrancy in open sea until full dependence on the host that feeds them, or on the accommodation that houses them. All intermediate possibilities that could be imagined seem to have been exhausted.*"]

Crustaceans show an extraordinary **diversity** for their **ways of life** and their **diets**. What is actually shared between the appearance of a sacculinid cirripede and a water flea, between a crab and a brine shrimp? As such, they rank among the most diverse groups

---

[1]) The original, 1996 text has been updated by Pierre Noël in March 2013; latest additions December 2013.

of the Animal Kingdom. A large number of habitats has been occupied by crustaceans; however, they are basically aquatic and mainly marine. Very few of them adapted to a life style outside the water. Some crustaceans are sessile (cirripedes); none is really able to fly.

Among arthropods, crustaceans fulfil the same role in the water that insects have in terrestrial and aerial habitats. However, their **numbers of species** are by far smaller, i.e., some 68 000 only (fossils not included), with the following distribution over the various classes (classification according to Forest, 2004; numbers adapted from Ahyong et al., 2011, and further literature):

Cephalocarida	13
Branchiopoda	±1180
Remipedia	27
Tantulocarida	36
Mystacocarida	13
Thecostraca	±1400
Copepoda	±16 000
Branchiura	±200
[Pentastomida	130]
Ostracoda	±7600
Malacostraca	±40 000

The number of **described species** of Crustacea increases with time. For instance, this number was successively estimated as 30 000 (Forest, 1969), 40 000 (Heywood & Watson, 1995), 52 000 (Monod & Laubier, 1996; Martin & Davis, 2001), and 68 000 (Brusca & Brusca, 2003). The average of new, valid species described every year is ca. 300. Furthermore, it is often considered that for groups of small size only half of the species actually present may have been described already (see Noodt, 1971 for copepods), which means that substantial numbers of species of crustaceans in nature still await scientific discovery and proper description. For amphipods, Bousfield (1978) stated that approximately 25 000 species would exist, whereas only 5000 species had by then been described. Furthermore, the growing application of molecular techniques is uncovering much cryptic speciation.

THE EXTREMES: SIZES AND ENVIRONMENTAL CONDITIONS

Size

Crustacean **adult size** varies from 0.1 mm (*Stygotantulus stocki*), less than 1 mm for copepods (0.18 mm), ostracodes, mystacocarids, and even peracarids such as isopods of the families Microcerberidae and Microparasellidae, or 1-2 mm (Isopoda: *Chauliodoniscus reyssi*: 1.3 mm, *Haploniscus unicornis*: 1.7 mm; *Carpias minutus*: 1.9-2.16 mm; Amphipoda: *Galapsiellus leleuporum*: 1-1.15 mm), to up to 65 cm (body length) for the Decapoda: *Homarus americanus*. These are selected minimal sizes as examples, but here are, as a contrast, some indications for maximal sizes that could be considered as exceptional within various groups:

Branchiopoda: 150 mm (*Triops*)
Cladocera: 18 mm (*Leptodora kindti*)
Copepoda: 312 mm (*Pennella balaenoptera*, without ovisac)
Cirripedia Thoracica Sessilia: 15-20 cm (*Austromegabalanus psittacus*)
Ostracoda: 25 mm (*Gigantocypris agassizi*)
Leptostraca: 37 mm (*Nebaliopsis typica*)
Mysidacea: 35 mm (*Gnathophausia ingens*)
Cumacea: 25 mm (*Diastylis goodsiri*)
Amphipoda: 20-32 mm (*Hirondellea gigas*); 50 mm (*Parargissa galatheae*); 100 mm (*Eurythenes gryllus*); 34 cm (*Alicella gigantea*) (Barnard & Ingram, 1986)
Isopoda: 50 cm (*Bathynomus giganteus*) (Lowry & Dempsey, 2006)
Euphausiacea: 93 mm (*Thysanopoda cornuta*)
Decapoda Dendrobranchiata: 32 cm (*Penaeus monodon*)
Decapoda Caridea: 22 cm (*Pasiphaea princeps*); 37 cm, 1.4 kg (*Macrobrachium carcinus*)
Decapoda Reptantia: **Astacidea**, 50 cm and 3.6-4 kg (*Astacopsis gouldi*, Tasmanian crayfish); **Nephropidea**, 65 cm and 20 kg (*Homarus americanus*) (Wolff, 1978); **Palinura**, 60 cm (*Sagmariasus verreauxi*); **Anomura**, 1.30 m leg span in *Paralithodes camtschaticus*; **Brachyura**, 40 cm width and 14 kg (*Pseudocarcinus gigas*); 1.20 m leg span (*Cyrtomaia* spp.); 2.50 m to 3.80 m leg span and carapace width 45 cm and 20 kg (*Macrocheira kaempferi*, Japanese giant spider crab)

Temperature

Temperatures of the habitats where crustaceans actually live range from the coldest waters (0°C and below) to 36°C (*Tigriopus*, a copepod), 44-45°C (*Thermosbaena mirabilis*), 42-48°C (*Potamocypris philotherma*), and 55°C (*Thermobathynella adami*) (Capart, 1951). Crabs of the family Bythograeidae inhabit **hydrothermal vents** in the Pacific and Atlantic oceans, between 2500 and 3500 m depth, and live there near **thermal chimneys** from which hot fluid is emitted at temperatures varying between 150 and 400°C. Their actual habitat would be located in a zone of warm water, probably 10-30°C, that exists between the hot water around the chimneys and the cold deep water of 1.5-2°C. The very common shrimps of the family Alvinocarididae can live for a short time at temperatures of about 40 to 50°C. However, the leptostracan *Dhalella caldariensis* occurs deep inside the **hot smoker vents**.

Altitude

Among aquatic crustaceans, Ostracoda are known from East Africa from habitats at 4000 m above sea level (*Eucypris neumanni, Eucypris latissima*, and *Trandosia jenkinsae*), at 3500-4000 m (*Gomphocythere angulata*) and at 4230-4540 m from Mount Kenya (Lindroth, 1953). Amphipoda have been reported from locations in the Andes (Ecuador) higher than 4000 m, for example, *Hyalella* aff. *curvispina* at 4100 m at Cotopaxi. Among terrestrial Amphipoda, an unidentified species of Talitridae was reported ("*Platorchestia platensis* [as *Orchestia platensis*]") (Tattersall, 1929) at some 2000 m on

Mount Tabwemasana, Espiritu Santo, Vanuatu. Even in France, several species of oniscid isopods are encountered as high as between 2500 m and 2800 m.

Deep sea

Occurrences of species of Crustacea in the deep sea have been reported down to great depths, and include:

Cirripedia: 6960 m
Tanaidacea: 7150 m (*Herpotanais kirkegaardi*), up to 6800 m (*Apseudes*) and 8300 m (*Neotanais hadalis*)
Cumacea: 7130 m
Isopoda Asellota: According to Birstein (1957) and Wolff (1962): *Rectisura brachycephala* [as *Storthyngura*]: 5670-5680 m; *Vanhoeffenura bicornis* [as *Storthyngura*]: 6156-6207 m; *Vanhoeffenura chelata* [as *Storthyngura*]: 5345-6860 m; *Rectisura tenuispinis* [as *Storthyngura*]: 7246 m; *Rectisura herculea* [as *Storthyngura*]: 6475-8100 m; *Rectisura vitjazi* [as *Storthyngura*]; *Fortimesus gigas* [as *Haplomesus*]: 7305-8430 m; and *Macrostylis galatheae*: 9820-10 000 m
For **Asellota**, Wolff (1962) reported on 356 species, of which 27 are encountered between 0 and 200 m, 133 between 200 and 2000 m, 183 between 2000 and 6000 m, and 13 between 6000 and 11 000 m, so a maximum in the abyssal, and very few in the hadal zones
Amphipoda (maximum depths): 5110 m (*Harpinia excavata*); 5207-6156 m (*Liljeborgia caeca*); 6475-6571 m (*Lepechinella ultraabyssalis*); 6660-6770 m (*Lepechinella wolffi*); 6960-7000 m (*Bathyschraderia magnifica*); 7210-7230 m (*Astyroides carinatus* and *Metaceradocoides vitjazi*); 7270 m (*Harpiniopsis spaercki*); 7290 m (*Bathyceradocus stephenseni*); 8000 m (*Astyra zenkevithchi*); 8300 m (*Abyssorchomene abyssorum* and *Princaxelia abyssalis*); 9120 m (*Cleonardo longipes* and *Eusirus fragilis*); 10 000 m (*Pardaliscoides longicaudatus*); 10 190 m (*Pseudotiron longicaudatus* and *Synopioides secundus*/*Halice macronyx*); 10 500 m (*Halice subquarta*)
Decapoda: **Brachyura** are represented down to 4300 m (*Ethusina abyssicola* and *Ethusina alba*). **Caridea** (Oplophoridae, Pasiphaeidae and Crangonidae) were found even deeper, and, up to 5500 m, some **Penaeoidea** (noticably *Plesiopenaeus armatus* and *Benthesicymus bartletti*), as well as **Anomura**: **Paguroidea** (*Parapagurus*) and **Galatheoidea** (*Munidopsis*)

Salinity

Some Anostraca, such as *Artemia* spp., are commonly found in salinities ranging from 45 tot 200 psu [practical salinity units], and can survive (and even live for some time) in supersaturated **brine** at 340 psu (Ben Naceur et al., 2012).

HABITATS OF CRUSTACEANS

While crustaceans occupy diverse **aquatic environments**, i.e., marine (from littoral to hadal) and both riverine and lacustrine **fresh waters**, there are also many paralittoral or amphibious species. In addition, the **colonization** of terrestrial habitats, though generally more or less moist, can be characterized as fully successful in the case of woodlice (Isopoda: Oniscidea) and in a few tropical Amphipoda of the family Talitridae. Also, a small number of Decapoda (*Gecarcinus*, *Gecarcoidea*, *Coenobita*, etc.) provide examples of terrestrial forms, albeit still depending on the **marine environment** for their **reproduction**. On the other hand, there are no really flying Crustacea: those few that are able to take off from water or ground contact, can only do so through jumps of limited reach and height, or following occasional, fortuitous circumstances.

In the oceans, crustaceans can be found existing at all depths, according to various life styles: planctonic, pelagic swimmers (i.e., nektonic), as benthic walkers or crawlers, burrowers, and as interstitially living organisms. Within the **meiobenthos**, copepods are the most important group, after Nematoda, and before Turbellaria, Gnathostomulida, Annelida, and Ostracoda.

A large number of crustaceans are **inquilines**, **commensals**, and **parasites**, as will be mentioned further below. In addition, many Crustacea live in **subterranean waters**. Each group, in fact, merits separate study for their specific ecological and morphological adaptations. In this general chapter, however, such an approach would be obviously far too comprehensive; the reader is consequently referred to the treatments of the various groups in the specific chapters in the systematic sections of the present series. As a consequence, we shall confine the current survey to a few, representative examples, viz., Copepoda (cf. Noodt, 1971), Amphipoda: Gammaridea (cf. Bousfield, 1978), and Isopoda and Tanaidacea (cf. Roman, 1970, 1979).

Copepoda. – With a possible littoral, and perhaps Mesozoic origin, Copepoda seem to have evolved along several different lines or **trends**, notably concerning:

- **reductions** in size (down to 0.1 mm), eyes, and pigmentation (troglobitic or abyssal forms), numbers of segments of appendages or of setae on the limbs, numbers of eggs per clutch (down to 1 or 2 only), and, perhaps also, in the number of larval stages;
- **morphological specializations**, especially involving the appendages (for instance, claws and suckers in some copepods parasitic on fish and invertebrates);
- **ecological specializations**: stenoecy (related with alimentary specializations), stenothermy, euryhalinity (colonization of brackish and fresh water), loss of natatory capacity, independence of a substratum (eupelagic forms), parthenogenesis, freely shedded eggs, resting eggs for overcoming periods of drought, development of different types of parasitism.

Amphipoda. – In Amphipoda, the classification of Gammaridea into families (e.g., Bousfield, 1978; but see Ahyong et al., 2011 for a recent overview) reveals an incredible degree of diversity. More than 20 superfamilies can be recognized, all characterized by their own set of specializations as regards **niches** and **habitats**. Some of the most noteworthy are as follows:

1. **Phoxocephaloidea**: marine, coastal and neritic to abyssal, burrowing, Austral regions.
2. **Lysianassoidea**: marine, mainly benthic, carnivorous, cleaners or inquilines, coastal to abyssal and hadal, cosmopolitans but mainly Arctic and Antarctic.
3. **Pontoporeioidea**: burrowing, marine, coastal, brackish waters or fresh waters, Holarctic temperate regions, south of the equator in the Americas, South Africa.
4. **Gammaroidea**: mainly brackish or fresh waters (70% of species are dulcaquicolous), epigeans, Northern Hemisphere (one-third of the known genera and one-quarter of the known species are endemic forms in Lake Baikal).

5. **Crangonyctoidea**: continental fresh waters, Holarctic and antiboreal, epigean and hypogean, absent from tropical regions.
6. **Niphargoidea**: exclusively hypogean, mostly European and West Asian, with radiations to Madagascar, Madeira, plus the centre and eastern parts of the U.S.A.
7. **Bogidielloidea**: exclusively hypogean, tropical regions, partly with much advanced morphological features, and (sub)vermiform.
8. **Eusiroidea**: free-swimming and epibenthic, marine (coastal and neritic), brackish and fresh waters, mainly northern Pacific and antiboreal region.
9. **Oedicerotoidea**: burrowing, marine (coastal and abyssal), fresh and brackish waters, cosmopolitan, noticeably diversified in cold waters of the Northern Hemisphere.
10. **Leucothoidea**: marine, crawling, "hanging", inquilines, commensals or even parasites (in fish: Laphystiopsidae), associated with hydrozoans (Stenothoidae), sponges (Leucothoidae, Anamixidae, Colomastigidae), tunicates (Leucothoidae), using vegetal remains (Plecostidae, Amphilodridae).
11. **Talitroidea**: benthic, littoral and coastal (marine and fresh water), and terrestrial; Talitridae being the only family to become entirely terrestrial; tropical and temperate regions, with a special diversification in the Southern Hemisphere.
12. **Stegocephaloidea**: marine, free swimmers and benthic, coastal and neritic, mainly from cold waters (Holarctic and antiboreal regions), Lafystiidae parasites in benthic fish.
13. **Synopioidea**: marine, pelagic and epibenthic (on soft substratum), coastal to abyssal, all oceans.
14. **Pardaliscoidea**: marine, pelagic in open seas (all oceans), and deep (down to hadal level), entering deep coastal waters (fjords).
15. **Liljeborgioidea**: marine, coastal benthic, brackish waters inquilines (polychaete tubes, sponges, corals, etc.: Sebidae, Liljeborgiidae); sometimes hypogean in fresh water (Salentinellidae).
16. **Dexaminoidea**: marine, coastal and abyssal, all oceans (mainly Pacific and antiboreal region), sometimes inquilines in tunicates (Dexaminidae *pro parte*).
17. **Ampeliscoidea**: marine, burrowing and tubicolous, coastal to abyssal, all oceans.
18. **Mephidippoidea**: marine, coastal and neritic, epibenthic and pelagic, mainly tropical to temperate.
19. **Melitoidea**: marine coastal, estuarine, cosmopolitan, some species abyssal; hypogean in brackish environments (Hadziidae).
20. **Corophioidea**: marine coastal, estuarine, living in burrows and building "nests" of agglomerated debris, sometimes mobile.

In addition to the suborder Gammaridea, three more suborders were recognized among Amphipoda in the classical system, i.e., that of Caprellidea (marine, littoral to abyssal, hanging to epibiontic, with one group, Cyamidae, ectoparasitic on Cetacea); that of Hyperiidea (pelagic and abyssopelagic, carnivorous or commensals of cnidarians, salps, etc.); and Ingolfiellidea (subterranean in fresh waters, brackish or marine littoral, and sometimes benthic abyssal). However, following Dahl (1977), also Bowman & Abele (1982) have included Ingolfiellidea in Gammaridea, while Myers & Lowry (2003) united Corophioidea with Caprellidea, thus creating the suborder Corophiidea. Hence, the currently accepted classification of Amphipoda comprises three suborders, Gammaridea, Corophiidea and Hyperiidea (cf. the chapter "Order Amphipoda Latreille, 1816" by D. Bellan-Santini, in one of the forthcoming volumes of the present series).

Isopoda. – The third example concerns Isopoda and Tanaidacea of the Tuléar reef formation in Madagascar (Roman, 1970, 1979), totalling approximately 190 species. The zone studied was divided into 20 geophysical facies. Next 17 ecological categories, each, of course, comprising one or more niches, were characterized, viz.:

1. Uncalcified animals (Porifera, Hydrozoa).
2. Calcified animals (Bryozoa, Scleractinia).
3. Uncalcified algae.
4. Sea grasses.
5. Crevices (fixed shells, e.g., the barnacle *Tetraclita*, and those of bivalve molluscs of the family Ostreidae, or wood with *Teredo* shipworms (Mollusca: Bivalvia), etc.).
6. Empty gastropod shells.
7. Soft bottoms.
8. Hard bottoms.
9. Animals on sediments (Porifera, Zoantharia).
10. Algae on soft bottoms.
11. Algae in or around crevices.
12. Sediments fixed by animals (Polychaeta).
13. Calcified free and mobile algae.
14. Calcified encrusting algae.
15. Calcified algae with branches and articulations.
16. Tide marks with slow desiccation.
17. Pebbles or boulders on sediment.

Within the framework of this complex of biotopes, morphological types have been defined in each suborder of tanaids and isopods. So, four types were recognized in the former suborder Flabellifera [the members of which are currently separated into two groups, placed, respectively, in Cymothoida and Sphaeromatidea] according to the general shape of their body, i.e., the **spindle** (with 10 subdivisions; families Cirolanidae, Corallanidae, Aegidae, Limnoriidae); the **sphere** (with 11 subdivisions; Sphaeromatidae); the **shield** (with four subdivisions; Cirolanidae and Sphaeromatidae); and the **truncated prism** (*Gibbosphaeroma*).

The components of the superfamily Cirolanoidea (4 families, 23 species) were subjected to a double classification, both ecological and according to the **geomorphology** of their actual habitat, including definitions of their specific preferences, as, e.g., for *Cirolana venusticauda* "sea weeds in crevices" and for *Paracirolana platysoma* "calcified algae with ramifications".

The members of the superfamily Sphaeromatoidea (3 families, 23 species) were similarly sorted, also with an indication of their ecological preferences:

– with a **biological determinant**: uncalcified animals (*Gibbosphaeroma*), oscula of sponges, leaves of marine sea grasses (*Paracilicaea teretron*), *Cymodoce* spp. (Polychaeta: Nereidae), etc.;
– with a **determinant relating to shelter**: tests of Tetraclitidae, Scleractinia, Bryozoa, old wood (*Sphaeroma terebrans*), large shells of *Conus* (*Sphaeroma sieboldii*), calcified algae forming free nodules (*Cymodoce bifida* and *Dynamenopsis* sp.);
– with an **algal determinant**: for example, algae in crevices (*Paracilicaea mossambica* juv., *Dynamenella octoloba*, *Cassidinidea quadricarinata*);
– with a **sedimentary determinant**: *Paradynamenella psammophila*, *Sphaeroma polydemum*, *Parasphaeroma granosum*, *Cilicaea splendida*, *Dynamene ramuscula*, *Paracilicaea* sp., *Dynamenella* sp.

THE OCEANIC ENVIRONMENT

Crustaceans are mostly marine animals, often benthic: even the so-called "swimming crabs" (Portunidae) are mainly depending on the presence of a substrate. Most forms are vagile but some are sessile, e.g., Cirripedia, parasitic Copepoda.

Species that can truly be characterized as "pelagic holoplanctonic", include only the numerous free-swimming Copepoda, some marine Cladocera, quite a number of Ostracoda, and Amphipoda: Hyperiidea, as well as many Decapoda (Penaeoidea, Caridea: Pasiphaeoidea and Oplophoroidea, etc.).

The marine habitats of Crustacea are so diverse that any attempt at sorting taxa will necessarily be artificial and represent no more than a coarse approach. Indeed no straightforward system, whether based on a function of bathymetry, or giving priority to the type of substratum, is fully satisfying. Hence, in the following sections a pragmatic and simplified arrangement has been chosen, in order to provide a somehow conveniently structured overview of the extreme variety of **ecological adaptations** encountered in crustaceans.

Coastal zone: littoral and neritic bottoms

HARD BOTTOMS

A series of habitats can be gathered here, albeit somewhat arbitrarily:

- **Rocks exposed to waves** constitute a rich environment, with vagile and active species such as various crabs (*Grapsus*, *Pachygrapsus*, *Plagusia*, *Goniopsis*, etc.) and isopods (*Ligia*). This also is the habitat where sessile (*Balanus*, *Chthamalus*, *Tetraclita*, etc.) or pedunculate cirripedes (*Pollicipes* [as *Mitella*]) are attached.
- Further down the **littoral zone**, many crustaceans hide under stones, in cavities, and also in the pools formed at low tide. Generally, these are small-sized species, or juveniles of which the adults are located deeper.
- **Non-rocky substrates** can be represented by either organic (wood, shells), or inorganic items (concrete or brick wharfs, plastic buoys, metal constructions). The role played by shells (Gastropoda, sometimes Bivalvia) as a home for many Paguroidea is well known. Calcareous tubes of serpulids can be inhabited by crustaceans, which are then "secondarily" tubicolous (the pagurid *Calcinus tubularis* and hyssurid isopods in tubes of serpulids, the amphipod *Americorophium spinicorne* in shipworm tubes), in contrast with those elaborating their own protective tube.
- **Coral reefs** are similar to hard substrata in many respects. However, this complex environment constitutes a highly specialized biome, exclusively in tropical seas. Scleractinian reefs shelter a **biocoenosis** as noticeable by its richness (abundance of individuals) as by its biodiversity (numbers of species) (Edmonson, 1946; Ekman, 1953; Wiens, 1962). Crustacea truly pullulate in coral reefs, where a full series of differentiated microbiotopes occurs: external slope, reef crest, internal reef flat with eelgrass beds, boulder zone, internal slope, lagoon, etc. The main constituents of this fauna, where most groups of crustaceans are represented, will not be detailed here: for instance, pontoniid commensal shrimps will be considered in the next pages.

- Many species of rocky locations occupy more or less dark habitats. They seek shelter under overhanging rocks, hide in superficial crevices, or live deep in truly sciaphilic habitats (submarine caves, tunnels, holes under coral reefs, etc.). In Madagascar, Pichon (1964) has shown that among three local species of spiny lobster, one (*Panulirus japonicus*) is really sciaphilic and restricted to cavites under domes of *Porites*, while the second one (*Panulirus versicolor*) has a less exclusive preference, and the third (*Panulirus ornatus*) is located in eelgrass beds with *Cymodoce* when young, but lives at the entrance of coral cavities when adult. The barnacle *Tetraclitella costata*, could also be classified among the infralittoral sciaphilic organisms (Vasseur, 1964), as do some Ostracoda (Reys, 1969).

Lignicolous and petricolous forms

Crustacea that have adapted to **wood boring** (or the firmer parts of some seaweeds), are scarce, as are those that could really dig into rocks (then mostly soft types of stone, but also rarely hard, such as basalt). Sometimes it is difficult to tell if a crustacean actually is the source of the cavity it occupies, or not. Any (natural) cavity can be colonized by a species looking for a shelter: the *Sphaeroma venustissimum* in Dakar are just occupying lodges drilled by *Petricolaria pholadiformis* (cf. Monod, 1947, 1952). On the other hand, Crustacea that perforate wood are not necessarily xylophagous, which term thus often is used abusively: though it is true that *Limnoria* eats the wood it is gnawing on, *Sphaeroma pentodon* in contrast, attacks the wood only to build itself a shelter — and that shelter then is only occasional, for the species prefers soft rocks.

Cirripedia. – Though it is sometimes difficult to separate symbiotic cirripedes from parasites associated with living tissues and from petricolous forms digging into calcified objects, we may still quote in this place:

- **Acrothoracica** or burrowing barnacles (Tomlinson, 1969) with:
  – Lithoglyptidae: *Lithoglyptes* (corals and molluscs), *Berndtia* (corals), *Weltneria* (corals, molluscs, limestone and clay or loam), *Kochlorine* (corals and molluscs), *Kochlorinopsis* (Bryozoa), *Balanodytes* (corals and barnacles);
  – Cryptophialidae: *Cryptophialus* (corals, molluscs, *Concholepas*, *Balanus*), *Australophialus* (Polyplacophora, Gastropoda, Bivalvia, *Austrominius*);
  – Trypetesidae: *Trypetesa* (gastropod shells occupied by Pagurids); in *Trypetesa lateralis*, the shell may be perforated from one side to the other and from inside to outside (Tomlinson, 1955, 1969, 1969a).

In addition, we may observe that some lodges of acrothoracicans or burrowing barnacles are known as fossils from the Carboniferous onward, i.e., in corals, oysters, and even rocks.

- Quite a number of **Thoracica: Balanomorpha** (*Pyrgoma*, *Creusia*, *Acasta*, etc.) are capable of penetrating into hard material: corals, antipatharians, gorgonians (Ross & Newman, 1973); in some instances, lepadomorphs are even found fully sunk into coral or limestone, for instance, representatives of the genus *Lithotrya*.

Isopoda. – Three families include perforating forms:

- **Sphaeromatidae**: some species dig galleries in wood, like *Sphaeroma terebrans*, a common species in tropical waters and well known for its attacks on mangrove roots, while the same can be mentioned for its congeners *Sphaeroma sieboldii* and *Sphaeroma retrolaeve*; other species, such as *Sphaeroma pentodon*, *Sphaeroma peruvianum*, *Gnorimosphaeroma oregonensis*, *Gnorimosphaeroma ovatum* [as *Eusphaeroma*], and even *Cymodocea japonica* (cf. Shiino, 1957) are suspected of similar activities. Reports stated that mangrove roots gave way more easily, to ultimately break down due to the settlement of oysters on them, when *Sphaeroma terebrans* was attacking those roots (Monod, 1931). In a general sense, these isopods have been considered either as deleterious, or as useful for Florida mangroves (Simberloff et al., 1978), the latter because their actions facilitate ramification of those mangrove roots. However, in cases like these in which such different aspects of the effects of the behaviour of animals are at issue, it is intrinsically impossible to formulate anything of a conclusive answer (compare Ribi, 1981).
- **Limnoriidae**: the genera *Limnoria* and *Chelura* (for the latter, see below, under Amphipoda) are the most well-known lignicolous crustaceans, and probably also the most notorious ones. In *Limnoria*, some 25 species have been described, classified into three subgenera. Although the actual **digestion of cellulose** by *Limnoria* spp. is still subject of discussion, there is no doubt that the animal feeds mostly upon pieces of wood. Some Limnoriidae were reported to attack the coverings of submarine telecommunication cables.
- **Cirolanidae**: a *Ceratolana papuae* was reported (Bowman, 1977) to dig long cylindrical galleries in mangroves in New Guinea, to the extent that one wondered if the "horns" of this species were actually used as files.

Amphipoda. – The representatives of the genus *Chelura* (family Cheluridae), with a few species, the most famous one of which undoubtedly is *Chelura terebrans*, seem to be invariably (and apparently obligatorily) associated with a *Limnoria*, whereas the isopod is capable of living alone. Its galleries are of similar size as those of *Limnoria*, though slightly longer (2.5 mm *vs* 1-2 mm). Two other genera of chelurids exist, *Tropichelura* (with *Tropichelura insulae*) and *Nippochelura* (*Nippochelura brevicaudata*). Finally, juveniles of *Peramphithoe humeralis* are able to make holes in the kelp *Macrocystis*, as do also isopods of the genus *Phycolimnoria*.

Decapoda. – Caridea: the petricolous snapping shrimp, *Alpheus saxidomus* lives on the Pacific coast of Costa Rica and is able to penetrate 10 cm into hard rock (basalt). Some Pylochelidae, Callianassidae, and Upogebiidae are also able to bore into hard rocks.

"OCCUPANTS" AND TUBICOLOUS FORMS

The category of borrowed and built shelters must be considered separately both from the burrows of burrowing crustaceans and the lodges or galleries of perforating crustaceans. Here not substrata but objects, more of less hollow, living or inert, are used by an individual for protection. These items may have a natural or artificial origin, they may have been simply appropriated and occupied, or secreted, or built (i.e., constructed). Intermediate conditions may occur, for instance, in the case of a natural but inanimate object (like a refitted shell) that was secreted by its original owner. When the shelter object is living, it is often impossible to distinguish inquilinism, commensalism, or even symbiosis.

Natural shelter objects. – There are many examples of Crustacea living inside natural shelter objects. In the following we give a representative, but certainly not exhaustive, selection.

Pagurotanais bouryi is a **tanaidacean** found in a shell from Cuba (Bouvier, 1918).

As far as **isopods** are concerned, Sphaeromatidae often occupy cavities of various origin, such as *Dynamene bidentata* in empty skeletons of *Perforatus perforatus* (as *Balanus*; Bourdon, 1964; Holdich, 1970), in which it is possible to find one male with a varying number of females (up to a maximum of 16). The same species also inhabits crevices on a rocky substratum. *Gnathia maxillaris* is also found in barnacle shells and in crevices. *Campecopea hirsuta* can live in empty *Chthamalus stellatus* walls (Panouse, 1940; Tétart, 1962) and in those of *Semibalanus balanoides*; Th. Monod personally found this species in *Amphibalanus amphitrite* in Banc d'Arguin, Mauritania. In contrast, Panouse (1940) found *Campecopea* only rarely in acorn barnacles, and then always close to *Lichina pygmaea*, which does not occur in Mauritania. Hyssurids, worm-like Anthuridea, occupy calcareous tubes of serpulid annelids (*Vermilia, Vermiliopsis, Salmacina, Hydroides*). *Eisothistos macrurus* was reported from Ischia (Italy) living in *Vermiliopsis infundibulum* tubes, and *Eisothistos pumilus* in tubes of *Salmacina dysteri*. *Eisothistos macrurus* might be able to enter the tube of an annelid and eat its primary occupant (Wägele, 1979, 1981). *Zenobia prismatica* (Valvifera) was reported from a *Pomatoceros* tube and *Cleantis tubicola* was found living in a hollow stem of a plant. Issei (1912) reported *Cleantis prismatica* [as *Zenobiana*] in *Posidonia* roots and rhizomes, stalks of Gramineae, the stem of a species of Lamiaceae, etc. In the English Channel, the species is present inside hollow parts of the common eelgrass, *Zostera*.

Amphipods use to find shelter in various kinds of hollow shells. Chevreux (1908), while examining *Buccinum undatum* inhabited by *Pagurus bernhardus*, reported as much as eight species of amphipods, some of which being rather rare (*Orchomene commensalis; Leucothoe incisa; Crassicorophium bonellii* [as *Corophium*]; *Caprella acanthifera*), the others representing common species (*Aora typica; Melita obtusata; Gammaropsis maculata* [as *Eurystheus maculatus*] and *Stenula rubrovittata* [as *Metopa*]). There is no possibility to properly distinguish between inquilines and commensals in cases like these. In South Australia, *Leucothoe spinicarpa* was found in sponges and tunicates. Another amphipod was reported to be clinging to the shell under which it found shelter. *Unciala* spp. have been reported from tubes of polychaetes. Other amphipods, such as *Siphonoecetes*, were found in shells of *Bittium, Rissoa* or *Dentalium*, a *Cyrtophium* in the stem of a grass (Poaceae) and *Apocorophium acutum* in a honeycombed worm tube (*Sabellaria*; Annelida: Polychaeta).

The use of empty gastropod shells by **Anomura: Paguroidea** is, of course, among the most widely known cases of "occupying" foreign objects. Yet, it should be noted

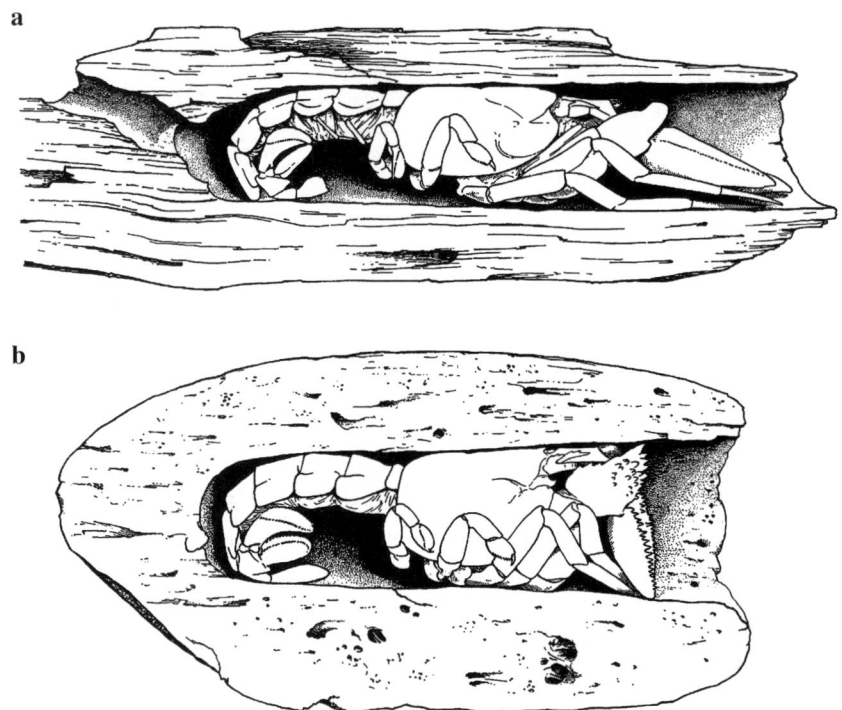

Fig. 26.1. a, *Pylocheles* (*Xylocheles*) *macrops* in a cavity in a piece of wood; b, *Pylocheles* (*Pylocheles*) *mortensenii* in a cavity in a piece of stone, i.e., pumice. [After Forest, 1987.]

that also other objects can be used by hermit crabs: wooden tubes (*Xylopagurus* spp.), shells of scaphopods (Scaphopoda: *Dentalium*) (*Pylopagurus*, *Parapagurus*), etc. The representatives of Pylochelidae (symmetrical pagurids) may inhabit pieces of wood, bamboo tubes, dentalid shells, pieces of pumice stone or soft limestone, depending on the various genera (fig. 26.1). *Porcellanopagurus* can adopt conical mollusc shells (*Patella*, *Siphonaria*) as a shelter, or even the shell of a bivalve (*Cardium*). Holes in a rock or in corals can be occupied (*Cancellus*) and Th. Monod has observed a small pagurid sheltered in an empty seed of a baobab on a beach in Senegal. *Coenobita* spp. are found in a variety of shells (Seurat, 1904), and in pieces of bamboo, or of coconuts, or even in a lamp glass...

Brachyura. In the family Dromiidae, *Hypoconcha* uses a bivalve shell in which it fits and to which it attaches, while *Conchoecetes*, remaining free in its borrowed shell, must hold on to that with its hind legs (André, 1937). Pinnotheridae are often associated with living molluscs, mainly bivalves but occasionally gastropods such as *Conus*. A male and a female are often found together within tubes of sedentary polychaetes, empty or not. In this last instance, the relationship with the worm is both offering a shelter and constitutes a case of actual commensalism. The most important tubicolous genera of such crabs are *Pinnixa*, *Pinnotheres* (Pinnotheridae), *Tritodymania* (Macrophthalmidae), and *Asthenognathus* (Varunidae), living with the polychaetes *Chaetopterus*, *Pectinaria*, *Amphitrite*, *Clymenella*, *Loimia*, *Arenicola*, etc.

Associations of **shrimp** are less common than those of crabs in this regard. Some may be mentioned nonetheless: *Alpheus lottini* in holes of *Lithophaga lithophaga*, *Alpheus dentipes* in spaces made by rock-boring mussels (Bivalvia: Mytilidae, genus *Lithophaga* [earlier as *Lithodomus*]) and in honeycombed worm tubes (*Sabellaria*), *Alpheus malleodigitus* in coral cavities, and *Alpheus simus*, a species living by pairs in similar cavities, but of which the opening is too small for them to get out.

Constructed or adapted shelters. – The shelters dealt with here are constructions, mostly tubular, in the proper sense, as well as lodges made in an organic substratum, such as Porifera tissues, either made, or adapted, and next occupied by Crustacea. Many crustaceans are able to build tubes or hollow "nests" used as permanent or temporary housing from various materials. An **ostracode** (*Cylindroleberis* sp.) is often cited as being capable of making itself a tube from agglomerated particles.

Many **tanaidaceans** are probably tubicolous, although most known specimens were collected in free condition, with the aid of a dredge. Buckle Ramìrez (1965) documented the example of *Heterotanais oerstedii*, euryhaline and littoral. This animal, whose **secretory glands** were already described by Blanc as early as 1884, constructs its tube from diatoms and various kinds of debris, and lines the inside with its own excrements that it disposes of in an appropriate way, viz., perpendicular to the axis of the tunnel. Mating takes place there, as well (fig. 26.2). In the Kerguelen Islands, the female of *Langitanais willemoesi* builds a nest of undefined shape with agglutinated materials (hydrozoans, debris from echinoderms, siliceous spicules from sponges, etc.) (Shiino, 1978). A female, with 3-51 juveniles as has been recorded, can be found inside such a lodge: it is (also) a nursery. Tubes made by *Tanais dulongii* [as *cavolinii*] have been described as well (Casabianca, 1966).

Cleantis prismatica (as *Zenobiana*; **Isopoda**: Valvifera) can inhabit a tube made from vegetal elements and sand grains. However, Maury (1925) stated that this kind of sand tube could not be a construction of the species itself: he once found this species in a the tube of a serpulid worm. Thus, *Cleantis* might rather be a "lodger" than the builder of its tube. Another species, *Cleantis phryganea* from Australia, inhabits hollow stems of plants, cut under a node, which thus make up a tube closed at one end.

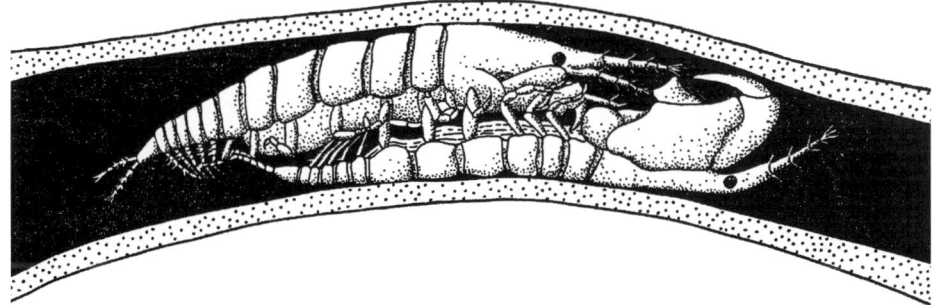

Fig. 26.2. Copulation of a pair of *Heterotanais oerstedii* inside the tube of the female. The male, recognizable by its large chela, is upside down, i.e., lying on its back. [After Gruner, 1979.]

While isopod constructors seem to be scarce, there are, in contrast, plenty to be found among **amphipods**. These either build rather amorphous masses, like a kind of purses or capsules, or well-formed tubes. Glands have been described in these animals that secrete a kind of glue through a dactylar pore. In this regard, we can mention the following species: *Ampelisca cristata* (pouches of mud), *Ampithoe rubricata*, *Aora typica*, *Cerapus tubularis* (erect tubes on leaves of seagrass), many species of *Corophium* and *Cymadusa*, *Ericthonius punctatus* [also as *Ericthonius brasiliensis* or *Cerapus brasiliensis*] (soft tubes made of mud, or tubes of muddy sand, fixed on seaweeds and hydrozoans, sometimes erect on those substrates), *Ericthonius pugnax* (tubes sticking on leaves of seagrass), *Haploops tubicola* (tubes of mud), *Jassa falcata* (tubes), *Leptocheirus pilosus*, *Microdeutopus gryllotalpa* (elongated tubes of mud in the grooves of a shell of *Cardium*), *Microprotopus maculatus*, *Parajassa pelagica* (tubes of mud), *Siphonoecetes* (nested tubes) (Ledoyer, 1969).

Barnard (1958) reported seven main species in California that foul ports and vessels, of which five are tubicolous: *Monocorophium acherusicum*, *Monocorophium insidiosum*, *Laticorophium baconi* [these three species all as *Corophium*], *Jassa falcata*, and *Ericthonius punctatus*. The work of Ennequist (1949) on amphipods from sandy bottoms of the Skagerrak, also concerns many tubicolous species (*Haploops tubicola*, *Autonoe longipes*, *Neohela monstrosa*, etc.) and their biology. In the Philippines, the spongiform construction of *Corophium shoemakeri* had earlier even been considered and described as a sponge, named *Oscarella malabonensis*. The phenomenon of amphipods arranging their home in sponges has been well documented (*Colomastix pusilla* in *Suberites domuncula*, *Suberites carnosus* and *Halichondria* sp.). Another amphipod, *Polycheria osborni*, as well as the juvenile *Polycheria antarctica*, are able to open an alveole in the wall of the tunicate *Amaroucium* (cf. Ennequist, 1949).

Within **Decapoda**, examples of shelters are provided by alpheid shrimp (Caridea): *Alpheus frontalis* is known to build its lodge with the aid of a blue alga, *Oscillatoria* (Cyanophyta), while *Alpheus pachychirus* can "stitch" with filaments from another species of Cyanophyta (*Plectonema*) (see fig. 26.3). Species of the genus *Trizocheles* (Pylochelidae) are generally associated with hexactinellid sponges, as is the stenopodidean shrimp *Spongicola venustus*. Finally, many stomatopods inhabit galleries (terriers with two openings), that are either constructed, or natural (Serène, 1954).

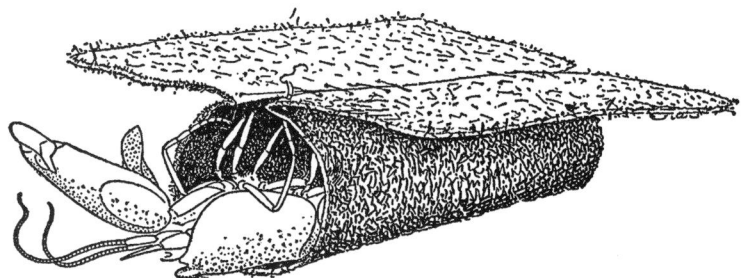

Fig. 26.3. *Alpheus pachychirus* shaping an algal mat into a tubular lodge by making a "seam" that holds the peripheral parts together. [After Cowles, in Schmitt, 1931.]

Soft sediments

Many groups of Crustacea include species that live inside **sandy beaches**. For example, from Arcachon, on the Atlantic coast of France, Salvat (1967) listed no less than 9 copepods, 1 mystacocarid, 1 tanaidacean, 4 isopods, and 12 amphipods from that habitat. Some 30 species belonging to 21 genera were reported from Banyuls-sur-Mer, on the Mediterranean coast of France (Bodiou & Chardy, 1973) as living in the sand of the beach. In the area of Tuléar (Madagascar), Pichon (1967) distinguished three zones:

1. **Supralittoral zone**, with *Talorchestia* sp., *Excirolana natalensis*, *Ocypode ceratophthalmus* and *Ocypode cordimanus*.
2. **Mediolittoral zone**, with *Excirolana orientalis*, *Urothoe* sp. and the same two *Ocypode* species as in the supralittoral zone.
3. **Infralittoral zone**, with *Albunea symmista*, *Albunea paretii*, *Hippa pacifica*, *Macrophthalmus grandidieri*, *Macrophthalmus convexus*, *Dotilla fenestrata*, *Euplax bosci*, etc.

The isopods present on beaches often are Cirolanidae, which are carnivorous and aggressive (*Eurydice*, *Excirolana*, *Pontogeloides*, etc.). Isopods occupy the medium level of the beach (Jones, 1971, 1974, 1979) between the supralittoral with *Ocypode*, *Tylos*, and Talitridae, and a lower zone with hippids. In general, Sphaeromatidae are mainly herbivorous and they are scarce on beaches. However, *Sphaeromopsis amathitis*, described from a beach in Kenya, seems a true burrower (Holdich & Jones, 1973). Salvat (1967) studied the biology of *Eurydice pulchra* and *Eurydice affinis*. These species appear to exhibit two phases of activity, a diurnal benthic one, and a nocturnal pelagic phase.

Many amphipods from various families (Haustoriidae, Phoxocephalidae, Dogielinotidae, Gammaridae, Oedicerotidae) that are adapted to an endopsammic life (Schellenberg, 1929; Dennel, 1933; Sameoto, 1969, 1969a; Bousfield, 1970) show two distinct adaptative types: the first is fusiform-elongate, with a developed rostrum or pseudo-rostrum, and the second is widened-fusiform, i.e., more or less truncate, and has a short rostrum. Some of these amphipods can "swim" in the sand, as also the isopods of the genus *Eurydice* do.

Within Decapoda, the Hippidea (*Hippa*, *Albunea*, *Emerita*, etc.) use to live at the lower level of the wave breaking zone, where they find their food: the animals almost completely sunk in the sand, facing the current, leaving nothing more above the sediment than their long and mobile antennules that collect alimentary particles brought in by the rising tide. Some species, like *Emerita analoga*, make up very dense populations, up to 560 ind./100 cm^2 and 22 000 ind./m^2 (Efford, 1965). Other species may rely on a different type of food: *Hippa testudinaria* is a predator, and also a detritivore (Thomassin, 1969).

Most Brachyura on beaches are burrowing. However, *Dotilla myctiroides* is able to build both a terrier that requires a resistant sand by the ebb tide, and a kind of shelter with a dome, a sort of "igloo" made from pellets, in which it sometimes seeks shelter during high tide (Tweedie, 1952).

Large Penaeidae from the neritic zone and also Crangonidae are able to burrow in soft sediments, where they stay by day. Their antennules, the tip of their rostrum and the animal's eyes are the only body parts visible. Processidae (ostrich shrimps) use to bury themselves completely, disappearing head first into the sand. At night, these shrimps wander in order to look for food, as well as to moult, mate, and lay their eggs.

BURROWERS

Where in the above we dealt with superficial, **horizontal burrowers** (Penaeidae, Crangonidae) that dig only shallowly into the sediment surface in order to cover themselves with sand, the crustaceans here at issue are those entering deep into the soft sediments, either sand, or mud. Their actions leave a **temporary hole** or a **permanent burrow** in the substrate and among them are crabs of the family Gecarcinidae, as well as the mud shrimp of the suborders Axiidea and Gebiidea (formerly collectively known as Thalassinidea) (Dworschak, 1983, 1987, 1988; Dworschak et al., 2012). Knowledge is scarce for burrows in benthic, neritic, bathyal or abyssal substrates of which the occupants are either unknown, or whose biology is known only in part. The Norway lobster, *Nephrops norvegicus* inhabits a burrow with two openings (Figuereido & Thomas, 1967). The Recent glypheid *Neoglyphea inopinata*, most probably spends part of its life in a **gallery** (Forest, 1981; Forest & de Saint-Laurent, 1989). Not surprisingly, the littoral species are better known, reason why we can treat those in more detail in the following. Yet, it should be made clear that between the superficial diggers and the deep burrowers, quite some intermediate cases do exist that can not easily be brought to either of those two categories.

Among Isopoda, the adult gnathiid *Paragnathia formica* occupies open cavities in the hardened mud of estuarine shores of the North Atlantic (Monod, 1926) and the cirolanid *Annina kumari* lives in galleries in the muddy bottoms of mangroves in Malaysia.

Quite a number of Amphipoda can burrow in sand or mud, but it is often impossible to say if they live in real holes or not. Burrowing methods and the adaptations of the animals' appendages may vary: these have been compared in detail for *Talitrus saltator*, *Bathyporeia sarsi* [as *robertsoni*] and *Corophium volutator* (cf. Schellenberg, 1929). Adult talitrids appear to hide in sand burrows by day and get out at night for feeding according to a necrophagous-detritivorous regime (Schellenberg, 1928, 1929).

Some brachyurans make extraordinary burrowers and live in dense populations on intertidal flats in the tropics. For instance, several species of Ocypodidae can live together in the same area but appear to be separated into distinct biotopes. For Tamsui (Taiwan), Takahasi (1935) gave the following scheme:

- **Sandy beaches**:
    - on dunes close to the high tide mark, *Ocypode cordimanus*; on the high tide mark, *Ocypode ceratophthalmus*;
    - intertidal zone, on the beach, *Scopimera globosa*, *Scopimera bitympana*; on fine sand, *Ilyoplax delsmani*, *Methypocoelis ceratophora* (as *Ilyoplax ceratophorus*), *Mictyris longicarpus*, *Uca lactea*;
    - in the algae zone, *Uca lactea*, *Scopimera globosa* (with *Sesarma haematocheir*).
- **Muddy beaches**:
    - on the beach, *Macrophthalmus dilatatus*, *Macrophthalmus japonicus*;
    - on clay and mud, *Uca arcuata*, *Uca formosensis*, *Uca vocans* [as *marionis*];
    - in the algae zone, *Uca arcuata*, *Uca formosensis*.

Takahasi (1935) described different types of galleries with the aid of **plaster castings**. Feeding in these species mostly occurs as a superficial scraping of the sandy-mud surface, and producing pellets of substrate that are ingested. The "tunnel" of *Mictyris* more or less corresponds to the "igloo" of *Dotilla* (cf. Tweedie, 1952).

In Djakarta, Verwey (1930) was able to observe the following zonation from the upper reaches of the beach down to the permanently submerged zone of the seawater:

1. *Episesarma mederi* [as *Sesarma taeniolata*].
2. *Uca lactea* [as *Uca consobrinus*].
3. *Uca (Australuca) signata* [as *Uca signatus*], *Ilyoplax delsmani*.
4. *Ilyoplax elegans, Uca urvillei, Paracleistostoma depressum*.

The diet of these species is mostly microphagous (vegetal debris, eggs and larvae of littoral animals, diatoms). There are many species of fiddler crabs (genus *Uca*) and one of them, *Uca tangeri* reaches Andalucia (southern Spain); its biology was studied by various authors such as Altevogt (1957, 1969), Von Hagen (1962), and Crane (1975). Fiddler crabs are often encountered in mangrove swamps with other decapods such as *Scylla, Sesarma, Metaplax,* and *Thalassina*.

The burrowing crabs in Djakarta were divided by Verwey (1930) into two groups, according to their **breathing physiology**: the "inhalers", which extrude water through exhalant openings and take it back by inhalant openings when oxygenated (*Sesarma, Ilyoplax, Macrophthalmus*), and the "non-inhalers", which can introduce water into their branchial chambers (*Grapsus, Uca, Ocypode*).

Several Decapoda are able to dig in the supra-littoral zone. This is the case for *Thalassina anomala*, whose galleries are crowned by mud chimneys and for the large species of *Cardisoma*, which are already semi-terrestrial. Many littoral mud shrimps (*Gebia, Upogebia, Callianassa*) inhabit galleries. The snapping shrimp *Alpheus djiboutensis* lives in association with the gobiid fish *Cryptocentrus cryptocentrus* (cf. Luther, 1958).

Stomatopoda use to stay in natural cavities or in selfmade tunnels: the large mantis shrimp *Lysiosquillina maculata* [as *Lysiosquilla*] occupies a gallery with two openings (Serène, 1954), and the spottail mantis shrimp *Squilla mantis* lives with the alpheid *Athanas amazone* (cf. Atkinson et al., 1997; Froglia & Atkinson, 1998).

THE INTERSTITIAL HABITAT

From 1930 onward (notably Hertzog, 1933; Delamare Deboutteville & Angelier, 1950; Ruffo & Delamare Deboutteville, 1952), extensive studies of the marine and freshwater **endopsammon** have occurred. This community appeared to be exceptionnally rich in Crustacea: Mystacocarida, Copepoda (mostly Harpacticoida), Ostracoda, Thermosbaenacea, Isopoda (Microcerberidae), Amphipoda, and Syncarida (Anaspidacea, especially Stygocarididae; Bathynellacea).

As a result of the comparable selective pressures, we can recognize many similar adaptations in the various groups of animals together constituting the **meiofauna**, presenting as many cases of the phenomenon of **convergent evolution**. Isopods can be considered a relevant example in this regard (Coineau, 1971). In this group, the body size is always very small (0.62-1.34 mm), the body as a whole is elongate, with a ratio width/length that may reach 1:11 up to 1:14; furthermore, there is a considerable flattening of the body, along with depigmentation, anophthalmy, muscular development related to crawling between sand grains, a distinct fragility, thigmotaxis, eurythermy, euryhalinity, microphagy, a reduction of the numbers of eggs produced, and an increase in the intramarsupial period of embryonic development.

Noodt (1974), however, questioned the importance of the role of an interstitial way of life for the evolution of the higher groups of Crustacea. In his view, the mesopsammic habitat, very ancient and relatively stable, and inhabited by epibenthic forms, would rather have led those forms to evolve at a slower pace, perhaps even **orthogenetic**, and with a tendency for **neoteny**. Such a scenario obviously would not be prone to yield large macro-evolutionary breaks, but it would have contributed to the possibilities of active **radiation**. If the habitat was not actually a *cul-de-sac*, it did make any chances for marked future evolution very limited. Consequently, the role of the littoral endopsammic environment has often been emphasized as providing a significant step in the penetration of underground fresh waters, i.e., the phreatic, hyporheic, and cavernicolous environments.

THE PLANT REALMS

Marine plants (both algae and seagrasses) harbour many species of crustaceans everywhere. Such "meadows" comprise a series of distinct microbiotopes. *Posidonia* beds represent complex habitats in which the leaves constitute a dominant element, while the thick webs of rhizomes shelter a specialized and rich fauna. Eelgrass beds in temperate regions with *Zostera*, *Cymodocea* and *Posidonia* already lodge many crustaceans (Amphipoda, Isopoda, majid Brachyura, hermit crabs, and shrimp), those in tropical areas are even much richer: first, with plants; second, with fauna. At Tuléar, Ledoyer (1967) listed 23 species of Caridea (10 families). The same Malagasy seagrass beds contained 13 species of Isopoda (cf. Roman, 1979) and 34 Amphipoda (16 families) (Ledoyer, 1967), while 85 species of Ostracoda were reported from *Posidonia* beds in the Provence (France) (Reys, 1964).

Seagrass beds are especially rich in caprellid amphipods: in the Dinard region (English Channel), Bertrand (1942) found *Caprella tuberculata*, *Caprella erethizon*, *Caprella acanthifera*, *Caprella linearis* and *Caprella fretensis* (with the isopod *Idotea neglecta* and the corophioid amphipod *Ampithoe rubricata*). In the Mediterranean, the mimetic *Idotea viridis* is both homochromic and homotypic on the *Posidonia* leaves where it hides.

These marine spermatophytes root by means of intricately intertwined, entangled rhizomes, which constitute a mass that is easy to penetrate for crustaceans and rich both in food and in opportunities for shelter. In this type of microbiotope, off Naples, Wägele (1979, 1981) found a series of Isopoda: Anthuridea, viz., Hyssuridae (*Kupellonura serritelson* and *Neohyssura spinicauda*), Anthuridae (*Apanthura corsica*) and Paranthuridae.

The notorious marine fouling on immersed objects, e.g., buoys, often includes abundant algae, among (and on) which crustaceans are present. In this habitat, the following species were found in California (Barnard, 1958): *Monocorophium acherusicum*, *Monocorophium insidiosum, Laticorophium baconi* [as *Corophium*], *Jassa falcata, Ericthonius punctatus* [as *Podocerus brasiliensis*] and *Stenothoe valida*.

Large algae (*Laminaria, Macrocystis, Umbricella, Durvillea*, etc.) have their own crustacean fauna. On the other hand, smaller, branched algae like *Cystoseira* also lodge a significant number of species (copepods, amphipods, isopods, and brachyurans; most often both adults and juveniles). On *Halimeda*, the crabs *Huenia proteus* and *Menaethius monoceros* are both mimetic and homochromic.

HYPERSALINE WATER BODIES

Crustaceans can be encountered in two types of **hypersaline waters**: in supralittoral pools on rocky shores that suffer strong evaporation, and in sheltered, shallow bays on arid coasts with no supply of fresh water (the salinity of the waters of Saint-Jean, Mauritania, can reach 80‰ salinity). Supralittoral puddles are the classical habitat in Europe of the harpacticoid copepod *Tigriopus*. The fauna of hypersaline bays does not differ significantly from that of the adjacent, "open" coasts.

THE ANCHIHALINE ENVIRONMENT

There exists a particular category of environments, often with a remarkable fauna, that comprises water bodies of variable salinity, generally supralittoral and hence close to the sea, but all the same cut off either completely, or at least largely from it. Holthuis (1963, 1973) mentioned that shrimp of different groups are predominantly red coloured in those places, and we cite some of these so-called **"anchihaline"** biotopes ("near the sea", Holthuis, 1973):

1. **Jameos del Agua, Lanzarote** (Canary Islands), a subterranean lake of marine water some 500 m away from the shore, with the blind anomuran, *Munidopsis polymorpha*, the mysidacean, *Heteromysoides cotti* and the amphipod *Parhyale fascigera* (cf. Fage & Monod, 1936), as well as the remipede *Morlockia ondinae* (cf. Schram, 1986; as *Speleonectes*).
2. **Anchihaline cave on Grand Bahama Island**, where the first species of the class Remipedia, *Speleonectes lucayensis* was found (Yager, 1981) deep in nearly anoxic water (0.0005 ppt dissolved oxygen).
3. **Supralittoral pools on Ascension Island**, where the shrimp *Procaris ascensionis* was discovered, the type of Procarididea, which lack chelate thoracic appendages, and a species of Atyidae, *Typhlatya rogersi* (cf. Chace & Manning, 1972).
4. **Galapagos, supralittoral crevices in lava**, where shrimps live (*Typhlatya galapagensis, Macrobrachium americanum*), together with amphipods (*Galapsiellus leleuporum*, blind, 1-1.5 mm, *Chelorchestia costaricana, Ampithoe* sp., *Cheiriphotis megacheles*), and the tanaidacean *Sinelobus stanfordi* [as *Tanais*] (Monod, 1970; Monod & Cals, 1970).
5. **Ras Muhamad, Sinai**, crevices, 150 m away from the sea, saline water (23.6 psu), with two shrimps: *Periclimenes pholeter* and *Calliasmata pholidota*.
6. **Maui Island, Hawaii**, cracks and intertidal basins in lava close to the sea, water of very variable salinity (10-30 psu): with the shrimps *Halocaridina rubra, Antecaridina lauensis, Calliasmata pholidota, Metabetaeus lohena, Alpheus* sp., *Palaemonella burnsi, Macrobrachium grandimanus*, and *Palaemon debilis*.

7. The shrimp *Ligur uveae* was found in **anchihaline stations** on Aldabra, the Moluccas, the Loyalty Islands, Fiji, the Funafuti atoll, and on the Philippines (Monod, 1968; Holthuis, 1973; Wear & Holthuis, 1977).
8. We may here add to this enumeration: *Metabetaeus minutus* from the **Tokelau Archipelago** and *Barbouria cubensis* from **Cuba**.

In concluding the section on **oceanic waters**, we may observe that some crustaceans serve as "indicators" for different kinds of water in the oceans: *Calanus hyperboreus*, *Metridia longa* (Arctic water), *Nematoscelis megalops*, *Thysanoessa longicaudata* (mixed, Arctic-Atlantic water), *Meganyctiphanes norvegica*, *Thysanoessa inermis*, *Nyctiphanes couchi* (Atlantic water), *Dosima fascicularis* [as *Lepas*], *Anomalocera patersoni* (Antarctic warm water).

Typically terrestrial organisms that dwell in a littoral habitat, might survive for some time in the plankton: this has been found for two specimens of woodlice (both *Tylos punctatus*) found "quite lively" in surface waters in the Bay of Newport, California (Menzies, 1952).

BRACKISH WATERS: ESTUARIES, LAGOONS AND MANGROVES

Brackish waters, and more precisely waters with **variable salinities**, make an environment that must be considered only somewhat autonomous, because in spite of the efforts made to determine accurate limits of the brackish realm, both in the direction of the sea, and towards the fresh waters of the adjoining land, such efforts are largely in vain. Both **estuaries** and **coastal lagoons** belong to this category.

Particular types of brackish **biocoenoses** can be defined by their **vegetation**, such as salt marshes with *Obione*, *Aster* and *Salicornia*, mud flats with *Spartina*, which develop where an active mud sedimentation occurs. In tropical zones, the mangroves occupy vast surfaces along many shores at those low latitudes and their faunas include many species of Crustacea. Among the latter are the swimming crabs (*Callinectes*, *Scylla*) and amphibious crabs (*Sesarma*, *Uca*) certainly as the most conspicuous. We may also note that Rhizocephala such as *Sacculina*, *Sesarmaxenos*, and *Ptychascus* can parasitize those brackish water crabs, thereby sometimes entering into fresh water (on *Sesarma*, *Uca*, *Macrophthalmus*, *Varuna*, *Eriocheir*).

Pneumatophores of *Avicennia* and the aerial roots of mangroves are often richly covered with acorn barnacles, the numbers of which may be correlated with seasonal variations in salinity, e.g., for the region of Lagos, Nigeria (Sandison & Hill, 1966). These roots are often attacked and drilled into by an isopod, *Sphaeroma terebrans* (*vide supra* in the section on Lignicolous forms).

Most species in mangroves spend their whole life in that same biotope. However, several decapods (various Penaeidae, *Macrobrachium*), shed their eggs in seawater, whereupon the juvenile stages enter estuaries and lagoons that function for them as nurseries. These prawns attain their sexual maturity in fresh water, but they go down the river again to shed their eggs at the mouth, in water of high(er) salinity.

Various authors (for instance Monod, 1925 and Fischer, 1940) ventured to conclude that the brackish fauna would not be primarily specific to the variable salinity of the environment, but rather linked to a favourable substrate, the mud, as well as to the trunks and aerial roots of the mangroves. We may indeed consider that the sessile fauna on the roots (barnacles, etc.) is epiphytic, while *Sphaeroma* spp. are endophytic, and some of the other crustaceans, like crabs, are limicolous, and form a community in the mud as well as on and around the bases of the trees. Yet, it would seem that further research will be necessary to solve this problem satisfactorily, since the general view of mangroves housing a limited number of species in abundantly large numbers of individuals might well have to be reconsidered as the study of this biotope intensifies, more in particular where species of small body size are concerned. For instance, Carvacho (1977), working on isopods in an Antillian mangrove, found six genera and seven species.

Deep benthos

BENTHOS ON SOFT SUBSTRATUM

Crustaceans, as already noted, are present at all depths, the troughs of the **hadal zone** included. Nevertheless, there is a steep decline in faunistic richness with the increase of depth. The "Challenger" Expedition could collect 190 species of Brachyura between 0 and 36 m, 75 at 36-180 m, 28 at 180-360 m, 21 between 360 and 900 m, 3 at 900-1800 m; and a mere 2 species between 1800 and 3600 m. **Physicochemical parameters** become more uniform with depth and the diversity of the littoral or neritic biotopes correlates with the relative monotony of the abyssal mud flats, which are only interrupted by large **geomorphological structures**, like the oceanic ridges and fields of basaltic lavas, guyots, and other sub-marine seamounts. Quite surprisingly, however, vegetal debris (wood, branches, bark, coconuts, and pieces of bamboo) may accumulate very deep, i.e., at the bottom of hadal troughs, and those plant remains appear to be occupied by many species of Mollusca, Polychaeta, and Crustacea (Wolff, 1980).

Without doubt, the **abyssal fauna** has originated from animals that migrated to deeper strata from the shallower littoral and bathyal zones (Bruun & Wolff, 1961). The periods when those movements took place can vary according to the various taxa. Such a historical reconstruction was tentatively drawn up for Decapoda Reptantia (cf. Beurlen, 1929), with the definition of three main periods of immigration to abyssal and hadal depths after a phase of **marine transgression**: (1) the Jurassic (Polychelidae, Homolodromiidae, *Munidopsis*), (2) the Cretaceous (Homolidae), and (3) the Eocene (Geryonidae and Majidae). Many archaic forms do exist at great depths, or are even strictly limited such habitats: Penaeidae, *Acanthocaris* and *Thaumastocheles* (Nephropidea), Polychelidae, *Uroptychus* (Chirostylidae), Homolodromiidae. The best adapted (or pre-adapted?) forms for living in the deep sea are often those with elongated legs, or similar spider-like morphological types, i.e., already possessing features that will facilitate life on the surface of mud. Another putatively pre-adapted group are the burrowers, for instance *Nephrops* and *Nephropsis*, that are able to dig galleries. Here we summarize the morphological particularities often present in deep-living Crustacea:

1. **Size** often large (Isopoda: *Bathynomus*, up to 50 cm; Amphipoda, up to 34 cm; many large Penaeidae, up to 30 cm; Brachyura: Geryonidae, *Macrocheira, Pseudocarcinus gigas*).
2. The **colour** red is common in Decapoda (in fact, this colour may be cryptic, for it appears as black under the aphotic conditions of the deep sea). However, many species are white or colourless, especially those that bury themselves in the sediment.
3. **Luminescence** is scarce in deep-water crustaceans. However, two pelagic shrimps, viz., *Heterocarpus dorsalis* and *Plesiopenaeus coruscans*, taken at 1010 m deep, were reported to produce clouds of bluish light.
4. Reduction of the **eyes**, as often found in benthic organisms.
5. Many forms are **spinose** (Isopoda: Asellota: Arcturidae; Anomura: Lithodidae) and shrimps have long antennular and antennal flagella (Penaeidae, Pandalidae, etc.).
6. There is a tendency, at least in Decapoda, for an increase in **egg size**, correlated with a reduction of the duration of the free-larval period.

Crustacea that live in the deep are represented by a few **eurybathic** groups and by many **stenobathic** forms, for both of which some examples are presented here. The eurybathic forms include *Verruca, Diastylis, Eudorella, Antarcturus*, and Lithodidae. The stenobathic catgory encompasses ostracodes (Benson, 1969), cirripedes (*Hexelasma, Megalasma*), tanaidaceans (*Typhlotanais, Sphyrapus, Cryptocope*), cumaceans (*Macrocylindrus, Bathycuma*, Platysympodidae, Procampilaspidae), isopods (quite a number of Asellota: *Munnopsis, Munnopsurus, Haploniscus, Nannoniscus*), amphipods (*Trischizostoma, Brugelia, Harpinia, Halirages, Byblis*), mysidaceans (*Pseudomysis, Parerythrops, Erythrops, Boreomysis, Acanthocaris*), decapods (Penaeoidea: *Amalopenaeus, Plesiopenaeus, Benthesicymus, Benthonectes*; Caridea: Oplophoridae, Glyphocrangonidae; Eryonidae: *Polycheles, Willemoesia*; Nephropidae: *Thaumastocheles, Nephropsis*; Axiidea: Axiidae; Anomura: Chirostyloidea (*Uroptychus*), Galatheoidea (*Munidopsis*), Paguroidea (Pylochelidae, Parapaguridae); Brachyura: many Homolodromiidae, Homolidae, Majoidea (*Platymaia*), and Geryonidae.

BENTHOS AROUND HYDROTHERMAL VENT SITES

Several hundreds of species have been reported in association with high temperature **hydrothermal vent sites** (up to 400°C) since their discovery in 1977 (Laubier, 1988, 1989). The ratio of endemics in these extreme biotopes is especially high (Tunnicliffe, 1998). The populations found thriving around such geological formations are based on **endosymbiosis** between various invertebrates from different groups (Vestimentifera, Bivalvia, Gastropoda) and chemo-autotrophic sulfo-oxydant bacteria. These bacteria receive their energy, at least in part, from the oxydation of reduced compounds ejected in the hydrothermal fluid, in particular hydrogen sulfide. Near such site, also archaeobacteria (Archaea) are abundant, forming bacterial mats on the substrate, and some of these bacteria can develop at temperatures between 100 and 110°C.

The faunas associated with these hydrothermal vent chimneys include a series of Crustacea, most of which are endemic in the environment, with its special physico-chemical properties. More than fifty have already been described from these realms, among which a leptostracan, *Dahlella caldariensis*; pedunculate cirripedes (*Neolepas zevinae*) as well as Cirripedia: Sessilia (*Eochionelasmus ohtai*); about thirty species of copepods, some of which are parasites of invertebrates, and one primitive calanid, genus *Isaacsicalanus*; one isopod, half a dozen Brachyura (Bythograeidae with the genera *Bythograea*, *Cyanagraea*, *Austinograea*, *Allograea*, *Gandalfus*, and *Segonzacia*); Galatheoidea of the genus *Munidopsis*; and alvinocaridid shrimps (a family close to Bresiliidae), including eight recently (1982 up to 2010) described genera: *Alvinocaridides*, *Alvinocaris*, *Chorocaris*, *Mirocaris*, *Nautilicaris*, *Opaepele*, *Rimicaris*, and *Shinkalcaris*.

Alvinocaridid shrimp and bythograeid crabs are well adpted to the extreme physico-chemical conditions of these habitats. The large crab *Cyanagraea* walks over the colonies of the polychaete *Alvinella* on which it feeds, and it has to deal with temperatures of 20 to 40°C or more when it predates on those worms, while in addition it experiences an acidic environment (pH between 2.5 and 3.5). Six species of *Bythograea* are now known (*Bythograea galapagensis*, *Bythograea intermedia*, *Bythograea laubieri*, *Bythograea microps*, *Bythograea thermydron* and *Bythograea vrijenhoeki*); together with *Segonzacia mesatlantica* these crabs feed mostly by grazing the distal ends of the gills of vestimentiferans. The shrimp *Rimicaris exoculata* is very common at sites on the Mid-Atlantic Ridge between 1000 and 3800 m. It is considered a primary consumer and it feeds on the bacterial mats that cover the substrates (basaltic lavas, polymetallic sulfure deposits on the chimneys); the species is not attracted by bait deposited in traps. The other species of Alvinocarididae would have a necrophagous or mixed alimentary regime (Segonzac et al., 1993); they stay in the periphery of the zones, i.e., there where the hydrothermal fluid diffuses, and are caught with baited traps.

Some Decapoda that usually live outside those hydrothermal zones are all the same found there, as well, like *Munidopsis* spp. (Munidopsidae). These are located close to the limit of the zone, where temperatures are only 1-2°C higher than the local temperature.

The large numbers of cirripedes observed at the sites in the West Pacific (the basin of North-Fidji, basin of Lau, south of Tonga) may be due to the large quantity of organo-mineral particles in suspension. **Panchronic forms**, i.e., "living fossils" like *Eochionelasmus ohtai* and *Neolepas zevinae*, would have migrated to the deep during the grand Cretaceous diversification, after having found a refuge near superficial hydrothermal sites.

The pelagic realm

The significant role crustaceans play in the pelagic fauna is well known, both as elements in the plankton (holo- or mesoplanktonic forms), and in the nekton. Most classes are represented in the pelagic: Branchiopoda (Cladocera), Ostracoda, a large number of Copepoda, Cirripedia (as larvae), and within Malacostraca: Hoplocarida (larvae), Mysida, Cumacea (occasionally near the shore), Isopoda, Amphipoda (mostly

Hyperiidea), Euphausiacea, and Decapoda (larvae and adults). Pelagic decapods present natatory adaptations such as multiplication of spines and setae ("Chaeto-planktonform"), a pronounced elongation of their body (*Lucifer*), or the attainment of flat, discoid shapes, e.g., phyllosomas. We need not dwell on this subject in detail here: the statements made above will be clear when consulting the various systematic chapters in the present series.

FRESH WATERS

Continental waters

Fresh water is often considered a **uniform environment** populated with species that show wide geographical distributions. This view is supported by the fact that small species (Cladocera, Copepoda, etc.) apparently are widely distributed, and that large rivers have extended basins that include the catchment areas of their tributaries. However, the true conditions of freshwater environments are difficult to generalize for the following reason: at a smaller scale, some fresh waters are effectively isolated, and some lakes are rich in endemic species: e.g., the Caspian Sea, Lake Baikal, and Lake Tanganyika. Hence, it is not surprising to find many purported archaic or relict forms in fresh waters, isolated by the geological evolution of the surface of the earth: phenomena related to earlier **glaciations**, the effects of **paleoclimatological conditions**, and **plate tectonics**.

Inhabiting large rivers has some biological consequences for non-microscopic Crustacea, for instance, the suppression of larval instars and the concurrent acquisition of (a degree of) direct development (crayfish, freshwater crabs). Adaptations like these are inevitable for the long-term continued existence of the species in its habitat.

In general, freshwater crustaceans are less diversified than those living in the sea. This is easily explained by their possible mode of colonization of the continental habitats, i.e., via **introgression** from the littoral. The occupation of the low-salinity environment may have followed several pathways: either by following the surface (estuaries, lagoons) in the case of large or medium-sized organisms, or through the subterranean route (interstitial environment, caves) for small or very small species. However, the remarkable diversification of freshwater crustaceans at specific and subspecific levels stands in great contrast to the relative uniformity of the taxa of higher rank.

Some groups have consistently remained marine, either in whole or nearly so. No freshwater forms exist for Cephalocarida, Mystacocarida, Cirripedia (with the exception of a few Rhizocephala found on freshwater crabs such as *Sesarmaxenos* on *Sesarma* and *Polyascus gregaria* on *Eriocheir*), Leptostraca, Euphausiacea, Stomatopoda, and Tanaidacea (with some exceptions, such as *Sinelobus stanfordi*, which is found up to 800 m altitude on St.-Helena, or *Nesotanais lacustris* on Rennell Island and the Solomon Islands).

In contrast, some groups are found in fresh waters only, which might argue in favour of an early separation from their marine ancestors: Spelaeogriphacea, Branchiopoda (Anostraca), Notostraca, Conchostraca, Bathynellacea (with only a few marine species, e.g., *Hexabathynella halophila* and others, see Camacho, 2003; Schram, 2008), Isopoda: Phreatoicidae, Thermosbaenacea (with, however, some brackish water species, or occurring in hypersaline waters, like *Halosbaena*), the crayfish (Astacidae, Cambaridae, Parastacidae), freshwater crabs (Potamidae), the galatheid genus *Aegla*, and many caridean shrimp (*Atya, Macrobrachium*). Brine shrimp (*Artemia*) appear to have specialized from fresh water environments to hypersaline conditions and have no marine relatives.

Yet it should be noted that in groups that have relatives in fresh water, the latter can be exceptions: Isopoda occurring in fresh waters are limited to Phreatoicidae, some Flabellifera [currently divided as Cymothoida and Sphaeromatidea], some Asellota, some Anthuridea (*Cyathura, Cruregens*), and rarely Valvifera, but no Gnathiidea at all. Amphipoda in fresh waters include no Hyperiidea nor Caprellidea, and only a few families of Gammaridea. Mysidacea [now as Mysida] comprise some freshwater genera (sometimes cavernicolous): *Mysis, Neomysis, Taphromysis, Limnomysis, Spelaeomysis, Stygiomysis, Katamysis*, etc. Bacesco (1940) noted that among 21 species in Romania, only eight are truly marine: *Katamysis warpachowskyi* was reported from the Danube, 150 km away from the sea, and so were *Limnomysis benedeni* at 400 km, and *Paramysis* (*Mesomysis*) *intermedia* at 600 km from the river mouth. Cumacea, though, are mostly marine; however, *Cumopsis goodsir* is still able to live in waters of low salinity (10‰) in some North Sea estuaries (Vader & Wolff, 1973). Freshwater crabs (Brachyura) are represented by some Grapsidae (*Sesarma* spp.), and a few other families such as Potamidae and Pseudothelphusidae. Freshwater caridean shrimps and prawns are more diverse, represented with Palaemonidae (*Macrobrachium, Palaemonetes*, etc.), Atyidae, Desmocarididae, Euryrhynchidae, Xiphocarididae, and some Alpheidae.

Accordingly, we can interpret this as the result of an effective "filtering" amongst crustaceans because of prevailing geophysical conditions, in a position to evolve into freshwater habitats. The "filters" imposed must be viewed as involving physiological factors, somewhat compartmented biotopes, and the frequent biological necessity to produce eggs resistant to desiccation, or, within Decapoda, to resort to an **abbreviated development**, with a suppression of free larval stages.

The genus *Jaera* offers an interesting example of a marine taxon that has locally entered fresh waters (Veuille, 1979). The group allegedly originated in the Miocene (ca. 20 Mya), and subsequently divided into two lineages. One such lineage involves commensals of *Sphaeroma serratum* (*Jaera hopeana*). The other gave off a marine Atlantic group ca. 15 Mya, and next, ca. 5.2 Mya, split on the one hand into a Mediterranean group of mostly brackish species (lagoons) with penetration into fresh waters in the Azores (*Jaera nordmanni guernei* and *Jaera nordica insulana*), and on another a Ponto-Caspian group inhabiting brackish or warm waters (Danube).

Cymothoida are mainly parasites or cavernicolous forms, which belong to the following families: Cirolanidae (*Typhlocirolana*, etc.), Cymothoidae, parasites of fish in South America (cf. Trilles, 1973: *Telotha, Paracymothoa, Artystone, Riggia, Philostomella, Lironeca, Asotana, Braga, Nerocila, Isonebula*), Corallanidae (*Austrorgathona*, a parasite of shrimps in Australia: Riek, 1953, 1967), and various Sphaeromatidae.

Scenarios for transition. – An **ecological classification** of freshwater and terrestrial Decapoda in the Antilles (Chace & Hobbs, 1969) is worth considering as far as possible **transitions** between marine and fresh waters are concerned:

1. **Marine species**: either climbing on rocks (*Grapsus, Plagusia*), or living on pebbles and various types of debris (*Petrolisthes, Pachygrapsus, Sesarma*), or being burrowers (*Uca, Sesarma, Aratus*).
2. **Marine species entering estuaries**: *Penaeus* spp., *Callinectes* spp.
3. **Freshwater species** staying in **contact with the sea** for their larval stages: *Atya, Macrobrachium, Sesarma*.
4. Species **remaining in fresh water** during their whole biological cycle:
   – **epigean species**: in phytothelmes of Bromeliaceae (*Metopaulias*), lakes and ponds (*Procambarus*), rivers (*Procambarus, Epilobocera, Pseudothelphusa, Sesarma*);
   – **subterranean species**: *Typhlatya, Troglocubanus*.
5. **Terrestrial species** with marine larvae: in shells of Gastropoda (*Coenobita*), burrowing in the sand (*Ocypode*) or in various coastal (littoral) substrates (*Sesarma, Cardisoma, Gecarcinus*).
6. **Species of brackish lagoons**: *Barbouria cubensis*.

From the above it seems clear, that transitions from the marine environment to freshwater and terrestrial habitats followed different pathways and will invariably have involved a number of successive stages. Also, **adaptations** of some taxa appear to have resulted in a total independence from the sea, whereas those in other groups still require at least some (often temporary) connection with waters of higher salinity.

Geographical distribution. – The geographical distribution of freshwater Crustacea is to be explained in the framework of **evolutionary history**, both with respect to **global geology**, and in regard of the evolution of the crustacean taxa concerned. Geology will often require an interpretation in connection with plate tectonics and the displacements of continents (or parts of continents) relative to each other. For more extensive and also more detailed information, reference is made in this place to papers by Monod (1975, general issues), Strenth (1976, North American Palaemonidae), Stock (1976a, Thermosbaenacea; and 1977a, Amphipoda: Gammaridea: Hadziidae), and Schminke (1974, Bathynellacea). [To our regret, acquiring a complete update of this section according to the current state of knowledge appeared not possible and would have involved a profound rewriting of the text. Nevertheless, we recommend that readers interested in modern approaches to geographical distribution of freshwater Crustacea consult the relevant sections in the various taxonomic chapters of these Treatise volumes; and the 16 papers within Balian et al. (2008). FRS, JCvVK, eds.]

Of course, various scenarios can be considered in each specific case. For instance, in order to arrive at a plausible reconstruction of the distribution of freshwater crustaceans on the Antilles (Stock, 1977, 1977a), the following models are possible:

1. **Random dispersal** from the continent or from nearby islands; probably applicable to species that have retained larvae that develop in seawater: shrimps like *Macrobrachium* and *Palaemon*, and crabs like *Sesarma*.

2. **Vicariance or fragmentation**: monophyletic island populations become separated through the mobility of land masses, e.g., parts of islands, influenced by plate tectonics, possibly from as early as the Mesozoic.
3. **Regression**: when the sea retreats from land masses that are being lifted up to altitudes above average sea level, marine littoral forms could have found an opportunity to penetrate into new, continental waters.

Clearly, in each specific case all available data should be considered together in order to arrive at plausible hypotheses with respect to that particular distribution. However, the various scenarios for individual taxa in the same region should not be contradictory: one area only has a single geological history, no matter how different an impact that may have had on different groups, all with their own possibilities and impossibilities.

Subterranean forms. – To conclude this section on crustaceans of continental waters, we can recognize that there is no doubt that **subterranean Crustacea** are from marine origin. However, features in their biology and of their geographical distribution have to be used to arrive at an acceptable scenario as to how they penetrated into fresh waters. In addition, the model should give a clue about the age of the transition, which, in any hypotheses involving a relict distribution, might well refer back to an ancient period of the Tertiary, or even to the Upper Cretaceous (Schminke, 1974; Stock, 1977a).

Thermal waters

Several crustaceans are found in thermal waters (Brues, 1924), which obviously requires a number of physiological adaptations. Random examples include:

1. **Copepoda**: *Cyclops quadricornis* (36°C).
2. **Ostracoda**: *Cypris balnearia* (40°C), *Darwinula malayica* (edge of a source at 45-50°C), *Potamocypris philotherma* (42-48°C, Nazare, Ethiopia).
3. **Syncarida**: *Thermobathynella adami* (up to 55°C, Zaïre).
4. **Thermosbaenacea**: *Thermosbaena mirabilis* (44-45°C).
5. **Isopoda**: *Thermosphaeroma* (four species, 27-43.8°C, Texas and Mexico); the phreatoicid *Hyperoedesipus plumosus* would have been collected in a "smoking" artesian well (cf. Nicholls, 1943-1944).
6. **Brachyura**: in Blanche Bay (New Britain, Bismarck Archipelago) Studer (1889) observed an *Ocypode* sp. living on and in littoral sands soaked with water of 50°C, and *Potamon fluviatile* was found in Algeria in waters of 36-40°C.

Subterranean waters

Though already impressive, the enumeration below represents only a concise review of the main groups of cavernicolous freshwater Crustacea:

1. **Cladocera**: some are troglobitic, for example *Ceriodaphnia cornuta* and *Daphnia ambigua* in caves on Cuba.

2. **Copepoda**: hypogean copepods are numerous, e.g., among Calanoida, in the family Diaptomidae (*Tropodiaptomus* (subgenus *Anadiaptomus*), *Microdiaptomus*); Cyclopoida, often very small, less than 0.5 mm (*Graeteriella*, *Speocyclops*, with the troglophilics, *Paracyclops*, *Diacyclops*, *Eucyclops*, etc.); Harpacticoida, with many genera, *Bryocamptus*, *Moraria*, *Elaphoidella*, *Parastenocaris*, etc., belonging to several families. Some of the hypogean copepods could be commensals, for instance *Attheyella pilosa* on the crayfish *Orconectes* (Kentucky, Indiana).
3. **Ostracoda**: hypogean ostracodes (*Cypridinopsis*, *Candona*, *Cryptocandona*, *Darwinula*, etc.) are generally more phreatobitic than actually cavernicolous. Some species are also commensals of Cambaridae in North America (*Cambarus*, *Procambarus*, *Orconectes*) where Entocytheridae have been found on their gills: *Entocythere*, *Sagittocythere*, *Uncinocythere*, *Dactylocythere*, and *Ankylocythere* (Rioja, 1942, 1943, 1949, 1955).
4. **Spelaeogriphacea**: one species (*Spelaeogriphus lepidops*), in a cave on the Table Mountain, Cape Town; another (*Potiicoara brasiliensis*), in the Bodoquena mountain range, Brazil. [From a total of four described species known worldwide.]
5. **Syncarida (Bathynellacea)**: Bathynellidae and Parabathynellidae, mesopsammic forms (with the exception of two species from Lake Baikal), are mostly phreatobitic: the first form discovered (Vejdovsky, 1880, *Bathynella natans*) came from a well in Prague (Czech Republic).
6. **Thermosbaenacea**: a monographic revision (Wagner, 1994) of this group, including morphology, taxonomy, phylogeny, and biogeography, has shown the presence of 34 species. These species are distributed into seven genera: *Thermosbaena*, *Monodella*, *Tethysbaena*, *Halosbaena*, *Limnosbaena*, *Tulumella*, and *Theosbaena*. The species of this group are always hypogean (endopsammic or cavernicolous) and in his monograph Wagner (1994) provides data on their partly Tethyan distribution and their habitats. (See also the chapter Thermosbaenacea in the present series.)
7. **Mysida**: many species are included in the genera *Antromysis* (Mexico, Antilles), *Spelaeomysis* (Mexico, Italy), *Stygiomysis* (Italy, Antilles), *Heteromysis* (Canary Islands), *Troglomysis* (in the area of former Yugoslavia), *Lepidops* (Zanzibar), etc.
8. **Isopoda**: **Anthuridae** include several hypogean species in the genera *Cruregens* (cf. Bowman, 1972; New Zealand) and *Cyathura* (Mexico and Cuba). Subterranean **Asellota** are often blind (Argano, 1977, 1979; Henry, 1978; Magniez, 1978, 1978a); they belong to four families: Asellidae (examples include *Asellus*, *Nipponasellus*, *Caecidotea* (= *Conasellus*), *Mancasellus*, *Lirceus*), Microparasellidae (*Microparasellus*, *Pseudoasellus*, *Microcharon*, *Angeliera*), Stenasellidae (*Stenasellus*, *Parastenasellus*, *Johannella*, *Magniezia*), and Stenetriidae (*Synasellus*, *Stygiasellus*). These asellotes are not fundamentaly "cavernicolous" but rather phreatobitic; they are able to colonize all types of waters, i.e., subterranean, karstic, hyporheic (waters in river sediments). **Cymothoida** include many hypogean genera, which all belong to the family Cirolanidae (*Antrolana*, *Ambolana*, *Cirolanides*, *Creaseriella*, *Faucheria* (cf. Bertrand, 1974), *Haptolana*, *Mexilana*, *Saharolana*, *Skotobaena*, *Specirolana*, *Sphaeromides*, *Troglocirolana*, *Typhlocirolana*). **Sphaeromatidea** are represented with at least three genera in

Sphaeromatidae (*Caecosphaeroma, Monolistra, Microlistra*). Phreatoicidae include a few hypogean species in the genera *Phreatomerus, Phreatoicus, Neophreatoicus* in New Zealand on the one hand, and *Nichollsia* in India, on the other.

Within **Oniscidea**, some 250 genera are cavernicolous; many belong to Trichoniscidae, and three forms are amphibious, or even (secondarily) aquatic: *Mexiconiscus laevis, Typhlotricholigioides aquaticus*, and *Cantabroniscus primitivus* (cf. Vandel, 1966).

9. **Amphipoda**: **Ingolfiellidea** (cf. Stock, 1977a) have a few phreatobitic or troglobitic species, for instance *Trogloleleupia* (formerly as *Ingolfiella*) in caves in central Africa, or *Ingolfiella fontinalis* (formerly also as *Gevgeliella*) in a spring on Bonaire. Gammaridea include a large number of genera with hypogean species: Gammaroidea (Typhlogammaridae, *Typhlogammarus*); Crangonyctoidea (Crangonyctidae, Neoniphargidae, *Crangonyx, Synurella, Stygobromus, Stygonectes, Pseudocrangonyx, Pseudoniphargus, Phreatoniphargus*); Niphargoidea (Niphargidae, *Niphargus*); Bogidielloidea (Bogidiellidae, *Bogidiella, Parabogidiella, Spelaeogammarus*); Eusiroidea (Calliopiidae, *Paraleptamphopus*); Liljeborgioidea (Sebidae, *Seborgia*); and Salentinellidae (*Salentinella*), Melitoidea (Hadziidae, *Hadzia, Metaniphargus, Meckelia, Teximeckelia*); Talitroidea (Hyalellidae, *Hyalella*).

10. **Astacidea** and **Anomura**: there are many hypogean species in North America: Astacidae (cf. Hobbs, Jr. et al., 1977) of the genera *Cambarus, Procambarus, Orconectes* and *Troglocambarus*). One Galatheoidea (*Munidopsis polymorpha*) lives in a marine cave in Lanzarote, Canary Islands, an another galatheoid, *Aegla cavernicola*, lives in a cave in Brazil.

11. **Brachyura** (cf. Delamare Deboutteville, 1976a, b; Cottarelli & Argano, 1977; Holthuis, 1979; Rodriguez, 1982): troglophilic crabs are common, for instance Grapsidae (*Sesarma*, etc.). A very few crabs are truly troglobitic: Potamidae (*Cerberusa caeca*, on Borneo), Gecarcinidae (*Adeleana chapmani*, slightly reduced eyes, Borneo), Pseudothelphusidae (*Typhlopseudothelphusa mocinoi, Typhlopseudothelphusa mitchelli* and *Typhlopseudothelphusa juberthiei*, blind, in Mexico and Guatemala), Trichodactylidae (*Trichodactylus mensabak*, blind, Mexico).

12. **Caridea**: cavernicolous shrimps are quite numerous, and the numbers of described species increases regularly. They belong to five families:
    - **Atyidae**: *Paratya* (New Zealand), *Typhlatya* (Cuba, Mexico, Galapagos Islands, Ascension), *Typhlopatsa* and *Parisia* (Madagascar), *Stygiocaris* (Australia), *Caridina* (*Caridina troglodytes*, New Ireland), *Antecaridina* (Fiji), *Troglocaris* (Europe), *Palaemonias* (U.S.A.), *Gallocaris* (endemic to France).
    - **Typhlocarididae**: *Typhlocaris* (Israel, Cyrenaica [= Libya], Italy).
    - **Palaemonidae**: *Cryphiops* and *Creaseria* (Mexico), *Palaemonetes* (U.S.A., Cuba), *Troglocubanus* (Cuba, Jamaica, Mexico), *Macrobrachium* (New Ireland, Assam, Mexico).
    - **Hippolytidae**: *Barbouria* (Cuba, etc.).
    - **Alpheidae**: *Alpheopsis* (Mexico).

As studies continue, the above list is likely to expand in the future.

Temporary waters

The ditches, pools and puddles, the temporarily drying ponds of temperate regions, like the lowland dhayas and gueltas of Saharian wadies, or the *Pfannen* (pans) of the Kalahari and of Namibia, all have a crustacean fauna adapted to prolonged periods of drought through the production of resting eggs (= diapausing eggs) that are resistant to desiccation. Anostraca, Notostraca, Cladocera, Ostracoda, and Copepoda are present in this type of environment. A syncaridan, *Koonunga cursor* (Australia), also has "resting" eggs, something only rarely encountered in Malacostraca.

Hypersaline water bodies

The classical biotope of *Artemia* spp. (Branchiopoda: Anostraca) comprises ponds and lakes of arid regions and deserts (see Sorgeloos et al., 1989) that obviously experience high evaporation. They can live in brines with a salinity of up to 340‰ and their encysted eggs, when ingested by flamingos, retain their faculty of development when they are extruded from the digestive tract of the bird, which permits considerably wide dispersal of this crustacean (Mac Donald, 1980).

Phytothelmes

The word phytothelme refers to small aquatic habitats on tropical plants such as inflorescences, tanks formed by leaves, and holes in trees. This last biotope may refer *sensu stricto* to the category of temporary waters and as such is comparable to biotopes like compost, humid soil, mosses. The fauna of these phytothelmes has been studied in particular on Bromeliaceae in tropical America, and also in South-East Asia. The results reveal rather rich faunas:

1. **Tropical American Bromeliaceae** (cf. Picado, 1913; Laessle, 1961; Hartnoll, 1964): Ostracoda: *Candonopsis anisitsi*, *Metacypris bromeliarum* and *Metacypris laesslei*, *Elpidium bromeliarum*. — Copepoda: *Tropocyclops prasinus*, *Elaphoidella sewelli*, *Ectocyclops phaleratus*. — Brachyura: the small grapsid species *Metopaulias depressus* lives in bromeliaceans of Jamaica up to 900 m altitude and reproduces on the plant itself, whereas other crabs of the same biotope, Grapsidae (*Sesarma bidentatum*, *Sesarma bromeliarum*, *Sesarma jarvisi*, *Sesarma verleyi*), Pseudothelphusidae (*Guinotia*, *Epilobocera*, *Pseudothelphusa*), or Trichodactylidae (*Trichodactylus*) reproduce in more extensive bodies of fresh water. Inside one *Aechmea paniculigera*, Laessle (1961) found no less than 30 *Metopaulias* (3 males, 2 females and 25 juveniles).
2. **South-East Asia (Indonesia)** — Inflorescences of *Zingiber macradenium*: the Copepoda *Epactophanes richardi* and *Phyllognathopus coecus* [as *Viguierella caeca*] — leaf tanks of *Colocasia*: the copepods *Attheyella ruttneri*, *Elaphoidella bromeliaecola*, *Elaphoidella elegans*, *Epactophanes richardi*, *Phyllognathopus coecus* [as *Viguierella caeca*], *Ectocyclops medius*, *Bryocyclops bogoriensis*; Brachyura: *Sesarma nodulifera* — leaf tanks of *Crinuma*: the copepod *Bryocyclops*

chappuisi — cultivated Bromeliaceae: the copepods *Bryocyclops bogoriensis, Bryocyclops anninae, Elaphoidella malayica, Elaphoidella bromeliaecola, Phyllognathopus coecus* [as *Viguierella caeca*] — on *Curtandra*: the copepods *Attheyella inopinata, Elaphoidella cornuta, Elaphoidella thienemanni, Epactophanes richardi, Phyllognathopus coecus* [as *Viguierella caeca*].
3. **Tree trunks, in Europe** — Ostracoda: *Candona pratensis, Candona compressa, Cyclocypris laevis*; Cladocera: *Alona affinis, Chydorus ovalis, Chydorus sphaericus*; Copepoda: *Tachidius brevicornis, Epactophanes richardi, Bryocamptus pygmaeus, Moraria arboricola, Diacyclops bisetosus, Diacyclops pulchellus*; Isopoda: *Asellus aquaticus* (one record from Denmark).

The tree trunk fauna is close to that of mosses. Some copepods, such as *Moraria arboricola*, are known from mosses as well as from tree trunks.

Muscicolous crustaceans

The habitat of muscicolous Crustacea, i.e., wet or at least humid mosses, is halfway between an aquatic and a terrestrial environment. Phytothelmes include free water, though sometimes temporarily; this is not the case in vegetation composed of mosses, such as *Sphagnum*, in which the habitat is only **moist**, but actually free water is absent. Muscicolous crustaceans are most often Copepoda: *Moraria* (*Moraria arboricola, Moraria monticola*, found up to 2800 m, *Moraria sphagnicola*), *Maraenobiotus, Epactophanes, Bryocamptus, Hypocamptus* (*Hypocamptus brehmi* up to 2400 m). Harding (1953) gave ranges for the hydrophily of copepods: *Moraria* and *Bryocamptus* are generally associated with damp or submerged mosses, but *Epactophanes* and *Maraenobiotus* were found in mosses that were only slightly moist; *Epactophanes muscicola* would avoid mosses that are really wet, and *Maraenobiotus vejdovskyi* is known from submerged plants as well as from mosses in forest litter, while the type-specimen of this species comes from mosses in Bohemia that get only wet from rain. Muscicolous Ostracoda are also known, for example in the genus *Mesocypris*. *Darwinula zimmeri* was found on Java among humid leaves of *Elatostemma*, and some actually terrestrial ostracodes exist as well, as is reported herein in the next section; some of these are humicolous, some even "arboricolous" (Menzel, 1923; Harding, 1955; Shornikov, 1980). Finally, a few Amphipoda are muscicolous and/or humicolous, like *Austroniphargus bryophilus* (Madagascar), *Talitrus* (*Talitroides*) *pacificus* (forest litter, Réunion, Comores, Madagascar), *Talitrus* (*Talitroides*) *topitotum* (forest litter, Indo-Pacific), and the anophthalmous *Hyale milloti*, living on damp mosses beside a waterfall on Moheli (Comores).

TERRESTRIAL HABITATS

In general, Crustacea have not been able to become independent of contact with water, either still requiring some form of direct contact, or at least needing the proximity of **accessible water**. This is in contrast to what we observe in many other Arthropoda. The extreme rarity or even complete absence of terrestrial Malacostraca in vast regions of the world has been explained by their competition with insects (Hexapoda; both as larvae and as imagines). This is in agreement with a relatively recent penetration of these crustaceans into fresh waters.

The relative failure of Recent malacostracans in conquering terrestrial habitats may be also explained in a physiological sense, i.e., through the persistence of a type of **respiration** that still requires an aquatic component, or at least a high level of **aerial humidity**: woodlice in the Sahara are only present near the sea shore, or on mountains.

In particular Oniscidea (and to a lesser extent the supralittoral Sphaeromatidae like *Campecopea*) present a series of biological adaptations ranging from those of the littoral *Ligia*, to the intricate features characterizing some woodlice of truly dry areas (Edney, 1954; Warburg, 1968). *Ligia simoni* is present at altitudes up to 1400 m in Colombia (Schmalfuss, 1978), and other *Ligia* spp. occupy similar habitats, although they belong to a primarily littoral genus but can yet reach tropical rain forests at considerable altitudes (*Ligia perkinsi* on Hawaii, *Ligia platycephala* in South America, *Ligia philoscoides* in Polynesia and *Ligia latissima* on New Caledonia). This case is especially interesting as an example of recent continental penetration by coastal forms facilitated by the presence of favourable **biotopes** and an available ecological **niche**.

Another example can be seen in the Azores where a species of *Jaera*, originally a genus of coastal, i.e., littoral marine Isopoda, colonizes fresh waters well up into the mountains, which might be possible in the absence of predators and competitors, for instance, the absence of gammarid Amphipoda (Veuille, 1979).

Generally speaking, only an extremely small group of Malacostraca has been successful in their attempts at making the transition from the water (whether the sea or fresh waters) to the land, i.e., in changing from water-dependent respiration to breathing air instead. Only one in fact, viz., the isopod family Oniscidae, can be considered fully adapted to a terrestrial life and to **aerial respiration**, the latter enabled by the presence of **pseudotracheae** in their pleopods. In many other instances (Anomura; Brachyura with the exception of Potamidae), the adaptation to terrestrial life is incomplete, for their larval development is still aquatic and necessitates regular migrations back to the sea for purposes of **reproduction**.

Several pathways can be imagined for the passage from the marine littoral to the terrestrial environment: from fresh water for ostracodes and potamids, from rocky coasts for grapsids, from muddy places and mangroves for gecarcinids, from sandy beaches for talitrids and ocypodids. Though obviously it is not easy to determine a relevant borderline between muscicolous or humicolous species, both still dependent on a high level of humidity, and the truly "terrestrial" forms, it is nonetheless possible to characterize terrestrial and semiterrestrial crustaceans in the following ways:

- **"Terrestrial" Copepoda**: These concern only incidental records: a copepodite IV of *Thermocyclops* sp. was reported from a site with wild bees in Betou (Congo) on 2 June 1980 (B. Dussart in litt., 9 September 1982) and the type of *Phyllognathopus camptoides*, a humicolous species, came from the cavity of a polydesmid diplopod (Bozic, 1965).

- **Terrestrial Ostracoda**: Many ostracodes are muscicolous or sphagnicolous; however, some terrestrial forms of Ostracoda are known. Litter from the forest of Knysna contained a *Mesocypris terrestris*, a humicolous species (Harding, 1943, 1955). It was obtained with a Berlese funnel, and found accompanied by a purely terrestrial microfauna (Acari, Arachnida, Insecta, Myriapoda, woodlice and talitrids). In the Salomon Islands, *Callistocypris zlotini* was found in humus, *Darwinula malayica* and *Terrestricypris arborea* in rotten wood, and all three species in the basis of an epiphytic fern at 2 m above the ground (Shornikov, 1980). In Indonesia, *Darwinula malayica* can allegedly be found down to 10 m deep in the soil (Menzel, 1923). Finally, two *Terrestricythere* spp. (*Terrestricythere ivanovae* and *Terrestricythere pratenesis*) were retrieved from supralittoral coastal and saline soils in Vladivostok (Shornikov, 1980).
- **Terrestrial Isopoda**: All terrestrial isopods are Oniscoidea, present everywhere in the world, except in the more arid parts of deserts and in polar regions, with the exception of an australian phreatoicoid, *Phreatoicopsis terricola* (whose true way of life remains enigmatic). The oniscoids occupy various environments, include 21 families that contain cavernicolous forms, and some genera live as commensals with ants. The best known myrmecophilic woodlouse is *Platyarthrus hoffmannseggii* (cf. Collinge, 1941; Brooks, 1942; Mathes & Strouhal, 1954). We may also mention *Platyarthrus acropyga*, an anophthalmous species that is commensal with the ant *Acropyga acutiventris*, *Cubaris granulatus* with *Bothroponera tesserinoda*, *Alloschizidium cottarellii*, also anophthalmous, and *Platyarthrus caudatus* with *Tetramorium brevicorne*.

  Some woodlice have a tendency to climb trees (Collinge, 1944). Vandel (1959: 135) formulated the hypothesis that rolling up into a ball, a behaviour frequently seen in woodlice, could represent an adaptation to dry climates, allowing the reduction of evaporation of water from the respiratory pleopods. This is a possible explanation in the case of Spelaeoniscidae, but also marine (*Campecopea*, etc.) and freshwater (*Skotobaena*) isopods are known to roll up themselves as an apparently regular part of their behaviour.
- **Terrestrial Amphipoda**: The only family having fully terrestrial species is Talitridae with 55 genera, mainly in the Indo-Pacific region on the Southern Hemisphere. Terrestrial amphipods are unknown in Europe and in most parts of Asia and the Americas (Grimmett, 1926; Williamson, 1951; Hurley, 1959, 1975; Bousfield, 1968, 1978). In actually terrestrial species, the following adaptations are present: a reduction of some structures (antennulae, palp of the maxillipedes, oostegites, pleopods), the development of the antennae, modifications of the male second gnathopods, the presence of lobate gills, a shortened urosome, and their jumping behaviour. Terrestrial amphipods have sometimes been reported from greenhouses in temperate zones: *Talitroides alluaudi*, *Talitroides hortulanus*, *Talitroides pacificus* [as *Talitrus*] (Calman, 1921; Stephensen, 1924).
- **Terrestrial Decapoda**: The atolls of Tokelau (Yaldwynn & Wodzicki, 1979) are inhabited by ten species of terrestrial decapods: four Anomura (three *Coenobita* spp. and the coconut crab, *Birgus latro*), six Brachyura, of which four Grapsidae (two *Geograpsus*, one *Metopograpsus*, one *Sesarma*) and two Gecarcinidae (two

Cardisoma). All these species have to return to the sea to allow their eggs to hatch. The coconut crab (Holthuis, 1959; Wiens, 1962; Cameron & Mecklunburg, 1973; Yaldwyn & Wodzicki, 1979) reaches 50 cm in length (Cuzent, 1884) and is able to climb a coconut tree with the greatest possible ease. Terrestrial Decapods sometimes undertake long mass migrations in order to reach the sea when their eggs are close to hatching: those of *Gecarcinus* or "tourlourous" in the Antilles, or of the famous "red crabs", *Gecarcoidea natalis* of Christmas Island are well known (Hicks et al., 1984).

"Flying" Crustacea

There exist no Crustacea that are really able to fly. Swarms of large numbers of the hyponeustonic copepod *Anomalocera patersoni* swimming close to the water-air interface have been reported to simulate rain on water. Another copepod, *Pontella mediterranea* is able to jump out of the water. Some cirripedes may accidentally be fixed on birds and be transported into air. This was reported for lepadids (*Lepas*: Targioni Tozetti, 1872; Giglioli, 1878; Roberts, 1948) and balanids (*Balanus*: Monod, in Ciurea et al., 1933).

COMMENSAL AND PARASITIC ASSOCIATIONS OF CRUSTACEA

The word **"associates"** corresponding to **"consortes"** (sing. "consors") of Pearse (1938) is used here (Hipeau-Jacquotte, 1972, 1974) in order to avoid a useless discussion about the terms phoresia, inquilinism, commensalism, mutualism, symbiosis, and parasitism. This subject, also, has been satisfactorily covered in contributions by, e.g., Caullery (1922), Brian (1931), Pearse (1938), Dales (1957), Noble (1964), and Monod (1976), and it has also been treated extensively in the chapter "Symbiosis and Parasitism in the Crustacea" in volume 3 of the present series (cf. Trilles & Hipeau-Jacquotte, 2012).

A large number of Crustacea is involved in **associations** of different types, and groups not including associating species are the following: Cephalocarida, Branchiopoda, Mystacocarida, Phyllocarida, Syncarida, Thermosbaenacea, Spelaeogriphacea, Cumacea, Tanaidacea, Euphausiacea, Reptantia: Scyllaridea and Astacidea, and Hippidea.

In Penaeoidea, the only instance reported would be a *Metapenaeopsis commensalis* found in Fiji inside a large sea anemone, but this was a single specimen and the association could have been accidental. On the other hand, the fact that some groups of Crustacea enter more often in associations than others, is well documented. Associating forms include many Copepoda, all Branchiura (invariably parasites), isopods in Cymothoida, and Caridea. In Nosy-Bé (Madagascar) for instance, 191 species of copepods were found on 275 marine invertebrates (Humes, 1963).

As far as Caridea are concerned, Alpheidae and specifically Pontoniinae, possess the larger number of consortes. There are some 300 species of Pontoniinae, about 150 of which are associated, with 36 genera out of a total of 44 (fig. 26.4). In Kenya, on a total of 67 species of shrimp, 37 (57%) are associated, viz., 3 with sponges, 29 with cnidarians, 2 with molluscs and 4 with echinoderms (Bruce, 1976).

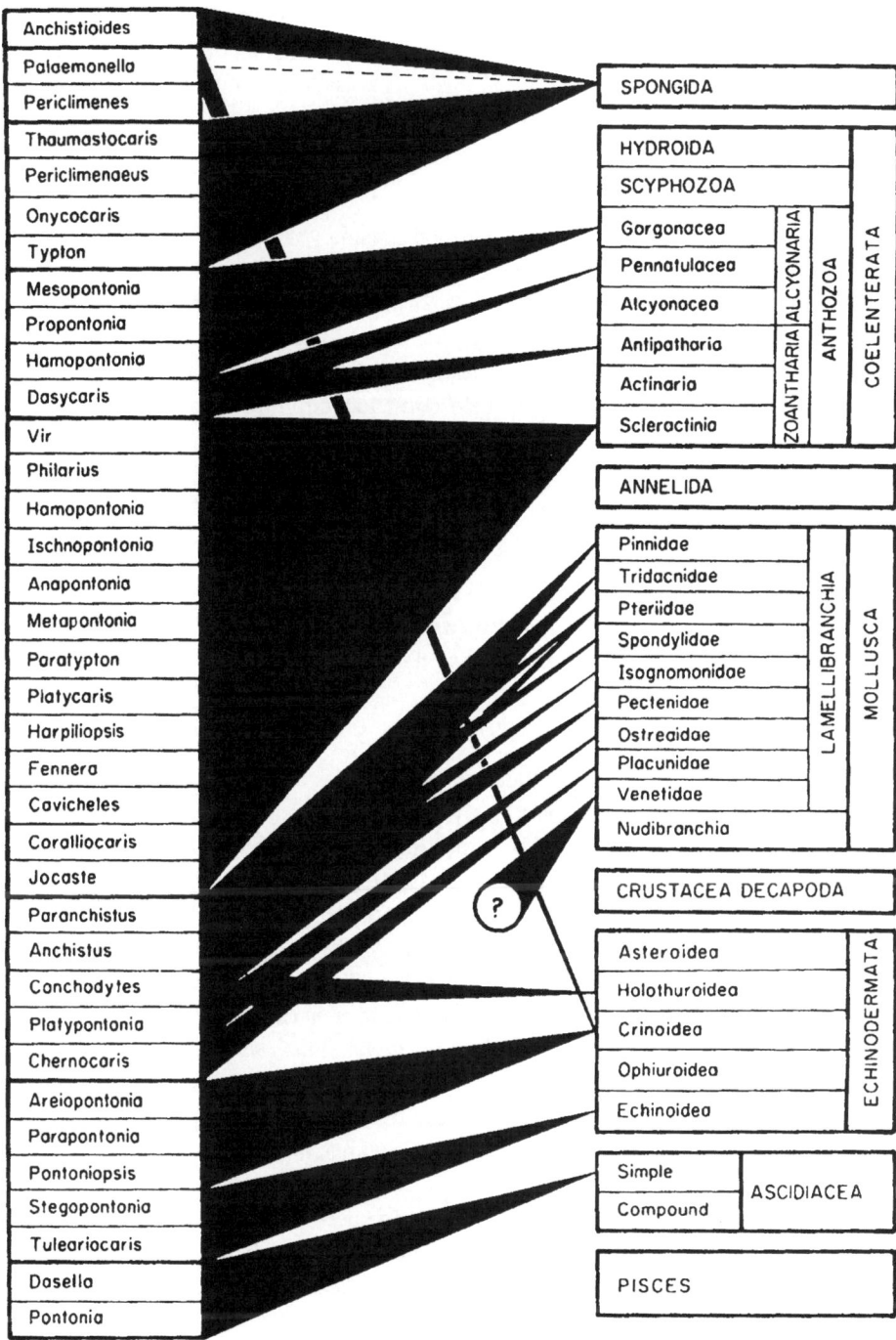

Fig. 26.4. Relationships between commensal genera of Indo-West Pacific Pontoniinae (except *Periclimenes*) and their hosts. [After Bruce, 1976.]

Various types of arrangements could be proposed in order to present an overview of the different phenomena encountered: according to type of association, by taxa of crustaceans, or by groups of associated organisms, i.e., **hosts**. This last mode of presentation has been adopted herein.

ALGAE

Young instars of two harpacticoid copepods, namely *Thalestris rhodymeniae* and *Diarthrodes feldmanni*, are able to make galleries in the fronds of various Rhodophyceae (*Rhodymenia, Rhodophyllis, Stenogramma, Polyneura, Cryptopleura, Nitophyllum*) where they induce the formation of **galls** and within which adults are found, with, occasionnaly, other organisms (Halacari, Nematoda). These gallicolous Copepoda were reported from Scotland and from the English Channel at Roscoff (Bocquet, 1953; Harding, 1954).

The copepod *Myzopontius australis* (Artotrogidae), found in Mauritius in the green alga *Codium arabicum* can be added here (Stock, 1966).

FORAMINIFERA

The only examples known so far are those of the cirripedes *Megatrema madreporarum* [as *Pyrgoma stockesi*] and *Amphibalanus amphitrite*, settled on *Miniacina miniacea*, as well as *Arcoscalpellum botellinae* on *Botellina pinnata*.

PORIFERA

In an important 1933 paper, Arndt described 464 cases of associations between Crustacea and Porifera, which encompass many groups of crustaceans: as an example, involving 11 genera with 21 species for **Amphipoda** alone. Many other examples have been reported since, in particular in **Copepoda**: *Hemicyclops perinsignis* with *Agelas* in Madagascar (Humes, 1973), *Asterocheres* and *Acontiophorus* with *Halichondria, Oscarella, Haliclona* and *Pericharax* in Mauritius (Stock, 1966), *Apodomyzon brevicorne* and *Apodomyzon longicorne* with *Haliclona* (cf. Stock, 1970), *Asterocheres parvus* and *Acontiophorus scutatus* with *Dysidea fragilis*, *Psilomyzon pauciseta* in *Ircinia, Tuphacheres micropus* in *Dysidea*, *Sestropontius bullifer* and *Pteropontius cristatus* in *Hymeniacidon, Cryptopontius capitalis* and *Cryptopontius minor* in *Petrosia, Entomopsyllus adriae* in *Verongia*, etc. Two families of Copepoda are exclusively parasites in sponges: Sponginticolidae with *Sponginticola* and Spongiocnizontidae with *Spongiocnizon* (Stock & Kleeton, 1963).

Many **Cirripedia: Thoracica**, such as *Acasta*, frequently embed in sponges.

Also, quite a number of **Tanaidacea** and **Isopoda** find refuge in the cavities of sponges: Gnathiidae, Sphaeromatidae, Cirolanidae, Aegidae (*Aega spongiophila* in *Euplectella aspergillum*).

Among **Amphipoda**, the examples of *Colomastix pusilla* in *Suberites domuncula* and of *Tritaeta gibbosa* in *Halichondria* and *Suberites* are well known (Bacesco & Mayer, 1960). *Anamixis pacifica* and *Anamixis linsleyi* were reported to live in marine sponges.

Sponges may also host **Paguroidea**, such as *Paguristes, Pagurus, Dardanus, Cheiroplatea, Pylocheles, Trizocheles*, or Stenopodidae, for instance *Spongicola venustus*, which lives as couples in *Euplectella*. Some Axiidae and Upogebiidae have a similar habitat. Other sponges (*Chelina, Halichondria, Halisarca*, etc.) can cover the carapace in many crabs of the taxa **Majidae** and **Dromiacea**. Finally, some *Cliona* can attack the tests of balanids.

A number of pontoniine shrimp (**Pontoniinae**: *Anchistioides, Thaumastocaris, Periclimenes, Periclimenaeus, Onychocaris, Typton*), as well as **Alpheidae** (*Synalpheus, Alpheus*), a few **Hippolytidae** (*Lysmata, Gelastocaris*, etc.) and **Disciadidae** (*Discias exul* on *Acarnus ternatus* in Kenya) are associated with sponges.

Hydrozoa

Some Copepoda (Lichomolgidae: *Macrochiron* spp. on *Aglaophenia, Gymnangium, Lithocarpus*, etc., and *Telestacicola* on *Aglaophenia*) and one shrimp (*Rapipontonia galene* on *Aglaophenia*) are associated with Hydroida. In Parang (Moluccas), the same host, *Aglaophenia cupressina*, is able to host as much as four different species of copepods.

One amphipod (*Metopa borealis*) and one isopod (*Astacilla longicornis*, fig. 26.5) were found associated with hydrozoans of the genus *Tubularia* (cf. Vader, 1972b), while there are also reports on Stenothoidae associated with Hydroida (*Metopa alderi* on the hydromedusae of *Tima bairdii*, Europe, and *Metopa borealis* on *Clytia*, Europe).

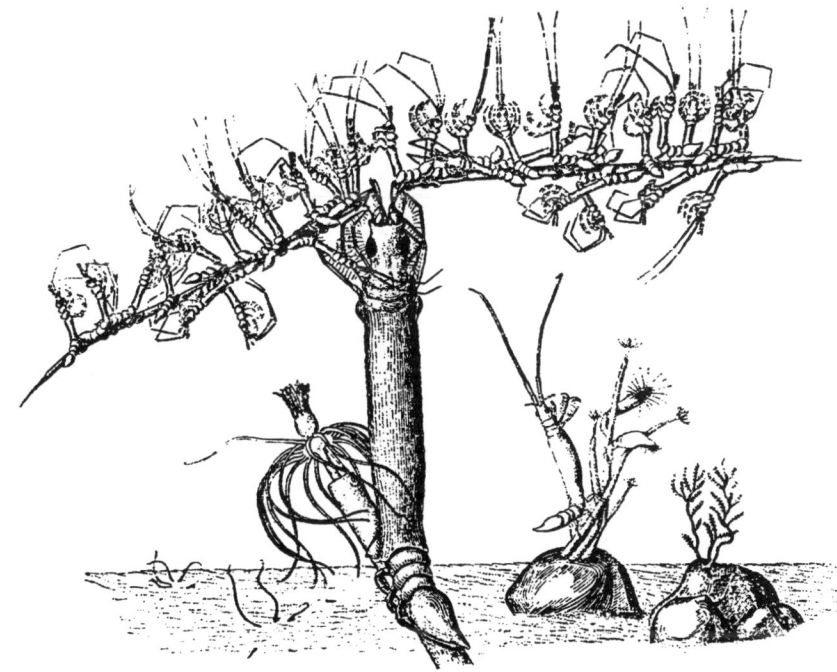

Fig. 26.5. A (female) specimen of *Astacilla longicornis* carrying its juveniles on its antennae, while itself holding on to the hydrozoan *Tubularia indivisa*. [After an original drawing by G. Thorson.]

Hydractinia echinata often covers the mollusc shell that various species of Paguroidea carry (André & Lamy, 1936).

Stylasterine corals (*Conopora, Crypthelia, Stylaster*) can have **galls** occupied by copepods like *Cecidomyzon, Oedomyzon, Cystomyzon,* and *Hammatimyzon* (Asterocheridae); the galls described by Zibrowius (1981) contain adult females that are definitively captive. Some Galatheoidea (*Eumunida*) would also live in association with Stylasterina (cf. de Saint-Laurent & Macpherson, 1990).

SIPHONOPHORIDA

The copepod *Nogagella siphonophoriae* is known from *Physophora hydrostatica* at Algiers (Rose, 1933), the cirripede *Dosima fascicularis* [as *Lepas*] from *Velella*, an hyperiid amphipod (*Primno* sp.) from the hind nectophores of *Abylopsis tetragona* and of *Sculeolaria chuni* (cf. Bowman, 1978), while the small crab *Planes minutus*, so frequent on various floating objects, was found on *Velella* and *Porpita*.

SCYPHOZOA

Associations between jellyfish and Crustacea are relatively scarce. However, the following groups are involved:

1. **Copepoda**: *Paramacrochiron japonicum* on *Thysanostoma thysanura*, *Paramacrochiron ennorense* on "jellyfish", *Paramacrochiron rhizostomae* on *Rhizostoma* sp., *Paramacrochiron sewelli* on *Lychnorhiza malayensis*, *Pseudomacrochiron stocki* on *Dactylometra quinquecirrata*, *Sewellochiron fidens* on *Cassiopea xamachana*, *Nitokra medusaea* on *Aurelia* sp. (Reddiah, 1968, 1969; Humes, 1969, 1970a).
2. **Cirripedia**: *Alepas* sp. (Monod, 1933).
3. **Amphipoda**: Associations between many species of Hyperiidea and jellyfish (*Cyanea, Chrysaora, Rhizostoma*, etc.) have been known for quite some time already. In contrast, associated Gammaridea are rather rare, and the only examples from this group are quoted here (cf. Vader, 1972b): Amphilochidae (*Gitanopsis* aff. *pusilla* on *Acromitus*, in India; and *Hoplopleon medusarum*, in South Africa); Acanthonotozomatidae (*Panoplea eblanae* on *Rhizostoma*, Europe); also Caprellidae (*Pariambus typicus* on *Rhizostoma*), which may, however, be accidental.
4. **Decapoda**: In Japan, three caridean shrimps were found on the jellyfish *Mastigias papua*: *Chlorotocella gracilis* (Pandalidae), *Latreutes anoplonyx* and *Latreutes mucronatus* (Hippolytidae) as well as *Ancylomenes holthuisi* [as *Periclimenes*] on *Cassiopea* (cf. Hayashi & Sadayashi, 1968; Bruce, 1972b). Some phyllosomas (larval Palinuroidea) were reported several times as associated with jellyfish, as well.

OCTOCORALLIA

Quite some copepods have been decribed from Octocorallia (Pennatulacea, Gorgonacea, Alcyonacea), especially Lichomolgidae, in which the genus *Lichomolgus* that

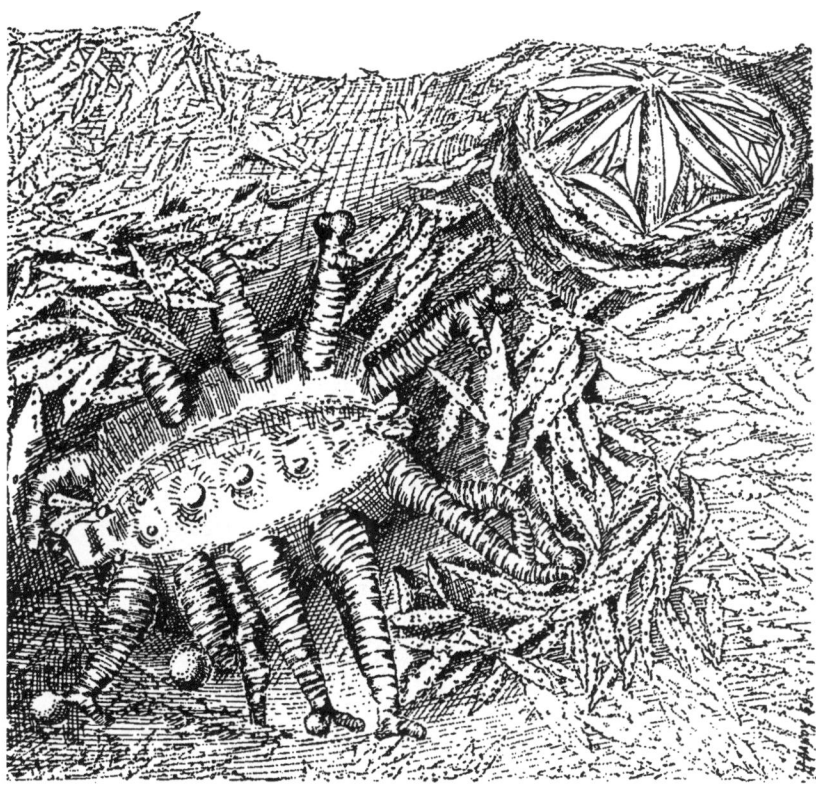

Fig. 26.6. A female *Linaresia mammillifera* (Copepoda: Lamippidae) *in situ* in a polyp of the gorgonian *Paramuricea clavata* (Anthozoa: Alcyonacea). [After Bouligand, 1960.]

comprised some forty species, has subsequently been split into approximately thirty new genera (Humes, 1957a, 1973a, b, c, 1974a, b, 1975, 1978; Bouligand, 1960, 1960a; Stock, 1963, 1975, 1978; Humes & Bruce, 1964; Humes & Ho, 1967, 1968a, b, c; Humes & Stock, 1973). Lichomolgids were reported:

1. In **Pennatulacea**: the genera *Doridicola, Pennatulicola, Lamippe*.
2. In **Gorgonacea**: with Lamippidae (genera *Lamippe, Magnippe, Isidicola, Linaresia*, etc.) (figs. 26.6, 26.7), the genus *Gasterocheres*, and many lichomolgids (*Doridicola* [as *Metaxymolgus*], *Paramolgus, Acanthomolgus*). Some of these copepods inhabit galls in the cortex of their host. An abyssal species, *Lamippe bouligandi*, is parasitic in an abyssal gorgonian of the North Atlantic, *Anthoptilum murrayi* (cf. Laubier, 1971).
3. In **Alcyonacea**: some Lamippidae and a large number of Lichomolgidae, the latter with some fifteen genera: *Anisomolgus, Doridicola* [as *Metaxymolgus*], *Acanthomolgus*, and *Paramolgus*, etc.

Cirripedes often settle on Octocorallia, for instance *Acasta* on Alcyonacea, *Octolasmis* on *Pennatula* and many genera (*Scalpellum, Smilium, Pachylasma, Balanus, Conopea,*

Acasta, Verruca) on the sea fan (Gorgonacea), while one Ascothoracida, genus *Gorgonolaureus*, lives in the cortex of one of them, *Paracis squamata*, and *Baccalaureus* parasitizes Zoantharia.

The sphaeromatid isopod *Cymodopsis gorgoniae* was described associated with an Octocorallia, and the crab, *Caphyra alcyoniophila*, lives embedded in an alcyonarian of Viet-Nam.

Majid crabs and species of the genera *Uroptychus* and *Galathea* are also sometimes associated with Octocorallia.

Various pontoniine shrimp are associated with Octocorallia as well (Bruce, 1970): *Mesopontonia*, *Balssia*, *Homodactylus*, *Neopontonides*, *Periclimenes*, and *Veleronia*, and so is a hippolytid, *Hippolyte commensalis*.

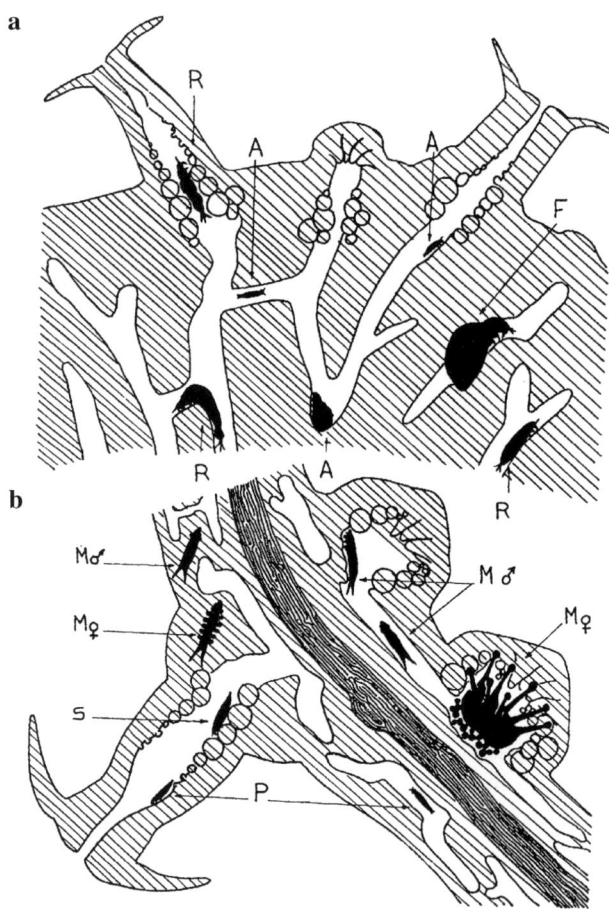

Fig. 26.7. Location of lamippid Copepoda in their hosts: a, in *Alcyonium palmatum*: A, *Lamippina aciculifera*; F, *Lamippella faurei*; R, *Enalcyonium rubicundum*; b, in *Paramuricea clavata*: M, *Linaresia mammillifera*; P, *Enalcyonium parvum*; S, *Enalcyonium setigerum*. [After Bouligand, 1960.]

ACTINIARIA

The various crustaceans associating with sea anemones can be enumerated as follows:

1. **Copepoda**: many genera and species of copepods live on and in Actiniaria. One very special instance is that of the calanoid *Ridgewayia fosshageni*, which is associated on the Atlantic coast of Panama with *Bartholomea annulata*, though without direct contact with the sea anemone. All other copepods in these associations are cyclopoids from several families: Lichomolgidae (*Doridicola* with *Actinia equina*, *Anemonia sulcata*; *Doridicola* [as *Metaxymolgus*] with *Radianthus*; *Paramolgus* with *Rhodactis*; *Aspidomolgus* with *Stoichactis* and *Homostichanthus*; *Indomolgus*, *Paranthessius*); Asterocheridae (*Asteropontius* with *Stoichactis* and *Condylactis*, *Asteropontides*); Antheacheridae (*Antheacheres duebenii* inducing galls in *Bolocera*; *Staurosoma parasiticum* in *Anemonia sulcata*), *Diaponticus* on *Anemonia sulcata*; finally, the enigmatic *Mesoglicola delagei* in *Corynactis viridis* (cf. Quidor, 1906; Bocquet & Stock, 1959; Haefelfinger & Laubier, 1965; Humes & Ho, 1967a; Humes, 1969a; Vader, 1970, 1970a; Humes & Smith, 1974).
2. The **ascothoracid cirripede**, *Petrarca bathyactidis* parasitic in *Bathyactis*.
3. **Amphipoda** are also more or less associated with Actiniaria, such as *Boeckosimus normani* [as *Onisimus*] with *Aristias neglectus*, *Allogaussia recondita* with a colonial Actiniaria from California (*Anthopleura elegantissima*), *Metopa solsbergi* with *Actinoloba dianthus* and *Elasmopus calliactis* with *Calliactis armillata* fixed on a shell of *Tonna* occupied by a pagurid crustacean in Hawaii (Edmondson, 1951). Vader (1983) published a review of the amphipods associated with Actiniaria: 22 species belonging to seven families, found on 20 species of sea anemones (eight families).
4. One **Mysida**, *Heteromysis actiniae*, lives among the tentacles of *Bartholomea annulata* (cf. Clarke, 1955).
5. The associations of **Paguroidea** and sea anemones are common along European coasts and well known: *Pagurus bernhardus* and *Dardanus arrosor* with *Calliactis parasitica*; *Pagurus prideaux* with *Adamsia palliata*, etc. In some instances the pagurid, without shell, is more or less covered by the sea anemone: *Paguropsis typica* or *Sympagurus pictus* make fine examples. We may also mention *Paracalliactis* associated with *Anapagurus* and *Parapagurus* (cf. Dechancé & Dufaure, 1959) or *Neoaiptasia commensali* living in Bombay with *Diogenes* and *Clibanarius* (Parulekar, 1969).
6. Some **Porcellanidae** can also be associated with large tropical sea anemones (*Stoichactis*); this is the case for *Petrolisthes maculatus* and *Neopetrolisthes oshimai* (cf. Gordon, 1960; Johnson, 1960).
7. One more exceptional case is that of the **crab**, *Mithraculus cinctimanus* found in the Antilles on the large sea anemone *Stoichactis helianthus* (cf. Manning, 1970; Patton, 1979). The fact that some small crabs can hold a sea anemone in each of their chelae, as *Lybia* does with *Triactis producta*, etc., *Polydectus cupulifer* with *Phellia* sp., or *Telmactis decora* and *Hepatus* sp. with *Actinoloba*, has now been known for a while.

8. Finally, some **shrimps** have to be mentioned here, for instance *Alpheus armatus* on *Bartholomea*, *Periclimenes brevicarpalis* on *Stoichactis* and *Actinodendron*, and *Periclimenes rathbunae* on *Stoichactis*, so as also *Thor amboinensis*. In Japan, Suzuki & Hayashi (1977) noted the hippolytid *Thor amboinensis* on *Parasicyonis*, *Radianthus* and *Stoichactis*, and the pontoniines *Hamopontonia* (on *Parasicyonis*), *Periclimenes brevicarpalis* (on *Parasicyonis* and *Radianthus*), *Periclimenes ornatus* (on *Parasicyonis*) and *Ancylomenes holthuisi* (on *Parasicyonis*, *Radianthus* and *Dofleinia*). The actual relationship with the sea anemone might be closer to parasitism than to commensalism, in the cases of *Periclimenes brevicarpalis*, *Periclimenes ornatus* and *Hamopontonia*.

Madreporaria [= Scleractinia] — corals

In the case of scleractinians, which essentially, together with hermatypic corals, constitute the builders of **coral reefs**, it is, in fact, the reef as a whole that forms a series of micro-habitats leading crustaceans to inquilinism or parasitism. Because of the wealth of data, we need to give here a concise review only of the groups entering into such types of **associations**. The fauna associated with corals is very rich: a single species, *Pocillopora damicornis*, studied in Australia, appeared to be connected with no less than 13 species of decapods: four swimming shrimp, three benthic shrimp, and six crabs (Patton, 1974), and for the area of Ceylon-Maldives Islands, 16 shrimp and 15 crabs (Garth, 1974). Despite the restrictions here imposed, the lists are nonetheless impressive:

1. **Copepoda**. The copepods inhabiting coral reefs belong to six families:
   – Asterocheridae, with *Asteropontius* (on *Pocillopora*, *Montipora*, *Stylophora*), *Bradypontius* (*Platygyra*), *Cholomyzon* (*Dendrophyllia*), *Monocheres* (*Pocillopora*), *Peltomyzon* (*Montastraea*), *Pteropontius* (*Echinopora*).
   – Clausiidae, with *Indoclausia* (*Montipora*), *Stockia* (*Favia*).
   – Corallovexiidae, with *Corallovexia* (*Montastraea*, *Acropora*, *Diploria*).
   – Lichomolgidae, with about 25 genera and 55 species (*Lichomolgus*, *Anchimolgus*, *Odontomolgus*, etc.) on many corals (*Pavona*, *Montipora*, *Psammocora*, *Porites*, *Favia*).
   – Pseudanthessiidae, with *Kombia* (*Psammocora* and *Porites*), *Rhynchomolgus* (*Psammocora*).
   – Xarifiidae, with two genera, *Xarifia* and *Orstomella* and about 25 species on *Acropora*, *Pocillopora*, *Favia*.
   About one hundred species of corals are known as **hosts** for copepods: a most remarkable diversity (Humes, 1960, 1962a, b, 1973c, 1974a, b, 1978, 1979; Humes & Ho, 1967, 1968c, d; Stock & Humes, 1969; Stock, 1975a).
2. **Cirripedia** (barnacles). Two groups of Cirripedia have species associated with corals: the order Thoracica, suborder Balanomorpha, and the order Acrothoracica or burrowing barnacles. Balanidae include only a few species associated with corals, for instance *Megabalanus stultus* on *Millepora*, *Hexacreusia durhami* [as

Armatobalanus] on *Porites, Tetraclita* sp. on *Heliopora*. The family Pyrgomatidae comprises symbionts or obligatory parasites in scleractinians (Hiro, 1938; Ross & Newman, 1973; Newman & Ladd, 1974); the family includes some 17 genera, among which *Cantellius, Hiroa, Creusia, Nobia, Pyrgoma, Savignium, Hoekia, Pyrgopsella, Boscia* (with *Boscia anglicum*, well known in the English Channel).

Acrothoracicans, the burrowing barnacles, often drill holes in the shells of molluscs (Tomlinson, 1969); some of them can inhabit corals, for example *Weltneria* in *Psammocora* and *Porites*, *Lithoglyptes* in *Acropora, Heliopora, Kochlorine* in *Acropora, Berndtia* in *Leptastraea* and *Psammocora, Balanodytes* in *Distichopora*, and *Cryptophialus* in *Acropora* and *Distichopora*.

3. **Brachyura**. Many crabs live in and on coral reefs. Some Majidae (species of *Paratymolus, Schizophrys*, and *Tylocarcinus*), one Parthenopidae (*Harrovia elegans*) and some Portunidae (*Thalamita pilumnoides, Thalamitoides quadridens*) are associated with the coralligenous substratum, and many Xanthidae are as well.

Among these, some crabs such as *Chlorodiella nigra, Phymodius ungulatus, Actaea superciliaris* are not found exclusively among corals, but other crabs like *Domecia hispida, Cymo andreossyi* and species of *Trapezia* inhabit Pocilloporidae and *Tetralia*, while *Cymo melanodactylus* and *Domecia glabra* dwell on Acroporidae. The most noticeable brachyurans certainly are those of the family Hapalocarcinidae, which induce true galls on madreporarians where they live. Some species like *Cryptochirus, Pseudocryptochirus* (fig. 26.8) or *Troglocarcinus* occupy simple lodges, from shallow holes to more deeply cut out cavities in the coral (*Turbinaria, Fungia, Oxypora*). Other species like *Hapalocarcinus marsupialis* or *Pseudohapalocarcinus ransoni* induce the formation of true galls, i.e., stimulate the coral host to grow and thereby build a lodge for them (on *Pocillopora, Seriatipora, Sideropora, Pavona*) (Shen, 1936; Fize & Serène, 1955, 1957; Fize, 1956; Garth, 1964, 1965, 1974; Garth & Hopkins, 1968). Hapalocarcinids from the Annam sea shore are described in detail and illustrated in colour in the monograph by Fize & Serène (1957).

4. **Caridea**. Several families of shrimp have species associated with corals (Bruce, 1972, 1975, 1976, 1976a, 1977, 1979; Garth, 1974, 1974a): many Pontoniinae (*Periclimenes madreporae, Periclimenes consobrinus, Periclimenella petitthouarsii, Ancylomenes magnificus, Harpilius lutescens, Vir, Paratypton, Platycaris, Coralliocaris, Jocastes*), and Alpheidae (*Synalpheus, Racilius, Alpheus*). The species *Paratypton siebenrocki* is highly modified: male and female live together firmly encysted in a lodge in a colony of *Acropora* (Bruce, 1969, 1977) (fig. 26.9); *Ctenopontonia cyphastreophila* lives in the furrow on the surface of the coral.

Other species are either flattened (*Platycaris*), or compressed (*Ischnopontonia lophos, Racilius compressus*). The subfamily Pontoniinae is particularly well represented among shrimps associated with Scleractinia, with 15 genera and 30 species found on 13 genera and 60 species of the corals.

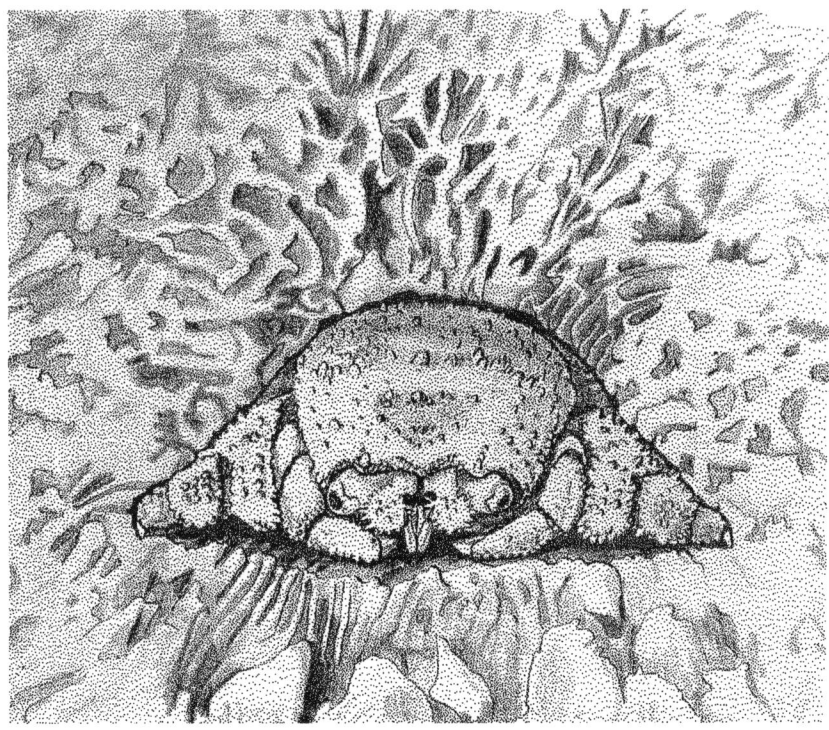

Fig. 26.8. The hapalocarcinid crab *Pseudocryptochirus crescentus* in the entrance of its lodge in the madreporarian *Pavona gigantea*. [After Garth & Hopkins, 1968.]

Antipatharia (black corals)

On these hosts (or substratum) we find the following crustaceans:

1. **Copepoda**: *Vahinius* spp. on *Cirripathes* and *Stichopathes*, *Paramolgus* spp. and *Thammolgus* on *Antipathes* (cf. Humes, 1967, 1969b, 1979a).
2. **Cirripedia**: Pedunculata: *Oxynaspis*, for example *Oxynaspis aurivillii* on a flagelliform antipatharian from the Marquesas Islands (Monod, 1979), and *Octolasmis*; also some Ascothoracida (*Laura gerardiae*, *Baccalaureus japonicus*, *Synagoga mira*).
3. **Decapoda: Caridea**: some Pontoniinae, like *Pontonides*, *Dasycaris*, *Periclimenes* (cf. De Ridder & Holthuis, 1979; Monod, 1979) and a Pandalidae, *Anachlorocurtis commensalis* on a species of *Antipathes* from Japan (Hayashi, 1975).
4. **Brachyura**: *Quadrella cyrenae* from the Marquesas Islands, on *Oxynaspis* and *Pontonides* (cf. Monod, 1979). A mimetic Majidae, genus *Xenocarcinus*, is also associated with an antipatharian coral.

Zoantharia [= Hexacorallia]

Many lichomolgid copepods of the genera *Temnologus*, *Doridicola* and *Pseudomolgus* were found on zoantharians (*Palythoa* spp.) from Madagascar. Some Ascothoracida

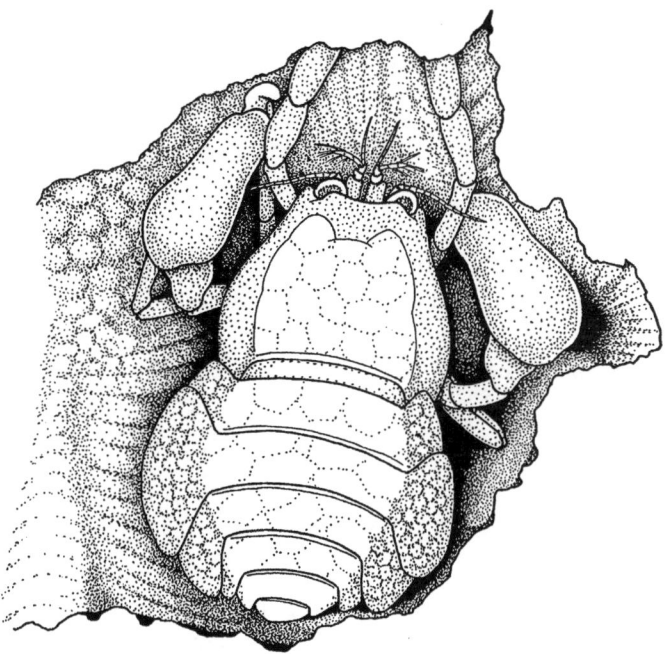

Fig. 26.9. An ovigerous female of *Paratypton siebenrocki*, *in situ* in a cyst of *Acropora*. [After Bruce, 1977.]

(*Baccalaureus maldivensis* and *Baccalaureus argalicornis*) live in *Palythoa* and the association of various zoantharians with Paguroidea, of which they partly cover the shell, is well documented: *Epizoanthus paguropsidis* and *Epizoanthus paguriphilus*, *Cyclactinia*, etc. *Diogenes ovatus* was found associated with *Palythoa senegambiensis* (cf. André & Lamy, 1939).

Turbellaria, Echiura, Sipunculida and Nemertea

Some incidental observations from miscellaneous groups of associates of Crustacea are here assembled:

1. **Copepods**: The lichomolgid copepod *Pseudanthessius latus* was found associated with the marine **polyclad turbellarian** *Cryptophallus magnus* (Illg, 1950).
2. Two **copepods** were found in the digestive tract of **Echiura** (*Goidelia* and *Echiurophilus*). Several crabs of the genus *Pinnixa* are commensal with Echiura in their galleries. For example, in California *Pinnixa franciscana* is associated with *Urechis caupo*; *Pinnixa lunzi* probably has a similar habitat (Glassell, 1937).
3. Two ectoparasitic **copepods** were found on the **sipunculidan** *Sipunculus nudus*: the clausidiid *Myzomolgus stupendus* and the catiniid *Catinia plana*. Both have, on the antennae, a disc for attachment on their host (Bocquet & Stock, 1957).
   Other copepods from the order **Poecilostomatoida** were described from on or in Sipunculida, for example *Ventriculina* and *Heliogabalus*, family Ventriculinidae.

4. Gallien (1935) described the **copepod** *Pseudanthessius nemertophilus* from the **nemertean** *Lineus longissimus* collected in the English Channel.

BRYOZOA

Barnacles are not uncommon on bryozoans with firm colonies; Pedunculata can use these as a site for fixation, for example *Scalpellum hispidum* on *Sarsiflustra abyssicola*, or *Scalpellum cancellatum* and *Scalpellum subalatum* on other calcified Bryozoa.

A burrowing acrothoracic cirripede, *Kochlorinopsis*, lives in *Discoporella umbellata*. Some instances are known of Bryozoa covering shells occupied by pagurids: *Conopeum commensale* in that way connected with *Pseudopagurus granulimanus*, *Cellepora senegambiensis* with *Pagurus alcocki* (fig. 26.10) and *Diogenes pugilator*, *Janaria mirabilis* with *Pagurus varians* (cf. André & Lamy, 1936; Monod & d'Hondt, 1978).

BRACHIOPODA

Some barnacles, for example *Balanus crenatus* and *Verruca stroemia*, occur on *Rynchonella psittacea*, as do specimens of an amphipod, *Aristias neglectus*, found in the branchial cavities of *Terebratulina caputserpentis* and of *Macandrevia cranium* (cf. Vader, 1970).

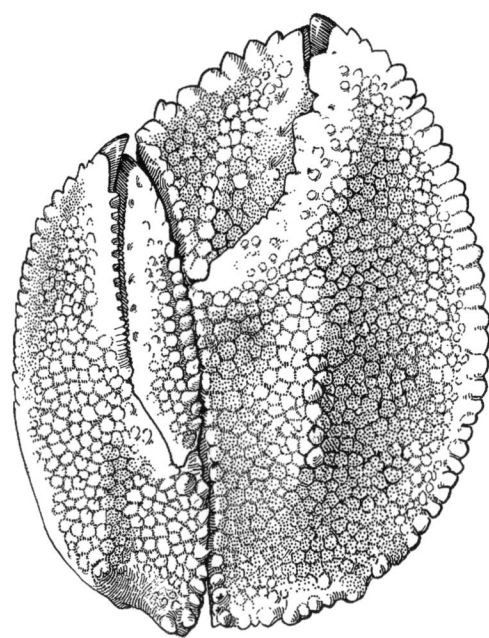

Fig. 26.10. *Pagurus alcocki*, dorsal view of the distal segments of its chelipeds: these modified segments together build an operculum that perfectly closes the entrance of its lodge in its host, a bryozoan. [After Forest & Ngoc-Ho, 1992.]

MOLLUSCA

Between the many species of the large phylum Mollusca and the likewise speciose, and hence large, arthropod superclassis Crustacea, we find, accordingly, large numbers of mutual associations. We can only indicate those in general terms within the framework of the following, necessarily concise, account.

1. **Copepoda**: About 70 genera of parasitic copepods are known to live on or in molluscs; they belong mostly to Cyclopoida and less frequently to Caligoida, Harpacticoida, Lernaeoida, or Monstrilloida, and never to Calanoida nor Notodelphyoida. Lichomolgidae is the most important family parasitic in molluscs, followed by Myicolidae, Clausidiidae, Mytilicolidae and even a few Ergasilidae and Ventriculinidae, as well as one Caligidae (on *Nautilus*).

    Genera on or in **Gastropoda: Prosobranchia** (*Epimolgus, Trochicola, Cerastocheres, Paranaeitis, Monstrilla, Myzotheridion*, etc.) are scarce. Much more genera are present on **Opisthobranchia** (*Lichomolgus, Doridicola, Anthessius, Artotrogus, Splanchnotrophus, Micrallecto, Nannalecto, Briarella*, etc., with either vagile (mantle cavity, gills), or endoparasitic species.

    Many genera are found on **Lamellibranchia**, internally in the intestine like *Mytilicola* or *Ostreicola*, or externally (*Modiolicola, Ostrincola, Pseudomyicola, Conchyliurus, Myocheres, Paranthessius,* and *Myicola*). Parasitic specificity is low: the genus *Paranthessius* is known from a dozen hosts (*Pecten, Mactra, Cardium, Chlamys, Panopea, Tivela,* and *Saxidomus*), whereas *Pseudomyicola* was found on species from a large number of families (Arcidae, Pinnidae, Pteriidae, Ostreidae, Spondylidae, Pectinidae, Mytilidae, Chlamidae, Diplodontidae, Cardiidae, Veneridae, Sanguinolariidae).

    In contrast, a single genus may have several species of parasites: *Tridacna* and *Hippopus* associated with *Anthessius solidus, Anthessius amicalis, Anthessius alatus, Anthessius discipedatus, Lichomolgus tridacnae* and *Lichomolgus hippopi*. Species of *Teredo* also are associated with Copepoda, whose morphology is often strongly modified (*Teredicola, Teredoika*).

    As far as **Cephalopoda** are concerned, only a few copepods have been reported so far: *Octopicola* spp. (on *Octopus*), some representatives of the subfamily Cholidyinae (Harpacticoida) as *Cholidya polypi* (on *Benthoctopus*), *Cholidyella* (on *Benthoctopus* and *Opisthoteuthis*), *Brescianiana* (on *Graneledone*), *Tripartisoma* (on *Pareledone*), *Avdeevia* (on *Megaleledone*), *Lichomolgus longicauda* (on *Sepia*), and a few forms belonging to genera ordinarily parasitic on fish, viz., *Pennella varians* (on *Loligo, Sepia* and *Moschites*) as well as the caligid of *Nautilus, Anchicaligus nautili*.

    Within **Polyplacophora** (formerly: Amphineura), copepods of the genus *Ischnochitonika* are parasites of chitons (Franz & Bullock, 1990). It seems that there is no copepod reported as associated with Monoplacophora or Scaphopoda.

2. **Cirripedia**: a few pedunculate cirripedes have been found on molluscs, like *Malacolepas conchicola* (on the Lamellibranchia *Cucullaea labiata* and *Venerupis mitis* in Japan), *Koleolepas* spp. (on shells of *Turbo* and *Fusinus* "with pagurids", in the

Pacific) and *Pagurolepas conchicola* (on gastropod shells "with pagurids", viz., *Fusinus, Bursa*, Buccinidae, in the Pacific and western Atlantic). The genus *Octolasmis* was found on a living *Murex*, a *Temnaspis* on *Nautilus*, and, of course, some *Lepas* on floating shells of *Spirula* and *Janthina*.

3. **Malacostraca. Isopoda** seem rarely associated with Mollusca and only some Sphaeromatidae can be cited here: *Dynamenella australis* is associated with limpets (South Africa), while in the western tropical Atlantic *Dynamenella perforata* was found on *Acanthopleura granulata* and *Chiton tuberculatus*, while the *Chiton* last-mentioned was also associated with *Dynamenella moorei, Dynamopsis dianae*, and *Exosphaeroma* spp. (Brattegard, 1968; Glynn, 1968).

More associations have been quoted for **Amphipoda** (cf. Stephensen, 1936; Vader, 1972a), for example *Calliopiella michaelseni* on limpet shells in South Africa, *Cardiophilus baeri*, commensal of *Didacna baeri* (Black Sea, Caspian Sea), *Hyale grandicornis* on Californian limpets (*Acmaea, Lottia*), *Hyale nilssoni* in connection with *Nucella lapillus* and *Patella vulgata* (Norway), *Hyale perieri* with *Patella* (Plymouth), *Parhyale hawaiensis* with *Chiton tuberculatus, Metopa glacialis* in *Modiolaria discors* and *Metopa groenlandica* in *Pandora glacialis* (cf. Stephensen, 1936). Many crabs of the family **Pinnotheridae** are associated with molluscs, especially Lamellibranchia (*Pinnotheres, Pinnixa, Opisthopus*), sometimes Gastropoda (*Conus, Megathura, Bullana, Navanax*).

Finally, many **shrimp** are commensals of molluscs, again in particular Lamellibranchia: some Alpheidae (*Athanas* spp.) and especially many Pontoniinae (*Anchistus, Paranchistus, Conchodytes, Platypontonia*). *Periclimenes imperator* would be associated with Nudibranchia. In this respect, an important study has been done by Hipeau-Jacquotte (1972, 1974) on Pontoniinae associated with Pinnidae in Tuléar (Madagascar).

ANNELIDA: POLYCHAETA

Representatives of this group of quite common animals in the marine environment have associations with various taxa of Crustacea:

1. **Copepoda**: at least a dozen families of copepods are associated with polychaetes, more often as external semi-parasites, sometimes as internal parasites with a strongly modified morphology. Here are some examples: Lichomolgidae (*Acanthomolgus* on *Serpula, Doridicola* on *Sabella, Myxomolgus* on *Myxicola, Nasomolgus* on *Sabellastarte, Sabelliphilus* on *Sabella* and *Spirographis, Serpuliphilus* on *Spirobranchus*), Pseudanthessiidae (*Pseudanthessius* on *Sabella, Hermodice*), Clausiidae (*Mesnilia* in the tubes of *Polydora, Rhodinicola* on various Maldanidae), Clausidiidae (*Hersiliodes* on *Clymene, Leiochone, Cotylomolgus* on *Lepidonotus*), Herpyllobiidae, especially on Aphroditiformia: Polynoidae (*Herpyllobius, Eurysilemum, Hedyphanella, Bradophila, Phallusiella, Saccolepis, Sarsilenium, Oestrella*), Serpulidicolidae (*Rhabdopus, Rhynchopus, Serpulidicola*), Eunicicolidae, Phyllodicolidae, Nereicolidae (*Nereicola* on several Nereidae, *Vectoriella* on some Paraonidae),

Anomopsyllidae for a species considered for a long time with doubtful affinities because apparently devoid of any buccal appendage (*Anomopsyllus pranizoides* on *Terebellides stroemii*, cf. Laubier, 1988), Gastrodelphyidae, Sabelliphilidae, Spiophanicolidae Monstrillidae (larva only).

Some genera are still of uncertain affinities, for example *Cotylomyzon* (on *Eupolyodontes*, *Amboine*), *Ophelicola* (abyssal Opheliidae, Bay of Biscay), as well as various extremely reduced internal parasites: *Xenocoeloma* (on *Polycirrus*), *Aphanodomus* (on *Amphitrite*), *Flabellicola* (on *Flabelligera*). Species of *Tychidion* would reportedly be occurring on *Lamellibrachia*, a genus of the family Siboglinidae (order Sabellida).

2. **Cirripedia**: Some cirripedes, like *Scalpellum* spp., can settle on the calcareous tubes of sedentary polychaetes (Monod, 1933).
3. **Stomatopoda**: A mantis shrimp from California, *Acanthosquilla digueti*, lives with a polynoid polychaete (*Lepidasthenia digueti*) in the gallery of a Hemichordata [acorn worm] (Coutière, 1905).
4. **Decapoda**: Tubes and galleries of large polychaetes can be inhabited by porcellanid anomurans or brachyuran crabs. Here are some examples from Brachyura: *Pinnixa* spp. (on *Chaetopterus*, *Clymenella*, *Pectinaria*), *Pinnaxodes*, *Pinnotheres* (with *Chaetopterus*), *Asthenognathus* (on *Amphitrite*), *Menippe*, etc.; and from Anomura: *Polyonyx* (with *Loimia*, *Chaetopterus*).

Pycnogonida

Some pedunculate cirripedes can settle on pycnogonids, for example *Weltnerium nymphocola* on *Boreonymphon robustum*, and *Scalpellum aduncum* on *Phoxichilidium fluminense*.

Xiphosura

The following barnacles were mentioned on horseshoe crabs: *Octolasmis warwicki* on *Tachypleus gigas*; and *Chelonibia patula*, *Megabalanus tintinnabulum* and even *Dosima fascicularis* on "*Limulus*".

Crustacea

Obviously, also associations exist between crustaceans and other species of this same group. We mention those in the following, specified by group of Crustacea.

Copepoda

In this group, the following associations with other Crustacea are noted:

1. **Tantulocarida**: These tiny crustaceans were discovered in 1975 (Becker, 1975) and live as ectoparasites on other Crustacea. The first species described, *Basipodella harpacticola*, was discovered on harpacticoid copepods. Several genera of Deoterthridae also live on Harpacticoida. *Deoterthron* was found on harpacticoids and on an ostracode (*Metavargula*) (Bradford & Hewitt, 1980).

2. **Cirripedia**: Many cases of fixation of the balanomorph *Conchoderma virgatum* on representatives of the copepod genus *Pennella*, itself parasitizing a fish (*Mola*, *Xiphias*, *Makaira*, *Diodon* or *Exocetus*) were reported by Monod (1933) and Dollfus (1948). *Conchoderma* is also found on other copepods, for instance *Philorthagoriscus serratus* and *Orthagoriscicola muricatus*, and also on *Pennella* rooted on cetaceans (*Balaenoptera* spp.).

Ostracoda

Choniostomatid copepods can parasitize ostracodes, for example *Sphaeronellopsis* on *Pseudophilomedes*, *Sarsiella*, *Parasterope* and *Metavargula* (cf. Monod, 1932; Kornicker & Bowman, 1939; Bowman & Kornicker, 1967, 1968). Some cryptoniscid Epicaridea (Isopoda) also parasitize Ostracoda, for instance *Cyproniscus cyprinidinae*.

Cirripedia

Many thoracican Cirripedia can be found fixed on other cirripedes (Monod, 1933), e.g., *Lepas anatifera* on *Conchoderma auritum*, *Dosima fascicularis* on *Lepas australis*, *Heteralepas quadrata* on *Lepas hilli*, *Conchoderma auritum* on *Coronula diadema*, *Ibla cumingi* on *Mitella mitella*, *Glyptelasma carinatum* on *Scalpellum giganteum* and *Octolasmis orthogonia*, *Glyptelasma gracilius* on *Scalpellum giganteum*, *Megalasma striatum* on *Scalpellum stearnsii*, *Chthamalus stellatus* on *Balanus crenatus*, *Balanus crenatus* on *Balanus tintinnabulum*. Arrangements like these are very common indeed.

A few Acrothoracica are present in shells of Cirripedia Sessilia, for instance *Cryptophialus* on *Balanus* and *Concholepas*, or *Australophialus* as a burrower on *Austrominius*.

A rhizocephalan belonging to the order Akentrogonida, *Chthamalophilus delagei*, is a parasite in *Chthamalus stellatus* (cf. Bocquet-Védrine, 1961).

A series of Isopoda: Cryptoniscoidea are hyperparasites of Rhizocephala: *Liriopsis* on *Septosaccus* and *Peltogaster*, *Euthylacus*, *Danalia* and *Perezia* on sacculinids of Brachyura (*Carcinus*, *Macropipus*, *Liocarcinus*, *Grapsus*, *Ostracotheres*).

Hemioniscus balani is a well known parasite in acorn barnacles and sometimes in Chthamalidae (cf. Prenant, 1924).

Stomatopoda

A few stomatopods seem to have associated crustaceans (endoecism). There is one example from Copepoda: *Hemicyclops acanthosquillae*, in Nosy-Bé, Madagascar (Humes, 1965), and a few others from caridean shrimp, genus *Athanas* (cf. Froglia & Atkinson, 1998; Hayashi, 2002).

Peracarida

A few species of the tantulocarid genus *Doryphallophora* live on isopods; some species of Deoterthridae (*Onceroxenus*) are parasites in tanaidaceans. [See also the chapter on Tantulocarida in the present series.]

Some copepods are known as associates of isopods: *Cithadius* on *Cyathura*, *Tisbe furcata*, *Dactylopusia neglecta*, *Laophonte* sp., and *Donsiella limnoriae* on *Limnoria lignorum*, and *Harrietella simulans* on *Limnoria tripunctata*; Choniostomatidae (*Sphaeronella*) live in the marsupia of several genera belonging to various groups, Serolidae, Gammaridea, Mysida. The genus *Paranicothoe* was found on two Bopyridae. The curious copepod *Rhizorhina ampeliscae*, family Herpyllobiidae, was found on *Ampelisca*.

Cirripedes can also settle on isopods, acorn barnacles on Sphaeromatidea (*Amphibalanus amphitrite* on *Sphaeroma walkeri*) or on Pedunculata (*Octolasmis bathynomi*, *Octolasmis aymonini*) on *Bathynomus* (cf. Monod, 1974).

Where Rhizocephala are concerned, *Duplorbis* is parasitic in isopods (*Duplorbis calathurae*, etc.). Some asellote Isopoda, like *Jaera hopeana* on *Sphaeroma serratum*, live as facultative commensals, or live as obligatory commensals, as *Iais* on *Sphaeroma* and *Exosphaeroma*.

Quite a series of Ostracoda live as commensals on marine isopods, for instance *Limnoria* (*Aspidoconcha*) on cavernicolous isopods, with the genus *Sphaeromicola* living on *Cirolana*, *Sphaeromides*, *Caecosphaeroma* and *Monolistra*; *Sphaeromicola dudichi* is marine and is found on the amphipod *Chelura terebrans* (cf. Roelofs, 1966).

Various Epicaridea [= Cymothoida: Bopyroidea] are parasites in Peracarida, for example *Cabirops* on *Pseudione diogeni* and *Pseudione fraissei* (other epicarids!), *Clypeoniscus* on species of *Idotea* (*Idotea balthica*, *Idotea pelagica*, *Idotea chelipes*) and on *Zenobia prismatica*, *Nannoniscus* on various Asellota, *Seroloniscus* on *Serolis*, *Astacilloechus* on *Astacilla*. Some others are also found on amphipods (*Podascon*) and on Cumacea (*Cumaoechus*). Some are hyperparasites, for instance *Cabirops* on *Bopyrus* or *Gnomoniscus* on *Podascon*. Dajidae (*Aspidophryxus*) are parasites in Mysida.

GLYPHEIDA

A parasitic copepod, *Nicothoe tumulosa*, was described from the gills of *Neoglyphea inopinata*.

ASTACIDEA, PALINURA AND ANOMURA

The associated crustaceans in these groups include:

1. **Ostracoda**: many species of ostracodes (*Entocythere*, *Ankylocythere*, *Uncinocythere*) are present on North American crayfish (*Cambarus*, *Procambarus*, *Orconectes*), some of which are cavernicolous.
2. **Cirripedia**: many thoracic cirripedes, both Pedunculata (often *Octolasmis*) and Sessilia, can be found fixed on lobsters, spiny lobsters, slipper lobsters; small species (*Octolasmis*) are especially common on sheltered places of the integument (ventral side, buccal region) and in the gill chambers, and also, for example, on the scaphognathite.
   Balanomorph cirripedes are already frequent on the shells occupied by hermit crabs, and exceptionnally also on the hermit crab itself: *Amphibalanus amphitrite*

Fig. 26.11. *Diogenes pugilator*: a large *Amphibalanus improvisus* and several juveniles are fixed on the left eyestalk of the hermit crab, and another small acorn barnacle (B) has settled on the propodus of the larger cheliped. [After Codreanu & Codreanu, 1959.]

on *Diogenes pugilator* (fig. 26.11). Rhizocephalans on Anomura are represented, on hermit crabs by the genera *Septosaccus*, *Peltogaster*, *Gemmosaccus*, and on Galatheoidea by the genus *Triangulus*.

3. **Copepoda**: These are present in the gill chamber and on the gills of large decapod crustaceans, and include species such as *Nicothoe astaci*, *Tisbe elongata*, *Sacodiscus ovalis* in the lobster *Homarus*, *Nicothoe* in Norway lobsters, *Atthyella pilosa* in American crayfish (genus *Orconectes*), *Atthyella trispinosa* and *Nitokra divaricata* in European crayfish; a choniostomatid (*Choniomyzon panuliri*) is found on *Panulirus homarus* in India. Copepods are also associated with Paguroidea, for example species that belong to the genera *Porcellidium* (on *Dardanus* spp.), *Paraidya* (also on *Dardanus* spp.), *Sunaristes* (*Sunaristes paguri* on *Pagurus bernhardus*, *Clibanarius erythropus*, *Diogenes pugilator*; *Sunaristes dardani* and *Sunaristes inaequalis* on *Dardanus* spp.), and *Caudella* on *Pagurus bernhardus*. *Hemicyclops*

spp. are found on *Upogebia*, *Callianassa* and *Axius*, *Clausidium* on *Callianassa* and, to conclude, *Halicyclops caridophilus* on *Thalassina* in Borneo.
4. **Syncarida**: The fact that syncarids can be associated with freshwater crayfish (Parastacidae) in their galleries is, in general, not well known. However, this is the case in *Allanaspides* associated with *Parastacoides* and also in *Micraspides* in relation to *Engaeus*.
5. **Amphipoda**: Some amphipods, like *Isaea*, are reported from lobsters (*Homarus*) or from spiny lobsters (Palinuridae), such as *Parapleustes commensalis* on *Panulirus interruptus* in California. A few amphipods were found with Paguroidea, but the presence of some of these, appears more likely as accidental. Others are indeed true commensals: *Stenula rubrovittata*, *Melita obtusata*, *Aora typica*, *Eurystheus maculatus*, *Gitanopsis paguri*, and *Gammaropsis nitida*.
6. **Isopoda**: Many Epicaridea are known from Anomura. They belong to the genera *Athelges*, *Parathelges*, *Pseudione*, *Pseudionella*, *Aporobopyrus*, *Pleurocrypta*, *Parione*, *Astralione*, and *Parionella*. More than 30 species of Bopyridae are known on 36 host species, for Porcellanidae only. *Danalia ypsilon* is directly attached on Galatheoidea and not through an intermediary rhizocephalan.
7. **Mysida**: Some mysids seem associated with Paguroidea, for instance *Gnathomysis gerlachei* with *Pagurus brevipes*. For *Heteromysis* spp., it is supposed they might play a cleaning role, while feeding on the faeces of the anomuran.
8. **Decapoda**: Some commensal decapods are known for *Callianassa californiensis* (cf. MacGinitie, 1934): the crabs *Scleroplax granulata*, *Pinnixa franciscana* and *Pinnixa schmitti*; there are also crabs associated in relation to the shrimp, *Betaeus ensenadensis*. Another alpheid, *Aretopsis amabilis* in the Seychelles, lives associated with the pagurids *Dardanus arrosor* and *Dardanus sanguinolentus*, and this one is supposed, in fact, to be coprophagous (Bruce, 1969a).

BRACHYURA

Quite some **Copepoda** live in the branchial chambers of crabs, for example *Cancrincola* (on *Cardisoma*, *Goniopsis*, *Sesarma*), *Antillesia* (on *Cardisoma*), or *Pholetischus* (on *Sesarma*). A few Choniostomatidae are also known from crabs: *Lecithomyzon* on *Carcinus maenas* or *Choniosphaera* on *Neptunus* and *Carcinus maenas*.

Various crustaceans live on large majid crabs that, with their spinose carapace, are rich in epizoonts and epiphyta. About 40 species of **amphipods** and some 28 species of **copepods** belonging to 15 genera and 9 families were identified on *Maja brachydactyla* [as *Maja squinado*] (cf. Jakubisiak, 1932). Only one of these species is probably species-specific (*Hemilaophonte janinae*) (Bertrand, 1942).

Many crabs can bear **cirripedes** that are either simply phoretic, or commensal: *Conchoderma virgatum* on *Neptunus pelagicus*; *Temnaspis tridens* on *Scalopidia spinosipes*; *Octolasmis* on *Geryon*, *Scylla*, *Callinectes*; *Poecilasma* on *Rochinia*; *Heteralepas* on *Percnon*.

In Grande-Ile (Louisiana), *Octolasmis muelleri* was found (Humes, 1941) on 19% of the males and 43% of the females of *Callinectes sapidus*, and also on *Callinectes ornatus*,

Menippe mercenaria, Panopeus herbstii, Calappa flammea and *Hepatus epheliticus*. The association of *Megabalanus decorus* with *Leptomithrax longipes* seems rather permanent.

Among **Rhizocephala**, many species parasitize brachyurans: primarily these are large numbers of sacculinids, like *Sacculina carcini* on *Carcinus, Portunus, Liocarcinus, Bathynectes, Pisa, Xantho*, but also *Ptychascus* on *Sesarma, Heterosaccus* on *Mithrax, Macrocoeloma, Charybdis* spp., *Microphrys; Sesarmaxenos* on *Sesarma; Loxothylacus* on *Xantho, Phymodius; Drepanorchis* on *Lybia, Portunus, Camphyra, Macropodia*.

Epicaridea of brachyurans are Bopyridae (*Cancricepon*), Entoniscidae (*Portunion, Cancrion, Pinnotherion, Priapon*) or Liriopsidae: *Danalia* on Hapalocarcinidae (*Danalia hapalocarcini* on *Hapalocarcinus marsupialis, Danalia* sp. on *Troglocarcinus*) are directly attached to the crab, and not through an intermediary rhizocephalan parasite. Some **mysids**, like *Antromysis anophelinae* and *Antromysis pectorum*, live in the galleries of *Cardisoma*.

DECAPODA: VARIOUS SHRIMP

As noted above, caridean shrimp are often associated with invertebrates from various taxa: sponges, cnidarians, echinoderms, and molluscs, to name the more obvious. They are also associated with other crustaceans. Some rare **copepods** are associated with penaeids or with caridean shrimp: *Hadrothoe crosnieri* is found on the carapace of *Aristeus virilis* in Madagascar, and *Paranicothoe* on *Plesionika ensis* and *Haliporoides triarthrus*. A large number of **Epicaridea** of the family Bopyridae live on the body or in the gill chambers of many Caridea (cf. Bourdon, 1968). Finally, the **isopods** *Tachaea picta* and *Tachaea caridophaga* [as *Austroongerthona*] are known to parasitize shrimp (*Macrobrachium, Paratya*).

ECHINODERMATA

Many species of echinoderms are associated with Crustacea from various groups, and in quite a large number of species. We cite here:

1. **Copepoda**: As with molluscs and annelids, within Crustacea the copepods are the more abundant parasites and commensals of echinoderms. They are represented by some 15 odd families: that number is an indication only, because the true systematic affinities of many internal parasites are difficult to establish.
   The **asteroids** are associated with many species of copepods, such as representatives of the genera *Astericola, Astroxynus, Leiocomes, Onychopygos, Stellicola, Stellicomes*, and *Synstellicola*.
   On **ophiuroids**, we find the genera *Asterocheres, Cancerillopsis, Doridicola, Pseudanthessius*, as well as some internal parasites, such as *Amphiurophilus, Arthrochordeumium, Codoba, Lernaeosaccus*, and *Ophioika*.
   Echinoids host a series of genera of cyclopoids (*Ascomyzon, Asterocheres, Calvocheres* (in spine galls), *Listinogaster, Macrochiron, Meomicola, Micropontius*, and *Pseudanthessius*), as well as an harpacticoid (*Porcellidium echinophilum*) and several internal parasites (*Dichelina phormosomae, Pionodesmotes phormosomae*).

Crinoids also have some associated Copepoda, such as *Enterognathus* spp., *Doridicola*, and *Kelleria*.

Copepods associated with **holothurians** are highly diverse, most of them are external, some internal (*Lichothuria, Lecanurius, Synapticola*); from the cyclopoids we list the following genera: *Allantogynus, Chauliolobion, Diogenella, Diogenidion, Nanaspis, Namakosiramia, Scambicornus*, and *Synaptiphilus*; the genera *Brychiopontius* and *Gomphopodaria* have been found on an abyssal holothurian of the genus *Oneirophanta*. Finally, a small number of harpacticoids such as *Mebis holothuriae, Tisbe holothuriae, Tisbe cucumariae* and *Tisbe japonica* are associated with sea cucumbers.

2. **Cirripedia**: Many Thoracica can settle on the spines of urchins or on crinoids. An exceptionnal case is that of *Balanus concavus pacificus*, which seems to be permanently fixed on the surface of *Dendraster excentricus* in California (Giltay, 1934). Ascothoracida are often parasites in echinoderms, for example: *Synagoga* in *Metacrinus*; *Ascothorax* in *Ophiocten, Amphiura, Ophionotus*; *Parascothorax* in *Ophiura*; *Dendrogaster* in *Echinaster, Solaster, Asterias, Leptasterias*; *Ulophysema* in *Pourtalesia, Echinocardium, Brissopsis*; and *Myriocladus* in some asteroids.

3. **Isopoda**: The asellote *Antias uniramea* is cited from cavities in the skin of a holothuroid (*Stichopus*), and a valviferan, *Colidotea rostrata* as living between the spines of the sea urchin *Strongylocentrotus purpuratus*. There are some other, analogous relationsips known, for example *Cirolana lineata* on the crinoid *Comanthus*.

4. **Amphipoda**: Associations between isopods and echinoderms are, in general, exceptional. However, this is different with amphipods (Vader, 1978), represented with 15 families and 23 genera as associates of Echinodermata: Amphilochidae (*Amphilochus*), Caprellidae (*Caprella* spp., *Pariambus*), Colomastigidae (*Colomastix*), Dexaminidae (*Tritaeta gibbosa*), Gammaridae (*Elasmopus, Melita obtusata*), Haustoriidae (*Urothoe*), Isaeidae (*Gammaropsis*), Ischyroceridae (*Ischyrocerus, Jassa*), Laphystiopsidae (*Laphystiopsis*), Liljeborgiidae (*Listriella*, currently known as *Idunella*), Lysianassidae (*Ambassia, Aristias, Aroui, Euonyx, Menigrates*), Paramphithoidae (*Epimeria*), Pleustidae (*Parapleustes*), Podoceridae (*Dulichia*), Stenothoidae (*Stenothoides*) and Syrrhoidae (*Pseudochiron*). The types of associations are discussed in detail by Vader (1978), but they are very variable and difficult to detail precisely. In fact, an array of situations seems to exist, ranging from accidental presence on a given host to actual parasitism, the latter being exceptional.

5. **Decapoda**: Where Anomura are concerned, we can cite Porcellanidae (for example, *Petrolisthes* on asteroids), Galatheoidea (*Galathea* and *Allogalathea* on genera such as *Comanthina, Comanthus*, and *Himerometra*).

Associations of Brachyura with echinoderms are known for *Eumedon, Zebrida* and *Dissodactylus* with sea urchins, while *Lissocarcinus* can be found in and on holothuroids, the pinnotherids, *Opisthopus* and *Pinnixa*, were found in holothurians, and *Pinnaxodes* as well as *Fabia*, in sea urchins. A special mention should be made of *Zebrida adamsii* since that is much more than a simple commensalism: in fact, this involves a "ravaging", i.e., truly destructive type of parasitism (Suzuki & Takeda, 1974).

Finally, a series of shrimps are associated with echinoderms, mainly pontoniines, and such with sea urchins (*Stegopontonia, Tuleariocaris, Periclimenes* spp.), crinoids (*Hippolyte, Palaemonella, Areiopontonza, Parapontonza, Pontoniopsis, Periclimenes* spp.), holothuroids (*Conchodytes, Periclimenes* spp.), and sometimes ophiurans (*Periclimenes*); some alpheids, genus *Athanas* are associated with echinoids (Jacquotte, 1964; Hipeau-Jacquotte, 1965).

Bruce (1982) published an important review of shrimps associated with echinoderms in the Indo-West Pacific: he made a census of some fifty species (34 Pontoniinae, 11 Alpheidae, 4 Gnathophyllidae, 1 Stenopodidae) on 71 host species. The pontoniine *Conchodytes tridacnae* is a commensal of the giant clam (*Tridacna*), and was also found in the cloacas of holothurians.

TUNICATA

A variety of Copepoda live on or in ascidians, for example *Clausia, Hemidoxyphium, Macrochiron, Lichomolgidium, Ascidioxymus*, and *Zygomolgus*. In particular the family Notodelphyidae includes many genera of ascidicolous copepods, whose morphology can be strongly modified: *Notodelphys, Doropygus, Pygodelphys, Mychophilus, Bonnierilla, Botachus, Ophioseides, Pholeterides, Brementia, Cochlodelphys, Sicyiodelphys, Anoplodelphys*, and *Syndelphys* (see Illg, 1958; Illg & Dudley, 1961; Lafargue & Laubier, 1968, 1968a, 1977, 1977a; Laubier & Lafargue, 1974).

The parasitic copepod *Gonophysema gullmarensis* presents a curious example of hermaphroditism. The genus *Sapphirina* is often associated with Cirripedia: Sessilia, which are themselves sometimes fixed to large, simple ascidians, while hyperiid Amphipoda frequently live in Thaliacea such as Salpida, Pyrosomatida or Doliolida. Finally, some shrimp are commensals of tunicates: Alpheidae (*Synalpheus*) or Pontoniinae: *Pontonia, Dasia, Sadella, Periclimenaeus* (fig. 26.12), etc.

PISCES

The following crustaceans are known as associates of fishes, both permanently or temporarily, and varying from mutually beneficial commensals to genuine parasites:

1. **Copepoda**: A very large number of copepods are fish parasites: within the family Belonidae (11 genera with 28 species), 23 species of parasitic copepods were found belonging to 11 genera in six families: Bomolochidae, Anthosomatidae, Lernaeolophidae, Caligidae, Ergasilidae, Lernaeidae (cf. Cressey & Collette, 1970). Other well-known families of Copepoda that include fish parasites are Dichelesthiidae, Chondracanthidae, Philichthyidae (these live in the mucus canals on the fish' head), and Lernaeopodidae. A list of parasitic copepods of fish from Italy, composed by Brian in 1906, could already enumerate no less than 180 species.

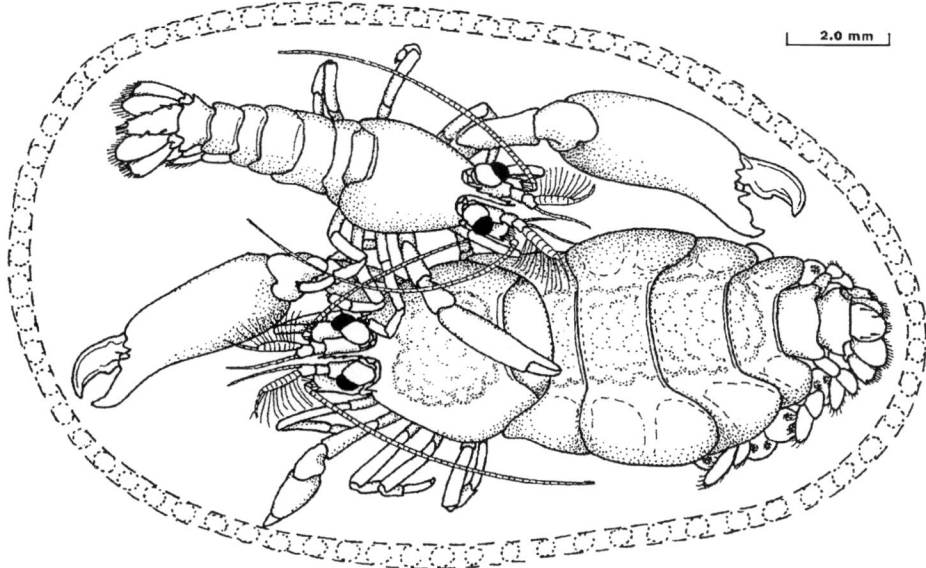

Fig. 26.12. A pair of *Periclimenaeus hecate* inside a compound ascidian (*Diplosoma*). [After Bruce, 1976.]

2. **Branchiura**: This group is completely parasitic: all four genera of Branchiura, mostly but not exclusively inhabiting fresh waters, are essentially, though not exclusively, parasites on fish.
3. **Ostracoda**: *Cypridina squamosa* was found on *Scorpaena scrofa* and could be haematophagous (Monod, 1923); *Cypridina parasitica* in the mouth of a smooth hammerhead (*Sphyrna zygaena*); and *Cypridina* sp. in the mucus canals of *Coryphaena hippurus* (cf. Brian, 1931).
4. **Cirripedia**: Some Lepadomorpha can settle on fish, for example *Anelasma squalicola* on *Etmopterus spinax* and *Cetorhinus maximus*; *Conchoderma virgatum* on *Xiphias*, *Mola mola*, *Diodon hystrix*, and *Gymnothorax*, which can also host *Conchoderma auritum* (cf. Monod, 1933); and one *Serranus lanceolatus* was found with an *Octolasmis warwicki*. From Balanomorpha, we can mention *Platylepas hexastylos ichthyophila* on *Lepisosteus*, an acorn barnacle on the pectoral fin of a *Xiphias gladius*, and *Balanus improvisus* on the mandible of a *Lucioperca* (cf. Ciurea et al., 1933).
5. **Amphipoda**: Associations between amphipods and fish are exceptionnal, with a few isolated cases such as *Trischizostoma nicaeense* and *Lafystius sturionis*, which do indeed have a relationship with fish.
6. **Isopoda**: Where examples of fish parasites are rare in Amphipoda, in contrast, they are numerous in isopods, among which the suborder Gnathiidea has larvae (pranizas) that live as parasites in fish, and where Cymothoida show long lists of forms leading from simple predation (haematophagy, cleaning of carcasses, etc.) to internal parasitism. The large family Cymothoidae is entirely parasitic, sometimes

at the body surface of fish, sometimes in the mouth or in the gill chamber. Their biology, their physiology, as well as their effect on the host are nowadays rather well known (Trilles, 1968; Romestand, 1978, 1979).

7. **Stenopodidea** and **Caridea**: A special type of association between crustaceans and fish is that of **cleaning shrimps**. These shrimp belong to the following families: Palaemonidae (Palaemoninae: *Palaemon adspersus* and *Palaemon elegans*; Pontoniinae: *Urocaridella, Periclimenes amethysteus, Periclimenes yucatanicus, Ancylomenes pedersoni, Ancylomenes holthuisi, Ancylomenes magnificus, Brachycarpus biunguiculatus*); Hippolytidae (*Lysmata amboinensis, Lysmata grabhami, Lysmata californica*, and *Lysmata seticaudata*); and Stenopodidae (*Stenopus hispidus, Stenopus spinosus*, and *Stenopus scutellatus*); most of these species have pretty colours (Limbaugh et al., 1961; Yaldwyn, 1964, 1966, 1968; Corredor, 1978; Östlund-Nilsson et al., 2005). Some crabs apparently are also cleaning fish, but this has been only rarely observed.

Amphibia

There is one parasitic copepod on Anura: *Lernaea cyprinacea* (= *Lernaea esocina* = *Lernaea ranae*) on *Dienictylus pyrrhogaster* in Japan, and on *Pelobates cultripes* in the Camargue, France (Delamare Deboutteville, 1958) and on *Rana clamitans* in Ohio, U.S.A. (Stunkard & Cable, 1931).

Reptilia

Five different genera of Cirripedia Sessilia can develop on marine turtles: *Chelonibia* spp., *Platylepas* spp., *Stomatolepas, Cylindrolepas*, and *Stephanolepas*, as well as the pedunculate cirripede *Conchoderma virgatum*. A small fauna of phoretic epizoonts can also be present on marine turtles, with tanaidaceans, amphipods as *Hyale grimaldii* and *Hyachelia tortugae*, and a small grapsid crab, a usual guest on floating objects, i.e., the Columbus crab, *Planes minutus*, cleaning its host from dead skin fragments and also active in the same sense in its anal area.

Aves

In spite of the great singularity of the scarce cases observed, we must recall here the association between Cirripedia and birds. However, for more specific information the reader is referred to the chapter "Symbiosis and Parasitism in the Crustacea" in volume 3 of the present series (see Trilles & Hipeau-Jacquotte, 2012).

Mammalia: Sirenia

The harpacticoid copepod, *Harpacticus pulex*, was recorded on a sirenian (*Trichechus manatus* in an aquarium in Miami) and a few cirripedes (*Platylepas* and *Chelonibia manati*) on *Trichechus*, the manatee, and *Platylepas* on a dugong (genus *Dugong*). The amphipod, *Cyamus rhytinae*, was found on Steller's sea cow, *Hydrodamalis gigas* (cf. Leung, 1967).

MAMMALIA: CETACEA

Cetaceans can be associated with many commensal or parasitic Crustacea. Two harpacticoid copepods, *Balaenophilus unisetus*, occurring on the baleens of *Balaenoptera* spp., and *Harpacticus pulex*, found on *Tursiops truncatus* in an aquarium in Miami, are known from cetaceans (Vervoort & Tranter, 1961; Banister & Grimoley, 1966). Also, the giant parasitic copepods of the genus *Pennella* (*Pennella balaenoptera*) have to be mentioned here.

Several cirripedes can settle on Cetacea as well. Species of Sessilia either are more or less superficial, like *Coronula*, or are deeply embedded in the skin, such as *Xenobalanus globicipitis*, *Tubicinella* or *Cryptolepas rachianecti*. Pedunculates are sometimes fixed on teeth or on baleens, or more often on specimens of *Coronula* spp. that are present on the host's skin. For instance, *Conchoderma* is rarely fixed on the skin of the host itself.

The cirolanid isopod, *Natatolana narica* [as *Cirolana*], was discovered in the nasal cavity of a dolphin, *Cephalorhynchus hectori*, in New Zealand (Bowman, 1971a). Finally, a family of amphipods, Cyamidae or whale lice, contains the largest number of crustacean parasites in cetaceans. Six genera and some 30 species are currently known: *Cyamus* (ca. 18 species), *Neocyamus* (1 species), *Platycyamus* (2 species), *Isocyamus* (4 species), *Syncyamus* (4 species), and *Scutocyamus* (2 species). *Cyamus* spp. do not have pelagic larvae, whence the question is: how do new infestations occur (breastfeeding, mating, occasional contacts?).

EPIZOONTS, COMMENSALS AND PARASITES OF CRUSTACEANS

As an appendix to the chapter on commensal and parasitic Crustacea, it may be useful to add some information on epizoonts, commensals and parasites of crustaceans.

Only on or in a single species of crab, e.g., *Carcinus maenas*, it is already possible to find the following life forms: Microsporidia: *Thelohania*, *Ormieresia*, and *Ameson*; Ciliata: *Phoretophrya*, *Synophrya*, *Epistylis*, and *Gymnodinioides*; Dinophyceae: *Hematodinium*, *Cecrinis*, and *Turniella*; Turbellaria: *Fecampia*; Trematoda: *Microphallus*; Nemertea: *Carcinonemertes* (on the eggs); Annelida: *Spirobranchus* [as *Pomatoceros*]; Bryozoa: *Electra pilosa*; Rhizocephala: *Sacculina carcini*; Epicaridea: *Portunion maenadis* (cf. Bourdon, 1965).

Within a single family of shrimp, the Penaeidae, the occurrence has been described of viruses, bacteria, Gregarina (*Porospora*, *Cephalolobus*, and *Uradiophora*), Microsporidia (*Nosema* and *Thelohania*), Ciliata (*Epistylis*), Trematoda (*Opecoeloides* and *Microcephalus*), Cestoda (*Prochristianella* and *Parachristianella*), Nematoda (*Contracaecum*), Crustacea (endoparasites; and also epizoonts: *Balanus* spp., *Lepas*).

The following, non-exhaustive list, mentions the main vegetal or animal organisms that crustaceans carry, or that can infest them:

Among **epiphyta**, the **algae** are usually present with various, and very different, species on the carapace of most Majidae. The following algae may be present on Copepoda: *Sti-*

geoclonium australense on *Lernaea*; Desmidiaceae on *Salmincola*; *Ectocarpus* and *Ceramium* on *Lernaeenicus* and, on the same species, a diatom, *Diploneis smithii* (Bacillariophyceae); while other diatoms may live on free-living Copepoda, for example *Protoraphis*, and on *Corycaeus*, *Pseudohimanthidium* (= *Sameioneis*).

Protista are well represented: **Ciliata** [= **Ciliophora**]: *Anophrys* (on *Carcinus*), *Anoplophrya* (*Homarus*, in intestine), *Carchesium*, *Chilodochona* (on *Ebalia*, *Macropipus*, *Liocarcinus*), *Cothurnia* (on *Ligia*, *Limnoria*), *Epistylis* (on many crustaceans), *Foettingeria* (on Caprellidae), *Folliculina*, *Microfolliculina*, *Folliculinopsis* (on *Limnoria*), *Gymnodinioides* (on Decapoda), *Lagenophrys* (on many Crustacea, including the crab *Metopaulias* from phytothelmes in Jamaica), *Myoschiston*, *Perezia* (on Copepoda), *Phoretophrya* (on *Nebalia*), *Polyspira* (on Decapoda), *Spirochona* (on Amphipoda, *Nebalia*), *Spirophrya* (on Copepoda), *Synophrya* (on *Macropipus*, *Liocarcinus*, *Carcinus*), *Trochilioides* (on *Pseudocaligus*), *Zoothamnium* (on Amphipoda). — **Acinetidae**: *Dendrosomides paguri*, *Ephelota* (on *Lernaeenicus*, *Caligus*, Caprellidea), *Ophyoderma* (on *Porcellana platycheles*, on Copepoda), *Paracineta homari* (on Paguroidea), *Rhabdophrya* (on Copepoda), *Thecacineta cypridinae*, *Tokophrya* (on Copepoda). — **Dinophyceae**: *Actinodinium* (on Copepoda: *Acartia*), *Blastodinium* (on Copepoda), *Chytriodinium* (on eggs of Copepoda), *Haematodinium* (on *Carcinus*, *Macropipus*, *Liocarcinus*), *Schizodinium* (on *Corycella*), *Syndinium* (on *Paracalanus*), *Typanodinium* (on eggs of Copepoda). — **Microsporidia**: *Ameson* (on *Carcinus*), *Coccospora* (on *Balanus*), *Inodosporus* (on *Palaemon*), *Octosporea* (on Amphipoda, *Daphnia*), *Ormieresia* (on *Carcinus*), *Pyrotheca* (on Copepoda), *Thelohania* (on many Crustacea). — **Gregarina**: *Cephaloidophora* (on Amphipoda), *Porospora* (on *Homarus*, *Macropipus*, *Liocarcinus*).

The group **Ellobiopsidae** [currently classified under the (sub)kingdom Chromista], with, e.g., *Ellobiopsis*, *Thalassomyces* [= *Amalocystis*], is parasitic in many groups of Crustacea: on Amphipoda (on *Eusirus*, *Parathemisto*, *Rhachotropis*), on Mysida (on *Gastrosaccus*, *Gnathophausia*, *Siriella*), on Euphausiacea (on *Euphausia*, *Meganyctiphanes*), and on Decapoda (on *Pasiphaea*, *Acanthephyra*, *Oplophorus*).

Among **Metazoa**, the following groups can be cited:

Hydrozoa: Various species are known as epizoonts of crustaceans, in particular *Clytia hemisphaerica* [as *Clytia johnstoni*] (fig. 26.13) and *Obelia geniculata* on parasitic Copepoda (on *Lernaeocera*, *Peniculus*, *Lernaeenicus*) (Dollfus, 1948; Delamare Deboutteville, 1951). An interesting example is that of the pelagic isopod, *Syscenus infelix*, of 24 mm body length, wearing a tuft of *Obelia longissima* of more than 100 mm (Vader et al., 1981).

Turbellaria: *Fecampia spiralis* in *Acanthoserolis schythei*, *Fecampia balanicola* and *Kronborgia caridicola* in a shrimp, and *Kronborgia amphipodicola* in an ampeliscid amphipod. *Temnocephala* is found on the gills of various Decapoda (e.g., *Paranephrops*, *Engaeus*) and on a species of Isopoda, *Phreatoicopsis terricola*.

Trematoda: Exceptionally adults, such as *Udonella* (on Copepoda: Caligidae), but generally larval forms such as *Hemiurus* (in Copepoda), *Halicometra* (in *Palaemon*), *Cymatocarpus* (in *Macropipus*, *Liocarcinus*).

Cestoda: Eutetrarhynchidae [earlier as Tetrarhynchidae] (in Brachyura and Paguroidea), *Echinobothrium* (in Paguroidea, *Crangon*, Amphipoda), *Calliobothrium* (in *Carcinus*).

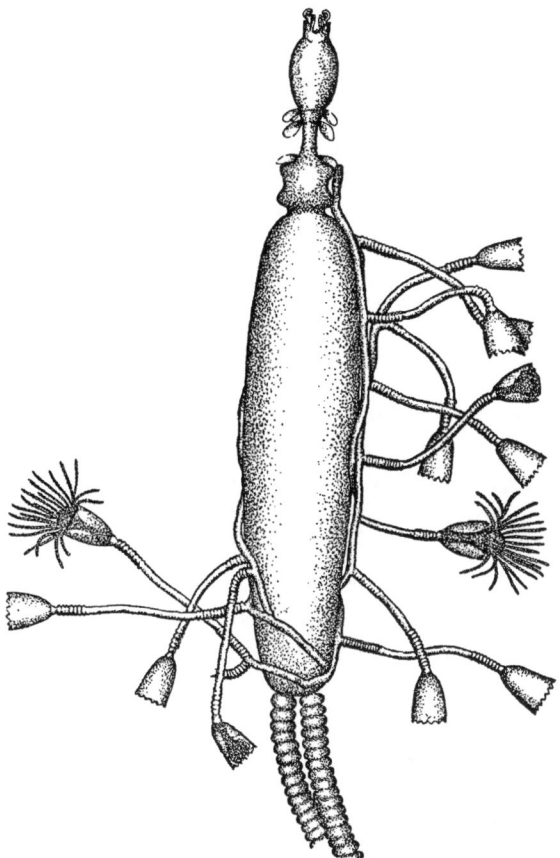

Fig. 26.13. A hydrozoan (*Clytia hemisphaerica* [as *Clytia johnstoni*]) on the parasitic copepod *Peniculus fistula*, itself a parasite on an individual of *Mullus surmuletus*. [After Delamare Deboutteville & Nunes, 1951.]

Nematoda: Very rare in Crustacea; larvae were present in the digestive tract of *Lernaeocera branchialis* (cf. Carpent, 1948), but these may have been ingested with the blood of the host.

Nematomorpha [a phylum also referred to as one out of four classes of a phylum Cephalorhyncha]: *Nectonema agile* is present in palaemonid prawns such as *Palaemon serratus* and *Palaemon elegans*.

Nemertea: *Carcinonemertes* on eggs of *Carcinus* and *Callinectes*; *Prostoma* on *Nautilograpsus*; and *Cephalothrix galatheae* on Galatheoidea.

Polychaeta: *Ophryotrocha geryonicola* in the gill chamber of Brachyura: Geryonidae.

Mollusca: *Mytilus edulis* were found on the shrimp *Pandalus montagui*, on *Upogebia* and on *Lernaeocera*; *Mytilus platensis* on the parasitic copepod *Trifur tortuosus*, and very young *Modiola* on *Achteinus*, another parasitic copepod.

Tardigrada: A marine tardigrade was found as a commensal of the isopod *Limnoria*.

Insecta: A small fly, *Lissocephala powelli* (Diptera), lays its eggs near the buccal appendages of several terrestrial crabs on Christmas Island; the crabs are: *Gecarcoidea natalis, Cardisoma hirtipes*, and *Sesarma obtusifrons*.

N.B.: *In the chapter "Symbiosis and Parasitism in the Crustacea", published in volume 3 of the present series (see Trilles & Hipeau-Jacquotte, 2012) more information is presented on the physiological and behavioural effects of parasites on the infected crustacean hosts.*

TO CONCLUDE

As we have seen in the aforegoing, crustaceans constitute a significant element in the earth's biosphere. They are found in every conceivable habitat, even the extreme or bizarre, in almost every corner of the world, even the remotest, and everywhere they make important members of the various biocoenoses. Though built to a common *Bauplan*, their diversity in overall body shape as well as in detailed anatomy is astonishing, and, moreover, they not infrequently populate their biotopes in vast numbers. Consequently, the role of Crustacea in all kinds of environments, by occupying a wide array of ecological niches, hardly can be underestimated. It is up to carcinologists to further expand and organize our knowledge of these remarkable animals.

APPENDIX I

Names of species of Crustacea mentioned in this chapter, with their authority and date, listed in alphabetical order [an occasional generic name that needs specification is also included]

ANIMALIA: CRUSTACEA
Abludomelita obtusata (Montagu, 1813)
Abyssorchomene abyssorum (Stebbing, 1888)
Acanthoserolis schythei (Lütken, 1858)
Acanthosquilla digueti [see *Alachosquilla digueti* (Coutière, 1905)]
Acontiophorus scutatus (Brady & Robertson, 1873)
Actaea superciliaris [see *Gaillardiellus superciliaris* (Odhner, 1925)]
Adeleana chapmani Holthuis, 1979
Aega spongiophila [see *Aegiochus spongiophila* (Semper, 1867)]
Aegiochus spongiophila (Semper, 1867)
Afrogitanopsis paguri (Myers, 1974)
Alachosquilla digueti (Coutière, 1905)
Albunea paretii Guérin-Méneville, 1853
Albunea symmista [see *Albunea symmysta* (Linnaeus, 1758)]
Albunea symmysta (Linnaeus, 1758)
Alcyonohippolyte commensalis (Kemp, 1925)
Alicella gigantea Chevreux, 1899
Allogaussia recondita (Stasek, 1858)
Alloschizidium cottarellii (Argano & Pesce, 1974)
Alona affinis (Leydig, 1860)
Alpheus armatus M. J. Rathbun, 1901
Alpheus djiboutensis De Man, 1909

Alpheus frontalis H. Milne Edwards, 1837
Alpheus laevis [see *Alpheus lottini* Guérin-Méneville, 1829 [in Guérin-Méneville, 1829-1838]]
Alpheus lottini Guérin-Méneville, 1829
Alpheus malleodigitus (Spence Bate, 1888)
Alpheus pachychirus Stimpson, 1860
Alpheus saxidomus Holthuis, 1980
Alpheus simus Guérin-Méneville, 1855-1856
Americorophium spinicorne (Stimpson, 1857)
Amigdoscalpellum hispidum (G. O. Sars, 1890)
Ampelisca cristata Holmes, 1908
Amphibalanus amphitrite (Darwin, 1854)
Ampithoe humeralis [see *Peramphithoe humeralis* (Stimpson, 1864)]
Ampithoe rubricata (Montagu, 1808)
Anachlorocurtis commensalis Hayashi, 1975
Anamixis linsleyi Barnard, 1955
Anamixis pacifica J. L. Barnard, 1955
Anchicaligus nautili Stebbing, 1900
Ancylomenes holthuisi (Bruce, 1969)
Ancylomenes magnificus (Bruce, 1979)
Ancylomenes pedersoni (Chace, 1958)
Anelasma squalicola Darwin, 1851
Annina kumari (Bowman, 1971)
Anomalocera patersoni Templeton, 1837
Anomopsyllus pranizoides G. O. Sars, 1921
Antecaridina lauensis (Edmondson, 1935)
Antheacheres duebenii M. Sars, 1857
Anthessius alatus Humes & Stock, 1965
Anthessius amicalis Humes & Stock, 1965
Anthessius discipedatus Humes, 1976
Anthessius solidus Humes & Stock, 1965
Antias uniramea [see *Halacarsantia uniramea* (Menzies & Miller, 1955)]
Antromysis anophelinae W. M. Tattersall, 1951
Antromysis peckorum Bowman, 1977
Antromysis pectorum [see *Antromysis peckorum* Bowman, 1977]
Aora typica Krøyer, 1845
Apanthura corsica Amar, 1953
Apocorophium acutum (Chevreux, 1908)
Apodomyzon brevicorne Stock, 1970
Apodomyzon longicorne Stock, 1970
Apohyale prevostii (H. Milne Edwards, 1830)
Arcoscalpellum botellinae (Barnard, 1924)
Aretopsis amabilis De Man, 1910
Aristeus coruscans [see *Plesiopenaeus coruscans* (Wood-Mason in Wood-Mason & Alcock, 1891)]
Aristeus virilis (Spence Bate, 1881)
Aristias neglectus Hansen, 1887
Armases roberti (H. Milne Edwards, 1853)
Armatobalanus durhami (Zullo, 1961) [see *Hexacreusia durhami* (Zullo, 1961)]
Artemia salina (Linnaeus, 1758)
Asellus aquaticus (Linnaeus, 1758)
Astacilla longicornis (Sowerby, 1806)

Astacopsis gouldi (Clark, 1936)
Asterocheres parvus Giesbrecht, 1897
Astyra zenkevitchi Birstein & M. Vinogradov, 1955
Astyra zenkevithchi [see *Astyra zenkevitchi* Birstein & M. Vinogradov, 1955]
Astyroides carinatus Birstein & Vinogradova, 1960
Attheyella (Chappuisiella) inopinata Chappuis, 1931
Attheyella (Chappuisiella) ruttneri Chappuis, 1931
Attheyella inopinata [see *Attheyella (Chappuisiella) inopinata* Chappuis, 1931]
Attheyella pilosa Chappuis, 1929
Attheyella ruttneri [see *Attheyella (Chappuisiella) ruttneri* Chappuis, 1931]
Atthyella pilosa [see *Canthocamptus pilosa* (Chappuis, 1929)]
Atthyella trispinosa [see *Canthocamptus trispinosus* Brady, 1880]
Austromegabalanus psittacus (Molina, 1788)
Austroniphargus bryophilus Monod, 1925
Austroongerthona picta [see *Tachaea picta* (Riek, 1967)]
Autonoe longipes (Lilljeborg, 1852)
Baccalaureus argalicornis Brattström, 1936
Baccalaureus japonicus Broch, 1929
Baccalaureus maldivensis maldivensis Pyefinch, 1934
Bagatus minutus [see *Carpias minutus* (Richardson, 1902)]
Balaenophilus unisetus P. O. Aurivillius, 1879
Balanus concavus pacificus Pilsbry, 1916
Balanus crenatus Bruguière, 1789
Barbouria cubensis (Von Martens, 1872)
Basipodella harpacticola Becker, 1975
Bathyceradocus stephenseni Pirlot, 1934
Bathynella natans Vejdovsky, 1882
Bathynomus giganteus A. Milne-Edwards, 1879
Bathyporeia robertsoni [see *Bathyporeia sarsi* Watkin, 1938]
Bathyporeia sarsi Watkin, 1938
Bathyschraderia magnifica Dahl, 1959
Benthesicymus bartletti S. I. Smith, 1882
Betaeus ensenadensis Glassell, 1938
Birgus latro (Linnaeus, 1767)
Boeckosimus normani (Sars, 1895)
Bopyrissa diogeni (Popov, 1929)
Bopyrissa fraissei (Carayon, 1943)
Boscia anglia [see *Megatrema anglicum* (G. B. Sowerby II, 1823)]
Botellina pinnata Pearcey, 1908
Brachycarpus biunguiculatus (Lucas, 1846)
Brincoxelia abyssalis [see *Princaxelia abyssalis* Dahl, 1959]
Bryocamptus pygmaeus (G. O. Sars, 1863)
Bryocyclops anninae (Menzel, 1926)
Bryocyclops bogoriensis (Menzel, 1926)
Bryocyclops chappuisi Kiefer, 1928
Bythograea microps de Saint Laurent, 1984
Bythograea thermydron Williams, 1980
Calanus hyperboreus Krøyer, 1838
Calappa flammea (J. F. W. Herbst, 1794)
Calcinus tubularis (Linnaeus, 1767)

Callianassa californiensis [see *Neotrypaea californiensis* (Dana, 1854)]
Calliasmata pholidota Holthuis, 1973
Callinectes ornatus Ordway, 1863
Callinectes sapidus M. J. Rathbun, 1896
Calliopiella michaelseni Schellenberg, 1925
Callistocypris zlotini Schornikov, 1980
Campecopea hirsuta (Montagu, 1804)
Candona compressa Koch, 1837
Candona pratensis Hartwig, 1901
Candonopsis anisitsi Daday, 1905
Cantabroniscus primitivus Vandel, 1965
Canthocamptus pilosa (Chappuis, 1929)
Canthocamptus trispinosus Brady, 1880
Canthocamptus vejdovskyi (Mrazek, 1893)
Caphyra alcyoniophila Monod, 1938
Caprella acanthifera Leach, 1814
Caprella erethizon Mayer, 1901
Caprella fretensis Stebbing, 1878
Caprella linearis (Linnaeus, 1767)
Caprella tuberculata Guérin, 1836
Carcinus maenas (Linnaeus, 1758)
Cardiophilus baeri G. O. Sars, 1896
Cardisoma hirtipes Dana, 1851
Caridina troglodytes Holthuis, 1978
Carpias minutus (Richardson, 1902)
Cassidinidea quadricarinata Pillai, 1954
Catinia plana Bocquet & Stock, 1957
Cephalothrix galatheae Dieck, 1874
Cerapus abditus [see *Ericthonius punctatus* (Spence Bate, 1857)]
Cerapus brasiliensis [see *Ericthonius punctatus* (Spence Bate, 1857)]
Cerapus tubularis Say, 1817
Ceratolana papuae Bowman, 1977
Cerberusa caeca Holthuis, 1979
Ceriodaphnia cornuta G. O. Sars, 1885
Chaenostoma boscii (Audouin, 1826)
Chauliodoniscus reyssi (Chardy, 1974)
Cheiriphotis megacheles (Giles, 1885)
Chelonibia manati Pilsbry, 1916
Chelonibia patula (Ranzani, 1818)
Chelorchestia costaricana (Stebbing, 1906)
Chelura terebrans Philippi, 1839
Chiromantes haematocheir (De Haan, 1833)
Chlorodiella nigra (Forskål, 1775)
Chlorotocella gracilis Balss, 1914
Cholidya polypi [see *Cholydiella polypi* (Farran, 1914)]
Cholydiella polypi (Farran, 1914)
Choniomyzon panuliri Pillai, 1962
Chthamalophilus delagei Bocquet-Védrine, 1957
Chthamalus stellatus (Poli, 1795)
Chydorus ovalis Kurz, 1875

Chydorus sphaericus (O. F. Müller, 1785)
Cilicaea splendida [authority not retrieved]
Cirolana lineata F. A. Potts, 1915
Cirolana narica [see *Natatolana narica* (Bowman, 1971)]
Cirolana venusticauda Stebbing, 1902
Cleantis phryganea (Hale, 1924)
Cleantis prismatica (Risso, 1826)
Cleantis tubicola (Thomson, 1885)
Cleonardo longipes Stebbing, 1888
Cleonardo longirostris [see *Cleonardo longipes* Stebbing, 1888]
Clibanarius erythropus (Latreille, 1818)
Colidotea rostrata (Benedict, 1898)
Colomastix pusilla Grube, 1861
Conchoderma auritum (Linnaeus, 1758)
Conchoderma virgatum (Spengler, 1789)
Conchodytes tridacnae Peters, 1852
Coronula diadema (Linnaeus, 1767)
Corophium shoemakeri Monod, 1955
Corophium uenoi Stephensen, 1938
Corophium volutator (Pallas, 1766)
Crassicorophium bonellii (H. Milne Edwards, 1830)
Cryptolepas rachianecti Dall, 1872
Cryptopontius capitalis (Giesbrecht, 1895)
Cryptopontius minor Stock, 1965
Ctenopontonia cyphastreophila Bruce, 1979
Cubaris granulatus Collinge, 1915
Cumopsis goodsir (Van Beneden, 1851)
Cyamus rhytinae (J. F. Brandt, 1846)
Cyclocypris laevis (O. F. Müller, 1776)
Cyclops quadricornis (Linnaeus, 1758)
Cymo andreossyi (Audouin, 1826)
Cymo melanodactylus De Haan, 1833
Cymodoce bifida Leach, 1818
Cymodopsis gorgoniae Baker, 1926
Cypridina parasitica [see *Vargula parasitica* (Wilson, 1913)]
Cypridina squamosa [see *Skogsbergia squamosa* (G. W. Müller, 1894)]
Cypris balnearia Moniez, 1939
Cyproniscus cypridinae (G. O. Sars, 1882)
Cyproniscus cyprinidinae [see *Cyproniscus cypridinae* (G. O. Sars, 1882)]
Dactylopusia neglecta (G. O. Sars, 1905)
Dahlella caldariensis Hessler, 1984
Danalia hapalocarcini Fize, 1955
Danalia ypsilon Smith, 1906-1907
Daphnia ambigua Scourfield, 1946
Dardanus arrosor (Herbst, 1796)
Dardanus sanguinolentus (Quoy & Gaimard, 1824)
Dardanus tinctor (Forskål, 1775)
Darwinula malayica Menzel, 1923
Darwinula zimmeri Menzel, 1916
Dhalella caldariensis Hessler, 1984

Diacyclops bisetosus (Rehberg, 1880)
Diacyclops nanus (G. O. Sars, 1863)
Diacyclops pulchellus [see *Diacyclops nanus* (G. O. Sars, 1863)]
Diarthrodes feldmanni Bocquet, 1953
Diastylis goodsiri (Bell, 1855)
Dichelina phormosomae Stephensen, 1933
Dies quadricarinatus [see *Cassidinidea quadricarinata* Pillai, 1954]
Diogenes pugilator (Roux, 1829)
Discias exul Kemp, 1920
Discias mvitae Bruce, 1976 [see *Discias exul* Kemp, 1920]
Domecia glabra Alcock, 1899
Domecia hispida Eydoux & Souleyet, 1842
Donsiella limnoriae Stephensen, 1936
Dosima fascicularis (Ellis & Solander, 1786)
Dotilla fenestrata Hilgendorf, 1869
Dotilla myctiroides (H. Milne Edwards, 1852)
Duplorbis calathurae Smith, 1906
Dynamene ramuscula Baker, 1908
Dynamenella australis [see *Ischyromene australis* (Richardson, 1906)]
Dynamenella moorei Richardson, 1905 [see *Dynamenella perforata* (Moore, 1901)]
Dynamenella octoloba [authority not retrieved]
Dynamenella perforata (Moore, 1901)
Dynamopsis dianae [see *Paradella dianae* (Menzies, 1962)]
Ectocyclops medius Kiefer, 1930
Ectocyclops phaleratus (Koch, 1838)
Eisothistos macrurus Wägele, 1979
Eisothistos pumilus Wägele, 1979
Elaphoidella bromeliaecola (Chappuis, 1928)
Elaphoidella cornuta Chappuis, 1931
Elaphoidella elegans Chappuis, 1931
Elaphoidella malayica (Chappuis, 1928)
Elaphoidella sewelli (Chappuis, 1928)
Elaphoidella thienemanni Chappuis, 1931
Elasmopus calliactis Edmondson, 1951
Elpidium bromeliarum Müller, 1880
Enalcyonium parvum [see *Lamippula parva* (Zulueta, 1908)]
Enalcyonium rubicundum Olsson, 1869
Enalcyonium setigerum (Zulueta, 1908)
Entomolepis adriae [see *Entomopsyllus adriae* (Eiselt, 1959)]
Entomopsyllus adriae (Eiselt, 1959)
Eochionelasmus ohtai Yamaguchi, 1990
Epactophanes muscicola (Richters, 1900)
Epactophanes richardi Mràzek, 1893
Episesarma mederi (H. Milne Edwards, 1853)
Ericthonius brasiliensis [see *Ericthonius punctatus* (Spence Bate, 1857)]
Ericthonius pugnax (Dana, 1852)
Ericthonius punctatus (Spence Bate, 1857)
Ethusina abyssicola S. I. Smith, 1884
Ethusina alba (Filhol, 1884)
Eucypris latissima Muller, 1898

Eucypris neumanni G. W. Muller, 1910
Euplax bosci [see *Chaenostoma boscii* (Audouin, 1826)]
Eurydice affinis H. J. Hansen, 1905
Eurydice pulchra Leach, 1815
Eurystheus maculatus [see *Gammaropsis maculata* (Johnston, 1828)]
Eurythenes gryllus Lichtenstein, 1822
Eusirus fragilis Birstein & M. Vinogradov, 1960
Eusphaeroma ovatum [see *Gnorimosphaeroma ovatum* (Gurjanova, 1933)]
Excirolana natalensis (Vanhoeffen, 1914)
Excirolana orientalis (Dana, 1853)
Exosphaeroma ramusculum [see *Dynamene ramuscula* Baker, 1908]
Fortimesus gigas (Birstein, 1960)
Gaillardiellus superciliaris (Odhner, 1925)
Galapsiellus leleuporum (Monod, 1970)
Gammaropsis anaculata (Johnston, 1828)
Gammaropsis nitida (Stimpson, 1853)
Gecarcoidea natalis (Pocock, 1888)
Geosesarma noduliferum (De Man, 1892)
Gigantocypris agassizi G. W. Müller, 1895
Gitanopsis paguri [see *Afrogitanopsis paguri* (Myers, 1974)]
Gitanopsis pusilla [see *Hourstonius pusilla* (K. H. Barnard, 1916)]
Glyptelasma carinatum (Hoek, 1883)
Glyptelasma gracilius (Pilsbry, 1907)
Gnathia maxillaris (Montagu, 1804)
Gnathomysis gerlachei [see *Heteromysis (Gnathomysis) gerlachei* (Bonnier & Pérez, 1902)]
Gnathophausia ingens (Dohrn, 1870)
Gnorimosphaeroma oregonensis (Dana, 1853)
Gnorimosphaeroma ovatum (Gurjanova, 1933)
Gomphocythere angresta [see *Gomphocythere angulata* Lowndes, 1932]
Gomphocythere angulata Lowndes, 1932
Gonophysema gullmarensis Bresciani & Lützen, 1960
Hadrothoe crosnieri Humes, 1975
Halacarsantia uniramea (Menzies & Miller, 1955)
Halice macronyx (Stebbing, 1888)
Halice subquarta Birstein & M. Vinogradov, 1960
Halicyclops caridophilus Humes, 1947
Haliporoides triarthrus Stebbing, 1914
Halocaridina rubra Holthuis, 1963
Hapalocarcinus marsupialis Simpson, 1859
Haplomesus gigas [see *Fortimesus gigas* (Birstein, 1960)]
Haploniscus unicornis Menzies, 1956
Haploops tubicola Lilljeborg, 1855
Harpacticus pulex Humes, 1964
Harpilius consobrinus De Man, 1902
Harpilius lutescens Dana, 1852
Harpinia excavata Chevreux, 1900
Harpinia spaercki [see *Harpiniopsis spaercki* (Dahl, 1959)]
Harpiniopsis spaercki (Dahl, 1959)
Harrietella simulans (T. Scott, 1894)
Harrovia elegans De Man, 1887

Hemicyclops acanthosquillae Humes, 1965
Hemicyclops perinsignis Humes, 1973
Hemilaophonte janinae Jakubisiak, 1932
Hepatus epheliticus (Linnaeus, 1763)
Herpotanais kirkegaardi Wolff, 1956
Heteralepas quadrata [see *Paralepas quadrata* (Aurivillius, 1894)]
Heterocarpus alphonsi [see *Heterocarpus dorsalis* Spence Bate, 1888]
Heterocarpus dorsalis Spence Bate, 1888
Heteromysis (*Gnathomysis*) *gerlachei* (Bonnier & Pérez, 1902)
Heteromysis (*Olivemysis*) *actiniae* Clarke, 1955
Heteromysoides cotti (Calman, 1932)
Heterotanais oerstedii (Krøyer, 1842)
Hexabathynella halophila Schminke, 1972
Hexacreusia durhami (Zullo, 1961)
Hippa pacifica (Dana, 1852)
Hippa testudinaria (J. F. W. Herbst, 1791)
Hippolysmata grabhami [see *Lysmata grabhami* (Gordon, 1935)]
Hippolyte commensalis [see *Alcyonohippolyte commensalis* (Kemp, 1925)]
Hirondellea gigas (Birstein & Vinogradov, 1955)
Homarus americanus H. Milne Edwards, 1837
Hoplopleon medusarum K. H. Barnard, 1932
Hourstonius pusilla (K. H. Barnard, 1916)
Huenia proteus De Haan, 1839
Hyachelia tortugae J. L. Barnard, 1967
Hyale grandicornis Krøyer, 1845
Hyale grimaldii Chevreux, 1891
Hyale perieri (Lucas, 1849)
Hyalella curvispina Shoemaker, 1942
Hymenopenaeus triarthrus [see *Haliporoides triarthrus* Stebbing, 1914]
Hyperoedesipus plumosus Nicholls & Milner, 1923
Hypocamptus brehmi (Van Douwe, 1922)
Ibla cumingi Darwin, 1851
Idotea balthica (Pallas, 1772)
Idotea chelipes (Pallas, 1766)
Idotea neglecta G. O. Sars, 1897
Idotea pelagica Leach, 1815
Idotea viridis [see *Idotea chelipes* (Pallas, 1766)]
Ilyoplax ceratophorus [see *Tmethypocoelis ceratophora* (Koelbel, 1897)]
Ilyoplax delsmani De Man, 1926
Ilyoplax elegans [see *Metaplax elegans* De Man, 1888]
Ingolfiella fontinalis Stock, 1977
Intersunaristes dardani (Humes & Ho, 1969)
Iphimedia eblanae Heller, 1867
Ischnopontonia lophos (Barnard, 1962)
Ischyromene australis (Richardson, 1906)
Jaera hopeana Costa, 1853
Jaera (*Jaera*) *nordmanni nordica insulana* Veuille, 1976
Jaera nordica insulana [see *Jaera* (*Jaera*) *nordmanni nordica insulana* Veuille, 1976]
Jaera nordmanni guernei Dollfus, 1889
Jassa falcata (Montagu, 1808)

Katamysis warpachowskyi G. O. Sars, 1877
Koonunga cursor Sayce, 1908
Kupellonura serritelson Wägele, 1981
Lamippe bouligandi Laubier, 1972
Lamippella faurei Bouligand & Delamare Deboutteville, 1959
Lamippina aciculifera (Zulueta, 1908)
Lamippula parva (Zulueta, 1908)
Langitanais willemoesi (Studer, 1883)
Laticorophium baconi (Shoemaker, 1934)
Latreutes anoplonyx Kemp, 1914
Latreutes mucronatus (Stimpson, 1860)
Laura gerardiae De Lacaze-Duthiers, 1865
Lembos longipes [see *Autonoe longipes* (Lilljeborg, 1852)]
Lepas anatifera Linnaeus, 1758
Lepas australis Darwin, 1851
Lepas fascicularis [see *Dosima fascicularis* (Ellis & Solander, 1786)]
Lepas hilli Leach, 1818
Lepechinella ultraabyssalis Birstein & Vinogradov, 1960
Lepechinella wolffi wolffi Dahl, 1959
Leptocheirus pilosus Zaddach, 1844
Leptodora kindti (Focke, 1844)
Leptomithrax longipes (Thomson, 1902)
Lernaea cyprinacea Linnaeus, 1758
Lernaea esocina [see *Lernaea cyprinacea* Linnaeus, 1758]
Lernaea ranae [see *Lernaea cyprinacea* Linnaeus, 1758]
Lernaeocera branchialis (Linnaeus, 1767)
Leucothoe incisa Robertson, 1892
Leucothoe spinicarpa (Abildgaard, 1789)
Leucothoides pacifica (Barnard, 1955)
Leucothoides Shoemaker, 1933 [accepted as *Anamixis* Stebbing, 1897]
Lichomolgus hippopi Humes, 1976
Lichomolgus longicauda (Claus, 1860)
Lichomolgus tridacnae Humes, 1972
Ligia perkinsi (Dollfus, 1900)
Ligia philoscoides Jackson, 1938
Ligia platycephala (Van Name, 1925)
Ligia simoni (Dollfus, 1893)
Ligur uveae [see *Parhippolyte uveae* Borradaile, 1900]
Liljeborgia caeca Birstein & Vinogradova, 1960
Limnomysis benedeni Czerniavsky, 1882
Limnoria lignorum (Rathke, 1799)
Limnoria tripunctata Menzies, 1951
Linaresia mammillifera Zulueta, 1908
Litoscalpellum giganteum (Gruvel, 1902)
Lysiosquilla maculata [see *Lysiosquillina maculata* (J. C. Fabricius, 1793)]
Lysiosquillina maculata (J. C. Fabricius, 1793)
Lysmata amboinensis (De Man, 1888)
Lysmata californica (Stimpson, 1866)
Lysmata grabhami (Gordon, 1935)
Lysmata seticaudata (Risso, 1816)

Macrobrachium americanum Spence Bate, 1868
Macrobrachium carcinus (Linnaeus, 1758)
Macrobrachium grandimanus (J. W. Randall, 1840)
Macrocheira kaempferi (Temminck, 1836)
Macrophthalmus convexus Stimpson, 1858
Macrophthalmus dilatatus De Haan, 1853
Macrophthalmus japonicus (De Haan, 1835)
Macrophthalmus grandidieri A. Milne-Edwards, 1867
Macrostylis galatheae Wolff, 1956
Maja brachydactyla Balss, 1922
Maja squinado (Herbst, 1788)
Malacolepas conchicola Hiro, 1933
Maraenobiotus vejdovskyi Màzek, 1893
Megabalanus decorus (Darwin, 1854)
Megabalanus stultus (Darwin, 1854)
Megabalanus tintinnabulum (Linnaeus, 1758)
Megalasma striatum Hoek, 1883
Meganyctiphanes norvegica (M. Sars, 1857)
Megatrema anglicum (G. B. Sowerby II, 1823)
Melita obtusata [see *Abludomelita obtusata* (Montagu, 1813)]
Menaethius monoceros (Latreille, 1825)
Menippe mercenaria (Say, 1818)
Mesocypris terrestris Harding, 1953
Mesoglicola delagei Quidor, 1906
Mesomysis intermedia [see *Paramysis intermedia* (Czerniavsky, 1882)]
Metabetaeus lohena A. H. Banner & D. M. Banner, 1960
Metabetaeus minutus (Whitelegge, 1897)
Metaceradocoides vitjazi Birstein & Vinogradov, 1960
Metacypris bromeliarum (Muller, 1881)
Metacypris laesslei Tressler, 1956
Metapenaeopsis commensalis Borradaile, 1898
Metaplax elegans De Man, 1888
Metis holothuriae (A. Milne-Edwards, 1891)
Metopa alderi (Spence Bate, 1857)
Metopa borealis G. O. Sars, 1882
Metopa glacialis (Krøyer, 1842)
Metopa groenlandica Hansen, 1888
Metopa solsbergi Schneider, 1884
Metopaulias depressus Rathbun, 1896
Metridia longa (Lubbock, 1854)
Mexiconiscus laevis (Rioja, 1956)
Microdeutopus gryllotalpa A. Costa, 1853
Microprotopus maculatus Norman, 1867
Mictyris longicarpus Latreille, 1806
Minyaspis aurivillii (Stebbing, 1900)
Mitella mitella [see *Pollicipes pollicipes* (Gmelin, 1790)]
Mithrax (Mithraculus) commensalis [see *Mithraculus cinctimanus* Stimpson, 1860]
Monocorophium acherusicum (Costa, 1853)
Monocorophium insidiosum (Crawford, 1937)
Moraria arboricola Scourfield, 1915

Moraria monticola (Menzel, 1912)
Moraria sphagnicola Gurney, 1930
Morlockia ondinae (Garcia-Valdecasas, 1984)
Munidopsis polymorpha Koelbel, 1892
Myzomolgus stupendus Bocquet & Stock, 1957
Myzopontius australis Nicholls, 1944
Natatolana narica (Bowman, 1971)
Nebaliopsis typica G. O. Sars, 1887
Nematoscelis megalops (G. O. Sars, 1883)
Neoglyphea inopinata Forest & de Saint-Laurent, 1975
Neohela monstrosa (Boeck, 1861)
Neohyssura spinicauda (Walker, 1901)
Neolepas zevinae Newman, 1979
Neopetrolisthes maculatus (H. Milne Edwards, 1837)
Neopetrolisthes oshimai [see *Neopetrolisthes maculatus* (H. Milne Edwards, 1837)]
Neotanais hadalis Wolff, 1956
Neotrypaea californiensis (Dana, 1854)
Nephrops norvegicus (Linnaeus, 1758)
Neptunus pelagicus [see *Portunus pelagicus* (Linnaeus, 1758)]
Nesotanais lacustris Shiino, 1966-1968
Nicothoe astaci Audouin & H. Milne Edwards, 1825-1826
Nicothoe tumulosa Cressey, 1976
Nippochelura brevicaudata (Shiino, 1948)
Nitokra divaricata Chappuis, 1923
Nitokra medusaea Humes, 1953
Nogagella siphonophoriae Rose, 1933
Nyctiphanes couchi (Bell, 1853)
Octolasmis aymonini Pilsbry, 1907
Octolasmis bathynomi [see *Temnaspis bathynomi* (Annandale, 1906)]
Octolasmis muelleri (Coker, 1902)
Octolasmis orthogonia (Darwin, 1851)
Octolasmis warwicki Gray, 1825
Ocypode ceratophthalmus (Pallas, 1772)
Ocypode cordimanus Latreille, 1818
Opecarcinus crescentus (Edmondson, 1925)
Orchomene commensalis Chevreux & Fage, 1925
Orchomenella abyssorum Stebbing, 1888 [see *Abyssorchomene abyssorum* (Stebbing, 1888)]
Orchomenella commensalis [see *Orchomene commensalis* Chevreux & Fage, 1925]
Orthagoriscicola muricatus (Krøyer, 1837)
Oxynaspis aurivillii [see *Minyaspis aurivillii* (Stebbing, 1900)]
Pagurapseudes bouryi [see *Pagurotanais bouryi* Bouvier, 1918]
Pagurolepas conchicola Stubbings, 1940
Paguropsis typica Henderson, 1888
Pagurotanais bouryi Bouvier, 1918
Pagurus alcocki (Balss, 1911)
Pagurus bernhardus (Linnaeus, 1758)
Pagurus brevipes [see *Dardanus tinctor* (Forskål, 1775)]
Pagurus prideaux Leach, 1815
Pagurus varians [see *Diogenes pugilator* (Roux, 1829)]
Palaemon adspersus Rathke, 1837

Palaemon debilis Dana, 1852
Palaemon elegans Rathke, 1837
Palaemonella burnsi Holthuis, 1973
Pandalus montagui Leach, 1814
Panopeus herbstii H. Milne Edwards, 1834
Panulirus homarus (Linnaeus, 1758)
Panulirus interruptus (J. W. Randall, 1840)
Panulirus japonicus (Von Siebold, 1824)
Panulirus ornatus (J. C. Fabricius, 1798)
Panulirus versicolor (Latreille, 1804)
Paracilicaea mossambica Barnard, 1914
Paracilicaea teretron Barnard, 1955
Paracirolana platysoma [neither combination nor authority retrieved]
Paracis squamata (Nutting, 1910)
Paracleistostoma depressum De Man, 1895
Paradella dianae (Menzies, 1962)
Paradynamenella psammophila [neither combination nor authority retrieved]
Paragnathia formica (Hesse, 1864)
Parajassa pelagica (Leach, 1814)
Paralepas quadrata (Aurivillius, 1894)
Paralithodes camtschaticus (Tilesius, 1815)
Paramacrochiron ennorense Reddiah, 1968
Paramacrochiron japonicum Humes, 1970
Paramacrochiron rhizostomae Reddiah, 1968
Paramacrochiron sewelli Reddiah, 1968
Paramithrax longipes [see *Leptomithrax longipes* (Thomson, 1902)]
Paramysis intermedia (Czerniavsky, 1882)
Parapagurus pictus [see *Sympagurus pictus* Smith, 1883]
Parapleustes commensalis Shoemaker, 1952
Parargissa galatheae J. L. Barnard, 1961
Parasphaeroma granosum [neither combination nor authority retrieved]
Paratypton siebenrocki Balss, 1914
Pardaliscoides longicaudatus Dahl, 1959
Parhippolyte uveae Borradaile, 1900
Parhyale fascigera Stebbing, 1897
Parhyale hawaiensis (Dana, 1853)
Pariambus typicus (Krøyer, 1844)
Pasiphaea princeps Smith, 1884
Penaeus monodon Fabricius, 1798
Peniculus fistula Von Nordmann, 1832
Pennella balaenoptera Koren & Danielssen, 1877
Pennella varians Steenstrup & Lütken, 1861
Peramphithoe humeralis (Stimpson, 1864)
Perforatus perforatus (Bruguière, 1789)
Periclimenaeus hecate (Nobili, 1904)
Periclimenella petitthouarsii (Audouin, 1826)
Periclimenes amethysteus (Risso, 1827)
Periclimenes brevicarpalis (Schenkel, 1902)
Periclimenes consobrinus (De Man, 1902) [see *Harpilius consobrinus* De Man, 1902]
Periclimenes holthuisi [see *Ancylomenes holthuisi* (Bruce, 1969)]

Periclimenes imperator A. J. Bruce, 1967
Periclimenes lutescens [see *Harpilius lutescens* Dana, 1852]
Periclimenes madreporae Bruce, 1969
Periclimenes magnificus [see *Ancylomenes magnificus* (Bruce, 1979)]
Periclimenes ornatus Bruce, 1969
Periclimenes pholeter Holthuis, 1973
Periclimenes rathbunae Schmitt, 1924
Periclimenes yucatanicus (Ives, 1891)
Petrarca bathyactidis Fowler, 1889
Petrolisthes maculatus [see *Neopetrolisthes maculatus* (H. Milne Edwards, 1837)]
Philorthagoriscus serratus (Krøyer, 1863)
Phreatoicopsis terricola Spencer & Hall, 1896
Phyllognathopus camptoides (Bozic, 1965)
Phyllognathopus coecus (Maupas, 1892)
Phymodius ungulatus (H. Milne Edwards, 1834)
Pilsbryscalpellum subalatum (Barnard, 1824)
Pinnixa franciscana Rathbun, 1918
Pinnixa lunzi Glassell, 1937
Pinnixa schmitti M. J. Rathbun, 1918
Pionodesmotes phormosomae Bonnier, 1898
Planes minutus (Linnaeus, 1758)
Platorchestia platensis (Krøyer, 1845)
Platyarthrus acropyga Chopra, 1924
Platyarthrus caudatus Aubert & Dollfus, 1890
Platyarthrus hoffmannseggii Brandt, 1833
Platylepas hexastylos ichthyophila Pilsbry, 1916
Pleonexes brevirostris (T. Scott & A. Scott)
Plesionika ensis (A. Milne-Edwards, 1881)
Plesiopenaeus armatus (Spence Bate, 1881)
Plesiopenaeus coruscans (Wood-Mason in Wood-Mason & Alcock, 1891)
Podoceropsis nitida [see *Gammaropsis nitida* (Stimpson, 1853)]
Podocerus brasiliensis [see *Ericthonius punctatus* (Spence Bate, 1857)]
Pollicipes pollicipes (Gmelin, 1790)
Polyascus gregaria (Okada & Miyashita, 1935)
Polycheria antarctica (Stebbing, 1875)
Polycheria osborni Calman, 1898
Polydectus cupulifer (Latreille, 1812) [in Milbert, 1812]
Pontella mediterranea (Claus, 1863)
Porcellana platycheles (Pennant, 1777)
Porcellidium echinophilum Humes & Gelerman, 1962
Portunus pelagicus (Linnaeus, 1758)
Potamocypris philotherma Rome, 1970
Potamon fluviatile (Herbst, 1785)
Potiicoara brasiliensis Pires, 1987
Princaxelia abyssalis Dahl, 1959
Procaris ascensionis Chace & Manning, 1972
Pseudanthessius latus Illg, 1950
Pseudanthessius nemertophilus Gallien, 1936
Pseudione diogeni [see *Bopyrissa diogeni* (Popov, 1929)]
Pseudione fraissei [see *Bopyrissa fraissei* (Carayon, 1943)]

Pseudocarcinus gigas Lamarck, 1818
Pseudocryptochirus crescentus Utinomi, 1944 [see *Opecarcinus crescentus* (Edmondson, 1925)]
Pseudohapalocarcinus ransoni Fize & Serène, 1956
Pseudomacrochiron stocki Reddiah, 1969
Pseudopagurus granulimanus (Miers, 1881)
Pseudotiron longicaudatus Pirlot, 1834
Psilomyzon pauciseta Stock, 1965
Pteropontius cristatus Giesbrecht, 1895
Pylocheles mortensenii Boas, 1926
Pyrgoma stockesi Gray, 1823 [see *Megatrema madreporarum* (Bosc, 1801)]
Quadrella cyrenae [see *Quadrella maculosa* Alcock, 1898]
Quadrella maculosa Alcock, 1898
Racilius compressus Paul'son, 1875
Rapipontonia galene (Holthuis, 1952)
Rectisura brachycephala (Birstein, 1957)
Rectisura herculea (Birstein, 1957)
Rectisura tenuispinis (Birstein, 1957)
Rectisura vitjazi (Birstein, 1957)
Rhizorhina ampeliscae Hansen, 1892
Ridgewayia fosshageni Humes & W. L. Smith, 1974
Rimicaris exoculata Williams & Rona, 1986
Rodriguezia mensabak (Cottarelli & Argano, 1977)
Ruffohyale milloti (Ruffo, 1958)
Sacculina carcini J. V. Thompson, 1836
Sacculina gregaria [see *Polyascus gregaria* (Okada & Miyashita, 1935)]
Sacodiscus ovalis (C. B. Wilson, 1944)
Sagmariasus verreauxi (H. Milne Edwards, 1851)
Scalopidia spinosipes Stimpson, 1858
Scalpellum aduncum [see *Weltnerium aduncum* (Aurivillius, 1894)]
Scalpellum cancellatum [see *Verum cancellatum* (Barnard, 1924)]
Scalpellum giganteum [see *Litoscalpellum giganteum* (Gruvel, 1902)]
Scalpellum hispidum [see *Amigdoscalpellum hispidum* (G. O. Sars, 1890)]
Scalpellum stearnsii Pilsbry, 1890
Scalpellum subalatum [see *Pilsbryscalpellum subalatum* (Barnard, 1824)]
Scleroplax granulata M. J. Rathbun, 1893
Scopimera bitympana Shen, 1930
Scopimera globosa De Haan, 1835
Semibalanus balanoides (Linnaeus, 1766)
Serolis scythei [see *Acanthoserolis schythei* (Lütken, 1858)]
Sesarma bidentatum Benedict, 1892
Sesarma bromeliarum [see *Armases roberti* (H. Milne Edwards, 1853)]
Sesarma haematocheir [see *Chiromantes haematocheir* (De Haan, 1833)]
Sesarma jarvisi Rathbun, 1914
Sesarma nodulifera Lenz, 1910
Sesarma obtusifrons Dana, 1851
Sesarma taeniolata [see *Episesarma mederi* (H. Milne Edwards, 1853)]
Sesarma verleyi Rathbun, 1914
Sestropontius bullifer Giesbrecht, 1899
Sewellochiron fidens Humes, 1969
Sinelobus stanfordi (Richardson, 1901)

Skogsbergia squamosa (G. W. Müller, 1894)
Spelaeogriphus lepidops Gordon, 1958
Speleonectes lucayensis Yager, 1981
Sphaeroma pentodon Richardson, 1904
Sphaeroma peruvianum Richardson, 1910
Sphaeroma polydemum [neither combination nor authority retrieved]
Sphaeroma retrolaeve Richardson, 1904
Sphaeroma serratum (Fabricius, 1787)
Sphaeroma sieboldii Dollfus, 1889
Sphaeroma terebrans Spence Bate, 1866
Sphaeroma venustissimum Monod, 1931
Sphaeroma walkeri Stebbing, 1905
Sphaeromonopsis amathitis [see *Sphaeromopsis amathitis* Holdich & Jones, 1973]
Sphaeromopsis amathitis Holdich & Jones, 1973
Spongicola venustus De Haan, 1844 [in De Haan, 1833-1850]
Staurosoma parasiticum Will, 1844
Stenopus hispidus (Olivier, 1811)
Stenopus scutellatus Rankin, 1898
Stenopus spinosus Risso, 1826
Stenothoe valida Dana, 1852
Stenula rubrovittata G. O. Sars, 1883
Stygotantulus stocki Boxshall & Huys, 1989
Stynocoides longicaudatus [see *Pseudotiron longicaudatus* Pirlot, 1834]
Sunaristes inaequalis Humes & Ho, 1969
Sunaristes paguri Hesse, 1867
Sympagurus pictus Smith, 1883
Synagoga mira Norman, 1888
Synopioides secundus Stebbing, 1888 [see *Halice macronyx* (Stebbing, 1888)]
Syscenus infelix Harger, 1880
Tachaea caridophaga (Riek, 1953)
Tachaea picta (Riek, 1967)
Tachidium brevicornis [see *Tachidius brevicornis* Lilljeborg, 1853]
Tachidius brevicornis Lilljeborg, 1853
Talitroides alluaudi (Chevreux, 1896)
Talitroides hortulanus (Caiman, 1912)
Talitroides pacificus Hurley, 1955
Talitroides topitotum (Burt, 1934)
Talitrus alluaudi [see *Talitroides alluaudi* (Chevreux, 1896)]
Talitrus saltator (Montagu, 1808)
Tanais cavolinii [see *Tanais dulongii* (Audouin, 1826)]
Tanais dulongii (Audouin, 1826)
Temnaspis bathynomi (Annandale, 1906)
Temnaspis tridens (Aurivillius, 1894)
Terrestricypris arborea Shornikov, 1980
Terrestricythere ivanovae Schornikov, 1969
Terrestricythere pratenesis Schornikov, 1980
Tetraclitella costata (Darwin, 1854)
Thalamita pilumnoides Borradaile, 1903
Thalamitoides quadridens A. Milne-Edwards, 1869
Thalassina anomala (Herbst, 1804)

Thalestris rhodymeniae (Brady, 1894)
Thermobathynella adami Capart, 1951
Thermosbaena mirabilis Monod, 1924
Thor amboinensis (De Man, 1888)
Thysanoessa inermis (Krøyer, 1846)
Thysanoessa longicaudata (Krøyer, 1846)
Thysanopoda cornuta Illig, 1905
Tisbe cucumariae Humes, 1957
Tisbe elongata (A. Scott, 1896)
Tisbe furcata (Baird, 1837)
Tisbe holothuriae Humes, 1957
Tisbe japonica Ho, 1982
Tmethypocoelis ceratophora (Koelbel, 1897)
Trandosia jenkinsae [neither combination nor authority retrieved]
Trichodactylus mensabak [see *Rodriguezia mensabak* (Cottarelli & Argano, 1977)]
Trifur tortuosus C. B. Wilson, 1917
Trischizostoma nicaeense (Costa, 1853)
Tritaeta gibbosa (Spence Bate, 1862)
Tropichelura insulae (Calman, 1910)
Tropocyclops prasinus (S. Fischer, 1860)
Trypetesa lateralis Tomlinson, 1953
Tuphacheres micropus Stock, 1965
Typhlatya galapagensis Monod & Cals, 1970
Typhlatya rogersi Chace & Manning, 1972
Tylos punctatus Holmes & Gay, 1909
Typhlopseudothelphusa juberthiei Delamare Debouteville, 1976
Typhlopseudothelphusa mitchelli Delamare Debouteville, 1976
Typhlopseudothelphusa mocinoi Rioja, 1952
Typhloschizidium cottarellii [see *Alloschizidium cottarellii* (Argano & Pesce, 1974)]
Typhlotricholigioides aquaticus Rioja, 1953
Uca (Australuca) bellator (White, 1847)
Uca (Australuca) signata (Hess, 1865)
Uca consobrinus [see *Uca lactea* (De Haan, 1835)]
Uca formosensis Rathbun, 1921
Uca (Gelasimus) vocans (Linnaeus, 1758)
Uca lactea (De Haan, 1835)
Uca marionis vocans [see *Uca (Gelasimus) vocans* (Linnaeus, 1758)]
Uca marionis Ward, 1928
Uca signatus [see *Uca (Australuca) signata* (Hess, 1865)]
Uca tangeri (Eydoux, 1835)
Uca (Tubuca) arcuata (De Haan, 1835)
Uca (Tubuca) urvillei (H. Milne Edwards, 1852)
Uca urvillei [see *Uca (Tubuca) urvillei* (H. Milne Edwards, 1852)]
Vanhoeffenura bicornis (Birstein, 1957)
Vanhoeffenura chelata (Birstein, 1957)
Vargula parasitica (Wilson, 1913)
Vermiliopsis infundibulum (Philippi, 1844)
Verruca stroemia (O. F. Müller, 1776)
Verum cancellatum (Barnard, 1924)
Viguierella caeca [see *Phyllognathopus coecus* (Maupas, 1892)]

Weltnerium aduncum (Aurivillius, 1894)
Weltnerium nymphocola (Hoek, 1883)
Xenobalanus globicipitis Steenstrup, 1851
Xylocheles macrops (Forest, 1987)
Zebrida adamsii White, 1847
Zenobia prismatica [see *Cleantis prismatica* (Risso, 1826)]
Zenobiana phryganea [see *Cleantis phryganea* (Hale, 1924)]

APPENDIX II
Names of species of non-Crustacea mentioned in this chapter, with their authority (and date, for animals), listed in alphabetical order per taxon

PROTISTA
[see also Sprague & Couchi, 1971]
    Dendrosomides paguri Collin, 1906
    Paracineta homari (Sand, 1899)
    Thecacineta cypridinae Collin, 1912

PLANTAE
    Diploneis smithii (Brébisson) Cleve [not: *Diplaneis*]
    Stigeoclonium australense K. Möbius
    Zingiber macradenium K. Schum.

ANIMALIA: NON-CRUSTACEA
PORIFERA:
    Acarnus ternatus Ridley, 1884
    Dysidea fragilis (Montagu, 1818)
    Euplectella aspergillum Owen, 1841
    Suberites domuncula (Olivi, 1792)

CNIDARIA [= COELENTERATA]:
HYDROZOA:
    Abylopsis tetragona (Otto, 1823)
    Aglaophenia cupressina Lamouroux, 1816
    Clytia johnstoni [see *Clytia hemisphaerica* (Linnaeus, 1767)]
    Clytia hemisphaerica (Linnaeus, 1767)
    Obelia geniculata (Linnaeus, 1758)
    Obelia longissima (Pallas, 1767)
    Physophora hydrostatica Forskål, 1775
    Sculeolaria chuni [see *Sulculeolaria chuni* (Lens & Van Riemsdijk, 1908)]
    Sulculeolaria chuni (Lens & Van Riemsdijk, 1908)
    Tima bairdii (Johnston, 1833)
    Tubularia indivisa Linnaeus, 1758

SCYPHOZOA:
    Cassiopea xamachana Bigelow, 1892
    Chrysaora quinquecirrha (Desor, 1848)
    Dactylometra quinquecirrata [see *Chrysaora quinquecirrha* (Desor, 1848)]
    Lychnorhiza malayensis Stiasny, 1920
    Mastigias papua (Lesson, 1830)
    Thysanostoma thysanura Haeckel, 1880

ANTHOZOA:
   Actinia equina (Linnaeus, 1758)
   Actinoloba dianthus de Blainville, 1830 [see *Metridium senile* (Linnaeus, 1761)]
   Adamsia palliata (Fabricius, 1779)
   Alcyonium palmatum Pallas, 1766
   Anemonia sulcata (Pennant, 1777)
   Anthopleura elegantissima (Brandt, 1835)
   Anthoptilum murrayi Kölliker, 1880
   Aristias neglectus Hansen, 1888
   Bartholomea annulata (Lesueur, 1817)
   Calliactis armillata Verrill, 1928
   Calliactis parasitica (Couch, 1842)
   Corynactis viridis Allman, 1846
   Metridium senile (Linnaeus, 1761)
   Neoaiptasia commensali Parulekar, 1969
   Paracis squamata (Nutting, 1910)
   Paramuricea clavata (Risso, 1826)
   Pavona gigantea Verrill, 1869
   Pocillopora damicornis (Linnaeus, 1758)
   Stichodactyla helianthus (Ellis, 1768)
   Stoichactis helianthus [see *Stichodactyla helianthus* (Ellis, 1768)]
   Triactis producta Klunzinger, 1877

PLATYHELMINTHES: TURBELLARIA:
   Fecampia balanicola Christensen & Hurley, 1977
   Fecampia spiralis [see *Kronborgia spiralis* (Baylis, 1949)]
   Kronborgia amphipodicola Christensen & Kanneworff, 1964
   Kronborgia caridicola Kanneworff & Christensen, 1966
   Kronborgia spiralis (Baylis, 1949)

BRYOZOA:
   Cellepora samboangensis Busk, 1884 [see *Celleporaria oculata* (Lamarck, 1816)]
   Cellepora senegambiensis [probably *error pro*: *Cellepora samboangensis* Busk, 1884]
   Celleporaria oculata (Lamarck, 1816)
   Conopeum commensale Kirkpatrick & Metzelaar, 1922
   Discoporella umbellata (Defrance, 1823)
   Janaria mirabilis Stechow, 1921
   Sarsiflustra abyssicola (G. O. Sars, 1872)

BRACHIOPODA:
   Hemithiris psittacea (Gmelin, 1790)
   Macandrevia cranium (O. F. Müller, 1776)
   Rynchonella psittacea [see *Hemithiris psittacea* (Gmelin, 1790)]
   Terebratulina caputserpentis [see *Terebratulina retusa* (Linnaeus, 1758)]
   Terebratulina retusa (Linnaeus, 1758)

NEMERTEA:
   Cephalothrix galatheae Dieck, 1874

NEMATOMORPHA:
   Nectonema agile Verrill, 1879

ECHIURA:
   Urechis caupo Fisher & MacGinitie, 1928

ANNELIDA:
   Lepidasthenia digueti Gravier, 1905
   Ophryotrocha geryonicola (Esmark, 1878)
   Terebellides stroemii (Sars, 1835)

MOLLUSCA:
   Acanthopleura granulata (Gmelin, 1791)
   Buccinum undatum Linnaeus, 1758
   Chiton tuberculatus [see *Chiton (Chiton) tuberculatus* Linnaeus, 1758; and/or *Chiton tuberculatus* Leach, 1852]
   Chiton tuberculatus Leach, 1852 [see *Leptochiton cancellatus* (Sowerby, 1840)]
   Chiton (Chiton) tuberculatus Linnaeus, 1758
   Cucullaea labiata (Lightfoot, 1786)
   Didacna baeri (Grimm, 1877)
   Irus mitis (Deshayes, 1854)
   Leptochiton cancellatus (Sowerby, 1840)
   Modiolaria discors [see *Musculus discors* (Linnaeus, 1767)]
   Musculus discors (Linnaeus, 1767)
   Mytilus edulis Linnaeus, 1758
   Mytilus edulis platensis d'Orbigny, 1842
   Mytilus platensis [see *Mytilus edulis platensis* d'Orbigny, 1842]
   Nucella lapillus (Linnaeus, 1758)
   Pandora glacialis Leach in Ross, 1819
   Patella vulgata Linnaeus, 1758
   Venerupis mitis [see *Irus mitis* (Deshayes, 1854)]

ARTHROPODA:
   ARACHNIDA: PYCNOGONIDA:
      Boreonymphon robustum (Bell, 1855)
      Pallenopsis fluminensis (Krøyer, 1844)
      Phoxichilidium fluminense [see *Pallenopsis fluminensis* (Krøyer, 1844)]

   ARACHNIDA: XIPHOSURA:
      Tachypleus gigas (O. F. Müller, 1785)

   INSECTA [= HEXAPODA]:
      Acropyga acutiventris Roger, 1862
      Bothroponera tesserinoda Forel, 1922 [also compare *Pachycondyla sulcata* Mayr var. *sulcatotesserinoda* Forel, 1900]
      Lissocephala powelli Carson & Wheeler, 1973
      Pachycondyla sulcata Mayr var. *sulcatotesserinoda* Forel, 1900
      Tetramorium brevicorne Bondroit, 1918

CHORDATA: CRANIATA:
   "PISCES":
      Cetorhinus maximus (Gunnerus, 1765)
      Coryphaena hippurus Linnaeus, 1758
      Cryptocentrus cryptocentrus (Valenciennes in Cuvier & Valenciennes, 1837)
      Diodon hystrix Linnaeus, 1758
      Epinephelus lanceolatus (Bloch, 1790)
      Etmopterus spinax (Linnaeus, 1758)
      Mola mola (Linnaeus, 1758)
      Mullus surmuletus Linnaeus, 1758

Scorpaena scrofa (Linnaeus, 1758)
Serranus lanceolatus [see *Epinephelus lanceolatus* (Bloch, 1790)]
Sphyrna zygaena (Linnaeus, 1758)
Xiphias gladius Linnaeus, 1758

AMPHIBIA:
Cynops pyrrhogaster (Boie, 1826)
Dienictylus pyrrhogaster [see *Cynops pyrrhogaster* (Boie, 1826)]
Pelobates cultripes (Cuvier, 1829)
Rana clamitans Latreille, 1801

MAMMALIA:
Cephalorhynchus hectori (P. J. van Beneden, 1881)
Hydrodamalis gigas (Zimmerman, 1780)
Trichechus manatus Linnaeus, 1758
Tursiops truncatus (Montagu, 1821)

BIBLIOGRAPHY

Note. — Please note that in this Bibliography series of "same author – same year" references start with the year as such in the first reference [0000], then with an "a" added for the second reference [0000a], then a "b" for the third reference [0000b], and so on.

ADAMOWICZ, S. J. & A. PURVIS, 2005. How many branchiopod crustacean species are there? Quantifying the components of underestimation. — Glob. Ecol. Biogeogr., **14** (5): 455-468.

AHYONG, S. T., J. K. LOWRY, M. ALONSO, R. N. BAMBER, G. A. BOXSHALL, P. CASTRO, S. GERKEN, G. S. KARAMAN, J. W. GOY, D. S. JONES, K. MELAND, D. C. ROGERS & J. SVAVARSSON, 2011. Subphylum Crustacea Brünnich, 1772. — In: Z.-Q. ZHANG (ed.), Animal biodiversity: an outline of higher-level classification and survey of taxonomic richness. Zootaxa, **3165**: 165-191.

ALTES, J., 1982. Les Liriopsidae. — Bull. Soc. Hist. nat. Afr. Nord, **69** (3-4): 3-35. [1981]

ALTEVOGT, R., 1955. Beobachtungen und Untersuchungen an indischen Winkerkrabben. — Z. Morph. Ökol. Tiere, **43**: 501-522.

— —, 1957. Untersuchungen zur Biologie, Ökologie und Physiologie indischer Winkerkrabben. — Z. Morph. Ökol. Tiere, **46**: 1-110.

— —, 1959. Ökologische und ethologische Studien an Europas einiger Winkerkrabbe *Uca tangeri* Eydoux. — Z. Morph. Ökol. Tiere, **49**: 123-146.

— —, 1969. An ethological reproductive isolation mechanism in sympatric species of *Uca* (Ocypodidae) of the eastern Pacific. — Forma et Functio, **3**: 238-248.

ANDRÉ, M., 1937. Relations entre la distribution géographique des écrevisses et celle de leurs parasites. — C. R. Soc. Biogéogr., **14** (120): 31-37.

— —, 1937a. Coquilles vides de Bivalves habitées par des Crustacés. — J. Conchyl., **81**: 72-81.

ANDRÉ, M. & E. LAMY, 1933. Crustacés xylophages et lithophages. — Bull. Inst. océanogr. Monaco, **626**: 1-21.

— — & — —, 1936. Colonies d'Hydraires ou de Bryozoaires fixées sur des coquillles à Pagures. — Bull. Soc. zool. Fr., **61**: 94-99.

— — & — —, 1939. Action des Pagures sur les coquilles qu'ils habitent. — J. Conchyl., **83**: 234-242.

ANKER, A., 2000. Taxonomical problems of the goby-associated species of *Alpheus* (Decapoda, Alpheidae). — I.O.P. Diving News, Japan, **11** (8): 2-7. [Text in Japanese, extensive abstract in English.]

— —, 2001. Taxonomie et évolution des Alpheidae (Crustacea: Decapoda). — In 2 volumes: 547 + 332 pp. (Thesis, Mus. natn. Hist. nat., Paris.)

ANKER, A. & T. M. ILIFFE, 2000. Description of *Bermudacaris harti*, a new genus, and species (Crustacea: Decapoda: Alpheidae) from anchialine caves of Bermuda. — Proc. biol. Soc. Wash., **113** (3): 761-775.

ARGANO, R., 1977. Asellota del Messico meridionale e Guatemala (Crustacea, Isopoda). — Accad. Naz. Lincei, **374** (quadr. 171) (III): 101-124.

— —, 1979. Isopodi (Crustacea: Isopoda). — In: Guide Animali Acque Interne Italiane, Verona, **5**: 1-65.

ARGANO, R. & G. L. PESCE, 1980. A cirolanid from subterranean waters of Turkey (Crustacea, Isopoda, Flabellifera). — Rev. suisse Zool., **87** (2): 439-444.

ARNDT, W., 1933. Die biologische Beziehungen zwischen Schwammen und Krebsen. — Mitt. zool. Mus. Berlin, **19**: 221-305.

ATES, R. M. L., 2003. A preliminary review of zoanthid-hermit crab symbioses (Cnidaria; Zoantharia/Crustacea; Paguridea). — Zool. Verh., Leiden, **345**: 41-48.

ATKINSON, R. J. A., C. FROGLIA, E. ARNERI & B. ANTOLINI, 1997. Observations on the burrows and burrowing behaviour of *Squilla mantis* (L.) (Crustacea: Stomatopoda). — Pubbl. Staz. zool. Napoli, (I, Mar. Ecol.) **18** (4): 337-359.

— —, — —, — — & — —, 1998. Observations on the burrows and burrowing behaviour of *Brachynotus gemmellari* and on the burrow of several other species occurring on *Squilla mantis* off Ancona, central Adriatic. — Scient. mar., **62** (1-2): 91-100.

AVDEEV, G. V., 1986. New harpacticoid copepods associated with Pacific cephalopods. — Crustaceana, **51**: 49-65.

BABA, K., 1979. Expédition Rumphius II (1975). Crustacés parasites, commensaux, etc. III. Galatheid Crustaceans (Decapoda, Anomura). — Bull. Mus. natn. Hist. nat., Paris, (A) (3), (4) **1**: 643-657.

BACESCO, M., 1940. Les Mysidacés des eaux roumaines. — Ann. Sc. Univ. Jassy, (2) **26** (2): 453-804.

BACESCO, M. & R. MAYER, 1960. Nouveaux cas de commensalisme. — Trav. Mus. Hist. nat. Grigore Antipa, **2**: 87-96.

BALIAN, E. V., C. LÉVÊQUE, H. SEGERS & K. MARTENS (eds.), 2008. Freshwater animal diversity assessment. — Hydrobiol., **595**: 1-301.

BANNISTER, J. L. & J. R. GRINDLEY, 1966. Notes on *Balaenophilus unisetus* P.O.C. Aurivillius, 1879, and its occurrence in the southern hemisphere (Copepoda, Harpacticoida). — Crustaceana, **10** (3): 296-302.

BARNARD, J. L., 1958. Amphipod crustaceans as fouling organisms. — Calif. Fish & Game, **44** (2): 161-170.

— —, 1959. Generic partition in the amphipod family Cheluridae, marine wood borers. — Pacific Nat., **1** (3-4): 3-12.

BARNARD, J. L. & C. L. INGRAM, 1986. The supergiant amphipod *Alicella gigantea* Chevreux from the North Pacific Gyre. — J. Crust. Biol., **6** (4): 825-839.

BARNARD, R. H., 1930. Scientific results of the Vernay-Lang Kalahari Expedition, March to September 1930. Crustacea. — Ann. Transvaal Mus., **16** (3): 483-492. [1935]

BECKER, J. H. & A. S. GRUTTER, 2004. Cleaner shrimp to clean. — Coral Reefs, **23**: 515-520.

BECKER, K. H., 1975. *Basipodella harpacticola* n. gen., n. sp. — Helgöl. wiss. Meeresunt., **27**: 96-100.

BEN NACEUR, H., A. BEN REJEB & M. S. ROMDHANE, 2012. Impacts of salinity, temperature, and pH on the morphology of *Artemia salina* (Branchiopoda: Anostraca) from Tunisia. — Zool. Stud., **51** (4): 453-462.

BENSON, R. H., 1969. Preliminary report in the study of abyssal ostracodes. — In: J. W. NEALE (ed.), The taxonomy, morphology and ecology of Recent Ostracoda, pp. 475-480. (Edinburgh.)

BERTRAND, H., 1942. Malacostracés de la région dinardaise (3e note). — Bull. Lab. marit. Dinard, **24**: 7-40.

BERTRAND, J.-Y., 1974. Recherches sur l'écologie de *Faucheria faucheri* (Crustacé, Cirolanidés). — Pp. 1-123. (Thesis, Univ. Paris 6, Paris.)

BIRSTEIN, J. A., 1957. Certain peculiarities of the ultra-abyssal fauna at the example of the genus *Storthyngura* (Isopoda, Asellota). — Zool. J. [Zool. Zhurn.], **36** (7): 961-985.

BLANC, H., 1884. Contribution à l'histoire naturelle des Asellotes Hétéropodes. Observations faites sur la *Tanais oerstedii* (Kroeyer). — Rev. suisse Zool., **1** (2): 189-258.

BLASTER, G. C., 1954. The biology and dispersal of *Mytilicola intestinalis* Steuer, a copepod parasite of mussels. — Fish. Invest., London, (2) **18** (6): 1-30.

BOCQUET, CH., 1952. Copépodes semi-parasites et parasites des Échinodermes de la région de Roscoff. Description de *Lichomolgus asterinae* n. sp. — Bull. Soc. zool. Fr., **77** (5-6): 495-504.

— —, 1953. Sur un Copépode Harpacticoïde mineur, *Diarthrodes feldmanni* n. sp. — Bull. Soc. zool. Fr., **78** (2-3): 101-105.

— —, 1963. Remarques morphologiques et systématiques. — Cah. Biol. mar., **4**: 65-79.

BOCQUET, CH. & J. H. STOCK, 1956. Copépodes parasites d'invertébrés des côtes de la Manche. I. *Endocheres obscurus*, nov. gen., n. sp., parasite de *Calliostoma zizyphinum*. — Arch. Zool. exp. gén., **93** (Notes et Rev., 3): 113-122.

— — & — —, 1956a. Copépodes parasites d'invertébrés des côtes de la Manche. II. Sur un Lichomolgide parasite des Gibbules, *Lichomolgus* (*Epimolgus*) *trochi*. — Arch. Zool. exp. gén., **94** (Notes et Rev., 1): 10-16.

— — & — —, 1957. Copépodes parasites d'invertébrés des côtes de France. I. Sur deux genres de la famille des Clausidiidae, commensaux de Mollusques, *Hersiliodes* Canu et *Conchyliurus* nov. gen. — Proc. Kon. ned. Akad. Wet., (C) **60** (2): 212-222.

— — & — —, 1957a. Copépodes parasites d'invertébrés des côtes de France. II, III. Notes taxonomiques et écologiques sur la famille des Mytilicolidae. — Proc. Kon. ned. Akad. Wet., (C) **60** (2): 223-229.

— — & — —, 1957b. Copépodes parasites d'invertébrés des côtes de France. IV. Le double parasitisme de *Sipunculus nudus* L. par *Myzomolgus stupendus* nov. gen., nov. sp. et *Catinia plana* nov. gen., nov. sp., Copépodes Cyclopoïdes très remarquables. — Proc. Kon. ned. Akad. Wet., (C) **60** (3): 410-431.

— — & — —, 1957c. Copépodes parasites d'invertébrés des côtes de France. V. Le genre *Synaptiphilus* Canu et Cuénot. — Proc. Kon. ned. Akad. Wet., (C) **60** (5): 679-695.

— — & — —, 1958. Copépodes parasites d'invertébrés des côtes de France. VI. Description de *Paranthessius myxicolae* nov. sp., Copépode semi-parasite du Sabellidae *Myxicola infundibulum* (Renier). — Proc. Kon. ned. Akad. Wet., (C) **61** (2): 243-253.

— — & — —, 1958a. Copépodes parasites d'invertébrés des côtes de France. VII. Caractères spécifiques et subspécifiques à l'intérieur du genre *Conchyliurus* Bocquet et Stock. — Proc. Kon. ned. Akad. Wet., (C) **61** (3): 308-324.

— — & — —, 1958b. Copépodes parasites d'invertébrés des côtes de la Manche. III. Sur deux espèces, jusqu'ici confondues, du genre *Anthessius*; description d'*Anthessius tessieri* n. sp. — Arch. Zool. exp. gén., **95** (Notes et Rev., 2): 99-112.

— — & — —, 1958c. Copépodes parasites d'invertébrés des côtes de la Manche. IV. Sur les trois genres synonymes de Copépodes Cyclopoïdes, *Leptinogaster* Pelseneer, *Strongylopleura* Pelseneer et *Myocheres* Wilson (Clausidiidae). — Arch. Zool. exp. gén., **96** (Notes et Rev., 2): 71-89.

— — & — —, 1958d. Copépodes parasites d'invertébrés des côtes de France. VIII. Le genre *Ischnurella* Pelseneer, synonyme de *Paranthessius* Claus (Cyclopoidea, Lichomolgidae). — Proc. Kon. ned. Akad. Wet., (C) **61** (6): 604-609.

— — & — —, 1959. Copépodes parasites d'invertébrés des côtes de la Manche. VI. Redescription de *Paranthessius anemoniae* Claus (Copepoda Cyclopoida) parasite d'*Anemonia sulcata* Pennant. — Arch. Zool. exp. gén., **98** (Notes et Rev., 1): 43-53.

— — & — —, 1959a. Sur la présence de "*Lichomolgus leptodermatus*" Gooding, copépode associé au pélécypode *Cardium crassum* Gmelin, sur les côtes françaises de la Manche. — Bull. Soc. Linn. Normandie, **9** (10): 119-120.

— — & — —, 1959b. Copépodes parasites d'invertébrés des côtes de France. X. Sur les espèces de *Paranthessius* (Cyclopoida, Lichomolgidae) du groupe des *Hermannella*, associés à des Pélécypodes. — Proc. Kon. ned. Akad. Wet., (C) **62** (3): 232-349.

— — & — —, 1959c. Copépodes parasites d'invertébrés des côtes de la Manche. V. Redescription de *Mesnilia cluthae* (Th. et A. Scott) (Copépode Cyclopoïde, famille des Clausiidae). — Arch. Zool. exp. gén., **97** (Notes et Rev., 1): 1-12.

— — & — —, 1960. Copépodes parasites d'invertébrés des côtes de la Manche. VII. Sur la présence d'*Octopicola superba* Humes, Lichomolgide associé à *Octopus*, le long des côtes de Bretagne. — Arch. Zool. exp. gén., **99** (Notes et Rev., 1): 1-7.

— — & — —, 1960a. Copépodes parasites d'invertébrés des côtes de France. XII. Étude de *Presynaptiphilus acrocnidae*, nov. gen., nov. sp., Copépode parasite de l'ophiure *Acrocnidae brachiata* (Montagu). — Proc. Kon. ned. Akad. Wet., (C) **63** (2): 220-229.

— — & — —, 1962. Copépodes parasites d'invertébrés des côtes de la Manche. IX. Cyclopoïdes associés à *Marthasterias glacialis* (L.). — Arch. Zool. exp. gén., **101** (Notes et Rev., 2): 79-91.

— — & — —, 1963. Some recent trends in work on parasitic Copepods. — Oceanogr. mar. Biol., annu. Rev., **1**: 289-300.

— — & — —, 1964. Copépodes parasites d'invertébrés des côtes de France. XVII. Le genre *Sabelliphilus* M. Sars, 1862 (Copépodes Cyclopoïdes, famille des Lichomolgidae). — Proc. Kon. ned. Akad. Wet., (C) **67** (3): 157-189.

BOCQUET, CH., J. H. STOCK & F. BENARD, 1959. Copépodes parasites d'invertébrés des côtes de France. IX. Description d'une nouvelle espèce remarquable de Lichomolgidae, *Heteranthessius scotti* n. sp. (Cyclopoida). — Proc. Kon. ned. Akad. Wet., (C) **62** (2): 111-118.

BOCQUET, CH., J. H. STOCK & G. KLEETON, 1963. Copépodes parasites d'invertébrés des côtes de la Manche. X. Cyclopoïdes Poecilostomes associés aux Annélides Polychètes de la région de Roscoff. — Arch. Zool. exp. gén., **102** (Notes et Rev., 1): 20-40.

— —, — — & — —, 1963a. Copépodes parasites d'invertébrés des côtes de la Manche. XI. Sur le développement de *Trochicola entericus* Dollfus, 1914, Copépode Cyclopoïde parasite de Trochidae. — Arch. Zool. exp. gén., **102** (Notes et Rev., 2): 49-68.

BOCQUET, CH., J. H. STOCK & F. LOUISE, 1963. Copépodes parasites d'invertébrés des côtes de France. XV. Le problème systématique d'*Asterocheres violaceus* (Claus) et d'*Asterocheres minutus* (Claus). — Proc. Kon. ned. Akad. Wet., (C) **66** (1): 37-53.

BOCQUET-VÉDRINE, J., 1961. Morphologie de *Chthamalophilus delagei* J. Bocquet-Védrine, Rhizocéphale parasite de *Chthamalus stellatus* (Poli). — Cah. Biol. mar., **2** (5): 455-593.

BODIOU, J.-Y. & P. CHARDY, 1973. Analyse en composantes principales du cycle annuel d'un peuplement de Copépodes Harpacticoïdes des sables fins infralittoraux de Banyuls-sur-Mer. — Mar. Biol., Berl., **20**: 27-34.

BORRADAILE, L. A., 1921. On the coral-gall prawn *Paratypton*. — Mem. Proc. Manchester lit. phil. Soc., **65** (11): 1-9.

BORUTZKY, E. V., 1947. Sur la faune des Copépodes Harpacticoïdes des microcontinents épiphytes. — Dokl. Akad. Nauk SSSR, **58** (9): 2105-2108.

BORZIC, B., 1965. Un nouveau *Phyllognathus* (Copépode Harpacticoïde) du Gabon. — Rev. Écol. Biol. Sol, **2**: 271-275.

BOSCHMA, H., 1937. The species of the genus *Sacculina* (Crustacea: Rhizocephala). — Zool. Meded., Leiden, **19**: 187-238.

— —, 1949. Ellobiopsidae. — Discovery Rep., **25**: 281-314.

— —, 1955. The described species of the family Sacculinidae. — Zool. Verh., Leiden, **27**: 1-74.

— —, 1959. Ellobiopsidae from tropical West Africa. — Atlantide Rep., **5**: 145-175.

BOTOSANEANU, L. & L. B. HOLTHUIS, 1970. Subterranean shrimps from Cuba (Crustacea: Decapoda Natantia). — Trav. Inst. Spéol. "Émile Racovitza", **9**: 121-133.

BOTOSANEANU, L. & J. H. STOCK, 1979. *Ambolana insula*, n. gen., n. sp., the first hypogean cirolanid isopod found in the Lesser Antilles. — Bijdr. Dierk., **49** (2): 227-233.

BOULIGAND, Y., 1960. Notes sur la famille des Lamippidae, première partie. — Crustaceana, **1** (3): 258-278.

— —, 1960a. Sur l'organisation des Lamippides, Copépodes parasites des octocoralliaires (première note). — Vie Milieu, **11** (3): 335-380.

BOURDON, R., 1963. Épicarides et Rhizocéphales de Roscoff. — Cah. Biol. mar., **4**: 413-434.

— —, 1964. Notes sur la biologie de "*Dynamene bidentata*" Adams (Isopode, Sphaeromatidae). — Bull. Acad. Soc. Lorraine Sci., Nancy, **4** (1): 155-162.

— —, 1967. Données complémentaires sur les Épicarides et les Rhizocéphales de Roscoff. — Bull. Acad. Soc. Lorraine Sci., Nancy, **6** (4): 279-286.

— —, 1968. Les Bopyridae des mers européennes. — Mém. Mus. natn. Hist. nat., Paris, (A, Zool.) **50** (2): 1-410.

— —, 1996. Les Bopyres des Porcellanes. — Bull. Mus. natn. Hist. nat., Paris, (3) **359** [Zool., **252**]: 165-245.

BOUSFIELD, E. L., 1968. Terrestrial adaptations in Crustacea: discussion. — Amer. Zool., **8** (3): 393-398.

— —, 1970. Adaptive radiation in sand-burrowing amphipod crustaceans. — Chesapeake Sci., **11** (3): 143-154.

— —, 1978. A revised classification and phylogeny of amphipod crustaceans. — Trans. R. Soc. Canada, (4) **16**: 343-390.

BOUSFIELD, E. L. & F. G. HOWARTH, 1976. The cavernicolous fauna of Hawaiian lava tubes. 8. Terrestrial Amphipoda (Talitridae) including a new genus and species with notes on its biology. — Pacific Insects, **17** (1): 144-154.

BOUVIER, E. L., 1918. Sur une petite collection de Crustacés de Cuba offerte au Muséum par M. de Boury. — Bull. Mus. natn. Hist. nat., Paris, **24**: 12-16.

BOWMAN, T. E., 1964. *Antrolana lira*, a new genus and species of troglobitic cirolanid isopod from Madison cave, Virginia. — Int. J. Speleol., **1** (1-2): 229-236.

— —, 1965. *Cyathura specus*, a new cave isopod from Cuba (Anthuridea, Anthuridae). — Stud. Fauna Curaçao Caribb. Isl., **22** (85): 88-97.

— —, 1971. *Excirolana kumari*, a new tubicolous isopod from Malaysia. — Crustaceana, **23** (1): 70-76.

— —, 1971a. *Cirolana narica*, n. sp., a New Zealand isopod (Crustacea) found in the nasal tract of the dolphin *Cephalorhynchus hectori*. — Beaufortia, **19** (252): 107-112.

— —, 1972. *Cithadius cyathurae*, a new genus and species of Tachidiidae (Copepoda: Harpacticoida) associated with the estuarine isopod, *Cyathura polita*. — Proc. biol. Soc. Wash., **85** (18): 249-254.

— —, 1975. A new genus and species of troglobitic cirolanid isopod from San Luis Potosi, Mexico. — Occas. Pap. Mus. Texas techn. Univ., **27**: 1-7.

— —, 1977. *Ceratolana papuae* a new genus and species of mangrove-boring cirolanid isopod from Papua, New Guinea. — Proc. biol. Soc. Wash., **90** (4): 819-825.

— —, 1977a. A review of the genus *Antromysis* (Crustacea: Mysidacea) including the new species from Jamaica and Oaxaca, Mexico, and a redescription and new records for *A. cenotensis*. — In: J. R. REDDELL (ed.), Studies on the caves and cave fauna of the Yucatan Peninsula. Assoc. Mexican Cave Stud. Bull., **6**: 27-38.

— —, 1978. Revision of the pelagic amphipod genus *Primno* (Hyperiidea: Phrosinidae). — Smithson. Contr. Zool., **275**: 1-23.

— —, 1981. *Thermosphaeroma milleri* and *T. smithi*, new sphaeromatid isopod crustaceans from hot spring in Chihuahua, Mexico, with a review of the genus. — J. Crust. Biol., **1** (1): 105-122.

BOWMAN, T. E. & L. G. ABELE, 1982. Classification of the recent Crustacea. — In: D. E. BLISS (ed.), The biology of Crustacea, **1**: 1-27. (Academic Press, New York, NY.)

BOWMAN, T. E. & L. S. KORNICKER, 1967. Two new crustaceans: the parasitic copepod *Sphaeronellopsis monothrix* (Choniostomatidae) and its myodocopid ostracod host *Parasterope pollex* (Cylindroleberidae) from the southern New England coast. — Proc. U.S. natl. Mus., **123** (3613): 1-28.

— — & — —, 1968. *Sphaeronellopsis hebe* (Copepoda, Choniostomatidae), a parasite of the ostracod, *Pseudophinomedes ferulanus*. — Crustaceana, **15** (2): 113-116.

BOXSHALL, G. A. & R. HUYS, 1990. New family of deep-sea planktonic copepods, the Paralubbockiidae (Copepoda: Poecilostomatoida). — Biol. Oceanogr., **6** (2): 163-173.

BRADFORD, J. M. & G. C. HEWITT, 1980. A new maxillopodan crustacean, parasitic on a myodocopid ostracod. — Crustaceana, **38**: 67-72.

BRATTEGARD, T., 1968. On an association between *Acanthopleura granulata* (Polyplacophora) and *Dynamene* spp. (Isopoda). — Sarsia, **32**: 11-20.

BRESCIANI, J. & J. LÜTZEN, 1960. *Gonophysema gullmarensis* (Copepoda parasitica). An anatomical and biological study of an endoparasite living in the ascidian *Ascidiella aspera*. I. Anatomy. — Cah. Biol. mar., **1** (2): 157-184.

— — & — —, 1961. *Gonophysema gullmarensis* (Copepoda parasitica). An anatomical and biological study of an endoparasite living in the ascidian *Ascidiella aspera*. II. Biology and development. — Cah. Biol. mar., **2** (4): 347-372.

— — & — —, 1994. Morphology and anatomy of *Avdeevia antarctica*, new genus, new species (Copepoda: Harpacticoida: Tisbidae), parasitic on an Antarctic cephalopod. — J. Crust. Biol., **14** (4): 744-751.

BRIAN, A., 1931. Il parasitismo fra gli animali marini. — Pp. 1-293. (Genova.)

BRIGHT, D. B. & CH.-L. HOGUE, 1972. A synopsis of the burrowing land crabs of the world and list of their arthropod symbionts and burrow associates. — Contrib. Sci. nat. Hist. Mus. Los Angeles Co., **220**: 1-58.

BROOKS, J. L., 1942. Notes on the ecology and occurrence in America of the myrmecophilous sowbug *Platyarthrus hoffmannseggi* Brandt. — Ecology, **23**: 427-437.

BROUSSEAU, D. J., K. KRIKSCIUN & J. BAGLIVO, 2003. Fiddler crab burrow usage by Asian shore crab. — Northeast. Nat., **10**: 415-420.

BRUCE, A. J., 1969. Notes on some Indo-Pacific Pontoniinae. XIV. Observations on *Paratypton siebenrocki* Balss. — Crustaceana, **17** (2): 171-186.

— —, 1969a. *Aretopsis amabilis* De Man, an alpheid shrimp commensal of pagurid crabs in the Seychelles Islands. — J. mar. biol. Ass. India, **11** (1-2): 175-181.

— —, 1970. Report on some commensal pontoniid shrimps (Crustacea Palaemonidae) associated with an Indo-Paific gorgonian host (Coelenterata: Gorgonacea). — J. Zool., Lond., **160**: 537-544.

— —, 1972. Shrimps that live with molluscs. — Sea Frontiers, **18** (4): 218-227.

— —, 1972a. An association between a pontoniinid shrimp and a rhizostomatous scyphozoan. — Crustaceana, **23** (3): 300-302.

— —, 1975. Coral reef shrimps and their colour patterns. — Endeavour, **34** (121): 23-27.

— —, 1976. Shrimps and prawns of coral reefs, with special reference to commensalism. — In: Biology and geology of coral reefs, (III, Biol.) **2**: 37-94.

— —, 1976a. *Discias mvitae* sp. nov., a new sponge associate from Kenya (Decapoda, Natantia, Disciadidae). — Crustaceana, **31** (2): 119-130.

— —, 1976b. Coral reef Caridea and "commensalism". — In: Symposium "Animal associates on coral reefs". Symp. Indo-Pac. trap. Reef Biology, Guam and Palau, June 23-July 5, 1974. Micronesia, **12** (1): 83-98.

— —, 1976c. Shrimps from Kenya. — Zool. Verh., Leiden, **145**: 1-72.

— —, 1977. Shrimps that live on corals. — Oceans, **1** (2): 70-75.

— —, 1979. *Ctenopontia cyphastreophila*, a new genus and species of coral associated Pontoniinae shrimp from Eniwetok Atoll. — Bull. mar. Sci., **29** (3): 423-435.

— —, 1979a. Notes on some Indo-Pacific Pontoniinae. XXXI. *Periclimenes magnificus* sp. nov., a coelenterate associate from the Capricorn Islands (Decapoda: Palaemonidae). — Crustaceana, (Suppl.) **5**: 195-208.

— —, 1982. The shrimps associated with Indo-west Pacific echinoderms, with the description of a new species in the genus *Periclimenes* Costa, 1844 (Crustacea: Pontoniinae). — Austral. Mus. Mem., **16**: 191-216.

BRUCE, C. & E. W. LINDGREN, 1969. *Harrietella simulans* (Copepoda: Harpacticoida) associated with *Limnoria tripunctata* (Isopoda) in North Carolina. — J. Elisha Mitchell Sci. Soc., **85** (2): 73-75.

BRUES, C. T., 1924. Observations on animal life in the thermal waters of Yellowstone Park, with a consideration of the thermal environment. — Proc. Amer. Acad. Arts Sci., **15**: 371-437.

BRUSCA, R. C. & G. BRUSCA, 2003. Invertebrates (2nd ed.). — Pp. 1-895. (Sinauer Associates, Sunderland, MA.)

BRUUN, A. & T. WOLFF, 1961. Abyssal benthic organisms: nature, origin, distribution and influence on sedimentation. — In: M. SEARS (ed.), Oceanography, pp. 391-397. (Amer. Ass. Advmt Sci.)

BUCKLE RAMIREZ, L. F., 1965. Untersuchungen über die Biologie von *Heterotanais oerstedi* (Crustacea, Tanaidacea). — Z. Morph. Ökol. Tiere, **55**: 714-782.

CALMAN, W. T., 1911. The life of Crustacea. — Pp. i-xvi + 1-280. (London.)

— —, 1919. Marine boring animals injurious to submerged structures. — Pp. 1-35. (Brit. Mus. (Nat. Hist.), London.)

— —, 1921. Notes on marine wood-boring animals. II. Crustacea. — Proc. zool. Soc. Lond., **1921**: 215-220.

CAMACHO, A., 2003. Historical biogeography of *Hexabathynella*, a cosmopolitan genus of groundwater Syncarida (Crustacea, Bathynellacea, Parabathynellidae). — Biol. J. Linn. Soc., Lond., **78**: 457-466.

CAMERON, J. & T. H. MECKLUNBURG, 1973. Aerial gas exchange in the coconut crab, *Birgus latro*, with some notes on *Gecarcoidea lalandei*. — Resp. Physiol., **19**: 245-261.

CAPART, A., 1951. *Thermobathynella adami*, gen. et sp. nov., Anaspidacé du Congo belge. — Bull. Inst. R. Sci. nat. Belgique, **27** (10): 1-4.

CAROLL, E., 1924. Sulla presenza della *Typhlocaris* (*T. salentina* n. sp.) in terra d'Otranto. Contributo alia conoscenza del genere. — Ann. Mus. Zool. R. Univ., Napoli, (N.S.) **5** (9): 1-20.

CARTON, Y., 1967. Description de *Nicothoe procircularis* n. sp. (Crustacea, Copepoda). Discussion sur la forme mâle. — Vidensk. Meddr Dansk naturh. Foren., **130**: 143-152.

— —, 1970. Le genre *Paranicothoe*, un nouveau représentant de la famille des Nicothoidae. — J. Parasit., (sect. 2) **56** (4): 47-48.

— —, 1970a. Description de *Paranicothoe* n. gen., un nouveau représentant de la famille des Nicothoidae. — Galathea Rep., **11**: 239-246.

CARTON, Y. & P. LÉCHER, 1964. Recherches sur l'anatomie interne et la biologie de *Selioides bocqueti* Carton, Copépode parasite de l'Aphroditidae *Scalisetosus assimilis* Mc'Intosh. — Bull. Soc. Linn. Normandie, **10** (4): 140-148.

CARVACHO, A., 1977. Isopodes de la mangrove de la Guadeloupe, Antilles Françaises. — Stud. Fauna Curaçao Caribb. Isl., **54** (174): 1-24.

CASABLANCA, M.-L. DE, 1966. Étude des conditions écologiques dans les étangs de la plaine orientale de la Corse et autécologie de l'Amphipode constructeur *Corophium insidiosum* Crawford. — Pp. 1-88. (Thesis, Univ. Marseille, Marseille.)

CAULLERY, M., 1922. Le parasitisme et la symbiose. — Pp. 1-399. (Paris.)

CAULLERY, M. & F. MESNIL, 1901. Recherches sur *Hemioniscus balani* Buchh., Épicaride parasite des Balanes. — Bull. Sci. France Belgique, **34**: 316-161.

CHACE, F. A., JR., 1943. Two new blind prawns from Cuba with a synopsis of the subterranean Caridea of America. — Proc. New Engl. zool. Club, **22**: 25-40.

— —, 1951. The number of species of decapod and stomatopod Crustacea. — J. Wash. Acad. Sci., **41** (11): 370-372.

CHACE, F. A., JR. & H. HOBBS, JR., 1969. The freshwater and terrestrial decapod crustaceans of the West Indies with special reference to Dominica. — Bull. U.S. natl. Mus., **292**: 1-258.

CHACE, F. A., JR. & R. B. MANNING, 1972. Two new caridean shrimps, one representing a new family, from marine pools on Ascension Island (Crustacea, Decapoda: Natantia). — Smithson. Contr. Zool., **131**: 1-18.

CHANGEUX, J.-P., 1960. Contribution à l'étude des animaux associés aux Holothuries. — Vie Milieu, **10**: 1-124.

CHEVREUX, ED., 1908. Sur les commensaux du Bernard l'Hermite. — Bull. Mus. natn. Hist. nat., Paris, **14**: 14-16.

CIUREA, J., TH. MONOD & G. DINULESCO, 1933. Présence d'un Cirripède operculé sur un Poisson dulçaquicole européen. — Bull. Inst. océanogr. Monaco, **615**: 1-32. [15 March 1933]

CLARKE, W. D., 1955. A new species of the genus *Heteromysis* (Crustacea, Mysidacea) from the Bahama Islands, commensal with a sea-anemone. — Amer. Mus. Novit., **1716**: 1-13.

CODREANU, R. & M. CODREANU, 1959. Données biologiques et statistiques sur un Pagure *Diogenes pugilator* (Roux) de la mer Noire, et ses Crustacés parasites. Essai d'analyse de ses caractères sexuels. — Univ. Iasi, Lucratile Ses. Stiintif. 15-17 sept. **1956**. Stat. Zool. mar. Agigea, pp. 315-348.

COINEAU, N., 1968. Contribution à l'étude de la faune interstitielle: Isopodes et Amphipodes. — Mém. Mus. natn. Hist. nat., Paris, (A, Zool.) **55**: 145-216.

— —, 1971. Les Isopodes interstitiels: documents sur leur écologie et leur biologie. — Mém. Mus. natn. Hist. nat., Paris, (A, Zool.) **64**: 1-170.

COLLINGE, W. E., 1941. Notes on the terrestrial Isopoda (woodlice), n° 1. — Northwest. Nat., **1941**: 123-127. [Observations on *Platyarthrus hoffmannseggii*: 123-124.]

— —, 1944. Notes on the terrestrial Isopoda (woodlice), n° 10. — Northwest. Nat., **1944**: 112-121. [On the climbing habits of some woodlice: 118-119.]

CONOVER, M. R., 1976. The influence of some symbionts on the shell-selection behaviour of the hermit crabs, *Pagurus pollicaris* and *Pagurus longicarpus*. — Anim. Behav., **24** (1): 191-194.

COTTARELLI, V. & R. ARGANO, 1977. *Trichodactylus* (*Rodrigueziu*) *mensabak* n. sp. (Crustacea, Decapoda, Brachyura), granchio cieco delle acque sotterranee del Chiapas (Messico). — Quad. Prob. Att. Sci. Cult., Accad. naz. Lincei, **171** (3): 207-212.

COUTIÈRE, H., 1905. Notes sur *Lysiosquilla digueti* n. sp. commensale d'un Polynoïdien et d'un Balanoglosse de Basse Californie. — Bull. Soc. Phil. Paris, **9** (VII): 174-179.

CRANE, J., 1975. Fiddler crabs of the world. Ocypodidae: genus *Uca*. — Pp. i-xxiii, 1-736. (Princeton Univ. Press, Princeton.)

CRAWFORD, G. I., 1937. A review of the amphipod genus *Corophium* with notes on the British species. — J. mar. biol. Ass. U.K., **21** (2): 589-630.

CRESSEY, R. F., 1976. *Nicothoe tumulosa*, a new siphonostome copepod parasitic on the unique decapod *Neoglyphea inopinata* Forest and Saint Laurent. — Proc. biol. Soc. Wash., **89** (7): 119-126.

CRESSEY, R. F. & B. B. COLLETTE, 1970. Copepods and needlefishes: a study in host-parasite relationships. — Fish. Bull., U.S., **68** (3): 347-432.

CRESSEY, R. F. & E. A. LACHNER, 1970. The parasitic copepod diet and life history of diskfishes (Echeneidae). — Copeia, **2**: 310-318.

CUZENT, G., 1884. Archipel des Paumotu (Paumotu-Tuamotu). — Bull. Soc. Acad. Brest, (2) **9** [1883-1884]: 49-90.

DAHL, E., 1977. The amphipod functional model and its bearing upon systematics and phylogeny. — Zool. Scr., **6**: 221-228.

DALES, R. PH., 1957. Interrelations of organisms. A. Commensalism. — In: J. W. HEDGPETH (ed.), Treatise on marine ecology and paleoecology, I, Ecology. Geol. Soc. Am. Mem., **67**: 391-412.

DANIEL, A., 1958. On *Platylepas indicus* n. sp., a new barnacle from the Madras coast of India. — Ann. Mag. nat. Hist., (13) **1**: 755-757.

DAVENPORT, J., 1994. A cleaning association between the oceanic crab *Planes minutus* and the loggerhead sea turtle *Caretta caretta*. — J. mar. biol. Ass. U.K., **74**: 735-737.

DE GRAVE, S., 1998. Pontoniinae (Decapoda, Caridea) associated with *Heliofungia actiniformis* (Scleractinia) from Hansa Bay, Papua New Guinea. — Belg. J. Zool., **128** (1): 13-22.

DE GRAVE, S. & C. H. J. M. FRANSEN, 2011. Carideorum catalogus: the recent species of the dendrobranchiate, stenopodidean, procarididean and caridean shrimps (Crustacea: Decapoda). — Zool. Meded., Leiden, **85** (9): 195-589.

DEBELIUS, H., 2001. Crustacea-guide of the world. Shrimps. Crabs. Lobsters. Mantis shrimps. Amphipods. — Pp. 1-321. (Editions Ikan, Frankfurt-a.-M.)

DECHANCÉ, M. & J.-P. DUFAURE, 1959. Une nouvelle association entre une Actinie et un Pagure. — C. R. Acad. Sci. Paris, **249**: 1566-1568.

DEFAYE, D., N. RABET & A. THIÉRY, 1998. Atlas et bibliographie des Crustacés Branchiopodes (Anostraca, Notostraca, Spinicaudata) de France métropolitaine. — Patrimoines nat., **32**: 1-61.

DELAMARE DEBOUTTEVILLE, C., 1950. Contribution à la connaissance des Copépodes du genre *Splanchnotrophus* Hancock et Norman, parasites de Mollusques. — Vie Milieu, **1** (1): 1-7.

— —, 1976. Intérêt biologique et écologique des crabes cavernicoles du Guatémala et du Mexique, appartenant au genre *Typhlopseudothelphusa* Rioja. — C. R. Acad. Sci. Paris, **283**: 837-840.

— —, 1976a. Sur la radiation évolutive des crabes du genre *Typhlopseudothelphusa* au Guatémala et au Mexique, avec description d'espèces nouvelles. — Ann. Spéléol., **31**: 115-129.

DELAMARE DEBOUTTEVILLE, C. & E. ANGELIER, 1950. Sur un type de Crustacé phréaticole nouveau: *Parabathynella fagei* n. sp. — C. R. Acad. Sci. Paris, **231**: 175-176. [10 juillet 1950]

DELAMARE DEBOUTTEVILLE, C. & L. LAUBIER, 1960. Les Phyllocolidae Delamare et Laubier, Copépodes parasites d'Annélides Polychètes Phyllodocides et leurs rapports avec les Copépodes annélidicoles. — C. R. Acad. Sci. Paris, **251**: 2231-2233.

DELAMARE DEBOUTTEVILLE, C., A. G. HUMES & J. PARIS, 1957. Sur le comportement d'*Octopicola superba* Humes, n. g., n. sp., parasite de la pieuvre *Octopus vulgaris* Lamarck. — C. R. Acad. Sci. Paris, **244**: 504-506.

DELAMARE DEBOUTTEVILLE, C. & L. P. NUNES, 1951. Étude de *Pennella remorae* Munay et remarques sur la biologie et la systématique des *Pennella* Oken. — Rev. Fac. Ciên. Lisboa, (2) (C) **1** (2): 341-352.

— — & — —, 1951a. Hydraires épizoïques sur les Copépodes parasites. — Vie Milieu, **2** (4): 421-432.

DENNELL, R., 1933. The habits and feeding mechanism of the amphipod *Haustorius arenarius* Slabber. — J. Linn. Soc. Lond., (Zool.) **38**: 363-388.

DESBRUYÈRES, D., P. CRASSOUS, J. GRASSLE, A. KHRIPOUNOFF, D. REYSS, M. RIO & M. VAN PRAËT, 1982. Données écologiques sur un nouveau site d'hydrothermalisme actif de la ride du Pacifique oriental. — C. R. Acad. Sci. Paris, (10) (3) **295**: 489-494.

DEXTER, D. M., 1977. Natural history of the Pan-American isopod *Excirolana braziliensis* (Crustacea: Malacostraca). — J. Zool., Lond., **183**: 103-109.

DOLLFUS, R. PH., 1948. Épizoïques (animaux et végétaux) sur des Copépodes parasites. Déformation pathologique d'un Copépode par une algue épizoïque. — Feuille Natural., (N.S.) **1948**: 23-27.

— —, 1958. *Xenobalanus globicipitis* Steenstrup (Cirripedia: Thoracica) accolé sur *Tursiops truncatus* (Montagu) à proximité de la côte nord du Maroc. — Bull. Inst. Pêches marit. Maroc, **16**: 55-59.

DOTY, M. S., 1957. Rocky intertidal surfaces. — In: J. W. HEDGPETH (ed.), Treatise on marine ecology and paleoecology, I, Ecology. Geol. Soc. Am. Mem., **67**: 535-585.

DWORSCHAK, P. C., 1983. The biology of *Upogebia pusilla* (Petagna) (Decapoda, Thalassinidea). 1. The burrows. — Pubbl. Staz. zool. Napoli, **4** (1): 19-43.

— —, 1987. The biology of *Upogebia pusilla* (Petagna) (Decapoda, Thalassinidea). 2. Environments and zonation. — Pubbl. Staz. zool. Napoli, **8** (4): 337-358.

— —, 1988. The biology of *Upogebia pusilla* (Petagna) (Decapoda, Thalassinidea). 3. Growth and production. — Pubbl. Staz. zool. Napoli, **9** (1): 51-77.

— —, 2000. Global diversity in the Thalassinidea (Decapoda). — J. Crust. Biol., **20** (special number 2): 238-245.

— —, 2001. The burrows of *Callianassa tyrrhena* (Petagna, 1792) (Decapoda: Thalassinidea). — Mar. Ecol., **22** (1-2): 155-166.

— —, 2005. Global diversity in the Thalassinidea (Decapoda): an update (1998-2004). — Nauplius, **13** (1): 57-63.

EDMONDSON, CH. H., 1946. Reef and shore fauna of Hawaii (2nd ed.) — Bernice P. Bishop Mus. Spec. Publ., **22**: i-iii, 1-381.

— —, 1951. Some central Pacific crustaceans. — Occas. Pap. B. P. Bishop Mus., **20** (13): 183-243.

EDNEY, E. B., 1954. Woodlice and the land habitat. — Biol. Rev., **29**: 185-219.

— —, 1960. Terrestrial adaptations. — In: T. H. WATERMAN (ed.), The physiology of Crustacea, **1**: 367-393.

EFFORD, I. E., 1965. Aggregation in the sand crab, *Emerita analoga* (Stimpson). — J. anim. Ecol., **34**: 63-75.

EKMAN, S., 1953. Zoogeography of the sea. — Pp. i-xiv, 1-417. (Sedgwick and Jackson, Ltd., London.)

EMERY, K. O. & R. E. STEVENSON, 1957. Estuaries and lagoons. — In: J. W. HEDGPETH (ed.), Treatise on marine ecology and paleoecology, I, Ecology. Geol. Soc. Am. Mem., **67**: 673-750.

ENNEQUIST, P., 1949. Studies on the soft-bottom amphipods of the Skagerak. — Zool. Bidr., Uppsala, **28**: 297-492.

ESTEVEZ, F. D. & J. L. SIMON, 1975. Systematics and ecology of *Sphaeroma* (Crustacea: Isopoda) in the mangrove habitats of Florida. — In: G. E. WALSH, S. C. SNEDAKER & H. J. TEAS (eds.), Proc. internat. Symp. Biol. Management Mangroves, 8-11 Oct. 1974, Univ. Florida, pp. 286-304. (Univ. Florida.)

EVINK, G. L., 1975. Macrobenthos comparisons in mangrove estuaries. — In: G. E. WALSH, S. C. SNEDAKER & H. J. TEAS (eds.), Proc. internat. Symp. Biol. Management Mangroves, 8-11 Oct. 1974, Univ. Florida, pp. 256-285. (Univ. Florida.)

FAGE, L. & TH. MONOD, 1936. La faune marine du Jameo de Agua, lac souterrain de l'île de Lanzarote (Canaries). — Arch. Zool. exp. gén., **78** (2): 97-113.

FARRAN, G. P., 1905. Occurrence of the floating barnacle, *Lepas fascicularis* (Ellis and Sol.). — Ann. Rept. Fish., Ireland, for **1902** and **1903**, Pt. II, app. VII: 209-210.

FIGUEIREDO, M. J. DE & H. J. THOMAS, 1967. *Nephrops norvegicus* (L., 1758) Leach, a review. — Oceanogr. mar. Biol., annu. Rev., **5**: 371-407.

FISCHER, P. H., 1940. Notes sur les peuplements littoraux d'Australie. III. Sur la faune de la mangrove australienne. — Mém. Soc. Biogéogr., Paris, **7**: 315-329.

FIZE, A., 1956. Observations biologiques sur les Hapalocarcinides. — Ann. Fac. Sci. Saïgon, **1956**: 1-30.

FIZE, A. & R. SERÈNE, 1955. Note préliminaire sur huit espèces nouvelles dont une d'un genre nouveau d'Hapalocarcinidae. — Bull. Soc. zool. Fr., **80** (5-6): 375-378.

— — & — —, 1957. Les Hapalocarcinides du Viet-Nam. — Mém. Inst. océanogr. Nhatrang, **10**: 1-202.

FOREST, J., 1969. Crustacés. — Encyclopædia Universalis, **4**: 179-185.

— —, 1981. Compte rendu et remarques générales/Report and general comments. — In: Résultats des Campagnes MUSORSTOM. **I**. Philippines (18-28 mars 1976). Mém. ORSTOM, **91**: 9-50.

— —, 1987. Les Pylochelidae ou "Pagures symétriques" (Crustacea: Coenobitoidea). — In: Résultats des Campagnes MUSORSTOM, **3**. Mém. Mus. natn. Hist. nat., Paris, **137**: 1-254.

— —, 2004. The Crustacea: definition, primitive forms, and classification. — In: J. FOREST & J. C. VON VAUPEL KLEIN (eds.), Treatise on zoology — anatomy, taxonomy, biology. The Crustacea, **1**: 3-12. (Koninklijke Brill, Leiden.)

FOREST, J. & N. NGOC-HO, 1992. Description de *Pagurus dartevellei* (Forest, 1958) (Crustacea, Decapoda, Paguridae). — Bull. Mus. natn. Hist. nat., Paris, (A) (1) (4) **14**: 217-227.

FOREST, J. & M. DE SAINT-LAURENT, 1989. Nouvelle contribution à la connaissance de *Neoglyphea inopinata* Forest et de Saint-Laurent, à propos de la femelle adulte. — In: J. FOREST (ed.), Résultats des Campagnes MUSORSTOM, **5**. Mém. Mus. natn. Hist. nat., Paris, (A) **144**: 75-92.

FRANZ, C. J. & R. C. BULLOCK, 1990. *Ischnochitonika lasalliana*, new genus, new species (Copepoda), a parasite of tropical western Atlantic chitons (Polyplacophora: Ischnochitonidae). — J. Crust. Biol., **10** (3): 544-549.

FROGLIA, C. & R. J. A. ATKINSON, 1998. Association between *Athanas amazone* (Decapoda: Alpheidae) and *Squilla mantis* (Stomatopoda: Squilidae). — J. Crust. Biol., **18** (3): 529-532.

GALLIEN, L., 1935. *Pseudanthessius nemertophilus* nov. sp., copépode commensal de *Lineus longissimus* Sowerby. — Bull. Soc. zool. Fr., **60**: 451-459.

GARTH, J. A., 1964. The Crustacea Decapopda (Brachyura and Anomura) of Eniwetok Atoll, Marshall Islands, with special reference to the obligate commensals of branching corals. — Micronesia, **1** (1-2): 137-144.

— —, 1965. The brachyuran decapod crustaceans of Clipperton Island. — Proc. Calif. Acad. Sci., (4) **33** (1): 1-46.

— —, 1974. Decapod crustaceans inhabiting reef-building corals of Ceylon and the Maldive Islands. — J. mar. biol. Ass. India, **15** (1): 195-212.

— —, 1974a. On the occurrence in the eastern tropical Pacific of Indo-West Pacific decapod crustaceans commensal with reef-building corals. — Proc. 2nd intern. Coral Reef Symp., **1**: 397-406.

GARTH, J. S. & T. A. HOPKINS, 1968. *Pseudocryptochirus crescentus* (Edmonson), a record crab of the corallicolous family Hapalocarcinidae (Crustacea: Decapoda) from the eastern Pacific with remarks on phragmosis, host specificity and distribution. — Bull. South. Calif. Acad. Sci., **67** (1): 40-48.

GERLACH, S. A., 1958. Die Mangrove Region tropischer Küsten als Lebensraum. — Z. Morph. Ökol. Tiere, **46**: 637-730.

GERSTAECKER, A., 1866-1879. Crustacea (Erste Hälfte). — Bronns Tierreich, **V** (1): 1-1320.

GIBSON, R. A., 1978. *Pseudohimanthidium pacificum*, an epizoic diatom new to the Florida Current. — J. Phycol., **14** (3): 371-373.

GIGLIOLI, E. H., 1876. Viaggio intorno al globo delle R. pirocorvetta italiana Magenta. — Pp. i-xxxviii, 1-1031. (Milano.)

GILTAY, L., 1934. Note sur l'association de *Balanus concavus pacificus* Pilsbry (Cirripède) et *Dendraster excentricus* (Eschscholtz) (Échinoderme). — Bull. Mus. roy. Hist. nat. Belgique, **10** (5): 1-7.

GLASSELL, S. A., 1937. *Pinnixa lunzi*, a new commensal crab from South Carolina, Charleston. — Charleston Mus. Leaflet, **9**: 1-8.

GOODING, R. V., 1957. On some Copepoda from Plymouth, mainly associated with invertebrates, including three new species. — J. mar. biol. Ass. U.K., **36**: 195-221.

— —, 1957a. "*Callianassa pugettensis*" (Decapoda, Anomura), type host of the copepod *Clausidium vancouverense* (Haddon). — Ann. Mag. nat. Hist., (12) **10**: 695-700.

GORDON, I., 1960. Additional notes on the porcellanid sea-anemone association. — Crustaceana, **1** (2): 166-167.

GOTTO, V., 1969. Marine animals partnerships and other associations. — Pp. 1-96. (London.)

GOUDEAU, M., 1977. Contribution à la biologie d'un Crustacé parasite: *Hemioniscus balani* Buchholz, isopode épicaride. Nutrition, mues et croissance de la femelle et des embryons. — Cah. Biol. mar., **18**: 201-242.

GOUDEY-PERRIÈRE, F., 1972. *Amphiurophilus amphiurae* (Hérouard), Crustacé Copépode endoparasite de l'ophiure *Amphipholis squamata* Della Chiaje (Échinoderme): étude du développement et données biologiques. — Pp. 1-75. (Thesis, doct. 3e cycle, Univ. Paris 6, Paris.)

— —, 1979. *Amphiurophilus amphiurae* (Hérouard), Crustacé Copépode parasite des bourses génitales de l'ophiure *Amphipholis squamata* Della Chiaje (Échinoderme): morphologie des adultes et étude des stades juvéniles. — Cah. Biol. mar., **20**: 201-230.

GRASSÉ, P.-P., 1952. Les Ellobiopsidae. — Traité de Zoologie, **1** (1): 1023-1030.

GREEN, J., 1968. The biology of estuarine animals. — Pp. i-x, 1-401. (London.)

GRIMMETT, R. E. R., 1926. Forest-floor covering and its life. — Trans. Proc. roy. Soc. New-Zealand, **56**: 623-640.

GUILLE, A. & L. LAUBIER, 1965. *Synaptiphilus cantacuzenei mixtus* ssp. nov., Copépode ectoparasite sur *Oerstergrenia digitata* à Banyuls-sur-Mer. — Crustaceana, **9** (2): 125-136.

HAEFELFINGER, H. R. & L. LAUBIER, 1965. Découverte en Méditerranée occidentale de *Mesoglicola delagei* Quidor, copépode parasite d'actinies. — Crustaceana, **9** (2): 210-212.

HAGEN, H.-O. VON, 1962. Freiland studien zur Sexual- und Fortpflanzungsbiologie von *Uca tangeri* in Andalusien. — Z. Morph. Ökol. Tiere, **51**: 611-725.

HANSEN, H. J., 1897. The Choniostomatidae, a family of Copepoda, parasitic on Crustacea Malacostraca and Ostracoda. — Pp. i-iii, 1-205. (Host & Son, Copenhagen.)

HARDING, J. P., 1953. The first known example of a terrestrial ostracod, *Mesocypris terrestris* sp. nov. — Ann. Natal Museum, **12** (3): 359-365.

— —, 1954. The copepod *Thalestris rhodymeniae* (Brady) and its nauplius, parasitic in the seaweed *Rhodymenia palmata* (L.) Grev. — Proc. zool. Soc. Lond., **124** (1): 153-161.

— —, 1955. The evolution of terrestrial habits in an ostracod. — In: Symp. on Organic Evolution. Bull. nat. Inst. Sci. India, **7**: 104-106.

HARTNOLL, R. G., 1964. The freshwater grapsid crabs of Jamaica. — Proc. Linn. Soc. Lond., (Zool.) **175** (2): 145-169.

HAYASHI, K.-I., 1975. *Anachlorocurtis commensalis* gen. nov., sp. nov. (Crustacea: Decapoda: Pandalidae), a new pandalid shrimp associated with antipatharian corals from central Japan. — Annot. zool. Japon., **48** (3): 172-182.

— —, 2002. A new species of the genus *Athanas* (Decapoda, Caridea, Alpheidae) living in the burrows of a mantis shrimp. — Crustaceana, **75** (3-4): 395-403.

HAYASHI, K. I. & M. SADAYASHI, 1968. Three caridean shrimps associated with a medusa from Tanabe Bay, Japan. — Publs Seto mar. biol. Lab., **16** (1): 11-19.

HEDGPETH, J. W., 1957. Sandy beaches. — In: J. W. HEDGPETH (ed.), Treatise on marine ecology and paleoecology, I, Ecology. Geol. Soc. Am. Mem., **67**: 587-608.
— —, 1957. Biological aspects. — In: J. W. HEDGPETH (ed.), Treatise on marine ecology and paleoecology, I, Ecology. Geol. Soc. Am. Mem., **67**: 693-729.
HEEGAARD, P., 1951. Antarctic parasitic copepods and an ascothoracid cirriped from brittle stars. — Vidensk. Meddr Dansk naturh. Foren., **113**: 171-190.
HENRY, J.-P., 1978. Observations sur les peuplements de Crustacés Aselloïdes des milieux souterrains. — Bull. Soc. zool. Fr., **103** (4): 491-497.
HERTZOG, L., 1933. *Bogidiella albertimagni* sp. nov., ein neuer Grundwasseramphipode aus der Rheinebene bei Strassburg. — Zool. Anz., **102** (9-10): 225-227.
HEYWOOD, V. H. & R. T. WATSON, 1995. Global biodiversity assessment. — Pp. 1-1152. (UNEP, Cambridge, U.K.)
HICKS, J., H. RUMPFF & H. YORSTON, 1984. Christmas crabs. — Pp. 1-76. (Christmas Isl. natural Hist. Assoc.)
HIPEAU-JACQUOTTE, R., 1964. Décapodes nageurs associés aux Échinodermes dans la région de Tuléar (sud-ouest de Madagascar). — Rec. Trav. Sta. mar. Endoume, Bull., **32** (48): 179-182.
— —, 1965. Un nouveau Décapode nageur (Pontoniinae) associé aux oursins dans la région de Tuléar: *Tulearocaris holthuisi* nov. gen. et nov. sp. — Rec. Trav. Sta. mar. Endoume, Bull., **37** (53): 247-259.
— —, 1972. Étude des crevettes Pontoniinae. — Pp. 1-212. (Thesis, Doct. Sci. nat., Univ. Aix-Marseille, **AO 4845**.)
— —, 1974. Étude des crevettes Pontoniinae (Palaemonidae) associées aux Mollusques Pinnidae à Tuléar (Madagascar). — Arch. Zool. exp. gén., (B) **115**: 359-386.
HIRO, F., 1933. Notes on two interesting pedunculate cirripeds, *Malacolepas conchicola* n. g., n. sp., and *Kaleolepas avis* (Hiro), with remarks on their systematic positions. — Mem. Coll. Sci., Kyoto Imp. Univ., (B) **8** (3): 233-247.
— —, 1937. Studies on the animals inhabiting reefs corals. I. *Hapalocarcinus* and *Cryptochirus*. — Palao trop. biol. Sta. Stud., **1**: 137-154.
— —, 1938. Studies on the animals inhabiting reefs corals. II. Cirripeds of the genera *Creusia* and *Pyrgoma*. — Palao trop. biol. Sta. Stud., **3**: 391-416.
— —, 1938a. A new coral inhabiting crab, *Pseudocryptochirus viridis* gen. n., sp. nov. (Hapalocarcinidae, Brachyura). — Zool. Mag., **50** (3): 149-151.
HO, J.-S., 1980. *Anchicaligus nautili* (Willey), a caligoid copepod parasitic in *Nautilus* on Palau, with discussion of *Caligulina* Heegaard, 1972. — Publs Seto mar. biol. Lab., **25** (1-4): 157-165.
— —, 1982. Copepods associated with echinoderms of the Sea of Japan. — Annu. Rep. Sado mar. biol. Sta., Niigata Univ., **12**: 33-61.
— —, 1984. Copepoda associated with sponges, cnidarians and tunicates of the Sea of Japan. — Annu. Rep. Sado mar. biol. Sta., Niigata Univ., **14**: 23-61.
— —, 1984a. New family of poecilostomatoid copepods (Spiophanicolidae) parasitic on polychaetes from southern California, with a phylogenetic analysis of nereicoliform families. — J. Crust. Biol., **4** (1): 134-146.
— —, 1990. A phylogenetic analysis of copepod orders. — J. Crust. Biol., **10**: 528-536.
— —, 1991. Phylogeny of Poecilostomatoida: a major order of symbiotic copepods. — In: Proceedings of the Fourth International Conference on Copepoda. Bull. Plankton Soc. Jap., (Spec. vol.) **1991**: 25-48.
HO, J.-S. & P. S. PERKINS, 1977. A new family of cyclopoid copepod (Namakosiramidae) parasitic on holothurians from southern California. — J. Parasit., **63** (2): 368-371.
HOBBS, H. H., JR., 1971. The entocytherid ostracods of Mexico and Cuba. — Smithson. Contr. Zool., **81**: 1-55.
— —, 1974. A checklist of the North and Middle American crayfishes (Decapoda: Astacidae and Cambaridae). — Smithson. Contr. Zool., **166** (3): 1-161.

— —, 1977. Cave inhabiting crayfishes of Chiapas, Mexico (Decapoda: Cambaridae). — Accad. naz. Lincei, **374** (171) (III): 1-161.

HOBBS, H. H., JR., H. H. HOBBS, III & M. A. DANIEL, 1977. A review of the troglobitic decapod crustaceans of the Americas. — Smithson. Contr. Zool., **244**: 1-183.

HOBBS, H. H., III & H. H. HOBBS, JR., 1976. On the troglobitic shrimps of the Yucatan Peninsula, Mexico (Decapoda: Atyidae and Palaemonidae). — Smithson. Contr. Zool., **240**: 1-23.

HOLDICH, D. M., 1970. Distribution and habitat preference of the Afro-European species of *Dynamene* (Crustacea: Isopoda). — J. nat. Hist., Lond., **4**: 429-438.

HOLDICH, D. M. & D. A. JONES, 1973. The systematics and ecology of a new genus of sand beach isopod (Sphaeromatidae) from Kenya. — J. Zool., Lond., **171**: 385-395.

HOLSINGER, J. R., 1978. Systematics of the subterranean amphipod genus *Stygobromus* (Crangonyctidae). Part 2: species of the eastern United States. — Smithson. Contr. Zool., **266**: 1-144.

HOLSINGER, J. R. & G. LONGLEY, 1980. The subterranean amphipod crustacean fauna of an artesian well in Texas. — Smithson. Contr. Zool., **308**: 1-62.

HOLTHUIS, L. B., 1947. On a small collection of isopod Crustacea from the greenhouse of the Royal Botanic Gardens, Kew. — Ann. Mag. nat. Hist., (11) **13**: 122-137.

— —, 1956. The troglobitic Atyidae of Madagascar (Crustacea, Decapoda, Natantia). — Mém. Inst. scient. Madagascar, (A) **11**: 97-140.

— —, 1956a. An enumeration of the Crustacea Decapoda Natantia inhabiting subterranean waters. — Vie Milieu, **7** (1): 43-76.

— —, 1959. Contributions to New Guinea carcinology. III. The occurrence of *Birgus latro* (L.) in Netherlands New Guinea (Crustacea, Decapoda, Paguridea). — Nova Guinea, (n. s.) **10** (2): 303-310.

— —, 1963. On red coloured shrimps (Decapoda, Caridea) from tropical land-locked saltwater pools. — Zool. Meded., Leiden, **38** (16): 261-270.

— —, 1973. Caridean shrimps found in land-locked saltwater pools at four Indo-West Pacific localities (Sinai Peninsula, Funafuti Atoll, Maui and Hawaii Islands) with the description of one new genus and four new species. — Zool. Verh., Leiden, **128**: 1-48.

— —, 1978. Zoological results of the British speleological expedition to Papua New Guinea, 1975. Cavernicolous shrimps (Crustacea: Decapoda: Natantia) from New Ireland and the Philippines. — Zool. Meded., Leiden, **53** (19): 209-224.

— —, 1979. Cavernicolous and terrestrial decapod Crustacea from northern Sarawak, Borneo. — Zool. Verh., Leiden, **171**: 1-47.

— —, 1980. *Alpheus saxidomus*, new species, a rock boring snapping shrimp from the Pacific coast of Costa-Rica, with notes on *Alpheus simus* Guérin-Méneville, 1856. — Zool. Meded., Leiden, **55** (4): 49-58.

HORA, S. L., 1933. A note on the bionomics of two estuarine crabs. — Proc. zool. Soc. London, **4**: 881-884.

HUMES, A. G., 1941. Notes on *Octolasmis muelleri* (Coker), a barnacle commensal on crabs. — Trans. Am. micr. Soc., **60** (1): 101-103.

— —, 1941a. A new harpacticoid copepod from the gill chambers of a marsh crab. — Proc. U.S. natl. Mus., **90** (3110): 379-386.

— —, 1947. A new cyclopoid from a Bornean crustacean. — Trans. Am. micr. Soc., **66** (3): 293-301.

— —, 1947a. A new harpacticoid copepod from Bornean crabs. — J. Wash. Acad. Sci., **37**: 170.

— —, 1949. A new copepod (Cyclopoida, Clausidiidae) parasitic on mud shrimps in Louisiana. — Trans. Am. micr. Soc., **68** (2): 93-103.

— —, 1953. *Ostrincola gracilis* C. B. Wilson, a parasite of marine pelecypods in Louisiana (Copepoda, Cyclopoida). — Tulane Stud. Zool., **1** (8): 99-107.

— —, 1954. *Mytilicola porrecta* n. sp. (Copepoda, Cyclopoida) from the intestine of marine pelecypods. — J. Parasit., **40** (2): 186-194.

——, 1956. *Pholetiscus rectiseta* n. sp., des cavités branchiales d'un crabe à Madagascar (Copepoda, Harpacticoida). — Mém. Inst. scient. Madagascar, (A) **11**: 79-84.
——, 1957. *Lamippe concinna* sp. n., a copepod parasitic on a West African pennatulid coelenterate. — Parasitol., **47** (3-4): 447-451.
——, 1957a. *Octopicola superba* n. g., n. sp., copépode cyclopoïde parasite d'un *Octopus* de la Méditerranée. — Vie Milieu, **8** (1): 1-8.
——, 1957b. Une nouvelle espèce de *Clausidium* (Copepoda, Cyclopoida) parasite d'une *Callianassa*, au Sénégal. — Bull. Inst. fr. Afr. noire, IFAN, (A) **19** (2): 485-488.
——, 1957c. The genus *Cancricola* (Copepoda, Harpacticoida) on the west coast of Africa. — Bull. Inst. fr. Afr. noire, IFAN, (A) **19**: 180-191.
——, 1957d. Deux Copépodes Harpacticoïdes nouveaux du genre *Tisbe*, parasite des Holothuries de la Méditerranée. — Vie Milieu, **8** (1): 9-22.
——, 1958. Copépodes parasites de Mollusques à Madagascar. — Mém. Inst. scient. Madagascar, (A) **2**: 286-342.
——, 1958a. Copepods parasites of molluscs. — 23rd annu. Meet. Amer. Malacol. Union, Bull., **24**: 1 p.
——, 1958b. *Antillesia cardisomae*, n. gen. and sp. (Copepoda, Harpacticoida) from the gill chambers of land crabs, with observations on the related genus *Cancrincola*. — J. Wash. Acad. Sci., **48** (3): 77-87.
——, 1960. New copepods from madreporarian corals. — Kiel. Meeresf., **16** (2): 229-235.
——, 1960a. The harpacticoid copepod *Sacodiscus* (= *Unicaltheuta*) *ovalis* (C. B. Wilson, 1944) and its copepodid stages. — Crustaceana, **1** (3): 279-294.
——, 1962. Eight new species of *Xarifia* (Copepoda, Cyclopoida), parasites of corals in Madagascar. — Bull. Mus. comp. Zool., **128** (2): 35-63.
——, 1962a. *Kombia angulata* n. gen., n. sp. (Copepoda, Cyclopoida) parasitic in a coral in Madagascar. — Crustaceana, **4** (1): 47-56.
——, 1963. New species of *Lichomolgus* (Copepoda, Cyclopoida) from sea anemones and nudibranchs in Madagascar. — Cah. ORSTOM, (Océanogr.) **6** [1964]: 59-130.
——, 1963a. *Octopicola stocki* n. sp. (Copepoda, Cyclopoida) associated with an *Octopus* in Madagascar. — Crustaceana, **5** (4): 271-280.
——, 1964. *Harpacticus pulex*, a new species of copepod from the skin of a porpoise and a manatee in Florida. — Bull. mar. Sci., **14** (4): 517-528.
——, 1965. New species of *Hemicyclops* (Copepoda, Cyclopoidea) from Madagascar. — Bull. Mus. comp. Zool., **134** (6): 159-259.
——, 1966. New species of *Macrochiron* (Copepoda, Cyclopoida) associated with hydroids in Madagascar. — Beaufortia, **14** (165): 1-28.
——, 1967. *Vahinius petax*, n. sp., a cyclopoid copepod parasitic in an antipatharian coelenterate in Madagascar. — Crustaceana, **13** (3): 233-242.
——, 1967a. Cyclopoid copepods of the genus *Paranthessius* associated with marine pelecypods in Chile. — Proc. U.S. natl. Mus., **124** (3628): 1-18.
——, 1967b. A new species of *Scambicornus* (Copepoda, Cyclopoida, Lichomolgidae) associated with a holothurian in Madagascar, with notes on several previously described species. — Beaufortia, **14** (173): 135-155.
——, 1968. Two new copepods (Cyclopoida, Lichomolgidae) from marine pelecypods in Madagascar. — Crustaceana, (Suppl.) **1**: 65-81.
——, 1968a. The cyclopoid copepod *Pseudomyicola spinosus* (Raffaele et Monticelli) from marine pelecypods, chiefly in Bermuda and the West Indies. — Beaufortia, **14** (178): 203-226.
——, 1968b. *Lecanurius rossmanianus*, a new cyclopoid copepod parasitic in holothurians. — Proc. biol. Soc. Wash., **81**: 179-190.
——, 1969. A cyclopoid copepod, *Sewellochiron fidens* n. gen., n. sp., associated with a medusa in Puerto-Rico. — Beaufortia, **16** (219): 171-183.

——, 1969a. *Aspidomolgus stoichactinus* n. g., n. sp. (Copepoda Cyclopoida) associated with an actiniarian in the West Indies. — Crustaceana, **16** (3): 225-242.

——, 1969b. Cyclopoid copepods associated with antipatharian coelenterates in Madagascar. — Zool. Meded., Leiden, **44** (1): 1-30.

——, 1969c. *Stellicola dentifer* n. sp. (Copepoda, Cyclopoida) associated with a starfish in Jamaica. — Breviora Mus. comp. Zool., **315**: 1-11.

——, 1969d. *Pseudanthessius pusillus*, n. sp., a cyclopoid copepod associated with a clypeasteroid echinoid in Madagascar. — Zool. Anz., **183** (3-4): 268-277.

——, 1970. A census of copepods associated with marine invertebrates in a tropical locality. — J. Parasit., **56** (4): 160-161.

——, 1970a. *Paramacrochiron japonicum* n. sp., a cyclopoid copepod associated with a medusa in Japan. — Publs Seto mar. biol. Lab., **18** (4): 223-232.

——, 1970b. Cyclopoid copepods of the genus *Paranthessius* associated with marine pelecypods in the West Indies. — Bull. mar. Sci., **20** (3): 605-625.

——, 1970c. *Clavisodalis heterocentroti* gen. and sp. n., a cyclopoid copepod parasitic on an echinid at Eniwetok Atoll. — J. Parasit., **56** (3): 575-583.

——, 1970d. *Stellicola acanthasteris* n. sp. (Copepoda, Cyclopoida) associated with the starfish *Acanthaster planci* (L.) at Eniwetok Atoll. — Publs Seto mar. biol. Lab., **17** (5): 329-338.

——, 1971. *Sunaristes* (Copepoda, Harpacticoida) associated with hermit crabs at Eniwetok Atoll. — Pacif. Sci., **25** (4): 529-532.

——, 1971a. Cyclopoid copepods (Stellicomitidae) parasitic on sea-stars from Madagascar and Eniwetok Atoll. — J. Parasit., **57** (6): 1330-1343.

——, 1972. *Sunaristes* and *Porcellidium* (Copepoda, Harpacticoida) associated with hermit crabs in New Caledonia. — Cah. ORSTOM, (Océanogr.) **10** (3): 263-266.

——, 1972a. *Pseudanthessius comanthi* n. sp. (Copepoda, Cyclopoida) associated with a crinoid at Eniwetok Atoll. — Pacif. Sci., **24** (4): 373-380.

——, 1973. *Hemicyclops perinsignis*, a new cyclopoid copepod from a sponge in Madagascar. — Proc. biol. Soc. Wash., **86** (25): 315-327.

——, 1973a. Cyclopoid copepods (Lichomolgidae) from octocorals at Eniwetok Atoll. — Beaufortia, **21** (282): 135-151.

——, 1973b. Cyclopoid copepods of the genus *Acanthomolgus* (Lichomolgidae) associated with gorgonians in Bermuda. — J. nat. Hist., Lond., **7**: 85-115.

——, 1973c. Cyclopoid copepods (Lichomolgidae) from fungiid corals in New Caledonia. — Zool. Anz., **190** (5-6): 312-333.

——, 1973d. Cyclopoid copepods associated with marine bivalve molluscs in New Caledonia. — Cah. ORSTOM, (Océanogr.) **11** (1): 3-25.

——, 1973e. *Tychidion guyanense* n. gen. n. sp. (Copepoda, Cyclopoidea) associated with an annelid off Guyana. — Zool. Meded., Leiden, **46** (4): 189-196.

——, 1973f. *Nanaspis* (Copepoda, Cyclopoida) parasitic on the holothurian *Thelenota ananas* (Jaeger) at Eniwetok Atoll. — J. Parasit., **59** (2): 384-395.

——, 1974. Cyclopoid copepods (Lichomolgidae) from gorgonians in Madagascar. — Proc. biol. Soc. Wash., **87** (37): 411-438.

——, 1974a. *Odontomolgus mundulus* n. sp. (Copepoda, Cyclopoida) associated with the scleractinian coral genus *Alveopora* in New Caledonia. — Trans. Am. micr. Soc., **93** (2): 153-162.

——, 1974b. Cyclopoid copepods associated with the coral genera *Favia*, *Favites*, *Platygyra* and *Merulina* in New Caledonia. — Pacif. Sci., **28** (4): 383-389.

——, 1974c. Cyclopoid copepods associated with opisthobranch molluscs in New Caledonia. — Crustaceana, **26** (3): 233-238.

——, 1974d. *Octopicola regalis* n. sp. (Copepoda, Cyclopoida, Lichomolgidae) associated with *Octopus cyaneus* from New Caledonia and Eniwetok Atoll. — Bull. mar. Sci., **24** (1): 76-85.

— —, 1974e. New cyclopoid copepods associated with an abyssal holothurian in the eastern North Atlantic. — J. nat. Hist., Lond., **8**: 101-117.

— —, 1975. Cyclopoid copepods (Lichomolgidae) associated with alcyonaceans in New Caledonia. — Smithson. Contr. Zool., **191**: 1-27.

— —, 1975a. Cyclopoid copepods associated with marine invertebrates in Mauritius. — J. Linn. Soc. Lond., **56** (8): 171-181.

— —, 1975b. *Hadrothoe crosnieri*, n. gen., n. sp. (Crustacea: Copepoda) from a penaeid shrimp (Crustacea: Decapoda) in Madagascar. — Zool. Anz., **195** (1-2): 21-34.

— —, 1975c. Cyclopoid copepods (Nanaspidae and Sabelliphilidae) associated with holothurians in New Caledonia. — Smithson. Contr. Zool., **202** (III): 1-41.

— —, 1976. Cyclopoid copepods associated with Tridacnidae (Mollusca: Bivalvia) in the Moluccas. — Proc. biol. Soc. Wash., **89** (43): 491-508.

— —, 1976a. Distribution and hosts of *Stellicola* (Copepoda, Cyclopoida) associated with *Linckia* (Asteroidea) in the Indo-West Pacific. — Beaufortia, **25** (321): 49-61.

— —, 1976b. Cyclopoid copepods associated with asteroid echinoderms in New Caledonia. — Smithson. Contr. Zool., **217** (III): 1-19.

— —, 1977. Pseudanthessid copepods (Cyclopoida) associated with crinoids and echinoids (Echinodermata) in the tropical western Pacific Ocean. — Smithson. Contr. Zool., **248**: 1-43.

— —, 1978. Cyclopoid copepods (Lichomolgidae) associated with hydroids in the tropical western Pacific Ocean. — Pacif. Sci., **31** (4): 335-353.

— —, 1978a. Lichomolgid copepods (Cyclopoida) with two new species of *Doridicola*, from sea pens (Pennatulacea) in Madagascar. — Trans. Am. micr. Soc., **97**: 524-539.

— —, 1978b. Lichomolgid copepods (Cyclopoida) associated with the fungiid corals (Scleractinia) in the Moluccas. — Smithson. Contr. Zool., **253**: 1-48.

— —, 1979. Coral-inhabiting copepods from the Moluccas, with a synopsis of Cyclopoidea associated with scleractinian corals. — Cah. Biol. mar., **20**: 77-107.

— —, 1979a. Poecilostome copepods associated with antipatharian coelenterates in the Moluccas. — Beaufortia, **28** (347): 113-120.

— —, 1987. Copepoda from deep-sea hydrothermal vents. — Bull. mar. Sci., **41**: 645-788.

— —, 1987a. A review of Copepoda associated with sea anemones and anemone-like forms (Cnidaria, Anthozoa). — Trans. Amer. phil. Soc., **72**: 1-120.

HUMES, A. G. & W. F. BRUCE, 1964. New lichomolgid copepods (Cyclopoida) associated with alcyonarians and madreporarians in Madagascar. — Cah. ORSTOM, (Océanogr.) **6**: 131-212.

HUMES, A. G. & R. F. CRESSEY, 1958. Copepod parasites of molluscs in West Africa. — Bull. Inst. fr. Afr. noire, IFAN, (A) **20** (3): 921-942.

— — & — —, 1958a. Four new species of lichomolgid copepods parasitic on West African starfishes. — Bull. Inst. fr. Afr. noire, IFAN, (A) **20** (2): 330-341.

— — & — —, 1958b. A new family containing two new genera of cyclopoid copepods parasitic on starfishes. — J. Parasit., **44** (4): 395-408.

— — & — —, 1960. Seasonal population changes and host relationships of *Myocheres major* (Williams), a cyclopoid copepod from pelecypods. — Crustaceana, **1** (4): 307-325.

HUMES, A. G. & P. A. GELERMAN, 1962. A new species of *Porcellidium* (Copepoda, Harpacticoida) from a sea urchin in Madagascar. — Crustaceana, **4** (4): 311-319.

HUMES, A. G. & G. HEDLER, 1972. New cyclopoid copepods associated with the ophiuroid genus *Amphioplus* on the eastern coast of the United States. — Trans. Am. micr. Soc., **91** (4): 539-555.

HUMES, A. G. & J.-S. HO, 1965. New species of the genus *Anthessius* (Copepoda, Cyclopoida) associated with molluscs in Madagascar. — Cah. ORSTOM, (Océanogr.) **3** (2): 79-113.

— — & — —, 1966. New lichomolgid copepods (Cyclopoida) from zoanthid coelenterates in Madagascar. — Cah. ORSTOM, (Océanogr.) **4** (2): 3-47.

— — & — —, 1967. New cyclopoid copepods associated with the alcyonarian coral *Tubipora musica* (Linnaeus) in Madagascar. — Proc. U.S. natl. Mus., **121** (3573): 1-24.

— — & — —, 1967a. Two new species of *Lichomolgus* (Copepoda, Cyclopoida) from an actiniarian in Madagascar. — Cah. ORSTOM, (Océanogr.) **5** (1): 3-21.

— — & — —, 1967b. New cyclopoid copepods associated with the coral *Psammocora contigua* (Esper) in Madagascar. — Proc. U.S. natl. Mus., **122** (3586): 1-37.

— — & — —, 1967c. New cyclopoid copepods associated with polychaete annelids in Madagascar. — Bull. Mus. comp. Zool., **135** (7): 377-413.

— — & — —, 1967d. New species of *Stellicola* (Copepoda, Cyclopoida) associated with starfishes in Madagascar, with a redescription of *S. caeruleus* (Stebbing, 1920). — Bull. Br. Mus. (nat. Hist.), (Zool.) **15** (5): 201-225.

— — & — —, 1968. Cyclopoid copepods of the genus *Lichomolgus* associated with octocorals of the family Alcyoniidae in Madagascar. — Proc. biol. Soc. Wash., **81**: 635-692.

— — & — —, 1968a. Cyclopoid copepods of the genus *Lichomolgus* associated with octocorals of the families Xeniidae, Nidaliidae and Telestidae in Madagascar. — Proc. biol. Soc. Wash., **81**: 693-749.

— — & — —, 1968b. Cyclopoid copepods of the genus *Lichomolgus* associated with octocorals of the family Nephtheidae in Madagascar. — Proc. U.S. natl. Mus., **125** (3661): 1-41.

— — & — —, 1968c. Lichomolgid copepods (Cyclopoida) associated with corals in Madagascar. — Bull. Mus. comp. Zool., **136** (10): 353-413.

— — & — —, 1968d. Xarifiid copepods (Cyclopoida) parasitic in corals in Madagascar. — Bull. Mus. comp. Zool., **136** (11): 415-459.

— — & — —, 1969. The genus *Sunaristes* (Copepoda, Harpacticoida) associated with hermit crabs in the western Indian Ocean. — Crustaceana, **17** (1): 1-18.

— — & — —, 1969a. Harpacticoid copepods of the genera *Porcellidium* and *Paraidya* associated with hermit crabs in Madagascar and Mauritius. — Crustaceana, **17** (2): 113-130.

— — & — —, 1969b. Cyclopoid copepods parasitic in holothurians in Madagascar. — J. Parasit., **55** (4): 877-894.

— — & — —, 1970. *Mytilicola fimbriatus* sp. n., a cyclopoid copepod parasitic in the marine pelecypod, *Arca decussata*, in Madagascar. — J. Parasit., **56** (3): 584-587.

— — & — —, 1970a. The genus *Diogenella* (Copepoda, Cyclopoida) parasitic in holothurians in the West Indies. — Crustaceana, **19** (1): 15-36.

— — & — —, 1970b. Cyclopoid copepods of the genus *Paranthessius* associated with crinoids in Madagascar. — Smithson. Contr. Zool., **54**: 1-29.

— — & — —, 1971. The genus *Diogenidium* (Copepoda, Cyclopoida) parasitic in holothurians in the West Indies. — Crustaceana, **20** (2): 171-191.

HUMES, A. G. & A. DE MARIA, 1969. The cyclopoid copepod genus *Macrochiron* from hydroids in Madagascar. — Beaufortia, **16** (216): 137-155.

HUMES, A. G. & W. L. SMITH, 1974. *Ridgewayia fosshageni* n. sp. (Copepoda: Calanoidea) associated with an actiniarian in Panama, with observations on the nature of the association. — Caribb. J. Sci., **14** (3-4): 125-139.

HUMES, A. G. & J. H. STOCK, 1965. Three new species of *Anthessius* (Copepoda, Cyclopoida, Myicolidae) associated with *Tridacna* from the Red Sea and Madagascar. — Israel South Red Sea Exped., 1962, Reports, **15**. Sea Fish. Res. Sta. Haifa, Bull., **40**: 49-74.

— — & — —, 1973. A revision of the family Lichomolgidae Kossmann, 1877, cyclopoid copepods mainly associated with marine invertebrates. — Smithson. Contr. Zool., **127**: i-v, 1-368.

HUMES, A. G. & R. D. TURNER, 1972. *Teredicola typicus* C. B. Wilson, 1942 (Copepoda, Cyclopoida) from ship worms in Australia, New Zealand and Japan. — Aust. J. mar. freshw. Res., **23**: 63-72.

HURLEY, D. E., 1959. Notes on the ecology and environmental adaptations of the terrestrial Amphipoda. — Pacif. Sci., **13** (2): 107-129.

— —, 1975. A possible subdivision of the terrestrial genus *Talitrus* (Crustacea Amphipoda; family Talitridae). — N.Z. oceanogr. Inst. Rec., **2** (14): 157-170.

HUTTON, R. F., 1962. Studies on parasites of *Penaeus duorarum*, the pink shrimp. — U.S. Publ. Health Serv., Gr. E-2008, Final Rept., pp. 1-27.

ILLG, P., 1949. A review of the copepod genus *Paranthessius* Claus. — Proc. U.S. natl. Mus., **99** (3245): 391-428.

— —, 1950. A new copepod, *Pseudanthessius latus* (Cyclopoida, Lichomolgidae) commensal with a marine flatworm. — J. Wash. Acad. Sci., **40** (4): 129-133.

— —, 1958. North American copepods of the family Notodelphyidae. — Proc. U.S. natl. Mus., **107** (3390): 463-649.

ILLG, P. & P. DUDLEY, 1961. Notodelphyid copepods from Banyuls-sur-Mer. — Vie Milieu, (Suppl.) **12**: 1-126.

ILLG, P. & A. G. HUMES, 1971. *Hemicoxiphium redactum*, a new cyclopoid copepod associated with an ascidian in Florida and North Carolina. — Proc. biol. Soc. Wash., **83** (49): 569-578.

ISSEL, R., 1912. Ricerche di etologia sull'Isopodo tubicolo *Zenobiana prismatica* (Risso). — Arch. Zool. exp. gén., **51**: 479-500.

JACQUOTTE, R., 1964. Notes de faunistique et de biologie marine de Madagascar. I. Sur l'association de quelques Crustacés avec des Cnidaires récifaux de la région de Tuléar (sud-ouest de Madagascar). — Rec. Trav. Sta. mar. Endoume, Bull., **32** (48): 175-178.

JAKUBISIAK, S., 1932. Sur les Harpacticoïdes hébergés par *Maia squinado*. — Bull. Soc. zool. Fr., **57**: 506-513.

JAUME, D. & F. BRÉHIER, 2005. A new species of *Typhlatya* (Crustacea: Decapoda: Atyidae) from anchialine caves on the French Mediterranean coast. — Zool. J. Linn. Soc., Lond., **144** (3): 387-414.

JOHNSON, D. S., 1960. On a porcelain crab, *Petrolisthes oshimai* (Miyake) from Christmas Island, Indian Ocean, with a note on the genus *Neopetrolisthes* Miyake. — Crustaceana, **1** (2): 164-165.

JONES, D. A., 1971. The systematics and ecology of some sand beach isopods (Crustacea: Eurydicidae) from the coast of Kenya. — J. Zool., Lond., **165**: 201-227.

— —, 1974. The systematics and ecology of some sand beach isopods (family Cirolanidae) from the coasts of Saudi Arabia. — Crustaceana, **26** (2): 201-211.

— —, 1976. The systematics and ecology of some isopods of the genus *Cirolana* (Cirolanidae) from the Indian Ocean region. — J. Zool., Lond., **178**: 209-222.

— —, 1979. The ecology of sandy beach in Penang, Malaysia, with special reference to *Excirolana orientalis* (Dana). — Est. coast. mar. Sci., **9**: 677-682.

KELLEY, L. S. & W. A. NEWMAN, 1974. The Indo-West Pacific genus *Pagurolepas* (Cirripedia, Poecilasmatidae) in Floridan waters. — Bull. mar. Sci., **24** (3): 628-637.

KIENER, A., 1966. Contribution à l'étude écologique et biologique des eaux saumâtres malgaches. — Vie Milieu, **16**: 1013-1149.

KING, CH. E. & L. S. KORNICKER, 1970. Ostracoda in Texas bays and lagoons: an ecological study. — Smithson. Contr. Zool., **24**: 1-92.

KLEETON, G. J., 1964. Sur la présence de *Pseudomyicola spinosus* (Raff. et Mont.) (Crustacea: Copepoda) dans l'Atlantique, avec une note sur la synonymie de *P. spinosus* et *P. glaber* Pearse. — Beaufortia, **11** (145): 171-177.

KOMAI, T. & M. SEGONZAC, 2003. Review of the hydrothermal vent shrimp genus *Mirocaris*, redescription of *M. fortunata* and reassessment of the taxonomic status of the family Alvinocarididae (Crustacea: Decapoda: Caridea). — Cah. Biol. mar., **44**: 199-215.

KORNICKER, L. S. & T. E. BOWMAN, 1969. *Sphaeronellopsis dikrothrix*, a new clioniostomatid copepod from the ostracod *Metavargula ampla*. — Crustaceana, **17** (3): 282-284.

KORSCHELT, E., 1933. Uber zwei parasitaire Cirripedien, *Chelonibia* und *Dendrogaster* nebst Angaben über die Beziehungen der Balanomorphen zu ihrer Unterlage. — Zool. Jahrb., (Syst.) **64** (1): 1-40.

KULBICKI, M. & C. ARNAL, 1999. Cleaning of fish ectoparasites by a Palaemonidae shrimp on soft bottoms in New Caledonia. — Cybium, **23** (1): 101-104.

LAESSLE, A. E., 1961. A micro-limnological study of Jamaican bromelids. — Ecology, **42**: 499-517.

LAFARGUE, F. & L. LAUBIER, 1968. *Cochlodelphys delamarei*, nouveau genre et nouvelle espèce de Copépode Notodelphyidae en Méditerranée occidentale. — C. R. Acad. Sci. Paris, **267**: 1375-1378.

— — & — —, 1968a. *Sicyodelphys bocqueti*, nouveau genre et nouvelle espèce de Copépode Notodelphyidae en Méditerranée occidentale. — C. R. Acad. Sci. Paris, **267**: 2163-2166.

— — & — —, 1977. Copépodes Notodelphyidae parasites de Didemnidae (Ascidies Aplousobranches) dans le golfe d'Eilat (Mer Rouge). — Arch. Zool. exp. gén., **118**: 173-196.

— — & — —, 1977a. Deux Copépodes Notodelphyidae nouveaux de la région de Singapour. — Arch. Zool. exp. gén., **119** (3): 479-486.

— — & — —, 1978. *Anoplodelphys* g. nov., Copépode Notodelphyidae parasite de Didemnidae (Ascidies Aplousobranches) en Méditerranée. — Crustaceana, **35** (3): 277-293.

LAUBIER, L., 1961. *Phyllodicola petiti* (Delamare et Laubier, 1960) et la famille des Phyllodicolidae, Copépodes parasites d'Annélides Polychètes en Méditerranée occidentale. — Crustaceana, **2** (3): 228-242.

— —, 1964. Présence de *Nereicola ovatus* Keferstein à Banyuls-sur-Mer. Données morphologiques nouvelles. — Bull. Mus. natn. Hist. nat., Paris, (2) **36** (5): 631-640.

— —, 1971. *Lamippe* (*L.*) *bouligandi* sp. nov., Copépode parasite d'Octocoralliaire de la mer du Labrador. — Crustaceana, **22** (3): 285-293.

— —, 1971a. *Rhodinicola thomassini* sp. n., un nouveau Copépode parasite d'Annélides Polychètes Maldanidae de l'Océan Indien. — Arch. Zool. exp. gén., **111** (4): 559-572.

— —, 1971b. Description du mâle du genre *Rhodinicola* Levinsen (Copépode Clausiidae). — Arch. Zool. exp. gén., **112**: 351-359.

— —, 1973. *Vectoriella ramosae* sp. n., un Copépode parasite d'Annélide Polychète en Méditerranée profonde. — Arch. Zool. exp. gén., **114** (1): 149-158.

— —, 1978. *Ophelicola drachi* gen. sp. n., un nouveau Copépode Cyclopoïde abyssal ectoparasite d'Annélides Polychètes Opheliidae. — Arch. Zool. exp. gén., **119** (1): 39-50.

— —, 1988. Écosystèmes benthiques profonds et chimiosynthèse bactérienne: sources hydrothermales et suintements froids. — In: M. DENIS (coord. ed.), Océanologie, actualité et prospective, pp. 61-99. (Centre d'Océanologie de Marseille.)

— —, 1988a. Le genre *Anomopsyllus* Sars, 1921, Copépode parasite d'Annélides Polychètes. *A. pranizoides* Sars, 1921 et *A. abyssorum* nov. sp. — Crustaceana, **55** (2): 180-192.

— —, 1989. Deep-sea ecosystems based on chemosynthetic processes: recent results on hydrothermal and cold seeps biological assemblages. — In: A. AYALA-CASTANARES ET AL. (eds.), Oceanography 1988, pp. 130-148. (Joint Oceanographic Assembly, Mexico, 1988, UNAM & CNCT Publ.)

LAUBIER, L. & Y. CARTON, 1974. Description de *Selioides guineensis* sp. n., Copépode Cyclopoïde parasite d'Aphroditidae. — Arch. Zool. exp. gén., **115**: 129-139.

LAUBIER, L. & F. LAFARGUE, 1974. Le genre *Brementia* Chatton et Brément, curieux Copépode Notodelphyidae ascidicole parasite de Didemnidae. — Crustaceana, **27** (3): 235-248.

LAUBIER, L. & D. REYSS, 1964. Sub-spéciation chez un Copépode parasite *Pseudomyicola spinosus* (Raff. et Mont.) et description de deux sous-espèces nouvelles. — Vie Milieu, (Suppl.) **17**: 291-308.

LAWRENCE, R. F., 1953. The biology of the cryptic fauna of forests. With special reference to the indigenous forests of South Africa. — Pp. 1-408. (Balkerian, Cape Town.)

LEBOUR, M. V., 1938. Decapod Crustacea associated with the ascidian *Herdmannia*. — Proc. zool. Soc. Lond., (B) **108** (4): 649-653. [1939]
LEDOYER, M., 1967. Amphipodes Gammariens des herbiers de Phanérogames marines de la région de Tuléar (République Malgache). Étude systématique et écologique. — Rec. Trav. Sta. mar. Endoume, (fasc. hors série), (Suppl.) **7**: 7-56.
— —, 1968. Amphipodes Gammariens de quelques biotopes de substrat meuble de la région de Tuléar. Étude systématique et écologique. — Ann. Univ. Madagascar, **6**: 15-62.
— —, 1968a. Les Caridea de la frondaison des herbiers de Phanérogames de la région de Tuléar (République Malgache). Étude systématique et écologique. — Ann. Univ. Madagascar, **6**: 63-121.
— —, 1969. Amphipodes tubicoles des feuilles des herbiers de Phanérogames marines de la région de Tuléar (Madagascar). — Rec. Trav. Sta. mar. Endoume, (fasc. hors série), (Suppl.) **9**: 79-182.
— —, 1970. Mysidacés des herbiers de Phanérogames marines de Tuléar (Madagascar). Étude systématique et écologique. — Rec. Trav. Sta. mar. Endoume, (fasc. hors série), (Suppl.) **10**: 223-227.
LEIGH-SHARPE, W. H., 1935. A list of British invertebrates with their characteristic parasitic and commensal Copepoda. — J. mar. biol. Ass. U.K., **20** (1): 47-48.
LELOUP, E., 1942. L'hydraire *Campanularia johnstoni* Alder et le mollusque *Mytilus edulis* Linné, épizoaires sur le crustacé *Pandalus montagui* Leach. — Bull. Mus. roy. Hist. nat. Belg., **18**: 1-4.
LEUNG, Y. M., 1967. An illustrated key to the species of whale-lice (Amphipoda, Cyamidae) ectoparasites of Cetacea, with a guide to the literature. — Crustaceana, **12** (3): 279-291.
— —, 1970. First record of the whale-louse genus *Syncyamus* (Cyamidae: Amphipoda) from the western Mediterranean, with notes on the biology of odontocete cyamids. — In: G. RILLERI, (ed.), Investigations on Cetacea, II, pp. 243-247.
LIMBAUGH, C., H. PEDERSON & F. A. CHACE, JR., 1961. Shrimps that clean fishes. — Bull. mar. Sci. Gulf Caribb., **11** (2): 237-257.
LINCOLN, R. J. & D. E. HENLEY, 1974. *Scutocyamus parvus*, a new genus and species of whalelouse (Amphipoda, Cyamidae) ectoparasitic on the North Atlantic white-beaked dolphin. — Bull. Br. Mus. (nat. Hist.), (Zool.) **27** (2): 59-64.
— — & — —, 1974a. Catalogue of the whalelice (Crustacea: Amphipoda: Cyamidae) in the collections of the British Museum (Natural History). — Bull. Br. Mus. (nat. Hist.), (Zool.) **27** (2): 65-72.
LINDROTH, S., 1953. Taxonomic and zoogeographical studies of the ostracod fauna in the inland waters of East Africa. — Zool. Bidr., Uppsala, **30**: 43-156.
LOWRY, J. K. & K. DEMPSEY, 2006. The giant deep-sea scavenger genus *Bathynomus* (Crustacea, Isopoda, Cirolanidae) in the Indo-West Pacific. — In: B. RICHER DE FORGES & J.-L. JUSTONE (eds.), Résultats des Campagnes MUSORSTOM, 24. Mém. Mus. natn. Hist. nat., Paris, **193**: 163-192.
LUTHER, W., 1958. Symbiose von Fischen (Gobiidae) mit einen Krebs (*Alpheus djiboutiensis*), im Roten Meer. — Z. Tierpsychol., **15**: 175-177.
MACGINITIE, G. E., 1934. The natural history of *Callianassa californiensis* Dana. — Am. Midl. Nat., **15** (2): 166-177.
MACNAE, W., 1967. Zonation within mangroves associated with estuaries in North Queensland. — In: G. H. LAUFF (ed.), Estuaries. Amer. Ass. Advmt Sci. Publ., **83**: 432-441.
MAGNIEZ, G., 1978. Les Sténasellides de France (Crustacés Isopodes Asellotes souterrains): faune ancienne et peuplements récents. — Bull. Soc. zool. Fr., **103** (3): 255-262.
— —, 1978a. Quelques problèmes biogéographiques, écologiques et biologiques de la vie souterraine. — Bull. Sci. Bourgogne, **81** (1): 21-35.
MANNING, R. B., 1970. *Mithrax (Mithraculus) commensalis*, a new West Indian spider crab (Decapoda, Majidae) commensal with a sea anemone. — Crustaceana, **19** (2): 157-160.

MARIN, I. N., A. ANKER, T. A. BRITAYEV & A. R. PALMER, 2005. Symbiosis between the alpheid shrimp, *Athanas ornithorhynchus* Banner and Banner, 1973 (Crustacea: Decapoda), and the brittle star, *Macrophiothrix longipeda* (Lamarck, 1816) (Echinodermata: Ophiuroidea). — Zool. Stud., **44** (2): 234-241.

MARQUES, G., 1978. Copepoda dulciaquicola *Copidodiaptomus steueri* (Brehm) casualmente encontrado na Sardena. — Garcia de Orta Sci. Zool., **7** (1-2): 7-10.

MARTIN, J. W. & G. E. DAVIS, 2001. An updated classification of the Recent Crustacea. — Nat. Hist. Mus. Los Angeles Count., (Sci. ser.) **39**: 1-124.

MARTÍNEZ GARCÍA, A., A. M. PALMERO, M. DEL CARMEN BRITO, J. NUNEZ & K. WORSAAE, 2009. Anchialine fauna of the Corona lava tube (Lanzarote, Canary Islands): diversity, endemism and distribution. — Mar. Biodiv., **39** (3): 169-182.

MATEUS, A. & E. MATEUS DE OLIVEIRA, 1978. Amphipodes endogés du Portugal. — Publ. Inst. Zool. Dr. A. Nobre, **142**: 11-26.

MATHES, I. & H. STROUHAL, 1954. Zur Ökologie und Biologie der Ameisenassel *Platyarthrus hoffmannseggii*. — Z. Morph. Ökol. Tiere, **43**: 82-93.

MAURY, A., 1925. Étude sur le comportement de *Zenobia prismatica* Risso. — Bull. Soc. Linn. Normandie, **7** (8): 89-92.

MCDERMOTT, J. J., J. D. WILLIAMS & C. B. BOYKO, 2010. The unwanted guests of hermits: a global review of the diversity and natural history of hermit crab parasites. — J. exp. mar. Biol. Ecol., **394**: 2-44.

MENZEL, R., 1923. Beitrage zur Kenntnis der Mikrofauna von Niederländisch-Ostindien. V. Moosbewohnende Ostracoden aus dem Urwald von Tjibodas. — Treubia, **3** (2): 193-196.

MENZIES, R. J., 1952. The occurrence of a terrestrial isopod in plankton. — Ecology, **33** (2): 303.

— —, 1957. The marine borer family Limnoriidae (Crustacea: Isopoda). — Bull. mar. Sci. Gulf Caribb., **7** (2): 101-200.

— —, 1959. The identification and distribution of the species of *Limnoria*. — In: D. L. RAY (ed.), Marine borers and fouling organisms, pp. 10-33. (Univ. Washington Press, Seatle, WA.)

MENZIES, R. J. & J. L. BARNARD, 1951. The isopodan genus *Iais* (Crustacea). — Bull. south. Calif. Acad. Sci., **50** (3): 136-151.

MILLER, M. A., 1968. Isopoda and Tanaidacea from buoys in coastal waters of the continental United States: Hawaii and the Bahamas. — Proc. U.S. natl. Mus., **125** (3652): 1-53.

MONOD, TH., 1923. Notes carcinologiques (parasites et commensaux). — Bull. Inst. océanogr. Monaco, **427**: 1-23.

— —, 1925. La région de la Basse Seule. Étude bionomique. — Trav. Stat. biol. Roscoff, **1925**: 1-74.

— —, 1926. Les Gnathiidae. Essai monographique. — Mém. Soc. Sci. nat. Maroc, **13**: 1-668.

— —, 1928. Sur quelques Copépodes parasites de Nudibranches. — Bull. Inst. océanogr. Monaco, **509**: 1-18.

— —, 1931. Sur quelques Crustacés aquatiques d'Afrique (Cameroun et Congo). — Rev. Zool. Bot. afr., **21** (1): 1-36.

— —, 1931a. Une association biologique multiple. — La Terre et la Vie, (n. s.) **11**: 691-693.

— —, 1932. Ueber drei indopazifische Cyprididen und zwei in Ostracoden lebende Krebstiere. — Zool. Anz., **98** (1-2): 1-8.

— —, 1933. Hôtes et supports chez les Cirripèdes thoraciques. — In: J. CIUREA, TH. MONOD & G. DINULESCO, Présence d'un Cirripède operculé sur un Poisson dulçaquicole européen. Bull. Inst. océanogr. Monaco, **615**: 1-32.

— —, 1934. Sur un Copépode parasite de *Trochus niloticus*. — Rec. Indian Mus., **36** (2): 213-218.

— —, 1936. *Conchoderma auritum* (L., 1767) Olfers, 1814 sur un *Ziphius* cf. *cavirostris* G. Cuvier, 1823. — Bull. Trav. Sta. Aquicult. Pêche Castiglione, **1937**: 205-210. [1938]

— —, 1940. Thermosbaenacea. — Bronns Tierreich, **5** (1) (4) (4): 1-24.

— —, 1942. Sur un *Sphaeroma* de Dakar. — Notes afr., **16**: 1-9. [October 1942]

——, 1968. Nouvelles captures du *Ligur uveae* (Borradaile) aux îles Loyalty (Crustacea, Decapoda). — Bull. Mus. natn. Hist. nat., Paris, (2) **40**: 772-778.

——, 1970. Sur quelques Crustacés Malacostracés des îles Galapagos récoltés par N. et J. Leleup (1964-1965). — Miss. zool. belg. Galapagos et Ecuador. Inst. Roy. Sci. nat. Belg., **2**: 11-53.

——, 1974. Sur un *Octolasmis* (Cirripedia) épizoaire sur un bathynome de Guyane. — Crustaceana, **26** (2): 219-222.

——, 1975. Sur la distribution de quelques Crustacés Malacostracés d'eau douce ou saumâtre. — Mém. Mus. natn. Hist. nat., Paris, (n. s.) (A, Zool.) **88**: 98-103.

——, 1976. Remarques sur quelques Cirolanides (Crustacés, Isopodes). — Bull. Mus. natn. Hist. nat., Paris, **31** (358) [Zool., **251**]: 133-161.

——, 1976a. Expédition Rumphius 11 (1975). Crustacés parasites, commensaux, etc. Introduction. — Bull. Mus. natn. Hist. nat., Paris, **391** [Zool., **273**]: 833-843.

——, 1977. 5. Tanaidacea. — In: La faune terrestre de l'île de Sainte Hélène, 4e partie. Ann. Mus. r. Afr. centr., (8) (Zool.) **220**: 457-465.

——, 1979. Crustacés associés à un Antipathaire des îles Marquises. — Cah. Indo-Pac., **1** (1): 1-23.

MONOD, TH. & PH. CALS, 1970. Sur une espèce nouvelle de crevette cavernicole *Typhlatya galapagensis*. — Miss. zool. belg. Galapagos et Ecuador (N. et J. Leleup, 1964-1965), **2**: 57-103.

MONOD, TH. & J. L. D'HONDT, 1978. A propos d'un échantillon de *Conopeum commensale* (Kirkpatrick et Metzelaar, 1922) (Bryozoa, Cheilostomata) trouvé dans un site archéologique mauritanien. — Bull. Inst. fr. Afr. noire, IFAN, (A) **40** (2): 423-427.

MONOD, TH. & R. PH. DOLLFUS, 1932. Les Copépodes parasites de Mollusques. — Ann. Parasitol., **10** (2): 129-204.

—— & ——, 1932a. Les Copépodes parasites de Mollusques. Premier supplément. — Ann. Parasitol., **10** (3): 295-299.

—— & ——, 1934. Les Copépodes parasites de Mollusques. Deuxième supplément. — Ann. Parasitol., **12** (4): 309-321.

MONOD, TH. & L. LAUBIER, 1996. Les crustacés dans la biosphère. — In: J. FOREST (ed.), Traité de zoologie, **7**, Crustacés (2): 91-166. (Masson, Paris.)

MONOD, TH., M. NICKLES & F. MOLL, 1952. Xylophages et pétricoles ouest-africains. — Inst. fr. Afr. noire, IFAN, **8**: 1-145.

MYERS, A. A., 1974. A new species of commensal amphipod from East Africa. — Crustaceana, **26** (1): 33-36.

MYERS, A. A. & J. K. LOWRY, 2003. A phylogeny and a new classification of the Corophiidea Leach, 1814 (Amphipoda). — J. Crust. Biol., **23** (2): 443-485.

NEALE, J. W., 1969. The taxonomy, morphology and ecology of recent Ostracoda. — Pp. i-viii, 1-553. (Edinburgh.)

NEWMAN, W. A., 1967. Shallow water versus deep-sea *Octolasmis* (Cirripedia: Thoracica). — Crustaceana, **12** (1): 13-32.

NEWMAN, W. A. & H. S. LADD, 1974. Origin of coral-inhabiting balanids (Cirripedia: Thoracica). — Verh. naturf. Ges. Basel, **84** (1): 381-396.

NG, P. K. L., D. GUINOT & P. J. F. DAVIE, 2008. Systema brachyurorum: Part I. An annotated checklist of extant brachyuran crabs of the world. — Raffles Bull. Zool., (Suppl.) **17**: 1-286.

NICHOLLS, G. E., 1943-1944. Phreatoicoidea. — Pap. R. Soc. Tasmania, **1942-1943**: 1-145, 34 figs. [1943] and 1-157, 80 figs. [1944].

NIZINSKI, M. S., 1989. Ecological distribution, demography and behavioural observations on *Periclimenes anthophilus*, an atypical symbiotic cleaner shrimp. — Bull. mar. Sci., **45** (1): 174-188.

NOBLE, E. R. & G. A. NOBLE, 1961. Parasitology. The biology of animal parasites. — Pp. 1-714. (London.)

NOODT, W., 1971. Ecology of the Copepoda. — Smithson. Contr. Zool., **76**: 97-102.

— —, 1974. Anpassung an interstitielle Bedingungen: ein Faktor in der Evolution höherer Taxa der Crustacea? — Faun.-ökol. Mitt., Kiel, **4**: 445-452.

OKADA, Y. K., 1932. Note on the parasitic copepod *Herpyllobius*. — Annot. zool. Japon., **13** (4): 407-413.

— —, 1938. Les Cirripèdes ascothoraciques. — Trav. Stat. zool. Wimereux, **13**: 489-512.

OSTLUND-NILSSON, S., J. H.-A. BECKER & G. E. NILSSON, 2005. Shrimps remove ectoparasites from fishes in temperate waters. — Biol. Lett., **1** (4): 454-456.

PALMER, J. D., 1975. Biological clocks in marine organisms: the control of physiological and behavioural tidal rhythms. — Pp. 1-65. (J. Wiley, New York.)

PANOUSE, J.-B., 1940. Notes biologiques sur les Sphaéromiens. 1. *Campecopea hirsuta* Montagu. — Bull. Soc. zool. Fr., **65**: 93-98.

PARULEKAR, A., 1969. *Neoaiptasia commensali* gen. et sp. nov.: an actiniarian commensal of hermit crabs. — J. Bombay nat. Hist. Soc., **66** (1): 57-62.

PATTON, W. K., 1967. Commensal Crustacea. — Proc. Symp. Crustacea, **3**: 1228-1244.

— —, 1974. Community structure among the animals inhabiting the coral *Pocillopora damicomis* at Heron Islands, Australia — In: W. B. VERNBERG (ed.), Symbiosis in the sea, pp. 219-243. (Belle W. Baruch Library in Marine Science, **2**.)

— —, 1979. On the association of the spider crab, *Mithrax* (*Mithraculus*) *cinctimanus* (Stimpson) with Jamaican sea anemones. — Crustaceana, (Suppl.) **5** [Studies on Decapoda]: 55-61.

PEARCE, J. B., 1962. Adaptations in symbiotic crabs of the family Pinnotheridae. — Biologist, **45**: 11-15.

PEARSE, A. S., 1913. On the habits of the crustacean found in *Chaetopterus* tubes at Woods Hole, Massachusetts. — Biol. Bull., Woods Hole, **24**: 102-114.

— —, 1938. Termes "parasite", "commensal", "symbiose", "consors". — J. Elisha Mitchell Sci. Soc., **54** (2): 1-195.

PERKINS, E. J., 1974. The biology of estuaries and coastal waters. — Pp. 1-678. (Academic Press, London and New York.)

PERTHUISOT, J. P., 1994. Les biocénoses du domaine paralique. — Patrimoines Naturels, **19**: 133-145.

PICADO, C., 1913. Les Broméliacées épiphytes considérées comme milieu biologique. — Bull. Sci. Fr. Belg., (7) **5** (47): 215-360.

PICHON, M., 1964. Contribution à l'étude de l'écologie et des méthodes de pêche des Palinuridae dans la région de Nosy-Bé (Madagascar). — Cah. ORSTOM, (Océanogr.) **11** (3): 71-101.

— —, 1967. Contribution à l'étude des peuplements de la zone intertidale sur sables fins et sables vaseux non fixés dans la région de Tuléar. — Rec. Trav. Sta. mar. Endoume, (fasc. hors série) (Suppl.) **7**: 57-100.

PILLAI, N. K., 1957. A new species of *Limnoria* from Kerala. — Bull. centr. Res. Inst., Trivandrum, (C) **5** (2): 149-157.

— —, 1959. *Choniomyzon* gen. nov. (Copepoda, Choniostomatidae) associated with *Panulirus*. — J. mar. biol. Ass. India, **4** (1): 95-99.

— —, 1963. Copepods associated with south Indian invertebrates. — Proc. Indian Acad. Sci., (B) **58** (4): 235-247.

PRECHT, H., 1935. Epizoen der Kieler Bucht. — Nov. Acta Leopold, **3** (15): 405-474.

PRENANT, M., 1924. *Hemioniscus balani* Buchholz, parasite accidentel de *Chthamalus stellatus* Ranz. — Bull. Soc. zool. Fr., **48**: 347-375.

QUIDOR, A., 1906. Sur *Mesoglicola delagei* (n. g., n. sp.) parasite de *Corynactis viridis*. — C. R. Acad. Sci., Paris, **143**: 613-616.

RANIZE, S., 1956. Descrizione di un interessante copepodo parasita di anellide raccolto fra le alghe nella scogliera di Oneglia. — Doriana, (Suppl. Ann. Mus. civ. Stor. nat. Genova) **67**: 1-7.

RAY, D. L. (ed.), 1959. Marine boring and fouling organisms. — Pp. i-xii, 1-1-536. (Seattle, WA.)
REDDIAH, K., 1968. Three new species of *Paramacrochiron* (Lichomolgidae) associated with medusae. — Crustaceana, (Suppl.) **1**: 193-209.
— —, 1969. *Pseudomacrochiron stocki* n. g., n. sp., a cyclopoid copepod associated with a medusa. — Crustaceana, **16** (1): 43-50.
REHM, A. & H. J. HUMM, 1973. *Sphaeroma terebrans*: a threat to the mangroves of southwestern Florida. — Science, N.Y., **182**: 173-174.
— — & — —, 1974. The effects of wood-boring isopod *Sphaeroma terebrans* on the mangrove communities in Florida. — Environ. Conserv., **3**: 42-57.
REMANE, A. & C. SCHLIEPER, 1958. Die biologie des Brackwassers. — Die Binnengewässer, **22**: 1-348.
REYNE, A., 1930. On the food habits of the coconut crab (*Birgus latro* L.) with notes on its distribution. — Arch. néerl. Zool., **3**: 283-320.
REYS, S., 1964. Note sur les Ostracodes des Phanérogames marines des côtes de Provence. — Rec. Trav. Sta. mar. Endoume, Bull., **32** (48): 183-202.
— —, 1969. Ostracodes des grottes sous-marines semi-obscures des côtes de Provence. — In: J. W. NEALE (ed.), The taxonomy, morphology and ecology of recent Ostracoda, pp. 330-333. (Edinburgh.)
RIBI, G., 1981. Does the wood boring isopod *Sphaeroma terebrans* benefit red mangroves (*Rhizophora mangle*)? — Bull. mar. Sci., **31** (4): 925-928.
RIDDER, CH. DE & L. B. HOLTHUIS, 1979. *Pontonides sympathes*, a new species of commensal shrimp (Crustacea: Decapoda: Pontoniinae) from Antipatharia in the Galapagos Islands. — Zool. Meded., Leiden, **54** (7): 101-110.
RIEK, E. F., 1953. A new corallanid isopod parasitic on freshwater prawns in Queensland. — Proc. Linn. Soc. N. S. Wales, **77** (5-6) [n° 363-364]: 259-261.
— —, 1967. A new corallanid isopod parasitic in Australian freshwater prawns. — Proc. Linn. Soc. N. S. Wales, **91** (3) [n° 411]: 176-178.
RIOJA, E., 1942. Estudios carcinológicos. XIII. Consideraciones y datos acerca del genero *Entocythere* (Crust. Ostracodos) y algunas de sus especies, con descripcion de una nueva. — An. Inst. Biol. Mexico, **13** (2): 685-697.
— —, 1943. Estudios carcinológicos. XIV. Nuevos datos acerca de los *Entocythere* (Crust. Ostracodos) de México. — An. Inst. Biol. México, **14** (2): 553-566.
— —, 1949. Estudios carcinológicos. XXI. Contribución al conocimiento de las especies del genera *Entocythere* de México. — An. Inst. Biol. México, **20** (1-2): 315-329.
— —, 1955. Estudios carcinológicos. XXXII. Primeros datos acerca de los especies del genera *Entocythere* (Crustaceos Ostracodos) de la isla de Cuba. — An. Inst. Biol. Mexico, **26** (1): 193-197.
ROBERTS, A., 1948. On a collection of birds and eggs from the Tristan da Cunha Islands made by John Kirby. — Ann. Transvaal Mus., **21** (1): 55-62.
RODRIGUEZ, G., 1982. Les crabes d'eau douce d'Amérique. Famille des Pseudotelphusidae. — Faune trop., **22**: 1-223.
— —, 1992. The freshwater crabs of America. Family Trichodactylidae and supplement to the family Pseudothelphusidae. — Faune trop., **31**: 1-189.
ROELOFS, J., 1966. Redescription de l'Ostracode marin *Sphaeromicola dudichi* Klie, 1938, et sa présence dans l'Atlantique. — Beaufortia, **13** (161): 207-212.
ROLLET, B., 1981. Bibliography on mangrove research 1600-1975. — Pp. i-xxviii, 1-479. (UNESCO.)
ROMAN, M.-L., 1970. Écologie et répartition de certains groupes d'Isopodes dans les divers biotopes de la région de Tuléar (sud-ouest de Madagascar). — Rec. Trav. Sta. mar. Endoume, (fasc. hors série) (Suppl.) **10**: 163-208.

— —, 1979. Tanaïdacés et Isopodes benthiques récifaux et littoraux du Sud-Ouest de Madagascar: autoécologie, synécologie et chorologie. — Pp. 1-427. (Thesis, Univ. Marseille III, Marseille.)

ROMESTAND, B., 1978. Étude écophysiologique des parasitoses à Cymothoida. — Pp. 1-284, i-xxxi. (Thesis, Univ. Montpellier, Montpellier.)

— —, 1979. Étude écophysiologique des parasitoses à Cymothoadiens. — Ann. Parasitol., **54** (4): 423-448.

ROSE, M., 1933. *Nogagella* n. g. *siphonophoriae* n. sp., Copépode Caligide parasite des Siphonophores. — Ann. Inst. océanogr., **13** (4): 119-133.

ROSS, A. & W. A. NEWMAN, 1973. Revision of the coral-inhabiting barnacles (Cirripedia: Balanidae). — Trans. San Diego Soc. nat. Hist., **17** (12): 137-173.

ROSS, D. M. & L. SUTTON, 1961. The association between the hermit crab *Dardanus arrosor* (Herbst) and the sea anemone *Calliactis parasitica* (Couch). — Ray Soc., Lond., (B, Biol. Sci.) **155**: 282-291.

ROUSSEL, D. J., 1971. Ecology and taxonomy of an epizoic diatom. — Pacif. Sci., **25** (3): 357-367.

RUFFO, S., 1958. Amphipodes terrestres et des eaux continentales de Madagascar, des Comores et de La Réunion. — Mém. Inst. scient. Madagascar, (A) **12**: 35-66.

— —, 1964. Studi sui Crostacei Anfipodi: LVIII. Un nuovo Ingolfiellide delle acque sotterranee dell'Africa di Sud Ouest. — Boll. Zool., **31** (2): 1019-1034.

— —, 1975. *Hyachelia tortugae* J. L. Barnard (Amphipoda, Hyalidae) nell'Oceano Atlantico. — Boll. Mus. civ. Stor. nat. Verona, **2**: 482.

RUFFO, S. & C. DELAMARE DEBOUTTEVILLE, 1952. Deux nouveaux amphipodes souterrains de France. *Salentinella angelieri* n. sp. et *Bogidiella chappuisi* n. sp. — C. R. Acad. Sci. Paris, **234**: 1636-1638. [16 avril 1952]

RUFFO, S. & A. VIGNA-TAGLIANTI, 1977. Secundo contributo alla conoscenza del genera *Bogidiella* in Messico e Guatemala (Crustacea, Amphipoda, Gammaridae). — Accad. naz. Lincei, **374** (171) (III): 125-172.

SAINT LAURENT, M. DE & E. MACPHERSON, 1990. Crustacea Decapoda: le genre *Eumunida* Smith, 1883 (Chirostylidae) dans les eaux néo-calédoniennes. — In: A. CROSNIER (ed.), Résultats des Campagnes MUSORSTOM, **6**. Mém. Mus. natn. Hist. nat., Paris, (A, Zool.) **145**: 227-288.

SALVAT, B., 1966. *Eurydice pulchra* Leach, 1815, *Eurydice affinis* H. J. Hansen, 1905. Taxonomie, éthologie, écologie, répartition verticale et cycle reproducteur. — Actes Soc. Linn. Bordeaux, (3) **103** (1): 1-77.

— —, 1967. La macrofaune carcinologique endogée des sédiments meubles intertidaux (Tanaïdacés, Isopodes et Amphipodes): éthologie, bionomie et cycles biologiques. — Mém. Mus. natn. Hist. nat., Paris, (n. s.) (A, Zool.) **45**: 1-275.

SAMEOTO, D. D., 1969. Comparative ecology, life histories and behaviour of intertidal sandburrowing amphipods (Crustacea: Haustoriidae) at Cape Cod. — J. Fish. Res. Bd Can., **26** (2): 361-388.

— —, 1969a. Some aspects of the ecology and life-history of three species of subtidal sandburrowing amphipods (Crustacea: Haustoriidae). — J. Fish. Res. Bd Can., **26** (5): 1321-1345.

SANDISON, E. E. & M. B. HILL, 1966. The distribution of *Balanus pallidus stutsburi* Darwin, *Gryphaea gasar* (Adanson) Dautzenberg, *Mercierella enigmatica* Fauvel and *Hydroides uncinata* (Philippi) in relation to salinity in Lagos Harbour and adjacent reefs. — J. anim. Ecol., **35**: 235-250.

SCHELLENBERG, A., 1928. Beobachtungen an dem Amphipoden *Talitrus saltator* (Mont.). — Zool. Anz., **79** (1/4): 78-82.

— —, 1929. Körperbau und Grabweise einiger Amphipoden. — Zool. Anz., **85** (5/8): 186-190.

SCHMALFUSS, H., 1978. *Ligia simoni*: a model for the evolution of terrestrial isopods. — Stuttg. Beitr. Naturk., (A) **317**: 1-5.

SCHMINKE, H. K., 1972. *Hexabathynella halophila*, gen. n., sp. n., und die Frage nach der marinen Abkunft der Bathynellacea (Crustacea, Malacostraca). — Mar. Biol., Berl., **15**: 282-287.

— —, 1974. Mesozoic intercontinental relationship as evidenced by bathynellid Crustacea (Syncarida, Malacostraca). — Syst. Zool., **23** (2): 157-164.

— —, 1978. Ökologische Aspekte der Gattungsgliederung der Familie Parabathynellidae (Bathynellacea, Malacostraca). — Z. zool. Syst. Entwickl., **11** (2): 154-160.

SCHMITT, W. L., 1931. Crustaceans. — In: Shelled invertebrates of the past and present. — Smithson. Sci. Ser., Wash., **10**: 87-248.

SCHRAM, F. R., 1986. Crustacea. — Pp. i-xiv, 1-606. (Oxford Univ. Press, New York.)

— —, 2008. Does biogeography have a future in a globalized world with globalized faunas? — Contr. Zool., **77** (2): 127-133.

SEGONZAC, M., M. DE SAINT-LAURENT & B. CASANOVA, 1993. L'énigme du comportement trophique des crevettes Alvinocarididae des sites hydrothermaux de la dorsale médio-atlantique. — Cah. Biol. mar., **34**: 535-571.

SEILACHER, A., 1969. Palaecology of boring barnacles. — In: Penetration of calcium carbonate substrates by lower plants and invertebrates (2nd ed.). Amer. Zool., **9** (3): 705-719.

SERÈNE, R., 1954. Observations biologiques sur les Stomatopodes. — Ann. Inst. océanogr., **29** (1): 1-94.

SEURAT, L., 1904. Observations biologiques sur les Cénobites (*Coenobita perlata* Edwards). — Bull. Mus. natn. Hist. nat., Paris, **10** (5): 238-242.

— —, 1905. Sur le crabe des cocotiers, *Birgus latro* L. — Bull. Mus. natn. Hist. nat., Paris, **11**: 146-147.

SHEN, C. J., 1936. Notes on the family Hapalocarcinidae (coral-infesting crabs) with description of two new species. — Hong Kong Nat., (Suppl.) **5**: 21-26.

SHIINO, S. M., 1950. The marine wood-boring crustaceans of Japan. I. Limnoriidae. — Wasman J. Biol., **8** (3): 333-358.

— —, 1957. The marine wood-boring crustaceans of Japan. II. Sphaeromidae and Cheluridae. — Wasman J. Biol., **15** (2): 161-197.

— —, 1968. A tanaid crustacean *Nesotanais lacustris* gen. and sp. nov., from Lake Tasano, Rennell Island. — Nat. Hist. Rennell Isl., Brit. Solomon Isl., **5** (58): 153-168.

— —, 1978. Tanaidacea collected by French scientists on board the survey-ship "Marion-Dufresne" in the regions around the Kerguelen Islands in 1972-1974-1975-1976. — Sci. Rept. Shima Marineland, **5**: 1-122.

SHOEMAKER, C. R., 1919. A new amphipod parasitic on a crinoid. — Proc. biol. Soc. Wash., **32**: 245.

— —, 1952. A new species of commensal amphipod from a spiny lobster. — Proc. U.S. natl. Mus., **102** (3299): 231-233.

SHORNIKOV, E. I., 1980. Ostracods in terrestrial biotopes. — Zool. J. [Zool. Zh.], **59** (9): 1306-1319.

SIMBERLOFF, D., B. J. BROWN & S. LOWRIE, 1978. Isopod and insect root borers — any benefit Floridan mangroves. — Science, N.Y., **201**: 630-632.

SORGELOOS, P., D. A. BENGTSON, W. DECLEIR & E. JASPERS (eds.), 1989. *Artemia* research and its applications, **3**. — Pp. i-xiv, 1-535. (Universa Press, Wetteren.)

SOUTY-GROSSET, C., D. M. HOLDICH, P. NOËL, J. D. REYNOLDS & P. HAFFNER (eds.), 2006. Atlas of crayfish in Europe. — Patrimoines nat., **64**: 1-187.

SPRAGUE, V. & J. COUCHI, 1971. An annotated list of protozoan parasites, hyperparasites, and commensals of decapod Crustacea. — J. Protozool., **18**: 526-537.

STASEK, C. R., 1958. A new species of *Allogaussia* (Amphipoda, Lysianassidae) found living within the gastrovascular cavity of the sea-anemone *Anthopleura elegantissima*. — J. Wash. Acad. Sci., **48**: 119-126.

STEPHENSEN, K., 1924. *Talitrus alluaudi* Chevreux. An Indo-Pacific terrestrial amphipod found in hot-houses in Copenhagen. — Vidensk. Meddr Dansk naturh. Foren., **78**: 197-199.

— —, 1933. Some new copepods, parasites of ophiurids and echinids. — Vidensk. Meddr Dansk naturh. Foren., **93**: 197-213.

— —, 1935. Some endoparasitic copepods found in echinids. — Vidensk. Meddr Dansk naturh. Foren., **98**: 223-228.

— —, 1936. Copepods found on *Limnoria lignorum*. — Kgl. Norske Vidensk. Selsk. Skr., **39**: 1-10. [1935]

STEPHENSEN, K. & G. THORSON, 1936. On the amphipod *Metopa groenlandica* H. J. Hansen found in the mantle cavity of the lamellibranchiate *Pandora glacialis* Leach in East Greenland. — Medd. Grønland, **118** (4): 1-7.

STOCK, J. H., 1950. Parasite or commensal? *Notodelphys weberi*, a new South African ascidicolid copepod. — Amsterdam Nat., **1** (2): 37-42.

— —, 1954. Redescription de *Tocochres cylindraceus* Pelseneer, 1929, copépode commensal de *Loripes lacteus*. — Beaufortia, **4** (38): 73-80.

— —, 1956. *Lichomolgus longicauda* (Claus, 1860), copepod parasite of *Sepia* in the North Sea. — Beaufortia, **5** (53): 117-120.

— —, 1959. Copepoda associated with Neapolitan invertebrates. — Pubbl. Staz. zool. Napoli, **31** (1): 43-58.

— —, 1959a. Copepoda associated with Neapolitan Mollusca. — Pubbl. Staz. zool. Napoli, **31** (1): 59-75.

— —, 1960. Sur quelques Copépodes associés aux Invertébrés des côtes du Roussillon. — Crustaceana, **1** (3): 218-257.

— —, 1962. Heremietkreeften en hun commensalen. — Natura, **49** (97): 69-71.

— —, 1963. Copépodes parasites d'Invertébrés des côtes de France. XVI. Description de *Pseudoclausia longiseta* n. sp. (Copépode Cyclopoïde, famille des Clausiidae). — Proc. Kon. Ned. Akad. Wet., (C) **66** (2): 139-152.

— —, 1964. Sur deux espèces d'*Anthessius* (Copepoda) des Indes orientales. — Zool. Meded., Leiden, **39**: 111-124.

— —, 1966. Cyclopoida Siphonostoma from Mauritius (Crustacea, Copepoda). — Beaufortia, **13** (159): 145-194.

— —, 1966a. On *Callocheres* Canu, 1893, and *Leptomyzon* Sars, 1915, two synonymous genera of Copepods. — Beaufortia, **13** (163): 221-239.

— —, 1966b. Copepods associated with invertebrates from the Gulf of Aqaba. 3. *Enterognathus lateripes* n. sp., a new endoparasite of Crinoida (Cyclopoida, Ascidicolidae). — Proc. Kon. Ned. Akad. Wet., (C) **69** (2): 211-216.

— —, 1967. Sur trois espèces de Copépodes synonymes ou confondues: *Asterocheres echinicola* (Norman), *A. parvus* Giesbrecht et *A. kervillei* Canu (Cyclopoida Siphonostoma). — Bull. zoöl. Mus., Univ. Amsterdam, **1** (4): 31-39. [1966]

— —, 1967a. Copepods associated with invertebrates from the Gulf of Aqaba. 3. The genus *Pseudanthessius* Claus, 1889 (Cyclopoida, Lichomolgidae). — Proc. Kon. Ned. Akad. Wet., (C) **70** (2): 232-248.

— —, 1967b. Copepods associated with invertebrates from the Gulf of Aqaba. 4. Two new Lichomolgidae associated with Crinoida. — Proc. Kon. Ned. Akad. Wet., (C) **70** (5): 569-578.

— —, 1967c. Report on the Notodelphyidae (Copepoda, Cyclopoida) of the Israel South Red Sea Expedition. — Bull. Sea Fish. Res. Sta., Haifa, **46**: 1-58.

— —, 1967d. *Mychophilus fallax* n. sp., a new vermiform copepod parasite of a Red Sea tunicate. — Bull. Sea Fish. Res. Sta., Haifa, **43**: 9-12.

— —, 1970. *Apodomyzon* n. gen., a highly transformed siphonostome cyclopoid copepod, parasitic in the sponge *Haliclona* from Roscoff. — Beaufortia, **18** (235): 141-160.

— —, 1971. *Micrallecto uncinata* n. gen., n. sp., a parasitic copepod from a remarkable host, the pteropod *Pneumoderma*. — Bull. zoöl. Mus., Univ. Amsterdam, **2** (9): 77-79.

— —, 1973. *Nannalecto fusii*, n. gen., n. sp., a copepod parasitic on the pteropod *Pneumodermopsis*. — Bull. zoöl. Mus., Univ. Amsterdam, **3** (4): 21-24.

— —, 1975. On twelve species of the genus *Acanthomolgus* (Copepoda Cyclopoida: Lichomogidae) associated with West Indian octocorals. — Bijdr. Dierk., **45** (2): 237-269.

— —, 1975a. *Peltomyzon rostratum*, n. gen., n. sp., a siphonostome cyclopoid copepod associated with the West Indian coral *Montastraea*. — Bull. zoöl. Mus., Univ. Amsterdam, **4** (14): 111-117.

— —, 1975b. Copepoda associated with West Indian Actiniaria and Corallimorpharia. — Stud. Fauna Curaçao Caribb. Isl., **48** (161): 88-118.

— —, 1976. A new member of the crustacean suborder Ingolfiellidea from Bonaire with a review of the entire suborder. — Stud. Fauna Curaçao Caribb. Isl., **50** (164): 56-75.

— —, 1976a. A new genus and two new species of the crustacean order Thermosbaenacea from the West Indies. — Bijdr. Dierk., **46** (1): 47-70.

— —, 1977. The zoogeography of the Crustacean suborder Ingolfiellidea with descripion of new West Indian taxa. — Stud. Fauna Curaçao Caribb. Isl., **55** (178): 131-146.

— —, 1977a. The taxonomy and zoogeography of the hadziid Amphipoda with emphasis on the West Indies taxa. — Stud. Fauna Curaçao Caribb. Isl., **55** (177): 1-130.

— —, 1978. *Magnippe caputmedusae* n. gen., n. sp., (Copepoda: Lamippidae), a highly transformed endoparasite in octocorals of the genus *Thesea* from the Gulf of Mexico. — Mem. Hourglass Cruises, **3** (5): 1-11.

— —, 1979. A new species of *Linaresia* (Copepoda: Lamippidae) endoparasitic in the octocoral *Placogorgia* from the Gulf of Mexico. — Mem. Hourglass Cruises, **5** (1): 1-7.

— —, 1979a. Serpulidicolidae, a new family of copepods associated with tubicolous polychaetes, with description of a new genus and species from the Gulf of Mexico. — Mem. Hourglass Cruises, **5** (2): 1-11.

— —, 1981. Association of Hydrocorallia Stylasterina with gall-inhabiting Copepoda Siphonostomatoidea from the south-west Pacific. — Bijdr. Dierk., **51** (2): 287-312.

— —, 1982. Description de *Cotylomyzon vervoorti* gen. et sp. nov., un Copépode Cyclopoïde très original, parasite d'un Polychète d'Amboine. — Neth. J. Zool., **32** (3): 364-373.

— —, 1988. Lamippidae (Copepoda: Siphonostomatoida) parasitic in *Alcyonium*. — J. mar. biol. Ass. U.K., **68**: 351-359.

— —, 1988a. A bizarre parasitic copepod (nereicoliform Poecilostomatoida) from the Great Barrier Reef. — Trop. Zool., **1**: 217-222.

STOCK, J. H. & A. G. HUMES, 1969. *Cholomyzon palpiferum*, n. gen., n. sp., a siphonostome cyclopoid copepod parasitic in the coral *Dendrophyllia* from Madagascar. — Crustaceana, **16** (1): 57-64.

STOCK, J. H., A. G. HUMES & R. V. GOODING, 1962. Copepoda associated with West Indian invertebrates. I. The genus *Nanaspis* (Siphonostoma, Nanaspidae). — Stud. Fauna Curaçao Caribb. Isl., **13**: 1-20, 47.

— —, — — & — —, 1963. Copepoda associated with West Indian invertebrates. II. Cancerillidae, Micropontiidae (Siphonostomata). — Stud. Fauna Curaçao Caribb. Isl., **15**: 1-23.

— —, — — & — —, 1963a. Copepoda associated with West Indian invertebrates. III. The genus *Anthessius* (Cyclopoidea, Myicolidae). — Stud. Fauna Curaçao Caribb. Isl., **17**: 1-37.

STOCK, J. H. & G. KLEETON, 1963. Copépodes associés aux Invertébrés des côtes du Roussillon. 4. Description de *Spongiocnizon petiti* gen. nov. sp. nov., Copépode spongicole remarquable. — Vie Milieu, (Suppl.) **17**: 325-336.

STOCK, J. H. & S. VAN DER SPOEL, 1976. *Pteroxena papillifera* n. gen., n. sp., an endoparasitic organism (Copepoda?) from the gymnosomatous pteropod *Notobranchaea*. — Bull. zoöl. Mus., Univ. Amsterdam, **5** (21): 177-180.

STODDART, D. R. & R. E. JOHANNES, 1978. Coral reefs: research and methods. — Pp. 1-581. (UNESCO.)
STORK, N. E., 1993. How many species are there? — Biodiversity and Conservation, **2**: 215-232.
STRENTH, N. E., 1976. A review of the systematics and zoogeography of the freshwater species of *Palaemonetes* Heller of North America (Crustacea Decapoda). — Smithson. Contr. Zool., **228**: 1-27.
STUDER, T., 1889. Die Forschungsreise S. M. S. "Gazelle" in den Jahren 1874 bis 1876 unter Kommando des Kapitän zur See Freiherrn von Schleinitz herausgegeben von dem Hydrographischen Amt des Reichs-Marine-Amts. — Vol. III, Zoologie und Geologie, pp. i-iv, 1-322. (E.S. Mittler, Berlin.)
STUNKARD, H. W. & R. M. CABLE, 1931. Note on a species of *Lernaea* parasitic in the larvae of *Rana clamitans*. — J. Parasit., **18**: 92-97.
SUZUKI, K. & K. I. HAYASHI, 1977. Five caridean shrimps associated with sea anemones in central Japan. — Publs Seto mar. biol. Lab., **24** (1-3): 193-208.
SUZUKI, K. & M. TAKEDA, 1974. On a parthenopid crab, *Zebrida adamsii* on the sea urchins from Suruga Bay, with a special reference to their parasitic relations. — Bull. nat. Sci. Mus., Tokyo, **17**: 287-296.
TAKAHASI, S., 1922. Habits of *Ilyoplax formosensis* Rathbun and one species of Ocypodidae. — Zool. Mag., **44** (529): 407-421.
— —, 1935. Ecological notes on the ocypodian crabs (Ocypodidae) in Formosa, Japan. — Annot. zool. Japon., **15** (1): 78-85.
TATTERSALL, W. M., 1929. A terrestrial amphipod from 6 000 feet on the New Hebrids. — Ann. Mag. nat. Hist., **10** (3): 96-97.
TÉTART, J., 1962. Étude morphologique de *Campecopea hirsuta* (Montagu). — Bull. Soc. Linn. Normandie, **10** (3): 158-164.
THIENEMANN, A., 1935. Die Tierwelt der tropischen Pflanzengewässer. — Arch. Hydrobiol., (Suppl.) **13** (Tropische Binnengewässer, **5**) (90): 1-91.
THOMASSIN, B., 1969. Identification, variabilité et écologie des Hippidae (Crustacea, Anomura) de la région de Tuléar, S.W. de Madagascar. — Rec. Trav. Sta. mar. Endoume, (fasc. hors série) (Suppl.) **9**: 135-177.
TOMLINSON, J. T., 1955. The morphology of an acrothoracean barnacle *Trypetesa lateralis*. — J. Morphol., **96** (1): 97-122.
— —, 1969. Shell-burrowing barnacles in penetrations of calcium carbonate substrates by lower plants and invertebrates. — Am. Zool., **9** (3) [2nd ed.]: 837-840.
— —, 1969a. The burrowing barnacles (Cirripedia: order Acrothoracica). — Bull. U.S. natl. Mus., **296**: 1-162.
TRILLES, J.-P., 1968. Recherches sur les Isopodes Cymothoidae des côtes françaises. — Pp. 1-793. (Thesis, Univ. Montpellier, Montpellier.)
— —, 1968a. Recherches sur les Isopodes Cymothoidae des côtes françaises. Systématique et faunistique. — Pp. 1-181. (2nd Thesis, Univ. Montpellier, Montpellier.)
— —, 1973. Notes documentaires sur les Isopodes Cymothoadiens parasites de poissons d'eau douce de l'Amérique du Sud. — Bull. Mus. natn. Hist. nat., Paris, **3** (114) [Zool., **88**]: 239-272.
TRILLES, J.-P. & R. HIPEAU-JACQUOTTE, 2012. Symbiosis and parasitism in the Crustacea. — In: J. FOREST, J. C. VON VAUPEL KLEIN, M. CHARMANTIER-DAURES & F. R. SCHRAM (eds.), Treatise on Zoology — anatomy, taxonomy, biology. The Crustacea, revised and updated from the Traité de Zoologie, **3**: 239-317. (Brill, Leiden.)
TUNNICLIFFE, V., 1988. Biogeography and evolution of hydrothermal vent fauna in the eastern Pacific Ocean. — Proc. Roy. Soc. Lond., (B) **233**: 347-366.
TWEEDIE, M. W. F., 1952. Two crabs of the sandy shore. — Malay. Nat. J., **7**: 3-10.

UDEKEM D'ACOZ, C. D', 2001. Description of *Pseudocoutierea wirtzi* sp. nov., a new cnidarian-associated pontoniine shrimp from Cape Verde Islands, with decalcified meral swellings in walking legs (Crustacea, Decapoda, Caridea). — Bull. Inst. Roy. Sci. nat. Belg., (Biologie) **70**: 69-90.

UTINOMI, H., 1938. Studies on the animals inhabiting reef corals. III. A revision of the family Hapalocarcinidae (Brachyura) with some remarks on their morphological peculiarities. — Palao trop. biol. Sta. Stud., **2** (4): 637-731.

— —, 1957. Studies on Cirripedia Acrothoracica. I. Biology and external morphology of the female of *Berndtia purpurea* Utinomi. — Publs Seto mar. biol. Lab., **6** (1): 1-26.

VADER, W., 1970. *Antheacheres duebeni* M. Sars, a copepod parasitic in the sea anemone, *Bolocera tuediae* (Johnston). — Sarsia, **43**: 99-106.

— —, 1970a. On the occurrence of a gallforming copepod in *Actinostola* spp. (Anthozoa). — Sarsia, **43**: 107-110.

— —, 1970b. The amphipod *Aristias neglectus* Hansen, found in association with brachiopods. — Sarsia, **43**: 13-14.

— —, 1971. De vlokreeft *Podoceropsis nitida*, een kostganger van heremietkreeften. — Levende Natuur., **74**: 134-137.

— —, 1972. Terrestrial Amphipoda collected in greenhouses in the Netherlands. — Zool. Bijdr., Leiden, **13**: 32-36.

— —, 1972a. Associations between amphipods and molluscs. A review of published records. — Sarsia, **48**: 13-18.

— —, 1972b. Associations between gammarid and caprellid amphipods and medusae. — Sarsia, **50**: 51-56.

— —, 1973. A bibliography of the Ellobiopsidae, 1959-1971, with a list of *Thalassomyces* species and their hosts. — Sarsia, **52**: 175-179.

— —, 1975. The sea anemone *Bolocera tuediae* and its copepod parasite, *Antheacheres duebeni*, in northern Norway. — Astarte, **8**: 37-39.

— —, 1978. Associations between amphipods and echinoderms. — Astarte, **11**: 123-134.

— —, 1983. Associations between amphipods (Crustacea: Amphipoda) and sea anemones (Anthozoa: Actiniaria). — Mem. Austr. Mus., **18**: 141-153.

VADER, W., D. H. JOHANNESSEN & B. O. CHRISTIANSEN, 1981. A pelagic isopod, *Syscenus infelix* overgrown with hydroids. — Fauna Norv., (A) **2**: 47-48.

VADER, W. & J. E. KANE, 1968. New hosts and distribution records of *Thalassomyces marsupii* Kane, an ellobiopsid parasite of amphipods. — Sarsia, **33**: 13-20.

VADER, W. & S. LONNING, 1973. Physiological adaptations in associated amphipods, a comparative study of tolerance to sea anemones in four species of Lysianassidae. — Sarsia, **53**: 29-40.

VADER, W. & W. J. WOLFF, 1973. The Cumacea of the estuarine coast of rivers Rhine, Meuse and Scheldt (Crustacea: Malacostraca). — Neth. J. Sea Res., **6** (3): 365-375.

VANDEL, A., 1959. Nouvelles recherches sur les Isopodes volvationnels exoantennés et la genèse de leurs coaptations. — Bull. Biol. Fr. Belg., **93** (2): 121-139.

— —, 1966. Sur l'existence d'Oniscoïdes menant une vie aquatique et sur le polyphylétisme des Isopodes terrestres. — Ann. Spéléol., **20**: 489-518. [1965]

VASSEUR, P., 1964. Contribution à l'étude bionomique des peuplements sciaphiles infralittoraux de substrat dur dans les récifs de Tuléar, Madagascar. — Rec. Trav. Sta. mar. Endoume, (fasc. hors série) (Suppl.) **2**: 1-77.

VERVOORT, W. & D. TRANTER, 1961. *Balaenophilus unisetus* P.O.C. Aurivillius (Copepoda Harpacticoida) from the southern hemisphere. — Crustaceana, **3** (1): 70-84.

VERWEY, J., 1930. Einiges über die Biologie ost-indischer Mangroven-krabben. — Treubia, **12**: 168-251.

VEUILLE, M., 1979. L'évolution du genre *Jaera* Leach (Isopodes: Asellotes) et ses rapports avec l'histoire de la Méditerranée. — Bijdr. Dierk., **49** (2): 195-217.

WÄGELE, J. W., 1979. Morphologische studien an *Eisothistos* mit Beschreibung von drei neuen Arten (Crustacea, Isopoda, Anthuridea). — Mitt. zool. Mus. Kiel, **1** (2): 1-19.

— —, 1981. Zur Phylogenie der Anthuridea (Crustacea, Isopoda) mit Beiträgen zur Lebensweise, Morphologie, Anatomie und Taxonomie. — Zoologica, **45** (2) [132]: 1-127.

WAGNER, H. P., 1994. A monographic review of the Thermosbaenacea (Crustacea: Peracarida). A study of their morphology, taxonomy, phylogeny and biogeography. — Zool. Verh., Leiden, **291**: 1-338.

WARBURG, M. R., 1968. Behavioral adaptations of terrestrial Isopods. — Am. Zool., **8**: 545-559.

WELLS, J. W., 1957. Coral reefs. — In: J. W. HEDGPETH (ed.), Treatise on marine ecology and paleoecology, I, Ecology. Geol. Soc. Am. Mem., **67**: 609-631.

WIENS, H. J., 1962. Atoll environment and ecology. — Pp. i-xxii, 1-522. (Yale Univ. Press, New Haven, CT.)

WIESER, W., 1963. Adaptations of two intertidal isopods. 11. Comparison between *Campecopea hirsuta* and *Naesa bidentata* (Sphaeromatidae). — J. mar. biol. Ass. U.K., **43**: 97-112.

WILLIAMS, A. A., 1980. A new crab family from the vicinity of submarine thermal vents on the Galapagos rift (Crustacea: Decapoda: Brachyura). — Proc. biol. Soc. Wash., **92** (2): 443-472.

WILLIAMS, J. D. & J. J. MCDERMOTT, 2004. Hermit crab biocoenoses: a worldwide review of the diversity and natural history of hermit crab associates. — J. exp. mar. Biol. Ecol., **305**: 1-128.

WILLIAMSON, D. I., 1951. On the mating and breeding of some semi-terrestrial amphipods. — Rep. Dove mar. Lab., (3) **12**: 49-62.

WILSON, M. S., 1957. Redescription of *Teredicola typica* C.B. Wilson (Crustacea: Copepoda). — Pacif. Sci., **11** (3): 265-274.

WIRTZ, P., 1995. Krebs-Seeanemonen-Symbiosen bei Madeira. — Natur u. Mus., **125** (5): 137-142.

— —, 1997. Crustacean symbionts of the sea anemone *Telmatactis cricoides* at Madeira and the Canary Islands. — J. Zool., Lond., **242**: 779-811.

— —, 1998. Caprellid (Crustacea) — holothurian (Echinodermata) associations in the Azores. — Arquipélago Life mar. Sci., **16** (A): 53-55.

WIRTZ, P. & C. D'UDEKEM D'ACOZ, 2001. Decapoda from Antipatharia, Gorgonaria and Bivalvia at the Cape Verde Islands. — Helgol. mar. Res., **55**: 112-115.

WIRTZ, P. & W. VADER, 1997. A new caprellid-starfish association: *Caprella acanthifera* s.l. (Crustacea: Amphipoda) on *Ophidiaster ophidianus* and *Hacelia attenuata* from the Azores. — Arquipélago Life mar. Sci., **14** (A): 17-22. [1996]

WOLFF, T., 1958. On the rare whale-louse *Platycyamus thompsoni* (Gosse) (Amphipoda, Cyamidae). — Vidensk. Meddr Dansk naturh. Foren., **120**: 1-14.

— —, 1960. Rankeføderne Conchoderma og *Coronula* pa hvaler. — Flora og Fauna, **66**: 1-8.

— —, 1962. The systematics and biology of bathyal and abyssal Isopoda Asellota. — Galathea Rep., **6**: 7-320.

— —, 1978. Maximum size of lobsters (*Homarus*) (Decapoda, Nephropsidae). — Crustaceana, **34** (1): 1-14.

— —, 1980. Animals associated with seagrass in the deep sea. — In: R. C. PHILLIPS & C. P. MCROY (eds.), Handbook of seagrass biology, pp. 199-224. (New York and London.)

WORMS, 2013. The World Register of Marine Species. — http://www.marinespecies.org

YAGER, J., 1981. Remipedia, a new class of Crustacea from a marine cave in the Bahamas. — J. Crust. Biol., **1**: 328-333.

YALDWYN, J. C., 1964. Pair association in the banded coral shrimp. — Aust. nat. Hist., **14** (9): 286.

— —, 1965. Crustacea of the arid inland. — Aust. nat. Hist., **15**: 132-136.

— —, 1966. Notes on the behaviour in captivity of a pair of banded shrimps, *Stenopus hispidus* (Olivier). — Aust. J. Zool., **13** (4): 377-389.

— —, 1968. Records of, and observations on, the coral shrimp genus *Stenopus* in Australia, New Zealand and the south-west Pacific. — Aust. J. Zool., **14** (4): 277-289.

YALDWYN, J. C. & K. WOZICKI, 1979. Systematics and ecology of the land crabs (Decapoda: Cenobitidae, Grapsidae and Gecarcinidae) of the Tokelau Islands, central Pacific. — Atoll Res. Bull., **235**: 1-53.

YONGE, C. M., 1963. The biology of coral reefs. — Adv. mar. Biol., **1**: 209-260.

ZHANG, Z. Q., 2011. Phylum Arthropoda Von Siebold, 1848. — In: Z.-Q. ZHANG (ed.), Animal biodiversity: an outline of higher-level classification and survey of taxonomic richness. Zootaxa, **4138**: 99-103.

ZIBROWIUS, H., 1981. Associations of Hydrocorallia Stylasterina with gall-inhabiting Copepoda Siphonostomatoidea from the south-west Pacific. Part I. On the stylasterine hosts, including two new species, *Stylaster papuensis* and *Crypthelia cryptotrema*. — Bijdr. Dierk., **51** (2): 268-286.

CHAPTER 27A

CRUSTACEANS AND MANKIND[1])

BY

HENRY-MICHEL CAUCHIE, THÉODORE MONOD (†) AND
LUCIEN LAUBIER (†)

Contents. – **Crustaceans in trophic networks. Edible crustaceans. Crustaceans as useful auxilliaries** – Crustaceans in traditional pharmacopeia – Crustaceans as sources of polymers or energy. **Detrimental impact of crustaceans on human activities** – Dike digging and damage to rice fields – Parasites or commensals of useful species – Invasive species – Biofouling. **Human pathologies linked to crustaceans** – Toxic decapods – Crustaceans as intermediate hosts or vectors of a variety of human parasites. **Crustaceans in mythology, art, history, and popular culture** – Zodiacal asterism – Legends, mythology, and representations. **Appendix. Bibliography.**

CRUSTACEANS IN TROPHIC NETWORKS

Crustaceans play a considerable **ecological role** in both marine and freshwater **trophic networks**. As well as constituting a significant link between **phytoplankton** and higher-level animals (Sommer & Stibor, 2002), they also make up a large part of the **detritivores**. In marine **ecosystems**, copepods are a key element of the marine trophic networks as they are the major link between the autotrophic and heterotrophic microorganisms and the higher trophic levels. Pelagic copepods are remarkably well adapted for the collection of phytoplanktonic cells. They are able to select their food items on the basis of their "taste" and reject inadequate particles (Moore et al., 1999; De Troch et al., 2012). Copepods are the dominant **secondary producers** and are very abundant in the upwelling and euphotic zones of the oceans (Sobrinho-Gonçalves et al., 2013). They have a special importance for humans, because they are a foodstuff for many species of fish, including those of **economic importance** (Beaugrand et al., 2003). By vertically migrating between surface waters and

---

[1]) The original chapter by Th. Monod & L. Laubier was updated July 2013 by H.-M. Cauchie; final additions March 2014.

deeper waters, copepods also play an important role in the **carbon transfer** in oceans (Frangoulis et al., 2005).

In the deeper zones, the size of the crustaceans increases globally and the feeding habits of the encountered species shift towards **detritivory** or **carnivory**. More than 125 species of crustaceans have been found around **hydrothermal vents** (Martin & Haney, 2005). Around the vents on the Mid-Atlantic Ridge, the shrimp *Rimicaris exoculata* (Decapoda, Caridea) is often the dominant secondary producer (Hügler et al., 2011). In such extreme environments, these shrimps live in association with the autotrophic bacteria upon which the productivity of the **epibiotic community** depends.

To a lesser extent, Euphausiacea play an analogue role regarding phytoplankton in pelagic zones. The name "krill" is of Norwegian origin and means "young fry of fish". Krill, however, mainly applies to species of Euphausiacea such as *Euphausia superba* (Antarctic krill), *Euphausia pacifica* (Pacific krill) and *Meganyctiphanes norvegica* (northern krill) (Nicol & Endo, 1999). In the North Atlantic, krill makes up virtually 100% of the diet of blue whales (*Balaenoptera musculus*), fin whales (*Balaenoptera physalus*) and sei whales (*Balaenoptera borealis*). A single blue whale can eat up to 735 kg of krill per day (Sigurjónsson & Víkingsson, 1997). They feed mainly on dense swarms of euphausiids (Croll et al., 2005). Other animals, such as squids, fish, seals, penguins, and petrels also actively participate in the exploitation of the krill biomass (Ish et al., 2004; Blanchet et al., 2013).

According to the most recent estimations using net-based assessments and electroacoustic techniques (Atkinson et al., 2009), krill stocks reached between 117 and 379 millions of metric tons in the Southern Ocean. Between 1995 and 2010, the commercial fishing catch of Antarctic krill reached levels of between 100 000 and 210 000 metric tons, i.e., less than 0.1% of the standing stocks.

In fresh water, crustaceans play a key role in pelagic zones, with cladocerans and copepods being major drivers of phytoplankton dynamics (Lampert & Sommer, 2007). In meso- and eutrophic lakes, the development of cladocerans in late spring leads to a temporary extinction of the algae due to the efficient grazing activity of these planktonic crustaceans (Lampert, 1978). This corresponds to a distinct "clear water phase" before the algae start to grow again in summer. Planktonic crustaceans in ponds and lakes are an important trophic link to the upper trophic levels composed of insect larvae such as *Chaoborus* (Diptera) and planktivorous fish including the stickleback (*Gasterosteus aculeatus*), European minnow (*Phoxinus phoxinus*), and many larval fish.

Cauchie (2002) has collected over 800 published estimates of annual production by crustaceans. Production appears to be highly variable, mainly depending on the intensity of **primary production**. On the whole, 60% of the crustacean production estimates fell between 0.1 and 5 $gC\,m^{-2}\,yr^{-1}$ in pelagic zones. Considering the primary production of seas and lakes to range on the whole from 1 to 200 $gC\,m^{-2}\,yr^{-1}$ (Nõges et al., 2011; Tekanova, 2012; Uitz et al., 2012; Matrai et al., 2013; among others), the **ecological efficiency** (the ratio of predator production to prey production) ranges from 2.5 to 10%. With crustaceans being the dominant primary consumers in these ecosystems, these figures are in line with the observation that 90-95% of the energy is lost with each transfer in the food chain.

EDIBLE CRUSTACEANS

Although the vast majority of crustaceans are theoretically edible by humans, only rather large and common species are generally collected. As a matter of fact, decapods (crabs, shrimps, lobsters, ...) constitute the vast majority of crustaceans eaten by humans.

There are, however, some exceptions. Stomatopods (*Squilla mantis* and *Erugosquilla massavensis*) are captured and consumed, notably in the Mediterranean area. According to FAO (2011), around 304 340 tons of stomatopods are caught annually, 37% of which are gathered on Chinese shores.

Barnacles (Maxillopoda, Cirripedia), particularly goose barnacles (*Pollicipes pollicipes*), are popular fare on the coasts of Spain, Portugal, and some parts of northern Africa. Where the local stocks are overexploited, we see an upcoming **fishery** in France to provide for export to those areas. *Pollicipes polymerus* and *Pollicipes elegans* are collected on the west coasts of North America and South America (from Mexico to Peru), respectively. In 2011, global production reached 292 tonnes per year (FAO, 2011). However, their harvesting led to severe decreases in local stocks, such as in Canada or Spain. In some areas, strategies for a **sustainable management** of this **living resource** have been developed (Molares & Freire, 2003). Barnacle culture is also envisaged for the future (López et al., 2010).

In the sand dunes of Edeyen Ubari (Fezzan, Libya), many small salted lakes harbour large quantities of *Artemia salina* (Anostraca). Local populations belonging to the Dauada[2]) tribes collect these *Artemia* and eat them, generally together with dipteran larvae from the family Ephydridae (H. Dumont, pers. comm.). These animals are collected using a fabric net and are sun-dried until they form a foul-smelling black paste with high nutritive value (Monod, 1969). Dees (1961) reported similar consumption of *Artemia* by Indians in the Great Salt Lake area of North America. In Laos, in rivers where species of *Phyllodiaptomus* (Copepoda) are abundant, local populations harvest them with nets and eat them (H. Dumont, pers. comm.). In Thailand, populations in the north-eastern part of the country cook fairy shrimp (Anostraca) for human consumption (Sanoamuang & Dumont, 2000).

Similarly to nuoc-mam, which is obtained from the fermentation of fish with sea salt, nuoc-tom is a sauce obtained by fermenting shrimps. Mam-tom designates a shrimp paste that is also obtained by fermentation. It is produced in Cambodia and South Vietnam.

Decapod flesh is sought-after not only for its taste but also for its **nutritional value**. Table I presents the composition of some crustaceans eaten in the U.K. and U.S.A. There are a growing number of studies promoting the consumption of crustaceans (Tou et al., 2007; Barrento et al., 2010; Marques et al., 2010; Özden & Erkan, 2011). Decapod flesh is indeed rich in ω-3 polyunsaturated fatty acids and is a source of protein with low fat content. **Cholesterol levels** range from low to medium. Tou et al. (2007) have also demonstrated that **antioxidant levels** in krill and shrimp are higher than in salmonid fish.

---

[2]) The name of the tribe comes from the word "dûd" meaning "worms". Literally, Dauada means "worm eaters" (Bellair-Baudier, 1949).

TABLE I

Composition of crustacean food items per 100 g of edible portions (Food Standards Agency, 2002; USDA, 2013)

	Water g	Protein g	Total fit g	PUFA g	Cholesterol mg	Carbohydrate g	Energy value kcal	kJ
Crab (boiled)	71	19.5	5.5	1.6	72	0	128	535
Lobster (boiled)	74.3	22.1	1.6	0.6	110	0	103	435
Prawns (boiled)	70	22.6	0.9	0.2	280	0	99	418
Shrimps (canned in brine, drained)	74.9	20.8	1.2	0.4	130	0	94	398
Crayfish (raw)	84.05	14.85	0.97	0.3	114	0	72	303.5

Despite the beneficial effects of consuming sea food, adverse allergic effects are regularly observed (Lopata et al., 2010). Tropomyosin, a major muscle protein found in decapods, is the best known **allergenic protein**. Furthermore, other proteins such as arginine kinase, myosin light chain, or sarcoplasmic calcium-binding protein are now known to cause **allergic reactions** in humans.

During the last four decades, the human consumption of **sea food** in general and of crustaceans in particular has gradually increased (Delgado, 2003) (fig. 27A.1). In 2011, freshwater **aquaculture** production reached 36.9 million tons. Crustaceans accounted for 6.5% of this production, i.e., 2.4 million tons (FAO, 2011). Penaeid shrimp and crabs (Brachyura) constituted about 60% of the total crustacean production from freshwater aquaculture (fig. 27A.2). Furthermore, a total of 2.7 million tons of crustaceans were

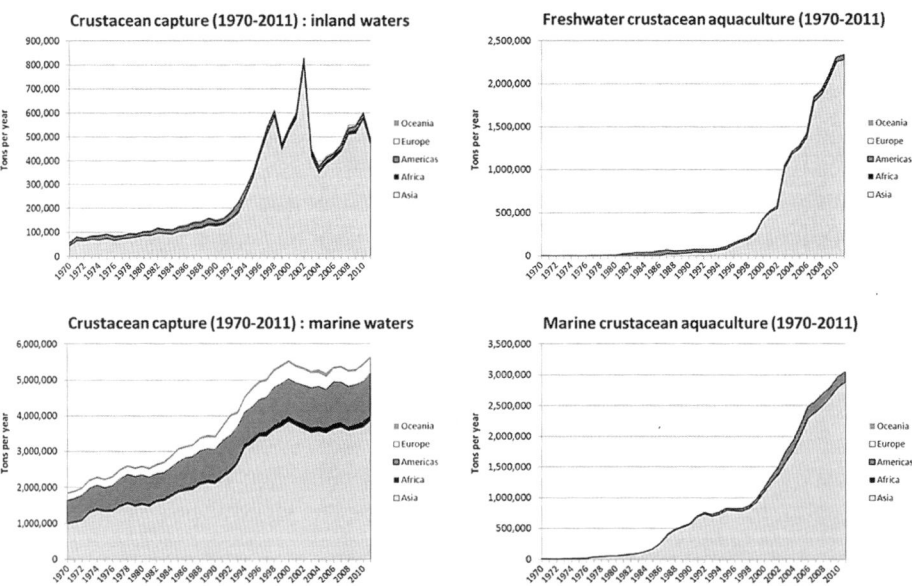

Fig. 27A.1. Global crustacean capture in natural environments and production in aquaculture.

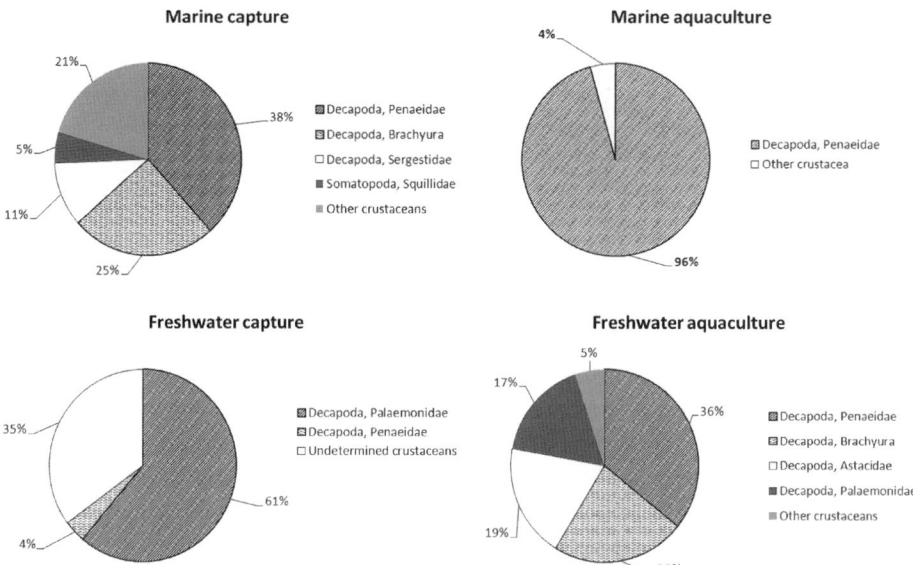

Fig. 27A.2. Proportions of the different crustaceans captured in natural environments and produced in aquaculture.

produced in brackish water aquaculture. Marine aquaculture alone yielded 0.7 million tons of crustaceans. Brackish and marine aquaculture productions were merged in fig. 27A.1 as production in salt waters. The Pacific white shrimp (*Litopenaeus vannamei*) is the major species raised in these waters (2.7 million tons in 2011) (fig. 27A.2).

Crustaceans are also increasingly being caught in natural environments (fig. 27A.1). In 2011, 5.7 million tons were captured in marine waters and 0.5 million tons in inland waters. Most were caught in Asia (68% in marine waters and 94% in inland waters). Similarly to what was observed for freshwater aquaculture, penaeid shrimp and crabs (Brachyura) were the predominant crustaceans caught in marine environments (fig. 27A.2). In fresh waters, shrimp belonging to the family Palaemonidae constituted more than 60% of the capture. Most species captured belonged to the genus *Macrobrachium*. In the FAO "freshwater capture" database, many species are categorized simply as "crustaceans" or "Natantia", which does not allow a precise attribution of the capture to taxonomic groups. This explains the high percentage of "undetermined crustaceans" in fig. 27A.2.

CRUSTACEANS AS USEFUL AUXILLIARIES

Crustaceans in traditional pharmacopeia

Many crustaceans have been used as **remedies** since antiquity and are still used in some countries. **Cures** based on crustaceans are described in "*De Materia Medica*" written around 60 AD by Pedanius Dioscorides (Book 2, chapter 12) as well as in "*Naturalis*

Historia" written around 78 AD by Pliny the Elder (books IX and XXXII). Dioscorides reported that ashes from freshwater crabs were a remedy against the bite of rabid dogs (Grmek & Guinot, 1965). This belief is also reported by Bacci (1586). Decapod **gastroliths** rich in calcium carbonate were, on the other hand, used to reduce excessive stomach acidity (Théodoridès, 1980). According to Hippocrates (5th-4th century BC) and Diocles of Carystus (4th century BC), the hard parts, flesh, and juice of crustaceans had diuretic properties. They were employed in the treatment of urinary track illness and also in gynaecology. During the 16th century, crabs were axiologically related to the chest and they were thought to cure tuberculosis, carcinomas and cancer in Europe (Grmek & Guinot, 1965). Woodlice were diuretic and used to treat scrofula and rheumatisms. As reported in many textbooks before the 19th century (Chevallier et al., 1829; Guibourt, 1851), they are the major ingredient in Morton's balsamic pills, used to treat chest diseases and coughs. Sachs von Löwenheim (1665) described in detail the considerable role of crayfish gastroliths in his large book dedicated to "crabs". He classified the effects of these *lapilli cancrorum* or *oculi cancrorum* according to the part of the body concerned: the head (headache, epilepsy, spasms, ...), chest (angina, pleurisy, emphysema, ...) heart, abdominal region, genital organs or skin.

Even in modern times, crustaceans are still used as remedies in traditional **pharmacopeia**. Alves & Alves (2011) have listed the **medicinal animals** used in popular medicine in Latin America, recording about 20 species of crustaceans used as remedies. Most species belong to the order Decapoda, with two major exceptions from the orders Stomatopoda (the mantis schrimp *Cloridopsis dubia*) and Isopoda (the pillbug *Armadillidium vulgare*). In Latin America, crustaceans are mainly recommended as remedies against asthma, bronchitis, epilepsy, irritation when milk teeth are growing, and haemorrhaging in women.

Crustaceans as sources of polymers or energy

Huge quantities of crab and shrimp shell waste are generated every year by the canned food industry. In the 1970s, this waste started to attract attention as a potential source of chitin for use in industry (Muzzarelli, 1977). **Chitin**, or β-1,4-N-acetyl-D-glucosamine, is indeed one of the major organic components of crustacean **exoskeletons** (Peter, 2005). This polysaccharide and its deacetylated derivative, **chitosan**, have very interesting properties including non-toxicity, biocompatibility, a fibrous structure, and abilities to adsorb metal ions, among others (Kurita, 2006). Thanks in particular to the amino groups, chitosan has been used for a large variety of applications, including the chelation of heavy metals (Gerente et al., 1999), organochlorinated compounds (Thomé et al., 1997), or industrial dyes (Juang et al., 1994). It has also been used for controlled drug delivery (Genta et al., 1999), wound healing (Muzzarelli et al., 1999), biomaterial building (Yilmaz, 2004), biodegradable surgical wire production (Foster et al., 2012), control of hypercholesterolaemia (Muzzarelli, 1999), antibacterial agents (Jia et al., 2001), and cosmetic additives (Leuba et al., 1991).

Crab and shrimp shells contain high levels of minerals (from 230 to 600 mg.g^{-1} dry weight) (Anderson et al., 1978; Andersen, 1991). This implies the use of drastic acid treatments that have an impact on the environment due to the production of large amounts of wastewater and the quality of the chitin extracted (No & Meyers, 1997), leading to the search for less mineralized sources of chitin as an alternative. Planktonic crustaceans appear as a potential alternative, since they have a reduced **mineral content** (from 100 to 200 mg.g^{-1} dry weight) (Cauchie et al., 1999). Large populations of cladocerans and copepods can be observed in **wastewater stabilization ponds**. Cauchie et al. (2002) have analysed the potential of using aerated **wastewater treatment** plants to produce chitin. They estimated a production rate of 1200 kg chitin per year for 60 000 m^2 ponds. Until now, however, ponds have not been exploited commercially for chitin or chitosan production. Instead, the ponds are sometimes used for producing food for fish in aquaculture (Prein, 1996) or biofuel feedstock (Kring et al., 2013), besides using planktonic crustaceans as a **water treatment agent** (Cauchie et al., 2000).

DETRIMENTAL IMPACT OF CRUSTACEANS ON HUMAN ACTIVITIES

Dike digging and damage to rice fields

Some crustaceans are considered **pests** because of their predation on fish or the impact of their **burrowing activities** on the stability of dikes (Rudnick et al., 2005). In Germany, the economic impact of the Chinese mitten crab *Eriocheir sinensis* has been estimated at 80 million Euros since 1912 (Gollasch et al., 2009). This estimate includes the impact on river bank erosion and losses in aquaculture and commercial fishing. The mud lobster *Thalassina anomala* also weakens levees and dikes with burrows that can reach 2 to 2.5 metres (Bhattacharya, 1999; Dubey et al., 2012). On the other hand, crabs and crayfish were reported to damage rice crops by feeding on young rice shoots (Foster & Harper, 2007).

The tadpole shrimp *Triops longicaudatus* is used as an efficient biological-control agent for mosquitoes in North American ponds (Fry et al., 1994). In rice fields, however, these branchiopods can cause severe damage by feeding directly on seedlings and uprooting seedlings (Tindall & Fothergill, 2012).

Parasites or commensals of useful species

The parasitic crustaceans that have an impact on fish aquaculture mainly belong to the families Caligidae and Ergasilidae (Copepoda), most often referred to as sea lice (Johnson et al., 2004). These parasites graze on the skin, gills, fins, or eyes causing ulceration or bleeding (Munday et al., 2003). Costello (2009) estimated the cost of sea lice infections in salmonid industry at around US$ 305 million annually around the world.

Exploited shellfish are also affected by crustaceans (see review by Bower et al., 1994). Molluscs such as oysters, mussels, clams, cockles or scallops are mainly affected by copepods from the genus *Mytilicola*. Infections by species of *Mytilicola* are also known as "**red worm disease**". Red worm prevalence can reach 100% in populations chronically affected by the parasite, in which case more than 30 copepods can be observed per mussel. There is, however, controversy on the real impact of *Mytilicola* on mussels. In a ten-year study, Davey (1989) suggested that red worms are living as commensals although *Mytilicola intestinalis* has been suspected of causing mass mortality among mussels (Korringa, 1951). The negative effects of another copepod species, *Pectinophilus ornatus*, on scallops are more obvious, with reduction in growth and fat content reported (Nagasawa & Nagata, 1992; Bower et al., 1994). Pea crabs (Decapoda, Pinnotheridae) are common parasites of bivalves. Reductions in growth rates and shell sizes have been observed in mussels parasitized by pea crabs (Bierbaum & Ferson, 1986). This has a significant impact on aquacultured mussels. Trottier et al. (2012) have estimated the production loss reaching US$ 2.16 million annually in New Zealand. The genus *Argulus* (Branchiura, Argulidae: fish lice) comprises another group of important ectoparasites in aquaculture. Sahoo et al. (2012) have estimated the production loss due this parasite at US$ 1428 per ha per year in farm ponds in India.

Invasive species

Biological invasion is a major driver of **biodiversity loss** in the world (Molnar et al., 2008). Many crustacean species are invasive, especially in marine ecosystems. Streftaris & Zenetos (2006) have included 15 crustaceans in the preliminary list of the 100 worst invaders in the Mediterranean Sea. These are mainly crabs (*Callinectes sapidus, Charybdis longicollis, Eriocheir sinensis, Percnon gibbesi, Portunus pelagicus*), shrimps (*Marsupenaeus japonicus, Melicertus hathor, Metapenaeus monoceros*), and the stomatopod *Erugosquilla massavensis*, with the major exception of the red worm *Mytilicola orientalis*. In fresh water, a famous example of native species being replaced by invaders concerns crayfish. Native European crayfish, such as *Astacus leptodactylus*, are threatened by the oomycete *Aphanomyces astaci* (crayfish plague) that is dispersed by invasive North American species including *Orconectes limosus, Procambarus clarkii*, and *Pacifastacus leniusculus*. The Ponto-Caspian invasive species in Europe are notably gammarids (*Chelicorophium robustum, Dikerogammarus bispinosus, Dikerogammarus villosus, Echinogammarus trichiatus*), and a mysid (*Hemimysis anomala*). The real socio-economic impact of biological invasion is difficult to assess, because it rarely affects an economic activity directly (Reinhardt et al., 2003), except in the case of aquaculture parasites, as discussed above (Lovell et al., 2006).

Biofouling

Objects submerged in the sea are rapidly covered by **epiphytes** and **sessile animals**. This phenomenon, known as "**biofouling**", has negative impact on human activities by increasing the fuel consumption of boats with colonized hulls and by increasing the need for structural maintenance of marine assets (Gollasch, 2002; Hewitt et al., 2009). Barnacles (*Balanus* spp., *Megabalanus* spp., and *Lepas anatifera*) are the main sessile crustaceans found on hulls but vagile crustaceans such as amphipods (among others *Corophium* spp., *Ericthonius* spp., *Jassa* spp., *Podocerus* spp.) can also be found on ship hulls (Feirrera et al., 2006). Decapods (*Panulirus laevicauda*) or stomatopod crustaceans (*Gonodactylaceus randalli*) can be found on oil platforms (Feirrera et al., 2006; Yeo et al., 2010). Hull fouling is thought to be a more important vector favouring the introduction of non-indigenous species in marine as well as freshwater ecosystems than transport in ballast water (Gollasch, 2002; Drake & Lodge, 2007). Biofouling has mainly been controlled by using antifouling paints such as the organotin tri-butyltin (TBT) (Minchin & Minchin, 1997). However, TBT was banned in 2003 due to its negative impact on aquatic ecosystems. Alternative antifouling substances that are less toxic to the environment are currently being explored (e.g., Lee et al., 2011).

HUMAN PATHOLOGIES LINKED TO CRUSTACEANS

Crustaceans are linked to **human health** through two aspects: firstly as toxic food items; and secondly as intermediate hosts for pathogenic bacteria or parasites.

Toxic decapods

Cases of poisoning after the consumption of crabs have been reported for many decades in the Pacific Ocean area. Human fatalities have occurred regularly, notably in the Philippines, Japan, Palau Islands, Fiji, Singapore, and Vanuatu (Alcala & Halstead, 1970; Ho et al., 2006; Asakawa et al., 2010). It is now known that these fatalities are due to the consumption of some crab species containing potent **neurotoxins** such as saxitoxin analogues and tetrodotoxins (Koyama et al., 1981; Daigo et al., 1985; Yasumura et al., 1986; Asakawa et al., 2010).

Crustaceans, mainly xanthid crabs, generally concentrate toxins from marine dinoflagellates, such as some species of *Alexandrium*, *Gymnodinium catenatum*, and *Pyrodinium bahamense*. Deeds et al. (2008) reviewed the saxitoxin concentrations and algal sources of crustacean contaminations. Besides true crabs such as *Metacarcinus magister* (earlier known as *Cancer magister*) and *Hemigrapsus oregonensis*, the hermit crabs of the genus *Pagurus*, lobsters such as *Homarus americanus*, or penaeid shrimp, were found to contain significant amounts of toxins.

Tetrodotoxins (TTX) are most often the dominant toxins in crabs. It is still unclear whether they are of endogenous or exogenous origin (Asakawa et al., 2010); TTX are notably produced by a wide array of bacterial genera including *Vibrio*, *Pseudomonas*, *Aeromonas*, *Alteromonas*, among others (Yu et al., 2004). It is plausible that crabs can accumulate toxins when eating contaminated items, such as mussels. Oikawa et al. (2002) observed that toxin concentrations in the viscera of the edible shore crab (*Telmessus acutidens*) increased linearly in relation to the amounts of contaminated mussels they eat. Chau et al. (2011) have started investigating the possible biosynthetic routes of TTX in non-bacterial organisms, but have not yet reached any definitive conclusions.

Crustaceans as intermediate hosts or vectors of a variety of human parasites

Dracunculiasis is a parasitic disease caused by *Dracunculus medinensis* (the Guinea worm). Guinea worms are spiruroid nematodes that utilize cyclopoid copepods as intermediate hosts (Boxhall & Defaye, 2008). Human infections occur when drinking water is consumed from stagnant sources containing contaminated copepods. In 1986, an estimated 3.5 million people were infected by Guinea worms in twenty countries in Asia and Africa (WHA, 1986). Dracunculiasis is currently moving towards global eradication (WHO, 2013), but remains notably present in South Sudan.

Paragonimiasis is an infectious, pulmonary disease caused by several trematode species of the genus *Paragonimus*. Humans can be contaminated when eating raw or undercooked infected crab or crayfish. Fifty-three species from 21 genera of freshwater crabs and crayfish have been identified by WHO (1995) as vectors of this disease. This list has been supplemented by new hosts discovered (among others by Sohn et al., 2009). The main species associated with paragonimiasis are decapods from the genera *Eriocheir*, *Potamon*, *Sinopotamon*, and *Cambaroides*. In 2005, it was estimated that more than 290 million people were at risk globally (Keiser & Utzinger, 2005). In the context of the growing human population, it is thought that the exponential growth of aquaculture will increase the prevalence of foodborne **trematodiasis** in the future.

The bacterium *Vibrio cholerae* is the causative agent of epidemic **cholera**. It provokes severe dehydrating diarrhoea, killing about 100 000 people worldwide each year according to the WHO. Huq et al. (1983) have demonstrated that *Vibrio cholerae* attaches to the surface of live copepods. As a consequence, transmission of *Vibrio cholerae* by drinking untreated surface water containing contaminated planktonic crustaceans appears significant in developing countries (Colwell & Huq, 1994). Colwell et al. (2003) demonstrated a 48% reduction in cholera occurrences following the filtration of surface water on 20 μm sieves in Bangladesh villages.

Diphyllobothriasis is a rare parasitic disease caused by cestodes (tapeworms) belonging to the genus *Diphyllobothrium*. Humans are generally infected after consuming raw fish (Dupouy-Camet & Peduzzi, 2004; Arizono et al., 2009). The life cycle of this parasite is as follows: immature eggs are produced by the adult tapeworm in the intestinal track of humans and pass out in faeces. The eggs mature in water and are ingested by calanoid and cyclopoid copepods, which are the first intermediate hosts. Copepods are then consumed by fish, the second intermediate host, and the parasite's larvae migrate into the fish flesh.

Finally, it can be noted that the blackfly *Simulium naevi*, which is the vector of *Onchocerca volvulus*, a parasitic nematode causing **onchocerciasis** (also called **river blindness**), lays its eggs on the legs of some African crabs, such as *Potamonautes* spp. (de Meillon, 1957). The link between crab population dynamics and the etiology of these diseases has, however, not yet been investigated.

CRUSTACEANS IN MYTHOLOGY, ART, HISTORY, AND POPULAR CULTURE

It is quite understandable that crustaceans, being mostly aquatic, rather inconspicuous animals, have played a lesser role in **history**, **art**, or **mythology** than vertebrates, primarily mammals, birds, and reptiles. Indeed, for a long time, they were considered simple "insects". Raban Maur (P.L., 111: 239) has even described them as simple "shells having legs" (*conchae crura habentes*). This does not mean, far from it, that Crustacea have been completely neglected in human cultures, as illustrated by the following remarks.

Zodiacal asterism

D'Arcy W. Thompson (1900) carried out an important study on "the Emblem of the Crab in relation to the sign Cancer". He highlighted the role of the zodiacal constellation Cancer in ancient social practices. Cancer was "*domus Lunae*" (the House of the Moon) and the Crab was associated with the Moon on coins from the cities of Consetia and Terrina[3]) during the Brutii era (around the 4th century BC) or with the lunar Diana of the Ephesians. Cancer was "*exaltatio Jovis*" (the Banner of Jove), and the Crab was peculiarly associated with the Bird of Jove in the coinage of Agrigentum[4]). Finally, Cancer is "*sedes Mercurii*" (the Seat of Mercury), the guardian of Mercury and the Crab is figured with the head of Hermes on coins of Aenus[5]).

Legends, mythology, and representations

Crustaceans are present in **classical antiquity**. In ancient Greek mythology, the queen of the Gods, Hera, sent a giant crab to bite the foot of Hercules during his fight against the nine-headed Lernaean Hydra. Hercules crushed the crab with his foot. Hera placed it in the sky for this service. Representations of Hercules pinched by the crab can be observed on ancient coins found at Phaestos, Crete (approximately 280 BC).

Descriptions of crustaceans appeared in natural history books such as "*Historia Animalium*" (343 BC) where Aristotle describes lobsters, crayfish, mantis shrimps, and crabs (book IV, chapter 2). In "*De natura animalium libri XVII*", Aelianus Claudius[6]) mentions the Red Sea crabs dedicated to Poseidon. A lot of information on crustaceans in the antique Mediterranean area is included in the book "*Die Antike Tierwelt*" by Keller (1913). This book notably mentions the presence of crustaceans such as spiny lobsters in Roman mosaics, such as those that can be seen at the Museo Archeologico Nazionale in Naples, Italy.

---

[3]) Consetia corresponds to the modern Consenza in Calabria (southern Italy). The old city of Terina was located near the modern city of Lamezia Terme.
[4]) Agrigentum corresponds to the modern city of Agrigento on the southern coast of Sicily.
[5]) Aenus corresponds to the modern city of Enez in Turkey.
[6]) Roman author and teacher of rhetoric (approx. 175-approx. 235 AD).

One of the most famous roles of crustaceans in folklore is the "Crab with a human face", popular in Japan (Neuville, 1938; André, 1939; Huard & Guinot, 1965). In Japan, the decapod *Heikeopsis japonica*, or "heikegani" in Japanese, can be found. Its shell bears a pattern resembling a human face, believed to be that of an angry samurai. The legend says that these crabs are reincarnations of the spirits of the samurais defeated at the Battle of Dan-no-ura in 1185. This story is depicted in the epic account of the struggle called "*Heike monogatori*" (Martin, 1993). In Japan, crabs also appear in children's stories (Roberts, 2009). In China, the samurai crab is known instead as "Guan Yu" crab, named after a famous general in the late Eastern Han Dynasty.

Before it was known that birds migrate, people in Celtic countries claimed that barnacle geese, *Branta leucopsis*, developed from goose barnacles, *Lepas anserifera* (cf. White, 1945). Similarities in colour and shape reinforced the legend. The species' name *anserifera* comes from the Latin roots *anser* for "goose" and *fera* for "bearing". Giraldus Cambrensis, a Welsh Monk, claimed in "*Topographia Hiberniae*" that barnacles were growing on trees as they were often found on driftwood. John Gerard represents a "goose-tree" in his "*Herball, or Generall Historie of Plantes*" (1597). In France, barnacles were first called "anatifères" then "anatifes", based on the Latin roots *anas* for "duck" and *fera* for "bearing". The French Renaissance poet Rémy Belleau portrays barnacles in his last work, "*Les Amours et nouveaux Eschanges des Pierres precieuses*" (1576):

Qui croiroit qu'une branche tender
Tombant dedans l'eau peust estendre
Ses fueilles en ailes d'oiseaux
Bois, escorces, nouveaux fruitages
S'emplumer en oysons sauvages
Naissant qui flottent sur les eaux?

[Who would believe that a tender branch
Falling in water can spread
Its leaves in bird's wings
Wood, bark, new fruits
Fledging into wild gosling
Born floating on water?]

Heron-Allen (1928) wrote a whole book on the legend of the barnacle.

The earliest representation of crustaceans in art can be traced back to ancient Egypt. Spiny lobsters were already present on an Egyptian painting from the XVIIIth Dynasty (approx. 1550-1292 BC). Crabs, probably from the genus *Potamon*, are represented in a bas-relief in the palace of Senacherib (705-681 BC). The Far East has been particularly rich in representations of crustaceans (paintings, bronzes, and ceramics).

Representations of Crustaceans are, on the contrary, rare during the Middle Ages (Charmantier, 2003). They come back in paintings during the Renaissance and peaked in 17th century Flemish and Dutch paintings. The following representative paintings can be cited: "*Crab*" by Albrecht Dürer (1495), "*Still Life with Nautilus Cup and Lobster*" by Jan Davidesz (1634) and "*Dog and cat in front of a lobster*" by Jan van Kessel, Junior (ca. 1700). Afterwards, crustaceans appeared more rarely during the 19th and 20th centuries: "*Still life with lobsters*" by Eugène Delacroix (1822), "*Two crabs*" by Vincent van Gogh (1885) and "*Lobster and cat*" by Pablo Picasso (1965) among others. Salvatore Dali created a sculpture called "Lobster phone" in 1936.

Captain Dabry de Thiersant (1826-1898) painted a series of watercolours showing China's crabs. Executed with realism, these watercolours are preserved at the National Museum of Natural History in Paris.

During the 19th and 20th centuries, crustaceans, mainly decapods, were also represented on stamps. Omori & Holthuis (2005) referenced nearly 1468 postage stamps depicting crustaceans.

Crustaceans play a minute role in literature, although several books contain the word "crab": "*The crab that played with the sea*" by Rudyard Kipling (1902)[7], "*The drummer-crab*" by Pierre Schoendoerffer (1976) or "*Crabwalk*" by Günter Grass (2002). In popular culture, crustaceans appear from time to time, such as in the B-series science-fiction film "*Attack of the Crab Monsters*" directed by R. Corman in 1953.

The role of crustaceans in, especially, the visual arts, is treated more extensively in the second part of this chapter 27, i.e., chapter 27B, in the present volume.

TO CONCLUDE

From the above it is evident that links between crustaceans and the various aspects of human societies are limited. Yet, in those fields where they do meet, like in the role of human food or as vectors of diseases, crustaceans often play a vital role as there they deeply affect human wellbeing and culture. The detailed study of crustaceans, therefore, is and remains a most relevant topic in biological science.

APPENDIX
Names of species mentioned in this chapter with their authority and date of publication, listed in alphabetical order for Crustacea and non-Crustacea (and subordinate taxa)

CRUSTACEA
Armadillidium vulgare (Latreille, 1804)
Artemia salina (Linnaeus, 1758)
Astacus leptodactylus Eschscholtz, 1823
Callinectes sapidus Rathbun, 1896
Cancer magister Dana, 1852 [currently as: *Metacarcinus magister* (Dana, 1852)]
Charybdis longicollis Leene, 1938
Chelicorophium robustum (G. O. Sars, 1895)
Cloridopsis dubia (H. Milne Edwards, 1837)
Dikerogammarus bispinosus Martynov, 1925
Dikerogammarus villosus (Sowinsky, 1894)
Echinogammarus trichiatus (Martynov, 1932)

---

[7]) The crab depicted in the story is, in fact, a limulid chelicerate!

Eriocheir sinensis H. Milne Edwards, 1853
Erugosquilla massavensis (Kossmann, 1880)
Euphausia pacifica Hansen, 1911
Euphausia superba Dana, 1852
Gonodactylaceus randalli (Manning, 1978)
Heikeopsis japonica (Von Siebold, 1824)
Hemigrapsus oregonensis (Dana, 1851)
Hemimysis anomala G. O. Sars, 1907
Homarus americanus H. Milne Edwards, 1837
Lepas anatifera Linnaeus, 1758
Lepas anserifera Linnaeus, 1767
Litopenaeus vannamei (Boone, 1931)
Marsupenaeus japonicus (Spence Bate, 1888)
Meganyctiphanes norvegica (M. Sars, 1857)
Melicertus hathor Burkenroad, 1959
Metacarcinus magister (Dana, 1852)
Metapenaeus monoceros (Fabricius, 1798)
Mytilicola intestinalis Steuer, 1902
Mytilicola orientalis Mori, 1935
Orconectes limosus (Rafinesque, 1817)
Pacifastacus leniusculus (Dana, 1852)
Panulirus laevicauda (Latreille, 1817)
Pectinophilus ornatus Nagasawa, Bresciani & Lützen, 1988
Percnon gibbesi (H. Milne Edwards, 1853)
Pollicipes elegans (Lesson, 1831)
Pollicipes pollicipes (Gmelin, 1790)
Pollicipes polymerus Sowerby, 1833
Portunus pelagicus (Linnaeus, 1758)
Procambarus clarkii (Girard, 1852)
Rimicaris exoculata Williams & Rona, 1986
Squilla mantis (Linnaeus, 1758)
Telmessus acutidens (Stimpson, 1848)
Thalassina anomala (Herbst, 1804)
Triops longicaudatus (LeConte, 1846)

NON-CRUSTACEA
BACTERIA
Vibrio cholerae Pacini, 1854

CHROMISTA [PLANTAE]
Gymnodinium catenatum Graham, 1943
Pyrodinium bahamense Plate, 1906

FUNGI
Aphanomyces astaci Schikora, 1906

ANIMALIA: NEMATODA
Dracunculus medinensis (Linnaeus, 1758)
Onchocerca volvulus Bickel, 1982

ANIMALIA: ARTHROPODA: HEXAPODA: DIPTERA
Simulium naevi [also cited as: *Simulium naevis* and *Simulium naevus*; authority and date not retrieved]

CHORDATA: VERTEBRATA
Balaenoptera borealis Lesson, 1828
Balaenoptera musculus (Linnaeus, 1758)
Balaenoptera physalus (Linnaeus, 1758)
Branta leucopsis (Bechstein, 1803)
Gasterosteus aculeatus Linnaeus, 1758
Phoxinus phoxinus (Linnaeus, 1758)

BIBLIOGRAPHY

ALCALA, A. C. & B. W. HALSTEAD, 1970. Human fatality due to ingestion of the crab *Demania* sp. in the Philippines. — Clin. Toxicol., **3**: 609-611.

ALVES, R. R. N. & H. N. ALVES, 2011. The faunal drugstore: animal-based remedies used in traditional medicines in Latin America. — J. Ethnobot. Ethnomed., **7** (9): 1-43.

ANDERSEN, S. O., 1991. Cuticular proteins from the shrimp, *Pandalus borealis*. — Comp. Biochem. Physiol., (B) **99**: 453-458.

ANDERSON, C. G., N. DE PABLO & C. R. ROMO, 1978. Antarctic krill (*Euphausia superba*) as a source of chitin and chitosan. — In: R. A. A. MUZZARELLI & E. R. PARISER (eds.), Proceedings of the First International Conference on Chitin and Chitosan, pp. 54-63. (MIT Sea Grant Program, Massachusetts.)

ANDRÉ, M., 1939. Un crabe japonais à face humaine. — Sci. natur., **1**: 169-170.

ARIZONO, N., M. YAMADA, F. NAKAMURA-UCHIYAMA & K. OHNISHI, 2009. Diphyllobothriasis associated with eating raw Pacific salmon. — Emerg. infect. Diseas., **15**: 866-870.

ASAKAWA, M., G. GOMEZ-DELAN, S. TSURUDA, M. SHIMOMURA, Y. SHIDA, S. TANIYAMA, M. BARTE-QUILANTANG & J. SHINDO, 2010. Toxicity assessment of the xanthid crab *Demania cultripes* from Cebu Island, Philippines. — J. Toxicol., **2010**: e172367.

ATKINSON, A., V. SIEGEL, E. A. PAKHOMOV, M. J. JESSOPP & V. LOEB, 2009. A re-appraisal of the total biomass and annual production of Antarctic krill. — Deep-Sea Res., (I, Oceanogr. Res. Pap.) **56**: 727-740.

BACCI, A., 1586. De venenis et antidotis prolegomena seu Communia praecepta ad humanam vitam tuendam saluberrima. — Pp. 1-83. (Impensis Ioannis Martinelli, Rome.)

BARRENTO, S., A. MARQUES, B. TEIXEIRA, R. MENDES, N. BANDARRA, P. VAZ-PIRES & M. L. NUNES, 2010. Chemical composition, cholesterol, fatty acid and amino acid in two populations of brown crab *Cancer pagurus*: ecological and human health implications. — J. Food Compos. Anal., **23**: 716-725.

BEAUGRAND, G., K. BRANDER, J. LINDLEY, S. SOUISSI & P. REID, 2003. Plankton effect on cod recruitment in the North Sea. — Nature, Lond., **426**: 661-664.

BHATTACHARYA, A., 1999. Embankments and their ecological impacts: a case study from the tropical low-lying coastal plains of the deltaic Sunderbans, India. — In: M. VOLLMER & H. GRANN (eds.), Large-scale constructions in coastal environments, pp. 171-180. (Springer, Berlin and Heidelberg.)

BIERBAUM, R. M. & S. FERSON, 1986. Do symbiotic pea crabs decrease growth rate in mussels? — Biol. Bull., Woods Hole, **170**: 51-61.

BLANCHET, M., M. BIUW, G. J. G. HOFMEYR, P. J. N. BRUYN, C. LYDERSEN & K. M. KOVACS, 2013. At-sea behaviour of three krill predators breeding at Bouvetøya — Antarctic fur seals, macaroni penguins and chinstrap penguins. — Mar. Ecol. Progr. Ser., **477**: 285-302.

BOWER, S. M., S. E. MCGLADDERY & I. M. PRICE, 1994. Synopsis of infectious diseases and parasites of commercially exploited shellfish. — Annu. Rev. Fish Diseas., **4**: 1-199.

BOXSHALL, G. A. & D. DEFAYE, 2008. Global diversity of copepods (Crustacea: Copepoda) in fresh water. — Hydrobiologia, **595**: 195-207.

CAUCHIE, H. M., 2002. Chitin production by arthropods in the hydrosphere. — Hydrobiologia, **470**: 63-96.

CAUCHIE, H. M., L. HOFFMANN & J. P. THOMÉ, 2000. Metazooplankton dynamics and secondary production of *Daphnia magna* (Crustacea) in an aerated waste stabilization pond. — J. Plankt. Res., **22**: 2263-2287.

CAUCHIE, H. M., M. F. JASPAR-VERSALI, L. HOFFMANN & J. P. THOMÉ, 1999. Analysis of the seasonal variation in biochemical composition of *Daphnia magna* Straus (Crustacea: Branchiopoda: Anomopoda) from an aerated wastewater stabilisation pond. — Ann. Limnol., **35**: 223-231.

CHARMANTIER, G., 2003. Crustaceans in art. — In: SICB 2003 Annual Meeting, Toronto, Jan. 4-8. [Oral pres.]

CHAU, R., J. A. KALAITZIS & B. A. NEILAN, 2011. On the origins and biosynthesis of tetrodotoxin. — Aquat. Toxicol., **104**: 61-72.

CHEVALLIER, A., A. RICHARD & J. A. GUILLEMIN, 1829. Dictionnaire des drogues simples et composées, ou Dictionnaire d'histoire naturelle médicale, de pharmacologie et de chimie pharmaceutique. — Vol. **4**, pp. 1-618. (Béchet Jeune (ed.), Paris.)

COLWELL, R. R. & A. HUQ, 1994. Environmental reservoir of *Vibrio cholerae*, the causative agent of cholera. — Ann. New York Acad. Sci., **740**: 44-54.

COLWELL, R. R., A. HUQ, M. S. ISLAM, K. M. A. AZIZ, M. YUNUS, N. H. KHAN ET AL., 2003. Reduction of cholera in Bangladeshi villages by simple filtration. — Proc. natn. Acad. Sci. U.S.A., **100**: 1051-1055.

COSTELLO, M. J., 2009. The global economic cost of sea lice to the salmonid farming industry. — J. Fish Diseas., **32**: 115-118.

CROLL, D. A., B. MARINOVIC, S. BENSON, F. P. CHAVEZ, N. BLACK, R. TERNULLO & B. R. TERSHY, 2005. From wind to whales: trophic links in a coastal upwelling system. — Mar. Ecol. Progr. Ser., **289**: 117-130.

DAIGO, K., A. UZU, O. ARAKAWA, T. NOGUCHI, H. SETO & K. HASHIMOTO, 1985. Isolation and some properties of neosaxitoxin from a xanthid crab *Zosimus aeneus*. — Bull. Jap. Soc. scient. Fish., **51**: 309-313.

DAVEY, J. T., 1989. *Mytilicola intestinalis* (Copepoda: Cyclopoida): a ten year survey of infected mussels in a Cornish estuary, 1978-1988. — J. mar. biol. Ass. U. K., **69**: 823-836.

DE MEILLON, B., 1957. Bionomics of the vectors of onchocerciasis in the Ethiopian geographical region. — Bull. World Health Org., **16**: 509-522.

DE TROCH, M., I. VERGAERDE, C. CNUDDE, P. VANORMELINGEN, W. VYVERMAN & M. VINCX, 2012. The taste of diatoms: the role of diatom growth phase characteristics and associated bacteria for benthic copepod grazing. — Aquat. microb. Ecol., **67**: 47-58.

DEEDS, J. R., J. H. LANDSBERG, S. M. ETHERIDGE, G. C. PITCHER & S. W. LONGAN, 2008. Non-traditional vectors for paralytic shellfish poisoning. — Mar. Drugs, **6**: 308-348.

DEES, L. T., 1961. Brine shrimp. — Fish. Leafl., **527**: 1-5. (Fish and Wildlife Service, U.S.)

DELGADO, C. L., 2003. Fish to 2020: supply and demand in changing global markets. — Pp. 1-226. (WorldFish, Penang.)

DRAKE, J. M. & D. M. LODGE, 2007. Hull fouling is a risk factor for intercontinental species exchange in aquatic ecosystems. — Aquat. Invas., **2**: 121-131.

DUBEY, S. K., A. CHOUDHURY, B. K. CHAND & R. K. TRIVEDI, 2012. Ecobiological study on burrowing mud lobster *Thalassina anomala* (Herbst, 1804) (Decapoda: Thalassinidae) in the intertidal mangrove mudflat of deltaic Sundarbans. — Explor. Anim. Med. Res., **2**: 70-75.

DUPOUY-CAMET, J. & R. PEDUZZI, 2004. Current situation of human diphyllobothriasis in Europe. — EuroSurv., **9** (5): i-ii, 1-467.
FAO, 2011. Fishery and aquaculture statistics — capture production. — FAO Yearbook of fishery and aquaculture statistics, pp. 1-594.
FERREIRA, C. E. L., J. E. A. GONÇALVES & R. COUTINHO, 2006. Ship hulls and oil platforms as potential vectors to marine species introduction. — J. coast. Res., (S1) **39**: 1340-1345.
FOOD STANDARDS AGENCY, 2002. McCance and Widdowson's The composition of foods (6th ed.). — Pp. 1-537. (The Royal Society of Chemistry, Cambridge.)
FOSTER, J. & D. HARPER, 2007. Status and ecosystem interactions of the invasive Louisianan red swamp crayfish *Procambarus clarkii* in East Africa. — In: F. GHERARDI (ed.), Biological invaders in inland waters: profiles, distribution, and threats, pp. 91-101. (Springer, Berlin and Dordrecht.)
FOSTER, L. J. & E. KARSTEN, 2012. A chitosan based, laser activated thin film surgical adhesive, 'SurgiLux': preparation and demonstration. — J. visual. Exp., **68**: e3527.
FRANGOULIS, C., E. CHRISTOU & J. HECQ, 2005. Comparison of marine copepod outfluxes: nature, rate, fate and role in the carbon and nitrogen cycles. — Adv. mar. Biol., **47**: 253-309.
FRY, L. L., M. S. MULLA & C. W. ADAMS, 1994. Field introductions and establishment of the tadpole shrimp, *Triops longicaudatus* (Notostraca: Triopsidae), a biological control agent of mosquitos. — Biol. Contr., **4**: 113-124.
GENTA, I., P. PERUGINI, F. PAVANETTO, T. MODENA, B. CONTI & R. A. MUZZARELLI, 1999. Microparticulate drug delivery systems. — In: P. JOLLES & R. A. MUZZARELLI (eds.), Chitin and chitinases, pp. 305-313. (Birkhäuser Verlag, Basel.)
GERENTE, C., Y. ANDRES & P. LECLOIREC, 1999. Uranium removal onto chitosan: competition with organic substances. — Environm. Technol., **20**: 515-521.
GOLLASCH, S., 2002. The importance of ship hull fouling as a vector of species introductions into the North Sea. — Biofoul., **18**: 105-121.
GOLLASCH, S., D. HAYDAR, D. MINCHIN, W. J. WOLFF & K. REISE, 2009. Introduced aquatic species of the North Sea coasts and adjacent brackish waters. — In: D. G. RILOV & D. J. CROOKS (eds.), Biological invasions in marine ecosystems, pp. 507-528. (Springer, Berlin and Heidelberg.)
GRMEK, M. D. & D. GUINOT, 1965. Les Crustacés dans la matière médicale européenne au XVe siècle. — Rev. Hist. Sci., (Applic.) **18**: 55-71.
GUIBOURT, N. J. B. G., 1851. Histoire naturelle des drogues simples ou Cours d'histoire naturelle. — Vol. **4**, pp. 1-70. (J. B. Baillière (ed.), Paris.)
HERON-ALLEN, E., 1928. Barnacles in nature and in myth. — Pp. 1-480. (Oxford University Press, London.)
HEWITT, C. L., S. GOLLASCH & D. MINCHIN, 2009. The vessel as a vector — biofouling, ballast water and sediments. — In: D. G. RILOV & D. J. CROOKS (eds.), Biological invasions in marine ecosystems, pp. 117-131. (Springer, Berlin and Heidelberg.)
HO, P. H., Y. H. TSAI, C. C. HWANG, P. A. HWANG, J. H. HWANG & D. F. HWANG, 2006. Paralytic toxins in four species of coral reef crabs from Kenting National Park in southern Taiwan. — Food Contr., **17**: 439-445.
HUARD, P. & D. GUINOT, 1965. Les Crabes de Chine dans une série d'aquarelles provenant de Dabry de Tiersant. — Vie Milieu, (Suppl.) **19**: 35-43.
HÜGLER, M., J. M. PETERSEN, N. DUBILIER, J. F. IMHOFF & S. M. SIEVERT, 2011. Pathways of carbon and energy metabolism of the epibiotic community associated with the deep-sea hydrothermal vent shrimp *Rimicaris exoculata*. — PLoS ONE, **6**: e16018.
HUQ, A., E. B. SMALL, P. A. WEST, M. I. HUQ, R. RAHMAN & R. R. COLWELL, 1983. Ecological relationships between *Vibrio cholerae* and planktonic crustacean copepods. — Appl. environm. Microbiol., **45**: 275-283.

ISH, T., E. J. DICK, P. V. SWITZER & M. MANGEL, 2004. Environment, krill and squid in the Monterey Bay: from fisheries to life histories and back again. — Deep-Sea Res., (II, Top. Stud. Oceanogr.) **51**: 849-862.

JIA, Z., D. SHEN & W. XU, 2001. Synthesis and antibacterial activities of quaternary ammonium salt of chitosan. — Carbohydr. Res., **333**: 1-6.

JOHNSON, S. C., J. W. TREASURER, S. BRAVO, K. NAGASAWA & Z. KABATA, 2004. A review of the impact of parasitic copepods on marine aquaculture. — Zool. Stud., **43**: 229-243.

KEISER, J. & J. UTZINGER, 2005. Emerging foodborne trematodiasis. — Emerg. infect. Diseas., **11**: 1507-1514.

KELLER, O., 1913. Die antieke Tierwelt. — Pp. 1-617. (Verlag Wilhelm Engelmann, Leipzig.)

KORRINGA, P., 1951. Le *Mytilicola intestinalis* Steuer (Copepoda – Parasitica) menace l'industrie moulière en Zélande. — Rev. Trav. Off. Pêch. marit., **17**: 9-13.

KOYAMA, K., T. NOGUCHI, Y. UEDA & K. HASHIMOTO, 1981. Occurrence of neosaxitoxin and other paralytic shellfish poisons in toxic crabs belonging to the family Xanthidae. — Bull. Jap. Soc. scient. Fish., **47**: 965.

KRING, S. A., X. XIA, S. E. POWERS & M. R. TWISS, 2013. Crustacean zooplankton in aerated wastewater treatment lagoons as a potential feedstock for biofuel. — Environm. Technol. [online preview; doi:10.1080/09593330.2013.795985]

KURITA, K., 2006. Chitin and chitosan: functional biopolymers from marine crustaceans. — Mar. Biotechnol., **8**: 203-226.

LAMPERT, W., 1978. Release of dissolved organic carbon by grazing zooplankton. — Limnol. Oceanogr., **23**: 831-834.

LAMPERT, W. & U. SOMMER, 2007. Limnoecology (2nd ed.). — Pp. 1-324. (Oxford University Press, Oxford.)

LEE, S., J. CHUNG, H. WON, D. LEE & Y. W. LEE, 2011. Analysis of antifouling agents after regulation of tributyltin compounds in Korea. — J. hazard. Mat., **185**: 1318-1325.

LEUBA, J. L., H. LINK, P. STOESSEL & J. L. VIRET, 1991. Cosmetic preparation containing chitosan. — Patent Nestec S.A. 428,882[5,057,542].

LOPATA, A. L., R. E. O'HEHIR & S. B. LEHRER, 2010. Shellfish allergy. — Clin. exp. Allerg., **40**: 850-858.

LÓPEZ, D. A., B. A. LÓPEZ, C. K. PHAM, E. J. ISIDRO & M. DE GIROLAMO, 2010. Barnacle culture: background, potential and challenges. — Aquacult. Res., **41**: e367-e375.

LOVELL, S. J., S. F. STONE & L. FERNANDEZ, 2006. The economic impacts of aquatic invasive species: a review of the literature. — Agric. Resour. Econ. Rev., **35**: 195-208.

MARQUES, A., B. TEIXERIA, S. BARRENTO, P. ANACLETO, M. L. CARVALHO & M. L. NUNES, 2010. Chemical composition of Atlantic spider crab *Maja brachydactyla*: human health implications. — J. Food Compos. Anal., **23**: 230-237.

MARTIN, J. W., 2003. The samurai crab. — Terra, **31**: 30-34.

MARTIN, J. W. & T. A. HANEY, 2005. Decapod crustaceans from hot vents and cold seeps. — Zool. J. Linn. Soc., **145**: 445-522.

MATRAI, P. A., E. OLSON, S. SUTTLES, V. HILL, L. A. CODISPOTI, B. LIGHT & M. STEELE, 2013. Synthesis of primary production in the Arctic Ocean: I. Surface waters, 1954-2007. — Progr. Oceanogr., **110**: 93-106.

MINCHIN, A. & D. MINCHIN, 1997. Dispersal of TBT from a fishing port determined using the dogwhelk *Nucella lapillus* as an indicator. — Environm. Technol., **18**: 1225-1234.

MOLARES, J. & J. FREIRE, 2003. Development and perspectives for community-based management of the goose barnacle (*Pollicipes pollicipes*) fisheries in Galicia (NW Spain). — Fish. Res., **65**: 485-492.

MOLNAR, J. L., R. L. GAMBOA, C. REVENGA & M. D. SPALDING, 2008. Assessing the global threat of invasive species to marine biodiversity. — Front. Ecol. Environm., **6**: 485-492.

MONOD, T., 1969. A propos du lac des Vers ou Bahr ed-Dûd (Libye). — Bull. Inst. fondam. Afr. noir., (A) **21**: 25-41.
MOORE, P. A., D. M. FIELDS & J. YEN, 1999. Physical constraints of chemoreception in foraging copepods. — Limnol. Oceanogr., **44**: 166-177.
MUNDAY, B. L., Y. SAWADA, T. CRIBB & C. J. HAYWARD, 2003. Diseases of tunas, *Thunnus* spp. — J. Fish Diseas., **26**: 187-206.
MUZZARELLI, R. A. A., 1977. Chitin. — (Pergammon Press, New York, NY.)
MUZZARELLI, R. A. [A.], 1999. Clinical and biochemical evaluation of chitosan for hypercholesterolemia and overweight control. — In: P. JOLLES & R. A. MUZZARELLI (eds.), Chitin and chitinases, pp. 293-304. (Birkhäuser Verlag, Basel.)
MUZZARELLI, R. A. [A.], M. MATTIOLI-BELMONTE, A. PUGNALONI & G. BIAGINI, 1999. Biochemistry histology and clinical uses of chitins and chitosans in wound healing. — In: P. JOLLES & R. A. [A.] MUZZARELLI (eds.), Chitin and chitinases, pp. 251-264. (Birkhäuser Verlag, Basel.)
NAGASAWA, K. & M. NAGATA, 1992. Effects of *Pectinophilus ornatus* (Copepoda) on the biomass of cultured Japanese scallop *Patinopecten yessoensis*. — J. Parasitol., **78**: 552-554.
NEUVILLE, H., 1938. Quelques remarques sur le Crabe dit «à face humaine» ou «des Samou-raïs» (*Dorippe japonica* von Siebold) et son rôle dans le folklore de l'Extrême-Orient. — Bull. Mus. natn. Hist. nat., **10**: 48-56.
NICOL, S. & Y. ENDO, 1999. Krill fisheries: development, management and ecosystem implications. — Aquat. liv. Resourc., **12**: 105-120.
NO, H. K. & S. P. MEYERS, 1997. Preparation of chitin and chitosan. — In: R. A. A. MUZZARELLI & M. G. PETER (eds.), Chitin handbook, pp. 475-489. (European Chitin Society, Grottammare.)
NÕGES, T., H. ARST, A. LAAS, T. KAUER, P. NÕGES & K. TOMING, 2011. Reconstructed long-term time series of phytoplankton primary production of a large shallow temperate lake: the basis to assess the carbon balance and its climate sensitivity. — Hydrobiologia, **667**: 205-222.
OIKAWA, H., M. SATOMI, S. WATABE & Y. YANO, 2005. Accumulation and depuration rates of paralytic shellfish poisoning toxins in the shore crab *Telmessus acutidens* by feeding toxic mussels under laboratory controlled conditions. — Toxicon, **45**: 163-169.
OMORI, M. & L. B. HOLTHUIS, 2005. Crustaceans on postage stamps from 1870 to and including 2002: revised articles for our paper in 2000 and addendum. — J. Tokyo Univ. mar. Sci. Technol., **1**: 1-39.
ÖZDEN, O. & N. ERKAN, 2011. A preliminary study of amino acid and mineral profiles of important and estimable 21 seafood species. — Brit. Food J., **113**: 457-469.
PETER, M. G., 2005. Chitin and chitosan from animal sources. — Biopolym. Online, doi: 10.1002/3527600035.bpol6015
PREIN, M., 1996. Wastewater-fed aquaculture in Germany: a summary. — Environm. Res. Forum, **5-6**: 155-160.
REINHARDT, F., M. HERLE, F. BASTIANSEN & B. STREIT, 2003. Economic impact of the spread of alien species in Germany. — Pp. 1-229. (Umweltbundesamt [Federal Environmental Agency], Berlin.)
ROBERTS, J., 2009. Japanese mythology A to Z (2nd ed.). — Pp. 1-138. (Chelsea House Publ., UK branch.)
RUDNICK, D. A., V. CHAN & V. H. RESH, 2005. Morphology and impacts of the burrows of the Chinese mitten crab, *Eriocheir sinensis* H. Milne Edwards (Decapoda, Grapsoidea), in south San Francisco Bay, California, USA. — Crustaceana, **78**: 787-807.
SACHS VON LÖWENHEIM, P. J., 1665. FAMMAPOAOFIA sive Gammarorum vulgo Cancrorum consideratio physico-philologico-historico-medico-chymica. — Pp. 1-21, 1-961. (Frankfurti et Lipsiae.)

SAHOO, P. K., HEMAPRASANTH, B. KAR, S. K. GARNAYAK & J. MOHANTY, 2012. Mixed infection of *Argulus japonicus* and *Argulus siamensis* (Branchiura, Argulidae) in carps (Pisces, Cyprinidae): loss estimation and a comparative invasive pattern study. — Crustaceana, **85**: 1449-1462.

SANOAMUANG, L. & H. J. DUMONT, 2000. Fairy shrimp: a delicacy in northeast Thailand. — Anostraca News, **8**: 3.

SIGURJÓNSSON, J. & G. A. VÍKINGSSON, 1997. Seasonal abundance of and estimated food consumption by cetaceans in Icelandic and adjacent waters. — J. northwest Atl. Fish. Sci., **22**: 271-287.

SOBRINHO-GONÇALVES, L., M. T. MOITA, S. GARRIDO & M. E. CUNHA, 2013. Environmental forcing on the interactions of plankton communities across a continental shelf in the Eastern Atlantic Upwelling System. — Hydrobiologia, **713**: 167-182.

SOHN, W. M., J. S. RYU, D. Y. MIN, H. O. SONG, H. J. RIM, Y. VONGHACHACK, D. BOUAKHASITH & V. BANOUVONG, 2009. *Indochinamon ou* (Crustacea: Potamidae) as a new second intermediate host for *Paragonimus harinasutai* in Luang Prabang Province, Lao PDR. — Korean J. Parasitol., **47**: 25-29.

SOMMER, U. & H. STIBOR, 2002. Copepoda – Cladocera – Tunicata: the role of three major mesozooplankton groups in pelagic food webs. — Ecol. Res., **17**: 161-174.

STREFTARIS, N. & A. ZENETOS, 2006. Alien marine species in the Mediterranean — the 100 'worst invasives' and their impact. — Medit. mar. Sci., **7**: 87-118.

TEKANOVA, E. V., 2012. The contribution of primary production to organic carbon content in Lake Onego. — Inland Wat. Biol., **5**: 328-332.

THÉODORIDÈS, J., 1980. Considerations on the medical use of marine invertebrates. — In: M. SEARS & D. MERRIMAN (eds.), Oceanography: the past, pp. 734-749. (Springer, New York, NY.)

THOMÉ, J. P., C. JEUNIAUX & M. WELTROWSKI, 1997. Application of chitosan for the elimination of organochlorine xenobiotics from wastewater. — In: M. F. A. GOOSEN (ed.), Applications of chitin and chitosan, pp. 309-331. (Technomic Publishing Company Inc., Lancaster.)

THOMPSON, D'ARCY W., 1900. XXII. — The emblem of the crab in relation to the sign Cancer. — Trans. Roy. Soc. Edinb., **39**: 603-611.

TINDALL, K. V. & K. FOTHERGILL, 2012. Review of a new pest of rice, tadpole shrimp (Notostraca: Triopsidae), in the midsouthern United States and a winter scouting method of rice fields for preplanting detection. — J. integr. Pest Managmt, **3**: B1-B5.

TOU, J. C., J. JACZYNSKI & Y. C. CHEN, 2007. Krill for human consumption: nutritional value and potential health benefits. — Nutrit. Rev., **65**: 63-77.

TROTTIER, O., D. WALKER & A. G. JEFFS, 2012. Impact of the parasitic pea crab *Pinnotheres novaezelandiae* on aquacultured New Zealand green-lipped mussels, *Perna canaliculus*. — Aquaculture, **344-349**: 23-28.

UITZ, J., D. STRAMSKI, B. GENTILI, F. D'ORTENZIO & H. CLAUSTRE, 2012. Estimates of phytoplankton class-specific and total primary production in the Mediterranean Sea from satellite ocean color observations. — Global biogeochem. Cycl., **26**: GB2024.

USDA, 2013. National Nutrient Database for Standard Reference, Release 25. — Nutrient Data Laboratory Home Page, http://www.ars.usda.gov/ba/bhnrc/ndl

WHA [WORLD HEALTH ASSEMBLY], 1986. Elimination of dracunculiasis: resolution of the 39th World Health Assembly. — Resolution no. **39.21**.

WHITE, B., 1945. Whale-hunting, the barnacle goose, and the date of the "Ancrene Riwle." Three notes on Old and Middle English. — Modern Language Rev., **40**: 205-207.

WHO [WORLD HEALTH ORGANIZATION], 1995. Control of foodborne trematode infections: report of a WHO study group. — Pp. 1-157. (WHO, Geneva.)

— —, 2013. Monthly report on dracunculiasis cases, January-February 2013. — Weekly epidemiol. Rec., **14** (88): 151-152.

YASUMURA, D., Y. OSHIMA & T. YASUMOTO, 1986. Tetrodotoxin and paralytic shellfish toxins in Philippine crabs. — Agric. biol. Chem., **50**: 593-598.

YEO, D. C. J., S. T. AHYONG, D. M. LODGE, P. K. L. NG, T. NARUSE & D. J. W. LANE, 2010. Semisubmersible oil platforms: understudied and potentially major vectors of biofouling-mediated invasions. — Biofouling, **26**: 179-186.

YILMAZ, E., 2004. Chitosan: a versatile biomaterial. — Adv. exp. Med. Biol., **553**: 59-68.

YU, C. F., P. H.-F. YU, P. L. CHAN, Q. YAN & P. K. WONG, 2004. Two novel species of tetrodotoxin-producing bacteria isolated from toxic marine puffer fishes. — Toxicon, **44**: 641-647.

CHAPTER 27B

CRUSTACEANS IN ART[1])

BY

GUY CHARMANTIER

Contents. – **Prologue and summary. Introduction. Chronology** – Prehistory – Ancient times – Middle Ages – Renaissance – Dutch Golden Age – The Enlightenment – More modern times. **Systematics. Symbolism** – Description – Mythology and religion – Allegories of water – Zodiacal signs: astrology – Symbolisms in the 17th century Dutch and Flemish art. **Portraits. Genre painting. Still life. Inconstancy. Conclusions. Post-script and acknowledgements. Appendix. Bibliography.**

PROLOGUE AND SUMMARY

Crustaceans have been represented by artists for their aesthetic or symbolic interest for 35 centuries or more. This review deals with the chronology, systematics, and symbolism of their artistic representations. The earliest known presence of crustaceans in art can be traced back to ancient Egypt and Assyria. Crustaceans were often present in Roman art from the 1st century BC to the 4th century AD, but they were seldom represented during the medieval period. Their occurrence in art increased from the 15th century and peaked in the 17th century Dutch and Flemish painting. Crustaceans have since then been moderately present in art (except during the Impressionist period, but they were one of the favourite animal subjects of Picasso) and in decorative arts. Mostly big, frequently captured, edible decapods such as crabs, lobsters, shrimp, and crayfish were pictured in art. In Roman times, crustaceans were represented as mythological symbols or as members of the marine fauna. Their use as astrological signs was frequent in the medieval period, and they often were symbolic parts of allegories of water or of mythological or religious scenes during the Renaissance. In 17th century Dutch and Flemish art, crustaceans were often an

---

[1]) Original section of Chapter 27, the manuscript for which was completed by the author in December 2010. Final additions March 2014.

element of genre paintings and still lifes either as part of the edible marine fauna in fish markets and kitchens, or as one of the components of food or banquet still lifes. If small and humble species indicate poverty, then crustaceans most often symbolized lavish food, feast, sensuality, luxury, and wealth, i.e., affluence. They have also been used as symbols of inconstancy.

INTRODUCTION

Since Palaeolithic times, over 35 000 years ago, human beings have created images of animals. Their representations thus predate the scientific interest in animal life that can be traced back to Aristotle's work (4th century BC), and the motivations behind them have been analysed in several studies (see for instance Cox & Povey, 1995: 11): "In all aspects of life, ..., animals had their part to play in the fulfilment of human needs, and have been represented by artists in commensurate fashion... Those themes which occur and recur throughout the history of animal representation are ultimately reducible to those most fundamental of human concerns: birth, sex, food, and death — those concerns which, in short, humans share with animals". To this list of preoccupations related to the human biological cycle, we can also add the aesthetic appeal of many animals to the viewer and potential artist. For these reasons, and also because they were more easily available, vertebrates, particularly mammals, birds, and reptiles have been more often represented than invertebrates. Among the latter, insects, aerial and beautiful, like butterflies, are frequently found in works of art. In contrast, crustaceans have been much less portrayed in art. Being mostly represented by small or medium-sized aquatic species, often difficult to capture or to observe, they have arisen less curiosity than terrestrial species, or than an aquatic group like fish (Jackson, 2012), which combines abundance of species, shape, and colour diversity, as well as being widely known for providing a large supply of food.

Even if crustaceans have generally triggered a comparatively modest interest among artists, their representations are numerous. The objective of this article is to offer an overview of the presence of crustaceans in art. This is a vast pursuit, which should be followed by more focused studies on selected works of art, and to which limits have voluntarily been set. First, artistic and scientific representations have to be distinguished. Until the Renaissance, separating one approach from the other is difficult. Even if precise representations of crustaceans, or of animals in general, can be found before that time, they are often linked with some sort of symbolism and it is difficult to describe them as scientific in the present meaning of the term. However, the scientific Renaissance, which followed the artistic Renaissance by approximately a century, was based on a renewed interest in nature, a consequence itself of overseas explorations and of the discovery of new parts of the earth. Direct observation of animals by scholars like Guillaume Rondelet (1507-1556), Ippolito Salviani (1514-1572), Conrad Gesner (1516-1565), Pierre Belon (1517-1564), and Ulysse Aldrovandi (1522-1605) lead to images whose main objective was to describe an animal (Grmek & Guinot, 1965a). This tendency increased from the 16th into the following centuries, particularly in the late 18th, e.g., the works of Linnaeus, and the 19th centuries during which the goal and interest of scientific description emerged with increasing sharpness. The resulting drawings and paintings, primarily aimed at description, will not be included in the present study. However, it is worth noting that, even with a descriptive aim, the advent of accurate observation and drawing was rather slow, as well shown by the careful analysis of mollusc and crab illustrations in pre-Linnaean times (Allmon, 2007). Another aspect of the use of crustaceans in illustrations that is not included in this study concerns postage stamps, a subject addressed by Holthuis (1967), who counted approximately 40 species, mostly decapods, used for that purpose. This number has increased later up to approximately 60 species (Monod & Laubier, 1996) then to 318 taxa, species or genus, identified from 1089 stamps by Omori & Holthuis (2000). The third limitation of the present article is that it is based mainly on western art, viewed in western sites and museums. Only brief mention will be made of crustaceans in art from other parts of the world, a topic that has been seldom explored (Harada, 1993).

To our knowledge, only a very limited number of references is available on this topic, a brief mention in a chapter on "Les Crustacés et l'homme" (Monod & Laubier, 1996) and a few articles devoted to particular aspects of the subject (Grmek & Guinot, 1965a, b; Huard & Guinot, 1965a, b; Chater, 1988; Holthuis, 1991). An abstract of the present article has been issued at the Second European Crustacean Conference held in Liège, Belgium (Charmantier, 1996). In this present overview, the questions of when (chronology), which (systematics) and why (symbolism) crustaceans were depicted in art will be addressed.

CHRONOLOGY

Prehistory

Palaeolithic images of animals represented almost exclusively terrestrial and aerial vertebrates (the salmons of the Magdalenian engravings of the Lorthet cave in France being an exception), with a large majority of mammals, and, to our knowledge, no crustaceans. The aquatic, especially marine, fauna was most probably first represented by fish, as exemplified by mudstone palettes in the form of a fish from early Egypt (4000-3600 BC, predynastic Nagada I, British Museum, London).

Ancient times

The oldest figure of a crustacean known at the time of this writing seems to concern spiny lobsters (*Panulirus penicillatus*?) in an Egyptian painting of the 18th Dynasty (1580-1350 BC) (Monod & Laubier, 1996). Several centuries later, crabs can be found among other marine animals in a scene depicting the transport of wood for the building of the palace of Sargon II, king of Assyria, at Khorsabad. The large alabaster reliefs, which covered the base of the brick walls, are dated from approximately 706 BC (see table I for dates, site of presentation/storage, and, when available, of the artist's name, for all works of art cited in text).

TABLE I
Choice of works of art with representations of crustaceans; MS, Manuscript

Artist/Site/MS	Date of work	Title	Support	Crustacean(s)	Location
Egypt	1580-1350 BC		Wall painting	1 *Panulirus penicillatus*	
Palace of Khorsabad (king Sargon II)	713-706 BC	Maritime Landscape: Transport of Cedar Wood from the Lebanon	Gypseous alabaster relief	2 or 3 crabs	Musée du Louvre, Paris
Acragas / Agrigente, Metya, Himera, Kos	5th-1st cent. BC	Diverse crustaceans	Silver coins	Potamid crabs, marine crabs, majid crabs, spiny and homarid lobsters, caridean shrimp, crayfish	Multiple (museums, private collections)
Pompei	1st cent. BC or AD	Marine Fauna	Mosaic	1 (penaeid?) shrimp, 1 spiny lobster	Museo Archeologico Nazionale, Naples
Pompei	1st cent. AD (before 79)	Cat Stealing a Bird	Mosaic	2 (penaeid?) shrimps	Museo Archeologico Nazionale, Naples
Pompei, Faun House	1st cent. AD (before 79)	Marine Fauna	Mosaic	1 spiny lobster	Museo Archeologico Nazionale, Naples
Pompei, Casa dei Vettii	1st cent. AD (before 79)	Cupid Riding the Crab	Wall painting	1 crab	Pompei
Herculaneum	1st cent. AD (before 79)	Shells, Lobster, Vase and Bird	Fresco	1 spiny lobster	Museo Archeologico Nazionale, Naples
Arles, France	Early 2nd cent. AD	Oil Lamp with Crab	Terracotta	1 crab (*Pachygrapsus marmoratus*?)	Musée de l'Arles Antique
Populonia, Triclinium	c. 100 AD	Marine Fauna	Mosaic	1 spiny lobster	British Museum, London
Carthage	1st-2nd cent. AD	Basket of Fish	Mosaic	1 spiny lobster	British Museum, London
Heraclitus, after Sosos of Pergamum	2nd cent. AD	The Unswept Floor	Mosaic	Fragments of spiny lobsters and shrimps (abdomen, legs)	Museo Laterano, Rome
—	3rd-2nd cent. BC	The Unswept Floor	Mosaic	Fragments of spiny lobster (cephalothorax)	Musei Vaticani, Rome

TABLE I
(Continued)

Artist/Site/MS	Date of work	Title	Support	Crustacean(s)	Location
Villa at Gargaresc, Tripolitania	150-225 AD	Crab	Mosaic	1 crab	British Museum, London
House of Venus	200 AD	Marine Fauna	Mosaic	Shrimps, spiny lobsters	Mactar
Houses at Hadrumete	2nd-3rd cent. AD	Fishing Scene	Mosaic	1 spiny lobster	Sousse Museum
House at Djemila	3rd cent. AD	Venus' Wash	Mosaic	1 crab	Archeological Museum, Djemila
Thermae of Themetra, Frigidarium	3rd cent. AD	The God Ocean	Mosaic	2 crustacean claws, several shrimps, spiny lobsters	Sousse Museum
Rome	4th cent. AD	Floor Decorated with Xenia	Mosaic	1 spiny lobster	Musei Vaticani, Rome
Rome, catacombs	Early cent. AD	Christian jewels	Stone rings, cameos	Elongated crustaceans (shrimp, lobster, crayfish?)	Cabinet des Médailles, Paris
Mimbres Valley, New Mexico	10th-12th cent.	Mogollon Pueblo Pot with "Water Bugs"		Ostracods?	
Kalila and Danuna, by Bidpai (Syria)	1200-1220	Heron Tryng to Carry a Crab to a Safe Site	Illumination	1 crab (10 legs)	Bibliothèque Nationale, Paris
Madeleine Church	c. 1130	Zodiac Signs on the Tympanum of the Narthex Interior Portal: the Apostles' Mission	Sculpture	1 crayfish (8 legs)	Vézelay, France
St Lazarus Cathedral	1120-1140	Zodiac Signs on the Tympanum of the Central Porch: Judgment Day	Sculpture	1 crayfish (8 legs)	Autun, France
Codex of Hildegarde of Bingen	c. 1180-1190	The Seasons	Illumination	2 half-visible crayfish or lobsters	State Library, Lucques, Italy
St Mary's Church	Date unknown betw. 14th-19th cent.	"Sol in Cancro", Zodiac Sign of Cancer	Stained glass	1 isopod (*Porcellio scaber*?)	Shrewsbury, U.K.

TABLE I
(Continued)

Artist/Site/MS	Date of work	Title	Support	Crustacean(s)	Location
St Austremoine Abbey	12th cent.	Zodiac Signs on the Apse: Cancer	Sculpture	1 crab (10 legs, 1 tail)	Issoire, France
Amiens Cathedral	1225-1236	Base of the Central Porch, Four-Leafed Patterns: Month of June	Sculpture	1 crab (10 legs, *Carcinus*?)	Amiens, France
Anthology of Iskandar at Chiraz, Persia	1410-11	Astrological Subjects	Illumination	1 crab (6 legs)	British Museum, London
Master of Catherine de Clèves (c. 1435-1460)	15th cent.	Hours of Catherine de Clèves: Saint Ambrosius	Illumination	1 crab (12 legs)	Pierpont Morgan Library, New York, NY
Limbourg Brothers (c. 1375/85-c. 1416)	1408-1416	Rich Hours of the Duke of Berry: Month of June	Illumination	1 crab (8 legs)	Musée Condé, Chantilly
—	1408-1416	Rich Hours of the Duke of Berry: the Anatomical Man	Illumination	2 crabs (8 legs)	Musée Condé, Chantilly
Barthélemy d'Eyck Enguerrand Quarton (1419-1466)	1440-1450	Book of Hours: Lobster	Illumination	1 *Homarus gammarus* (10 legs)	Pierpont Morgan Library, New York, NY
Fastolf Master	c. 1440-1450	Book of Hours: Month of June	Illumination	1 crayfish (10 legs)	Bodleian Library, Oxford
Giovani di Francia	1420-1473/85	Last Supper of Christ	Wall painting	Crayfish	Several churches in Alps of northern Italy (Vito di Cadore, Feltre, Villapiana, Porcen, Servo, Sospirolo, Lentiai, Cusighe, Cenceníghe, Taibon, Vallada Agordina, Belluno, Castion, Cadore, Arsie, ...)
Iseppo da Cividal	?				
Giovanni da Mel	1480-c. 1549				
Paris Bordon	1500-1571				
Antonio Vassilacchi	1556-1629				
Cesare Vecellio	1521-1601				
Alessandro Maganza	1556->1630				
Tomaso Vecellio	1587-1629				
Francesco Frigimelica	1560->1649				
Giovanbattista Volpato	1633-1706				
Girolamo Pellegrini	1624->1700				
Anonymous	1492	Mass-book of Jean de Foix	Illumination	1 crayfish (8 legs)	Bibliothèque Nationale, Paris

TABLE I
(Continued)

Artist/Site/MS	Date of work	Title	Support	Crustacean(s)	Location
Albrecht Dürer (1471-1528)	1494	De Predestinatione, Fool in crab's walk	Woodcut	1 crayfish	Germanisches National Museum, Nürnberg
—	1495	Crab	Quill and brush on paper	1 crab (*Eriphia* sp.)	Museum Boymans-Van Beuningen, Rotterdam
—	1495	Lobster	Quill and brush on paper	1 *Homarus gammarus*	Staatliche Museen Preussischer Kulturbesitz, Berlin
Alart Du Hameel (c. 1449-c. 1506)	Early 16th cent.	Saint Christopher	Burin engraving	1 crayfish or lobster	Musée du Louvre, Paris
Johann von Kaub (?)	c. 1500	Hontus sanitatis	Engraving	1 homarid lobster	Bibliothèque Nationale, Paris
Hieronymus van Aeken Bosch (c. 1450-1516)	c. 1503-05	Triptych of The Garden of Delights. Central section	Oil on wood	1 branchiopod-like animal, lobster shell with people	Museum of Prado, Madrid
Frans Crabbe van Espleghem (c. 1480-1553)	c. 1522-25	The Virgin and Child Sitting in a Yard	Burin engraving	1 crab as signature	Musée du Louvre, Paris
—	Mid 16th cent.	Christ in Front of Caiaphas	Etching	1 crab as signature	Musée du Louvre, Paris
—	Mid 16th cent.	The Entombment of Christ	Etching	1 crayfish as signature	Musée du Louvre, Paris
—	Mid 16th cent.	The Resurrection	Etching	1 crayfish as signature	Musée du Louvre, Paris
Bernard Palissy (c. 1510/1589-90)	Late 16th-early 17th cent.	Plate with Snake	Ceramic	2 crayfish	Musée du Louvre, Paris
—	Late 16th-early 17th cent.	Plate with Men and Women	Ceramic	1 crayfish	Musée du Louvre, Paris
—	Late 16th-early 17th cent.	Water	Ceramic	1 crayfish	Musée du Louvre, Paris
Giorgio Vasari (1511-1574)	1556-59	Elements: Allegory of Water	Wall painting	1 spiny lobster	Come 1st apartments, Florence

TABLE I
(Continued)

Artist/Site/MS	Date of work	Title	Support	Crustacean(s)	Location
Giuseppe Arcimboldo (1527/30-1593)	1566	Aqua	Oil on canvas	1 hoplocarid, 1 shrimp, 1 crayfish, 1 *Cancer pagurus*	Kunsthistorisches Museum, Vienna
Vincenzo Campi (1536-1591)	Late 16th cent.	Christ in the House of Mary and Martha	Oil on wood	1 *Homarus gammarus*	Galleria Estense, Modena
Pieter Pourbus (1523/24-1584)	16th cent.	Triptych of The Bruges Fishmongers. Center section: Calling of Four Disciples and Miraculous Draught	Oil on wood	1 crab	Musée des Beaux-Arts, Brussels
Georg Hoefnagel (1542-1600)	1592	Archetypa Studiaque Patris	Etching	1 crayfish	Private collection
Pieter Bruegel the Elder (c. 1525-1569)	1556	Big Fish Eat Little Fish	Ink drawing, brush and pen	1 crayfish, 1 crab	Graphische Sammlung Albertina, Vienna
—	1562	The Fall of Rebel Angels	Oil on wood	1 shrimp with fish tail, crab claws on a vial	Musée des Beaux-Arts, Brussels
Pieter van der Heyden (c. 1530-1575)	1557	Big Fish Eat Little Fish (see Bruegel, 1556)	Engraving	1 crayfish, 1 crab	Bibliothèque Royale Albert 1er, Brussels
Jan Bruegel (1568-1625)	c. 1600	Water	Oil on canvas	1 hoplocarid, 1 spiny lobster, 2 *Homarus gammarus*, 3 majid crabs, 3 crabs	Musée du Louvre, Paris
—	1618	Taste	Oil on wood	1 *Homarus gammarus*	Museum of Prado, Madrid
Georg Flegel (1566-1638)	Early 17th cent.	Preparation of Meal	Oil on canvas	1 crayfish	Kunstmuseum, Basel
Jacob van Es (c. 1590-1666)	17th cent.	Still Life with Squirrel	Oil on canvas	1 *Homarus gammarus*	Musée des Beaux-Arts, Rennes
Balthasar van der Ast (1593/94-1657)	1620/21	Still Life with Fruits and Flowers	Oil on panel	1 hermit crab	Rijksmuseum, Amsterdam

TABLE I
(Continued)

Artist/Site/MS	Date of work	Title	Support	Crustacean(s)	Location
—	1622	Still Life with Fruit Basket	Oil on wood	2 crayfish, 2 shrimps	North Carolina Art Museum, Raleigh, NC
—	1623	Still Life with Fruits, Flowers and Shells	Oil on wood	1 hermit crab	Musée des Beaux-Arts, Lille
Osias Beert (c. 1580-1624)	1622	Still Life	Oil on canvas	1 *Homarus gammarus*	Musée des Beaux-Arts, Brussels
Abraham van Beyeren (1620/21-1690)	Mid/late 17th cent.	Still Life with Lobster	Oil on canvas	1 *Homarus gammarus*	Kunsthaus, Zürich
—	Mid/late 17th cent.	Still Life	Oil on canvas	1 *Cancer pagurus*	Rijksmuseum, Amsterdam
—	Mid/late 17th cent.	Banquet Still Life	Oil on canvas	1 *Cancer pagurus*	Mauritshuis, The Hague
—	Mid/late 17th cent.	Still Life with Fish and Stone-bottle	Oil on canvas	1 *Cancer pagurus*	Musée des Beaux-Arts, Brussels
—	Mid/late 17th cent.	Still Life with Chicken and Lobster	Oil on canvas	1 *Homarus gammarus*	Ashmolean Museum, Oxford
—	Mid/late 17th cent.	Still Life with Lobster	Oil on canvas	1 *Homarus gammarus*	Ashmolean Museum, Oxford
—	Mid/late 17th cent.	Still Life with Fish	Oil on canvas	1 *Homarus gammarus*, 1 *Cancer pagurus*	Kunst Museum, Copenhagen
—	Mid/late 17th cent.	Still Life with Fish	Oil on canvas	1 *Cancer pagurus*	Musée du Louvre, Paris
—	1652	Still Life with Lobster	Oil on canvas	2 shrimps, 1 *Homarus gammarus*	Alte Pinakothek, Munich
—	1654	Still Life with Fruits and Lobster	Oil on wood	1 shrimp, 1 *Homarus gammarus*	Museum Boymans-Van Beuningen, Rotterdam
—	post-1654	Still Life with Fruits and Lobster	Oil on wood	1 shrimp, 1 *Homarus gammarus*	Historisches Museum, Frankfurt-am-Main

TABLE I
(Continued)

Artist/Site/MS	Date of work	Title	Support	Crustacean(s)	Location
—	post-1654	Still Life with Fruits and Lobster	Oil on wood	1 shrimp, 1 *Homarus gammarus*	Museum, Leipzig
—	1667	Sumptuous Still Life	Oil on canvas	1 *Homarus gammarus*	Los Angeles County Museum of Art, CA
Pieter Claesz (1597/98-1660/61)	1641	Still Life with Ewer and Lobster	Oil on wood	1 *Homarus gammarus*	Private collection
—	1644	Still Life with Glass of White Wine and Crab	Oil on wood	1 *Cancer pagurus*	Musée des Beaux-Arts, Strasbourg
Willem Claesz. Heda (1593/94-1680/82)	1648	Breakfast with a Crab	Oil on canvas	1 *Cancer pagurus*	Hermitage Museum, Saint Petersburg
—	1658	Still Life with Pie, Silver Ewer and Crab	Oil on canvas	1 *Cancer pagurus*	Frans Hals Museum, Haarlem
Jacob Van Es (1590-1666)	17th cent.	Still Life with Squirrel	Oil on canvas	1 *Homarus gammarus*	Musée des Beaux-Arts, Rennes
Frans Francken II (1581-1642)	c. 1620	Abdication of Emperor Charles V, Brussels, 25 October 1555	Oil on canvas	1 *Homarus gammarus*	Rijksmuseum, Amsterdam
Pieter Gallis (1633-1697)	17th cent.	Still Life	Oil on canvas	1 *Homarus gammarus*	Ashmolean Museum, Oxford
Nicolas van Gelder (1623/36-1677)	1667	Still Life with Blue Tablecloth	Oil on canvas	1 *Homarus gammarus*, 4 shrimps	Private collection
Frans Hals (1580/85-1666)	1624-1627	Banquet of the Officers of the St George Civic Guard	Oil on canvas	1 *Cancer pagurus*	Frans Hals Museum, Haarlem
Jan Davidz. de Heem (1606-1683/84)	1634	Still Life with Lobster and Nautilus Cup	Oil on canvas	1 *Homarus gammarus*	Staatsgalerie, Stuttgart
—	1648	Still Life with Lobster	Oil on canvas	1 *Homarus gammarus*, 2 crabs, shrimps	Museum Boymans-Van Beuningen, Rotterdam
—	1648-49	Still Life with Fruits and Lobster	Oil on canvas	2 shrimps, 1 *Homarus gammarus*, 2 crayfish	Staatliche Museen, Berlin

TABLE I
(Continued)

Artist/Site/MS	Date of work	Title	Support	Crustacean(s)	Location
—	17th cent.	Still Life with Fruits and Lobster	Oil on canvas	1 *Homarus gammarus*	Musée des Beaux-Arts, Brussels
—	17th cent.	Still Life with Fruits and Lobster	Oil on canvas	1 *Homarus gammarus*	Museum of Art, Toledo, OH
—	17th cent.	Still Life	Oil on canvas	1 shrimp	Musée du Louvre, Paris
Willem van Aelst (1625/6-1686?)	1661?	Breakfast with Nautilus and Crabs	Oil on canvas	2 crabs	Staatliches Museum, Schwerin
Cornelis de Heem (1631-1695)	1659	Still Life	Oil on wood	2 shrimps, 2 crayfish	Musée Fabre, Montpellier
Willem Kalf (1619-1693)	c. 1653	Still Life with the Drinking Horn of the Saint Sebastian Archers' Guild, Lobster and Glasses	Oil on canvas	1 *Homarus gammarus*	National Gallery, London
Jan van Kessel (1626-1679)	1664-66	Europe	Oil on copper	1 *Homarus gammarus*, 5 crabs (incl. 1 ovigerous majid)	Alte Pinakothek, Munich
	1664-66	Africa	Oil on copper	5 crabs (incl. 2 majids), 1 hermit crab	Alte Pinakothek, Munich
	1664-66	Asia	Oil on copper	1 *Homarus gammarus*, 2 hermit crabs (?)	Alte Pinakothek, Munich
—	17th cent.	Still Life	Oil on copper	4 terrestrial isopods (1 *Armadillidium*, 1 *Philoscia*)	Ashmolean Museum, Oxford
Adriaen van Nieulandt (1587-1658)	1616	Kitchen Scene	Oil on canvas	2 *Homarus gammarus*, 1 *Cancer pagurus*	Herzog-Anton-Ulrich Museum, Braunschweig
Adriaen van Ostade (1610-1685)	1672	The Fishwife	Oil on canvas	1 *Cancer pagurus*	Rijksmuseum, Amsterdam
Clara Peeters (1594-?)	17th cent.	Still Life	Oil on canvas	4 shrimps	Ashmolean Museum, Oxford
—	17th cent.	Still Life	Oil on canvas	1 *Homarus gammarus*, 1 *Cancer pagurus*, shrimps, small crabs	Private collection

TABLE I
(Continued)

Artist/Site/MS	Date of work	Title	Support	Crustacean(s)	Location
Pieter de Ring (c. 1615/1660)	17th cent.	Still Life with a Golden Goblet	Oil on canvas	1 cirripede on lobster, 1 *Homarus gammarus*, 1 crab	Rijksmuseum, Amsterdam
Peter Paul Rubens (1577-1640), Jan Bruegel (1568-1625)	1600	Banquet of Acheloos	Oil on canvas	1 *Homarus gammarus*	Metropolitan Museum of Art, New York, NY
Peter Paul Rubens (1577-1640)	1619/20	The Miraculous Draught of Fishes	Black chalk, oil and pen on paper	2 crabs	National Gallery, London
Salomon van Ruysdael (1600/03-1670)	1666	Fish	Oil on canvas	1 crab	H. Wetzlar collection, Amsterdam
Pieter Cornelisz van Ryck (1568-1628?)	1604	Kitchen Scene	Oil on canvas	Several *Homarus gammarus* and crabs	Herzog-Anton-Ulrich Museum, Braunschweig
Dieck Sauts (17th cent.)	17th cent.	Still Life with Oyster and Crabs	Oil on canvas	2 crabs	Ashmolean Museum, Oxford
Michiel Simons (?-1673)	1649	Still Life with Bowl of Strawberries and Crab	Oil on wood	1 crab, 3 shrimps	St Luc Gallery, Vienna
Frans Snyders (1579-1657), Cornelis de Vos (1584-1651)	1st half 17th cent.	The Larder	Oil on canvas	1 *Homarus gammarus*	Musée des Beaux-Arts, Brussels
Frans Snyders (1579-1657)	c. 1610	Kitchen-maid with Provisions	Oil on canvas	1 *Homarus gammarus*	Wallraf-Richartz Museum, Cologne
—	1st half 17th cent.	The Fishmongers	Oil on canvas	5 crabs (incl. 1 *Cancer pagurus*), 2 *Homarus gammarus*	Musée du Louvre, Paris
—	1st half 17th cent.	Still Life	Oil on canvas	1 *Homarus gammarus*	Musée de Valenciennes
—	1st half 17th cent.	Fish Stall	Oil on canvas	1 *Homarus gammarus*, 3 crabs (*Carcinus maenas*)	Musée des Beaux-Arts, Brussels

TABLE I
(Continued)

Artist/Site/MS	Date of work	Title	Support	Crustacean(s)	Location
—	1st half 17th cent.	Still Life with Roe Deer	Oil on canvas	1 *Homarus gammarus*	Musée des Beaux-Arts, Brussels
—	1st half 17th cent.	Still Life with Lobster and Fruits	Oil on canvas	1 *Homarus gammarus*	Staatliche Museen, Berlin
Adriaen van Utrecht (1599-1652)	1st half 17th cent.	Still Life with Lobster	Oil on canvas	1 *Homarus gammarus*	Ashmolean Museum, Oxford
Constantin Verhout (1638-1667)	17th cent.	Still Life	Oil on canvas	1 crab, 3 shrimps	Ashmolean Museum, Oxford
Joris van Son (1623-1667)	1660	Still Life with Lobster	Oil on canvas	1 *Homarus gammarus*	Private collections
Jeremias van Winghe (1578-1645)	1615	Kitchen Still Life	Oil on wood	1 *Homarus gammarus*, approx. 10 crabs	Sotheby's Monaco, 1986
Jan Ficht (1611-1661)	17th cent.	Still Life with Lobster	Oil on canvas	1 *Homarus gammarus*	Museum Boymans-Van Beuningen, Rotterdam
Otto Marseus van Schrieck (c. 1619-1678)	1644-47?	Underwood with Convolvulus, Snake, Frog and Crab	Oil on canvas	Crabs	Musée des Beaux-Arts, Quimper
Christiaan Luyckx (1623-1653)	1640s	Still Life	Oil on canvas	1 *Homarus gammarus*	Private collection
Abraham Susenier (1620-1666)	1666	Still Life	Oil on canvas	1 *Homarus gammarus*	Private collection
Abraham van Calraet (1642-1722)	17th cent.	Man, Woman and Child	Oil on canvas	1 *Homarus gammarus*	Dordrechts Museum, Dordrecht, Netherlands
Pieter van Overschee (17th cent.)	1660?	Still Life with Lobster	Oil on canvas	1 *Homarus gammarus*	Private collection
Andries de Koninck (?-1659)	17th cent.	Still Life of Fruit and Lobster with Parrot	Oil on canvas	1 *Homarus gammarus*	Private collection
Dutch Anonymous	17th cent.	Back to the Harbour	Tapestry	1 *Homarus gammarus*	Castle of Cheverny, France

TABLE I
(Continued)

Artist/Site/MS	Date of work	Title	Support	Crustacean(s)	Location
Abraham Janssens (1567/76-1632) [attributed to]	c. 1617	Inconstancy	Oil on canvas	1 *Homarus gammarus*	Statens Museum for Kunst, Copenhagen
Jacques Linard (1597-1647)	1634	Still Life with Nautilus Cup and Lobster	Oil on canvas	1 *Homarus gammarus*	Musée du Louvre, Paris
Willem Jansz Blaeu (1571-1638)	17th cent.	Celestial Globe: Cancer	Globe	1 crab	Museu de Marinha, Lisbon
Giovan Battista Recco (1615-1660)	1653	Lobster and seafood	Oil on canvas	1 spiny lobster, 1 slipper lobster (*Scyllarus* sp.), 1 crab	Private collection
Giuseppe Recco (1634-1695)	c. 1680	The Riches of the Sea with Neptune and two Sea Nymphs	Oil on canvas	2 spiny lobsters	Art Gallery of South Australia, Adelaide
Baltasar Gomez Figueira (1604-1674)	1645	Oranges, Onions, Fish and Crab	Oil on canvas	1 crab, 2 shrimps	Musée du Louvre, Paris
Bartolomé Esteban Murillo (1617/18-1682)	1645-1655	Young Beggar	Oil on canvas	Bits of shrimps	Musée du Louvre, Paris
Evaristo Baschenis (1617-1677)	17th cent.	Still Life with Musical Instruments	Oil on canvas	1 crayfish	Wallraf-Richartz Museum, Cologne
Giuseppe Recco (1634-1695), Luca Giordano (1634-1705)	c. 1680	The Riches of the Sea with Neptune and two Sea Nymphs	Oil on canvas	5 spiny lobsters	Art Gallery of South Australia, Adelaide
Giaccomo Ceruti (1698-1767)	1720-25	Still Life with Lobster and Crab	Oil on canvas	1 *Cancer pagurus*, 1 *Homarus gammarus*	
—	1736-38	Still Life with Lobster	Oil on canvas	1 *Homarus gammarus*	Coll. S. Lodi, Campione
—	1736-38	Still Life with Hunting Bag	Oil on canvas	1 *Homarus gammarus*	Staatliche Kunstammlungen, Kassel
Charles Le Brun (1619-1690)	1669	The Elements: Water	Tapestry	2 *Homarus gammarus*	Mobilier National, Paris

TABLE I
(Continued)

Artist/Site/MS	Date of work	Title	Support	Crustacean(s)	Location
Albert Eckhout (1610-1666), Georg Marcgraf (1610-1644)	1637-1644	Le Cheval Rayé (Anciennes Indes)	Tapestry	1 hoplocarid, 1 spiny lobster, 6 crabs	Grandmaster's Palace, Malta
Gobelin factory	1687-1730	Le Cheval Rayé (Nouvelles Indes)			Mobilier National, Paris
—	1740-1768	Le Combat des Animaux (Anc. Indes, Nouv. Indes)	Tapestry	1 crab	Mobilier National, Paris
—	—	Le Chasseur Indien (Anc. Indes, Nouv. Indes)	Tapestry	1 hoplocarid, 1 shrimp, 1 slipper lobster, 7 crabs/1 hoplocarid, 1 spiny lobster, 1 slipper lobster, 7 crabs	Mobilier National, Paris
—	—	Le Roi Porté (Anc. Indes) La Négresse portée (Nouv. Indes)	Tapestry	4 crabs 3 crabs	Mobilier National, Paris
Charles Collins (c. 1680-1738)	1738	Lobster on a Delft Dish	Oil on canvas	1 *Homarus gammarus*	Tate Gallery, London
Justus Juncker (1703-1767)	c. 1750	Still Life with Trout	Oil on canvas	Several crayfish	Wallraf-Richartz Museum, Köln
Daniel Chodowiecki (1726-1801)	1774	Elementarwerke für die Jugend ind ihre Freunde	Engraving	1 homarid lobster	Colonial Williamsburg, VA, Foundation
Ludwigsburg Factory, Germany (1756-1824)	1765-1775	Tureen	Porcelain	1 crayfish	Art Museum, St Louis, MO
Anne Vallayer-Coster (1744-1818)	1817	Vase, Lobsters, Fruits and Game	Oil on canvas	1 *Homarus gammarus*	Musée du Louvre, Paris
Eugène Delacroix (1798-1863)	1826	Still Life with Lobsters	Oil on canvas	2 *Homarus gammarus*	Musée du Louvre, Paris
James Ensor (1860-1949)	1888	Still Life with Fish and Shells	Oil on canvas	2 *Cancer pagurus*, 2 *Homarus gammarus*	The Art Institute, Chicago, IL

TABLE I
(Continued)

Artist/Site/MS	Date of work	Title	Support	Crustacean(s)	Location
Edouard Manet (1832-1883)	c. 1860-65	Shrimps and shell	Oil on canvas	Shrimps	Van Gogh Museum, Amsterdam
Vincent van Gogh (1853-1890)	1889	Crab on its Back	Oil on canvas	1 *Cancer pagurus*	British Museum, London
C. F. Hancock (1807-1881), Hancocks & Co., jewellers, Est. 1849	1853	Silver Crayfish Salt Cellar	Cast silver	1 crayfish	British Museum, London
French Service Rousseau re-issued by Joy & Levellié	Designed 1866 1903-1913	Plate with Crayfish	Earthenware	1 penaeid shrimp	British Museum, London
Filippo De Pisis (1896-1956)	1929	Still Life with Shrimps and Shells	Oil on canvas	Approx. 20 shrimps	Galleria Civica d'Arte Moderna e Cotemporana, Torino
Salvador Dali (1904-1989)	1936	Lobster Telephone	Mixed media, incl. steel, plaster, rubber, resin, paper	1 homarid lobster	Tate Modern Gallery, London
Jürg Kreienbühl (1932-2007)	1960s	Interior with Lobsters	Oil on canvas	5 lobsters, 3 shrimps, 4 other decapods	A. Blondel collection, MNHN, Paris
Paul Klee (1879-1940)	1940	Woodlouse in the Enclosure	Pastel	1 asellid isopod	Kunstmuseum, Bern
—	1940	Assel	Gouache	1 asellid isopod	Rolf Bürgi private collection, Bern
Max Ernst (1891-1976)	1926	The Crab	Oil on canvas	2 crabs	Musée d'Art Moderne, Brussels
Pablo Picasso (1881-1973)	1936	The Lobster	Sugar lift aquatint	1 spiny lobster	Victoria & Albert Museum, London
—	1941	Boy with a Lobster	Oil on canvas	1 spiny lobster	Musée Picasso, Paris
—	1948	Lobster and Bottle	Oil on canvas	1 homarid lobster	Private collection

TABLE I
(Continued)

Artist/Site/MS	Date of work	Title	Support	Crustacean(s)	Location
—	1962	Still Life with Cat and Lobster	Oil on canvas	1 homarid lobster, 2 other lobsters	Open-Air Museum, Hakone
—	1965	Cat and Lobster on a Beach	Oil on canvas	1 homarid lobster	Galerie Louise Leiris, Paris
Jean Hélion (1904-1987)	1975-76	Market with Lobsters	Oil on canvas	*Homarus gammarus*	Musée de Vannes
Peter Lörincz (1938-)	1995	Cathedral	Watercolour	14 peneid shrimps	Bibliothèque, Gignac
Gilles Dulis (contemp.)		Vélocité	Paint on glass	1 majid crab	?
Jeff Koons (1955-)	2003		Oil on canvas	1 homarid lobster	?
—	2003	Lobster	Polychromed aluminum	1 homarid lobster	Private collection. Exhib. Versailles 2008, Serpentine Gall. London 2009

Crustaceans were certainly represented in Greek art, particularly in "*xenia*"[2]), which had an influence on Roman art (see below). However, Greek works of art that dealt with animals represented mostly vertebrates, infrequently insects among the invertebrates (Bodson, 2008), but relatively few representations of crustaceans have reached us. A notable exception is found in Phoenician and Greek coins (Metya, Acragas / Agrigente, Himeria, Kos, etc.) from the 5th-1st centuries BC, on which different decapods, crabs, majid crabs, shrimp, spiny and homarid lobsters, and crayfish are illustrated (Delorme & Roux, 1987; Monod & Laubier, 1996; Charbonneau-Lassay, 2006; D. Guinot, H. Laufer, pers. comm.). Almost all of these crustaceans belong to marine species, except for the crayfish and the frequent representation of freshwater potamid crabs on coins from Acragas / Agrigente. The first historical occurrence of numerous artistic figures of crustaceans is noted in Roman times, straddling the shift from the pre-Christian to the Christian era. Over five centuries, extending from the end of the 2nd century BC to the 4th century AD, the marine fauna from the Mediterranean Sea becomes a regular fixture in Roman art. Fish, involving numerous species, are the most frequently represented in these works, along with molluscs, octopus, jellyfish, and crustaceans (mostly spiny lobsters, shrimps, and crabs). Such works have been found in all the territories controlled by the Romans, particularly in Italy (with the striking case seen in all the preserved works in Herculaneum and Pompei) and in North Africa. Crustaceans can be found on three main types of media. Mosaics, which are probably the best known among them (Charles-Picard, 1978), have usually retained the colours of the various species. Along with a depiction of crustaceans, viewed laterally or dorsally, they frequently stage a struggle between an octopus and a spiny lobster, a scene often found in Pompei (fig. 27B.1). Mural paintings also present scenes that sometimes include crustaceans. In association with fruits, flowers, bread, fish, ewers, platters, pitchers, they constitute *xenia*, resembling a category of painting later called **'still life'** (Bryson, 1995). According to Sterling (1985), the Greeks probably painted the first still lifes, particularly during the 3rd and 2nd centuries BC, but all of them have disappeared during the course of two millennia. They are only known through ancient descriptions. However, the style was transmitted to the Romans who used them to decorate their villas in Rome, Pompei, Herculaneum and in all the Roman empire from North Africa to the Rhine. Along with still lifes, some xenia depict everyday or symbolic scenes similar to the **'genre painting'** of later centuries (see below in section on Symbolism). Xenia appear mostly as mural paintings, but their style is also found in some mosaics. The crustacean species illustrated in the xenia are the same as in mosaics in general. In addition to the descriptive mosaics and the xenia, crustaceans have also been used by Romans to decorate objects, usually ordinary objects of a household such as plates, pitchers, oil lamps, etc. The oil lamp illustrated in fig. 27B.2 dates from the early 2nd century AD, and includes in its circular part a medallion circling a crab drawn in relief. It was found and probably manufactured near Arles, in southern France. The quadrangular shape of the cephalothorax and the aspect and proportions of this part of the body and of the pereiopods suggest that this crab was a representation of

---

[2]) A category of painting similar to "still life"; see further below.

Fig. 27B.1. Marine fauna, including a fight between an octopus and a spiny lobster. Note the shrimp at the upper part of the picture. Faun House, Pompei (before 79 AD). Soprintendenza Speciale per i Beni Archeologici di Napoli e Pompei. [Reproduced with permission.]

Pachygrapsus marmoratus, a grapsid common on the rocky parts of the shore which, in Roman times, was much closer to Arles than it is today.

Finally, crustaceans were more abundant in Roman art while relatively scarce during the Greek period. The curiosity of Greek philosophers for the world and their attempts at a rational, pre-scientific explanation of its operation might have influenced Grecian art. For instance, Aristotle's work on animal classification might have been reflected in the artistic depictions crafted by his fellow countrymen. However, though Greek pre-xenia images of biological objects have been found (Sterling, 1985), Greek art in essence was more humanistic than naturalistic. The Roman contribution to science was by comparison much weaker but their artistic rendering of animal life shows their naturalistic interest. Behind their reference to food gifts, the xenia also perhaps attest of the rationalism and epicurean

Fig. 27B.2. Oil lamp decorated with a crab, presumably *Pachygrapsus marmoratus*; early 2nd century AD. Potter mark on the back side. Inscription: LHOSCRI. Terracotta, beige paste, brown varnish; 10.9 × 7.9 × 3.1 cm. Musée de l'Arles Antique, France, inv. FAN.91.00.2044, © M. Lacanaud. [Reproduced with permission.]

hedonism that was underlying the contemplation, and soon-to-come consumption, of marine life (see on this subject Sterling, 1985: 9-15; Bryson, 1995: 16-59).

During approximately a millennium following the Roman era, the frequency of crustaceans decreased dramatically in western art, with only a few exceptions. Early Christian rings and other jewelry (from the 2nd century onwards) were ornamented with fish, the basic Christian animal symbol, few of them carrying an image of a crustacean (see Symbolism). The jewelry of the Merovingians, from the 5th-8th centuries in Western Europe, also sometimes are decorated with images of, or are shaped as, crustaceans that, while crudely represented, are nonetheless recognizable through their segmentation and appendages (Charbonneau-Lassay, 2006).

Until the 12th century and the foundation of European universities, the greatest part of the intellectual effort in western countries was oriented toward religious aspects accompanying the initial rise and subsequent prevalence of the Christian religion. The situation was different in other parts of the world. Figures of crustaceans on paintings, bronzes, and ceramics were numerous in the Far East (Monod & Laubier, 1996). In the Middle East, the expansion of Islam was accompanied during a period extending from the 8th to the 11th-12th centuries by a keen interest in sciences and art. Crustaceans had their part in depictions of animals in wall paintings, ceramics, and books (Gray, 1977). In America, an unusual representation of crustaceans during this period has been reported on a Pueblo pot made by the Mogollon people who lived in the Mimbres Valley area of New Mexico during the 10th-12th centuries. The pot is decorated with 'water bugs' that look like ostracodes, perhaps *Chlamydotheca* sp., which live in fresh waters of Central America (Arnold, 1982; J. Vannier, pers. comm.).

Middle Ages

As stated above, the occidental Medieval period was poor in animal representations; the art of this time was almost entirely oriented towards religious matters. During the medieval period, the theme of zodiacal signs became common, and crustaceans (crabs, or less often crayfish) were then used almost exclusively, but frequently, to represent the sign of Cancer in books or on sculptures in Romanesque churches (figs. 27B.3, 27B.4). Zodiacal crustaceans are also present on stained-glass windows. Chater (1988) and Hopkin (2003) report the presence of an isopod (*Porcellio scaber*?) illustrating the 'Sol in Cancro',

Fig. 27B.3. Sculpture of the zodiacal Cancer sign; 12th century. Apse of the Saint-Austremoine Abbey, Issoire, France. [Photograph G. Charmantier.]

Fig. 27B.4. Sculpture of the zodiacal Cancer sign, presumably *Carcinus maenas*. Note the sculpted mower as additional symbol of June; 13th century. Porch of the Notre Dame cathedral, Amiens, France. [Photograph G. Charmantier.]

probably referring to the crab *Cancer*, on a stained glass at St Mary's church, Shrewsbury, England. A collection of stained glass from the 14th-19th centuries is displayed in that church, but the exact age of this particular illustration is not known.

In the following four centuries that cover the Romanesque and Gothic periods, and still with an almost exclusively religious inspiration, mural representations and altarpieces, tapestries, and books were the supports of artistic expression. Books, and among them the so-called Books of Hours, which contained compilations of prayers for lay persons, are particularly interesting because the text of some or all of their pages (depending on the wealth status of their owner) was 'illuminated', i.e., illustrated, with painted and/or gold-embossed drawings. After scholars such as St Thomas d'Aquino or Roger Bacon had reconciled theological thought with the experience of nature (Sterling, 1985), the art of illumination included more images taken from nature in the 14th and 15th centuries (Baron, 1981; Avril & Reynaud, 1993; Dupont & Gnudi, 1994; Pächt, 1997).

Beginning in the 14th century, artists illustrated several parts of those pages with living forms, mainly plants but also animals, and including a few crustaceans, which were then represented in the illuminations. As noted above, the species belong to the decapods, with one of the earliest a very acute image of the clawed lobster *Homarus gammarus* painted in a medallion by Barthélemy d'Eyck and Enguerrand Quarton in 1440-1450 (Avril & Reynaud, 1993). They painted the lobster with its natural bluish colour against a contrasting red and orange paving, on which projected shades of the animal and perspective in the decorative motif give a sense of relief. In most cases, crabs, close in shape to *Cancer pagurus*, are included among zodiacal signs as in the Rich Hours of the Duke of Berry [*Très Riches Heures du Duc de Berry*] executed by the Limbourg brothers in 1408-1416 (Pognon, 1979). A particular case is represented by the presence of crayfish in scenes of the Last Supper of Christ painted between the 13th and 16th century on large murals of several churches of the Italian Alps (Claut, 2000; Romeri, 2000); their possible symbolism is discussed below.

Renaissance

At the transition from the end of the medieval period to the Renaissance, the understanding of the world by western people changed profoundly following the rediscovery of the Antique sciences and arts, and the discovery of the New World. Mankind then started its long realization that it inhabits the world and does not rule it. The translation of this trend in art was a renewed interest for nature and a progressive transition toward new forms of pictorial art. In more than one way, this approach had connections with Greek and Roman art, but with the addition of new physical surfaces for the paintings (wood, canvas), which then led to the establishment of different kinds of representations during the next centuries. The distinction made by Sterling (1985) between 'megalography' and 'rhopography' is now largely accepted (Bryson, 1995). Megalography is the depiction of those things in the world that are great: legends, religious matters, battles, historical crises; Rhopography (from Greek *rhopos* [ῥῶπος], trivial objects, small wares, trifles) is the depiction of unimportant things, "the unassuming material base of life that "importance" constantly overlooks" (Bryson, 1995: 61). From the former to the latter, a hierarchical system of artistic representation was progressively constituted, from **historical depictions** (**religious** or not), to **portraits**, **landscapes**, **genre painting**, and the **still life**. It is interesting to note that the presence of humans tends generally to decrease in this ranking, and the presence of animals tends to increase. Crustaceans, humble representatives among the 'artistic' animal crowd, were thus bound to be subjects for rhopography. In fact, they are mostly found in still lifes and genre painting.

Before addressing them, the case of the German artist Albrecht Dürer is interesting to consider as his life (1471-1528) spanned the transition to the Renaissance. Within one year, he used crustaceans in very contrasting ways. In an early commission, he illustrated through woodcuts made in 1494 the work by Sebastian Brant "Das Narrenschiff — The Ship of Fools" published from 1494 to 1520. One of those illustrations, "De Predestinatione", depicts a fool symbolically carried on the back of a big crayfish (see below). One year later, in 1495, one of his "Tierstudien" depicts a crab (*Eriphia* sp.?) that fills the entire painting in a way that conveys detailed and careful observation, announcing, as it were, the illustrations of Gesner and other 16th century naturalists.

The rediscovery of the Roman xenia led Italians (in particular Michelangelo da Caravaggio), Flemish painters, and then Spanish artists such as Sanchez Cotan and Francisco de Zurbaran with their *'bodegones'*, towards naturalism. Flower and fruit paintings flourished, and animals were progressively added to those as tokens of realism. These animals were logically mostly terrestrial (insects and reptiles) and only a few crustaceans can be found among them. This is particularly so in different works by Georg Hoefnagel (16th century), who transformed his flower paintings into albums of botany and zoology, as did Jan van Kessel one century later (Chater, 1988).

Other contemporary artists used crustaceans in original ways. Hieronymus [or: Jheronimus, or: Jeroen] Bosch mingled parts of crustaceans with those of other animals in his works to put together monsters. Inspired by him, Alart Du Hameel, also from 's-Hertogenbosch, put a homarid lobster in his early 16th century engraving of Saint Christopher. In the same period, an original and apparently unique use of crustaceans is that of Frans Crabbe van Espleghem, from Malines (Jean-Richard, 1997); using a pun on his name, the so-called 'Crayfish Master' signed several of his etchings with a drawing of a crab or a crayfish. The version of the proverb 'Big fish eat little fish' (1556) by Pieter Bruegel the Elder, in the manner of Bosch, presented a striking engraving of a gigantic stranded fish cut open and regurgitating tens of smaller fish (Tolnay & Bianconi, 1968). Comparatively small crayfish and crabs seem about to feast on them (Robert-Jones & Robert-Jones, 1997). Other Bosch-inspired works of Bruegel the Elder also include the creation of fantastic animals from crustacean parts, as in "The Fall of Rebel Angels" (1562). Giuseppe Arcimboldo used unlikely combinations of objects, usually flowers or vegetables, to paint satirical human portraits. One of them, "Aqua", dating from 1566, is a collection of aquatic species comprising mainly marine animals including molluscs, fish, and mammals, and at least four carefully painted crustaceans: a shrimp and a crayfish, one *Cancer pagurus* and, unusually, an hoplocarid stomatopod, perhaps *Squilla mantis*. This painting offers one of the few examples of the presence of (a) crustacean(s) in a portrait.

In a very different style, Bernard Palissy, in the late 16th-early 17th century, frequently used crustaceans to decorate his elegant ceramics, mostly plates and trays. His favourite species included marine shelled molluscs and an array of species living close to or in fresh water: snakes and frogs, fish and crayfish. The latter are very finely moulded, covered in a green-brownish, close-to-natural colour, and they can be identified either as *Astacus astacus* or *Austropotamobius pallipes*.

Dutch Golden Age

Afterwards, the presence of crustaceans in art reached a second peak (the first dating back to the Romans) during the 17th-century so-called Golden Age of Dutch and Flemish painting (Guratzsch, 1980; Schama, 1991; Bryson, 1995). During this period, genre painting and still lifes became widely accepted as respectable, if somehow minor, forms of art. One of their interests for the next centuries' connoisseurs lies in the wealth of information on everyday life they both retain and transmit. Their depictions extend from trivial matters to the behavioural or metaphysical aspects these images could imply. As Flanders and the Netherlands possess a long sea shore and have developed strong connections with the sea through fishing and trade, it is not surprising that sea life, especially as a source of food, was familiar to their inhabitants including the artists, hence the profusion of marine animals in their art. The importance of the Dutch and Flemish painting of the 17th century in terms of quality, its abundance, and symbolism, both in the general history of art and within the special scope of this chapter, warrants a special treatment, and we will deal with it in the Symbolism section.

A particular aspect of still lifes that reflected the mounting interest for natural history was paintings of **'curiosity rooms'** that became fashionable in the 17th century. These depicted accumulations of objects, including animals and among them crustaceans, or often represented other paintings of these objects collected by wealthy or powerful members of the society (Schama, 1991; Schneider, 1994). In other countries, a large number of still lifes and genre painting works were also executed during the 17th century, often under the Dutch and Flemish style influence. In Spain, the tradition of painting 'bodegones' was continued, in particular by Diego Velasquez. An intermediate between genre painting and still life, the 'bodegon' depicted shop, market or kitchen scenes and associated human figures to objects. However, few crustaceans can be observed in the bodegones. One example is a painting made in 1645 by Baltasar Gomez Figueira, a Portuguese painter trained in Spain, which associates a fish, onions, brightly coloured oranges, cooked shrimps and crab (Sterling, 1985). In France, the middle of the 17th century was marked by the foundation in 1648 of a powerful Academy aimed at controlling the expression of art, which was considered a tool of propaganda by the political power (Sterling, 1985). Rhopography was consequently partly replaced by megalography in which crustaceans had a small place. Two homarid lobsters were, for instance, represented among other marine animals in a pompous "Tapestry of the Elements: Water" woven in 1669 under the direction of Le Brun, co-organizer of the Academy with Colbert, and head of the Gobelins factory.

During the 17th then the 18th century, the naturalistic trends apparent in painting affected other forms of art, in particular the manufacture of tapestries. Genres scenes were appealing to prospective buyers, and several tapestries made in western countries such as the Netherlands or France include crustaceans. A particular case is the two series of Gobelin tapestries made in Paris at the order of King Louis XIV and depicting Brazilian people, landscapes and fauna, including numerous crustaceans (Holthuis, 1991; see below in the Symbolism section).

The Enlightenment

The influence of Dutch and Flemish painting was noticeable from the 17th to the following centuries and it will probably endure into the 21st century. For instance, in the British Isles, Charles Collins painted in 1738 his "Lobster on a Delft Dish" in a sober and simple setting reminiscent of the early Dutch school. In Italy and Spain, the Northern influence was present in the work of different artists, as in a still life painted in 1720-25 by Giaccomo Ceruti in which two crustaceans, *Homarus gammarus* and *Cancer pagurus*, are precisely depicted, the former having its claws tied up with string. Whether this was a cook's precaution or common practice in lobster fisheries, as with the current elastic bands, remains to be explored. In France, Jacques Linard was directly influenced by the Dutch style in the 17th century. Later, Jean-Baptiste Siméon Chardin conveyed the tradition of Dutch-inspired still lifes and a member of his school, Anne Vallayer-Coster, included a homarid lobster in at least one of her works in 1817. Even in the 20th century, Jean Hélion continued the trend with his "Saga des Homards" from 1975-76. In North America, despite a marked interest for naturalistic subjects evident from genre scenes and landscapes to still lifes, to our knowledge, in Audubon's work (1827-1838) very few if any crustaceans appear (Walker & Prown, 1987; Tyler, 1990). Without further contributing to the confusion that puts *Limulus* among crustaceans, a specimen of this chelicerate was depicted among other marine animals by John White in his "Indians fishing" (c. 1585, British Museum, London). Still in North America, crustaceans do not appear to have been represented by the Inuit, although their art is largely inspired by the animal kingdom (Swinton, 1972).

The appeal of genre painting and still life decreased in the 19th century, while landscapes and portraits became fashionable. Eugène Delacroix, fond as he was of animal painting, reserved his skills almost exclusively to the depiction of vertebrates, particularly mammals. In a rare still life, painted in 1826, he offers a curious composition of two red homarid lobsters among dead game and a gun, in the foreground of a Constable-inspired landscape. As Delacroix in his paintings, François Pompon (1855-1933) was also almost exclusively inspired by mammals in his numerous sculptures of animals. James Ensor, a Belgian artist inspired by Dutch painters in his still lifes, often represented fish. Among these and elements of tableware, he depicted in 1888 a crab (*Cancer pagurus*) and two homarid lobsters, in a rare comparison of their colours before and after boiling.

More modern times

On the whole, crustaceans were only scarcely incorporated into the art of the 19th and 20th centuries. Few of them are found in the works of the Impressionists, but Manet included shrimps in a still life in the 1860s and Van Gogh represented a cancrid crab in 1889. During the 20th century, Max Ernst, offered in 1926 a rare painting of two crabs seemingly active and moving, one of them fading away in what looks like sediment. A darker atmosphere hangs over the two paintings of woodlice from 1940 by Paul Klee (Chater, 1988; Grohmann, 1954, cited in Chater, 1988), and on the 1960s' image by Jürg Kreienbühl of several species of lobsters preserved as dry exoskeletons or in alcohol in a general atmosphere of disorder and decay. The 20th century artist who included most crustaceans in his work is Picasso, particularly during the years he spent in the 1940s on the French Riviera in Nice and Vallauris. Shrimp, spiny and clawed lobsters, well recognizable despite Picasso's play with their shape, were among his favourite crustaceans as artistic subjects, as well reportedly as seafood (Cox & Povey, 1995). Among contemporary artists, Jean Helion also expressed an interest for edible crustaceans in the "Saga des Homards" series painted in 1975-76 on Belle-Ile, as well as Peter Lörincz (penaeid shrimps associated with the cathedral of Lodève, France) and Gilles Dulis (humorous picture of a majid crab riding a multiple tandem in "Vélocité").

In another aspect of art, crustaceans were also used in the decorative arts, especially from the 17th and 18th centuries. The Royal Palace in Amsterdam makes notable use of carved lobsters and crabs in the marble-clad decorations of the main reception hall and adjacent corridors. Examples also abound of the use of crustaceans to decorate tableware, for example a crayfish on top of a porcelain tureen (made in 1765-1775 at the Ludwigsburg factory, Germany), a silver crayfish salt-cellar (1853, by C. F. Hancock, London), shrimp on glazed earthenware plates (French service Rousseau, designed in 1866 and re-issued by Joy and Levellié in 1903-1913).

During the 20th century, crustaceans were sometimes used in widely discussed forms of art that we gather here together under the kitsch denomination, at the same time evoking high levels of popularity and criticism. One could argue that Salvador Dali's 1936 "Lobster Telephone" might belong to that category: the surrealist artist put a homarid lobster to a highly unusual use by covering a phone receiver by a bright orange-coloured model of the crustacean. This type of use has continued in contemporary times with crustaceans, mostly big edible decapods, applied on countless pieces of tableware, plates, trays, mugs, silverware, or used as toys, etc. Kitsch has also been used recently for Jeff Koon's Popeye and lobster series and for a huge lobster (2.46 m long and weighing 48 kg) for which discussions and arguments started not primarily from the crustacean itself but from the locations chosen for its exhibition, at the Versailles Castle in 2008 and in the Serpentine Gallery, Kensington Gardens, Hyde Park, London in 2009. Koon said that he was attempting an art that does not alienate anyone. We won't go further in that discussion, but popular it was, and these lobsters started heated controversy. Worth mentioning, if beyond the scope of this article, is also the contemporary presence of crustaceans in advertisement, either to promote their sales or as part of another message.

SYSTEMATICS

An overview of the systematics of crustaceans used in works of art throughout the centuries, that we have examined or either read, or heard of (table I; summary in table II), reveals very consistent trends: these were mostly large and edible species belonging to marine Decapoda. Freshwater decapods, mainly crayfish, are less frequently represented. Among the most illustrated groups are homarid lobsters (approximately 20%), crabs (36%, of which 14% are cancrid crabs), crayfish (8%), spiny lobsters (10%), other decapods, mainly shrimp (20%). Non-decapod Malacostraca, e.g., isopods and stomatopods, were seldom represented, and non-Malacostraca seem only represented by branchiopods, as well as ostracodes and other maxillopodans.

The dominant groups or species of crustaceans represented by artists changed with time as will attest a few examples. During the Roman period, the most frequent groups were Mediterranean species such as spiny lobsters and to a lesser degree, crabs and shrimp. Medieval artists concentrated mainly on crabs, crayfish, and homarid lobsters.

TABLE II

Systematics in works of art: percentage of each systematic group as represented.
Total number of works: 136; total number of crustaceans: 294

Non-MALACOSTRACA (Ostracoda, Branchiopoda, Maxillopoda, Cirripedia): 1.0%
MALACOSTRACA: 99.0%
  HOPLOCARIDA, Stomatopoda: 1.4%
  EUMALACOSTRACA: 97.6%
    Peracarida, Isopoda: 2.4%
    Eucarida, Decapoda: 95.2%
      Penaeoidea: 2.1% (10.8% of "shrimps")
                                                     "shrimps": 19.5%
    Caridea: 17.4% (89.2% of "shrimps")
    Reptantia: 75.7%
      Palinura: 9.8% (32.6% of "lobsters")
      ("spiny lobsters")
      Astacidea: 28.0% [including 7.7% of "crayfish"
                              20.3% of "homarid lobsters"
                              (67.4% of "lobsters")]
    Anomura: 2.0%
    ("hermit crabs")
    Brachyura: 35.9% (including 13.6% of "*Cancer*"
    ("crabs")   6.8% of "majid crabs"
                       79.6% of "other crabs")

Seventeenth-century Dutch and Flemish painters were mostly inspired by the Atlantic fauna: among the representations listed in table I, the most frequent species belong to crabs (37%, of which 24% are *Cancer pagurus*) and lobsters (34%, of which 97% are *Homarus gammarus*) with few shrimps (15%), hermit crabs (5%), and crayfish (4%).

SYMBOLISM

Why were crustaceans used in art? What were their functions and the symbolic messages they carried if any? These questions reflect wider questions concerning the use of animals in art.

Objects represented in art possess a literal sense and a symbolic sense (Hall, 1996). At certain historical periods, e.g., during medieval times, the latter corresponded almost exclusively to a **religious meaning** related to faith and morals. Other periods, e.g., the post-Renaissance centuries, were permeated by **humanist** and **naturalist** erudition. Artists had then a wider view of the world around them and the represented objects were laden, coded with symbols of contemporary culture, economy, politics, and ethics. The decoding of works of art, given understanding of the way of thinking which accompanied their production could give insights into the contemporary social history, cultural, political and religious concepts, ideology, and sexuality. This is particularly evident in **genre scenes** and **still life paintings**. Works of art can thus be approached with pleasure and thought for their aesthetic and symbolic dimensions.

As with other subjects of artistic interest, the animals that have triggered the attention of artists carry their load of symbolism. Animal depictions in prehistoric times are still being debated, but of utmost importance is the fact that animals were the first subjects represented artistically by humans, which underlines the tightness of their relationships. Interpreted from the perspective of our present knowledge and sensibility, the motivations behind ancient animal artistic representations might be related to exploitation (animal-food/-game/-carrier/-general helper), religion (animal-god/-gift to god), contemplation (animal-aesthetics), and relationships (animal-companion). The different perceptions that humans have of animals also reflect their own self-awareness of the surrounding world. This perception focused for the most part on the closely-related mammals, but a few thoughts can also be applied to crustaceans. These invertebrate creatures that, as we have seen, were comparatively rare in art, nevertheless carry a diverse load of symbolism.

Description

As with other forms of animal life, analysing the ways crustaceans were described may help understand their **symbolism**. Given the systematic biases that have prevailed for their choice by artists, crustaceans have some stereotyped and simple features that help evaluate the quality of their description, e.g., the ten pereiopods of the decapods, the aspect of the cephalothorax in brachyurans and macrurans, the number of segments of the pleon in macrurans. The lowest level of accuracy in depiction occurred during the medieval period in Western countries. The number of pairs of pereiopods of decapods, mainly crabs and crayfish, may vary from 3 to 6. Even when this number is right, the shape of these crustaceans is deeply modified (fig. 27B.3) with a frog- or human-like head, fingered-legs, and a tail often associated with the devil. The identification of the animal is often only made possible by the presence of claws. Clearly, such representations most probably were executed without direct observation of the animal; they were not accurate depictions but were vehicles of symbolism. An exception linked to the shift from the Romanesque to the Gothic period is illustrated by a well-observed crab sculpted at Amiens, France (fig. 27B.4, and see further comments below). In contrast, one of the goals of many Roman artists was a precise presentation of the crustaceans depicted in their works. It seems likely that they used direct observation of marine life available from close-by Mediterranean fishermen. This is in agreement with the almost strictly Mediterranean origin of the fauna they depict. Their spiny lobsters (*Palinurus elephas*?) are generally accurate down to the design of their antennae or of the number and shape of their maxillipeds, pereiopods, and uropods (fig. 27B.1). Shrimp depicted by Romans belong to carideans (*Palaemon* sp., *Palaemonetes* sp.) or penaeids (*Penaeus kerathurus*). The precise proportions of the specimen of *Pachygrapsus marmoratus* adorning an oil lamp have already been discussed (fig. 27B.2). However, the Romans not only depicted the shape of these animals, but also their colours (retained on wall paintings and, mainly, on mosaics; this is particularly true for richly-coloured fish) and their movements and behaviour. In addition to reminding the viewer of the Romans' taste for violent struggle, the classic lobster and octopus fight scene is enhanced by the well-observed and correctly rendered attitudes of both animals (fig. 27B.1). However, a stylistic tendency is evident in the work of some artists of the 2nd and later centuries AD, in which the realistic depiction of animals gives way to a certain amount of impressionism or didactism (Charles-Picard, 1978). Some Roman artists such as Heraclitus, following Greek artists as Sosos of Pergamum, have also represented trompe l'oeil 'Unswept floors' in which broken objects, detritus, litter, are scattered on a mosaic floor. Sometimes, the shade of these objects has even been added for increased realism. Although generally viewed as a symbol of disorder (Bryson, 1995), such pieces depict the objects, as fragments of carapaces or pereiopods of crustaceans, with such precision as to allow the identification of the order or genus. A comparison with stomach contents and the present studies they elicit is here de rigueur.

A comparable concern for accuracy started to reappear at the end of the Gothic period and, mainly, during the Renaissance. A striking example of the shift between the dominance of religious symbolism during the Romanesque period and the rising interest for observation in Gothic times is offered by the comparison between two above-cited representations of crabs. In the Romanesque church of Saint Austremoine in Issoire (France) dating from the 12th century, the crab is barely recognizable, while it is finely and precisely crafted, and probably made from observation of *Carcinus maenas*, in the Gothic cathedral of Amiens built in the 13th and 14th centuries (figs. 27B.3 and 27B.4). Starting in the 14th century, i.e., earlier than the Renaissance, depiction of macruran-shaped crustaceans 'evolved' into identifiable crayfish or lobsters, and crabs were often undeniable *Cancer pagurus*. This trend became apparent in some of the 14th- and 15th-century illuminations and books of hours, with ample variations between artists. For instance, the difference in accuracy is obvious between the well-rendered lobster illuminated by Barthélémy d'Eyck and Enguerrand Quarton in 1440-1450 and the approximate, if recognizable, *Cancer pagurus* shape painted by the Limbourg brothers in the Rich Hours of the Duke of Berry in 1408-1416. Another meaningful example of the transition between symbolism and observation arises at the Renaissance from the comparison of the 1494 crayfish carrying a fool and the 1495 crab painting, both by Albrecht Dürer. But the quest for precision was well established in the 16th century, with for instance the detailed study of various crustaceans by Jan Bruegel's "Water", Arcimboldo's "Aqua", and the 17th-century Dutch and Flemish artists gave accurate depictions of crustaceans in their genre and still life paintings. As with the Romans, their abundance is certainly linked to the proximity of the North Sea, and to the large part that seafood played in the local diet and its ready availability, particularly from the beach ['strand' in Dutch] of Scheveningen close to The Hague. These artists depicted a moderately large number of species (table I), most often with a great attention to details. This is particularly evident when rarely used species are depicted, as in the still life renditions by Balthasar van der Ast in which gastropod shells, finely painted, host a hermit crab, one of the rare examples of artistic use of anomurans. Rare examples of isopods can be found in a still life by Jan van Kessel (Sutton, 1972, cited by Chater, 1988), two of them identified as *Armadillidium* sp. and *Philoscia* sp. by Chater (1988). Perhaps one of the most striking examples of the attention to details of 17th-century Dutch painters is given by Pieter de Ring's "Still Life with a Golden Goblet", in which a meticulously depicted homarid lobster (*Homarus gammarus*) carries a balanoid cirripede on its cephalothorax. That specimens of *Homarus* can carry cirripedes on their cuticle during their long intermoult phase is well known by today's scientists, but it required a keen eye in the 17th century to spot one of these epibionts.

In the following centuries, the artistic depiction of crustaceans generally decreased in accuracy, and for instance the number of segments in the pleon in lobsters was not Picasso's primary concern. However, among the 32 sugar-lift aquatint prints he made from 1936 for an album published in 1942 to illustrate Buffon's *"Histoire Naturelle"* (Cox & Povey, 1995), one was devoted to depicting a spiny lobster. The beast, seen from one side, has only three pereiopods, the most posterior one seemingly attached to the first pleomere, but the general aspect of the lobster is rendered in a naturalistic way that was uncommon for Picasso. This work's originality lies in the presentation of the crustacean in its habitat, under water, surrounded by aquatic plants and two fish. This is a rare occurrence since crustaceans are most generally presented out of water, alive or dead, if not boiled as attested by their colour. Only the Romans on their mosaics have consistently attempted to present marine life, including crustaceans, alive in their medium, but the lack of perspective makes their works much less lively than Picasso's.

An expressive interest can lead from art to the illustration of systematic studies. This tendency resulted in an abundant production of scientific drawings and paintings, especially during the 19th century. Hoefnagel's work is intermediate on the pathway between art and science. A similar situation is found in the Gobelin tapestries made from Marcgraf's observations and Eckhout's paintings during the Dutch colonial rule of northeastern Brazil (1624-1654) when Prince Johan Maurits was governor-general (1637-1644). These works and tapestries and their history have been precisely described by Holthuis (1991). Two series of Gobelin tapestries were made: "Les Anciennes Indes" (1687-1730) and "Les Nouvelles Indes" (1740-1768). At least ten species of decapods, mainly crabs, are depicted in these tapestries. The 'curiosity room' paintings of the 17th and 18th centuries also belong to these works halfway between art and science. They were aimed at the description of natural history artifacts and their exhibition also had the advantage of demonstrating the wealth and culture of their owner. As animals hidden below the water surface and sometimes deep in the sea, crustaceans were true 'curiosities' and their presence is consistent in most of these works. Jan van Kessel for instance depicted several decapods among many other animals, with a clear attempt at accuracy, in his numerous panels assembled in 1664-66 to illustrate "Europe", "Africa" and "Asia".

Mythology and religion

Crustaceans have sometimes been associated with mythological and religious scenes. In the 706 BC Assyrian scene cited in the Chronology section, one of the crabs depicted in the relief bears a scorpion-like tail. Is it an error of the artist, although the animal is clearly immersed in water? This composite beast might hypothetically be a reminder of an Assyrian god of war, Ninurta, represented as a scorpion on a stone, a Kudurra (= 'boundary stone', South Babylonia, c. 1100 BC) used indeed as a boundary stone to delineate a piece of land.

The Greek/Roman god of the oceans, Poseidon/Neptune, was often represented by the Romans in association with symbols of marine life, most often seaweeds and fish, and occasionally crustaceans. One example is a mosaic from the Themetra thermae (3rd century AD) picturing the head of the god, whose hair is topped by two horn-like crustacean claws. The connection of the goddess Aphrodite/Venus with aquatic life, at her birth, was also common. This tradition was later retained, from the 14th-15th centuries, when the rediscovery of art from antiquity wielded a strong influence on the Renaissance artists. For example, in a wall painting by Giorgio Vasari at the Palazzio Vechio in Florence dating from 1556-59, part of an "Allegory of Water" (see below) is devoted to the birth of Venus. A spiny lobster is visible among the presents (fish, erect gastropod, pearls) offered to her. The offering of gifts from the sea is also one of the themes of the "Banquet of Acheloos" painted by Peter Paul Rubens and Jan Bruegel in 1600. Despite the freshwater affinity of this Greek river divinity, the offerings are mainly of marine origin. Among them, carried by a man emerging from the water, is a live lobster whose right claw is humorously located very close to the ample naked behind of a siren. In a very similar setting painted in 1620 by the Flemish painter Frans Francken II representing the "Abdication of Emperor Charles V", the siren is much safer since she is herself carrying the offered lobster.

Thus, this lively approach for dealing with the gods did not imply soberness for painters of that time, but it contrasted with the seriousness that accompanied earlier religious Christian scenes, where the symbolic presence of fish was frequent while crustaceans had limited opportunities to participate. However, during the early period of development of Christianity, a few representations of crustaceans can be found, particularly on engraved ring-stones or cameos of the early centuries AD. Charbonneau-Lassay (2006) cites several of these jewels, some of them originating from the Foggini collection: the elongated crustaceans seem to fight other marine creatures, protecting a fish that is peacefully swimming below. As an interpretation, Lassay (2006) proposes that, thanks to its carapace, the well-protected crustacean might represent the invulnerability of Christ defending the human souls. However, in later representations, the crustacean religious symbolism was generally darker. For example, at a time of growing internal conflict in Christianity leading to the reform movement, Dürer's illustration from 1494 of Sebastian Brant's "Das Narrenschiff — The Ship of Fools" shows a crayfish carrying a fool, an illustration of human follies. In a similar fashion, Alart Du Hameel's early 16th century depiction of Saint Christopher crossing the river also gives a crayfish or lobster a bad role as, in its attempt to pinch the saint's leg, the crustacean is one of the antagonistic forces aimed at the failure of the crossing. A similar negative role is again attributed to crustaceans in the "Fall of Rebel Angels" by Pieter Bruegel the Elder (1562). The painting represents the fight of St Michael against evil, symbolizing the confrontation between right and wrong, on which side crustacean-like creatures are exhibiting prominent claws. In a triptych of the "Bruges Fishmongers" made by Pieter Pourbus in the second half of the 16th century, the central panel depicts the miraculous draught. Only fish are represented in the net brought on board the boat where Christ and the fishermen stand. A single crab is a humble and apparently neutral witness of the scene from the sandy beach. However, Rubens' "Miraculous Draught of Fishes" (1619/20) shows a few crabs among the fish caught in the net on Lake Tiberiade.

Christ stands on the prow of a boat, with fishermen gathered around in stormy weather. The catch that symbolizes saved souls is usually represented by fish in such scenes as in the previous work. Here, the two crabs, which apparently have the same role, reach a symbolic artistic pinnacle since they are equalled to human souls. Another of Rubens' message might be that even crab-looking, clawed, aggressive, awkwardly moving souls might be saved.

A particular case of crustaceans represented in a religious context is their presence among the food present on the table of the Last Supper of Christ, frequently encountered in several churches of the Italian Alps (Claut, 2000; Romeri, 2000). They were painted on murals between the 13th and 16th century, coloured either in natural dark or in red. Their symbolic meaning has led to opposing interpretations. In some instances, they are oriented toward Judas, often the dark ones, and they might be a symbol of evil or sin. The sideward gait of the crayfish, often associated with hypocrisy (see paragraph on Inconstancy), would sustain this interpretation. However, the change of colour of a crustacean when cooked is often associated to the Resurrection of Christ: within this view, the red crayfishes would carry a positive symbol linked to eternal life, in addition to their simple enjoyable taste as food (Rigaux, 2000; Young, 2000).

In line with the above paragraph may be the metaphysical preoccupations of the Swiss artist Paul Klee who, a few months before his death in 1940, painted two pictures of woodlice, "Assjel im Gehege" ("Woodlouse in the Enclosure") and "Assel". These works have been much commented upon (review in Chater, 1988). The isopods are represented as black, fishbone-like structures on a bright pink and orange background. The general pattern is that of railings, and two other grates are added in the first painting. These works have been interpreted in two different ways. As they probably reflect the thoughts of a man about to die, one can see in the animal meandering between grates the difficult but not necessarily unsuccessful passage from life to death and after-death ("stricture and free passage, threat and hope", according to Grohmann, 1954, cited by Chater, 1988). The opposite view is that the crustaceans are "harbingers of doom" and that they represent "the insidious forces that were slowly sapping his own vital juices and preparing him for eternal confinement" (Verdi, 1984, cited by Chater, 1988).

Therefore, crustaceans have often been associated with evil symbols. As creatures living under water, often bearing claws, moving in unusual and unpredictable ways, they were mysterious and unsettling beasts for artists of different centuries. However, albeit not often, they also have been used as positive reminders of Resurrection, a symbol probably related to their spectacular change toward a bright red colour after cooking.

Allegories of water

As mainly aquatic-living species, crustaceans may symbolize water. As one of the four primal elements, water is frequently personified by a river god carrying an overturned urn from which water flows into expanses of water, river, lakes, or the sea where aquatic animals are depicted. Not surprisingly, crustaceans have frequently been used in allegories of water along other creatures such as molluscs, fish, and dolphins.

Many above-cited Roman mosaics, whose primary objectives were demonstrative or mythological, also carry a symbolism, even of reverence, to water. The aquatic element was tremendously important for Roman civilization: the sea was used as a way of transportation for goods and armies, and as a source of food; freshwater was transported over long distances through aqueducts, and it was eliminated after use through drainage systems; thermal water was enjoyed in thermae. For the Romans, crustaceans thus took their part in the celebration of water.

The rediscovery of Roman art was certainly not foreign to the renewed interest in water of the 16th-18th centuries. I have already cited Vasari's (1556-59) "Allegory of Water" and Rubens' and Bruegel the Elder's (1600) "Banquet of Acheloos" (see Symbolism, and Mythology and Religion). The title of "Aqua", by Giuseppe Arcimboldo (see Chronology) is a symbol in itself. Besides his ceramic plates rich in freshwater fauna, Bernard Palissy also crafted in the late 16th century a painting-like ceramic picture celebrating "l'Eau". In it, a young woman pours water from two jars; its flow carries stylized water-living creatures including a crayfish or lobster down to a river which opens to the sea.

The same subject was interpreted in a very different manner at the same period in "Water", a painting by Jan Bruegel. In a mostly terrestrial landscape, water flows from mollusc shells held by a young woman and an old man. A waterfall sends water into what appears to be a lake (and might have been the sea in the artist's intent) since it feeds a running stream in the left foreground of the painting. On and along the stream's banks, an array of animals is displayed: shore birds, fish, molluscs and several species of crustaceans, young children and a cupid, and even monkeys — all of them precisely depicted. Crustaceans are represented by several species of lobsters, crabs including majids, homolid anomurans, stomatopods, etc. This naturalistic work would be worth a thorough systematic study. The "Tapestry of the Elements: Water" of Le Brun, dating from 1669, also gathers the allegorical images symbolizing water, and includes two homarid lobsters, but its main object was, not surprisingly at that time in France, to celebrate the power of the monarch Louis XIV.

Zodiacal signs — Astrology

Cancer is the fourth sign of the zodiac, associated with the month of June/July (actually, 22 June to 22 July; Thibaud, 1994; Hall, 1996). According to the Greek legend, a crab sent by Hera/Juno bit the foot of Hercules as he was fighting the Lernaean hydra, and the goddess placed the crustacean among the zodiacal signs as a token of appreciation (Hall, 1996; Monod & Laubier, 1996). Illustrations of the zodiac with crustaceans, crabs essentially, have been and still are extremely numerous, from the medieval period to contemporary horoscopes. Therefore, only a few examples of the variety of crustaceans used in astrology will be given in the present study. In the Chronology section, we have attested the presence of crustaceans as church sculptures (crab of the abbey of Issoire, France, 12th century, fig. 27B.3; crab of the cathedral of Amiens, France, 13th-14th century, fig. 27B.4), on stained-glass windows (St Mary's Church at Shrewsbury, England; Chater, 1988), as illustrations of books (Handbook of Astronomy and Comput, Metz, 820-840), or as illuminations in books of hours (Rich Hours of the Duke of Berry, by the Limbourg brothers, 1408-1416). During the same period, a crab was figured in a Persian astrological text from the 1410-11 work "Anthology of Iskandar" (Gray, 1977). Zodiacal crustaceans also adorn terrestrial maps and maps of the sky, nautical charts, and globes.

What all these pictures have in common is that the crustacean is viewed as a symbol, and it is depicted for this purpose and most often without much attention to its morphology. The Persian crab just mentioned has six pereiopods, and the Limbourgs' crab, although recognizable as a *Cancer*, has only eight. However, the crab from Issoire shows the right number of pereiopods, but as already discussed its overall shape is far from accurate. Even today, a look at the number of pereiopods of the countless crustaceans that decorate horoscopes can bring surprises. However, even during the Gothic period, some artists offered an accurate description of crustaceans. An example is given by the crab sculpted on the porch of the cathedral of Amiens in the 13th-14th century with such precision that it is recognizable as *Carcinus maenas* (fig. 27B.4).

If crabs constitute the most frequent symbol of the Cancer zodiacal sign, other species can also play this role. The isopod reported by Chater (1988) and Hopkin (2003) in Shrewsbury is certainly exceptional, but crayfish (or clawed lobsters in some cases?) appear quite commonly in the zodiac, with the same problems as the crabs regarding the accuracy of their morphology. An example is given by the presence of a crayfish in a book of hours illuminated in France for an English purchaser by the Fastolf Master in 1440-1450. The apparently indiscriminate use of a crab or a crayfish (or even a woodlouse) by the artists is an indication that they were able to mentally, perhaps subconsciously, associate them into the single systematic entity of the crustaceans. This process had been made manifest by Aristotle approximately eighteen centuries earlier.

Typical activities such as mowing or haymaking were the traditional symbols for June, but these were often represented by a companion image of the crab or crayfish, as in the case of the Limbourgs' and Falstof Master's illustrations, in the 1492 mass-book of Jean de Foix, or at Amiens (fig. 27B.4). This raises the question of the association of crustaceans with June, and apparently this stems from the often backward or sideward gait of crustaceans, be they crabs or crayfish. Their peculiar way of walking was viewed as a symbol of the backward shift in day-length, which begins to decrease in the Northern Hemisphere at the end of June. During medieval times, the Cancer sign, which is a lunar sign, was also considered a symbol of primitive aquatic life: somehow frightening because of its association with water and sediment. It was represented at the centre of the zodiac, in close association with the Lion, a solar sign. Taken together, the two signs, from the moon to the sun, were the symbols of cyclic transformation of old into new creatures, of birth and rebirth, both at the physical and spiritual levels (Thibaud, 1994).

Symbolisms in the 17th century Dutch and Flemish art

As underlined in the section Chronology, the so-called "Golden Age" of painting in the Netherlands and Flanders, in the 17th century, constitutes one of those periods in human intellectual history when several factors, seemingly coordinated and occurring over a short span of time, affect history, politics, sociology, human relations, and vision of the world, and result in bursts of wisdom, perception, and cognition in philosophy, science, or art. Thus, the sudden and short-lived expansion and blossoming of the Dutch and Flemish art, and the place of crustaceans within it, have historical roots that have been studied (see Guratzsch, 1980; Sterling, 1985; Schama, 1991; Grimm, 1992; Schneider, 1994; Bryson, 1995; Mai et al., 1997; Elbert-Schifferer, 1999) and that will be summarized hereafter.

The Dutch and Flemish provinces had strong political links with the Duchy of Burgundy in the 14th-16th centuries and they were an important part of the Habsburg empire of Emperor Charles V (1515-1555). During the 16th century, the southern provinces benefited from an increase in their economic development at the centre of which was the fishing trade (particularly herring), and shipyard industries based at the harbour of Antwerp. Agriculture and weaving also played a significant role in the good fortune of the region. However, upon the death of Charles V, his son Philip II attempted to rule the empire from Spain and a war soon followed that would last approximately eighty years. At the same time, religious confrontations between advocates of reform and counter-reform became chronic. A major political event occurred in 1609 with a truce and a *de facto* acknowledgement of a Dutch republic. The treaty of Münster in 1648 separated the Netherlands from the southern province of Flanders that remained under Spanish control. Calvinism and Catholicism were, respectively, predominant north and south of the new border. Economic consequences of these events were extremely important. Following the 1609 truce, and despite the ongoing outbursts of war, the Netherlands enjoyed fast-growing economic prosperity based on maritime trade, shipping, its colonial possessions in the East and West Indies, the development of banks and stock exchange in Amsterdam, and the industriousness of its people. The northern cities of Amsterdam, Haarlem, Leyden (Leiden), and The Hague (Den Haag) became centres of activity and of prosperity, whereas in the south, Antwerp (Antwerpen) entered into a phase of recession that worsened after 1640. The economic activity was accompanied by the reinforcement of professional associations called guilds, and wealth separated the society into classes among which the middle class 'bourgeoisie' ranks were growing fast. As Bryson (1995: 98) puts it, the Netherlands were then "the first European society to experience the problem of massive oversupply", since the country "became the richest nation the Western world had yet seen". Managing this sudden influx of money was not easy.

Channeling it back as investment had its limitations. Spending it to improve one's personal way of life was something new to these rather austere, hard-working, Calvinist people, and it took a few decades for this behavioural and moral adjustment to firmly settle. As brilliantly discussed by Bryson (1995: 96):

> "... with industrialization the old discourse of ethics, which measures wealth according to moral conduct, is obliged to co-exist alongside a newer perception which substitutes for the term "luxury" the term "affluence". "Luxury" can never shed its ties to its medieval past, to the idea of psychomache, the battle of the soul against the deadly sins, *luxuria, superbia, vanagloria, voluptas, cupiditas*. "Affluence" is neutral on this subject: it transfers the phenomenon of surplus wealth from ethics to aesthetics. Affluence assumes that expenditure is not a matter of morals but of style."

The middle class found a stimulating, pleasant and ethical way of spending its money in art, especially the kind of art reflecting their way of daily life, with which they took to decorate the interior of their homes. Hence the increasing differentiation between portrait, landscape, genre painting, and still life. While narcissism was certainly not absent from this process, it also reveals a genuine interest for, and openness to, the surrounding world and a preoccupation for the meaning of life. Ironically, by the time the Dutch society had adjusted to 'affluence' and its consequences, recession hit in the 1660s with the competition of other growing economic powers, Britain and France in particular. Of course, orders to artists were among the first to suffer from the situation and they decreased sharply.

The Golden Age of Dutch and Flemish painting had thus lasted for half a century only, but its influence would be felt for centuries to come. The influences on Dutch and Flemish art were multiple as we have seen in the Chronology section, from Roman xenia to Burgundy, Italian, and for a lesser part Spanish art. Painters settled their studios in the most active cities cited above. In particular, The Hague was known for its numerous paintings of fish displays. This early specialization of the cities (flowers in Haarlem, etc.) tended to decrease with the number and travels of the artists. Crustaceans were abundantly represented in genre and still life paintings. These two types of artistic expressions will be analysed separately after a mention of the use of crustaceans in portraits (see list of works of art in table I).

Portraits

As already noted, crustaceans seldom appear in portraits and this remains true in the Netherlands. A notable exception is offered by the "Banquet of the Officers of the St George Civic Guard" painted in 1624-27 by Frans Hals. The painter, born in Antwerp, worked in Haarlem where he became one of the greatest Dutch painters of portraits both of individuals, and of groups such as members of guilds. Here, he portrayed an assembly of important citizens of Haarlem, in a manner close to genre painting as if the figures were acting in a theater. People, not food, are at the centre of the scene, but a crab, *Cancer pagurus*, is visible on the table. Its brownish colour is in close harmony with that of the clothes of the closest men. Along with oysters, also visible, and wine, humoristically required by the front guest who looks at the viewer, the crab will be part of the seafood shared by florid-faced men who seem to enjoy life.

Genre painting

This form of art attracts the viewer by its depiction of the daily life of various people on the streets, at work, at home. Genre and still life painting constitute a rich source of thought and information for the viewer and/or the scholar in almost any field. Different sub-categories of genre painting can be distinguished within the scope of this article.

Banquets like that cited above might for a part be treated here. However, **market scenes**, through the wealth of shapes and colours of the goods offered for sale, and through the more lively and diverse activity of people, were a primary subject matter of genre painting. The growing success of **agriculture** and **cattle farming** led to paintings of inland markets or butcher shops, where crustaceans were of course usually absent. An exception can be found in Frans Snyders' and Cornelis de Vos' "Larder" (first half of the 17th century), where lobsters are set beside other rich food such as peacock, swan, deer, and fruits. *Homarus gammarus* was thus in the 17th century a symbol of luxury. Much more interesting are the depictions of **fish markets**, which were numerous since the sea was never far away and the fishermen provided a significant part of the people's diet. Thus, the painted fish market scenes are numerous, by artists such as Adriaen van Ostade, Salomon van Ruysdael, Frans Snyders, and most of them include crustaceans alongside fish. A side remark concerning all the market scenes is that seafood is consistently presented uncooked and, of course, in the absence of any preserving device such as ice or even salt. Smoked herrings are the exception. As crustaceans and fish, even gutted as they often appear, spoil fast, they had to be sold and eaten quickly except during the cold season.

In the "Fishwife" painted by the Haarlem-born artist Adriaen van Ostade in 1672, a small brightly lighted fish-stall takes up the main part of the painting. Stopped in a very natural movement of rest while scaling a fish, probably a cod, the fishwife looks at the viewer. Other fish include a salmon and a few plaice. A crab, *Cancer pagurus*, lies on its back in front of the stall. A child looks (mischievously?) at other individuals engaged in probably symbolic actions: exchange of money, a game? The painter did not forget the influence of the church, well lighted albeit in the background, on daily life. At the other end of a diagonal originating from the bell tower, the influence of religion is also conveyed by the mass, if not the number, of fish, themselves a powerful Christian symbol. Thus, the single crab seems the only — no lobster here — luxury of a local market.

A similar fish market scene was treated differently by Frans Snyders in "The Fishmongers", painted in the first half of the 17th century. The stage, a large stall-table occupying the front of the painting, is set on the wharves of what appears in the background as a large fortified harbour, perhaps Antwerp where the painter was born and where he lived. At each end of the table, a man and a woman are hard at work, staring intently down at their task. A heap of dead or dying marine animals is piled on and below the table: fish, particularly a huge sturgeon, turtles, mammals (porpoise, otter, seal), oysters, and crustaceans including majids and *Cancer irroratus*, deep-sea anomurans and *Homarus gammarus*. The crustaceans, some of them lying on their backs, participate to the ambivalent feeling of the scene. Even if one or two lobsters are still alive, these animals have been killed for the satisfaction of people, probably rich, who will eat them as delicacies. The sense of doom hanging over the stall is enhanced by the woman's frown and the man's cleaver cutting into the red flesh of a fish, but is further increased by the almost audible howl of a still-surviving seal (probably chosen for its expressive face) in a last attempt to escape. In contrast with Van Ostade's quiet and peaceful "Fishwife", the 'luxury' crustaceans such as lobsters along with the other animals depicted in Snyders' violent painting, tell a story of greed and death and of insatiability of mankind.

Accumulations of food, usually more varied than in the fish markets, were also represented in another type of genre painting, the **'kitchen scenes'**. The abundance of food (the "glutted kitchen") "... was an ancient theme in Netherlands culture extending back to the great medieval guild feasts of Flanders and the country kermissen (carnivals)" (Schama, 1991: 152). Human figures are generally still present in these paintings, sometimes in the background, which brings these works closer to a still life. While the market scenes were presented in the open air, the kitchen scenes are generally set in dark rooms with small or no openings. In the "Kitchen Scene" from Adriaen van Nieulandt (1616), luxurious food includes two *Homarus gammarus* and one *Cancer pagurus*, various rich fowl (peacock, turkey, swan), game (deer, boar), and fruits. This kitchen clearly belongs to a well-to-do household, perhaps in Amsterdam where the Antwerp-born painter worked and died. Meat and shellfish in abundance are usual symbols of banquets and carnal pleasure (Schneider, 1994; Bryson, 1995). They are frequently associated to sensual and erotic allusions, such as the cleavage of the young woman and her plucking of poultry. This activity was linked to the German popular 'vogelen' (derived from 'Vogel', bird), literally and crudely 'to f...' (Schneider, 1994). A cruder allusion is the presence of elongated phallus-like vegetables, as here the asparagus, or skewers in many other representations. In this painting, crustaceans were thus associated to sensuality and lust, whose symbols covered three quarters of the surface of the painting. But on the left side, one quarter of the work was used to balance this materialistic atmosphere with counter-symbols of Christian morality. As often in these paintings, a biblical scene (not yet precisely interpreted: Schneider, 1994) is represented in the background. The accumulation of bread on the left might be a Christian reminder of the miracle of the loaves and the fallen metal vases above the bread are a clear, classic signal about the precariousness of life. This painting is thus organized around the call for pleasure (represented among other symbols by the crustaceans) and the warning of virtue, as are other similar scenes by Pieter Cornelisz van Rijck and Frans Snyders.

STILL LIFE

A human presence, which was an important part of a genre painting, disappears in a still life, except for the viewer who contemplates objects, usually food set on a table. These works are also laden with symbolism, in which crustaceans are frequently present and sometimes play a prominent role. As the objects are viewed from a closer range, they, including crustaceans, appear bigger and are thus sometimes very precisely depicted. Crustaceans, live or dead, were usually painted with their natural colours in genre painting. In a still life, food is intended to be used soon, and it sometimes is half eaten. Consequently, the crustaceans are depicted after boiling, i.e., in a reddish colour that was often used by the artists to focus the viewer's eyes.

The styles of the still lifes produced followed the evolution of the Dutch society from austerity to affluence. We will present them in that order, although the dates of the different works may overlap, as the earlier style lasted and resurfaced on occasion.

The theme of the **'served tables'** was first illustrated by the 'Ontbijtjes' or light meals, usually breakfast, that became common during the first half of the 'Golden Period' of the 17th century. One classic example is the "Glass of White Wine and Crab" by Pieter Claesz (1644; fig. 27B.5). The painter was born in Burgsteinfurt, Westphalia but worked

Fig. 27B.5. Pieter Claesz, *Still Life with Glass of White Wine and Crab*, 1644. Oil on wood. Musée des Beaux-Arts de Strasbourg. Photograph E. Bacher. [Reproduced with permission.]

and died in Haarlem. He was one of the main promoters of the 'monochrome' still life, well exemplified by the general light-brown shade of the painting. The colour of *Cancer pagurus*, as that of the orange, enlightens the picture. The crab is finely depicted, down to its antennae, eyestalks, maxillipeds and pereiopod setae. Along with bread, a few hazelnuts, the fruit, wine and beer, it will constitute the main course of a light but delightful meal. However, the overall impression presented is not a home of poor people, but nevertheless an austere, probably Calvinist household. The silver plates, the decorated knife handle, the sweet orange, the pressed tablecloth are indicators of wealth and careful housekeeping. However, even this frugal meal is enjoyable, and its consumer must be reminded of life's shortness and precarious course, hence the presence of several counter-symbols: crumpled tablecloth, cracked nut, plate precariously placed at the edge of the table, fallen and empty glass. Watches, symbolizing the passing of time, have also been used in many other instances. On the whole, the crab thus is here a symbol of a reasonably pleasant, balanced life.

In a second period, around the mid-century, the affluence of their possessors became more obvious in the paintings. For instance Willem Claesz Heda, in Haarlem, retained the monochrome tonality and the general presentation of the table. But, in a setting like his "Still Life with Pie, Silver Ewer and Crab" (1658), the austerity is partly gone. The food, which includes a *Cancer irroratus*, is the same as in Pieter Claesz's preceding work. Beside the silver plates, a silver ewer and salt grinder, a silver-mounted *Nautilus* shell (symbol of overseas exchanges) and a fine leather box, are prominently displayed. Even with a few counter-symbols (fallen shell, crumpled cloth), this picture speaks of mounting wealth. Contemporary artists such as Abraham van Beyeren (who worked in different cities, among which The Hague, Delft, Amsterdam) tended to increase the richness of the setting and of the food through the addition of elaborate silver or golden utensils, of ham, various fruits, etc...., to meals which, from hearty 'ontbijtjes', were becoming true banquets. Hence the designation of 'banketje' or 'pronkstilleven' or 'luxury banquet-pieces' given to these paintings. The mounting abundance of food was sometimes compensated by a growing disorder of the table, as if the carefully-working painter was hinting at a criticism aimed at the rich owner of his work, since it sometimes looked "... as though the meal had been consumed by bears" (Bryson, 1995: 122). The monochrome tonality was replaced by brighter colours. Interestingly, the as yet almost ubiquitous *Cancer irroratus* now more often represented lying on its back, was often replaced by a more luxurious red *Homarus gammarus*. The appeal of this kind of art was such that still-life paintings were very abundantly produced. Abraham van Beyeren made many of them, along with such contemporary artists as Balthasar van der Ast, Osias Beert, Jacob van Es, Cornelis de Heem, Pieter de Ring (see table I).

Finally, during the second half of the 17th century, the banquet-pieces reached an apex in luxury and homarid lobsters became a staple of these works, not only for the taste and sensuality they represent but also for their appealing shape and their bright-red colour. The two masters of this style were Jan Davidsz de Heem (born in Utrecht, he worked in Leyden, Utrecht and Antwerp, where he died) and especially Willem Kalf (born in Rotterdam, worked in Paris, Rotterdam and Amsterdam, where he died). For instance, in a 1649 "Still Life with Fruits and Lobster", De Heem puts a lobster and two smaller crayfish beside a pile of diverse fruits, vegetal leaves, and an ornate mother-of-pearl and golden metal ewer. The abundance of fruits, very expensive items in 17th-century Holland, and the lobster are proofs of wealth that wants to be obviously, even ostentatiously, expressed. Counter-symbols were notably absent from these works, except for the odd watch in some of them. Other contemporary painters as Pieter de Ring produced similar works in which lobsters were the brightest objects. The techniques used to express their colour and texture are detailed in Wallert (1999). While Willem Kalf was also addressing his work to the very rich, he did so in an elaborate and elegant way. His lavish "Still Life with the Drinking Horn of the Saint Sebastian Archers' Guild, Lobster and Glasses" (1653) (fig. 27B.6) attests of his mastery. Although not reverting to the monochrome severity of Pieter Claesz, he achieves a paradoxical luxurious austerity through several means. The number of objects is severely reduced, but each is chosen for its representation of richness and beauty, from the thick edge of the table, the richly-decorated silver setting of the horn, the Venetian glass in the background and the decorated Dutch glass in the foreground, the Persian rug, to the silver plate and tray, and to the lobster. This is certainly one of the most accomplished representations of *Homarus gammarus* ever done by an artist. Claws, cephalothorax, abdominal segments, pereiopods, antennae, eyestalks, are carefully and exactly depicted. The flexion of the pleon, which is not folded as usual, but extended on the telson and uropods, is used by Kalf to study its reflection in the gleaming silver. Because,

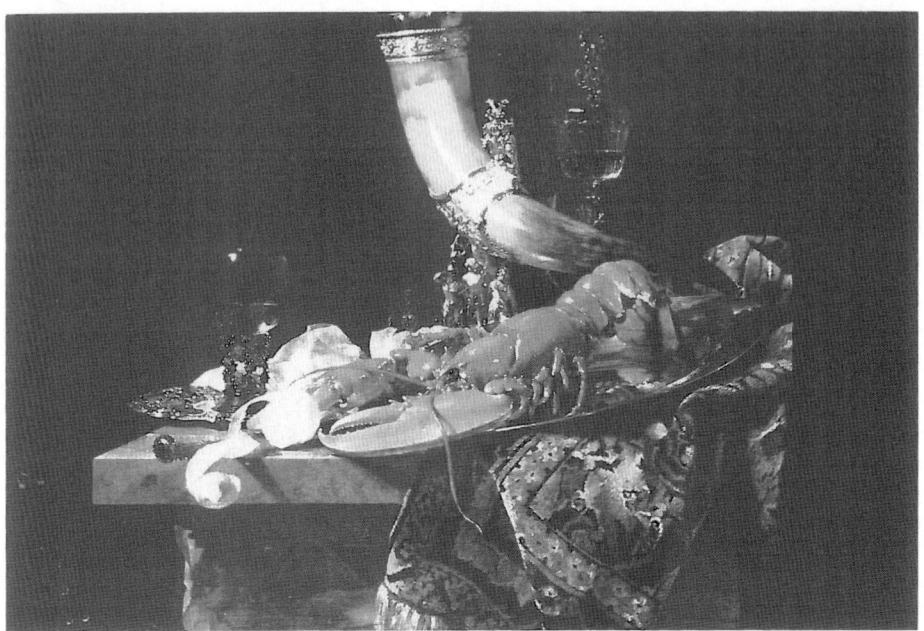

Fig. 27B.6. Willem Kalf, *Still Life with the Drinking Horn of the Saint Sebastian Archers' Guild, Lobster and Glasses*, c. 1653. Oil on canvas. National Gallery, London. [Reproduced with permission.]

as underlined by Schneider (1994) and Bryson (1995), the emphasis in Kalf's work is not any more on food, but on the crafted objects, the quality of which we have cited, and on the play of light on them. The black background increases the dramatic effect of the strong light that illuminates the table, and which becomes the theme of Kalf's art.

In summary, Willem Kalf paints "...illusions of wealth, to sell to the merchants of Amsterdam...His work touches on the deep-seated vein of fantasy, the one which sends crowds queuing up to look at...the Mona Lisa — or the rocks sent back from the moon" (Bryson, 1995: 127). It is quite striking that a lobster is part of such a dream. Although the usual interpretation of comparable works points to the taste for luxury of their owners, a different interpretation was given by the German poet Conrad von Würzburg (cited in Cox & Povey, 1995) who equated the lobster to Christ: as the dead boiled lobster was brighter in its red colour than when alive, so Christ escaped the misery of life and regained the splendor of his divinity through death. The validity of this interpretation can be questioned since Kalf did not include any counter-symbol in his painting, except for the crumpled rug and the position of a few objects on the table edge. It is worth noting that the appeal of the still life banquets among the Dutch fast-growing bourgeoisie was such that many painters (table I) competed for this market and the number of these art pieces is such that they are beyond a comprehensive survey. In most of them, the crustacean is represented by a lone lobster.

The celebration of luxury, ostentation and earthly preoccupation was balanced at the same period by the Counter-Reformation that inspired different schools of painting, particularly in catholic parts of Europe. We have already mentioned (Symbolism, Mythology, and Religion) the "Miraculous Draught of Fishes" painted by Rubens c. 1630, in which crustaceans symbolize saved souls, in a Flemish and Spanish environment quite different from the Dutch atmosphere. And, during the years of Van Beijeren's, De Heem's, and Kalf's activity, Murillo in Spain was painting the "Young Beggar" (1645-1655). He also used a sombre background to focus on a sunlit young boy, seated on the ground, dressed in rags, looking distressed, with most of his hair probably lost to scabies. Beside his dirty feet, the remains of a poor meal lay on the ground under the form of bits of shrimps. Far from the glory of lobsters, these humble crustaceans are here symbols of destitution and renunciation.

INCONSTANCY

The backward or sideward, somehow unpredictable gait of decapod crustaceans has triggered the interest of humans, probably throughout the centuries. This distinctive feature could not go unnoticed of people who lived in times when natural events were deciphered and interpreted as expressions of a misunderstood magical world. Moulting, and the particular behaviour associated with it, may also have been a strange process for observers. We have seen that their gait linked crabs and crayfish in the zodiac with the declining daylight of the month of June/July. In addition, the backward or sideward walk of crabs and lobsters earned them to symbolize changeableness, even fickleness, in a word, 'inconstancy'. One early example of this use is found in the Casa dei Vittii in Pompei, thus before 79 AD on a wall painting entitled "Cupid Riding the Crab". The two actors of the tale are painted in monochrome ochre on a black background, so as to attract the viewer's eyes. One character is a crab, relatively large-sized, apparently on the move, its eyestalks eagerly pointed forward, its ten pereiopods active. We can note again that, for a Roman artist, dealing with a symbol does not forbid accuracy, at least for the number of walking legs. Riding on his back is a Cupid, with wings and arrow, attempting to lead the crab with a string attached to the anterior part of the animal. But which figure is the leader here? The answer is in the backward stumble of the Cupid, who tries not to fall down following an unexpected forward move of the crab. What will happen next can only be guessed, depending on the whimsical crab, and Cupid, i.e., love, will keep moving in an inconstant way.

Fourteen centuries later, the fool on a crayfish back drawn by Dürer in 1494 illustrates a similar trend to support the commentary written in the 1511 edition of Sebastian Brant's "Das Narrenschiff — The Ship of Fools": "Who, lacking merit, seeks a prize / And on a fragile reed relies / His plans go backward crab-wise". The deep meaning of this sentence is associated with the controversy over the doctrine of predestination discussed by Luther, that is beyond the scope of this text; but the symbolism of inconstancy / going backward is very clear, even if the 'crab' is a crayfish in that case.

Sixteen hundred years after the Roman Cupid on his crab, and following Albrecht Dürer by a little more than a century, an anonymous Dutch artist of the 17th century, probably

Fig. 27B.7. Attributed to Abraham Janssens, *Inconstancy*, c. 1617. Oil on canvas. Statens Museum for Kunst, Copenhagen. [Reproduced with permission.]

Abraham Janssens, used a lobster with again a similar intent in the 'Inconstancy' (fig. 27B.7). It shows a woman whose presence is doubly enhanced by the use of four colors in respective clear-sombre contrasts: white skin – dark blue dress – pale yellow drapery – black background. As an allegory of inconstancy, this work is crammed of related symbols: the feminine gender of the woman, pointed by a prejudiced supposedly male artist; her youth, which will not last; her fleeting beauty attested by her face, her bare arms and breasts; she is wearing a crown, a symbol of the instability of power, and holding a quarter of moon, which is forever changing; and finally, she is grasping the tip of the telson of a live homarid lobster, which is trying to escape. In itself, and through its movement (and perhaps later moulting), the lobster is thus well related to inconstancy.

CONCLUSIONS

Crustaceans have been included in artistic representations for approximately at least over 3500 years. Judging from their number, they were much less considered than vertebrates or insects and their presence in art has often been discreet. However, their role became prominent during two historical times, the Roman period and the 17th-century Dutch and Flemish 'Golden Age'. With only few exceptions, big, frequently captured, edible decapods were pictured in art, with a distinct preference for crabs, lobsters, shrimps, and crayfish.

The motivations underlying their representations are found in the human concerns, related to the biological needs of feeding and species perpetuation, and to the metaphysical aspirations linked to the yearning for beauty, the need to understanding the world and the consciousness of death. It is thus not surprising that crustaceans, as other animals, have been associated to, and sometimes taken as symbols of, food, sex, aesthetics, religion. Some of them were appreciated for their shape and colour, the same and others triggered the interest of naturalists. If their association with food, water and the zodiac appear logical, some of their other symbolic and occasionally antagonistic stands are more surprising. The opposition is striking between the crabs as saved souls of Rubens, and the sensuality and lust symbolized by the lobsters of Van Nieulandt. The difference in attitude toward life is well illustrated by the use of *Cancer* and *Homarus* by Claesz and Kalf. And not many symbols can illustrate the contrast between luxury and poverty as well as the blazing lobster of Kalf and the dull shrimp fragments of Murillo. Perhaps most touching is the symbolic link drawn between crustaceans and inconstancy, which reaches to the very basis of human frailty, the unpredictable course of a lifetime, and the finite nature of life.

POST-SCRIPT AND ACKNOWLEDGEMENTS

This article has been written by a scientist whose research field is the adaptive ecophysiology of developing crustaceans and fish. Science has its way of transforming animals into sources of samples, drops of blood, tissue sections, DNA sequences and numerical data. Biologists though do not, or should not, forget that they are working on organisms. Sooner or later, someone working in organismal biology, or integrative biology, has to pause and consider the animals (or plants) he/she has been working on for a few or many years, in a really integrated way, i.e., one that attempts to consider living forms in all their dimensions. And then comes the realization, known but not fully realized that others, including artists, also have their say about life. This was the start of the story leading to this article, followed by a long search for crustaceans in art, with the continuous help (in this and all parts of the process) of my wife and daughters, Mireille (companion of numerous museum visits), Anne and Isabelle (who kept hunting crustaceans in art over the years). This preamble makes also clear that the author, aside from a strong interest in history of art, has no particular training in that field, which may contribute to explain, if not justify, the gaps and perhaps awkwardness sometimes present in the text. It appears also clearly from the considerations in the above that the amplitude of the subject demands by necessity future work to fully open the field, i.e., extend it to other cultures, find other symbolisms, etc. — all hopefully based on collaboration between scientists and art historians.

The help of many people has been instrumental both in the preparation and publication of this study, and I wish to extend my sincere and appreciative thanks to them all. Its first public appearance, in an oral form, was as a talk given at the Second European Crustacean Conference held in Liège, Belgium in September 1996. Dr. André Péqueux was kind enough to accept it, and to convince the other organizers that it might entertain scientific delegates. The talk was later presented in March 2001 in the "Séminaire d'Anthropozoologie" organized by Prof. François Poplin at The Muséum National d'Histoire Naturelle (MNHN) in Paris, and at the January 2003 SICB meeting in Toronto, thanks to Dr. Stacia Sower, University of New Hampshire.

The late Dr. Raymond B. Manning, Senior Zoologist at the National Museum of Natural History, Smithsonian Institution, who attended the 1996 Liège meeting, was the first to encourage me to write and publish a manuscript on the topic presented at the conference, and I am indebted to him for this crucial impetus. Several crustaceologist colleagues made persistent efforts to have the manuscript published and to find funds for its printing, particularly Dr. Raymond B. Manning, and also Paul Clark, Natural History Museum, London (whose help to improve the English language was also appreciated), along with Dr. Joel W. Martin, former President of the Crustacean Society (TCS), Natural History Museum, Los Angeles, Dr. Jens T. Høeg, then President of TCS, University of Copenhagen, Dr. David K. Camp, Editor of TCS, Florida Marine Research Institute, Saint Petersburg, and his daughter Allison Ben David, Museum of Fine Art, Saint Petersburg, FL. More urgent professional matters linked to research and teaching in aquatic ecophysiology (science versus art) delayed the completion of the manuscript until a fruitful and decisive conversation in early 2010 in Utrecht with Dr. J. C. von Vaupel Klein, formerly at the University of Leiden. As Editor of the Treatise on Zoology, Crustacea (along with the late Prof. Jacques Forest, MNHN, and Advisory Editor Prof. Frederick R. Schram, formerly at the University of Amsterdam, who further improved the English of the manuscript), he accepted the publication of this study as a chapter in Volume 4, for which I warmly thank him and Koninklijke Brill Academic Publishers, Leiden. I am grateful for the efficient and kind help from Mr. Michiel Thijssen, M.Sc., Senior Acquisitions Editor Biology & History of Science at Brill.

I also received advice, references, and photographs from the following people: Dr. Valérie Boudier, Paris; Mr. Chauvet, Musée Picasso, Paris; Prof. Marit E. Christiansen, University of Oslo; Ms. Teresa E. Cinquantaquattro, Sopritendente, and Ms. Alessandra Villone, Archivist, Soprintendenza Speciale per i Beni Archeologici di Napoli e Pompei; Ms. Louise Engell Behrensdorff, Consultant, Copydan Billed Kunst, Copenhagen; Prof. Danièle Guinot, Muséum National d'Histoire Naturelle (MNHN), Paris; Dr. Christophe Haond, Montpellier; Prof. L. B. Holthuis (†), Nationaal Natuurhistorisch Museum [currently: Naturalis Biodiversity Center], Leiden; Mr. Daragh Kenny, National Gallery Picture Library, London; Dr. Wulf Kobusch, Universität Bielefeld; Prof. Hans Laufer, Univer-

sity of Connecticut, Storrs; Ms. Lorène Linarès-Henry, Librarian, Musée de l'Arles Antique; Dr. Pierre Noël, MNHN, Paris; Ms. Catherine Paulus, Assitant Curator, Musée des Beaux-Arts de Strasbourg; Prof. François Poplin, MNHN, Paris; Mr. Michel Richard, Réunion des Musées Nationaux, Paris; Dr. Luciana Romeri, Université de Caen; Dr. Guiomar Rotllant, IRTA, Tarragona; Dr. Michelle de Saint-Laurent (†), MNHN, Paris; Mr. Jean-Pierre Samoyault, Conservateur général du Patrimoine, Manufacture des Gobelins, Paris; Mr. Claude Sintes, Curator, Musée de l'Arles Antique; Dr. Pierre Thuet and Prof. Jean-Paul Trilles, Université Montpellier 2; Prof. J.-P. Truchot, Arcachon; Dr. Jean Vannier, University of Lyon; Mlle Renée Zuza, Mobilier National, Paris. To all, my sincere thanks.

APPENDIX

Species names mentioned in this chapter with author and date

MALACOSTRACA
HOPLOCARIDA
   STOMATOPODA
      Squilla mantis (L., 1758)
EUMALACOSTRACA
  PERACARIDA
   ISOPODA
      Porcellio scaber Latreille, 1804
  EUCARIDA
   DECAPODA
    DENDROBRANCHIATA
      Penaeus kerathurus (Forskål, 1775)
    ASTACIDEA
      Astacus astacus (L., 1758)
      Austropotamobius pallipes (Lereboullet, 1858)
      Homarus gammarus (L., 1758)
    PALINURA
      Palinurus elephas (Fabricius, 1787)
      Panulirus penicillatus (Olivier, 1791)
    BRACHYURA
      Cancer irroratus Say, 1817
      Cancer pagurus L., 1758
      Carcinus maenas (L., 1758)
      Pachygrapsus marmoratus (Fabricius, 1787)

BIBLIOGRAPHY

ALLMON, W. D., 2007. The evolution of accuracy in natural history illustration: reversal of printed illustrations of snails and crabs in pre-Linnaean works suggests indifference to morphological detail. — Archives of Natural History, **34**: 174-191.
ARNOLD, D. L., 1982. Pueblo pottery. — National Geographic, **162**: 593-605.

AVRIL, F. & N. REYNAUD, 1993. Quand la peinture était dans les livres. — Pp. 1-48. (Flammarion, Paris.)
BARON, F., 1981. Les fastes du Gothique. Le siècle de Charles V. — Pp. 1-456. (Réunion des Musées Nationaux, Paris.)
BODSON, L., 2008. Les connaissances zoologiques de l'Antiquité grecque et romaine: aperçu de leurs spécificités fondamentales et de leur actualité. — Soc. Roy. Cercles Naturalistes Belgique, **2008**: 1-17.
BRYSON, N., 1995. Looking a the overlooked: four essays on still life painting. — Pp. 1-192. (Reaktion Books, London.)
CHARBONNEAU-LASSAY, L., 2006. Le bestiaire du Christ. — Pp. 1-997. (Albin Michel, Paris.)
CHARLES-PICARD, G., 1978. L'âge d'or de la mosaïque romaine en Afrique du Nord. — Dossiers de l'Archéologie, **31**: 12-31.
CHARMANTIER, G., 1996. Crustaceans in art. — In: Second European Crustacean Conference, Liège, Belgium, Abstracts, p. 155.
CHATER, A. U., 1988. Woodlice in the cultural conciousness of modern Europe. — Isopoda, **2**: 21-39.
CLAUT, S., 2000. Iconografia eucaristica nell'Alto Veneto. — In: L. ROMERI (ed.), I gamberi alla tavola des Signore. Civis, (Suppl.) **16**: 63-82.
COX, N. & D. POVEY, 1995. A Picasso bestiary. — Pp. 1-208. (Academy Editions, London.)
DELORME, J. & C. ROUX, 1987. Guide illustré de la faune aquatique dans l'art grec. — Pp. 1-176. (APDCA, Juan-Les-Pins.)
DUPONT, J. & C. GNUDI, 1994. La peinture Gothique. — Pp. 1-215. (Skira, Geneva.)
ELBERT-SCHIFFERER, S., 1999. Natures mortes. — Pp. 1-420. (Citadelles & Mazenod, Paris.)
GRAY, B., 1977. La peinture Persane. — Pp. 1-191. (Skira-Flammarion, Geneva, Paris.)
GRIMM, C., 1992. Natures mortes. — Pp. 1-251. (Herscher, Paris.)
GRMEK, M. D. & D. GUINOT, 1965a. Les crabes chez Ulysse Aldrovandi: un aperçu critique sur la carcinologie du XVIe siècle. — Vie et Milieu, (Suppl.) **19**: 45-64.
— — & — —, 1965b. Les Crustacés dans la matière médicale européenne au XVIe siècle. — Revue d'Histoire des Sciences et de leurs Applications, **18**: 55-71.
GURATZSCH, H., 1980. L'âge d'or de la peinture flamande et hollandaise. — Pp. 1-303. (VNU Books International, Paris.)
HALL, J., 1996. Dictionary of subjects and symbols in art. — Pp. 1-349. (John Murray, London.)
HARADA, E., 1993. Carcinology in classical Japanese works. — In: F. R. SCHRAM (ser. ed.), F. TRUESDALE (ed.), History of carcinology. Crustacean Issues, **8**: 243-258. (A. A. Balkema, Rotterdam.)
HOLTHUIS, L. B., 1967. Schaaldieren (Crustacea) afgebeeld of postzegels. — Zool. Bijdr., **8**: 1-21.
— —, 1991. Marcgraf's (1648) Brazilian Crustacea. — Zool. Verhand., **268**: 1-123.
HOPKIN, S., 2003. Woodlice, chiselbobs and sow-bugs. — British Wildlife, **14**: 381-387.
HUARD, P. & D. GUINOT, 1965a. Les crabes de Chine dans une série d'aquarelles provenant de Dabry de Thiersaut. — Vie et Milieu, (Suppl.) **19**: 35-43.
— — & — —, 1965b. Les crabes de Chine dans une série d'aquarelles provenant de Dabry de Thiersaut. — Bulletin de l'Ecole Française d'Extrême-Orient, **52**: 551-557, pls. LXXV-LXXIX.
JACKSON, C. E., 2012. Fish in art. — Pp. 1-248. (Reaktion Books Ltd., London.)
JEAN-RICHARD, P., 1997. Graveurs en taille-douce des Anciens Pays-Bas. — Pp. 1-175. (Réunion des Musées Nationaux, Paris.)
MAI, E., S. EBERT-SCHIFFERER & P. BEUSEN, 1997. L'art gourmand. — Pp. 1-383. (Crédit Communal, Bruxelles.)
MONOD, TH. & L. LAUBIER, 1996. Les Crustacés et l'Homme. — In: P. P. GRASSÉ (ser. ed.), Traité de zoologie, **7** (2), J. FOREST (ed.), Crustacés: généralités et systématique, pp. 166-186. (Masson, Paris.)

OMORI, M. & L. B. HOLTHUIS, 2000. Crustaceans on postage stamps from 1870-1997. — Rep. Tokyo Univ. Fisher., **35**: 1-89.

PÄCHT, O., 1997. L'Enluminure Médiévale. — Pp. 1-224. (Macula, Paris.)

POGNON, E., 1979. Les Très Riches Heures du Duc de Berry. — Pp. 1-123. (Seghers, Paris.)

RIGAUX, D., 2000. La Cène aux écrevisses: une image spécifique des Alpes italiennes. — In: L. ROMERI (ed.), I gamberi alla tavola des Signore. Civis, (Suppl.) **16**: 11-28.

ROBERT-JONES, P. & F. ROBERT-JONES, 1997. Pierre Brugel l'Ancien. — Pp. 1-352. (Flammarion, Paris.)

ROMERI, L., 2000. I gamberi alla tavola des Signore. — Civis, (Suppl.) **16**: 1-95.

SCHAMA, S., 1991. The embarrassment of riches. An interpretation of Dutch culture in the Golden Age. — Pp. 1-698. (Fontana Press, HarperCollins, London.)

SCHNEIDER, N., 1994. Les natures mortes. — Pp. 1-215. (Benedikt Taschen Verlag Gmbh, Cologne.)

STERLING, C., 1985. La nature morte, de l'Antiquité au XXe siècle. — Pp. 1-163. (Macula, Paris.)

SWINTON, G., 1972. Sculpture of the Eskimo. — Pp. 1-255. (The Canadian Publishers, McClelland and Stewart, Toronto.)

THIBAUD, R.-J., 1994. Dictionnaire de l'art Roman. — Pp. 1-350. (Dervy, Paris.)

TOLNAY, C. DE & P. BIANCONI, 1968. Tout l'oeuvre peint de Bruegel l'Ancien. — Pp. 1-120. (Flammarion, Paris.)

TYLER, R., 1990. American canvas. The art, eye and spirit of pioneer artists. — Pp. 1-208. (Portland House, Crown Publishers, New York.)

WALKER, J. & J. D. PROWN, 1987. La peinture Américaine. Des origines à l'Armory Show. — Pp. 1-134. (Skira, Geneva.)

WALLERT, A., 1999. Still lifes: techniques and style. The examination of paintings from the Rijksmuseum. — Pp. 1-112. (Rijksmuseum, Amsterdam.)

YOUNG, C. C., 2000. Variations on a theme: crayfish at the Last Supper. — In: L. ROMERI (ed.), I gamberi alla tavola des Signore. Civis, (Suppl.) **16**: 45-56.

CHAPTER 54

ORDERS LOPHOGASTRIDA BOAS, 1883, STYGIOMYSIDA TCHINDONOVA, 1981, AND MYSIDA BOAS, 1883 (ALSO KNOWN COLLECTIVELY AS MYSIDACEA)[1])

BY

KARL J. WITTMANN, ANTONIO P. ARIANI AND JEAN-PAUL LAGARDÈRE

Contents. – **Introduction** – Outline of the orders – Historical outline – Terminology and definitions. **External morphology** – Habitus – Size of adults – Carapace – Cephalothorax – Pleon – Telson – Integument and colour – Cephalic appendages – Thoracic appendages – Pleonal appendages. **Internal morphology** – Musculature – Nervous system – Sensory organs – Digestive system and digestion – Circulatory system – Respiratory system – Reproductive system – Excretory system and excretion – Endocrine organs. **Reproduction and sexuality** – Sexual dimorphism – Intersexuality – Sex ratio – Mating and oviposition – Fecundity – Incubation – Adjustment of reproductive parameters. **Development and moulting** – Marsupial development – Moulting and growth – Regeneration – Life cycle. **Ecology and ethology** – Habitat and distribution – Locomotion, orientation, and taxis – Migration – Social aggregation – Grooming – Food and feeding – Trophic interactions – Symbiotic associations – Parasites. **Ecological and economic importance** – Contribution to biodiversity – Impact on ecosystems – Importance in fisheries. **Phylogeny and biogeography** – Fossil record – Phylogeny – Biogeography. **Systematics** – Guidelines to classification – Classification – Keys to the families, subfamilies, and tribes of the Lophogastrida, Stygiomysida, and Mysida. **Appendix. Acknowledgements. Bibliography.**

INTRODUCTION

The unity or disunity of the **Mysidacea** as competing hypotheses is currently among the central controversies in phylogenetic research on malacostracan Crustacea. In the first

---

[1]) Revised and updated from the treatise by H. Nouvel (†), J.-P. Casanova & J.-P. Lagardère (1999), November 2013; latest additions February 2014.

volume of the present series, Forest (2004) noted that the traditional systematics, which outline the unity of the order Mysidacea as a container for the suborders Mysida and Lophogastrida, is under discussion. As an alternative hypothesis he noted a potential placement of the Mysida outside the Peracarida, in this case he envisaged a potential approach to the Eucarida. Ultimately, however he inclined towards the traditional hypothesis and scheduled the Mysidacea to be treated at order level in the present series. Meanwhile, new **morphological evidence** (Wirkner & Richter, 2007, 2010) would seem to favour the traditional hypothesis, whereas new **genetic evidence** (Spears et al., 2005; Meland & Willassen, 2007) points to the alternative hypothesis. Jenner et al. (2009), in turn, argued that any previously published evidence is unable to solve the problem. In fact, both competing hypotheses are strongly supported by data, and no definite decisions can yet be seen to emerge. As a provisional option within a controversial environment, the Lophogastrida, Stygiomysida, and Mysida are treated here as separate orders, as discussed below in 'Outline of history' and 'Phylogeny'.

Outline of the orders

The Lophogastrida, Stygiomysida, and Mysida have a great number of characters in common, stimulating decade-long discussions on the extent to which these characters are plesiomorphic within the Malacostraca and whether these taxa are to be kept separately or to be united in the traditional order Mysidacea (see below, 'Outline of history' and 'Phylogeny'). Together they represent shrimp-like (figs. 54.1-54.5) Eumalacostraca, at least some of them pertaining to the Peracarida. The well-developed **carapace** covers most of the **cephalothorax**, or even the entire cephalothorax in certain Lophogastrida (*Gnathophausia*, *Chalaraspidum*; fig. 54.2A, C). It is fused dorsally with the cephalic region and at most four anterior thoracic somites, and projects freely above some of the posterior ones (exceptions below, 'Carapace'). The free space below the carapace contributes to a **carapace cavity**. Dorsally, the carapace is often divided by a distinct, transverse **cervical sulcus** (figs. 54.4, 54.5, 54.7E, F, 54.9G), in lateral view located more or less above the mandibles. Laterally it is mostly not fused with the thoracic pleurites, leaving space for the lateral portions of the carapace cavity. **Respiratory tissue** is provided by the wall of the carapace cavity (fig. 54.31D; Mysida and Stygiomysida) or by **gills** on the coxae of the thoracic appendages (figs. 54.6E, 54.16C, E, G; Lophogastrida). The **respiratory current** is to a minor extent driven by a small leaf-like, first thoracic epipod, but mainly by the movements of the thoracic exopods. Paired compound **eyes** are normally well developed, moveable by **eyestalks**. Stalks and/or cornea are occasionally fused and/or reduced, particularly in species from deep-water and subterranean habitats. Eight pairs of **thoracopods** are present (figs. 54.16, 54.17), their exopods with a basal plate and with a setose multi-segmented flagellum. The exopods serve for swimming and support respiratory currents, in many species also **filtration currents**. Exopods 1, 2, and 8 may be reduced or absent in certain taxa (figs. 54.16A, B, 54.17A, B). The anterior 1-2, in certain taxa up to 3-4, pairs of thoracic endopods are mostly specialized as maxillipeds (figs. 54.15M, N, 54.16A-C, 54.17A-C) or 'gnathopods' (fig. 54.18H). The remaining endopods were previously often termed 'pereiopods' (e.g., Tattersall & Tattersall, 1951) but are equipped for manipulations such as grasping food rather than for walking, although some crawling does occur. As in most Eumalacostraca, the **gonopores** are located on the coxae of the sixth thoracopods in females, and of the eighth thoracopods in males (fig. 54.20). The **uropods** are well developed, biramous, and flattened. Together with the **telson** they form a strong **tail fan** (fig. 54.22) suitable to support **body flipping** for sudden **escape movements**. The **brood pouch** or **marsupium** (fig. 54.19) is formed by 1-7 pairs of **oostegites** on the posterior thoracic appendages. The young pass through an **embryonic** and two non-feeding **larval stages** within the marsupium. They are liberated as miniature 'adults' equipped with all appendages and capable of self-sustained life, but lacking secondary sexual characteristics.

The **Lophogastrida** (figs. 54.1A, 54.2) are characterized by external gills (fig. 54.6E, G), seven pairs of oostegites (fig. 54.19A), and well developed pleopods (natatory in both sexes). With the **Stygiomysida** (figs. 54.1B, C, 54.3) they share the absence of **statocysts** in the tail fan and the absence of well-developed tubular **penes**. The Stygiomysida have 4-7 pairs of oostegites (fig. 54.19D), bilobate male gonopores (fig. 54.20B; Wittmann, 2013b), biramous pleopods (fig. 54.17G-K) with largely, but never entirely, reduced rami in both sexes, and spinose inner lobes on the sympods of the uropods (fig. 54.22F-H). They share the absence of gills with the **Mysida** (figs. 54.1D, E, 54.4, 54.5). This order is characterized by 1-7 pairs (mostly 2-3) of functional oostegites, in most taxa also by the presence of statocysts in the endopods of the uropods (figs. 54.22K-O, 54.26, 54.27), and well-developed tubular penes (fig. 54.20C, E-G); with few exceptions the pleopods are reduced to small, unsegmented rods or plates in females (fig. 54.21F, H), in part also in males (fig. 54.21R-U).

Historical outline

An excellent overview of the **classification history** of Crustacea is given in the third volume of the present series (Monod & Forest, 2012), mainly dealing with class to order levels. The present outline gives complementary information about the difficult birth, the rise, and the potential decline or renaissance of the order Mysidacea Boas, 1883.

The first mysid taxa. — The first description of a species of the Mysida in today's understanding was given by O. F. Müller (1776) for *Cancer flexuosus*, defined as "*Macrourus pedibus pectoris duplici serie, abdominis membranis, quinque utrinque, branchialibus*". Today it is known as *Praunus flexuosus*, a common near-shore species from coastal waters of the north-east Atlantic and the Baltic. Shortly thereafter, Slabber (1778) published the first realistic drawing of a mysid species, based on the knowledge of that time correctly identified by him as *Cancer*, today known as *Mesopodopsis slabberi*. A few years later, Fabricius (1780) established a number of new taxa — *Cancer pedatus*, *Cancer bipes*, and *Cancer oculatus* — from waters off Greenland. All his types have been lost. These taxa experienced quite different fates: *Cancer pedatus* remained of obscure identity (according to Holmquist, 1958, possibly any euphausiacean), *Cancer bipes* became the well-known

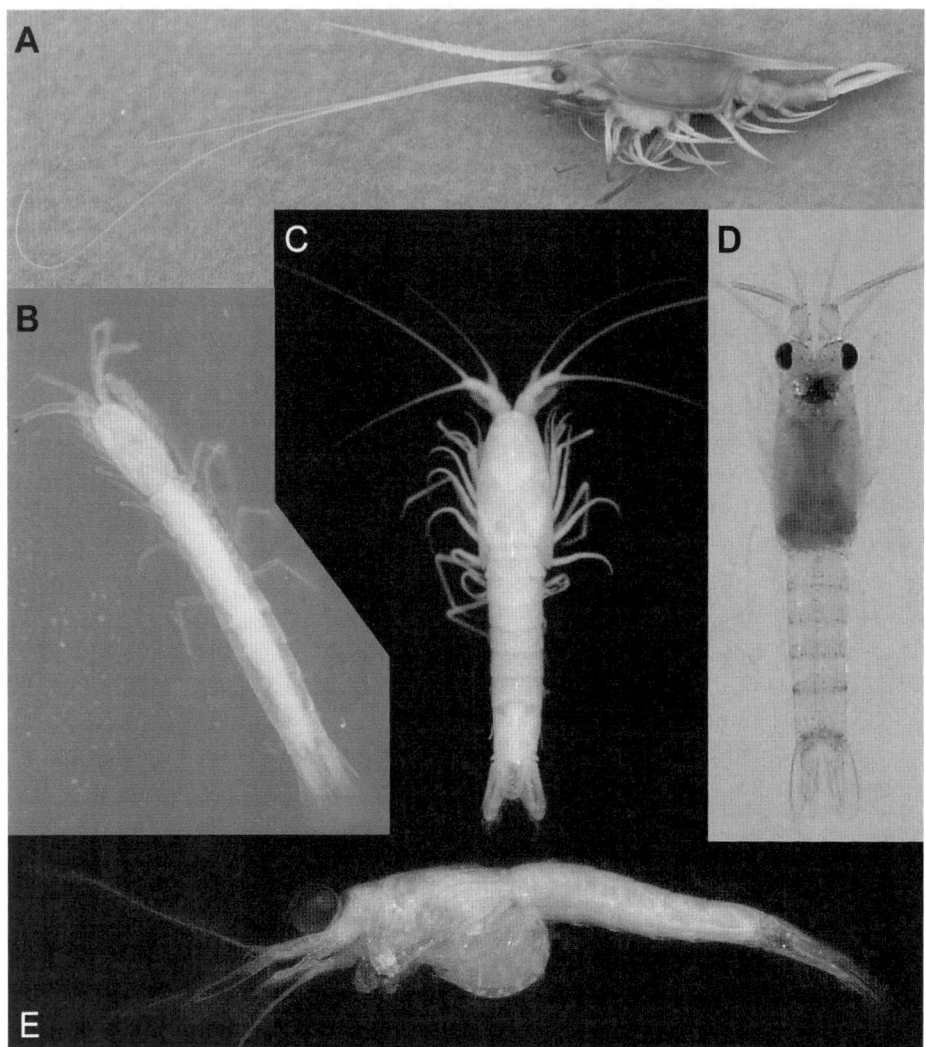

Fig. 54.1. Diversity of Lophogastrida (A), Stygiomysida (B, C), and Mysida (D, E). A, *Gnathophausia zoea* Willemoës-Suhm, 1873; B, *Stygiomysis hydruntina* Caroli, 1937; C, *Spelaeomysis bottazzii* Caroli, 1924; D, *Heteromysis wirtzi* Wittmann, 2008; E, *Hemimysis lamornae mediterranea* Băcescu, 1937. [A, photo José Antonio González; B, after Inguscio, 1998; C, E, photo Antonio P. Ariani; D, after Wittmann, 2008 (photo Peter Wirtz).]

leptostracan *Nebalia bipes*, and *Cancer oculatus* became the now well-known mysid *Mysis oculata*, widely distributed in Arctic waters.

***Establishment of* Mysis**. – Explicitly referring to *Cancer pedatus*, Latreille (1802) established the genus *Mysis*, and associated *Cancer oculatus* and *Cancer bipes* with his new genus. The **etymology** of '*Mysis*' was not indicated, so it remained unclear which, if any, of several potential meanings was intended by him. No later authors commented on the fact that Agassiz (1843: 19) and similarly also Costa (1847: 3, in footnote) had indicated for the genus *Mysis* "μύσις, oculorum conniventia" (transl.: *"mysis*, closing of eyes"), obviously referring to the Greek noun μύσις (*musis* = *mýsis* = the pressing together of lips or eyelids; Hentschel & Wagner, 1976). Nonetheless, the meaning is obscure in the present context and these etymological notes appear to simply be literal translations.

Inversion of **Mysis**. – Leach (1815) redefined the **morphological concept** of the genus *Mysis* by adding just the **brood pouch** — not mentioned by previous authors — as a diagnostic character: *"Ad feminae abdominis basin est uterus externus e membranis duobus concavis falvuliformibus eformatus, quo pulli nuper ex ovo exclusi vivunt, crescunt"*. He restricted *Mysis* to three species rearranged by him, one of which, *Mysis Fabricii*, is today considered a junior synonym of *Mysis oculata*. From that time onwards most authors emphasized *Mysis* as a mysid in today's understanding. In 1830, Leach described a new genus and species, *Megalophthalmus Fabricianus*, and referred *Cancer pedatus* to the new taxon. In conclusion, by his combined papers from 1815 and 1830, he had inverted the precedence explicitly indicated by Latreille (1802) for *Cancer pedatus* over *Cancer oculatus* as type species of the genus *Mysis*. No types are known for *Megalophthalmus Fabricianus*, which remained essentially unrecognizable; nonetheless, it was tentatively referred to different *Mysis* species by various authors (Holmquist, 1958).

Further establishment and crisis. – The redefined *Mysis* became **type genus** of the family Mysidae Haworth, 1825, and later also of the suborder Mysida Boas, 1883, the order Mysidacea Boas, 1883, and also of additional, today poorly known and outdated higher taxa, such as the suborder (as "Tribus") Mysidea Dana, 1852, and the superorder Mysiformida Czerniavsky, 1882. Despite being originally based on a taxon of obscure identity, the *Mysis*-based taxa and terminologies proposed by various authors (Haworth, 1825; Burmeister, 1837; Dana, 1852; Czerniavsky, 1882; Boas, 1883; Norman, 1892; Calman, 1904; Hansen, 1910) remained collectively stable for many decades. They then came under risk of collapsing due to the **priority rules** explicitly established with the **nomenclatural code** in the 20th century. Tattersall & Tattersall (1951) tried to circumvent the problem by claiming that *"C. pedatus* has not been identified with certainty, but is probably a synonym of *C. oculatus"*. However, their attempted solution is poorly compatible with the species descriptions given by Fabricius (1780) and, in addition would have required a **reversal of precedence**. Holmquist (1958) proposed as an effective solution that the International Commission on Zoological Nomenclature (ICZN) should use its plenary power to suppress the taxon *Cancer pedatus* Fabricius, 1780, and to fix *Cancer oculatus* Fabricius, 1780, as the type species of *Mysis*. Finally, these proposals were adopted by the ICZN (1959) together with a package of additional rules necessary in this context.

Schizopoda. – In 1817, Latreille established the Schizopodes (later termed Schizopoda) for reception of mysids and leptostracans, characterized by a well-developed carapace and biramous thoracopods. This system was developed further by Latreille (1825), H. Milne Edwards (1837), Dana (1852), and others. Then including the Euphausiacea, the Schizopoda became a widely acknowledged standard for many decades in various variants, including (Czerniavsky, 1882) or excluding (G. O. Sars, 1885a; Zimmer, 1909; Illig, 1930) the leptostracans.

Lophogastrida. – The age of the great oceanographic expeditions yielded the first lophogastrids. Almost eight decades after the first description of a mysid, the first lophogastrid was described by Dana (1852) as *Eucopia australis*. Based on this species, he established the Eucopidae (now Eucopiidae) as a new family within penaeid decapods by defining "*Carapax non rostratus, fronte integro. Pedes thoracici elongato-palpigeri, palpis natatoriis. Maxillipedes 2di 3tii et pedes 1mi monodactyli et suprehensiles*". A few years later, M. Sars (1857) established the genus *Lophogaster* with the species *Lophogaster typicus*, a benthopelagic crustacean with wide distribution in the eastern Atlantic and Mediterranean. The generic name is derived from the Greek λόφος (= tuft, mane) and γαστήρ (= stomach). In 1870, his son G. O. Sars used this genus to define the family Lophogastridae within the Schizopoda. Based on considerations on the homology of crustacean appendages, Boas (1883) aborted the Schizopoda and placed the Lophogastridae (without Eucopiidae) together with the Mysidae, both distant from the Euphausiacea, at suborder rank within the order **Mysidacea**, established by him. In fact, Lophogastrida and Mysida share a great number of morphological characters, as shown above in 'Outline of the orders'.

Stygiomysida. – The first representative of this order, *Spelaeomysis bottazzii*, was discovered in subterranean waters of Apulia (south-eastern Italy) and briefly described by Caroli (1924) as representative of a new mysidacean genus. Almost at the same time, a second genus and species, *Lepidophthalmus servatus*, was described by Fage (1924, 1925) from subterranean waters of Zanzibar (south-eastern Africa) as a representative of the new family **Lepidophthalmidae**. However, the generic name created by him is a junior homonym of a decapod genus. Zimmer (1927) noted this **homonymy** and created the replacement name *Lepidops*. Clarke (1961a) noted that this last name was again a junior homonym of a decapod genus and created the second replacement name *Lepidomysis*. Finally, Ingle (1972) withdrew this last name in favour of its senior synonym, *Spelaeomysis* Caroli, 1924. The invalid family name Lepidophthalmidae was twice replaced by family names based on junior homonyms of decapods, first by Lepidopidae Stammer, 1933, and then by Lepidopsidae Villalobos, 1951, and finally became the still valid name **Lepidomysidae** Clarke, 1961a. Meanwhile, Caroli (1937) had discovered a highly aberrant mysidacean at the same groundwater localities as in 1924. He named it *Stygiomysis hydruntina*, by deriving the new genus name from the infernal river Στυξ of the ancient Greek and Roman mythologies. Based on this genus he defined a new family, **Stygiomysidae**. His discussion pointed to the considerably **reduced carapace** (figs. 54.1B, 54.3A; shorter than in Petalophthalmidae), the thoracopods 2-4 resembling **gnathopods** (fig. 54.17C, D; as in Eucopiidae), the peculiar morphology of the **uropodal sympods** (fig. 54.22F; reminiscent of those of the stomatopods), eyes reduced to plates, missing statocysts, and probably seven [actually four] pairs of oostegites as in [unlike] *Spelaeomysis*.

Disputes on the Mysidacea. – Accompanied by a number of additional modifications, the unity of the Mysidacea within the superorder Peracarida was acknowledged by most authors up to the 2000s (Norman, 1892; Calman, 1904; Holt & Tattersall, 1905;

Hansen, 1910; Tchindonova, 1981; Nouvel et al., 1999). Nonetheless, Watling (1981, 1983, 1999), Schram (1984), Dahl (1992), Kobusch (1999), and Martin & Davis (2001) had meanwhile developed a variety of alternative phylogenetic reconstructions by using various methods of **morphological analysis**. Their conclusions had little in common among each other, apart from doubting the unity of the order Mysidacea and placing the Mysida and Lophogastrida as separate orders in various positions within or outside the Peracarida, respectively. Nonetheless, more recent detailed morphological analysis suggested a renaissance of the unity of the Mysidacea within the Peracarida (De Jong-Moreau & J.-P. Casanova, 2001; Richter & Scholtz, 2001; Wirkner & Richter, 2007, 2010).

Recent **genetic analyses** pointed again to a disunity of the Mysidacea. From rDNA sequence analysis, Jarman et al. (2000), Spears et al. (2005), and Meland & Willassen (2007) obtained the Lophogastrida mostly within the Peracarida, but the Mysida elsewhere within the Eumalacostraca, in any case distant from the Lophogastrida. In their analysis, Meland & Willassen (2007) included also two species of Stygiomysida, formerly considered a suborder of the Mysida, and obtained the Stygiomysida closer to the Lophogastrida.

Jenner et al. (2009) compared and tested different methods of **phylogenetic reconstruction** and concluded that the published evidence could not satisfactorily resolve the relationships between different higher taxa within the Eumalacostraca. This uncertainty is reflected by the above-discussed pendular movements of mysidacean classification in the 1980-2000s. Irrespective of a potential unity or disunity of the Mysidacea, Richter (2003) concluded a **monophyly** of the Lophogastrida from apomorphic characters of the mouthparts, and Wittmann (2013b) a monophyly of the Mysida based on the armature of the male genital pores. Unless new evidence is presented, keeping the Lophogastrida, Stygiomysida, and Mysida as different taxa at whatever level promises more stable systematics than any kind of fusion.

Terminology and definitions

General terminology. – The present terminology essentially follows Tattersall & Tattersall (1951), with emendations by Băcescu (1940, 1954) and Ariani & Wittmann (2000). For reasons of brevity, '**mysids**' is here often used for Mysida, '**lophogastrids**' for Lophogastrida, '**stygiomysids**' for Stygiomysida, and '**mysidaceans**' for the conglomeration of Lophogastrida, Stygiomysida, and Mysida. The appendages along with their segmental and setation patterns are distinguished as in figs. 54.10, 54.14, 54.18, 54.20, 54.21. In this context, 'multi-segmented' means more than two-segmented. Features of the foregut are termed according to De Jong-Moreau & J.-P. Casanova (2001); synonyms used by Kobusch (1998) are given in parentheses in the legend of fig. 54.29. Tagmata and body segmentation are termed according to Gruner & Scholtz (2004) in the first volume of the present series.

The terminology of stages within the **moult cycle** is given by Charmantier-Daures & Vernet (2004) in the first volume of the present series. They essentially follow the system of Drach (Drach & Tchernigovtzeff, 1967), but use the term '**diecdysis**' instead of the misleading 'intermoult period'. This last term was previously more commonly used in the mysid literature (Cuzin-Roudy & Tchernigovtzeff, 1985). **Embryonic** and **larval stages** (**nauplioids, postnauplioids**) in the marsupium are distinguished according to Wittmann (1981a).

Taxonomy. – The families and subfamilies of the Lophogastrida, Stygiomysida, and Mysida are acknowledged essentially according to Meland & Willassen (2007); for taxonomic modifications of their scheme see 'Systematics' below. Wherever convenient, taxa quoted by previous authors are cited in the currently valid form, for reasons of brevity in most cases without additional indication of the originally quoted taxa.

Classification and metric measurements. – Literature data were readjusted, transformed and/or recalculated if they differed from the following scheme: **Body size** (mm) is the distance from the anterior margin of the carapace to the end of the telson without spines. **Egg size** (mm) is expressed as the geometric mean of apparent length and width of the mostly sub-ellipsoid bodies. Degree of **endemism** is expressed according to Myers & De Grave (2000). Salinity (S) is expressed as a dimensionless equivalent of electric conductivity, and the Venice System was used for **salinity classification** of water bodies (Por, 1972). The suffixes '-haline' and '-saline' are used for water bodies only, '-halophilic' and '-halobious' for biota (similar to Ziemann & Schulz, 2011).

Cuticle structures. – Beyond the classical distinction of setae and spines (Tattersall & Tattersall, 1951), the '**fringes**' (fig. 54.9C-E) are non-sensory cuticle structures found in cumaceans by Klepal & Kastner (1980) and introduced for the study of mysids by Wittmann & Ariani (1998). Similarly, Wittmann (1985) introduced the numbers and arrangement of small pores (fig. 54.9F-H) on the carapace as diagnostic characters of certain taxa of Mysida. '**Fenestra paracornealis**' (Ariani & Wittmann, 2000) is a pale, pigment free, rounded and slightly elevated spot (fig. 54.12D) dorsally on the eyestalks near the cornea.

Thoracic endopods. – For the eight pairs of endopods (limbs), Tattersall & Tattersall (1951) and many others distinguished two pairs of **maxillipeds** versus six pairs of **pereiopods**. In contrast, Bowman (1977) and other, mainly American authors, emphasized the second endopod as a pereiopod, thus counting seven pairs of pereiopods. Bowman (1973) and Bowman et al. (1984) applied this scheme also for the two families of the Stygiomysida, i.e., the Stygiomysidae and the Lepidomysidae, irrespective of different numbers of maxilliped-like endopods. The different terminologies resulted in inconsistent numbering of pereiopods, persisting up to the present in the literature. Gruner & Scholtz (2004) located this problem widely in the malacostracan literature and therefore proposed to use only the term '**thoracopods**'. We avoid 'pereiopods' also because the etymology as 'walking legs' only marginally fits the behaviour of mysidaceans. The various taxa sometimes use the thoracic endopods 3-8 for crawling, but among a greater diversity of functions, mainly to form a filtration basket, to grasp and manipulate food, and/or to hold onto the substrate.

The slenderness of thoracic endopods is evaluated by the ratio (R: Băcescu, 1940) between the length and maximum width of the merus (*sensu* Tattersall & Tattersall, 1951; this is the carpus *sensu* Băcescu, 1954) for certain endopods, particularly the sixth (R_6)

or the seventh (R_7) (arrows in fig. 54.18M; Ariani & Wittmann, 2000). The '**propodal formula**' shows, according to Ariani & Wittmann (2000), the variability in the number of carpopropodal segments along the series of third to eighth thoracic endopods: ranges are reported as "a-b" or "b-a" according to the prevailing frequency, and parentheses indicate rare or exceptional values. For instance, the carpopropodus of thoracopods 3-8 in *Diamysis lagunaris* is described with 3-2 (4), 2-3, 2, 2, 2, and 2 (3) segments, respectively. '**Paradactylary setae**' (fig. 54.18O; Băcescu, 1940) are the usually two pairs of setae positioned at the end of the propodus, and flanking the dactylus at its base, often showing a peculiar morphology. Below the terminal claw 0-3 (mostly one) '**subungulary setae**' (fig. 54.18O) may be present on the dactylus.

Features of the pleon. – Uniramous pleopods are identified according to W. M. Tattersall (1951) as derivates of endopods (fig. 54.21S) if they show outwards (= laterally) directed lobes (= **exites** = **pseudobranchial lobes**) or rudiments of such lobes. The term '**scutellum paracaudale**' (fig. 54.22O) is used for the latero-dorsal plates protruding from the sixth pleonite and flanking the telson (Ariani & Wittmann, 2000). Voicu (1974, 1981) defined the '**statolith formula**' according to the arrangement of pore groups and numbers of pores within each group, in semicircular series at the surface of the statoliths (static bodies), starting with the most caudal pores (a + b + c + d ...; Schlacher et al., 1992). In the variant by Wittmann (1992a) this includes also variation (in parentheses) and total numbers of pores, for example $2 + 3 + (4\text{-}10) + (2\text{-}4) + (3\text{-}6) = 18\text{-}24$ in *Mesopodopsis aegyptia*.

EXTERNAL MORPHOLOGY

Habitus

As typical for Malacostraca, the Lophogastrida, Stygiomysida, and Mysida — together exposed in fig. 54.1 — share the '**caridoid facies**'. According to Calman (1909), this facies is characterized by the presence of three **tagmata**, which are represented by a five-segmented head (not counting a likely present ocular somite) fused with an eight-segmented thorax to form a cephalothorax, and by a six-segmented pleon. The following features complete the 'caridoid' picture of a typical member in these orders: moveable **stalked eyes**, biramous **antennulae** and **antennae**, a well-developed **antennal scale**, a carapace enveloping the thorax, biramous thoracopods, mostly biramous male pleopods, a ventrally flexing pleon capable of **tail flipping**, and a strong tail fan. In addition, the adult females share a large brood pouch, whose large size strongly modifies the body shape below the thorax. Both sexes of these orders also share a normally well-developed carapace, dorsally adhering to all cephalic somites plus the anterior 1-4 thoracic somites, and, with certain exceptions, projecting freely over some of the more posterior somites.

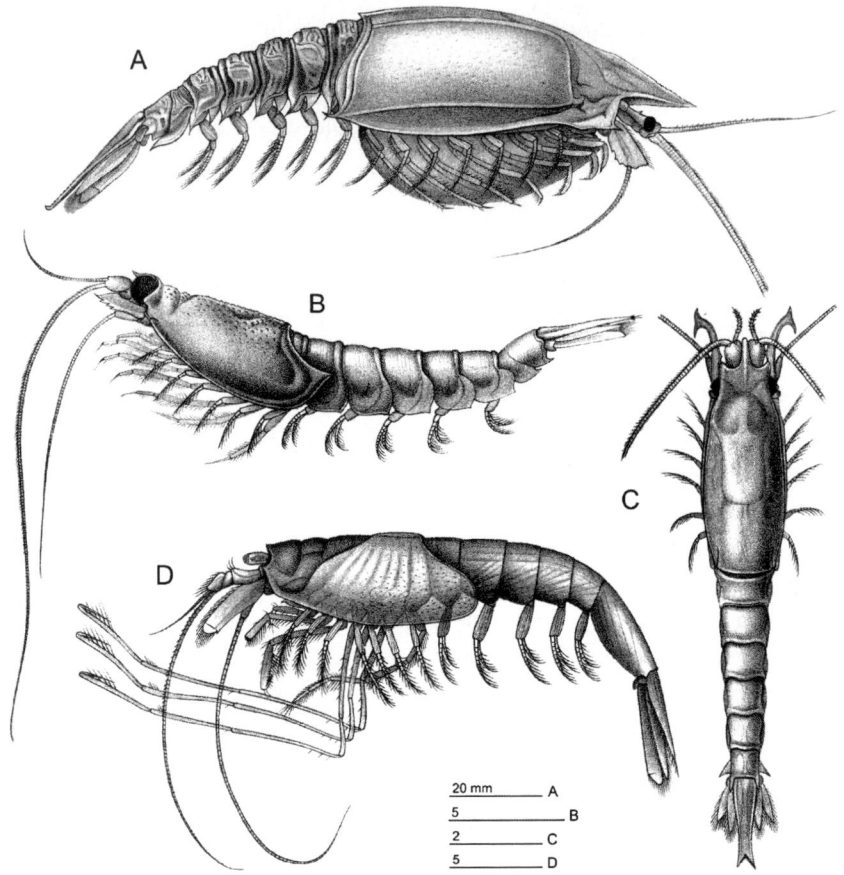

Fig. 54.2. Habitus of females in the Lophogastrida families Gnathophausiidae (A), Lophogastridae (B, C), and Eucopiidae (D). A, *Gnathophausia ingens* (Dohrn, 1870); B, *Lophogaster typicus* M. Sars, 1857; C, *Ceratolepis hamata* G. O. Sars, 1883; D, *Eucopia australis* Dana, 1852. [After G. O. Sars, 1885a.]

Lophogastrida (fig. 54.2). – The Gnathophausiidae show a comparatively rigid, keeled carapace. In several species, particularly in *Gnathophausia zoea*, the carapace bears an extremely long, toothed rostrum and a posterior spear-like prolongation (fig. 54.6C, D). As a striking feature, the Eucopiidae show strong, subchelate thoracic endopods 2-4, strongly contrasting with the very thin, elongate, weakly subchelate endopods 5-7 (fig. 54.2D).

Stygiomysida (fig. 54.3). – All species show at least some degree of **eye reduction**, indicative of their typically **subterranean** mode of life. Within this order, only *Spelaeomysis* shows a relatively 'normal' mysidacean habitus, the exception being a terminally rounded, subtriangular to semi-circular, anterior projection (fig. 54.3B) from the penultimate thoracic somite. This large, striking flap, also termed a 'scale', overlaps with the postero-median margin of the carapace from behind. There is no such scale in *Stygiomysis*, which is characterized by a **vermiform body** (figs. 54.1B, 54.3A) and a strongly shortened,

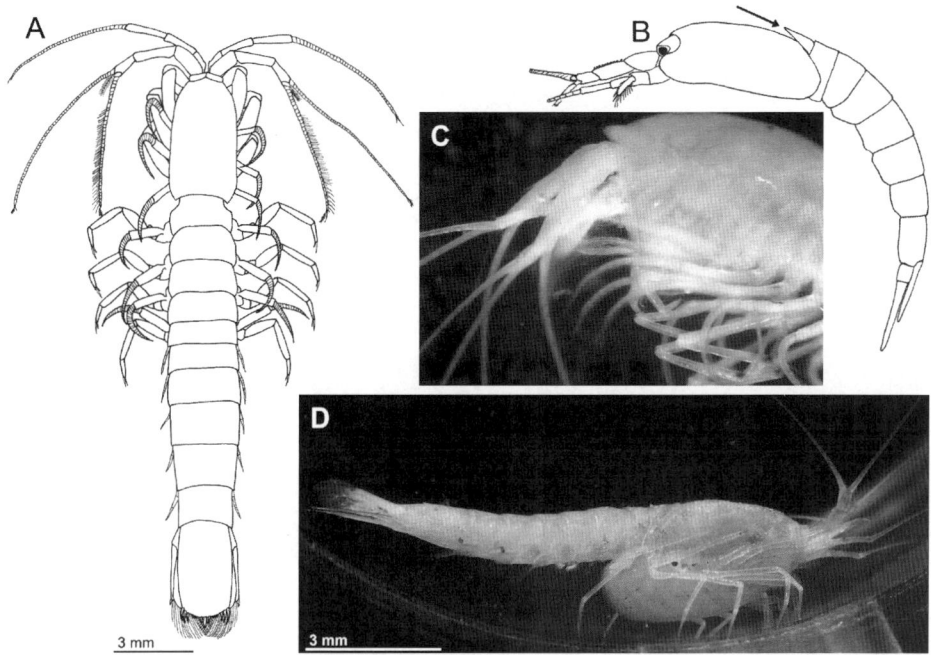

Fig. 54.3. Habitus of the Stygiomysida families Stygiomysidae (A) and Lepidomysidae (B-D). A, *Stygiomysis cokei* Kallmeyer & Carpenter, 1996, dorsal; B, lateral aspect of *Spelaeomysis cardisomae* Bowman, 1973, most appendages omitted, arrow shows anterior lobe of ultimate thoracic tergite; C, 'face' of *Spelaeomysis bottazzii* Caroli, 1924, in lateral view; D, lateral aspect of laboratory-kept *Spelaeomysis bottazzii* bearing nauplioid larvae. [A, after Kallmeyer & Carpenter, 1996; B, after Bowman, 1973; C, after Inguscio, 1998; D, photo Antonio P. Ariani.]

reduced carapace, posteriorly not extending beyond the fusion zone with the four anterior thoracomeres. According to *in vivo* observations by Inguscio (1998), the slender body form enables this highly specialized, **stygophilic** animal to perform very quick movements with abrupt direction changes of almost 90°. We interpret this as being advantageous for locomotion through the 'tunnel'-network of meso-interstitial groundwater habitats. Correspondingly, we consider the absence of a free posterior margin of the carapace in *Stygiomysis*, as well as the protection of the free margin by a flap of the penultimate thoracomere in *Spelaeomysis*, as adaptations to prevent hooking the carapace on the wall of narrow 'tunnels' upon backwards movements. Such movements appear necessary at least when the animal reaches a dead end.

Mysida (figs. 54.4, 54.5). – Most species show a fully 'normal' mysidacean habitus. There are varying degrees of **eye modifications**, mostly eye reductions, in most subterranean and also in many marine, meso- to bathypelagic species. Some holopelagic forms show an elongate trunk, either by elongation of certain thoracomeres (fig. 54.5G) or of the pleon (see below, 'Cephalothorax'). In several pelagic genera, particularly *Caesaromysis* (fig. 54.12C) and *Echinomysis*, the carapace and most pleomeres bear great numbers of long spiniform processes, giving these animals a hedgehog-like appearance. Unlike all

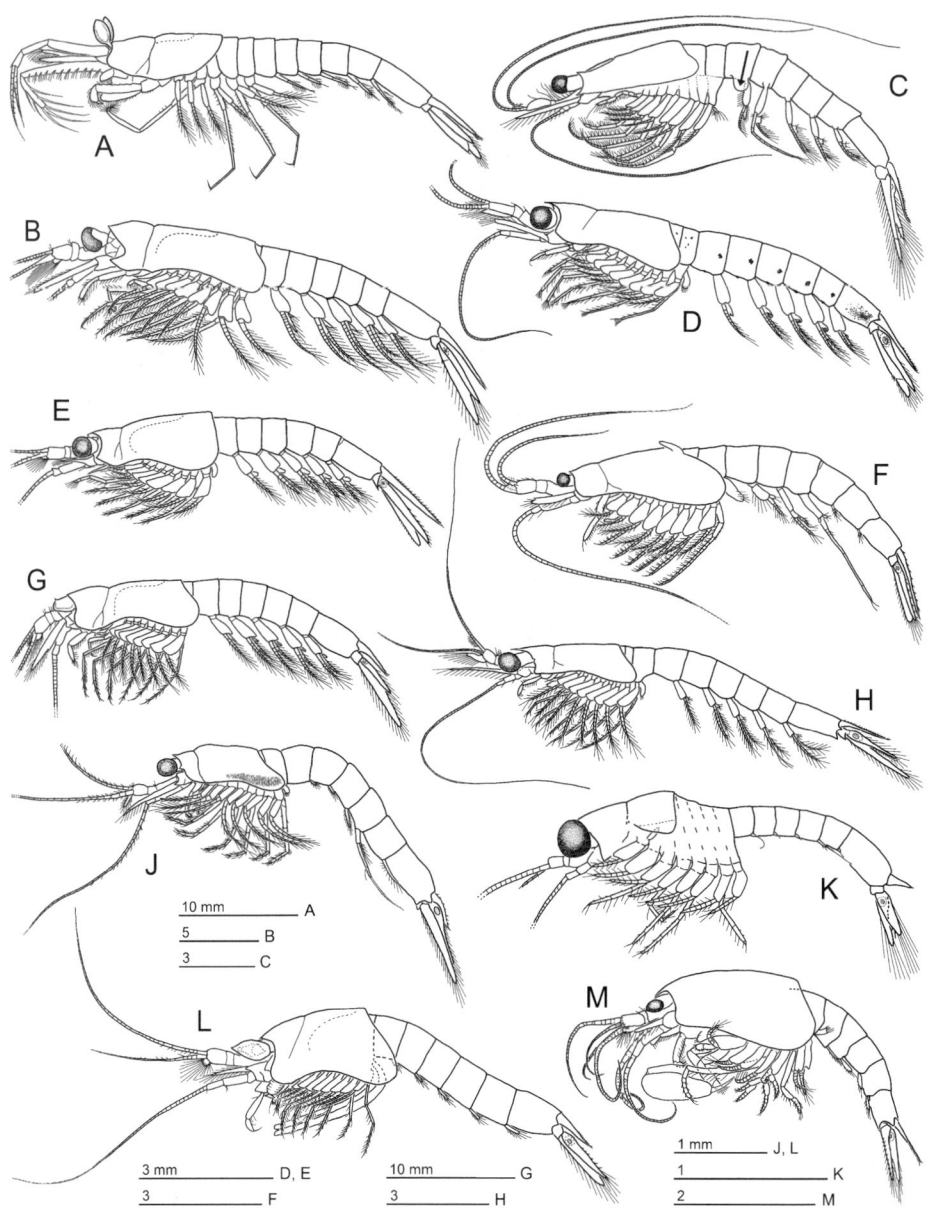

Fig. 54.4. Habitus of males in the Mysida families Petalophthalmidae (A) and Mysidae (B-M), the latter represented by all ten of its acknowledged subfamilies: Boreomysinae (B), Rhopalophthalminae (C), Siriellinae (D), Gastrosaccinae (E, F), Erythropinae (G), Leptomysinae (H), Mysinae (J), Palaumysinae (K), Mysidellinae (L), and Heteromysinae (M). A, *Petalophthalmus armiger* Willemoës-Suhm, 1875; B, *Boreomysis obtusata* G. O. Sars, 1884; C, *Rhopalophthalmus egregius* Hansen, 1910, arrow points to the pleural plate of the first pleomere; D, *Siriella thompsonii* (H. Milne Edwards, 1837); E, *Anchialina truncata* (G. O. Sars, 1883); F, *Haplostylus lobatus* (H. Nouvel, 1951); G, *Amblyopsoides crozetii* (Willemoës-Suhm in G. O. Sars, 1884);

remaining Mysida, several genera of the Gastrosaccinae are capable of digging and seemingly have a posteriorly elongated marsupium (fig. 54.5F). Actually, the elongation is due to a pair of large pleurites projecting from the first pleomere and supporting the posterior part of the marsupium from the side and from behind. This support may serve in **brood protection** when the mothers bury themselves up to 5 cm into the sediment.

Size of adults

Mysida. – Typical adult **body lengths** of epipelagic and coastal species are 3-8 (range 1-17) mm in **tropical**, 6-15 (3-50) mm in **temperate**, and 10-30 (4-77) mm in **Arctic zones**. The increase of **body size** with increasing latitudes is discussed in 'Growth' below. **Bathypelagic species** are 8-35 (6-85) mm long. Except in a few species, adult males are on average smaller than females, mainly due to earlier attainment of sexual maturity, shorter life span, and/or smaller **moult increments** after attainment of **adulthood** (Murano, 1964b, 1999b; Clutter & Theilacker, 1971; Hakala, 1978; Hanamura, 1999). The largest Mysida species, *Birsteiniamysis inermis* from the boreal North Atlantic, reaches up to 85 mm (Mauchline & Murano, 1977). This species is characterized by strong size variations between different geographical localities. Marked seasonal size differences are observed in many species. For instance, off Arcachon (France), adult *Schistomysis spiritus* are larger in winter than in autumn. Seasonal differences in relative dimensions (**cyclomorphosis**) of appendages were also described; on this basis, Wittmann (1992b) identified *Siriella adriatica* as the **overwintering generation** of *Siriella gracilipes*. Adult sizes of mysids typically decrease with increasing temperature (i.e., warmer season or climate) due to earlier attainment of sexual maturity and reduced longevity, despite the fact that daily growth rates increase with increasing temperature, e.g., in *Neomysis intermedia* (cf. Toda et al., 1983a, 1984).

Lophogastrida and Stygiomysida. – Body sizes of adults range from 6 to 39 mm in Lophogastridae, 27-66 mm in Eucopiidae, and 40-351 mm in Gnathophausiidae. The maximum size was found by Clarke (1961b) in *Gnathophausia ingens* (fig. 54.2A). According to J.-P. Casanova (1996a), the females of *Paralophogaster glaber* showed maximum sizes of 21.5 or 33 mm when collected near the Indonesian islands Kai and Tanimbar, respectively, despite a relatively small distance of only 300 km between these islands. Among adult Stygiomysida, the Lepidomysidae measure 3-11 mm, the Stygiomysidae 5-21 mm.

H, *Leptomysis gracilis* (G. O. Sars, 1864); J, *Diamysis cymodoceae* Wittmann & Ariani, 2012; K, *Palaumysis philippinensis* Hanamura & Kase, 2002; L, *Mysidella typhlops* G. O. Sars, 1872; M, *Heteromysis harpaxoides* Băcescu & Bruce, 1980. [A, B, D, E, G, after G. O. Sars, 1885a; C, after O. S. Tattersall, 1952; F, after Tattersall & Tattersall, 1951; H, L, after G. O. Sars, 1879; J, after Wittmann & Ariani, 2012a; K, after Hanamura & Kase, 2002; M, after Daneliya, 2012; A-M, modified.]

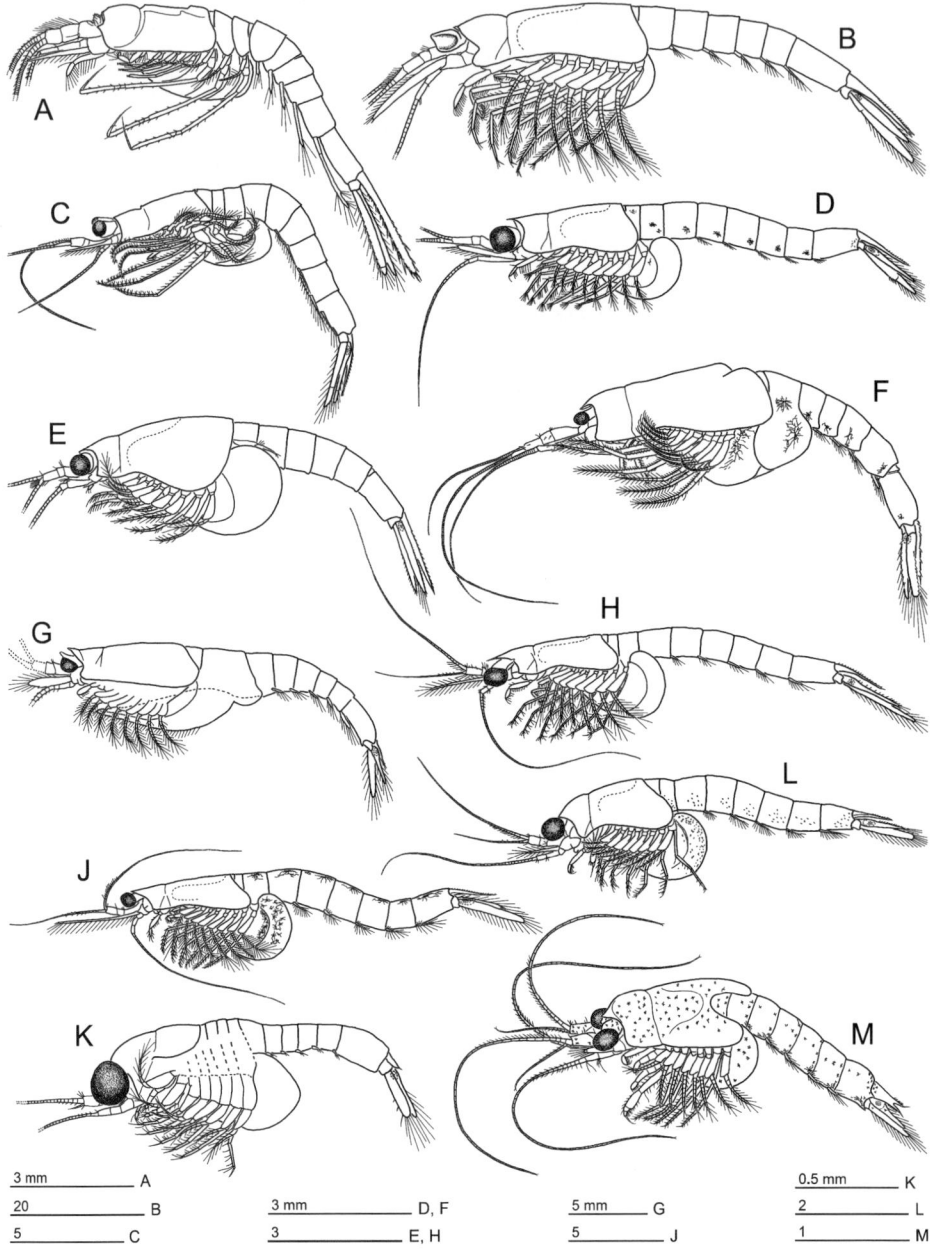

Fig. 54.5. Habitus of females in the Mysida families Petalophthalmidae (A) and Mysidae (B-M), the latter represented by all ten of its acknowledged subfamilies: Boreomysinae (B), Rhopalophthalminae (C), Siriellinae (D), Gastrosaccinae (E, F), Erythropinae (G), Leptomysinae (H), Mysinae (J), Palaumysinae (K), Mysidellinae (L), and Heteromysinae (M). A, *Hansenomysis falklandica* O. S. Tattersall, 1955; B, *Birsteiniamysis inermis* (Willemoës-Suhm, 1874); C, *Rhopalophthalmus tartessicus* Vilas-Fernández, Drake & Sorbe, 2008; D, *Siriella thompsonii* (H. Milne Edwards, 1837);

Carapace

Gross structure and formation. – In the Lophogastrida, Stygiomysida, and Mysida, the carapace adheres to the cephalothorax from the **tergite** of the antennal somite backwards over the region of the mouthparts and over 1-4 anterior thoracic somites. Along its margins the carapace — during larval growth — drives backwards, like a wedge, those **hemi-tergites** and **pleurites** belonging to somites not directly contributing to **carapace formation** (B. Casanova, 1987, 1991), at least in Lophogastrida and Mysida (figs. 54.6A, B, 54.8D). The **epimeron**, a membranous structure lining the carapace inside, is laterally linked with the exoskeleton of the cephalothorax (fig. 54.8A-C). The remaining posterior thoracomeres may be covered by the carapace, but all of them are free and distinct. During **larval development** the posterior progression of the carapace leads to dorsal opening of the tergites of the cephalic somites and, in Gnathophausiidae, of half of the first thoracic somite. The **cephalic tergites** plus the anterior 3.5 thoracic tergites become opened in Lophogastridae, Eucopiidae, and Mysidae (B. Casanova et al., 2002). The larval genesis of the carapace has not yet been studied in Petalophthalmidae, Lepidomysidae, and Stygiomysidae.

Rostrum and analogous structures. – Anteriorly, the carapace is produced into a frontal shield (**rostrum**) which may be pointed (figs. 54.2A, 54.6C, D, F), rounded (fig. 54.7D, H), or rarely absent. This shield represents a prolongation of the antennal tergite (fig. 54.6A, B; B. Casanova, 1991). In several genera of Mysinae (**Mysida**: Mysidae), 1-2 **subrostral processes** project anteriorly from the **antennal somite**, emerging between the ocular **symphysis** and the carapace. In the mostly Ponto-Caspian genera *Paramysis*, *Caspiomysis*, *Katamysis*, and in others, the subrostral process extends anteriorly beyond the true rostrum if there is one. In several species, most strikingly in *Paramysis ullskyi* and *Paramysis inflata*, there is a well-formed subrostrum, whereas a true rostrum is completely missing. Particularly in the Gastrosaccinae, one may find large suborbital median processes emerging between the ocular and the antennular symphyses. Such species may show up to three rostrum-like anterior projections (fig. 54.7D).

In the **Lophogastrida** genus *Gnathophausia* there is a strong triquetrous rostrum with serrated edges (figs. 54.1A, 54.6C, D). In the Lophogastridae and Gnathophausiidae, the rostrum usually shows oblique longitudinal ridges, often elongated by strong spines (in certain species of *Gnathophausia*). These spines are most strongly developed in juveniles and may be serrated (fig. 54.6C). The Eucopiidae have either no rostrum or a mostly short, anteriorly well-rounded rostrum with bare margins. In addition, there is always a distinct,

---

E, *Anchialina truncata* (G. O. Sars, 1883); F, *Gastrosaccus spinifer* (Goës, 1864); G, *Longithorax fuscus* Hansen, 1908, thoracic endopods not shown; H, *Leptomysis gracilis* (G. O. Sars, 1864); J, *Praunus flexuosus* (O. F. Müller, 1776); K, *Palaumysis simonae* Băcescu & Iliffe, 1986; L, *Mysidella typica* G. O. Sars, 1872; M, *Heteromysis wirtzi* Wittmann, 2008. [A, after O. S. Tattersall, 1955; B, D, E, after G. O. Sars, 1885a; C, after Vilas-Fernández et al., 2008; F, after Wittmann et al., 2011; G, after Tattersall & Tattersall, 1951; H, J, L, after G. O. Sars, 1879; K, after Hanamura & Kase, 2002; M, from the illustration by Wittmann on the cover of Crustaceana, vol. **86**, 2013; A-M, modified.]

anteriorly directed process, commonly termed '**pseudorostrum**'. This process emerges below the ocular symphysis and probably pertains to the **antennular somite**, *ergo* not to the antennal somite as in the Lophogastridae and Gnathophausiidae (fig. 54.6A, B).

External structures of the carapace. – The carapace is mostly thin and membranous, or somewhat calcified only in certain Lophogastridae (figs. 54.2B, 54.6F) and Gnathophausiidae (figs. 54.2A, 54.6C, D). The carapace of the Lophogastrida and Stygiomysida, and in most species of Mysida, shows a distinct transverse furrow (= **cervical sulcus**; figs. 54.2B, 54.7E, F, 54.9F, G) approximately above the mandibles. This furrow is completely missing in certain taxa of Heteromysinae. In some species of *Lophogaster* the surface of the carapace may be granulose (fig. 54.6F) or uniformly finely carved (Fage, 1942; J.-P. Casanova, 1993, 1997). In the Mysinae genus *Diamysis*, the morphology and distribution of '**fringes**' (as defined above; figs. 54.7C, 54.9C-E) on the carapace are of great value in defining different species (Wittmann & Ariani, 1998, 2012a, b; Ariani & Wittmann, 2000). Fringes of spectacular length are found (Wittmann & Ariani, 2012a) in *Diamysis cymodoceae*. Similarly, submedian lobes (fig. 54.7E) and fringe-like processes (not fringes *sensu* Klepal & Kastner, 1980) from the terminal margin of the carapace show a large diversity, which is useful for species definitions in the genera *Gastrosaccus* and *Haplostylus* (Gastrosaccinae). The posterior margin of the carapace is generally emarginated, leaving 1-2 (range 0.5-3) ultimate thoracic somites exposed. In this manner, the carapace forms two backwards directed lobes covering most of the thorax in lateral view. The ultimate thoracic somite, however, remains laterally uncovered in many species of Siriellinae (figs. 54.4D, 54.5D); even more somites of the elongated thorax in *Longithorax* (Erythropinae) remain uncovered (fig. 54.5G).

The branchiostegal carapace. – The Lophogastrida clearly show a **branchiostegal carapace** (fig. 54.6E) in that the **carapace cavity** serves to protect the **gills**, which arise from the coxae of the thoracic appendages (figs. 54.6E, G, 54.16C, E, G). Nonetheless, one of the typically four gills of a given thoracopod is often directed inwards to the ventral midline of the body (fig. 54.6E), i.e., away from the carapace cavity. Some of the posterior gills may also project laterally out from the carapace cavity in large specimens of Eucopiidae (fig. 54.6G). According to Wirkner & Richter (2013), the inner lining of the carapace duplicature in *Lophogaster typicus* shows a similar network of **haemolymph channels** as in the carapace of the Mysida discussed below. *Lophogaster* is well equipped with branchiae, raising the question about what the function of this lining might be: in analogy to ultrastructural findings of Kikuchi & Matsumasa (1993) in the estuarine tanaid *Sinelobus stanfordi*, Wirkner & Richter (2013) proposed a mainly **osmoregulatory function** of this lining. A certain osmoregulatory capability is generally found in waterborne respiratory tissue. In *Sinelobus stanfordi*, the ultrastructural features of the branchial versus branchiostegal epithelia point to different functions in osmoregulatory ion transport, in each case in addition to the respiratory function (Kikuchi & Matsumasa, 1993). Based on these considerations and on the strictly euhaline distribution of *Lophogaster*, it is premature to venture proposing a major osmoregulatory versus a minor respiratory

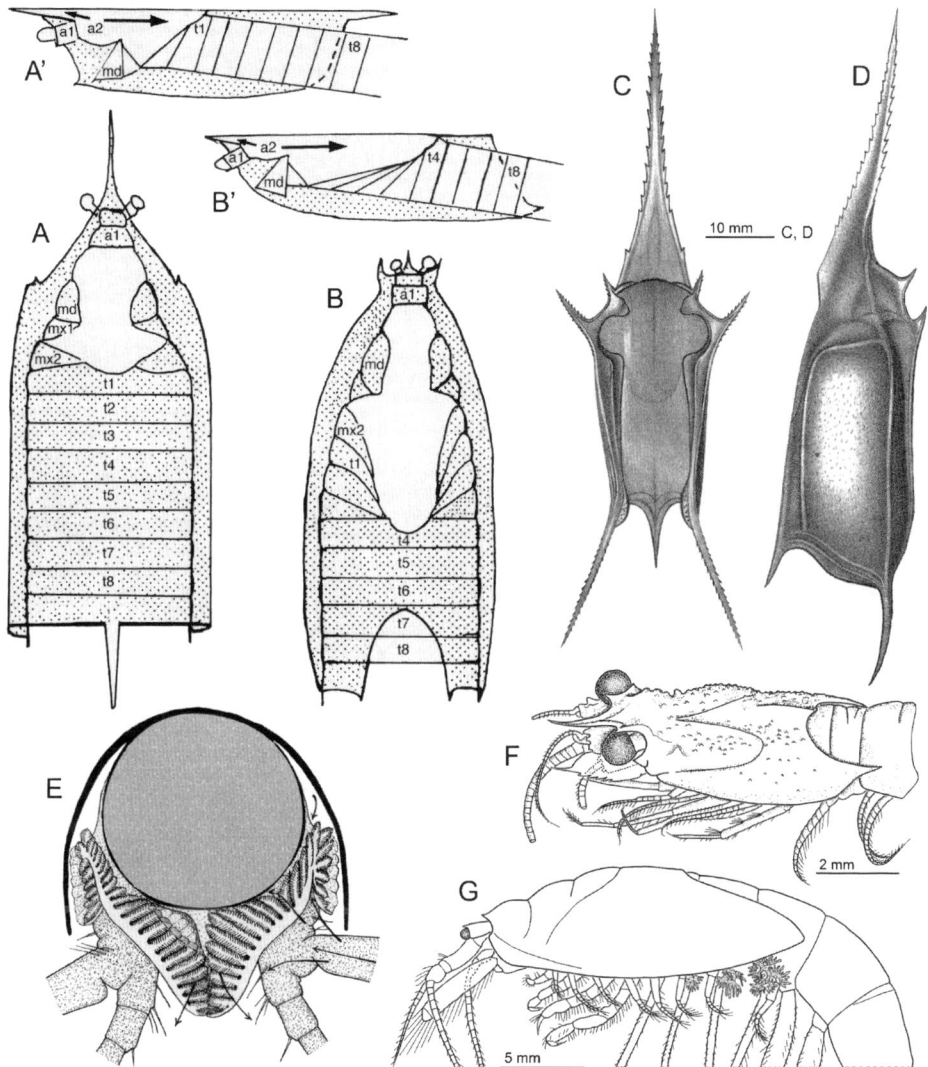

Fig. 54.6. The carapace of the Lophogastrida families Gnathophausiidae (A, C, D), Lophogastridae (B, E, F), and Eucopiidae (G). A, A', schematic outline of the extension of the carapace and the epimeres (dotted) in *Gnathophausia* Willemoës-Suhm, 1873, dorsal (A) and lateral (A') aspects; B, B', the same for *Lophogaster* M. Sars, 1857; C, detached carapace of a juvenile *Gnathophausia ingens* (Dohrn, 1870), inner ventral face; D, same species as before, carapace of a more advanced juvenile, lateral; E, cross section between two consecutive thoracic appendages in *Lophogaster typicus* M. Sars, 1857, showing the arrangement of branchiae; arrows indicate respiratory water flow; F, *Lophogaster pacificus* Băcescu, 1985, dorso-lateral; G, *Eucopia sculpticauda* Faxon, 1893, note that the carapace leaves some of the posterior gills exposed in lateral view. Abbreviations indicate somites of the cephalothorax: a1, antennula; a2, antenna; md, mandible; mx1, maxillula; mx2, maxilla; t1 to t8, thoracomeres 1 to 8. [A, B, after B. Casanova, 1991; C, D, after G. O. Sars, 1885a; E, after Manton, 1928a, and Nouvel et al., 1999; F, after Băcescu, 1985; G, after Tattersall & Tattersall, 1951; E-G, modified.]

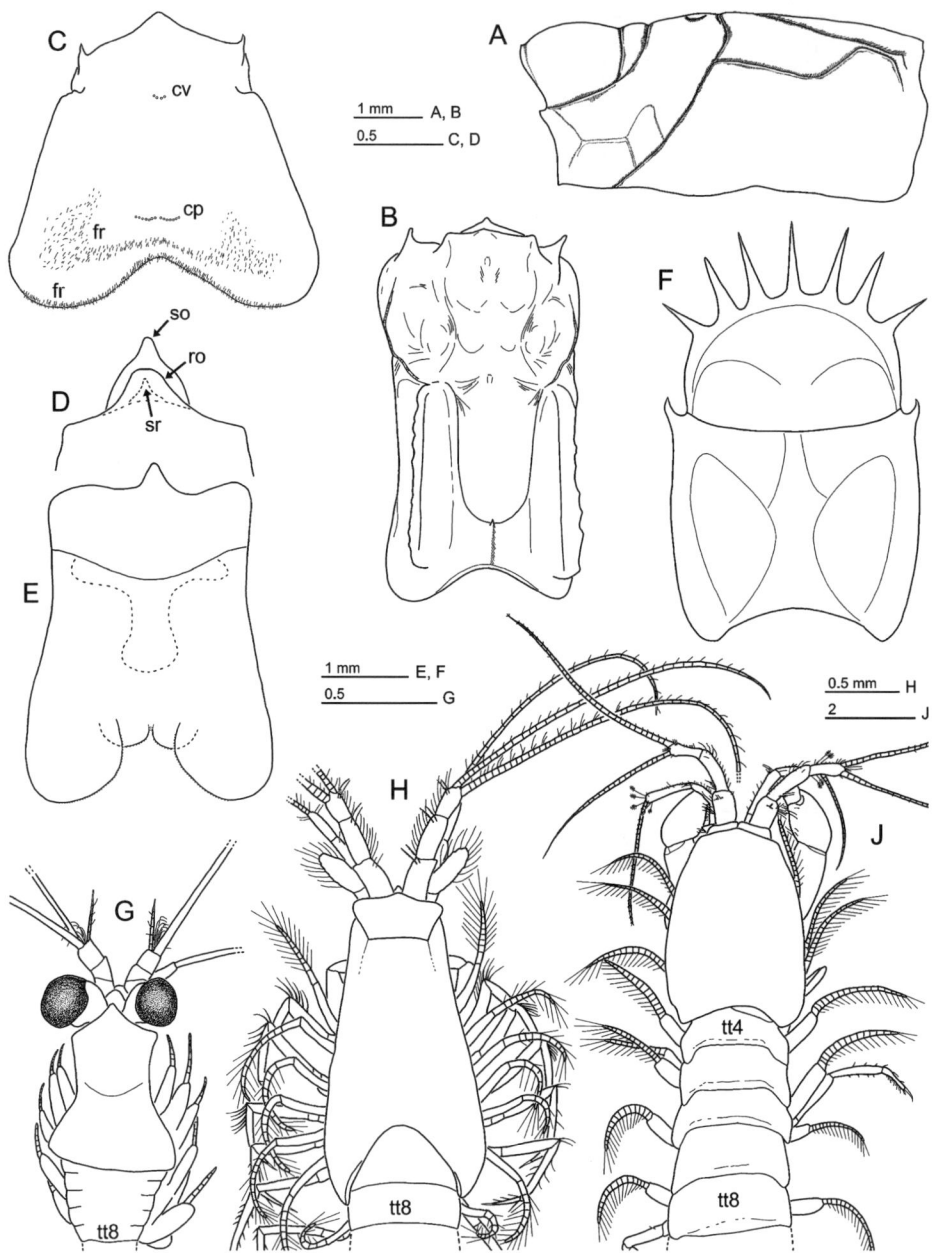

Fig. 54.7. The carapace of the Mysida (A-G) and the Stygiomysida (H, J): examples for the families Petalophthalmidae (A, B), Mysidae (C-G), Lepidomysidae (H), and Stygiomysidae (J). A, *Hansenomysis abyssalis* Lagardère, 1983, lateral; B, *Hansenomysis nouveli* Lagardère, 1983, dorsal; C, *Diamysis bahirensis* (G. O. Sars, 1877), carapace expanded (flattened) on slide, dorsal; D, rostrum and analogous processes in female of *Gastrosaccus roscoffensis* Băcescu, 1970; E, male of previous species, carapace flattened as in (C), dorsal; F, *Chunomysis diadema* Holt & Tattersall, 1905,

function of the branchiostegal epithelia. A definite conclusion will require appropriate physiological data.

The respiratory carapace (fig. 54.31D). – The Stygiomysida and Mysida lack gills, but the inner surface of the freely projecting parts of the carapace, i.e., the wall of the carapace cavity, is lined with **respiratory tissue**. Wägele (1994) and Kobusch (1999) termed this a '**respiratory carapace**'. The respiratory lining is most striking in certain Gastrosaccinae. As an example of an opposite development, in the stygobiotic Stygiomysidae the posterior parts of the carapace are reduced, and the lateral carapace cavities are shortened (fig. 54.3A); this is not the case in their also stygobiotic relatives, the Lepidomysidae (fig. 54.3B). The reduction of the respiratory tissue together with the carapace in the former are interpreted above ('Habitus') as a consequence of acquiring a **vermiform body**, adaptive for locomotion in underground habitats. A posteriorly shortened, but less reduced carapace cavity is found in the tropical, marine, cave-dwelling genus *Palaumysis* (Mysidae: Palaumysinae; figs. 54.5K, 54.7G). These mysids have large, functional eyes and a comparatively stout body. The degree of carapace reduction varies between the various species of this genus (Hanamura & Kase, 2003). We conclude, that the reduced respiratory surface in *Palaumysis* is a consequence of **dwarfing**. In most species of this genus, the females grow to about the lower limit of body length (1 mm) required for successful incubation of their relatively large larvae (see below, 'Egg size').

Cephalothorax

Metameric patterns. – According to Hansen (1925) the unpaired mobile elements ("one" in *Gnathophausia*, two in *Boreomysis* and *Mysis*) in front of the cephalic region may represent ocular and antennular somites, perhaps comparable to those in stomatopods. Actually, examination of the internal cuticular organization of the cephalothorax showed that tergites of the ocular as well as the antennular somites (fig. 54.8) are present in *Gnathophausia zoea*, yet the tergite of only the **antennal somite** contributes to **carapace formation** (B. Casanova, 1987, 1991). The posterior thoracic somites, not attached to the carapace, are dorsally, laterally, and particularly ventrally well distinguished (figs. 54.6A, B, 54.8A, C, D) in Lophogastrida and Mysida. They show a similar development except for a strong elongation of certain thoracic somites in a few essentially holopelagic genera belonging to the family Mysidae: the first somite in *Arachnomysis* or the eighth somite in *Longithorax* (fig. 54.5G) and *Gymnerythrops*.

---

dorsal; G-J, different degrees of coverage of the cephalothorax by the carapace, as exemplified by *Palaumysis philippinensis* Hanamura & Kase, 2002 (G, most setae omitted), *Spelaeomysis longipes* (Pillai & Mariamma, 1964) (H) and *Stygiomysis aemete* Wagner, 1992 (J), anterior body regions in dorsal view. Abbreviations: cp, cardial pore group; cv, cervical pore group; fr, fringes; ro, rostral process; so, suborbital process; sr, subrostral process; tt4, tt8, thoracic tergites. [A, B, after Lagardère, 1983; C, after Wittmann & Ariani, 2012a; D, E, after Wittmann et al., 2011; F, after Tattersall & Tattersall, 1951; G, after Hanamura & Kase, 2002; H, after Pillai & Mariamma, 1964; J, after Wagner, 1992; A-J, modified.]

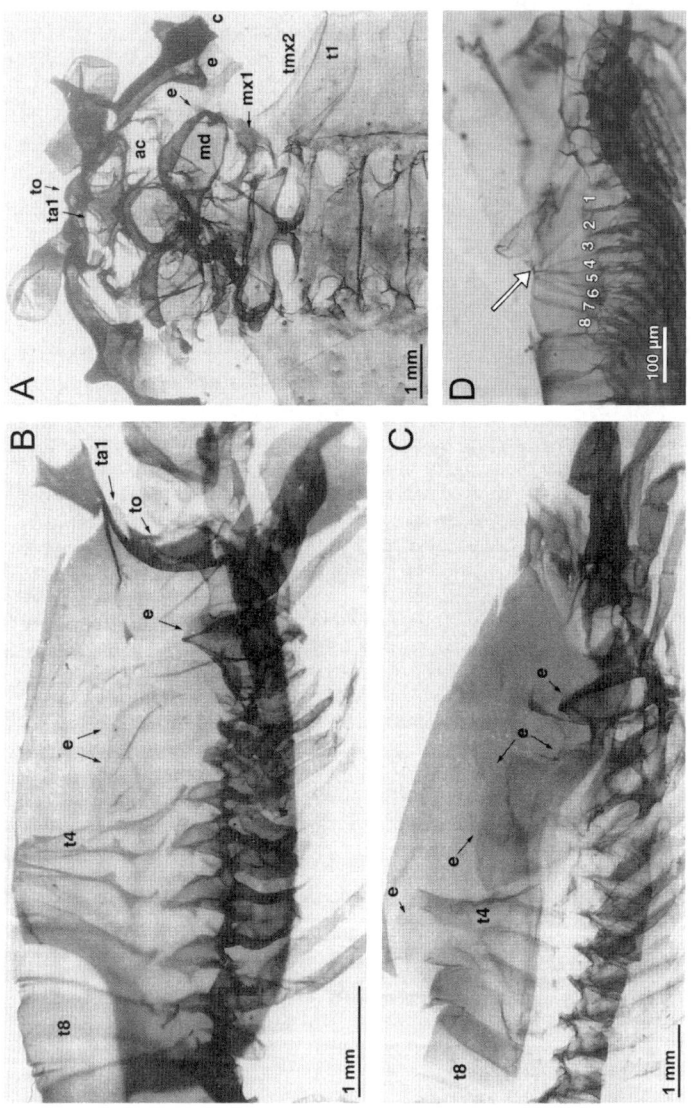

Sternal projections. – Most Lophogastrida and certain Mysida show mostly unpaired median projections from thoracic sternites, often equipped with scales or hairs. The presence and structure of these **sternal projections** commonly differ between age stages and/or sexes. In those species with distinct sternal projections in adult males (figs. 54.15N, 54.20F, 54.35A, D, E), one typically finds some projections also in juveniles (fig. 54.35C) of both sexes, but no such projections or only rudiments in adult females (figs. 54.15M, 54.35B). Less frequently, the situation is inverse: several species of Mysidae show large foliaceous sternal processes, projecting from the ultimate thoracomere in adult females only. These unpaired processes probably contribute to the posterior closure of the marsupium (see below, 'Oostegites').

Pleon

Unlike Isopoda and Amphipoda, and certain taxa within the Decapoda, the pleon is never reduced. Rather, it is always well developed and capable of emergency **tail flipping** in the Lophogastrida, Stygiomysida, and Mysida. The five anterior pleonites are generally similar in size; only the sixth is typically distinctly longer.

Lophogastrida. – Unlike in the family Eucopiidae (figs. 54.2D, 54.19A), all **pleonites** in the Lophogastridae (fig. 54.2B) and Gnathophausiidae (fig. 54.2A) are flanked by distinctly projecting, rounded or terminally pointed **pleural plates**. In these families the sixth pleonite shows a more or less distinct transversal groove imitating an additional segmental border (fig. 54.2A, B). This insinuates that the sixth pleonite may have originated from the fusion of two ancestral somites as still found in the Leptostraca. Interestingly, the ultimate pleonite lacks appendages in the Leptostraca, whereas its posterior margin bears the uropods in the Lophogastrida. According to Manton (1928a) the embryonic development and the arrangement of the musculature favour the hypothesis of two fused somites. Finally, the sternites show non-dimorphic, median, spiniform processes in *Lophogaster*.

Stygiomysida. – The pleura of the pleon lack conspicuous plate-like extensions in both families of this order. The sympods of pleopods 3-5 are bilaterally connected by a series of transverse, membrane-like **sternal lamellae**. These lamellae may incorporate the sympods (*Stygiomysis holthuisi*; fig. 54.17L; Gordon, 1960) or leave the sympods separate (*Spelaeomysis cardisomae*; fig. 54.17M; Bowman, 1973).

Fig. 54.8. Internal cuticular organisation of the cephalothorax in Lophogastrida (A, B) and Mysida (C, D). A, *Gnathophausia zoea* Willemoës-Suhm, 1873, dorsal view after median exposure of the tergites; B, C, *Lophogaster typicus* M. Sars, 1857 (B) and *Boreomysis microps* G. O. Sars, 1883 (C), lateral view after removal of the right body half; D, thorax of the postnauplioid stage of *Praunus flexuosus* (O. F. Müller, 1776) in lateral view, arrow points to the posterior suture of the carapace at the junction between the numbered thoracic somites 3 and 4. Abbreviations: ac, arthrodial cavities; c, carapace; e, epimeres; md, mandible; mxl, maxillula; ta1, antennular tergite; tmx2, maxillary tergite; to, ocular tergite; t1-t8, thoracic somites 1-8. [A-C, after B. Casanova, 1991; D, after B. Casanova et al., 2002, with permission by "© Canadian Science Publishing or its licensors", Ottawa.]

Mysida. – Conspicuously projecting pleural plates on the pleon are rare in this order. The females of Gastrosaccinae show **pleural plates** on the first pleonites, supporting the marsupium from the side and from behind (fig. 54.5F). Unlike in other genera of this subfamily, in *Archaeomysis* such plates are also present in males, but smaller. In certain species of this genus, both sexes show additional plates decreasing in size from pleomere 2 to 5. Also *Rhopalophthalmus* (Rhopalophthalminae) shows ventrally projecting pleural plates on the first pleonites, but only in the males (arrow in fig. 54.4C). In several species of Mysinae, e.g., most distinctly in *Paramysis portzicensis*, the pleonites show transversal folds simulating double somites. In a number of Mysinae (Mysidae) the telson is flanked by latero-dorsal plates (**scutella paracaudalia**; fig. 54.22O), protruding from the sixth pleonite. The scutella are important for the diagnosis of *Diamysis* species (Ariani & Wittmann, 2000). In embryos of Mysida, rudiments of a seventh pleon somite are indicated by the appearance of separate sixth and seventh ganglia during the development of *Hemimysis* and *Mesopodopsis* (Manton, 1928b; Nair, 1939).

Telson

The trunk ends in a dorsoventrally flattened **telson** (fig. 54.22D), ventrally bearing the anus (fig. 54.22F) in sub-basal position. Apart from this, the telson shows a great variability in form and armature between the various taxa from order down to species level. It may be linguiform, semi-ellipsoidal, triangular, but mostly trapezoidal, often with a more or less incised terminal margin.

Lophogastrida (fig. 54.22A-E). – The Eucopiidae and Lophogastridae show a roughly linguiform telson whose margins are armed with spines at least on the terminal half. The telson of *Eucopia sculpticauda* is exceptional by two subapical constrictions. In most species of *Lophogaster* and *Paralophogaster* the terminal margin of the telson shows a small, transverse, median portion between a pair of large apical spines; this portion bears a few laminae with barbed setae in-between (fig. 54.22A). The telson of the Gnathophausiidae has two long, longitudinal keels on the dorsal surface. Its small, crescent-shaped terminal part is separated by a strong subapical constriction (fig. 54.22D).

Stygiomysida (fig. 54.22F-H). – All taxa show a comparatively short, linguiform to trapezoid telson. At least the terminal margin, mostly also part of the lateral margins (usually their distal parts) are armed with spines. In several species of *Spelaeomysis*, one or more barbed setae are present on either side of the medio-apical spine (fig. 54.22H).

Mysida (fig. 54.22J-O). – The terminal incision, if present, of the telson is generally armed with spine-like laminae, less frequently with (additional) spines, or again less frequently there are only bare margins. The lateral margins are generally armed with spines all along, or only along part of their length, but they may be completely smooth in a few taxa. Fine hairs may be present in the space between these spines, particularly in Mysinae, e.g., the genera *Paramysis* and *Neomysis*. The tip of the telson is often equipped with a pair of plumose or barbed setae, particularly in Siriellinae (fig. 54.22L). Such setae insert on the bottom of the **telson cleft** in a number of Leptomysinae genera, e.g.,

Doxomysis, Prionomysis, and *Tenagomysis*. Rarely there may be a brush of plumose setae on the ventral face of the telson, as in the Leptomysinae genera *Notomysis* from coastal waters of India and Australia, and the closely related *Antichthomysis* from Tasmania. In the Rhopalophthalminae, the terminal margin of the telson is always armed with four large, microserrated spines. In the Petalophthalminae genera *Petalophthalmus* (fig. 54.22J), *Parapetalophthalmus*, and *Pseudopetalophthalmus*, the median portion of the terminal margin of the telson is lined with spine-like laminae, and the lateral portions by a number of barbed (microserrated) setae (spines).

Integument and colour

Cuticular structures. – The body surface of the Lophogastrida, Stygiomysida, and Mysida appears generally smooth, not considering structures <2 µm. Nonetheless, in certain taxa, scales, hairs, spines, or burls may be present on limited regions of the body. In a few Mysida, such as *Acanthomysis longicornis* and *Leptomysis gracilis*, most of the cuticle is covered by small scales, giving the entire body a hispid appearance. In most taxa, however, such scales show a more restricted distribution on the body wall and/or the appendages (fig. 54.9J, K). In the males of certain *Diamysis* species, part of the carapace may be covered by fringes, i.e., hair-like **non-sensory cuticular structures**; these fringes are strikingly long and densely arranged in *Diamysis cymodoceae* (fig. 54.9C-E; Wittmann & Ariani, 1998, 2012a, b; Ariani & Wittmann, 2000).

In addition to the **integumental sensilla** that occur on various appendages, all pelagic crustaceans examined by Mauchline (1977) after treatment with potassium hydroxide showed small '**pores**' (*sensu lato*; 1-20 µm) related to sense organs or to glands. These pores are present on the body wall and in most taxa also on the carapace. Mauchline (1977) found a median group of four larger pores dorsally near the posterior margin of the carapace in the lophogastrid *Eucopia sculpticauda*. Each lateral pair is connected by series of smaller pores, together forming a **compound organ** with a thin central area (fig. 54.9A). According to Mauchline (1977) this organ may be analogous to the compound organs with unknown function on the carapace of the nebaliacean *Nebaliopsis typica*. Mauchline (1977) and Wittmann (1985 and later) noted small pores (1-4 µm) on the dorsal face of the carapace, the pleon, and on dorsally uncovered thoracomeres in many species of Mysidae (Mysida). These pores show species-specific distribution patterns and are of taxonomic value. The Leptomysinae and Mysinae often bear a dense group of pores near the median, posterior margin of the carapace (fig. 54.9H); often there is also a transverse row in **cardial position** (fig. 54.9B, C, F), in most cases arranged in two symmetrical subgroups; and slightly less frequently a mostly small group in **cervical position** (fig. 54.9C, F, G). These groups, especially the posterior group, also occur in certain Erythropinae and Heteromysinae, but otherwise numbers and distribution of pores are more variable in these subfamilies. SEM-observations in *Diamysis bacescui* revealed two types of pores: the cardial and cervical pores resemble plant stomata due to their lip-like margins, whereas the remaining, more widely scattered ones are generally smaller and have a more simple, rounded opening (Wittmann & Ariani, 1998).

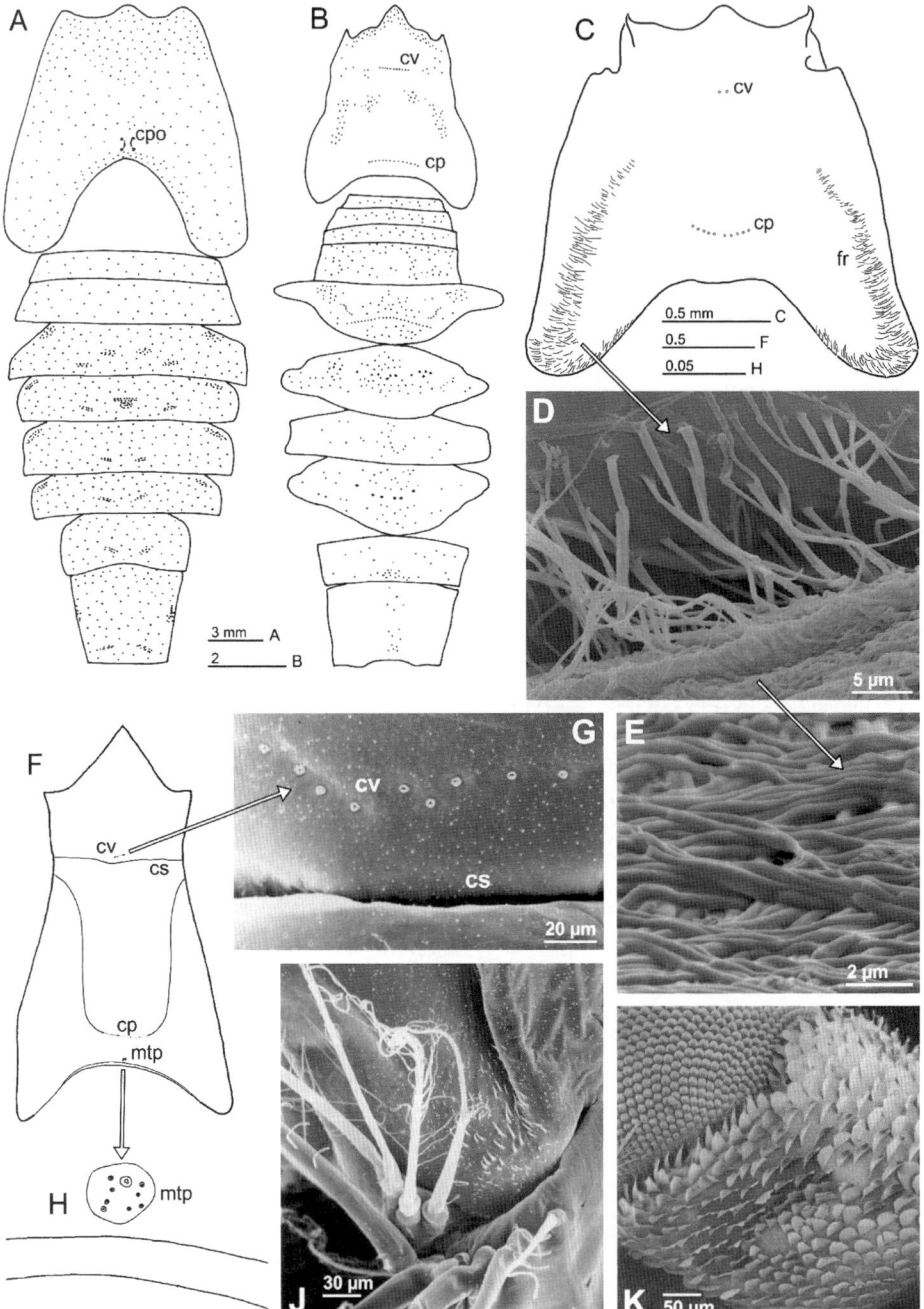

Fig. 54.9. Cuticle structures on the carapace and the tergites of the trunk in Lophogastrida (A) and Mysida (B-K), dorsal views only. A, B, pores (*sensu lato*) in the integument of the carapace, several posterior thoracomeres and all pleomeres after treatment with potassium hydroxide (KOH), *Eucopia sculpticauda* Faxon, 1893 (A) and *Praunus flexuosus* (O. F. Müller, 1776) (B); C, pores and

Colour (fig. 54.1). – For the nature of crustacean pigments and chromatophores, see the first volume of the present series (Noël & Chassard-Bouchaud, 2004). Most bathypelagic mysids are more or less uniformly red with purple, orange, violet, or rarely nigrescent (*Longithorax fuscus*) tinge. The benthopelagic and littoral species are rarely colourless or uniformly coloured. Most show a more or less dense cover by ramifications of **chromatophores** containing black, brown, or red pigment. There may also be substances giving white, yellow, or blue reflections. The patterns and action of the chromatophores produce a mostly spotted aspect, varying with illumination and colour patterns of the substrate. Nonetheless, the colour patterns also show specific components useful for visual distinction of species. Many species exhibit a longitudinal dorsal stripe, which may be darker or, in contrast, lighter than the rest of the body. Numbers, distribution and, in certain cases, colour of the chromatophores may be of taxonomic value. Thus, the striking and constant difference in colour (red and black, respectively) of the dorsal chromatophores gave first hints to the splitting of *Diamysis lagunaris* from *Diamysis mesohalobia*, two taxa that were long taxonomically confused (Ariani & Wittmann, 2000: fig. 2D-F). The semi-hypogean species *Diamysis camassai* shows a darker **pigmentation** in summer compared to winter and to its epigean congeners, probably to avoid light damage upon migration from dark to light parts of its habitat (Ariani & Wittmann, 2002). Certain pigment cells are found in the **hypodermis**, others on or close to the surface of **internal organs** such as those pertaining to the nervous system, stomach, and gonads (Keeble & Gamble, 1904; Degner, 1912a, b). Often the chromatophores appear strongly ramified, as those present on the telson and oostegites in several species of *Diamysis* and *Paramysis*. It is noteworthy that Costa (1847) already remarked on the ramifications visible on the oostegites in his description of *Acanthomysis longicornis* from the Gulf of Naples, but oddly considered them as pertaining to the circulatory system. The cornea varies from red to brown or black, rarely blue (*Mesopodopsis slabberi*), often with metallic bronze or golden reflection.

Important components of the **body colour** are often not related to **pigment cells**. Such colours are due to stomach and intestine contents, to the colour of eggs in the ovaries and/or in the marsupium, and to yellow to orange-red **fat bodies** (oil globules) visible through the more or less transparent body wall. In certain species the entire cuticle (*Heteromysis*) or only that of setae (*Praunus*) may show a violet tinge. In coastal Mysidae, the eggs or their yolk are mostly pale to yellow or orange, but also red, green, blue, or violet. In *Leptomysis*

---

fringes on the carapace of male *Diamysis cymodoceae* Wittmann & Ariani, 2012, carapace expanded (flattened) on slide, dorsal; D, detail of (C) showing the fringes; E, detail from another specimen, like (D), showing a dense fur formed by fringes; F, carapace with three major pore groups as typical for species of *Leptomysis* G. O. Sars, 1869; G, cervical pore group in female *Leptomysis mediterranea* G. O. Sars, 1877; H, mid-terminal pore group located on a slight elevation directly in front of the carapace margin in male *Leptomysis posidoniae* Wittmann, 1986; J, scales and setae on the basis of the terminal segment of the antennular peduncle in female *Leptomysis buergii* Băcescu, 1966; K, scales covering the eyestalk in a female *Leptomysis posidoniae*. Abbreviations: cp, cardial pore group; cpo, compound organ *sensu* Mauchline (1977); cs, cervical sulcus; cv, cervical pore group; fr, fringes; mtp, medio-terminal pore group. [A, B, after Mauchline, 1977; C-E, after Wittmann & Ariani, 2012a; F-K, after Wittmann 1986a, b; A-C, F, H, modified.]

the colour of yolk appears to be related to the habitats in which the species live (Wittmann, 1981a). The eggs of several *Heteromysis* species are brilliantly green (visible through the transparent carapace in fig. 54.1D), strongly contrasting with the predominantly orange-red body colour of the incubating mothers (Wittmann, 2008; Daneliya, 2012).

Cephalic appendages

Antennula (figs. 54.10A, B, 54.11A, B, 54.12A-G). – The **antennules** show a three-segmented **peduncle** with two unequal, multi-segmented **flagella** equipped with a variety of sensory setae. The inner flagellum is generally shorter than the outer in Lophogastrida (fig. 54.1A) and Mysida (figs. 54.4, 54.5), but not so in the Stygiomysida family Lepidomysidae (fig. 54.3A). Compared to females, the male flagella are mostly larger and more robust, often with different sets of sensory setae; these differences are most striking in *Hansenomysis* (Petalophthalmidae: Hansenomysinae). The males of *Mesopodopsis* and *Surinamysis* (Mysidae: Mysinae) show a long, unsegmented lobe resembling a third, **rudimentary flagellum**. In most Mysidae males, the distal segment of the peduncle bears a conical or subcylindrical (rarely sub-globular), inwards and/or forward to backwards directed process with a brush of sensory setae (fig. 54.12C, E, G). This process is commonly termed **appendix masculina** (see below, 'Sexual dimorphism'). In males of certain Heteromysinae this appendix may be very short, rarely absent, but always shows the brush of setae typical for males in the family Mysidae (see below, 'Sensory setae'). The **antennular peduncle** is usually more stout in males (fig. 54.12E) than in females (fig. 54.12F). In most species it shows additional dimorphic elements regarding lobes, spines and/or setae.

Antenna (figs. 54.10C-G, 54.11B, D, 54.12A-D, H-L). – The **antennal sympod** consists of three often poorly discernible segments. In the Mysida, the **end sac** (fig. 54.12L) of the **antennal gland** is generally large and may be confused with a segment. The sympod bears a plate-like **exopod** (= antennal scale) and a multi-segmented **endopod**. The endopod shows a somewhat robust trunk (= **antennal peduncle**) composed of 3-4 segments, with a long multi-segmented flagellum. The **antennal scale** varies greatly in form and size between the taxa (fig. 54.12A-D, H-L). It may be unsegmented or show a

Fig. 54.10. Antennae and mouthparts in the Lophogastrida families Gnathophausiidae (A, C, G-K, M, O), Eucopiidae (B, D, L, N), and Lophogastridae (E, F). A, antennula with eye and rostrum in *Gnathophausia longispina* G. O. Sars, 1883; B, the same for *Eucopia australis* Dana, 1852; C, antenna of *Gnathophausia ingens* (Dohrn, 1870); D, antenna of *Eucopia australis*; E, antenna of *Lophogaster typicus* M. Sars, 1857; F, antenna of *Ceratolepis hamata* G. O. Sars, 1883; G, antenna of *Gnathophausia longispina* G. O. Sars, 1883; H, labrum and mandibles in *Gnathophausia longispina*; J, detail of (H) showing the masticatory surface of the mandibular trunk, posterior face; K, labium (paragnaths) of *Lophogaster typicus*, anterior face; L, maxillula of *Eucopia australis*; M, maxillula of *Gnathophausia longispina*; N, maxilla of *Eucopia australis*; O, maxilla of *Gnathophausia longispina*. Abbreviations: ba, basis; cx, coxa; ed, endite; end, endopod; ex, exopod; lo, luminescent organ; pcx, praecoxa; pl, retroverted palp. [A-J, L-O, modified after G. O. Sars, 1885a; K, original drawing of an adult male with 20 mm body length from Hjeltefjord (Norway).]

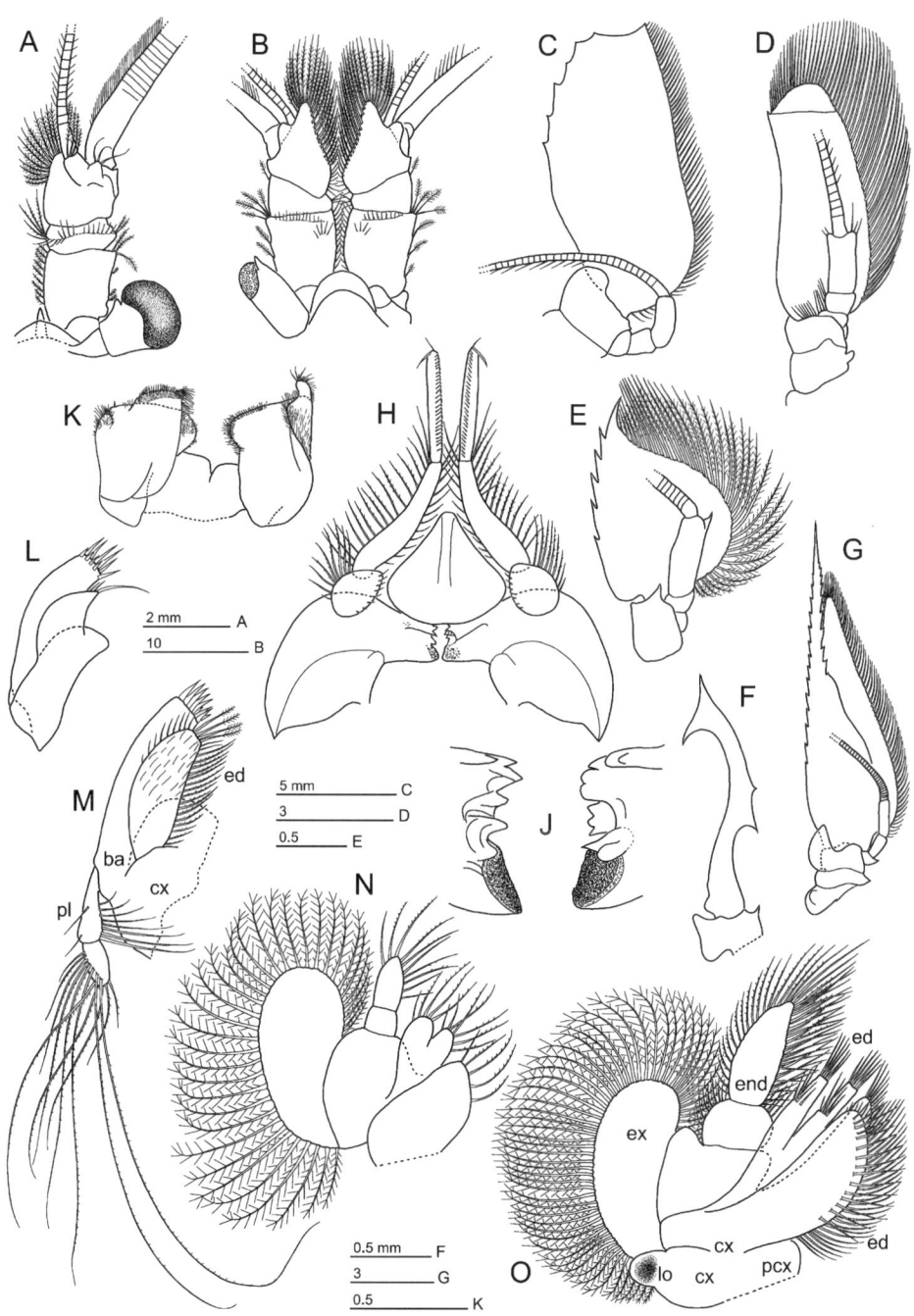

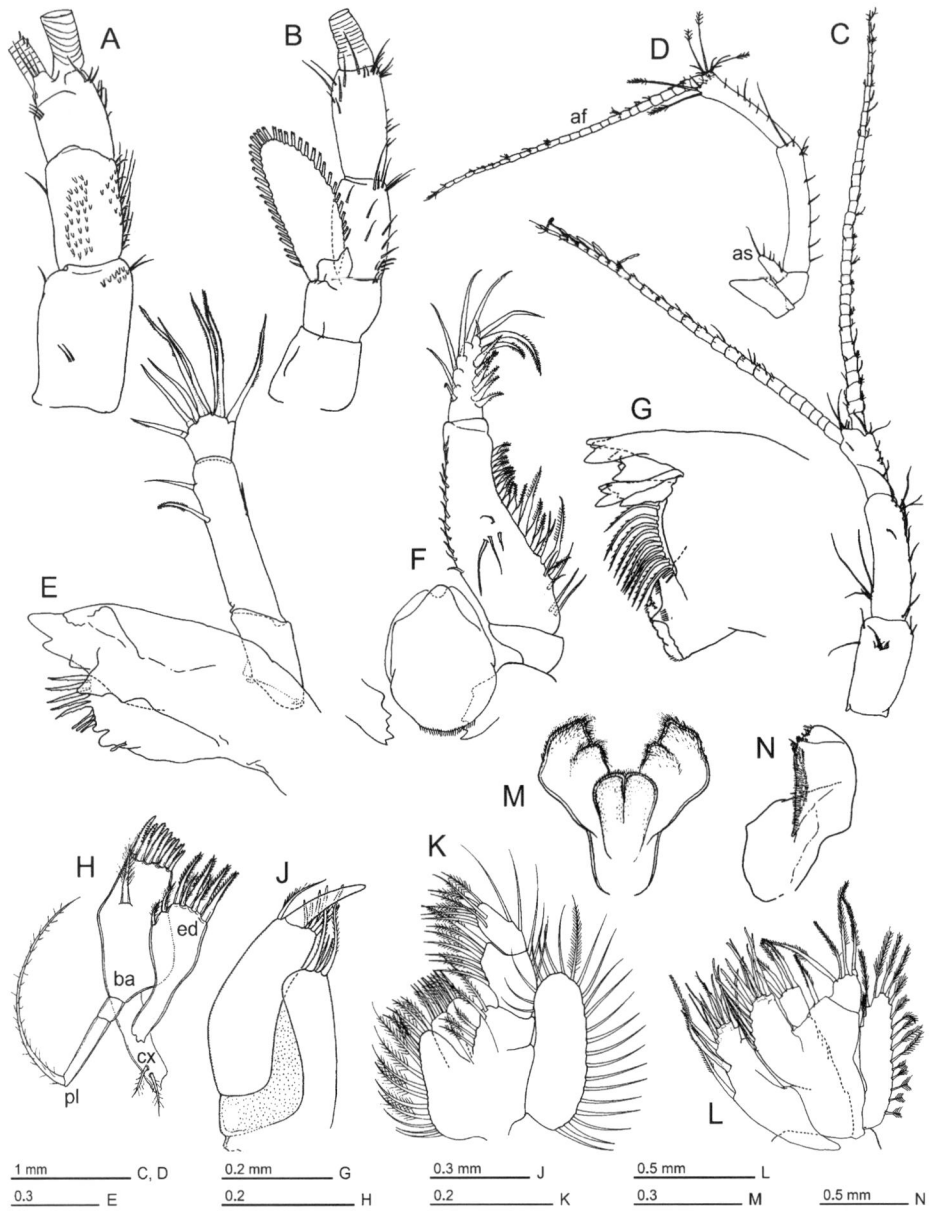

Fig. 54.11. Antennae and mouthparts in the Stygiomysida families Lepidomysidae (A, B, F-H, K, M) and Stygiomysidae (C-E, J, L, N). A, antennula of male *Spelaeomysis cardisomae* Bowman, 1973; B, antenna of the same male as in (A); C, antennula of female *Stygiomysis aemete* Wagner, 1992; D, antenna of the same female as in (C); E, right mandible of the same female as in (C); F, left mandible with labrum *in situ*, *Spelaeomysis cardisomae*; G, masticatory portion of the left mandible in *Spelaeomysis bottazzii* Caroli, 1924; H, maxillula in *Spelaeomysis longipes* (Pillai & Mariamma, 1964); J, maxillula in *Stygiomysis hydruntina* (Pillai & Mariamma, 1964); K, maxilla in *Spelaeomysis cochinensis* Panampunnayil & Viswakumar, 1991; L, maxilla in

small terminal segment separated by a transverse or oblique articulation. Its borders are mostly lined by plumose setae; nonetheless, the outer border may be (in part) smooth, serrated, or spinose. The outer border is slightly dimorphic in *Eucopia australis*. In the lophogastrid *Ceratolepis hamata* the antennal scale is curiously modified, resembling a fishhook (fig. 54.10F). Within the Stygiomysida it is small in *Spelaeomysis* (figs. 54.7J, 54.11B) and only vestigial (fig. 54.11D) in *Stygiomysis*. It is vestigial also in a few Mysida, for example miniaturized and scale-like in *Palaumysis*, reduced to a small spine in *Arachnomysis* and *Chunomysis*, and small, styliform in *Caesaromysis* (fig. 54.12C).

Mouthfield (figs. 54.13A, 54.14A). – The **mouth** is delimited anteriorly by the unpaired **labrum**, posteriorly by the paired lobes of the **labium**, and laterally by three pairs of appendages, i.e., the **mandibles**, **maxillules**, and **maxillae**. More caudally, the area of the mouthparts in Mysida is closed by an anteriorly directed lobe (fig. 54.15M) originating from the first thoracic sternite. De Jong-Moreau et al. (2001) found that gut content analyses generally agreed well with the morphology of **peri-oral structures** in Mysidae species, reflecting the different feeding types: saprophagous (*Bacescomysis abyssalis*), phytophagous (*Birsteiniamysis inermis*), omnivorous (*Hemimysis speluncola*), and carnivorous (*Siriella armata*). Remarkably, their results suggest that a large **processus molaris** is not always associated with herbivorous feeding as was previously generally assumed (Mauchline, 1980).

Labrum (figs. 54.10H, 54.11F, 54.13B, F, 54.14A, C, 54.15A-C). – The **labrum** has the form of a ventrally protruding hump, anteriorly mostly rounded and posteriorly truncate. However, in the Siriellinae, Gastrosaccinae, Mysidellinae, and certain Mysinae it extends anteriorly in a median spiniform process (fig. 54.15C). The posterior margin is formed by two **asymmetrical lobes** bearing setae, small spines, cuticular ridges, and/or tooth-like structures. In the Lophogastridae, Gnathophausiidae, Boreomysinae, and certain Erythropinae, the right lobe forms a kind of inwards-bent tooth. In *Gnathophausia longispina* this tooth bears stiff setae that appear apically frayed (fig. 54.13B). As a diagnostic character of the Mysidellinae, the ventral face of the labrum is posteriorly produced into two strongly unequal lobes. In *Lophogaster typicus*, the external face of the labrum is occupied by numerous pores associated with **glandular units** involved in coating and digesting food; accordingly, the labrum plays a role not only in the mechanical but also in the chemical breakdown of food (De Jong et al., 2002).

Mandible (figs. 54.10H, J, 54.11E-G, 54.13A, E, 54.14, 54.15D). – The **asymmetric mandibles** are composed of a trunk and a three-segmented palp, whereby the proximal segment is the shortest. The inner margin of the trunk shows four domains, presented here in a series from distal to proximal (fig. 54.14C): the **pars incisiva** (processus incisivus)

Stygiomysis aemete; M, labium (paragnaths) in *Spelaeomysis longipes*; N, left paragnath in *Stygiomysis aemete*. Abbreviations: af, antennal flagellum; as, antennal scale; ba, basis; cx, coxa; ed, endite; pl, palp. [A, B, F, after Bowman, 1973; C-E, L, N, after Wagner, 1992; G, after Ariani, 1982; H, M, after Pillai & Mariamma, 1964; J, after Gordon, 1960; K, after Panampunnayil & Viswakumar, 1991; E, K, N, modified.]

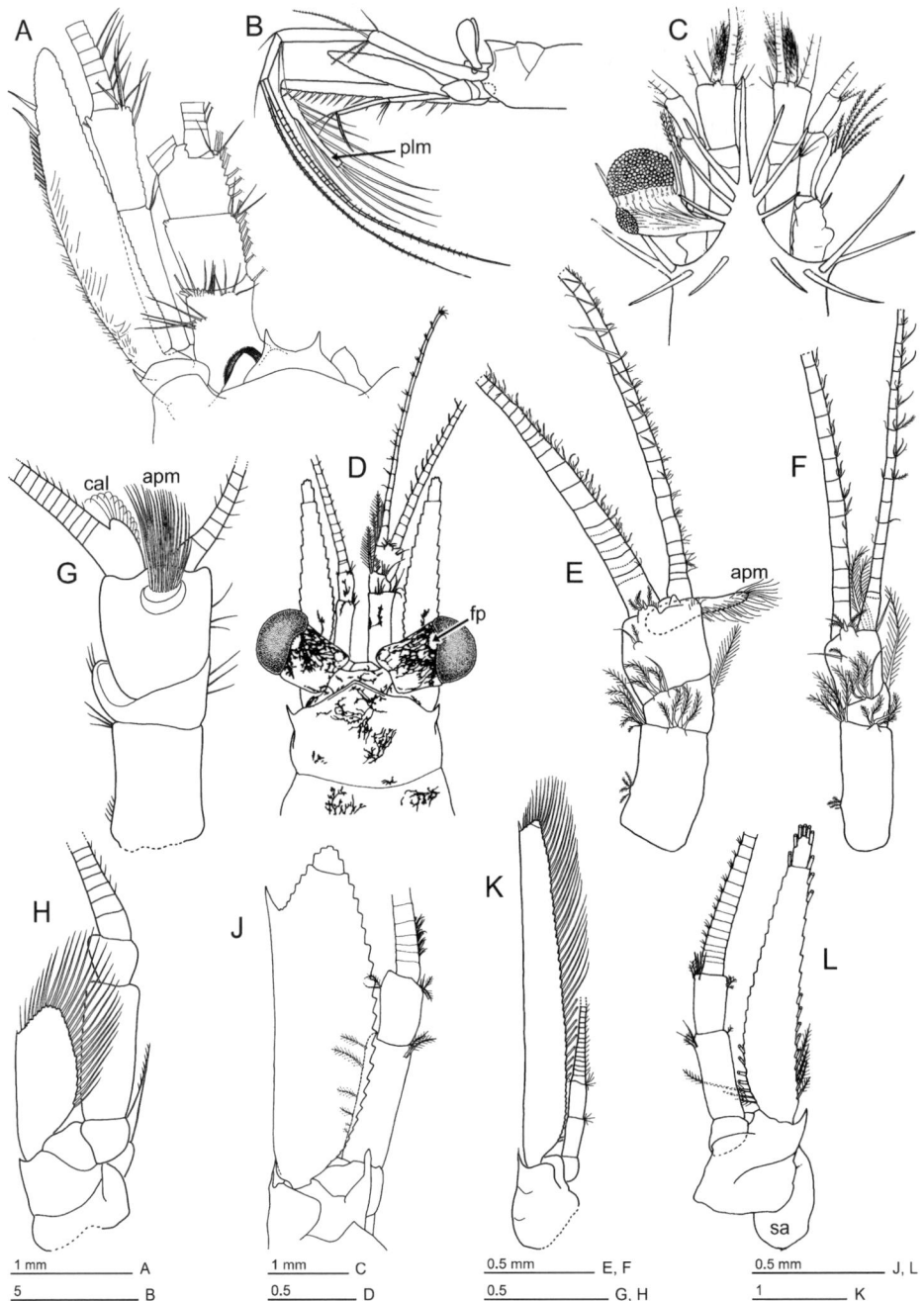

Fig. 54.12. Antennae in the Mysida families Petalophthalmidae (A, B) and Mysidae (C-L). A, eyeplate and left antennula and antenna of female *Hansenomysis pseudofyllae* Lagardère, 1983, dorsal; B, cephalic region of female *Petalophthalmus armiger* Willemoës-Suhm, 1875, lateral; C, cephalic region of male *Caesaromysis hispida* Ortmann, 1893, dorsal; D, cephalic region of female

and the **digitus mobilis** (lacinia), each with several strong, mostly smooth teeth; the **pars centralis** (processus incisivus accessorius, spine row) usually with serrated or spinose teeth; and the **pars molaris** (processus molaris) with a more or less strong grinding surface. The equipment of the domains with teeth or grinding lamellae varies between the left and right mandibles. Size and strength of the partes molares vary with the feeding habits of the various species (Mauchline, 1980). Certain domains may be missing in some genera. Particularly, the right digitus mobilis is often rudimentary or missing. The inner margin of the trunk is rectilinear and forms a broad cutting blade in *Mysidella*. The **mandibular palp** commonly shows moderate variations. In several species of *Petalophthalmus*, however, it is strongly enlarged and prehensile with a serrated inner margin of the median segment (fig. 54.12B). In Lophogastrida, the right mandible shows no digitus mobilis (lacinia; Richter, 2003), and the left mandible bears a semicircular incisory process and a furrow that receives the strongly developed left paragnath of the labium. According to B. Casanova et al. (2002) the gnathal part of the mandibles is prolonged by portions of the pleurites and tergites of the **mandibular somite** during larval development in Lophogastrida, Mysida, Euphausiacea, and certain Decapoda.

Labium (figs. 54.10K, 54.11M, N, 54.13C, D, 54.14A, 54.15H). – The **labium** of the Mysidae is generally formed by two roughly symmetrical **paragnaths** that are more or less connected at their base. The labium of *Thalassomysis* is aberrant by the strongly **asymmetrical paragnaths** showing a very large common basis; and by the right paragnath showing a cavity to receive part of the labrum. In the Stygiomysida genus *Stygiomysis* the paragnaths are long (fig. 54.11N) and apically widely separated. The left paragnath is more strongly developed than the right (fig. 54.10K) in most, probably all, Lophogastrida (Clarke, 1961b; Richter, 2003). Its anterior and inner faces are lined by tubercles and ridges, forming a **masticatory structure** together with the concave posterior margin of the corresponding mandible (Clarke, 1961b). The well-studied paragnaths of *Gnathophausia childressi* (fig. 54.13C, D) show a strong musculature comparable to that of the maxillipeds. They possibly actively contribute to mastication, like the mandibles with which they are connected. The presence of **molar structures** on the left paragnath (fig. 54.10K), stronger than on the right one, supports this hypothesis (J.-P. Casanova, 1996b).

Diamysis lagunaris Ariani & Wittmann, 2000, dorsal; E, antennula of male *Diamysis fluviatilis* Wittmann & Ariani, 2012, dorsal; F, antennula in female *Diamysis fluviatilis*; G, antennula of male *Anchialina truncata* (G. O. Sars, 1883), ventral; H, antenna of male *Anchialina truncata*, dorsal; J, antenna of *Siriella castellabatensis* Ariani & Spagnuolo, 1976; K, antennal scale of *Praunus flexuosus* (O. F. Müller, 1776); L, antenna of *Diamysis mesohalobia* Ariani & Wittmann, 2000. Abbreviations: apm, appendix masculina; cal, callynophore; fp, fenestra paracornealis; plm, palpus mandibularis; sa, sacculus (end sac) of the antennal gland. [A, after Lagardère, 1983; B, after Tattersall & Tattersall, 1951; C, after O. S. Tattersall, 1955; D, L, after Ariani & Wittmann, 2000; E, F, after Wittmann & Ariani, 2012b; G, H, after G. O. Sars, 1885a; J, after Ariani & Spagnuolo, 1976; K, after G. O. Sars, 1879; A, B, D-L, modified.]

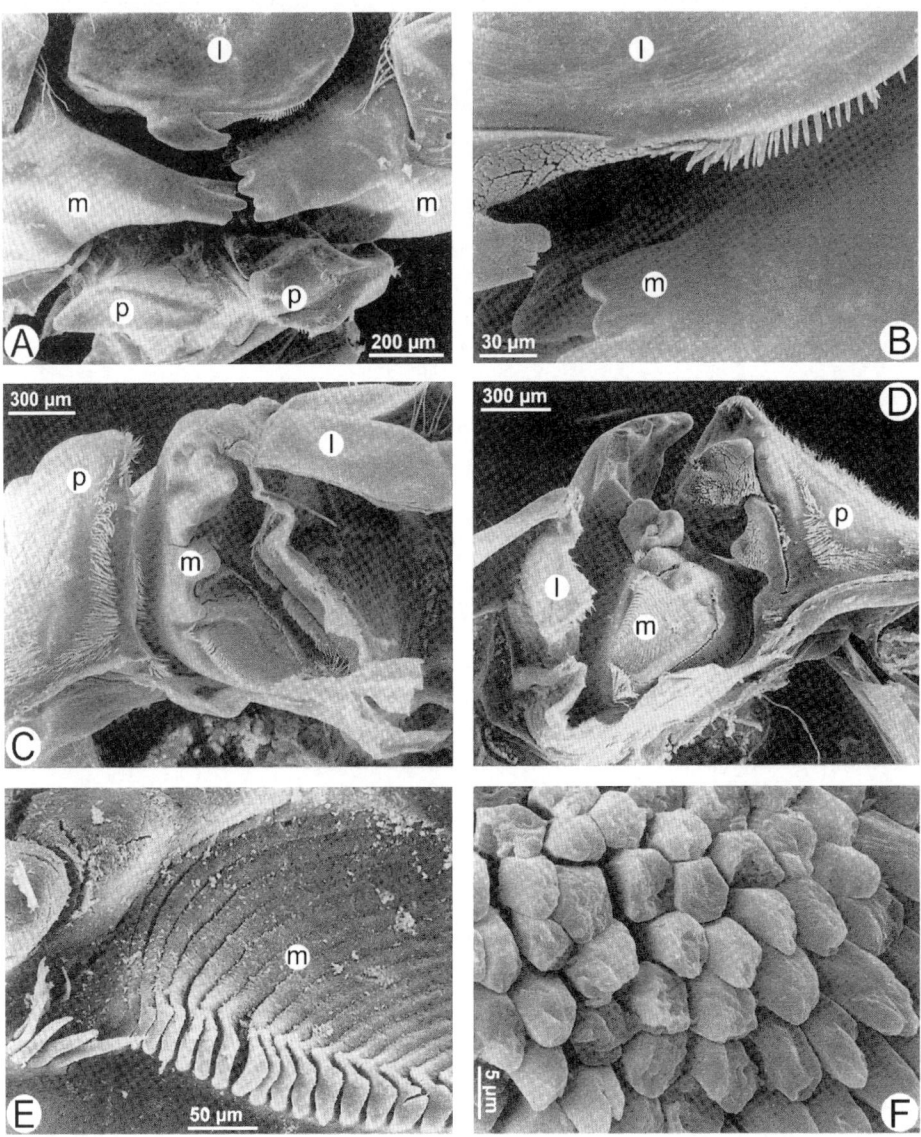

Fig. 54.13. Structures of the mouthfield in the Lophogastrida species *Gnathophausia longispina* G. O. Sars, 1883 (A-E) and *Gnathophausia childressi* Casanova, 1996 (F). A, mouth area in ventral view; B, detail of labrum; C, D, inner lateral view on medio-sagittal sections of mouth area, showing coaptation of the right (C) and left (D) paragnaths; E, F, details of grinding surfaces of mandibles (E) and labrum (F). Abbreviations: l, labrum; m, mandibles; p, paragnaths. [A, C-F, after J.-P. Casanova, 1996b; B; after Nouvel et al., 1999.]

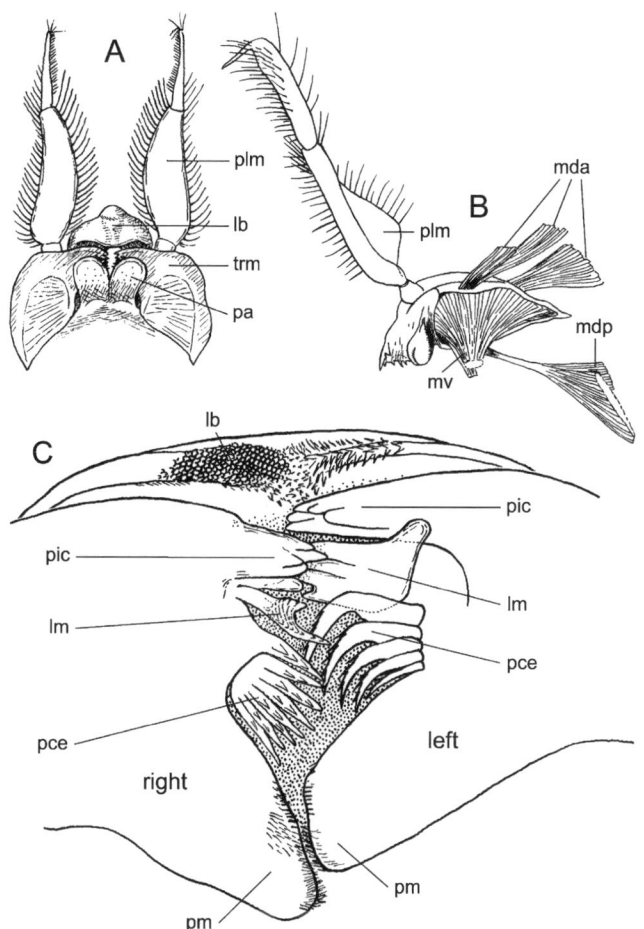

Fig. 54.14. Structures of the mouthfield in Mysidae (Mysida). A, *Birsteiniamysis inermis* (Willemoës-Suhm, 1874), mouthparts *sensu stricto* in ventral view; B, *Neomysis rayii* (Murdoch, 1885), musculature of right mandible in inner lateral view; C, *Hemimysis lamornae* (Couch, 1856), details of labrum and masticatory edges in ventral view. Abbreviations: lb, labrum; lm, lacinia mobilis; mda, mdp and mv, dorsal-anterior (mda), dorsal-posterior (mdp), and ventral (mv) muscles, respectively; pa, paragnaths (labium); pce, pars centralis (processus incisivus accessorius, spine row); pic, pars incisiva (processus incisivus); plm, palpus mandibularis; pm, pars molaris (processus molaris); trm, mandibular trunk (endite). [A, after G. O. Sars, 1885b; B, after Snodgrass, 1952; C, modified after Cannon & Manton, 1927a.]

Maxillula (figs. 54.10L, M, 54.11H, J, 54.15E). – According to the interpretation by Hansen (1925) the **maxillula** is formed by three segments (praecoxa, coxa, basis) of the sympod. Coxa and basis each bear an endite with setae and spines. The Gnathophausiidae (fig. 54.10M) and the Lepidomysidae (fig. 54.11H) each show a small, two-segmented, **retroverted palp** that may be interpreted as a vestigial endopod. In the Mysidae and certain Gnathophausiidae, there is a small lamelliform, setose lobe (**pseudoexopodite**;

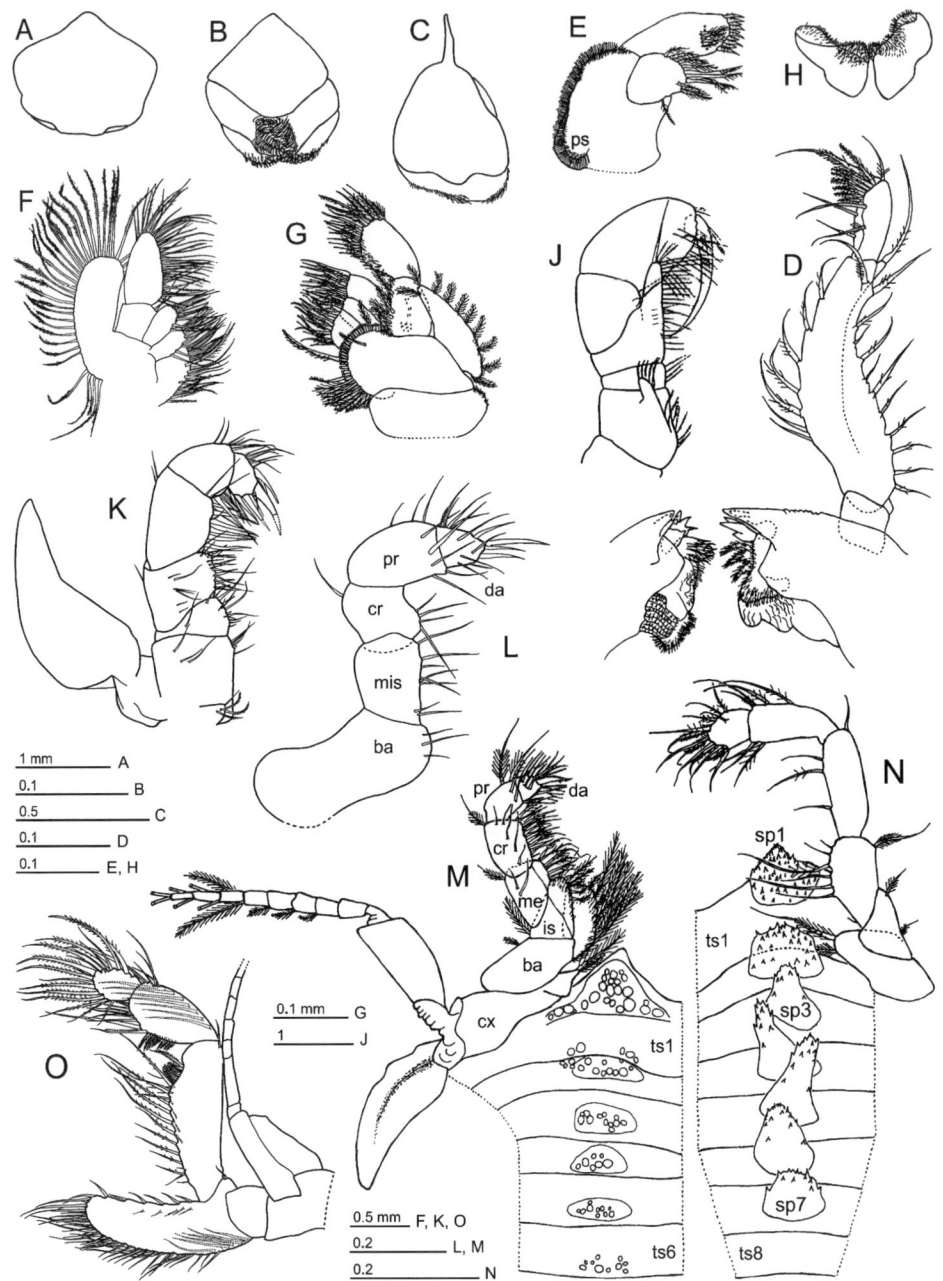

Fig. 54.15. Mouthparts and maxillipeds in the Mysida families Petalophthalmidae (A, F, J, K, O) and Mysidae (B-E, G, H, L-N). A, labrum of *Hansenomysis atlantica* Lagardère, 1983, ventral; B, labrum of *Heteromysis arianii* Wittmann, 2000, oblique ventral view; C, labrum of *Gastrosaccus roscoffensis* Băcescu, 1970, slightly oblique ventral view; D, mandibles of *Heteromysis arianii*, caudal aspect; E, maxillula of *Heteromysis arianii*, caudal; F, maxilla of *Hansenomysis atlantica*, caudal; G, maxilla of *Heteromysis arianii*, frontal aspect; H, labium (paragnaths) of

fig. 54.15E), folded from the base, considered by Hansen (1925) to be an exite of the praecoxa.

Maxilla (figs. 54.10N, O, 54.11K, L, 54.15F, G). – The second **maxilla** resembles a series of leaves with a common basis. It shows a quite uniform structure with the sympod consisting of three more or less distinct segments: the coxa with a simple endite, and the basis with a bilobate endite. These endites bear dense rows of mostly plumose setae along their margins. The endopod (= **maxillary palp**) is reduced to two foliaceous segments; the terminal margin of its distal segment is well equipped with setae. In addition, it is armed with spines or teeth in various taxa. The exopod is usually represented by a rounded plate with setae along the outer margin. However, it is reduced or completely absent (*Metamysidopsis*) in a few species. On the outer margin of the coxa of the Gnathophausiidae, a red papilla (fig. 54.10O) marks the common orifice of two glands producing a **luminescent secretion**.

Thoracic appendages

General structure of the thoracopods (figs. 54.16, 54.17A-F, 54.18). – All taxa have eight pairs of thoracic appendages. The anterior two pairs of endopods are normally modified as **maxillipeds** (fig. 54.15M, N). Only the Eucopiidae (fig. 54.19A, B) and the Stygiomysidae (fig. 54.17D) show four pairs of maxillipeds. It is a matter of definition whether the prehensile third endopods, termed gnathopods (fig. 54.18H), of many Heteromysinae (Mysidae) are to be counted as maxillipeds.

The thoracopods were classically described as consisting of a one-segmented sympod with an endopod and a natatory exopod. The endopod was schematized as five main articles: a short ischium, elongate merus and carpus, elongate propodus (often secondarily subdivided), and a short dactylus with claw. The main knee was seen between carpus and propodus (Băcescu, 1954). The exopod appears composed of a basal plate and a multi-articulate flagellum bearing plumose setae.

Based on the comparative anatomy of the Malacostraca and on the study of the musculature (principally the flexor and extensor dactyli), Hansen (1925) proposed a modified concept for the mysidaceans (fig. 54.18A-G): firstly, the sympod shows a more or less distinct subdivision in **basis** and **coxa**, sometimes also a praecoxa. Secondly, the main **articulation of the leg** is placed between **merus** and **carpus**. Proximally to

---

Heteromysis arianii; J, first thoracic endopod (maxilliped 1) of *Petalophthalmus armiger* Willemoës-Suhm, 1875, caudal; K, first thoracic endopod with epipod in *Hansenomysis atlantica*, caudal; L, first thoracic endopod of *Mysidopsis gibbosa* G. O. Sars, 1864; M, first thoracopod (caudal aspect) with thoracic sternites 1-6 (ventral) in female *Heteromysis arianii*; N, second thoracic endopod (caudal) with thoracic sternites 1-8 (ventral) in male *Heteromysis arianii*; O, second thoracopod of *Hansenomysis pseudofyllae* Lagardère, 1983, caudal. Abbreviations: ba, basis; cr, carpus; cx, coxa; da, dactylus; is, ischium; mis, merischium; pr, propodus; ps, pseudexopodite; sp1, sp3, sp7, processes from sternites 1, 3, or 7; ts1, ts6, ts8, thoracic sternites 1, 6, and 8, respectively. [A, F, K, O, after Lagardère, 1983; B, D, E, G, H, M, N, after Wittmann, 2000; C, after Wittmann et al., 2011; J, L, after Tattersall & Tattersall, 1951; A, E, G, J-L, O, modified.]

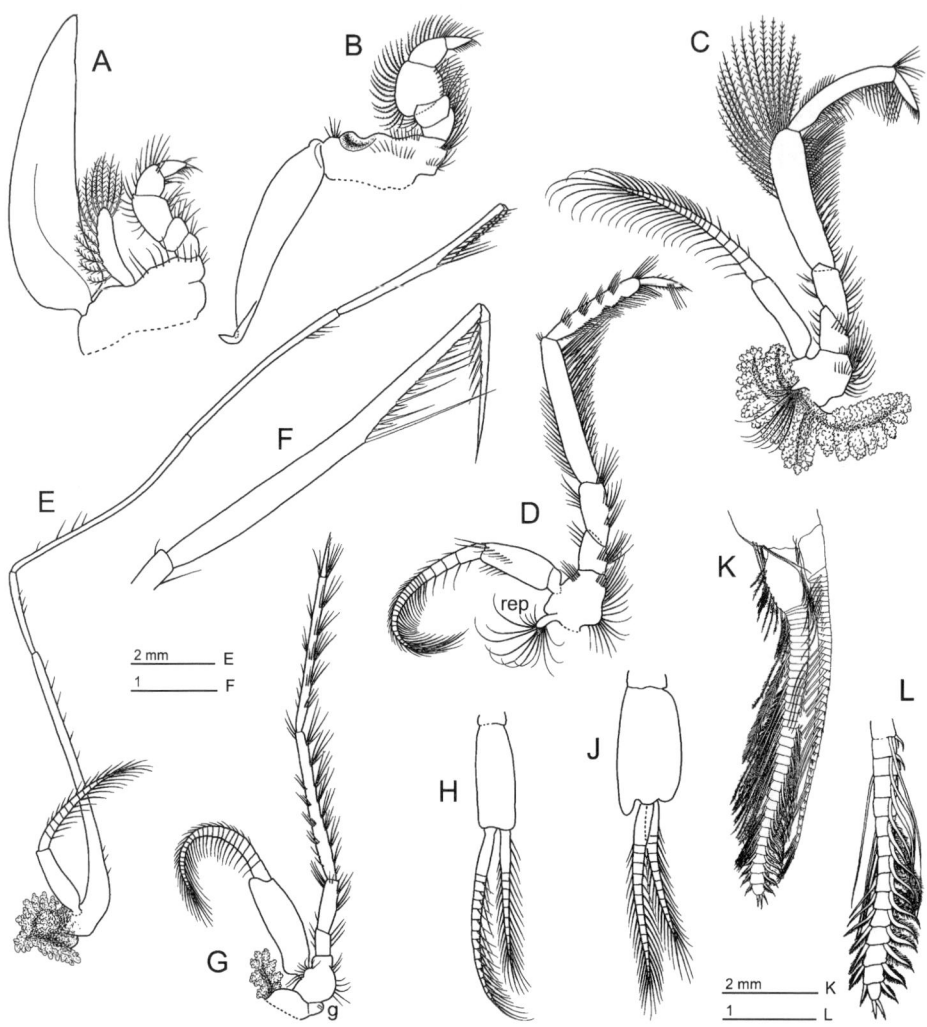

Fig. 54.16. Thoracopods and pleopods in the Lophogastrida families Eucopiidae (A, E, F, H, J) and Gnathophausiidae (B-D, G, K, L). A, first thoracopod of *Eucopia australis* Dana, 1852; B, the same for *Gnathophausia longispina* G. O. Sars, 1883; C, second thoracopod of *Gnathophausia longispina*; D, third thoracopod of male in *Gnathophausia longispina*, branchiae removed to render rudimentary epipod (rep) visible; E, fifth thoracopod in *Eucopia australis*; F, terminal portion of endopod as in (E); G, eighth thoracopod with male gonopore (g) in *Gnathophausia longispina*; H, pleopod of *Eucopia australis* female; J, pleopod of *Eucopia australis* male; K, second male pleopod in *Gnathophausia zoea* Willemoës-Suhm, 1873; L, terminal portion of endopod as in (K). [A-J, modified after G. O. Sars, 1885a; K, L, after Lagardère & Nouvel, 1980a; K, modified.]

this articulation there are three articles: praeischium, ischium, and merus. Distally there are as a general rule (e.g., in *Siriella*; fig. 54.18B) the carpus, the propodus, and the dactylus with claw. In many Erythropinae the carpus is separated from the (one- to multi-segmented) propodus by an oblique segmental border (fig. 54.18D). Depending on genera and higher taxa, the carpus and the propodus may be distinct, or may be fused to form a **carpopropodus** (fig. 54.18F, G), which, then, may be subdivided by secondary articulations. In case of only one subdivision, its significance may be established based on the course and insertion of the musculature. This scheme was adopted by most subsequent authors (e.g., Tattersall & Tattersall, 1951). Nonetheless, the concept of a fused and then subdivided carpopropodus should not be used in an over-schematized manner. There is a great diversity between taxa, particularly within the order Mysida. In many Mysinae and Leptomysinae, particularly in *Paramysis* (fig. 54.19E), the first segment after the knee may be identified as carpus due to different size, development, and setation patterns compared to the remaining propodal segments.

First thoracopod (figs. 54.15M, 54.16A, B, 54.17A, B). – The endopod is pediform only in the Carboniferous Lophogastrida Peachocarididae (Schram, 1986). In extant taxa it is modified as a **maxilliped** throughout. It is more or less shortened, sturdy, compressed, formed by 5-7 articles and a terminal claw. The two distal articles are broadened and flattened. They are turned to the sagittal plane of the body, thus forming with the endites of the proximal articles a scissor-like structure. In certain species of *Boreomysis* the distal article forms a subchela together with a bulge of the preceding article. Well-developed endites may be present on 1-4 basal articles of the endopod. Occasionally, some of the basal articles may be fused. The basis may show a great transversal extension due to a large distance between exopod and endopod. In most taxa the sympod bears a **foliaceous epipod** penetrating into the carapace cavity. Most Mysida show a typically small endite inserting subterminally on the inner anterior face of the basis (fig. 54.15M). In this order, the **exopod** is generally well developed, with basal plate and a mostly eight-segmented flagellum (fig. 54.15M): in *Mysimenzies* (Mysida: Erythropinae) the exopod is reduced to a lamina, in the Lophogastrida to an unsegmented blade (fig. 54.16A) or rod (or may be vestigial; fig. 54.16B), in the Stygiomysida to a leaf-like blade (fig. 54.17A, B). The first exopod is completely reduced in the Petalophthalmidae (Mysida).

Second thoracopod (figs. 54.15N, 54.16C). – The endopod normally shows less strong modifications as a **maxilliped** compared to that of the first thoracopod. Shape and degree of modification vary according to the various taxa. At family level, this endopod is least maxilliped-like, more closely resembling the subsequent thoracopods in the Gnathophausiidae (fig. 54.16C). Basal endites are present in several taxa, e.g., *Ceratomysis*. The distal article of the endopod is often flat, with brushes of setae and spines. The second endopod may show a **prehensile structure** interpreted as suitable for grasping or for brushing. In the Stygiomysida it is prehensile. The second **exopod** is well developed, natatory in this order, as in the Lophogastrida, and in most Mysida, but missing in Hansenomysinae (Mysida: Petalophthalmidae).

Third to eighth thoracopods (figs. 54.16D-G, 54.17C-F, 54.18, 54.19B, E). – These appendages are almost always well developed. The **natatory exopods** consist of a **basal plate** and a multi-segmented, setose **flagellum**. Differences at specific level concern mainly segmental numbers of the flagellum and the outer corner (rounded or spiniform) of

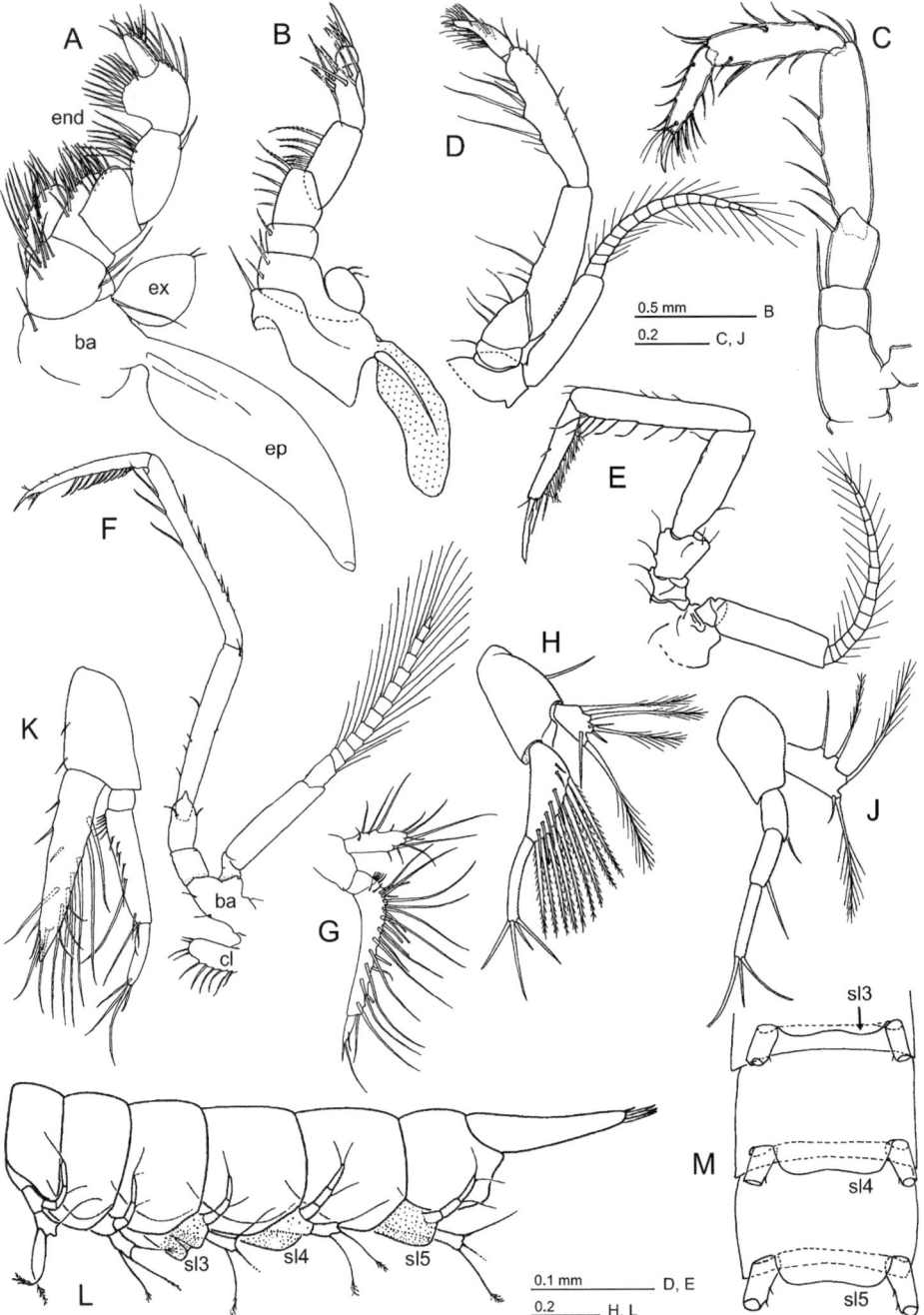

Fig. 54.17. Thoracopods, pleopods, and pleon in the Stygiomysida families Lepidomysidae (A, C, F, G, K, M) and Stygiomysidae (B, D, E, H, J, L). A, first thoracopod in *Spelaeomysis olivae* Bowman, 1973; B, first thoracopod in *Stygiomysis holthuisi* (Gordon, 1958); C, third thoracic endopod in *Spelaeomysis longipes* (Pillai & Mariamma, 1964); D, fourth thoracopod in *Stygiomysis aemete*

the basal plate of the exopods (Ariani & Wittmann, 2000). As an exception, exopod 8 is missing in the lophogastrid *Ceratolepis hamata*.

The **endopods** are typically **pediform**. Their length may be subequal to conspicuously unequal. In Mysidae, the length typically increases from endopod 3 caudally up to endopod 6 or 7, and may decrease behind. The third and fourth endopods are modified as **maxillipeds** (gnathopods) in the Eucopiidae (fig. 54.19B) and the Stygiomysidae (fig. 54.17D), similarly the third but not the fourth in Lepidomysidae (fig. 54.17C). In certain Heteromysinae (Mysidae) the third endopods are considered to be **gnathopods**, because they are more robust than the posterior endopods and because their strong claw folds up to act as a prehensile subchela (fig. 54.18H). The **eighth endopods** are rudimentary in *Rhopalophthalmus* and even more strongly reduced in *Pseudomysidetes*. In contrast, they are strong and prehensile in *Hyperiimysis*. The Eucopiidae show a very particular structure of the thoracopods (fig. 54.2D): endopods 2-4 are short but strong and subchelate (fig. 54.19A, B), endopods 5-7 are very long, thin and subchelate (fig. 54.16E, F), whereby endopod 8 is almost normal and slender but without claw.

Important variations regard numbers and structure of the distal, i.e., the 'tarsal', articles of the endopods: carpus and propodus distinct or fused to a **carpopropodus** or, in turn, the carpopropodus subdivided into a few up to many articles (fig. 54.18B-G; cf. above). In the Mysida genus *Diamysis*, the numbers of carpopropodal segments in the series of third to eight thoracic endopods show some variability (2-3, exceptionally 3-4); nonetheless, the patterns of segment numbers are within certain limits characteristic of different (sub)species. The **dactylus** of the Mysida may be normal (fig. 54.18H-M) or more or less distinctly fused with the **claw**, or may be rudimentary, as in the last thoracopod of *Paramysis pontica*. In certain cases there is no claw. In addition to the normally present claw, the dactylus may bear several setae (fig. 54.18K, O). Those below the claw are termed **subungulary setae** (fig. 54.18O); they may be long, occasionally microserrated, and may resemble a second claw, particularly if the true claw is thin, needle- or seta-like. On the terminal margin of the **propodus** there are often 1-2 (mainly two) pairs of **paradactylary setae** (fig. 54.18O) flanking the dactylus. Some of these setae are often microserrated and/or may show series of thin secondary projections together forming comb-like structures. Some of the latter have been interpreted as cleaning setae (see below, 'Grooming').

Branchiae (figs. 54.6E, 54.16C, E, G, 54.19B). – Arborescent **gills** at the coxae of thoracopods are typical of the order Lophogastrida. Hansen (1925) and Tattersall &

Wagner, 1992; E, eighth thoracopod in *Stygiomysis aemete*; F, eighth thoracopod in male *Spelaeomysis olivae*; G, second pleopod in male *Spelaeomysis olivae*; H, second pleopod in male *Stygiomysis holthuisi*; J, second pleopod in female *Stygiomysis holthuisi*; K, third pleopod in male *Spelaeomysis cardisomae* Bowman, 1973; L, pleon of female *Stygiomysis holthuisi* in lateral view; M, pleomeres 3-5 in male *Spelaeomysis cardisomae*, ventral. Abbreviations: ba, basis of sympod; cl, coxal lobe; end, endopod; ep, epipod; ex, exopod; sl3 to sl5, sternal lamellae from pleomeres 3-5. [A, F, G, K, M, after Bowman, 1973; B, H, J, L, after Gordon, 1960; C, after Pillai & Mariamma, 1964; D, E, after Wagner, 1992; D, E, G, modified.]

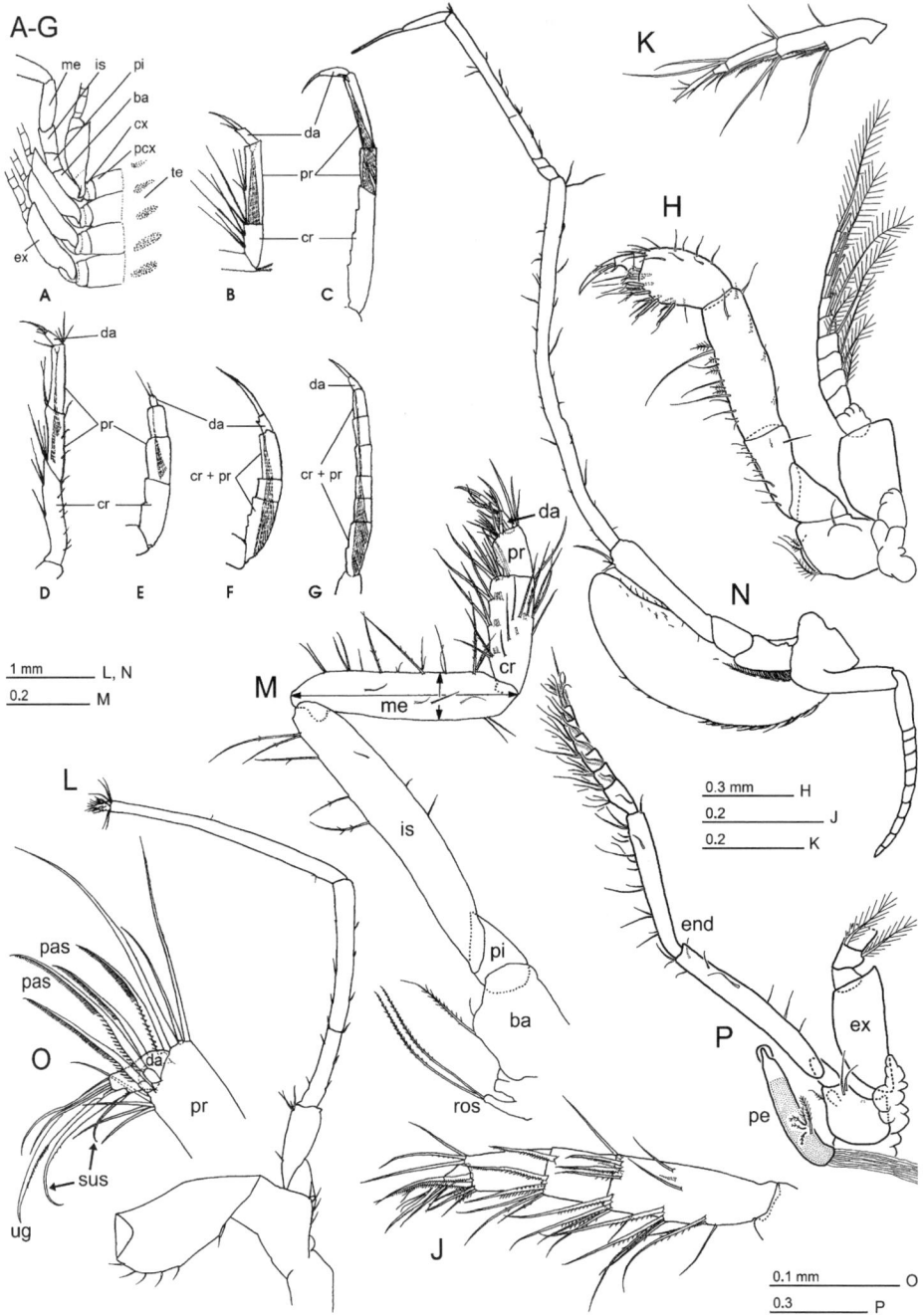

Fig. 54.18. Structure of the thoracopods 3-8 in the Mysida families Mysidae (A-K, M, O) and Petalophthalmidae (L, N). A-G, endopod structure according to interpretation of Hansen (1925, setae partly omitted). A, *Praunus* Leach, 1814, lateral view on thorax showing bases of endopods 2-5; B-G, distal portions of posterior endopod in *Siriella* Dana, 1850 (B), *Boreomysis* G. O. Sars, 1869 (C),

Tattersall (1951) interpreted these branchiae as epipods (quoted by them as pre-epipods), i.e., as **podobranchiae**. If any, there are normally four gills per thoracopod. The main gill is often directed medially and may surround the leg from behind to meet the remaining three gills that are arranged laterally below the lateral projection of the carapace. The lophogastrid gills are always found on thoracopods 3-7, and are generally more strongly developed in males than in females. There is an additional **rudimentary gill** on thoracopod 8 in *Gnathophausia* (fig. 54.16G). *Eucopia crassicornis* from oxygen-poor waters in the Canal of Mozambique shows more strongly developed branchiae compared with the congeneric species from waters with normal oxygen levels (J.-P. Casanova, 1997). There are no branchiae in the Stygiomysida and Mysida; however, small humps with thin cuticle in comparable position may possibly represent rudiments of branchiae in a few forms of the latter order.

Oostegites (fig. 54.19). – The brood lamellae are attached at the basis of the inner (= medial) margins of the thoracic appendages, together forming a brood pouch (**marsupium**) in females. The Lophogastrida, one family (Lepidomysidae) of the Stygiomysida, and the more primitive (sub)families of the Mysida show a total of seven pairs of oostegites on thoracopods 2-8 (fig. 54.19A, D). The Stygiomysidae, as the second family of Stygiomysida, have only four pairs of oostegites in unique position at thoracopods 4-7 (Bowman et al., 1984). Within the order Mysida the **numbers of oostegites** show a clear tendency for reduction: seven pairs in Petalophthalmidae and Boreomysinae, four pairs in *Thalassomysis*, and only 1-3 pairs (or 2-4, including vestigial ones) in the remaining taxa. In Mysida, the oostegites arise from the posterior thoracopods and are largest on thoracopod 8. Fully developed oostegites bear rows of plumose setae at least along the lower margins. Not considering these setae, the oostegites may represent simple plates or may be equipped with motile or immotile ventilation (fig. 54.19E) or **cleaning lobes**. These lobes are furnished with long, microserrated or barbed setae. Functionally similar lobes may also arise from thoracic sternites (fig. 54.19F, G). In certain taxa of Mysida, unpaired foliaceous sternal lobes from the eighth thoracomere (fig. 54.19G), or paired pleural plates from the first pleomere (fig. 54.5F), may contribute to the posterior closure of the marsupium.

Amblyops G. O. Sars, 1872 (D), *Anchialina* Norman & Scott, 1906 (E, endopod 5), *Mysidopsis* G. O. Sars, 1864 (F), and *Praunus* (G); H, third thoracopod in female *Heteromysis dardani* Wittmann, 2008; J, 'tarsus' of third endopod in *Diamysis camassai* Ariani & Wittmann, 2002; K, 'tarsus' of fourth endopod in *Ischiomysis telmatactiphila* Wittmann, 2013; L, fifth endopod with oostegite in adult female *Hansenomysis nouveli* Lagardère, 1983; M, sixth endopod with rudimentary oostegite in adult female *Diamysis camassai*, arrows indicate mode of measurement of merus length and width; N, sixth thoracopod in *Hansenomysis abyssalis* Lagardère, 1983, setae of exopod omitted; O, tip of eighth thoracic endopod in male *Diamysis mesohalobia* Ariani & Wittmann, 2000; P, eighth thoracopod with penis in *Heteromysis wirtzi*, terminal portion of exopod omitted. Abbreviations: ba, basis; cr, carpus; cx, coxa; da, dactylus; ex, exopod; is, ischium; me, merus; pas, paradactylary setae; pcx, praecoxa; pe, penis; pi, praeischium; pr, propodus; sus, subungulary seta; te, tergite; ug, unguis. [A-G, after Hansen, 1925, as arranged by Nouvel et al., 1999, again modified; H, P, after Wittmann, 2008; J, M, after Ariani & Wittmann, 2002; K, after Wittmann, 2013a; L, N, after Lagardère, 1983; O, after Ariani & Wittmann, 2000; A-G, K, L, N-P, modified.]

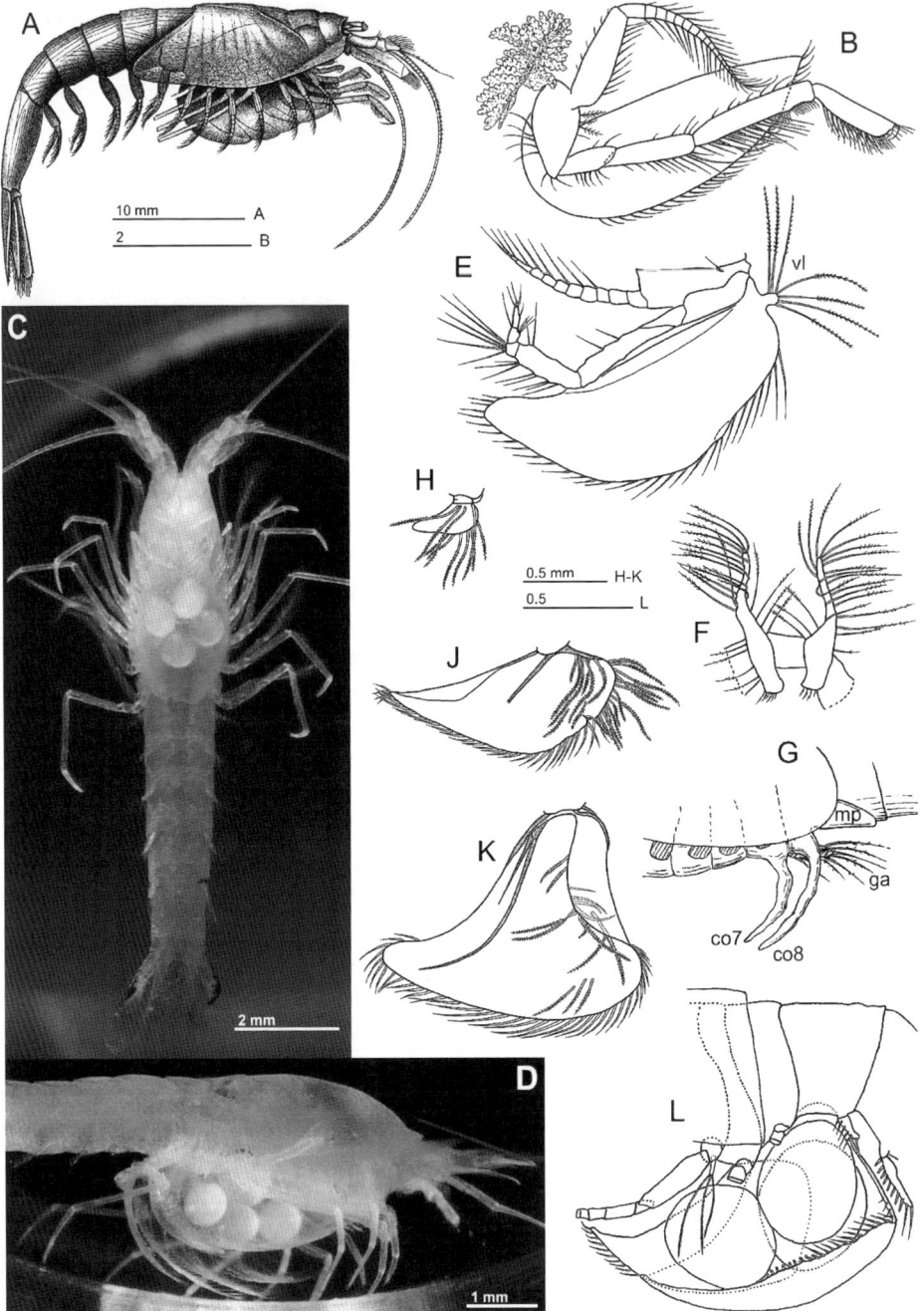

Fig. 54.19. Oostegites and associated structures in the Lophogastrida (A, B), Stygiomysida (C, D) and Mysida (E-L). A, *Eucopia australis* Dana, 1852, lateral, distal portions of thoracic endopods 5-8 omitted; B, same species, third thoracic endopod with gills and oostegite, lateral; C, *Spelaeomysis bottazzii* Caroli, 1924, incubating 4 eggs, *in vivo*, ventral; D, same species, living specimen

In addition, many species show hairy lamellae or **ventilation plates** on the bases of oostegite-bearing thoracopods, suggesting that these structures represent rudimentary (parts of) oostegites. The oostegites are generally reduced during the repose of the ovaries in Lophogastrida (J.-P. Casanova, 1977) and Stygiomysida (Ariani & Wittmann, 2010), as is the case in most remaining orders of Peracarida. A **reduction of the oostegites** at the end of the reproductive period was reported by Nouvel & Nouvel (1939) for the mysid *Praunus flexuosus* from north-eastern Atlantic coasts. Nonetheless, in most Mysida the oostegites are generally not reduced at the moult after release of the young (Ariani & Wittmann, 2000).

External male genitalia (figs. 54.4L, 54.16G, 54.18P, 54.20). – In the males of **Mysida**, with the exception of the Rhopalophthalminae, there is a well-developed **tubular penis** (fig. 54.20C, E-G) arising from the **coxa** of each **eighth thoracopod**. The **gonopore** is (sub)terminal in posterior position. It is typically flanked by two or more setose lobes forming a **closing apparatus** (**valves**; fig. 54.20C, D, G). The penes may be very large in several genera (*Mysidetes, Pseudomysidetes, Mysidella*; Lagardère & Nouvel, 1980b; Wittmann, 2013b). In certain taxa, the normally slightly bent, subcylindrical penes can be enlarged by large **terminal lobes** (*Ischiomysis* in fig. 54.20G, certain *Heteromysis*; Wittmann, 2013a) or by longitudinal wing-like blades (*Paramysis*; Daneliya et al., 2007; Wittmann & Ariani, 2011). The penes of *Mysifaun erigens* may be expanded to giant size by **erectile tissue** (fig. 54.20E; Wittmann, 1996). A strange situation was found by Wittmann (2013b) in two species of *Rhopalophthalmus* (Rhopalophthalminae; fig. 54.20D): there is no penis, but the genital opening is flanked by two small lobes distally on the inner margin of the strongly enlarged coxa of the eighth thoracopods. The **vas deferens** is terminally widened in a large **seminal vesicle** occupying most of the space within this coxa. Endopod 8 is vestigial in males, more strongly reduced compared to the also reduced endopod in females. Only the males bear a rounded **pleural plate** (arrow in fig. 54.4C) ventrally projecting from the first pleomere, and show modified second pleopods (fig. 54.21C). The close spatial vicinity of these uncommon structures suggests some role in the unknown process of mating in this subfamily.

Both genera (families) of **Stygiomysida** have no penes. The male **gonopore** is between an anterior setose and a posterior bare lobe on the inner margin of the (not conspicuously enlarged) coxa of the eighth thoracopods. These lobes are of about equal size (fig. 54.20B)

---

demonstrating its individual oostegites by keeping them straddled, lateral; E, eighth thoracopod of *Paramysis pontica* Băcescu, 1940, note setose ventilation lobe (vl) caudally on oostegite; F, 'ventilation lobes' of fourth thoracic sternites in *Gastrosaccus widhalmi* (Czerniavsky, 1882); G, special processes from thoracic sternites 7 and 8, probably involved in incubating the young of *Neomysis integer* (Leach, 1814), lateral; H-K, series of oostegites 1-3 in *Leptomysis lingvura marioni* Gourret, 1888, inner lateral view; L, right half of brood pouch in *Diamysis camassai* Ariani & Wittmann, 2002, inner lateral view, part of appendages and setae omitted. Abbreviations: co7, co8, 'cotyledons' from seventh and eighth sternites, respectively; ga, grooming appendage; mp, median sternal plate; vl, ventilation lobe. [A, B, after G. O. Sars, 1885a; C, D, photo Antonio P. Ariani, part of background cleaned with electronic tools; E, after Băcescu, 1954; F, after Băcescu, 1940; G, after Nouvel et al., 1999; H-K, after Wittmann, 1982; L, after Ariani & Wittmann, 2002; B, E, F, H-K, modified.]

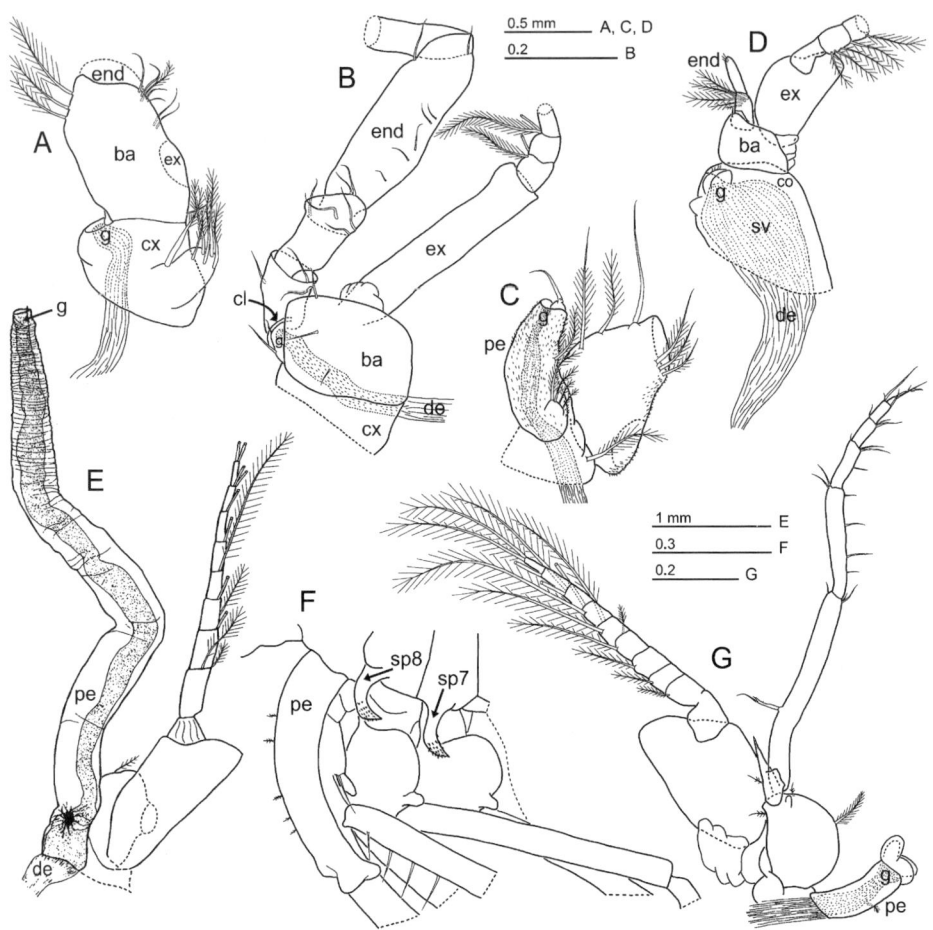

Fig. 54.20. External male genitalia as derivates of the eighth thoracopods in the Lophogastrida (A), Stygiomysida (B), and Mysida (C-G). A, *Eucopia unguiculata* (Willemoës-Suhm, 1875); B, *Stygiomysis hydruntina* Caroli, 1937; C, *Pseudomma armatum* Hansen, 1913; D, *Rhopalophthalmus terranatalis* O. S. Tattersall, 1957; E, *Mysifaun erigens* Wittmann, 1996; F, *Heteromysis armoricana* H. Nouvel, 1940 (setae partly omitted); G, *Ischiomysis telmatactiphila* Wittmann, 2013. Abbreviations: ba, basis; cx, coxa; de, ductus ejaculatorius; end, (insertion of) endopod; ex, (insertion of) exopod; g, gonopore; pe, penis; sp7, sp8, processes from sternites 7 and 8; sv, seminal vesicle. [A-D, after Wittmann, 2013b; E, after Wittmann, 1996; F, after Nouvel, 1940 (scale corrected); G, after Wittmann, 2013a; A-G, modified.]

in *Stygiomysis*, whereas the anterior lobe (fig. 54.17F) is distinctly larger than the posterior one in *Spelaeomysis*. Also this pair of lobes forms a **closing apparatus** (**valve**) similar to the more diverse lobes on the penis in Mysida. The **Lophogastrida** males show only a slot-like orifice, without lobes, on the coxae of the eighth thoracopods (fig. 54.20A). The gonopore is on top of an anvil-like elevation in *Paralophogaster*, or of a dome-shaped elevation in *Gnathophausia*. There is no distinct elevation in two species of *Eucopia* and one *Lophogaster* examined by Wittmann (2013b).

Pleonal appendages

PLEOPODS

A pair of pleopods inserts latero-ventrally on each of the first to fifth somites of the pleon. The **pleopods** represent uniramous or biramous, more or less natatory to reduced, rod-like appendages, showing consistent differences between the various taxa from order down to genus level.

Lophogastrida (fig. 54.16H-L). – The lophogastrids have a large, flat shaft (sympod) formed by a short coxa and a much longer basis, the latter bearing two natatory, multi-segmented rami furnished with plumose setae. The sexual differentiations of the male pleopods are weak. This regards modifications of the second endopod (fig. 54.16K, L) in Gnathophausiidae (Lagardère & Nouvel, 1980a), and an average larger size of male pleopods (fig. 54.16J) compared to that in females (fig. 54.16H) in Eucopiidae (Tattersall & Tattersall, 1951).

Stygiomysida (fig. 54.17G-L). – All pleopods are biramous, but are nonetheless reduced in both sexes. The reduction regards the number of segments: sympod and endopod each are unsegmented, and the exopod reduced to a few segments. The **exopod** of the second **male pleopod** is slightly modified and has only two segments in *Stygiomysis hydruntina* (fig. 54.17H) or three in *Spelaeomysis bottazzii* (fig. 54.17G). In females of both species, the same exopod shows one additional articulation compared to the respective males.

Mysida (fig. 54.21). – The pleopods are more or less rudimentary, rod- or platelet-like in females (but biramous in females of *Archaeomysis*). The pleopods of males are typically less reduced, but there is at least some reduction in the endopod of the first pair. Several subfamilies of the Mysidae have one, rarely more, modified **male pleopods**, which may play some role in mating (probably not directly involved in sperm transfer). With some variations according to the genera, the male modifications regard mainly the second exopod in Rhopalophthalminae (fig. 54.21C), the third in Gastrosaccinae (fig. 54.21K-M), and the fourth in Mysinae (fig. 54.21O-Q) and Leptomysinae (fig. 54.21N). In certain genera of Gastrosaccinae the third male pleopods show very complex modifications (fig. 54.21K, L). In males of most Siriellinae the endopods of pleopods 2-5 have spirally coiled exites (= **pseudobranchial lobes**; fig. 54.21G). In many species of Siriellinae modified setae are present on both rami of the fourth pair of male pleopods, occasionally also on the third pair, exceptionally on the endopod of the second pair. In the Erythropinae, the male pleopods are mostly not or only weakly modified, rarely with a strongly modified seta as in *Parapseudomma calloplura* (fig. 54.21J); if any modification, then mostly regarding one or more of **endopods** 2-5, or again less frequently of **exopods** 2-5 (Nouvel & Lagardère, 1976). In males of certain Boreomysinae, there are small modifications of the third exopod (fig. 54.21D), rarely of the second plus third exopods. In Mysidellinae, Heteromysinae (except *Harmelinella*), and Palaumysinae, all pleopods are rudimentary in both sexes, although several taxa show some small modifications exclusive of males (fig. 54.21R-U).

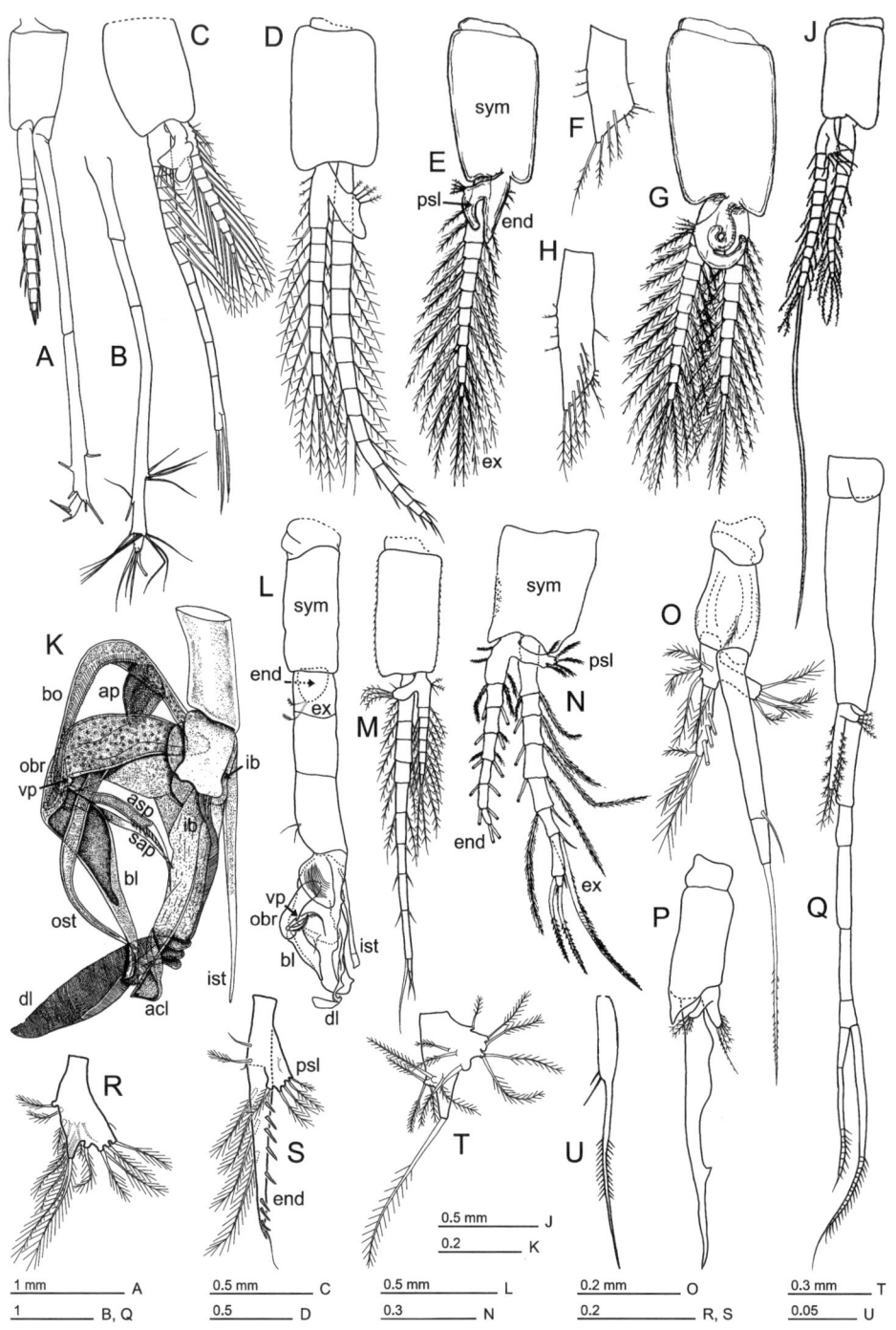

In *Palaumysis* the fourth pleopod bears a large apical, spine-like modified seta in males only (fig. 54.21U; Hanamura & Kase, 2003). The males of several species (subgenera) of *Heteromysis* are characterized by modified setae and/or spines on some of pleopods 2-5 (Wittmann, 2008; Price & Heard, 2011). In contrast to the remaining Heteromysinae, the males of *Harmelinella* show a uniramous and sub-segmented third pleopod, which is elongate and apically modified (Ledoyer, 1989).

UROPODS

The paired **uropods** insert latero-terminally on the sixth pleonite (figs. 54.1C, D, 54.2C, 54.3A, 54.19C, 54.22A-D, F-N). They consist of a short unsegmented **sympod** bearing two dorso-ventrally flattened rami, together with the unpaired telson constituting the usually strong **tail fan**. Considered an exclusive feature among the three orders, the sympod of the Stygiomysida shows a spinose, backwards-directed **inner lobe**. In the family Stygiomysidae this lobe is strong, reaching nearly to the tip of both rami of the uropod; its set of large spines extends even further (fig. 54.22F). A distinct, but much shorter spinose lobe is present in the Lepidomysidae (fig. 54.22G, H). The **exopod** is subdivided in its distal half by a **transverse suture** in certain Lophogastrida (Eucopiidae and Gnathophausiidae; fig. 54.22C, D) and Stygiomysida (Lepidomysidae; fig. 54.22G, H). Within the Mysida it is subdivided in the Petalophthalmidae (fig. 54.22J) and two subfamilies of the Mysidae: the Siriellinae and Rhopalophthalminae (fig. 54.22K, L). There is an incomplete, **subbasal suture** in Boreomysinae. The exopod is undivided in the

Fig. 54.21. Pleopods in the Mysida families Petalophthalmidae (A, B) and Mysidae (C-U), the latter represented by all ten of its acknowledged subfamilies: Rhopalophthalminae (C), Boreomysinae (D), Siriellinae (E-H), Erythropinae (J), Gastrosaccinae (K-M), Leptomysinae (N), Mysinae (O-Q), Heteromysinae (R, S), Mysidellinae (T), and Palaumysinae (U). A, fifth pleopod of male *Hansenomysis abyssalis* Lagardère, 1983, setae in part truncated; B, the same for female *Hansenomysis nouveli* Lagardère, 1983; C, second male pleopod in *Rhopalophthalmus egregius* Hansen, 1910; D, third male pleopod in *Boreomysis megalops* G. O. Sars, 1872; E-H, *Siriella clausii* G. O. Sars, 1877, first pleopods of male (E) and female (F), second pleopod of male (G), and fifth pleopod of female (H); J, fourth male pleopod in *Parapseudomma calloplura* (Holt & Tattersall, 1905); K, strongly modified terminal portion of exopod of third pleopod in terminal male of *Coifmanniella johnsoni* (W. M. Tattersall, 1937); L, third pleopod of subterminal male of *Chlamydopleon aculeatum* Ortmann, 1893; M-Q, fourth male pleopods in *Paranchialina angusta* (G. O. Sars, 1883) (M), *Leptomysis heterophila* Wittmann, 1986 (N), *Diamysis cymodoceae* Wittmann & Ariani, 2012 (O), *Limnomysis benedeni* Czerniavsky, 1882 (P), and *Mysis mixta* Lilljeborg, 1852 (Q), respectively; R, S, first (R) and second (S) male pleopods of *Heteromysis wirtzi* Wittmann, 2008; T, first male pleopod in *Mysidella biscayensis* Lagardère & Nouvel, 1980; U, third male pleopod in *Palaumysis pilifera* Hanamura & Kase, 2003. Abbreviations: acl, accessory lobe; ap, apophysis; asp, apical spine; bl, blade; bo, bow; dl, distal lobe; end, endopod; ex, exopod; ib, inner branch; ist, inner stylet; obr, outer branch; ost, outer stylet; psl, pseudobranchial lobe (exite); sap, subapical spine; sym, sympod; vp, ventral process. [A, B, after Lagardère, 1983; C, after Murano, 1988; D, Q, after G. O. Sars, 1879; E-H, after G. O. Sars, 1877; J, after Tattersall & Tattersall, 1951; K, after Băcescu, 1968b; L, after Wittmann, 2009; M, after G. O. Sars, 1885a; N, after Wittmann, 1986a; O, after Wittmann & Ariani, 2012a; P, after G. O. Sars, 1893; R, S, after Wittmann, 2008; T, after Lagardère & Nouvel, 1980b; U, after Hanamura & Kase, 2003.] [A, D, J-N setae omitted; A-Q, T, U, modified.]

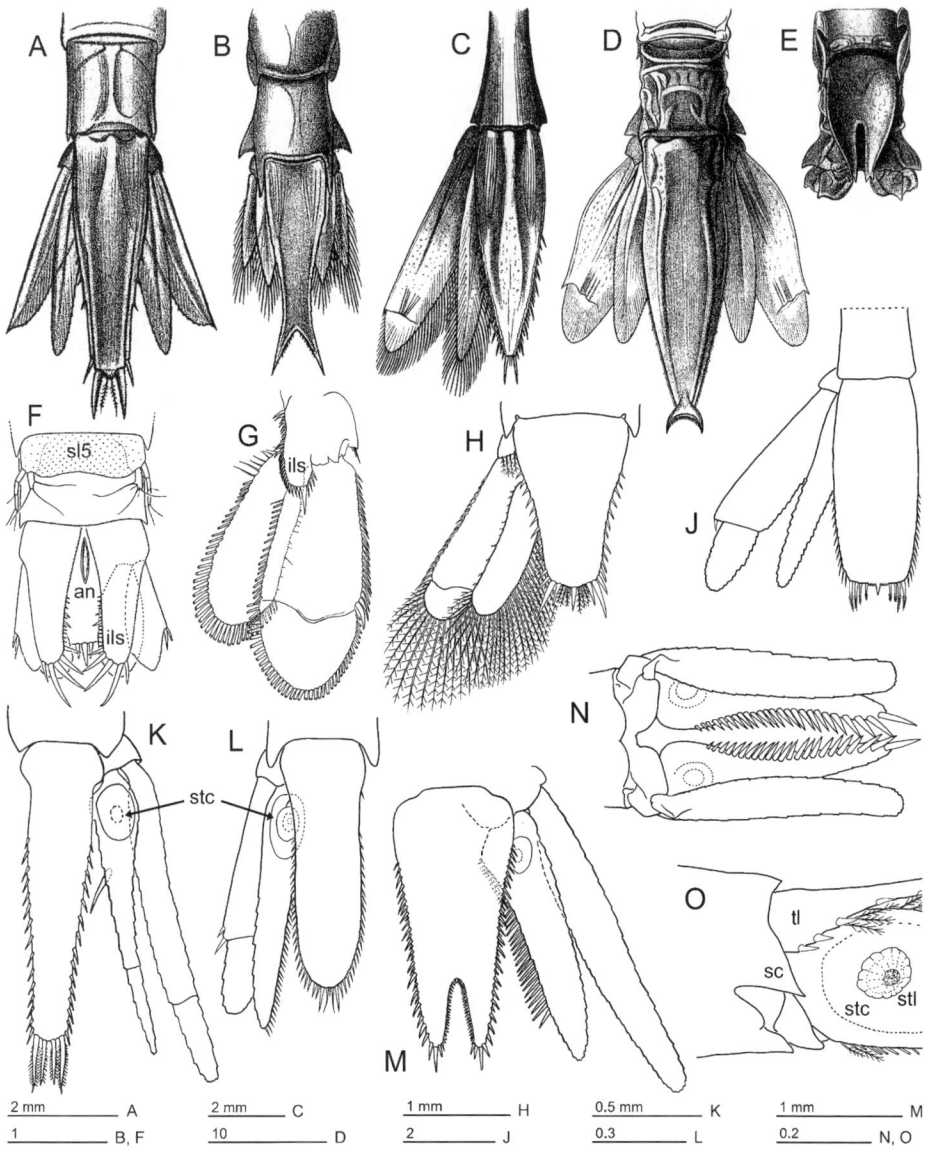

Fig. 54.22. Uropods, telson, and associated structures in the Lophogastrida families Lophogastridae (A, B), Eucopiidae (C), and Gnathophausiidae (D, E); in the Stygiomysida families Stygiomysidae (F) and Lepidomysidae (G, H); and in the Mysida families Petalophthalmidae (J) and Mysidae (K-O); dorsal view unless stated otherwise. A, *Lophogaster typicus* M. Sars, 1857; B, *Ceratolepis hamata* G. O. Sars, 1883; C, *Eucopia australis* Dana, 1852; D, *Gnathophausia ingens* (Dohrn, 1870); E, *Gnathophausia ingens*, epimeron from the sixth pleonite, ventral; F, appendages of the pleomeres 5 and 6 in *Stygiomysis holthuisi* (Gordon, 1958), ventral; G, *Spelaeomysis cardisomae* Bowman, 1973, uropods, ventral; H, *Spelaeomysis bottazzii* Caroli, 1924; J, *Petalophthalmus armiger* Willemoës-Suhm, 1875; K, *Rhopalophthalmus longicauda* O. S. Tattersall, 1957; L, *Siriella gracilis* Dana, 1852; M, *Mysidetes intermedia* O. S. Tattersall, 1955; N, *Mysidopsis robustispina*

remaining subfamilies of the Mysidae (fig. 54.22M, N). Its margins may be entirely lined by setae, may be smooth, or (in part) armed with spines or spine-like setae. The **endopod** has a well-developed transverse suture only in the Rhopalophthalminae; this suture is at about one quarter endopod length from the tip. This subfamily is exceptional in having both rami of the uropods subdivided (fig. 54.22K). In Mysidae, the apical portions of the endopod are mostly narrower than those of the exopod (fig. 54.22K, L). The basal portions of the endopod are swollen to accommodate the usually large, globular **statocyst** (fig. 54.22K-O; see below, 'Sensory organs').

INTERNAL MORPHOLOGY

The main internal organs of the Mysida are located in the cephalothorax as shown in fig. 54.23 for females (A) and males (B).

Musculature

Wollner (1924) described the complex musculature responsible for the movements of antennae, eyestalks, and mouthparts. The body **musculature** is also quite complex and composed of a system of metamerically arranged muscles: dorsal, latero-dorsal, ventral, and superficial-ventral muscles, plus basal **extensors** and **flexors** of the appendages. Detailed descriptions are given by Daniel (1928, 1933) for the lophogastrids *Gnathophausia zoea* and *Lophogaster typicus*, and for the mysid *Praunus flexuosus*. His meticulous figures are reproduced and discussed in the review by Mauchline (1980) (see also Manton, 1928a; Needham, 1937; Mayrat, 1955, 1956b). The musculature inserts partly on a non-chitinous **internal skeleton** including an **endosternite**, tendons, and columns (Debaisieux, 1954). According to Nath (1974), the cavernicolous Stygiomysida *Spelaeomysis longipes* shows a simplified musculature probably due to reduced swimming activity and the resulting atrophy of muscles otherwise used by epigean species for swimming.

Nervous system

Neural chain. – The gross morphology of the nervous system (fig. 54.24) of the Mysida and Lophogastrida was described so far by only few authors (G. O. Sars, 1867, 1885a; Illig, 1912). Our knowledge is still rudimentary compared with that about the Decapoda (Harzsch et al., 2012). The **cerebral ganglia** or brain are differentiated into **protocerebrum**, **deutocerebrum**, and **tritocerebrum**, with a **post-oesophageal**

Brattegard, 1969, ventral; O, *Diamysis bahirensis* (G. O. Sars, 1877), terminal portion of sixth pleonite, lateral. Abbreviations: an, anus; ils, inner lobe from sympod of uropods; sc, scutellum paracaudale; sI5, sternal lamellae from pleonite 5; stc, statocyst; stl, statolith; tl, telson. [A-E, J, L, after G. O. Sars, 1885a; F, after Gordon, 1960; G, after Bowman, 1973; H, after Ariani, 1981c; K, after O. S. Tattersall, 1957; M, after O. S. Tattersall, 1955; N, after Brattegard, 1973; O, original, adult female from the El Bahira lagoon at the Mediterranean coast of Tunisia.] [A, D, J, M, N setae (partly) omitted; F-N, modified.]

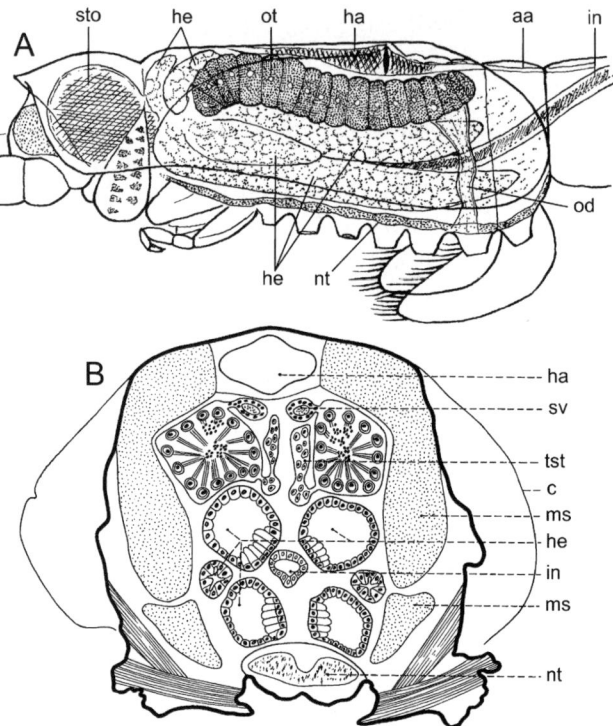

Fig. 54.23. Localization of the main internal organs in Mysida, schematic. A, female *Mysis relicta* Lovén, 1862, sagittal view; B, transverse section through thorax of male *Praunus flexuosus* (O. F. Müller, 1776). Abbreviations: aa, abdominal aorta; c, carapace; ha, heart; he, hepatopancreas; in, intestine; ms, muscles; nt, nerve tract; od, oviduct; ot, ovarian tubes; sto, stomach; tst, testicles; sv, seminal vesicle. [After Nouvel et al., 1999, based on G. O. Sars, 1867 (A) and Labat, 1961 (B).]

commissure, and are located above and in front of the mouth. The paired optical, antennular, and antennal nerves depart from the anterior portion of the cephalic mass; this mass is connected by the **peri-oesophageal connectives** with the ventral **ganglion chain**. Within this chain the paired **ganglia** are tightly interconnected medially. These connectives are generally distinct. A primitive situation is found in the mysid *Birsteiniamysis inermis* (fig. 54.24C): the ganglion masses related to the mouthparts are well separated with short connectives; their chain is continued by eight ganglion masses related to the thoracopods, plus six masses for the pleon. These elements are generally more condensed in the other species that have been studied. In the mysid *Mysis relicta* (fig. 54.24B), the eleven post-cerebral masses of the cephalothorax are reduced to ten by fusion of the two anterior ones. These ten masses are still discernible within the long band of ganglia. In the lophogastrid *Eucopia* (fig. 54.24A), the first pair of **thoracic ganglia** is more or less integrated into the single extended ganglion mass innervating the mouthparts; moreover, the ganglia of the eighth pair are smaller and more tightly interconnected, thus resembling the ganglion masses of the pleon. This resemblance is underlined by the visual aspect of the **neural chain**, which shows a distinct change between thoracomeres 7 and 8: short distances

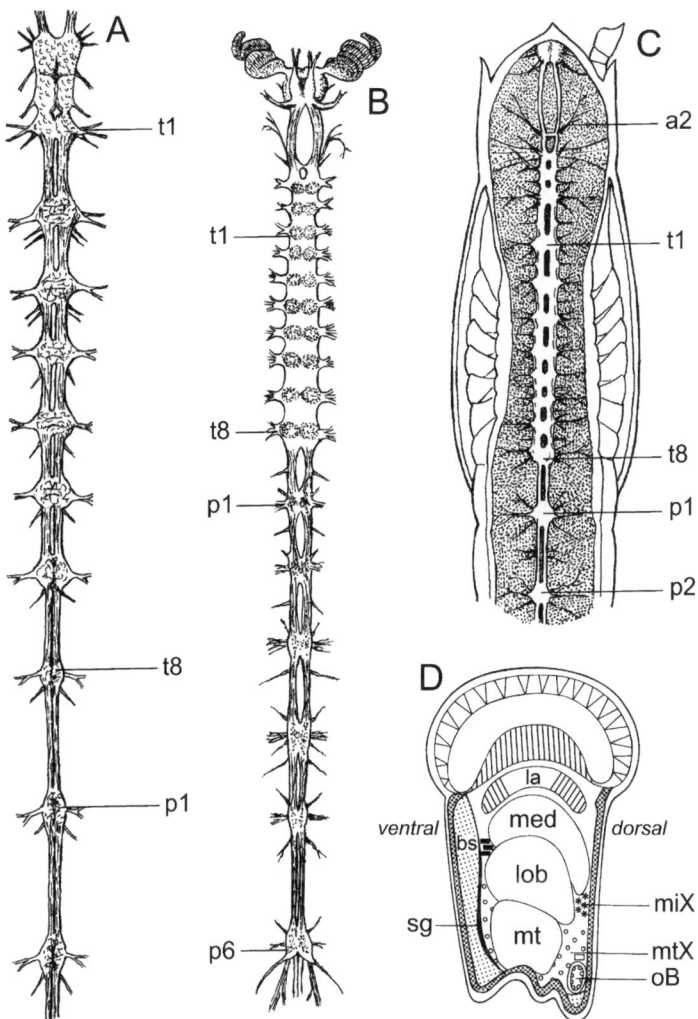

Fig. 54.24. The nervous system in Lophogastrida (A) and Mysida (B-D). A-C, neural chain in *Eucopia sculpticauda* Faxon, 1893 (A), *Mysis relicta* Lovén, 1862 (B), and *Birsteiniamysis inermis* (Willemoës-Suhm, 1874) (C), symbols indicate targets innervated by respective ganglia; D, neuropils and neurosecretory centres in eyestalk of *Siriella armata* (H. Milne Edwards, 1837), strongly schematic dorsoventral section, the graphic symbols indicate different types of neurosecretory cells. Abbreviations: a2, antennal nerve; bs, blood sinus; la, lamina; lob, lobula; med, medulla; miX, medulla interna and externa X-organ; mt, medulla terminalis; mtX, medulla terminalis X-organ; oB, organ of Bellonci; p1-p6, ganglia innervating pleomeres 1-6; sg, sinus gland; t1-t8, ganglia innervating thoracomeres 1-8. [A, C, modified after G. O. Sars, 1885a, originally named *Gnathophausia longispina* (A) or *Boreomysis scyphops* (C), respectively; B, after G. O. Sars, 1867; D, after Cuzin-Roudy & Saleuddin, 1985, with permission by "© Canadian Science Publishing or its licensors", Ottawa.]

between the pairs of ganglia and a distinct space between left and right neural chains in front of this limit, but longer distances between consecutive ganglia and medially tightly adjoining neural chains behind. Such differences between thorax and pleon are generally found in Lophogastrida. This explains the error by G. O. Sars (1885a: pl. X fig. 12), who mistook the ganglia of the ultimate thoracomere as pertaining to the first pleomere. An additional error by G. O. Sars (1885a: pl. VIII fig. 19) regards the nervous system of *Gnathophausia*, where one among the eight thoracic ganglia is missing and the first one is figured as pertaining to the mouthparts despite being actually well separated. Exactly this error justifies reproducing the anterior part of his drawing, although it must be attributed to a different genus, namely *Eucopia*, with which it shows an astonishing coincidence (fig. 54.24A). In both genera, the ultimate ganglion mass of the pleon is more strongly developed compared with the preceding one, probably as a consequence of the fusion of the sixth and seventh pleomeres.

Brain and associated organs. – The histology of the **brain**, the ganglia in the eyestalks, and associated **sense organs** and **endocrine glands** were studied by Hanström (1947, 1948), Mayrat (1956a), Hogstad (1969), Elofsson & Dahl (1970), Strausfeld & Nässel (1981), Cuzin-Roudy & Saleuddin (1985), and Elofsson & Hagberg (1986). Only a short summary is given here beyond the gross structures of the neural chain mentioned above:

The **pons protocerebralis** with the central body (**corpus centrale**) of the **protocerebrum** are both homogeneous in Lophogastrida, but divided into glomeruli in Mysida. The glomerular **paracentral lobe** is broadly connected with the central body and shows strong commissures. A median **frontal organ** is known only from the lophogastrid *Eucopia* (Hanström, 1947). It is inside the frontal plate and may attain a considerable size; each of several cell pairs forms an acidophilic plaquette. No **naupliar eye** is present in Lophogastrida, Stygiomysida, or Mysida. In the **deutocerebrum**, the layer of ganglion cells contains a group of cells related to the **olfactory lobes**, more strongly developed in Lophogastrida than in Mysida. The antennular fibres end in the glomeruli of the olfactory and the parolfactory lobes; a large unpaired **glomerulus** appears to be a peculiarity of the Lophogastrida and to have no equivalent in other crustaceans.

Optic ganglia. – The optic ganglia are located in the **eyestalks** in all taxa (fig. 54.24D). They include three columnar **optic neuropils** and the **medulla terminalis** as prolongations of the protocerebrum. As in insects, the three neuropils are currently termed — in inward-directed series — **lamina**, **medulla**, and **lobula** (Strausfeld & Nässel, 1981; Harzsch et al., 2012), whereas previously in a more crustacean-focused approach as **lamina ganglionaris**, **medulla externa**, and **medulla interna** (Elofsson & Dahl, 1970; Cuzin-Roudy & Saleuddin, 1985; Nouvel et al., 1999). The **sinus gland** is a discoid thickening of the **medulla terminalis** at the level of a large blood sinus in the lophogastrids *Eucopia* and *Gnathophausia* and in the mysids *Boreomysis* and *Mysis*. In these genera it is separated from the neuropil by a layer of cellular bodies, but not so in the mysid genus *Praunus*. The sinus gland of the mysid *Siriella armata* extends externo-ventrally over medulla terminalis and lobula (fig. 54.24D; Cuzin-Roudy & Saleuddin, 1985). Like most malacostracans, the lophogastrids and mysids also have two **chiasmata** in the eyestalk, the distal one between lamina and medulla, the proximal one between medulla and lobula. The proximal

chiasma may be indistinct in reduced eyes due to the tendency of the proximal **neuropils** to withdraw into the brain (Elofsson & Dahl, 1970).

Using immunohistochemical methods, Benzid et al. (2006) found different signals of the neurotransmitter **serotonin** in the lamina after a dark to light **illumination change** versus a light to dark change in the diurnally migrating mysid *Hemimysis speluncola*. No such difference was recorded in the non-migrating *Leptomysis lingvura*. The authors concluded that the lamina can function as a **photoreception signal** integrator during phases of illumination change.

Organ of Bellonci. – This organ, also termed **'sensory papilla X-organ' (SPX-organ)**, is found in most crustaceans (Hallberg & Chaigneau, 2004). In Lophogastrida and Mysida, it is located directly under the dorsal cuticle near the base of the eyestalk (figs. 54.24D, 54.25E) or, if present, near the base of eyestalk papillae (Hanström, 1947; Mayrat, 1956a; Dahl & von Mecklenburg, 1969; Hogstad, 1969; Kauri & Dahl, 1975; Cuzin-Roudy & Saleuddin, 1985). Dahl & von Mecklenburg (1969) described the **organ of Bellonci** as a group of cells and a vesicle surrounded by a connective tissue sheath near the base of the eyestalk papilla in *Boreomysis arctica* and assumed **neurosecretory** and **neurohaemal functions**. Re-examination by Kauri & Dahl (1975) revealed sensory neurons with ramifying cilia branches protruding into the cavity, thus pointing rather to a **sensory function**. The organ is distally connected with a nerve coming from the **eyestalk papilla**, and proximally with a nerve entering the medulla terminalis. In most species of the family Mysidae, the eyestalk papillae show a great morphological diversity up to reduction or complete absence (fig. 54.25). For example in *Siriella armata*, the organ of Bellonci represents a single cavity near the base of the eyestalk (fig. 54.24D; Cuzin-Roudy & Saleuddin, 1985). This cavity is lined by sensory cells that, with their axon-like elongations, extend to the **medulla terminalis**. The cavity is probably analogous to the structure Stammer (1936) interpreted as a rudiment of lenses in the cavernicolous, anophthalmic mysid *Troglomysis vjetrenicensis*. A chemosensory or else a photosensitive function of the organ of Bellonci is assumed for most crustaceans (Hallberg & Chaigneau, 2004), but its precise functions still remain unknown for Lophogastrida and Mysida. Despite its homonymous designation as 'X-organ' by many authors, the organ of Bellonci is not identical with the neurosecretory organs of the eyestalk, termed "ME-MI X-organ" and "MT X-organ" by Cuzin-Roudy & Saleuddin (1985) (fig. 54.24D; see also below, 'Endocrine organs').

Sensory organs

EYES AND VISION

Gross morphology of the eyes (fig. 54.25). – The **compound eyes** are each borne by a cylindrical or almost conical, more or less elongate mobile **eyestalk**. The **cornea** is generally globular, hemispheric, occasionally reniform. *Birsteiniamysis inermis* has quite particular eyes: cup-shaped, with about 3000 ommatidia (Elofsson & Hallberg, 1977, as *Boreomysis scyphops*). In several bathypelagic Mysidae, e.g., in *Euchaetomera* and

Caesaromysis, and in epipelagic *Carnegieomysis*, the cornea is divided into a frontal and a **lateral cornea**, more or less separated, sometimes widely separate (fig. 54.25G); it is in part formed by two types of visual elements with probably different visual function. The latero-posterior ommatidia are much larger than the frontal ones in the **subdivided cornea** of *Paraleptomysis* (fig. 54.25H). Although pertaining to different subfamilies, Leptomysinae and Mysinae, respectively, the genera *Dioptromysis* and *Kainommatomysis* have in common a very large, specialized latero-posterior **lens** (fig. 54.25F), separated from the **frontal cornea**. According to Nilsson & Modlin (1994), in *Dioptromysis paucispinosa* this lens consists of a large facet over an equally enlarged **crystalline cone**, projecting an image onto a specialized **retina** with 120 densely packed, extremely narrow **rhabdomes** (eye structure by Land, 2004, reproduced in fig. 6.22 of the present series). The centre of this zone performs six times better than the normal eye and covers a visual field of 15-20° with large binocular overlap, in good analogy with a pair of binoculars. With appropriate eyestalk movements the acute zone can to a limited extent face forwards and upwards, according to the interpretation by Nilsson & Modlin (1994) in favourable direction for spotting prey. For this species, however, and also for its Red Sea counterpart *Kainommatomysis schieckei*, there is no published evidence for a **predatory mode of life**. A few specimens of the latter species were studied by one of us (KJW) in the Gulf of Sinai and the Gulf of Aqaba. Besides eye morphology, they shared with the Caribbean *Dioptromysis paucispinosa* an **epibenthic lifestyle** during daytime and stomachs that contained diatoms and remains of other microalgae, but no animal remains. Thus, the eco-physiological context of the binocular-like eye modifications remains obscure.

Structure of the cornea. – As in the Euphausiacea, the compound eyes of Mysida are of the **refracting superposition type** (Strausfeld & Nässel, 1981; Nilsson & Modlin, 1994; Land, 2004). The cornea is formed by juxtaposed **ommatidia** showing (Parker, 1891; Chun, 1896; Mayrat, 1956a; Elofsson & Hallberg, 1977; Hallberg, 1977) similar numbers of cells as in decapods. Each **ommatidium** contains a vaulted **lens**, representing a transparent part of the cuticle secreted by two **corneagenous cells**. Above there is a **crystalline cone** with its large basis turned to the cornea, comprising two compound halves. Among the four **crystalline cells** only two secrete each a half cone into their cytoplasm. A short axial crystalline tract, formed by four crystalline cells, penetrates a short distance. The **retinula** contains eight (*Mysidopsis*, *Praunus*, *Siriella*) or only seven (*Erythrops*, *Neomysis*) **photoreceptor cells** (Hallberg, 1977; Strausfeld & Nässel, 1981), one of which is the most voluminous, while another, the accessory cell, penetrates into the axis. All except the latter are prolonged by a **nerve fibre**. The relatively short rhabdome is connected with the crystalline tract by a crystalline formation, the **epirhabdome** (absent in *Erythrops*); the latter consists of four branches which surround the accessory retinula cell and reunite on the top of the rhabdome (Hallberg, 1977). The **rhabdome** appears to be formed by eight chitinous, serrate, intricate **rhabdomeres**, secreted by the **retinula cells**. It is fused and banded as in decapods and euphausiaceans (Strausfeld & Nässel, 1981). **Screening pigment cells** surround the crystalline part and the proximal portions of the retinula cells, and contain a **melanin pigment** that becomes particularly dense around the rhabdome. The granules of the distal and proximal reflecting pigment cells probably contain **pteridine**. There is a wide clear zone between the cones and the rhabdomes. Rays entering from many facets can be brought into focus in this zone. **Pigment migration** for **dark and light adaptation** in the compound eye is similar to that described for decapods (Hallberg et al., 1980). The migration ranges of the screening pigments are relatively small, always leaving a clear zone between distal and proximal pigment cells. The proximal part of the crystalline cones is more pointed in the light- versus the dark-adapted eye.

The **numbers of ommatidia** vary considerably between the various species of Mysida and Lophogastrida, according to Mauchline (1980) without recognizable, consistent reasons. By contrast, differences between populations of the same species may be explained by the different light intensities they experience in nature. According to J.-P. Casanova (1977), the lophogastrid *Eucopia unguiculata* shows degenerate eyes, but the Atlantic specimens have larger eyes with more ommatidia compared to Mediterranean specimens sampled by him in the same depth but at lower **light intensity**.

Vision. – According to Beeton (1959), the mysid *Mysis diluviana* is sensitive to wave lengths of 515 and 395 nm, probably in correspondence with two different **visual pigments**. Lindström & Nilsson (1988) argued that geographically isolated populations of "*Mysis relicta*" may show different **spectral sensitivity** and tolerance to light intensity, but actually this may reflect differences between separate species described much later by Audzijonyte & Väinölä (2005, 2006): *Mysis salemaai* from brackish waters of the Baltic versus *Mysis relicta* from freshwater lakes in Scandinavia. Laboratory experiments on seven Baltic species of mysids by Lindström (2000) showed that the eye spectral sensitivity differs in accordance with wavelength-shifts in light transmittance of the water bodies from which the respective test specimens were taken. The eyes of *Mysis relicta* from Lake Pääjärvi (Finland) responded to near-infrared light, their spectral sensitivity being among the most red-shifted known (Lindström & Meyer-Rochow, 1987; Lindström, 2000). The experiments of Moeller & Case (1995) on the lophogastrid *Gnathophausia ingens* indicated differences in visual function, particularly regarding **circadian rhythms** of visual sensibility, between juveniles and adults. This is related to the shallower habitat of the juveniles.

According to Hallberg et al. (1980), the screening pigments of *Neomysis integer* need about 20 minutes to migrate upon **light-adaptation**, but more than one hour upon **dark-adaptation**. The position of the pigment in light-adapted eyes varies with light intensity. Those authors observed no cyclic diurnal pigment migrations in animals kept either permanently in total darkness or under continuous illumination. Lindström (1992, 2000) found that the eyes of *Mysis relicta* are easily damaged by strong light intensity; particularly if the specimens were forced from the dark deep waters of freshwater lakes to the bright daytime light at the surface upon sampling. Feldman et al. (2010) reported higher concentrations of retinol and other **retinoids** in a *Mysis relicta* population from the same (above quoted) dark, deep, freshwater Lake Pääjärvi, compared to animals from more shallow waters in the Baltic Sea. This points to a larger storage of chromophores or precursors for **dark regeneration** of visual pigment. Suddenly exposing the lake animals

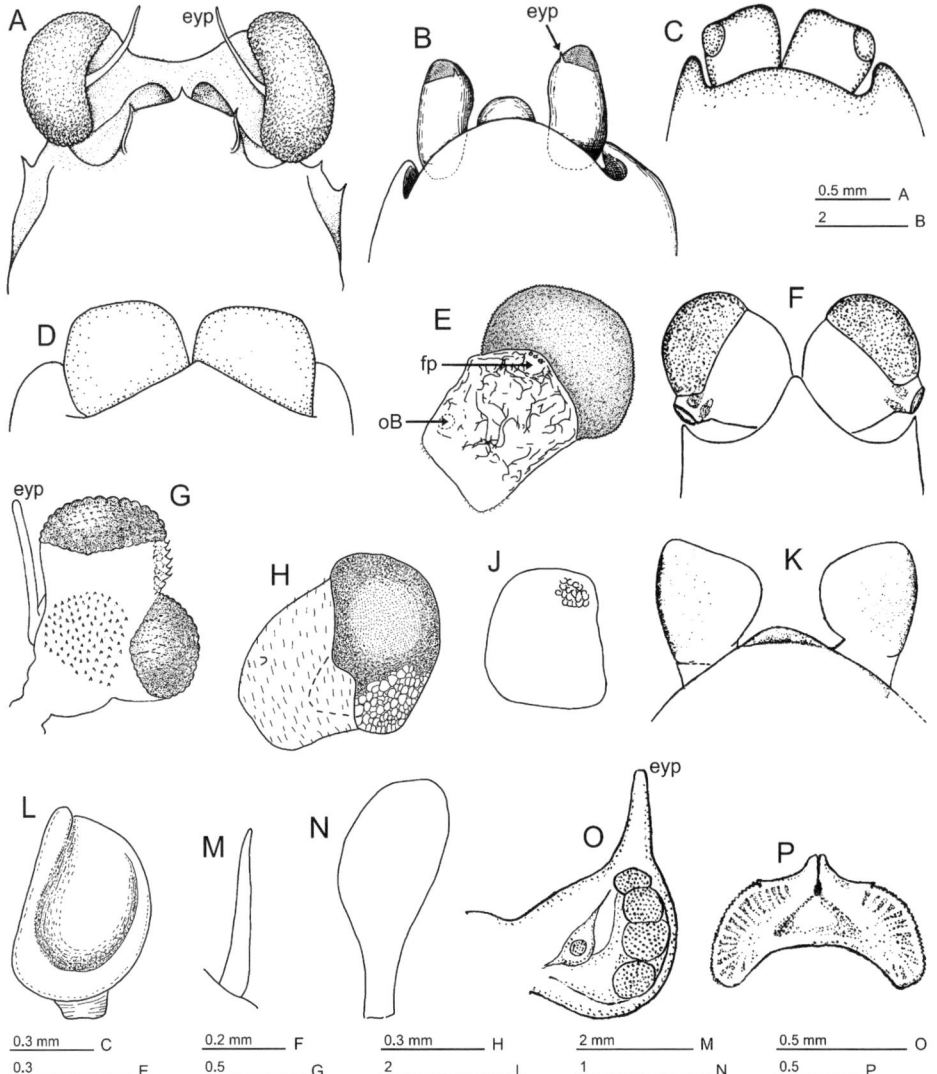

Fig. 54.25. The diversity of eye modifications in Lophogastrida (A, B), Stygiomysida (C, D), and Mysida (E-P). A, *Paralophogaster foresti* Băcescu, 1981; B, *Eucopia major* Hansen, 1910; C, *Spelaeomysis cochinensis* Panampunnayil & Viswakumar, 1991; D, *Spelaeomysis olivae* Bowman, 1973; E, *Diamysis lagunaris* Ariani & Wittmann, 2000; F, *Dioptromysis perspicillata* Zimmer, 1915; G, *Euchaetomera zurstrasseni* (Illig, 1906); H, *Paraleptomysis dimorpha* Wittmann, 1986, female; J, *Heteromysoides cotti* (Calman, 1932); K, *Antromysis cubanica* Băcescu & Orghidan, 1971; L, *Birsteiniamysis inermis* (Willemoës-Suhm, 1874); M, *Ceratomysis spinosa* Faxon, 1893; N, *Petalophthalmus armiger* Willemoës-Suhm, 1875; O, *Dactylerythrops dactylops* Holt & Tattersall, 1905; P, *Pseudomma affine* G. O. Sars, 1870. Abbreviations: eyp, eyestalk papilla; fp, fenestra paracornealis; oB, organ of Bellonci. [A, after Băcescu, 1981; B, after Nouvel, 1943; C, after Panampunnayil & Viswakumar, 1991, permission granted by Springer Verlag, Vienna; D, after Bowman, 1973; E, after Wittmann & Ariani, 2012a; F, after Zimmer,

to stronger light could induce massive pigment activation and resulting **photoreceptor damage**. Similar laboratory results were obtained by Attramadal et al. (1985): the marine deep-water (mesopelagic) mysid *Boreomysis megalops* showed pathological changes in eye morphology as well as abnormal vertical zonation behaviour after exposure to surface daylight. These findings on *Mysis* and *Boreomysis* raise strong questions about earlier works on vision and behaviour of deep-water species if the experimental specimens were exposed to bright light upon sampling and/or experimentation.

Eye reduction (fig. 54.25C, D, J-P). – The cornea is often reduced to a small **number of ommatidia** with often rudimentary structure, and the pigment may have disappeared completely in deep-water and cavernicolous species, clearly in relation to the poor light conditions of their habitat. The cavernicolous mysid *Heteromysoides cotti* (fig. 54.25J) from marine lavatunnels at Lanzarote (Canary Islands) has small eyes with at most 30 ommatidia. According to an electron microscopic study by Meyer-Rochow & Juberthie-Jupeau (1987), these ommatidia show signs of degeneration; the eyes are probably not equipped with any **polarization sensitivity** and incapable of **form vision**, but may be able to distinguish different light intensities and possibly also the direction of light. The loss of **visual abilities** appears to be counterbalanced by more effective **tactile** and **chemosensory abilities** in subterranean mysids (cf. Crouau, 1981).

The various species are not affected in the same way by the light conditions of their environment, and factors associated with phyletic lineages probably play some role. Thus, in the lophogastrid genus *Eucopia*, most species show a rather small cornea (fig. 54.25B). This is less evident in *Eucopia sculpticauda*, despite its preference for deep waters. This species is considered to be the most primitive in its genus as indicated by morphological, anatomical, and molecular characters (J.-P. Casanova et al., 1998); it is positioned near species of Gnathophausiidae, which all have normally developed eyes. More or less pronounced **eye reductions** have been observed in about one tenth of the known species. This is typically correlated with eyestalk modifications (Stammer, 1936; Zharkova, 1970; Nath et al., 1972a).

The generally subterranean stygiomysid genus *Spelaeomysis* shows various degrees of eye reduction, both between adults of different species and during the course of marsupial development. In adults of most species the eyes are reduced to tile-like plates (fused to a single median **eye plate** in *Spelaeomysis longipes*) without any trace of pigment, but *Spelaeomysis cardisomae* has distally widened, latero-distally well-pigmented (fig. 54.3B) eyestalks with only a few ommatidia (Bowman, 1973). In *Spelaeomysis nuniezi* only the postnauplioid larvae show almost normal eyes (Ortiz et al., 2005), but the adults have the eyes reduced to sub-quadrangular plates completely lacking cornea and pigment (Băcescu & Orghidan, 1971). Also in *Spelaeomysis bottazzii*, both adults and juveniles (fig. 54.38K)

1915; G, after Illig, 1930; H, after Wittmann, 1986c; J, after Calman, 1932; K, after Băcescu & Orghidan, 1971; L, after Hansen, 1908; M, after Faxon, 1895; N, after W. M. Tattersall, 1925 (scale derived from Tattersall & Tattersall, 1951); O, after Holt & Tattersall, 1905; P, after G. O. Sars, 1870; A, D-O, modified.]

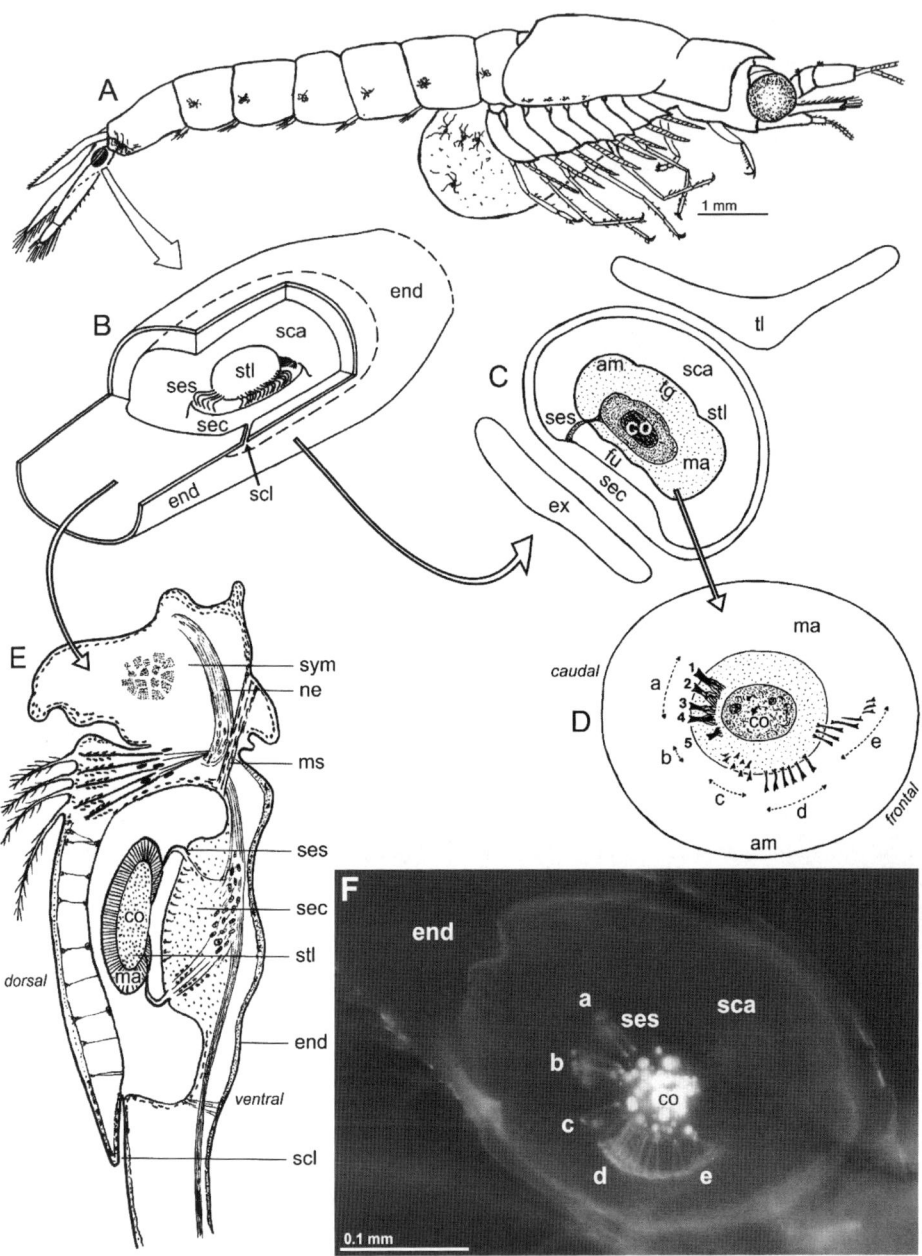

Fig. 54.26. Statocyst allocation and structure in Mysidae (Mysida). A, allocation of the statocyst in uropodal endopods of female *Siriella gracilipes* H. Nouvel, 1942, lateral view (setae partly omitted), right statolith indicated as black ellipsoid; B, schema of statocyst in *Neomysis integer* (Leach, 1814); C, schematic cross section through tail fan of *Siriella clausii* G. O. Sars, 1877, at height of statolith; D, groups (a-e) of non-sensory apical portions of sensory setae entering left statolith of *Siriella clausii*, ventral view; E, semi-schematic longitudinal section through uropod of

have the eyes reduced to unpigmented plates, but the postnauplioids are still showing almost cylindrical stalks with pale cornea (fig. 54.38J; Ariani & Wittman, 2010).

Eyestalks nearly or completely fused to a single median **eye plate** are also found in certain Mysida, such as the Mysidae *Pseudomma* (fig. 54.25P), *Parapseudomma*, *Calyptomma*, and *Michthyops*. Occasionally the eye plate may be hidden by the frontal plate as in the Petalophthalmidae *Hansenomysis* and *Bacescomysis*.

Additional eye-related structures. – The eyestalks are often furnished with a finger-like, mostly lateral **sensory papilla** of variable size (fig. 54.25A, B, G, O). In the lophogastrid *Paralophogaster foresti*, this papilla is remarkably large, often exceeding the diameter of the cornea (fig. 54.25A; Băcescu, 1981). A small basal ganglion, connected with the optic centres, sends its prolongations to this organ, which in certain taxa may be reduced from a papilla to a sensory pore. In the mysid *Boreomysis* a nerve from the papilla innervates the organ of Bellonci (see above). The detailed function of the **eyestalk papillae** is still unknown. A **fenestra paracornealis**, i.e., a pale, pigment-free, rounded and slightly elevated spot on the eyestalks near the cornea, has been described in certain species of the mysid genus *Diamysis* (figs. 54.12D, 54.25E; Ariani & Wittmann, 2000). An additional one is associated with the organ of Bellonci, located dorsally near the inner basal corner of the eyestalks. The visibility of these features depends on the state of preservation (may be invisible in old museum specimens), as well as on the pigment richness of the eyestalk surface. Branches of eyestalk chromatophores may surround but do not extend over these fenestrae, often resulting in a striking colour contrast to the surrounding parts of the eyestalk. A series of a few separate ommatidia may be present along the distal margin of the fenestra (fig. 54.25E). After bleaching in Swan-medium, a ganglion mass becomes visible below the fenestra paracornealis; this mass is similar in size and structure to that of the **organ of Bellonci**. Both fenestrae clearly favour light penetration to sensory organs. Prolonged bleaching revealed that the ganglion mass associated with the fenestra paracornealis is present in many (possibly all) *Diamysis* species, but that parts of it are often located below the cornea.

STATIC ORGANS

Statocyst structure (fig. 54.26). – Statocysts are present only in the family Mysidae (G. O. Sars, 1867; Bethe, 1895; Debaisieux, 1947, 1949a; Espeel, 1985; Schlacher et al., 1992; Ariani et al., 1993), where they are located at the base of the endopods of the uropods. They consist of a large vesicle formed as an invagination of the integument.

Praunus flexuosus (O. F. Müller, 1776); F, basis of right uropodal endopod in *Diamysis mesohalobia* Ariani & Wittmann, 2000, fresh preparation in ambient water 55 min after moult, core (co) showing incipient mineralization, (a-e) groups of sensory setae, dorsal dark-field aspect. Abbreviations: am, ambitus; co, core; end, endopod of uropod; ex, exopod of uropod; fu, fundus; ma, mantle; ms, muscle; ne, nerv; sca, statocyst cavity; scl, statocyst slit; sec, sensory cushion; ses, sensory setae; stl, statolith; sym, sympod of uropods; tg, tegmen; tl, telson. [A, after Wittmann et al., 1993, permission granted by John Wiley & Sons, Ltd., Hoboken; B, after Espeel, 1985; C, D, after Schlacher et al., 1992; E, after Debaisieux, 1949a; F, after Ariani et al., 1982; B, C, modified; D, permission granted by Springer Verlag, Vienna.]

The cavity is entirely lined by cuticle and contains ambient water. It is connected with the outside milieu through the **statocyst slit**. Its cuticle is entirely shed upon each moult. A small muscular bundle, extending from the external wall of the uropod to the **statocyst**, apparently adjusts the tension in the latter. The ventral wall of the cavity bears a **sensory cushion**: an arch of **sensory setae** penetrating with their non-sensory distal portions from below into a usually large, complex statolith. A few comparatively large setae emerge from the posterior part of this arch, and more, mostly smaller ones, from the outer anterior part. The setae are organized in serial groups along this arch, indicated by (a-e) in fig. 54.26D, F. The setae are shed at moulting together with the cuticle. According to Espeel (1986), the sensitivity of the sensory cells of the old setae is maintained until the moment of ecdysis.

Statolith composition and structure (fig. 54.27A-G). – In most Mysidae, each **statolith** consists (Wittmann et al., 1993) of two distinct parts. The first is the **core** (nucleus). Usually it contains both organic and mineral materials. It is rarely exclusively organic as erroneously believed in early works on species with fluorite statoliths. The second is the **mantle**, almost entirely mineralized with **fluorite** (CaF_2; fig. 54.27D, E). In a relatively small number of species, the mantle is mineralized with calcium carbonate ($CaCO_3$) in the metastable crystal phase of **vaterite** (fig. 54.27B, F; Ariani et al., 1993). Vaterite statoliths have so far been found in only seven genera of the subfamily Mysinae. Carbonate mysid statoliths in the stable crystal phase of **calcite** (fig. 54.27C, G), however, are known from Miocene deposits of eastern Europe, which was covered at that time by the extensive brackish basin of the Paratethys (Voicu, 1974, 1981). The finding of calcite in all examined Paratethyan fossils and in a few aged museum specimens of (otherwise vaterite-bearing) extant species, suggests that the calcite content is normally produced *post mortem* by spontaneous phase transformation of the originally present metastable vaterite to the stable calcite during fossilization and/or long-term storage (Wittmann et al., 1993).

Unlike in extant fluorite statoliths (fig. 54.27D), the core is mostly not covered ventrally by the mantle in both Recent and fossil carbonate statoliths (fig. 54.27B, C). Note also the strong similarity between the statoliths of the extant *Paramysis lacustris* (fig. 54.27B) and those obtained from Miocene Paratethys sediments (fig. 54.27C). Any type of carbonate statoliths may show a central, non-mineralized cavity within the core. The relative size of the central cavity tends to increase with individual size of the statolith and may open ventrally to the exterior as a hole in about central position. This hole, named **hilum** (hi), is still rather small in fig. 54.27B, but figures of much more prominent hila are available in Wittmann et al. (1993). The entry points of the apical parts of the sensory setae into the statoliths mark a semi-circular series of pores with diameters decreasing (in ventral view) in clockwise or anticlockwise order, according to left (fig. 54.27A, D) or right (fig. 54.27B, C) statoliths, respectively. These pores appear to be arranged in distinct groups, thus allowing to derive a characteristic **statolith formula** (Voicu, 1974), such as $2+3+1+d+e = n$, where d and e indicate groups with variable numbers of pores, and n the total number of pores (Wittmann, 1992a). Extensive investigations by Schlacher et al. (1992), however, showed that only the first 2-3 groups of pores may have some taxonomic value, although even the configuration of these groups is rarely exclusive for a single genus.

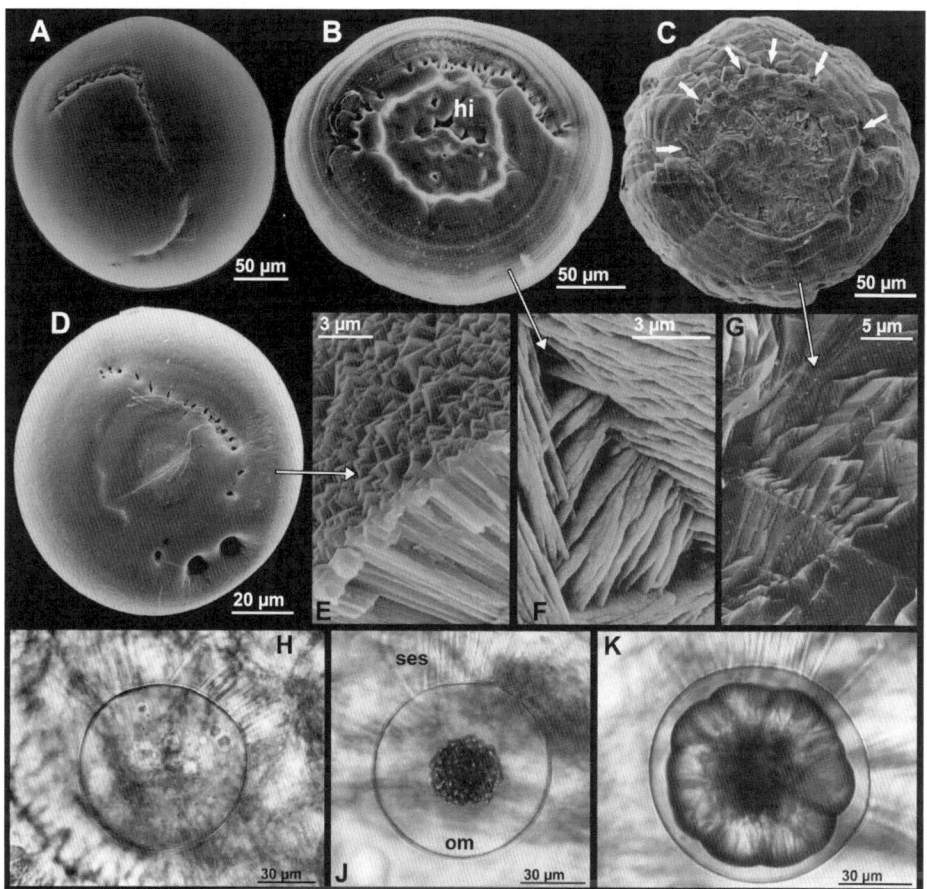

Fig. 54.27. Structure (A-D), mineral composition (E-G), and formation (H-K) of statoliths in recent (A, B, D-F, H-K) and fossil (C, G) Mysidae (Mysida). A-D, ventral aspects of statoliths *in toto*, SEM-images; A, non-mineral (organic) left statolith of *Rhopalophthalmus terranatalis* O. S. Tattersall, 1957; B, right vaterite statolith in *Paramysis lacustris* (Czerniavsky, 1882); C, right fossil calcite statolith from Upper Miocene deposits of the brackish Paratethys, short arrows point to selected pores along pore arch of lith; D, left fluorite statolith of *Haplostylus magnilobatus* (Băcescu & Schiecke, 1974); E-G, SEM-images of mineral structures from outer surface (E, F) and/or artificial fracture surfaces (E, G), respectively; E, cubic crystal habits of fluorite in *Mysidium integrum* W. M. Tattersall, 1951; F, needle-like vaterite aggregates in *Schistomysis assimilis* (G. O. Sars, 1877); G, rhombohedral crystal habits of calcite in fossil statolith as in (C); H-K, vaterite mineralization of organic matrix in time series of 12, 48, and 180 min, respectively, after moult in *Diamysis mesohalobia* Ariani & Wittmann, 2000, diascopic dark-field aspects of fresh preparations in ambient water. Abbreviations: hi, hilum; om, organic matrix; ses, sensory setae. [A, after Schlacher et al., 1992, permission granted by Springer Verlag, Vienna; B, D, photo Thomas Schlacher & Karl J. Wittmann; C, E-G, after Wittmann et al., 1993, permission granted by John Wiley & Sons, Ltd., Hoboken; H-K, after Ariani et al., 1982.]

A further type of static bodies is non-crystalline, **organic statoliths** (fig. 54.27A), peculiar to the primitive subfamilies Boreomysinae and Rhopalophthalminae (Ariani et al., 1993).

A study at worldwide scale (Ariani et al., 1993) revealed correlations between **crystallographic characteristics** of statoliths, and the ecology and biogeography of mysids: **fluorite precipitation** apparently prevails in the sea. Vaterite was mainly found in fresh and brackish waters, **non-crystalline** statoliths predominantly in the deep sea (except for the coastal Rhopalophthalminae). The **vaterite precipitation** in most species of *Diamysis* and *Paramysis*, which today inhabit the Mediterranean and Black Sea basins, was considered indicative of a Paratethyan origin: the **ancestors** of these taxa are thought (Ariani, 1981a; Wittmann & Ariani, 2011) to have been drained into the Mediterranean as a consequence of the Mediterranean salinity crisis (Hsü et al., 1977).

Statolith formation (fig. 54.27H-K). – Unlike the statoliths of Decapoda, the **static bodies** of the Mysidae (Mysida) are endogenous and renewed upon each moult. As a phylogenetic-ontogenetic parallel, an **organic matrix** is formed before the start of the **mineralization** process, as shown by the laboratory study of Ariani et al. (1982) on the formation of vaterite statoliths in *Diamysis*. In these experiments, the statoliths were formed within only a few hours after moulting (fig. 54.27H-K). The organic matrix consists of acid **mucopolysaccharides** and **glycoproteins** (Ariani et al., 1982) in the vaterite statoliths of *Diamysis*, but mainly of sulphated mucopolysaccharides (Espeel, 1987) in the fluorite statoliths of *Neomysis*. Wittmann & Ariani (1996) determined the concentration factors of ions, particularly fluoride (F^-), invested into the renewal of the statolith. A typical fluorite statolith contains about 180 000 times the amount of fluorine present in the seawater contained in the statocyst. So far, detailed sources and **uptake of ions** for **statolith formation** remain obscure. Ariani et al. (1999) experimentally investigated the role of the **organic matrix** for the renewal of vaterite statoliths at moulting in a species of *Diamysis*. This involved using an inhibitor (acetazolamide) of carbonic anhydrase or a generically toxic substance (ethanol) for comparison. Under both these conditions the mineral precipitation was not stopped, but in a few cases there was *in vivo* precipitation of calcite instead of vaterite, possibly correlated to failure of, or damage to, the organic frame that mediates the mineralization process.

Statocyst function. – Early authors attributed auditory functions to the statocyst, consequently erroneously describing it as "otocyst". They were not completely wrong, as sensitivity to vibrations is not excluded (Cohen, 1955). Delage (1887) was the first to suggest that the **statocyst** is sensitive to the action exerted by the weight (actually also the inertia) of the statolith and that the perceptions by the sensory setae are determinants of locomotor reflexes for orientation. Today, the statocyst is generally seen as an **equilibrium organ** for stabilization of body position and for directional swimming. Laboratory studies by Schöne (1954) for decapods and by Neil (1975a, b) for the mysid *Praunus flexuosus* demonstrated that the right and left statocysts co-operate in controlling the **eyestalk movements**. This shows that a major function of mysid statocysts is stabilization of the **visual field**, as is generally known for static organs in visually oriented animals, also in humans. In line with this, Mysinae with well-developed eyes generally have large statoliths, whereas species with reduced eyes have on the average smaller ones, and those without any visual elements again smaller ones (Wittmann et al., 1990). Correspondingly, the diameter of fluorite statoliths is generally larger in species of Mysinae from light-exposed, epipelagic and coastal habitats as compared to darker, deep-water habitats; the epipelagic species of Siriellinae have mostly larger statoliths compared to the neritic species. Together with landmarks perceived by the visual system, the statocysts are also involved in the perception of water currents, helping to guide the horizontal migration of *Hemimysis speluncola* back into dark submarine caves at daybreak (Passelaigue & Bourdillon, 1986; Passelaigue, 1989).

SENSORY SETAE

Diversity of sensory setae. – Many types of sensory setae (fig. 54.28) in Mysida and Lophogastrida are located on various parts of the integument (Debaisieux, 1947, 1949a, b; Crouau, 1978a, 1981, 1989; Johansson & Hallberg, 1992; Johansson et al., 1996). They are particularly abundant on the antennulae, antennae, and mouthparts, as well as on the locomotory appendages. Some are plumose and may be important for hovering, others are smooth or barbed and may have diverse, in some cases still hypothetical, functions, particularly as **aesthetascs** (olfactory hairs) or as tactile organs (**mechanoreceptors**). Crouau (1981) found 18 different types of setae on the antennulae and antennae in laboratory specimens of the tropical, subterranean, anophthalmic mysid *Antromysis juberthiei*. In 1987, the same author reported only 12 different types on the mouthparts and thoracic endopods of the same species, yet with greater morphological differences between these types (fig. 54.28). In 1989, he grouped these 12 types in the categories of plumose (fig. 54.28F-H), serrulate (M-R), pappose (J), small (K), and pushing (L) setae.

Mechanoreceptors. – According to Crouau (1978b, 1979), the **mechanoreceptors** on the antennula of *Antromysis juberthiei* respond to tactile and hydrodynamic stimuli, including waterborne vibrations and turbulences. Hallberg & Chaigneau (2004) interpreted the **perception of vibrations** as a kind of hearing, thus as a useful adaptation for localization of prey by these subterranean, anophthalmic mysids. Crouau (1982) presented a detailed hypothesis on how the initially mechanical signal becomes processed into an electric signal in these sensilla: **reception** of the stimulus in the external parts of the sensillum, **transmission** by the distal and middle regions of the outer dendritic segment, mechanical **amplification** by the proximal region, and finally **transduction** of the mechanical signal into an electric signal at the junction of the outer and inner dendritic segments (see figs. 7.12, 7.13 in volume 1 of the present series: Hallberg & Chaigneau, 2004).

Chemoreceptors. – Compared with mechanoreceptors, the diversity of **chemoreceptors** (fig. 54.28S-U) is generally greater on the flagella of the antennulae in Mysida and Lophogastrida (Juberthie-Jupeau & Crouau, 1977; Crouau, 1978a; Hallberg et al., 1992; Johansson & Hallberg, 1992; Johansson et al., 1996). Juberthie-Jupeau & Crouau (1977)

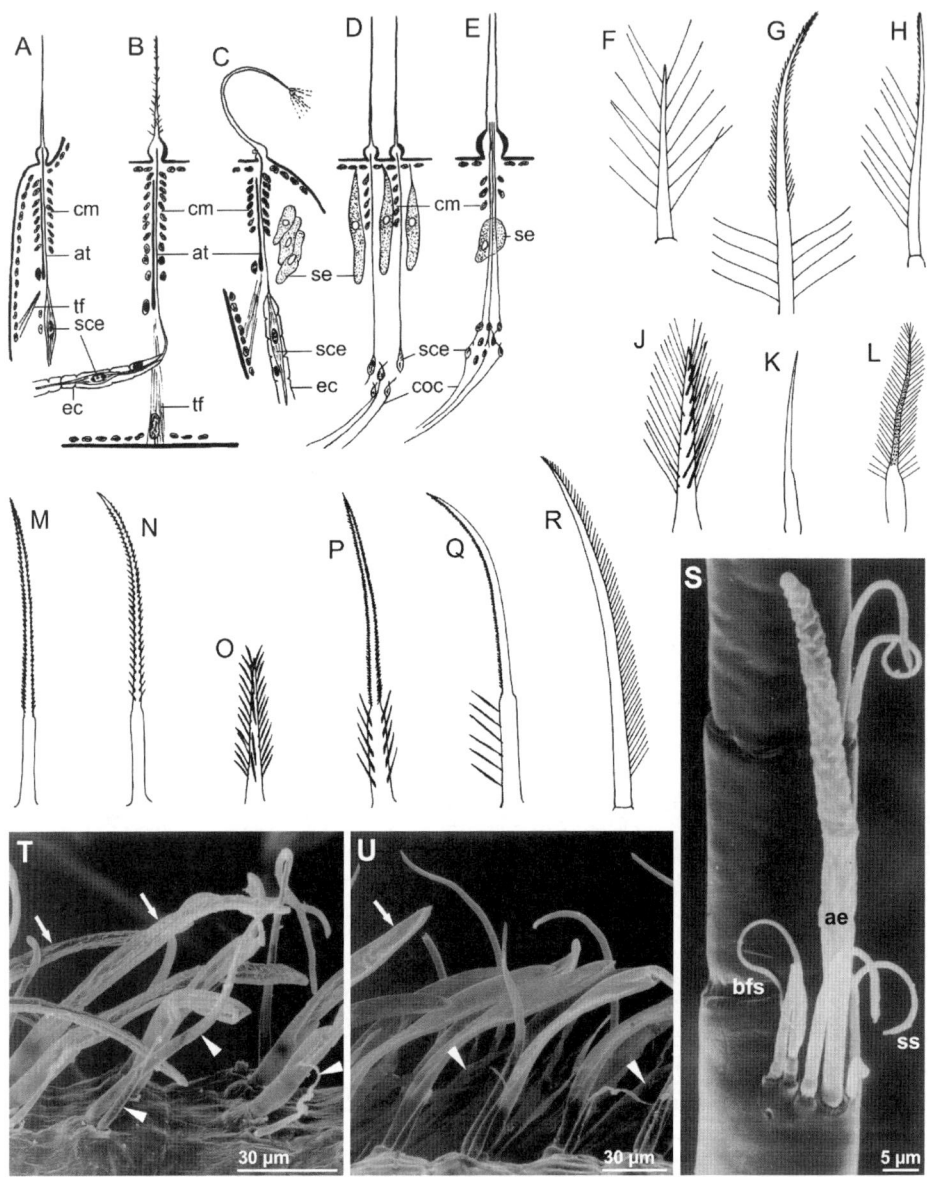

Fig. 54.28. Sensory setae in Mysida (A-S) and Lophogastrida (T, U). A-E, Interpretative scheme for various sensory setae of the mysid *Praunus flexuosus* (O. F. Müller, 1776), using terminology of Debaisieux (1949a). A, armed seta from sympod of antennula; B, armed seta dorsally on uropods; C, sensory seta from statocyst; D, unarmed simple seta from appendix masculina of antennula; E, olfactory seta; F-R, twelve types of setae from mouthparts and thoracic endopods of the tropical subterranean, anophthalmic mysid *Antromysis juberthiei* Băcescu & Orghidan, 1977: in terminology of Crouau (1989) these are plumose (F-H), pappose (J), small (K), pushing (L), and serrulate (M-R) setae, respectively; S, aesthetasc (ae) and companion setae on antennula of *Antromysis juberthiei*; T, U, setae on outer antennular flagellum of female (T) and male (U) *Lophogaster typicus* M. Sars,

gave a detailed description of an antennular **aesthetasc** of *Antromysis juberthiei*, including a schematic representation, which is reproduced in fig. 7.19 in volume 1 of the present series (Hallberg & Chaigneau, 2004). During the moulting process of **antennal sensilla** in the mysid *Neomysis integer*, the dendritic segments of the old sensory cells remain continuous with the dendritic outer segments of the new sensillum; ensuring that the sensitivity of the receptors is maintained until **ecdysis** (Guse, 1980). Olfactory abilities are important in the search for food, detection of predators, recognition of conspecifics, and many other purposes.

Male-specific sets of setae. – Clearly for the detection and pursuit of potential mating partners (Guse, 1983b; Hallberg et al., 1992; Johansson & Hallberg, 1992; Johansson et al., 1996), the males of most Mysidae have a dense brush of setae on the **appendix masculina** (fig. 54.12E, G), which emerges ventrally from the terminal segment of the antennular peduncle. Rarely, e.g., in certain species of *Boreomysis* and *Heteromysis*, this brush may arise (almost) directly from the terminal segment of the peduncle. In each mode of setae implantation, the sensilla-bearing parts are moveable by a special muscular arrangement (Johansson & Hallberg, 1992). The Petalophthalmidae and certain Mysidae lack an appendix masculina; instead their males show a thickened basal portion of the outer **antennular flagellum** furnished with dense rows of **aesthetascs**. Some Boreomysinae and Gastrosaccinae (fig. 54.12G) have an appendix masculina together with a modified outer flagellum of the male antennula. In certain Mysidae, a basally thickened, setose outer flagellum is present in both sexes, but usually less strongly in females. Lowry (1986) emphasized the particular modifications of the outer antennular flagella as a distinct sensory organ, termed '**callynophore**' by him, found in many amphipods and in certain mysids (fig. 54.12G), isopods, and decapods. He considered the function of these structures to be detection of receptive females by chemoreception in those species in which they occur only in reproductive males, or to be food detection where they occur in both sexes.

OTHER SENSORY ORGANS

The Petalophthalmidae genera *Hansenomysis* and *Bacescomysis* show a deep depression close to the base of the dorsal face of the proximal segment of the antennula. The opening is partly covered by a fleshy lamella and was initially interpreted as a rudimentary eye in these otherwise anophthalmic genera. Due to its position similar to that in decapods, this depression has also been considered a statocyst, but there are neither any endogenous nor exogenous static bodies. According to O. S. Tattersall (1961) the presence of small, rounded, stainable areas point rather to a chemosensory function. In slightly more anterior position on the same segment of the antennula, *Mysimenzies* (Mysidae: Erythropinae) has

---

1857, arrows indicate aesthetascs, arrowheads indicate companion setae (T) or typical male sensilla (U). Abbreviations: ae, aesthetasc; at, axial tigelle; bfs, bifid seta; cm, chitinogenic matrix; coc, connective cell; ec, ensheathing cell; sce, sensory cell; se, secretory cell; ss, simple seta; tf, tonic fibrils. [A-E, after Debaisieux, 1949a; F-R, after Crouau, 1987; S, after Crouau, 1989, permission granted by "© Canadian Science Publishing or its licensors", Ottawa; T, U, after Johansson et al., 1996; K, L, modified.]

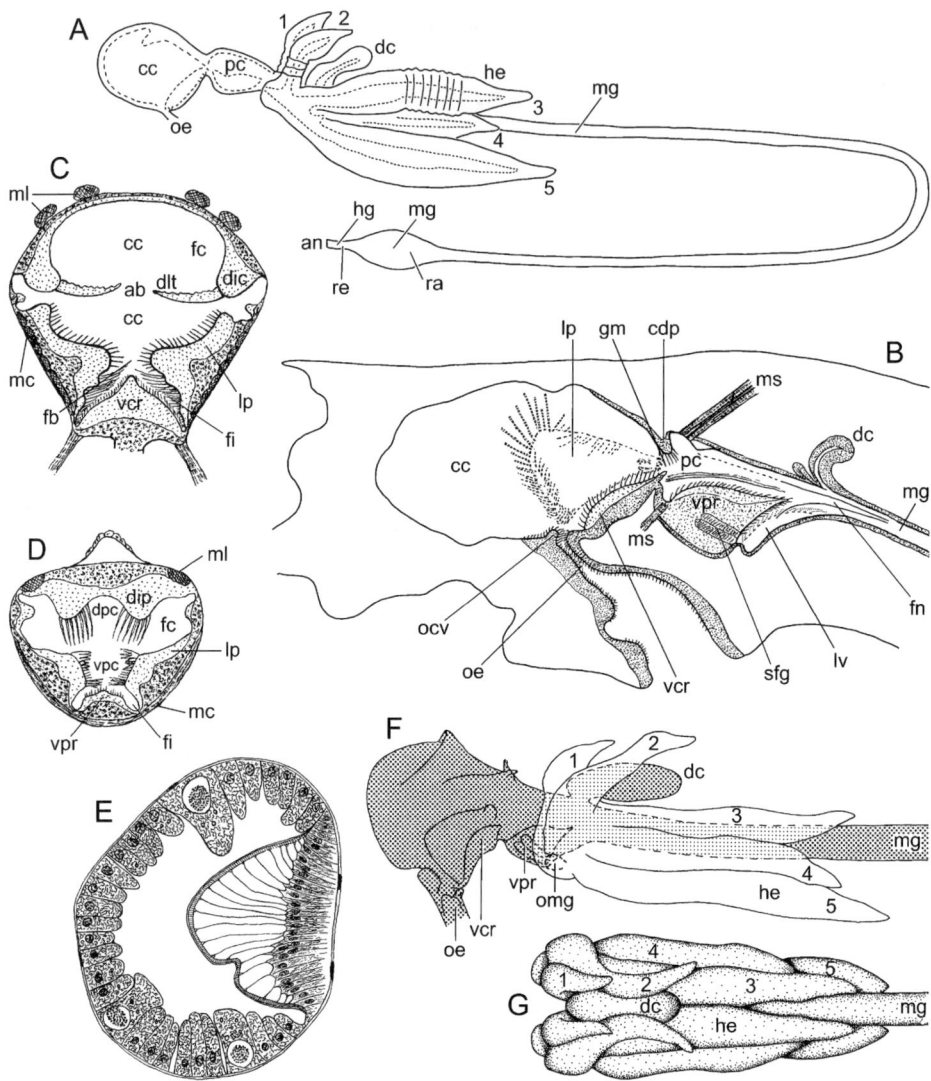

Fig. 54.29. Digestive system of the Mysidae (Mysida) (semi-schematic; A, B, F, simplified). A, entire digestive tract of *Mysis stenolepis* S. I. Smith, 1873; B-E, fore- to midgut of *Praunus flexuosus* (O. F. Müller, 1776), inner semi-transparent lateral view on right half of anterior region (B) and cross sections (C-E) through posterior part of cardiac chamber (C), approximately mid-region of pyloric chamber (D), and ventral lobe of hepatopancreas (E); F, G, *Neomysis integer* (Leach, 1814), midgut glands in relation to foregut in lateral (F) and dorsal (G) view. Abbreviations: 1-5, numbers indicating the series of lobe pairs of hepatopancreas; ab, alimentary belt; an, anus; cc, cardiac chamber (cardia); cdp, cardiac dorsal piece (superomedianum); dc, dorsal caecum (dorsal diverticulum); dic, dorso-lateral infolding of cardia; dip, dorso-lateral infolding of pylorus; dlt, dorso-lateral teeth; dpc, dorsal pyloric chamber; fb, filtration belt; fc, food channel; fi, filtration channel; fn, funnel (lamina dorsalis posterior); gm, gastric mill; he, hepatopancreas; hg, hindgut; lp, lateralia (lateral plates); lv, lamella ventralis; mc, circular muscle; mg, midgut; ml, longitudinal muscle; ms, muscle; ocv, oesophageal

a **sensory fossette**, whereas its close relative *Marumomysis* merely has a setose elevation (Băcescu, 1971a; Murano, 1999a). So far, no detailed knowledge is available about the function of the various **pores** related to sense organs and to glands, present on eyestalks, appendages, carapace, and body wall (see also above, 'Integument and colour').

Digestive system and digestion

History and gross structure. – The anatomy and structure of the **digestive system** (figs. 54.29, 54.30) were studied in various species of Lophogastrida (Siewing, 1953, 1956; Oshel & Steele, 1988; De Jong, 1996; De Jong & B. Casanova, 1997; De Jong & J.-P. Casanova, 1997; De Jong-Moreau et al., 2000; De Jong et al., 2002), Stygiomysida (Nath & Pillai, 1972a), and Mysida (Van Beneden, 1861; G. O. Sars, 1867; Gelderd, 1909; Illig, 1912; Molloy, 1958; Haffer, 1965; Nath & Pillai, 1976; Friesen et al., 1986b; Storch, 1989; Metillo & Ritz, 1994; Kobusch, 1998; De Jong-Moreau et al., 2000). All three orders are covered by the studies of Kobusch (1999) and De Jong-Moreau & J.-P. Casanova (2001) on the structure and evolution of the foregut. The early studies were greatly improved upon the availability of electron microscopic techniques from the 1980s onwards. The digestive system of these orders is of similar complexity as in decapods (Ceccaldi, 2006) and like these shows three major parts (fig. 54.29A): the anterior intestine or foregut (**stomodaeum**), consisting of oesophagus and stomach; the intermediate intestine or midgut (**mesenteron**) composed of the midgut tube, **hepatopancreas** (midgut gland) and potential additional midgut caeca; and the posterior intestine or hindgut (**proctodaeum**). The **foregut** and **hindgut** are of ectodermal origin and entirely lined by cuticle, which is renewed at each moult, whereas the **midgut** is of endodermal origin and without chitin lining. The digestive tube is equipped with parietal circular and longitudinal muscles. Some muscle bundles, particularly important at the height of the stomach, are connected with the exoskeleton.

Foregut (figs. 54.29, 54.30). – The transversely cleft mouth is followed by a laterally compressed, conical vestibular cavity, with the oesophagus departing from its bottom. The **oesophagus** is nearly vertically (dorsally) directed in all Lophogastrida and Stygiomysida so far examined (fig. 54.30A-C). It is forward inclined in most Mysida (fig. 54.29B), but vertically directed in certain Mysidae (*Mesopodopsis slabberi*) and Petalophthalmidae (*Ceratomysis* and *Petalophthalmus*). The ingestion of large food particles, already more or less split by the mandibles and maxillae, is facilitated by **peristaltic movements** of the oesophagus. According to Kobusch (1998) a total of up to four ventral, lateral, and/or dorsal setose infoldings protrude into the oesophagus of the Mysida. These folds, termed '**cardio-oesophageal valves**', prevent the reflux of food into the oesophagus.

---

cardiac valve (valvula dorsalis oesophagi); oe, oesophagus; omg, outpocketing of the midgut; pc, pyloric chamber (pylorus); ra, rectal ampoule; re, rectum; sfg, secondary filter groove; vcr, ventral cardiac ridge (inferomedianum anterius); vpc, ventral pyloric chamber; vpr, ventral pyloric ridge (inferomedianum posterius). [A, after Friesen et al., 1986b; B-E, after Gelderd, 1909; F, G, after Kobusch, 1998; A, B, F, G, modified.]

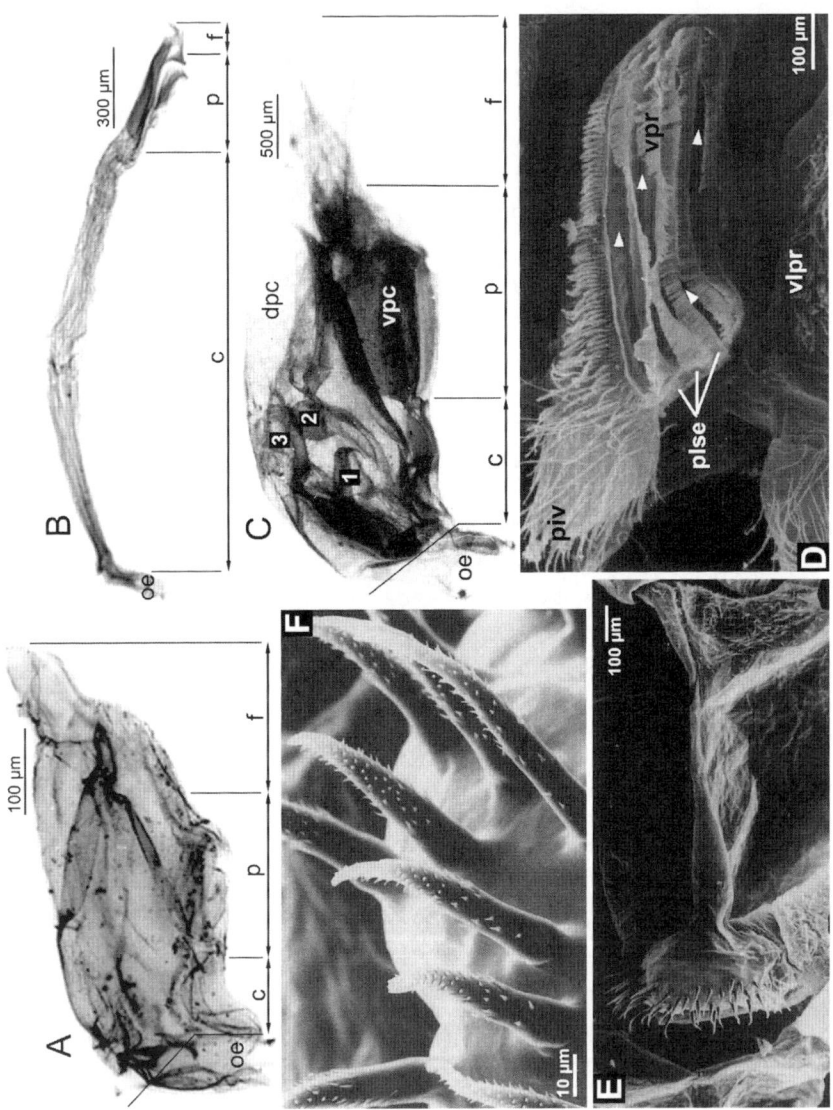

There may be some discrepancies in definition with respect to De Jong-Moreau & J.-P. Casanova (2001), who recognized only the dorsal cardio-oesophageal valve (fig. 54.29B) as being typical of all Mysidae, whereas they found only the ventral ones in all Stygiomysida as well as in most Petalophthalmidae and Lophogastrida; those authors found no valves in *Petalophthalmus armiger* (Petalophthalmidae). In certain species of *Lophogaster* (Lophogastridae), the valves are also missing (De Jong, 1996) or reduced to a wide, extremely flattened fold (Kobusch, 1999). The inner walls of the oesophagus are always equipped with setae that face the stomach in order to promote the passage of food and to impede reflux.

The oesophagus leads to the **stomach**, which in all taxa has a complex internal structure formed by ridges and folds of different form and size, equipped with setae and/or spines (fig. 54.30D-F). The stomach is stout and bulbous in most Mysida (fig. 54.29B), but extremely slender in certain Stygiomysida (Stygiomysidae; fig. 54.30B), and of intermediate shape in the remaining Stygiomysida (Lepidomysidae; fig. 54.30A), in the Lophogastrida (fig. 54.30C), and in certain Mysida (*Mesopodopsis slabberi*). The stomach is functionally subdivided into an anterior **cardiac chamber**, a posterior **pyloric chamber**, and the **funnel** (figs. 54.29B, 54.30A-C). Both chambers are in turn subdivided into a dorsal region or '**alimentary belt**' for storage, mechanical breakdown by mastication, and transportation of the ingested material, and a ventral region or '**filtration belt**' formed by the primary cardiac filters and the more sophisticated secondary pyloric filters for separation of fine particles and fluids. The dorsal alimentary belt of most Lophogastridae (*Ceratolepis*, *Chalaraspidum*, and *Lophogaster*, but not so in *Paralophogaster*) is strongly developed and forms a peculiar storage bag.

In the Mysida, the **cardiac chamber** is well separated from the **pyloric chamber** by a dorsal **gastric mill** at its posterior end (fig. 54.29B). In contrast, the gastric mill is in more anterior position within the cardiac chamber in Lophogastrida and Stygiomysida. In this case, the two chambers are separated only by a pair of ventral cardio-pyloric valves. In all three orders the gastric mill is generally composed of a median dorsal tooth (fig. 54.30E, F) and 1-2 pairs of lateral teeth (figs. 54.29C, 54.30C). Any of these teeth may be reduced to simple folds or ridges depending on the taxon. The teeth, ridges, and folds may show complex armatures of spinules and/or secondary teeth (Kobusch, 1998, 1999). The lophogastrid genus *Lophogaster* has a modified gastric mill with the first pair of lateral

---

Fig. 54.30. Foregut of the Stygiomysida (A, B) and Lophogastrida (C-F); light microscopy photographs in lateral view (A-C) and SEM-micrographs of internal details (D-F). A-C, note the great differences in longitudinal extension of cardiac (c), pyloric (p), and funnel (f) regions of foregut; A, *Spelaeomysis quinterensis* (Villalobos, 1951); B, *Stygiomysis major* Bowman, 1976; C, *Eucopia sculpticauda* Faxon, 1893; D, right side of ventral pyloric ridge in *Eucopia australis* Dana, 1852, arrowheads indicate lateral grooves; E, antero-dorsal cardiac tooth in *Gnathophausia ingens* (Dohrn, 1870); F, detail of (E) showing spines of cardiac tooth. Abbreviations: 1, first lateral teeth; 2, second lateral teeth; 3, antero-dorsal cardiac tooth; dpc, dorsal pyloric chamber; oe, oesophagus; piv, pyloro-intestinal valve; plse, lateral rows of plumose setae; vlpr, right ventro-lateral pyloric ridge; vpc, ventral pyloric chamber; vpr, ventral pyloric ridge. [A, B, after De Jong-Moreau & J.-P. Casanova, 2001; C, D, after De Jong & B. Casanova, 1997; E, F, after De Jong & J.-P. Casanova, 1997, permission granted by "© Canadian Science Publishing or its licensors", Ottawa; A-D, modified.]

teeth resembling internal mandibuliform appendages and with ventro-lateral cardiac plates bearing spines reinforcing the gastric armature (De Jong, 1996). Within the lophogastrid genus *Eucopia*, only *Eucopia sculpticauda* shows a dorsal tooth; the remaining species conserve only a simple fold representing a rudiment of the basis of this tooth. Within the Stygiomysida, *Spelaeomysis* shows a normal equipment consisting of a dorsal tooth and a pair of lateral teeth; by contrast, the gastric mill of *Stygiomysis* is reduced to a vestigial dorsal tooth in the anterior part of the extremely elongate cardiac chamber.

The primary **cardiac filter** is formed by a pair of ventro-lateral longitudinal folds fringed by long setae; these folds cover a ventro-lateral furrow (**filtration channel**) on both sides of a median ventral fold (fig. 54.29C, D). In *Stygiomysis major* (Stygiomysida), however, the cardiac folds lack setae and, therefore, lack their filter function (De Jong-Moreau & J.-P. Casanova, 2001). The dorsal **pyloric chamber** is a simple canal for the transport, through the funnel to the midgut, of the large particles and faecal material that did not pass through the filters. The dorsal pyloric region is elongated by the tube-like **funnel** (fig. 54.29B), which penetrates lengthwise into the lumen of the caudally adjacent midgut. The funnel is in open connection with the interior of the midgut by a longitudinal ventral fissure (Storch, 1989). As a derivate of the foregut, the funnel is of ectodermal origin, even though penetrating inside the endodermal midgut.

A pair of longitudinal folds, fringed by long setae, separates the dorsal pyloric region from the ventral one (fig. 54.29D). The secondary **pyloric filter** shows principally the same configuration as the primary filter. A pair of ventro-lateral folds separates the ventro-lateral furrows at each side from the median ventral part. The median ventral ridge represents the most strongly projecting structure of the pyloric chamber. Laterally, it bears several rows of setae, each row covering a furrow. These setae can form a very fine filtration net. In the Stygiomysida and Mysida, there are usually 2-3 setal rows (fig. 54.29B) depending on the taxon, but five rows are present in *Ceratomysis* (Mysida: Petalophthalmidae). The Lophogastrida generally show greater numbers of rows, 4-5 in *Lophogaster* and 5-7 in *Eucopia* (fig. 54.30D) and *Gnathophausia*. Intraspecific variations of these numbers are correlated with differences in body size. The ventro-lateral folds bear numerous rows with rasp-like setae. The ventral and ventro-lateral valves (fig. 54.30D) are prolonged towards the intestine.

Midgut (fig. 54.29A, F, G). – Normally, the **midgut** starts dorsally at the posterior pyloric chamber. However, its insertion is shifted ventrally in certain Lophogastrida (Lophogastridae: *Ceratolepis*, *Chalaraspidum*, *Lophogaster*), probably in relation to the large, dorsally overlying storage bag in these genera. The Mysida are equipped with a comparatively short, unpaired **dorsal diverticulum** or a **dorsal caecum** (fig. 54.29A, F, G), or a pair of very small diverticula at the junction between stomach and midgut. The unpaired dorsal diverticulum of the Stygiomysida species *Spelaeomysis longipes* is extremely long, extending to the anterior end of the stomach (Nath & Pillai, 1976). According to Mauchline (1980), the dorsal diverticula secrete the **peritrophic membrane** into the intestine. However, Friesen et al. (1986b) interpreted the potential function of the dorsal caecum in the mysid *Mysis stenolepis* more cautiously: "... the zymogene-like granules and large amounts of rough endoplasmic reticulum suggest a secretory role". No such diverticula or caeca are present in Lophogastrida (De Jong-Moreau et al., 2000). Posterior **midgut caeca** are so far known only from the freshwater subterranean Stygiomysida *Spelaeomysis longipes* and are used for calcium storage (see below, 'Moulting').

The hepatopancreatic ducts join the midgut at the level of the ventro-lateral folds via a pair of latero-posterior orifices (fig. 54.29F). The Mysida generally have five pairs of **hepatic caeca** within the thorax. Two among the five pairs are very short and directed obliquely upwards, whereas the remaining three are posteriorly directed along the intestine (fig. 54.29A, F, G). Among these last three pairs, the upper and the lower pairs extend along most of the thorax; the median pair is shorter and closest to the intestine walls. The stygiomysid *Spelaeomysis longipes* has a total of only three pairs, of which the shortest one turns posteriorly and the two longer pairs extend dorsally along the intestine (Nath & Pillai, 1972a). In the Lophogastrida, the **hepatopancreas** consists of only two large and irregularly shaped caeca, one on each side of the intestine (De Jong-Moreau et al., 2000).

After approaching the sternites, the intestine runs through the dorsal lacuna of the pleon. Friesen et al. (1986b) studied the ultrastructure of the intestine and the hepatic caeca of *Mysis stenolepis*. These authors reported the main cell types known from decapods (Ceccaldi, 2006): E- or S-cells (embryonic or **stem cells**, respectively), F (**fibrillary cells**; supply enzymes for extracellular digestion), B (**blister-like cells**, implicated in intracellular digestion), and R (**resorptive cells**; for lipoprotein metabolism). De Jong-Moreau et al. (2000) studied two species of Lophogastrida plus three species of Mysida and found marked differences in hepatopancreas structure between the two orders. They identified only B- and R-cells but no F-cells in their material, and doubted that the "F-cells" claimed by Friesen et al. (1986b) for *Mysis stenolepis* were true F-cells. The posterior part of the midgut forms an ampoule-like structure (fig. 54.29A).

Hindgut (fig. 54.29A). – The rectum is very short and restricted to the sixth pleonite; it is lined by a chitinous cuticle. The anus opens ventrally at the base of the telson (fig. 54.22F).

Digestion. – The first phase of **digestion** takes place at the level of the external mouthparts, mainly by breaking the food in pieces and grinding it. In *Lophogaster typicus*, secretions of the labrum probably help to coat and digest the food (De Jong et al., 2002). The second main phase takes place in the stomach. This involves (1) **mechanical digestion** by mastication, (2) **chemical digestion** by digestive enzymes, (3) **separation of fluids and fine particles** from large particles by **filters**, and (4) **transport of fluids** and fine particles to the hepatopancreas and of large particles to the intestine for subsequent defecation through the rectum. In *Lophogaster*, the reduced size of the intestine and its ventral junction with the stomach at the secondary filter do not permit the transport of large particles via the intestine. Large, non-reducible particles are probably regurgitated in *Lophogaster*, where this phenomenon is additionally facilitated by the (near) absence of the cardio-oesophageal valve.

The final phases of the **digestive cycle** take place in the **hepatic caeca** or hepatopancreas, namely (1) an extra- and intracellular **enzymatic digestion** of the pyloric filtrate, and (2) the **production of digestive enzymes** for the phases of digestion in the stomach. In addition, peculiar spines, each with an apical pore, are present in the lophogastrid *Eucopia sculpticauda* on each of the first lateral teeth of the gastric mill. This suggests a secretory role of the subjacent stomachal epithelium (De Jong & B. Casanova, 1997).

Zimmer (1932) and Molloy (1958) observed **antiperistalsis**, which may be continuous or may vary with gut content, depending on species in Mysida. Fox (1952) observed anal and oral intake of water in various crustaceans, including the mysids *Hemimysis lamornae* and *Siriella armata*. He concluded that oral and anal drinking is needed to stretch the gut wall muscles until they can contract by antiperistalsis.

Circulatory system

The **circulatory system** (fig. 54.31A, C-E) was studied in several species of Mysida by Van Beneden (1861), G. O. Sars (1867), and particularly by Delage (1883), Claus (1876, 1884), and Wirkner & Richter (2007). Mayrat (1956b) re-described the cephalic aorta and its branches. Gadzikiewicz (1905) provided several histological details of the heart. Many characters of the circulatory system described by Siewing (1953, 1956) for the Lophogastrida species *Eucopia sculpticauda* and *Eucopia unguiculata* are reminiscent of those in Mysida. A slightly more simple vascular system was found by Belman & Childress (1976) in the lophogastrid *Gnathophausia ingens*. Major progress was achieved by Wirkner & Richter (2007), who applied the corrosion casting method in combination with computer-aided 3D reconstruction on three species of mysids and on the lophogastrid *Lophogaster typicus*. So far no details are available about the circulatory system of the Stygiomysida. A detailed overview of crustacean circulatory systems is available in the second volume of the present series (Mayrat et al., 2006).

CIRCULATORY SYSTEM OF THE MYSIDA

General features of the circulatory system (fig. 54.31C-E). – According to Delage (1883), Mayrat (1956b), and Wirkner & Richter (2007), this system is characterized by a contractile **tubular heart** that extends dorsally from thoracomere 1 to at least thoracomere 7, in part back to the anterior part of pleomere 1. It shows two pairs of **ostia** within the inflated median part, close to the border of thoracomeres 3 and 4: the posterior pair is more ventrally located than the anterior one (fig. 54.31D). Due to its uppermost position within the thorax, the heart is, on its ventral face, connected with the underlying dorsal diaphragm. Its innervation was studied by Alexandrowicz (1955) in *Praunus flexuosus*. There are at least three systems of nervous elements: a local ganglion system with its cell bodies on the dorsal wall and its axons innervating muscle fibres; a pair of cardiac nerves connecting with the central nervous system; finally, a particular innervation in the valves of the arteries leading away from the heart.

Anterior arteries (fig. 54.31C). – The **anterior aorta**, equipped with a valve near its origin, lengthens the heart anteriorly. It widens to a **myoarterial formation** (accessory pumping structure) attached to the posterior wall of the stomach, continues over the stomach, and then bends ventrally at the anterior margin of the stomach, where it gives off a sac-like widening. Then it continues anteriorly as an **ophthalmic artery**, which bifurcates to the eyes. Ventrally it emits a small upper cephalic artery plus a pair of **cephalic arteries** that ramify in the brain and anastomose in a plexus from which the antennular arteries depart. Here, the aorta shows a second myoarterial formation and emits a pair of **antennal arteries** and finally opens into the labral haemocoel. In its anterior region, the heart emits a pair of backwards-curved cardiac (hepatic) arteries that supply musculature, ovaries, and midgut glands (hepatopancreas). More caudally and ventrally, three small, unpaired arteries penetrate the dorsal diaphragm and branch into the viscera of the body cavity.

Descending artery (fig. 54.31C). – An additional, very important artery originates at the posterior ventral region of the heart. This is the **sternal artery** or **descending artery** (equipped with valves), which descends almost vertically in the posterior part of the seventh somite, passes aside of the intestine, then turns anteriorly along the sternal walls and the nervous chain up to the mouth area. Before bending anteriorly, the sternal artery emits, depending on the taxon, 1-2 branches which supply the appendages of the 2-3 posterior thoracic somites with lateral arteries, while the main trunk, which crosses the ventral nervous chain, supplies the appendages of the more anterior somites at least up to the maxillae. The sternal artery also supplies the nervous chain and the sternal portions of the thorax.

Abdominal artery (fig. 54.31C). – The heart is prolonged posteriorly by an **abdominal artery**. The latter is also equipped with valves at its point of origin, runs dorsally along the digestive tube, sends ventrally an arteriole to the ultimate thoracic somite and to each of the five anterior pleonites and their appendages, as well as a lateral arteriole to each of these latter. The abdominal artery bifurcates upon reaching the ultimate pleonite: one branch ends in the telson; the other, smaller one turns anteriorly after having sent arterioles into the rami of the uropods.

Lacunae and pericardium (fig. 54.31D, E). – When leaving the arteries, the blood drains into the **lacunae**, of which the principal one is located between the thoracic viscera. Two other important lacunae extend into the pleon: one into the medio-dorsal flanks of the intestine, the other into the ventral region accompanying the nervous chain. The blood that circulated in the secondary lacunae of the head and its appendages as well as the eyes, returns in part into the thoracic lacunae, in part into the lateral duplicatures of the carapace, where it proceeds along the lower (= inner) margin of the carapace (fig. 54.31D). The blood coming from the lacunae of the pleon returns partly into the **pericardium** (fig. 54.31E), partly into the large **thoracic lacuna**. From the latter, some of the sanguine flux enters the lacunae of the appendages, the remainder joins the torrent coming from the base of the eyes, and flows also to the lower margin of the carapace. All the blood passing through the respiratory lining in the duplicature of the carapace flows up to the dorsal region of the carapace, and flows into the pericardium through a short canal. Finally, the blood circulating in the seven posterior pairs of thoracic appendages also returns to the pericardium through seven pairs of **podo-pericardial sinuses** (fig. 54.31E).

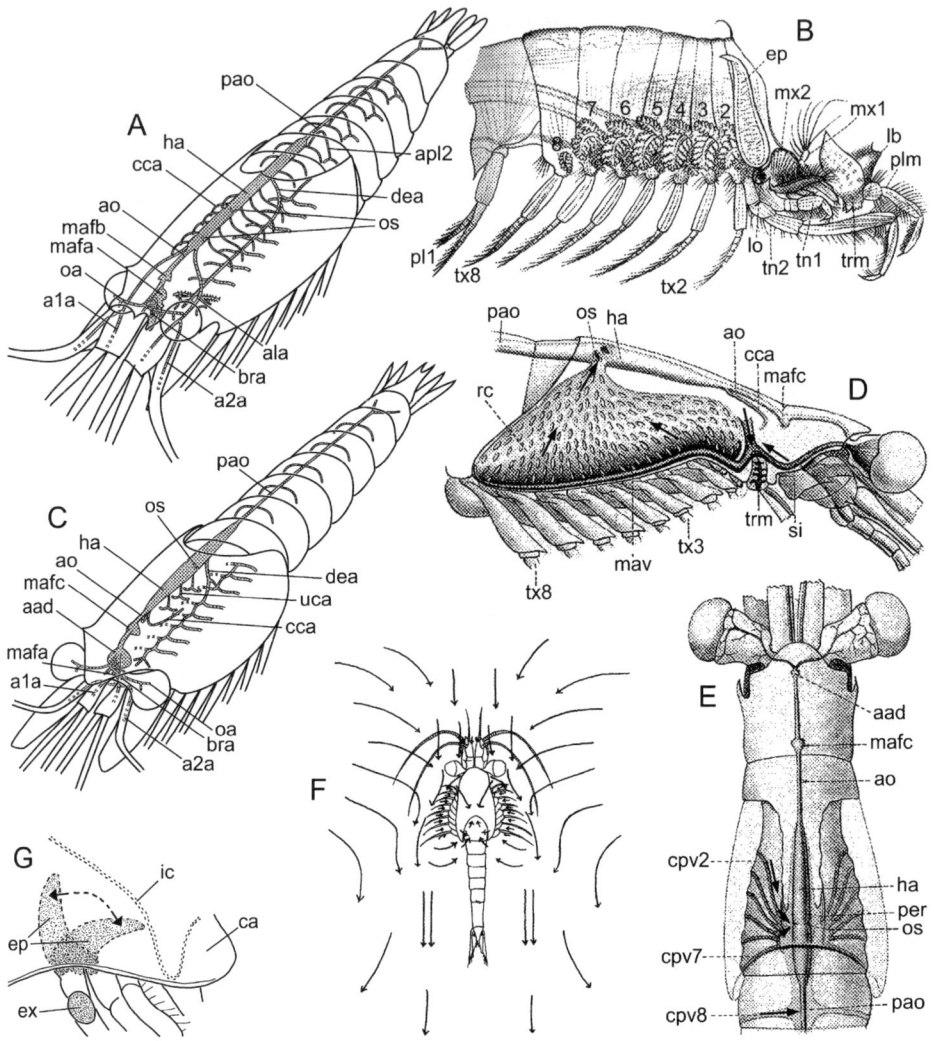

Fig. 54.31. Circulatory and respiratory systems of the Lophogastrida (A, B) and Mysida (C-G). A, C, interpretative schemes for heart and arterial system in *Lophogaster typicus* M. Sars, 1857 (A) and *Boreomysis arctica* (Krøyer, 1861) (C); B, series of branchiae on thoracopods 2-8 in *Gnathophausia longispina* G. O. Sars, 1883, carapace removed, lateral; D, E, semi-schematic presentations of afferent lacunae and sinuses in *Praunus flexuosus* (O. F. Müller, 1776) showing lacunae of respiratory carapace (D, lateral) and vessels returning from thoracopods (E, dorsal, overlying tissues removed), arrows indicate direction of flow; F, natatory and feeding currents, also supporting respiration, produced by a freely swimming *Hemimysis lamornae* (Couch, 1856); G, same species as before, arrows show movements of right epipod in respiratory chamber, lateral. Abbreviations: a1a, first antennal artery; a2a, second antennal artery; aad, anterior stomach aorta dilation; ala, anterior lateral artery; ao, anterior aorta; apl1-apl5, arteries for the pleopods 1-5; bra, brain artery; ca, carapace; cca, cardiac artery; cpv2-cpv8, cruro-pericardial vessels from thoracopods 2-8; dea, descending artery; ep, epipod; ex, base of thoracic exopod; ha, heart; ic, insertion of the carapace; lb, labrum; lo, luminescent organ; mafa-mafc, myoarterial formations a-c; mav, marginal afferent vessel of the

Circulatory system of the Lophogastrida

This system (fig. 54.31A) differs from that of the Mysida essentially by an additional (third), much more anterior pair of **ostia**, and by a greater number and different arrangement of **cardiac arteries**, whereby the posterior-most artery extends into the first pleomere. The antennae are supplied by a pair of anterior **lateral arteries** that branch as the first pair of arteries directly off the heart. The **podo-pericardial sinuses** of thoracopods 2-7 fuse with sinuses from the respective gills and then extend to the pericardium (Wirkner & Richter, 2007).

Respiratory system

Lophogastrida (fig. 54.31B). – Gas exchange takes place mainly in the gills, which arise as epipods from the coxae of thoracopods 2-7. The importance of the gills for respiration is underlined by their strong development in *Eucopia crassicornis*, described from oxygen-poor waters in the Canal of Mozambique (J.-P. Casanova, 1997). **Gills** are always present irrespective of body size and cuticle thickness, as exemplified by comparison between the 'giant' *Gnathophausia* and the small *Paralophogaster*. The **respiratory current** is driven by the movements of the exopods of the natatory thoracopods and by the back and forth movements of the **epipod** of the first thoracic appendages. According to Childress (1971), in *Gnathophausia ingens* the **exopod of the maxilla** also contributes by ventilation movements.

Mysida (fig. 54.31D, F, G). – Gas exchange takes place mainly through the inner face of the **carapace duplicature** according to the classical interpretation (e.g., Delage, 1883; Tattersall & Tattersall, 1951; Kobusch, 1999). This conclusion is essentially based on anatomical evidence, particularly on the characteristic network of **haemolymph channels** below the lining of the inner carapace wall (fig. 54.31D). In the space between the duplicature and the lateral wall of the thorax, the water driven by the beats of the epipod (fig. 54.31G) pertaining to the first thoracopod, circulates quite busily (Cannon & Manton, 1927a; Attramadal, 1981). The reduction of the respiratory carapace surface in the cave-dwelling *Palaumysis* is interpreted above (under 'Carapace') as a consequence of dwarfing. Due to the absence of respiratory tissue it appears unlikely that the laminar (e.g., in Mysinae, Leptomysinae) or curled (in males of most Siriellinae) **pseudobranchial lobes** on the basis of the endopods of the pleopods (fig. 54.21G) have some respiratory function in the Mysidae.

Stygiomysida. – This order shares the **respiratory carapace** with the Mysida (Wägele, 1994; Kobusch, 1999). Almost all species of Stygiomysida show a subterranean mode

---

carapace; mx1, maxillula; mx2, maxilla; oa, optical artery; os, ostium; pao, posterior aorta; per, pericard; pl1, first pleopod; plm, palpus mandibularis; rc, respiratory carapace; si, sinus; tn1-tn2, thoracic endopods 1, 2; trm, mandibular trunk; tx2-tx8, thoracic exopods 2-8; uca, unpaired cardiac artery. [A, C, after Wirkner & Richter, 2007 (figures modified and kindly made available by C. Wirkner); B, after Sars, 1885a; D, E, after Delage, 1883; F, G, after Cannon & Manton, 1927a; A, C-G, modified.]

of life. Among the two genera constituting this order, only *Stygiomysis* has a caudally strongly shortened carapace with reduced surface of the carapace cavity (figs. 54.3A, 54.7J; cf. above, 'Carapace'). The species of this genus are comparatively large (5-21 mm) and at least *Stygiomysis hydruntina* is capable of performing relatively quick movements (Inguscio, 1998). Therefore, neither dwarfing nor a strongly reduced basic metabolism are available as potential explanations for reduction of the respiratory carapace.

Reproductive system

As in most Malacostraca, the Lophogastrida, Stygiomysida, and Mysida are gonochoristic, apart from rare effects of intersexuality. The secondary sexual characteristics are acquired successively upon a number of moults. Unlike males, the females of certain taxa may adopt a resting stage at the end of the reproductive period (see below, 'Adjustment of reproductive parameters'). Hermaphroditism and reversal of sex were never observed, not counting partial reversal undergone by intersexes (see below, 'Intersexuality').

Female genital apparatus. – According to G. O. Sars (1867) and Holmquist (1959), the **female genital apparatus** (fig. 54.32) of the mysid *Mysis relicta* contains two parallel, joined cylindrical tubes, connected by a sacciform median bridge representing a true **ovary**. This portion of the apparatus is located between the digestive tube and the bottom of the pericardium. It extends from the region of the stomach to the posterior margin of the second pleonite (J.-P. Casanova, 1977) in mature females of the lophogastrid *Eucopia unguiculata*, and to the margin of only the first pleonite (Mauchline, 1980) in the mysid *Praunus flexuosus*. A quite large **oviduct** departs from the posterior portion of each tube. The only process occurring inside the oviduct is **oocyte** growth. According to Nair (1939), the mysid *Mesopodopsis orientalis* shows two pairs of parallel, longitudinal, differently developed tubes. The ovary opens on the right and on the left into a first pair of shorter thin tubes which lead with their anterior end into a second pair of longer and thicker tubes. The swollen terminal portion of the oviducts is glandular, with secretion taking place shortly before oviposition. Observed through the transparent **ovarian tubes**, the yolk of the oocytes may appear colourless or quite often vividly yellow, red, green, blue, or violet. As in all Malacostraca, the **female genital orifice** is located at the base of the sixth pair of thoracic appendages. The eggs are relatively large (see 'Egg size' below) and have a distinctly eccentric nucleus. **Fertilization** takes place after **oviposition** in the marsupium. So far no **receptacula seminis** have been found in any females of the three orders.

Male genital apparatus (fig. 54.33). – The **male genital apparatus** was first described by G. O. Sars (1867) for the mysids *Mysis relicta* and *Praunus inermis* as a horseshoe-like tube bearing about twenty testicular lobes, with two **deferent canals** departing from the ends of that tube. Later observations showed that this concept was erroneous. Labat (1961) observed for *Praunus flexuosus* that two independent, parallel, elongated **testes** are located in the medio-dorsal anterior region of the thorax, between the heart and the hepatic lobes, and behind the stomach. Along each **testis**, 5-6 efferent canals end in large cysts, most of the latter in lateral arrangement, only the anterior pair in frontal position. These cysts were

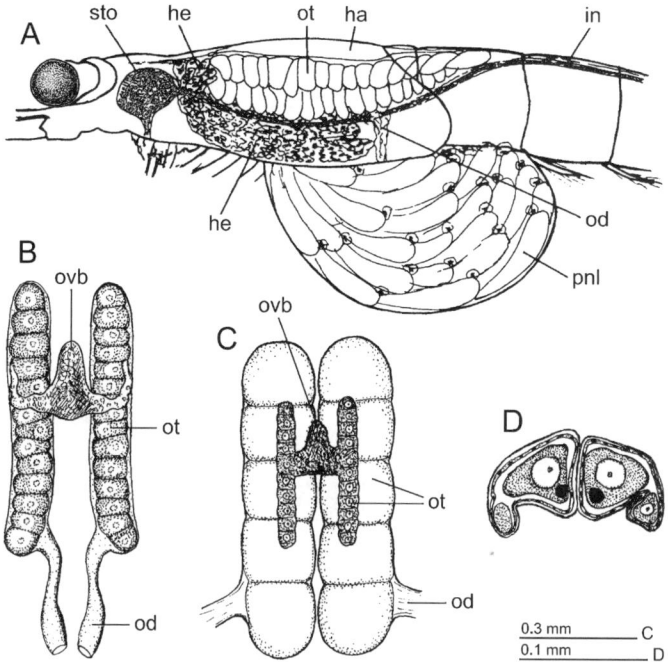

Fig. 54.32. Female genital apparatus of the Mysida. A, its position in body of *Praunus flexuosus* (O. F. Müller, 1776), lateral view; B, ovary and oviducts in young female *Mysis relicta* Lovén, 1862, dorsal view; C, ovary of *Mesopodopsis orientalis* (W. M. Tattersall, 1908), ventral face; D, transversal section through the latter, at connection with the two pairs of longitudinal tubes. Abbreviations: ha, heart; he, hepatopancreas; in, intestine; od, oviduct; ot, ovarian tubes; ovb, ovarian bridge; pnl, postnauplioid larvae in marsupium (highly schematic); sto, stomach. [A, modified after Mauchline, 1980; B, after G. O. Sars, 1867; C, D, after Nair, 1939.]

previously considered as true testes. They themselves communicate with two large tubes, which are independent of each other and located dorsally with respect to the testes. These tubes should be precisely named **seminal vesicles**, at least for their anterior, blind-ending parts, which are joined, curved down to the ventral parts of the body, and show a glandular epithelium that secretes mucus and other substances. Each seminal vesicle is prolonged posteriorly by a deferent canal that swells, forming an ampulla prior to ending at the **male genital orifice** close to the tip of the respective penis.

The **spermatogonia** grow in the swellings of the testicular chain and then pass through the cysts, where they undergo meiotic divisions and **spermatogenesis**. In each cyst, all elements are in the same meiotic state as found in the symmetrical cyst. In *Praunus flexuosus*, the ripe **spermatozoa** are released into the seminal vesicles in groups of 3, 4, or 5 (Labat, 1962). Each group is surrounded by the mucous substance secreted by the glandular epithelium of the seminal vesicle. This forms a type of **spermatophore**. The spermatophores accumulate in the deferent canals of the terminal ampulla. *Leptomysis lingvura* produces a much larger spermatophore: a sack-like, distally narrowing, sticky, soft membrane containing a large sperm mass (fig. 54.36G; Wittmann, 1982). The structure

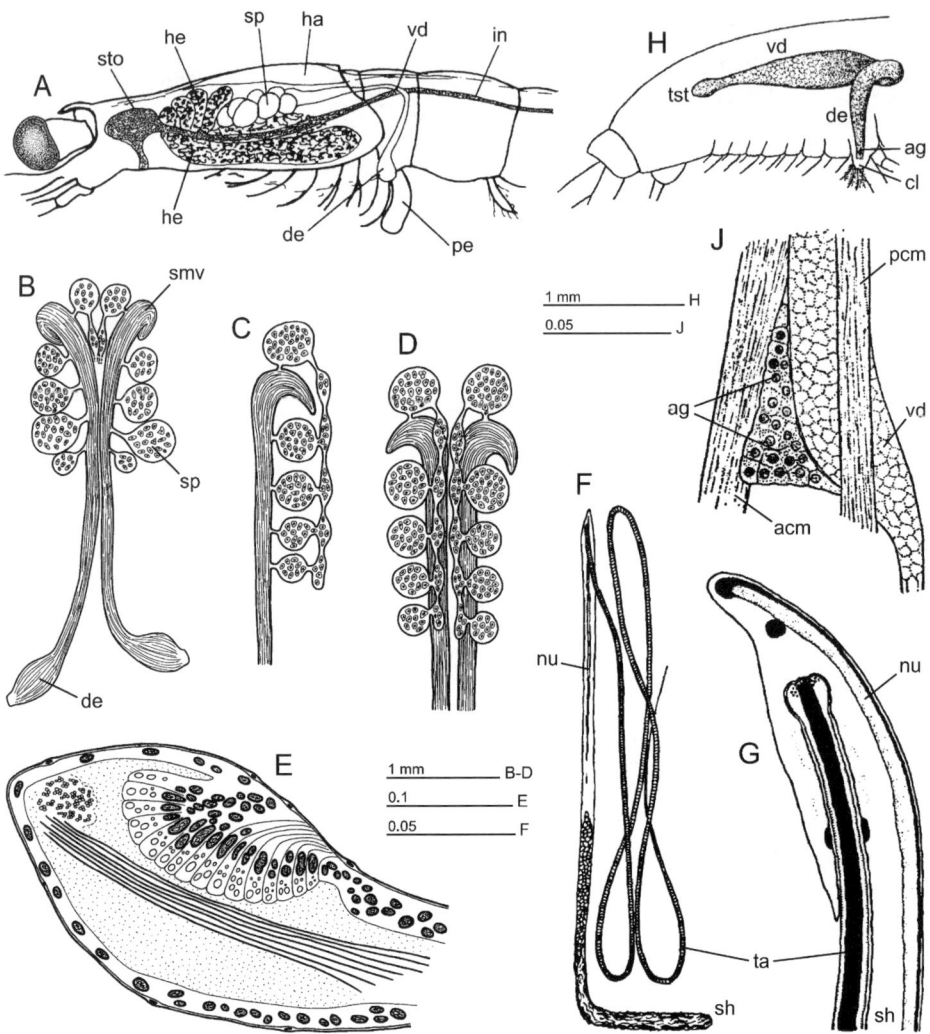

Fig. 54.33. Male genital apparatus of the Mysida (A-G) and Stygiomysida (H, J). A, its position in body of *Praunus flexuosus* (O. F. Müller, 1776), lateral view; B, genital apparatus of same species in dorsal view, object slightly expanded, only beginning of anterior portion of fine testicular canals shown; C, anterior portion of (B) in lateral view; D, anterior portion of (B) in ventral view; E, section through anterior end of seminal vesicle of *Praunus flexuosus*; F, spermatozoon of *Mysis oculata* (Fabricius, 1780); G, anterior portion of spermatozoon of *Praunus inermis* (Rathke, 1843) showing insertion of tail, schematic; H, position of genital apparatus in *Spelaeomysis longipes* (Pillai & Mariamma, 1964); J, androgenic gland with juxtaposed structures in *Spelaeomysis longipes*. Abbreviations: acm, anterior coxal muscle; ag, androgenic gland; cl, coxal lobe; de, ductus ejaculatorius (ejaculatory canal); ha, heart; he, hepatopancreas; in, intestine; nu, nucleus; pcm, posterior coxal muscle; pe, penis; sh, shaft; smv, seminal vesicle; sp, spermatidic pouch; sto, stomach; ta, tail; tst, testis; vd, vas deferens. [A, after Mauchline, 1980; B-E, after Labat, 1961; F, after Retzius, 1909; G, after Fain-Maurel et al., 1975, permission granted by Elsevier B.V., Amsterdam; H, J, after Nath et al., 1972b; A, F, H, J, modified.]

of the spermatozoa of the Mysida corresponds to that of the Amphipoda and Isopoda. They are whip-like, oblong, and immotile, composed of a long and thin shaft with an even longer, transversely striated tail (flagellum) (fig. 54.33F; Fain-Maurel et al., 1975; Wittmann, 1982). They are 0.5-1.3 mm long, including the long flagellum, and thus may be visible already with low-power microscopy through the wall of the ductus ejaculatorius (Wittmann, 2013b).

In the mysids *Praunus flexuosus* and *Paramysis nouveli* (cf. Juchault, 1963; Meusy, 1963), a pair of **androgenic glands** is present in the last thoracomere, close to the end of the external wall of the deferent canals. An identical arrangement can be found in the androgenic glands of the cavernicolous Lepidomysidae *Spelaeomysis longipes*, but their structure appears to indicate a weaker secretory activity (fig. 54.33J; Nath et al., 1972b).

Finally, mention should be made of the valuable survey by Kasaoka (1974) on the male genital apparatus in the mysids *Archaeomysis grebnitzkii* and *Neomysis awatschensis*. The structure of the external male genitalia is treated above in the section 'Thoracic appendages'.

Excretory system and excretion

Excretion takes place in **segmental glands** (fig. 54.34). All Lophogastrida and Mysida have a pair of **antennal glands** consisting of a **sacculus** and a more or less long and irregular **secretory canal** that swells into a **vesicle** at its terminal portion. Only the sacculus located in the praecoxa of the antenna, shows excretory function. The entire gland is of mesodermal origin, except for the ectodermal **nephrostome** located on the latero-ventral face of the antennal coxa (Grobben, 1881; Cannon & Manton, 1927b; Vogt, 1932, 1933, 1935a).

Mysida (fig. 54.34A, B). – Tchindonova (1981) found hypertrophic **antennal glands** in eight genera of Erythropinae (Mysida) characterized by eye plates with reduced visual elements. In contrast, normally developed antennal glands are present in genera with well-developed eyes. Vogt (1932) found mesodermal cells, termed by him 'athrocytes' (fig. 54.34A; i.e., not 'arthrocytes'), at the base of the thoracopods in Mysida and concluded that they possibly represent storage cells and do not contribute to excretion. Without indication of details or literature sources, Tattersall & Tattersall (1951) did not exclude a certain contribution by these cells to excretion. Studies on other crustacean groups by Hosfeld & Schminke (1997) showed that such cells possibly represent **podocytes**, i.e., typical excretory cells.

Lophogastrida (fig. 54.34C). – **Antennal glands** as well as **maxillary glands** are present in this order. Both types of excretory organs have principally the same structure. In *Lophogaster*, the excretory canal is straight, with the **nephrostome** placed at the end of the first segment of the maxillae (Cannon & Manton, 1927b). Such glands were also described for *Eucopia* by Siewing (1956). According to Calman (1909) the maxillary glands produce a **bioluminescent secretion** in Gnathophausiidae (luminescent organs in figs. 54.10O, 54.31B). Frank et al. (1984) reported that *Gnathophausia ingens* loses its luminescent

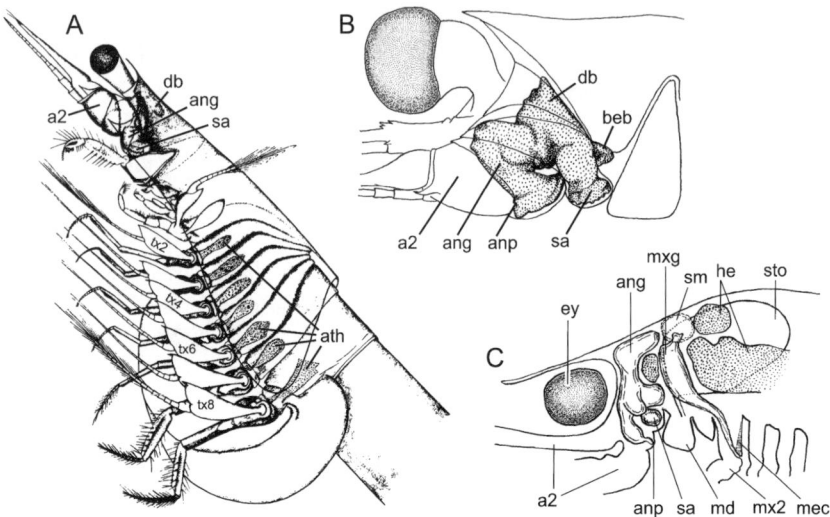

Fig. 54.34. Excretory system in Mysida (A, B) and Lophogastrida (C). A, antennal gland and series of athrocytes in cephalothorax of *Praunus flexuosus* (O. F. Müller, 1776); B, left antennal gland in *Mysis relicta* Lovén, 1862; C, reconstruction of excretory organs on left side of *Lophogaster typicus* M. Sars, 1857. Abbreviations: a2, antenna; ang, antennal gland; anp, antennal nephroporus; ath, athrocytes; beb, blind ending bladder; db, dorsal bladder; ey, eye; he, hepatopancreas; md, mandible; mec, mesodermal cells; mx2, maxilla; mxg, maxillary gland; sa, sacculus (end sac) of the antennal gland; sm, sacculus (end sac) of the maxillary gland; sto, stomach; tx2-tx8, thoracic exopods 2-8.
[A, after Vogt, 1932; B, after Vogt, 1933; C, after Cannon & Manton, 1927b; A-C, modified.]

capacity in the laboratory when fed a diet restricted to tissues from non-bioluminescent animals and rapidly regains this capacity after ingesting certain luminescent prey. This would indicate that luminescence would not originate from the species itself, but rather be dependent upon components of its food.

Endocrine organs

Endocrine organs of the eyestalk. – In the Mysida and Lophogastrida, **neurohormones** involved in the endocrine control of moulting and reproduction are produced in the eyestalks, as in Decapoda, rather than more directly in the brain as in Amphipoda and Isopoda. The **sinus gland** in the eyestalk and the **frontal organ** in the frontal plate were already described above, together with the nervous system of the lophogastrid *Eucopia*. The sinus gland was localized and described in detail within the eyestalk (fig. 54.24D) of the Mysida genera *Mysis*, *Paramysis*, and *Siriella* by Gabe (1953), Hogstad (1969), and Cuzin-Roudy & Saleuddin (1985). These later authors described the "medulla externa – medulla interna X-organ (ME-MI X-organ)" and "medulla terminalis X-organ (MT X-organ)" as **neurosecretory** and **neurohaemal organs** involved in moulting and reproduction. Cauterization of the MI-ME X-organ showed no effect when performed after apolysis (segregation of the old cuticle), but inhibited moulting when performed before apolysis

(Cuzin-Roudy & Saleuddin, 1985) in both sexes of *Siriella armata*. In the latter case, secondary vitellogenesis ceased and the breeding females could not stop incubation — and their marsupial young could not perform the second larval moult. Kulakovskii (1969, 1971) described neurosecretory cells of the eyestalk involved in the hormonal control of the chromatophores in *Mysis oculata*. Previously assumed neurosecretory and neurohaemal functions of the **organ of Bellonci** (SPX-organ) in the eyestalk appear currently doubtful (see above, 'Nervous system').

Y-organ. – The "lateral organ" described by Vogt (1935b) for the first maxillary segment of the mysid *Gastrosaccus spinifer* was listed by Charmantier-Daures & Charmantier (2006) as a **Y-organ**. Gabe (1953, 1956) identified a Y-organ in the mysid genera *Siriella*, *Praunus*, and *Paramysis*, where it is located in the second maxillary segment and innervated by the sub-oesophageal ganglion. It is comparable to the Y-organ of the decapod crustaceans (Lachaise et al., 1993) and to the moulting glands of insects (Covi et al., 2012), and it probably plays a role for the secretion of the **moulting hormone**.

REPRODUCTION AND SEXUALITY

Sexual dimorphism

Beyond the gonads or primary sex organs (cf. above, 'Reproductive system') and the accessory sex organs (penes and oostegites; cf. 'Thoracic appendages') all taxa show at least some secondary sexual traits, which together constitute a generally strong sexual dimorphism.

Lophogastrida. – In this order, **sternal projections** are generally present on the posterior thoracomeres of juveniles. These projections become more conspicuous with increasing body length, and form one — sometimes two — strong peaks in adult males (fig. 54.35A). In contrast, these projections are reduced in females with approaching sexual maturity, at the same time when granular areas appear at their position. These areas bear long barbed setae (fig. 54.35B) penetrating the space between eggs (embryos) or larvae (Fage, 1936, 1940, 1941, 1942; Nouvel, 1942a, 1943). In *Eucopia*, **sexual dimorphisms** may concern the antennular peduncles, the antennal scales (also in *Gnathophausia*), and in a very constant manner the pleopods, the thoracic exopods, and the gills. The latter are more strongly developed in the males (fig. 54.35L) than in the females (fig. 54.35K). In the Lophogastrida, the male-specific sensilla appear to be generally less restricted to certain portions of the antennula than in the Mysida (Johansson et al., 1996): All along the outer antennular flagellum of male *Lophogaster typicus*, there are male-specific aesthetascs (fig. 54.28U) in addition to those also present in females (fig. 54.28T). The pleopods of the Lophogastrida are well developed and natatory, not or only weakly modified in both sexes (see above, 'Pleopods').

Stygiomysida. – The exopod of the second male pleopod is slightly modified and generally shows more setae, but has always one segment less compared to that of adult females (fig. 54.17H versus 54.17J): in *Spelaeomysis* the male shows only three segments instead of four (Pesce, 1976a), and in *Stygiomysis* it shows only two instead of three segments (Bowman, 1976).

Mysida. – Diverse sexual differences, sometimes quite particular ones, may concern the formation of the frontal plate, antennulae, antennal scale, certain thoracic and pleonal appendages, as well as the armature of telson and uropods, particularly the form of their spines. In certain cases the pigmentation also varies according to sex and state of sexual maturity.

The males of most taxa have a stronger, stouter antennula with greater diversity and numbers of sensory setae. In adult males of the family Mysidae, the **appendix masculina** (see above, 'Antennula'), less frequently termed '**processus masculinus**' or '**lobus masculinus**', is covered by a dense brush of sensory setae (fig. 54.12C, E, G). The setae are lacking or are very sparse in subadult males. Their full presence marks the attainment of the adult stage. The male lobe varies greatly between the taxa. It is very long in the Siriellinae, but may be reduced to a small setose ridge or hump in other subfamilies. In certain Boreomysinae and Heteromysinae it represents only a brush of setae ventrally on the terminal segment of the antennular peduncle.

As a diagnostic feature of the subfamily Gastrosaccinae (Mysida: Mysidae), the first pleomere of the females shows **pleural plates** (fig. 54.5F) supporting the marsupium. Such plates are generally absent in males; if present, they are smaller than in females. Curiously, there is also an inverse case characteristic at subfamily level within the Mysidae: in all species of Rhopalophthalminae, the males show a pair of well-projecting, rounded pleural plates (arrow in fig. 54.4C) from the first pleomere, whereas such plates are completely absent in the females (O. S. Tattersall, 1957; Panampunnayil & Biju, 2006; Vilas-Fernández et al., 2008; Hanamura et al., 2011). The proximity of these plates to the specially **modified genital structures** in males of this subfamily suggest some unknown role in mating (Wittmann, 2013b).

Compared to the Lophogastrida, **sternal projections** from the thoracomeres are much less common and are scattered over various taxa in the Mysida. In certain species of *Heteromysis*, such projections are well developed and covered by spines or teeth (fig. 54.15N) in males, whereas they are absent or small, hump-like (fig. 54.15M), never with spines or teeth, in females (Nouvel, 1940; Băcescu, 1968a; Wittmann, 2000, 2008). Many Atlantic and Pontomediterranean species of *Paramysis* show lobe-like to long sickle-shaped projections in adult males (fig. 54.35D, E). These projections are shorter and more hump-like in juveniles of both sexes (fig. 54.35C). In analogy to the Lophogastrida, such projections are missing in adult females (Labat, 1953; Wittmann & Ariani, 2011).

The **thoracic endopods** are often more strongly developed and may bear different sets of setae in the males: in *Heteromysis* and closely related genera, the 'tarsus' (i.e., carpopropodus plus dactylus) of the third male endopod (fig. 54.35G) is typically stouter and more prehensile than in the females (fig. 54.35F). Depending on the species of *Surinamysis*, at least some among endopods 5-8 are longer, with particular sets of setae

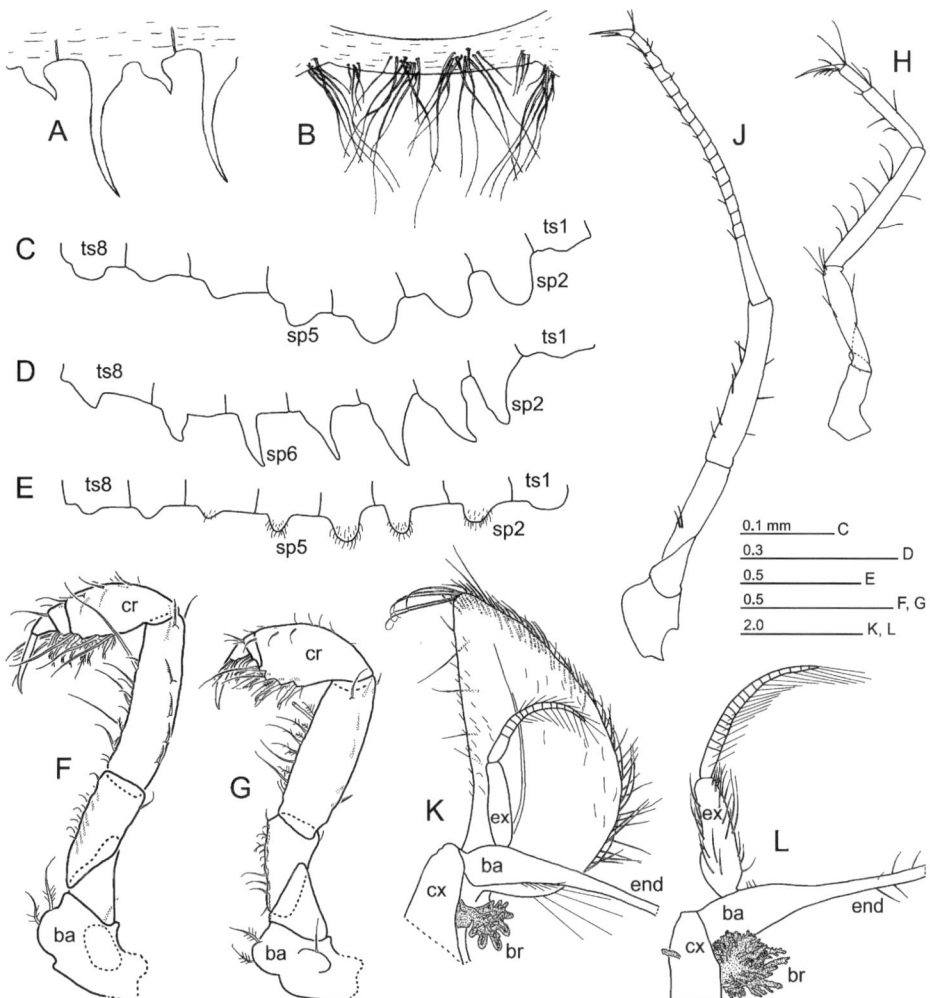

Fig. 54.35. Sexual dimorphism of thoracic sternites and thoracopods in Lophogastrida (A, B, K, L) and Mysida (C-J). A, B, equipment of some thoracic sternites in *Lophogaster typicus* M. Sars, 1857, for male (A) in lateral view and for female (B) in frontal view; C-E, series of thoracic sternites 1-8 in lateral view, for juvenile (C) *Paramysis pontica* Băcescu, 1940, an adult male (D) of same species, and an adult male (E) of *Paramysis bakuensis* G. O. Sars, 1895; F, G, third thoracic endopods in female (F, frontal aspect) versus slightly smaller male (G, caudal aspect) *Heteromysis wirtzi* Wittmann, 2008; H, J, seventh thoracic endopods in female (H) versus male (J) *Surinamysis merista* (Bowman, 1980); K, L, size of gills and exopods of seventh thoracopods in female (K) versus male (L) *Eucopia unguiculata* (Willemoës-Suhm, 1875). Abbreviations: ba, basis; br, branchiae; cr, carpus; cx, coxa; end, endopod; ex, exopod; sp2-sp6, sternal processes after thoracomeres 2-6; ts1-ts8, thoracic sternites 1-8. [A, B, after Fage, 1942; C-E, after Wittmann & Ariani, 2011; F, G, after Wittmann, 2008; H, J, after Bowman, 1980; K, L, after Nouvel, 1942a; A-L, modified.]

or with other modifications in males (fig. 54.35J) compared to those of females (fig. 54.35H). For female-specific structures associated with the marsupium see above, under 'Oostegites'.

The **male pleopods** (fig. 54.21) are well developed or generally less reduced compared to those of the females. Among the ten subfamilies of the Mysidae, strong modifications mark the male pleopods of the Rhopalophthalminae, Gastrosaccinae, Mysinae, Leptomysinae, and less strongly also Siriellinae (at the level of **pseudobranchial lobes**); such modifications are also less frequent and, if present, weaker in the Erythropinae. Male adulthood in many taxa is also marked by fully developed **modified pleopods**, besides penes and appendix masculina. Only small modifications, if any, are found on the well-developed male pleopods of the Boreomysinae. In Palaumysinae, Heteromysinae, and Mysidellinae, the pleopods are strongly reduced in both sexes, without or with only small, yet particular, modifications in the male.

Intersexuality

Although much more rarely than in several other crustacean orders, particularly amphipods, abnormal sexuality situations — referred to as **intersexuality** — have been observed also in the Mysida:

Masculinization. – **Masculinized females** were occasionally found in samples from brackish and marine coastal waters of the north-east Atlantic and the Baltic: on several occasions but always with low incidence in populations of the mysid *Neomysis integer* and so far in only one specimen of *Gastrosaccus spinifer* (Kinne, 1955; Holmquist, 1957; Hough et al., 1992; Mees et al., 1995). The masculinized adult females of *Neomysis integer* exhibited a normal marsupium containing embryos or nauplioid larvae, but also elongated fourth pleopods otherwise typical of adult males. Histological examination by Hough et al. (1992) indicated fully **functional ovaries**; in addition, there was no evidence of any testicular tissue or of parasitism. According to those authors, the observed intersexuality might reflect a rare genetic abnormality.

Feminization. – As an opposite phenomenon, **feminized males** were found in field samples by Yamashita et al. (2001). Among 9282 specimens of the mysid *Orientomysis mitsukurii* sampled in Sendai Bay, Japan, there were seven feminized males with empty marsupium, six of which with a brood pouch smaller than that of normal adult females. These intersexes had no ovaries but contained **testicular tissue** with spermatozoa. They showed elongated fourth pleopods, but no externally visible genital papillae. Intersexuality had an incidence of 2.5% among adults of the winter generation, but only at a station polluted with nonylphenol and bisphenol-A, known as xeno-oestrogens in vertebrates.

Apart from xenobiotics, feminizing effects on males may also be induced by ellobiopsid parasites (see below, 'Parasites').

Sex ratio

The details of **sex determination** are so far unknown in Lophogastrida, Stygiomysida, and Mysida. For sex verification see above, 'Reproductive system'. The only experimental work has been done by Ortega-Salas et al. (2008), who obtained sex ratios of 2.5-3.0 in favour of females in a laboratory culture of the mysid *Mysidopsis californica* under semi-controlled conditions.

In samples from natural populations the **sex ratio** may be near 1.0, but more often the females tend to outnumber the males. Mauchline (1980) reviewed the data on sex ratios in mysids and lophogastrids, and attributed the strong variations observed in natural populations to differences between sexes regarding age at attainment of maturity, longevity, habitat, and bathymetric plus regional distribution. Some more recent publications compiled here support the previous findings on Mysida:

Hanamura (1999) observed strong variations without a clear seasonal trend in the incidence of females of *Archaeomysis articulata* from marine sandy beaches in Japan. In this population the size at maturity varied with seasons in both sexes, and the **overwintering males** seemed to mature and to die earlier than the females. Nonetheless, the sex ratio was near 1.0 when the data were pooled over the main **breeding season**.

Grabe (1989) found sex ratios near 1.0 at most sampling dates for *Rhopalophthalmus tattersallae* from the Arabian Gulf [= Persian Gulf]. San Vicente & Sorbe (2003) obtained 1.0-1.8 in samples of five species of *Schistomysis* from the south-eastern Bay of Biscay (north-eastern Atlantic). Gergs et al. (2008) found an extremely large ratio of 34.9 in May versus only 1.0 in July in a non-indigenous freshwater population of *Limnomysis benedeni* in Lake Constance near the confluence of the Rhine River (western-central Europe). In a non-indigenous population of *Hemimysis anomala* in River Trent in England, Nunn & Cowx (2012) obtained 1.5-3.3 in May-June but only 0.5 in June-July, and argued this to be related to **differential mortality**.

Mating and oviposition

Mating (fig. 54.36). – The process of mating has so far been observed only in a few Mysidae (Mysida) species, namely by Nouvel (1937, 1940) in *Praunus flexuosus* and *Heteromysis armoricana*, by Nair (1939) in *Mesopodopsis orientalis*, by Labat (1954) in *Paramysis nouveli*, by Murano (1964b) in *Neomysis intermedia*, by Clutter (1969) and Clutter & Theilacker (1971) in *Metamysidopsis elongata*, and by Wittmann (1982) in *Leptomysis lingvura*. These species have a pair of well-developed, **tubular penes** in common. **Mating** occurs during the night, shortly after the moult of the female, but only if the female has a new egg clutch ready in the **ovarian tubes** or if this clutch is already freshly extruded into the brood pouch. There is no **precopula** comparable to that in gammarids. There is also no evidence pointing to **sperm competition** or **agonistic behaviour**. The males seem to distinguish the females ready for mating at a certain distance, probably due to **chemotactic substances** emitted by the latter (Clutter, 1969; Wittmann, 1982). In the above-listed species, the approach of the sexes proceeds very fast. In most species, the male approaches the female from below, and by rotating its own (fig. 54.36B) or the female's (fig. 54.36E) body usually adopts a head-to-tail, ventral to ventral position (fig. 54.36B, C, E), in which the sperm is deposited into the marsupium. The sperm becomes transferred by the well-developed tubular penes, probably without direct

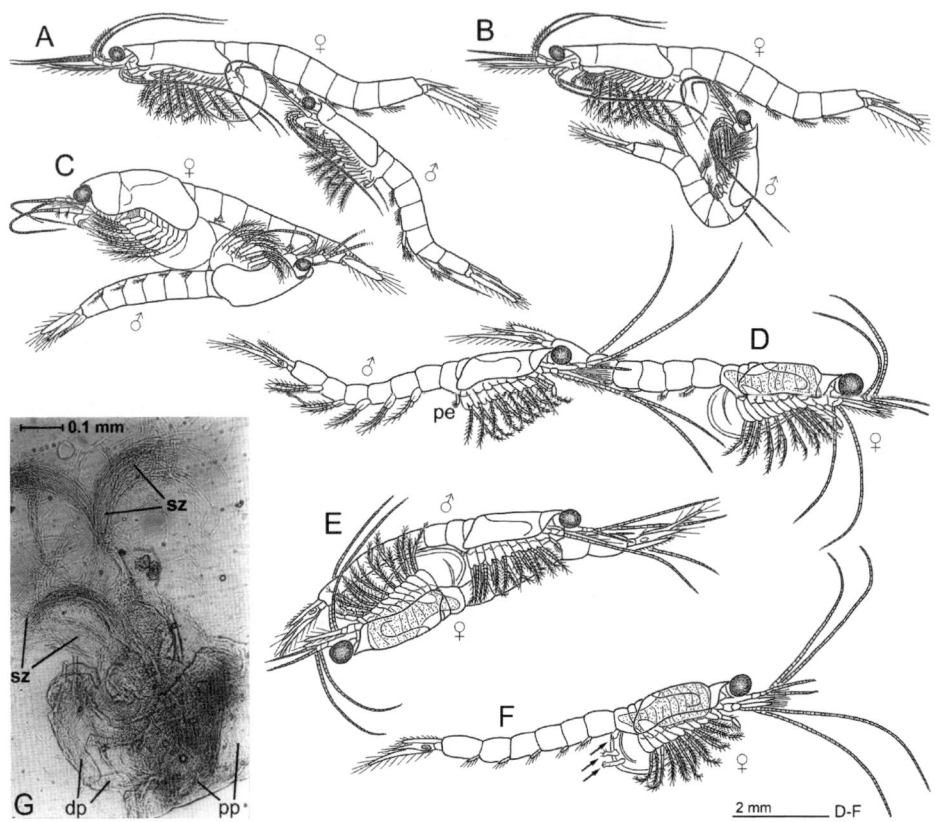

Fig. 54.36. Mating in Mysidae (Mysida). A, B, initialization (A) and main phase (B) of copulation in *Praunus flexuosus* (O. F. Müller, 1776); C, copulation in *Heteromysis armoricana* Nouvel, 1940; D, E, initialization (D) and main phase (E) of copulation in *Leptomysis lingvura* (G. O. Sars, 1868); F, female of *Leptomysis lingvura* shortly after copulation with several males, arrows point to spermatophores fixed on outside of brood pouch; G, spermatophore of *Leptomysis lingvura*. Abbreviations: dp, distal portion of spermatophore; pe, penis; pp, proximal portion of spermatophore; sz, spermatozoa. [A, B, after Nouvel, 1937; C, after Nouvel, 1940; D-G, after Wittmann, 1982; A-F, modified; G, permission granted by Elsevier Verlag, Munich.]

intervention of pleopods. In *Praunus flexuosus* (fig. 54.36B) and *Paramysis nouveli*, the modified fourth pleopods support the male in keeping the proper position during copulation (Nouvel, 1937; Labat, 1954). This could be a key to understand why **modified male pleopods** are present in most species of Mysidae. The mating animals are in contact for less than one minute; the male leaves the female immediately after **sperm transfer**, which takes only a fraction of a second. In general, the female is impregnated only once per egg clutch and, following this, flees from other approaching males.

However, females of the swarming species *Leptomysis lingvura* usually mate with a number of different males. In this case, the sperm is transferred in a comparatively large **spermatophore**. This is clearly advantageous for fast mating — one second or less in this species (Wittmann, 1982). Normally, the sperm is transferred directly into the brood pouch. After experimentally induced copulation with several males, however, some spermatophores were fixed on the outside wall of the brood pouch (fig. 54.36F). The outside displacement of spermatophores may possibly be due to less suitable mating responses and/or rejection of additional males by the females after preceding successful copulation (Wittmann, 1982).

Without mating the brood becomes lost within a few days (Nouvel, 1940; Murano, 1964b; Wittmann, 1982). A new sperm transfer appears necessary for each brood clutch; as far as known, there is no **receptaculum seminis**.

The sexual morphological peculiarities of certain mysids in the Antarctic are interpreted (Wittmann, 1996) as being adaptive for males. They possibly have a chance to reproduce only once in their lifetime during a very short period. Supporting this concept, the males of *Mysis relicta* die shortly after **copulation** in a subarctic, oligotrophic freshwater lake in Finland (Hakala, 1978). If copulation proceeds as rapidly as in the above-quoted species from temperate climates (Nouvel, 1940; Wittmann, 1982), it would be crucial for males to detect receptive females by using their antennular sense organs and to introduce their penes into the brood pouch as fast as possible. Here, **giant penes** (as in *Mysifaun erigens* and several species of *Mysidetes*) may be useful to gain precedence in mating, or may even favour higher precision in sperm transfer (Wittmann, 1996). In order to transfer genes to the next generation, **fitness for survival** may be less important than **fitness for reproduction**: in fact, the presence of these giant penes may be inconvenient for normal activities such as swimming. A partial solution to these conflicting adaption patterns may be **erectile penes** (fig. 54.20E), so far known (Wittmann, 1996) from only one mysid species, *Mysifaun erigens*. Erection is generally a rare phenomenon in crustaceans, but a common component of the **reproductive behaviour** in Stomatopoda (Caldwell, 1991).

Oviposition. – The process of **oviposition** was observed in the same species as above for mating. Moult of the female and subsequent deposition of unfertilized eggs may already start at dawn (Ariani et al., 1982; Wittmann, 1982), but proceeds mainly during the night. The female isolates herself, becomes immobile, and then shows flexions of the body. Simultaneously to the right and to the left, the eggs are slowly extruded into the **oviducts**, which dilate somewhat. In *Neomysis integer* and *Leptomysis lingvura*, each oviduct extrudes a small transparent **egg sac** close to the first pair of oostegites (Kinne, 1955; Wittmann, 1981a). The eggs are then discharged into the swelling egg sacs until each egg sac fills a lateral half of the brood pouch and contains half the brood. The delicate, deciduous egg sacs are often overlooked in peracarids (Johnson et al., 2001), suggesting that they may be possibly more common also in mysids. With or without egg sacs, the eggs are extruded into the brood pouch through the **gonopores** on the coxae of the sixth thoracopods. Here, they become fecundated and then immediately covered by a chorion consisting of substances secreted by the terminal portions of the oviduct. The eggs become larger upon fertilization and continue to swell during the course of embryonic development.

Fecundity

EGG SIZE

Biological limits of egg size. – The Lophogastrida, Stygiomysida, and Mysida share in common with the Amphipoda the release of young as miniature adults from the brood pouch. The young do not feed before release, which may be the main reason for a large lower limit of **egg size** of about 0.2 mm in Amphipoda and 0.3 mm in Mysida (considering only fertilized eggs; Steele & Steele, 1975; Wittmann, 1984; Johnson et al., 2001). In contrast, copepods, barnacles, isopods, cumaceans, and decapods have free-living, feeding larval stages and, accordingly, produce generally smaller eggs (with a larger range of sizes in decapods). As a curious consequence of **large minimum egg size**, the postnauplioid larvae of three *Palaumysis* species from tropical marine caves are too large to be entirely covered by the marsupial plates of their small mothers, whose body length is only 1.3-2.3 mm (Hanamura & Kase, 2002, 2003). **Diameters of fertilized eggs** range from 0.3-1.8 mm in Mysida (Mauchline, 1973b; Wittmann, 1984). Typical values from temperate climates are 0.4-0.8 mm. Mauchline (1973b) listed 0.8-4.0 mm for seven comparatively large species of Lophogastrida. This range is unrepresentative for the entire order as long as data for the small-sized species of *Paralophogaster* are missing. Only one report on egg size is available for the Stygiomysida: Ariani & Wittmann (2010) found an average of 0.68 mm in *Spelaeomysis bottazzii*. When comparing different species, egg diameters generally increase with increasing parental body size, increasing latitude, or decreasing average ambient temperature (Wittmann, 1984). There are smaller differences or even none upon comparison of egg sizes between different individuals or populations belonging to the same species.

Seasonal variations of egg size. – In several species from temperate (subtropical to boreal) climates, the **winter eggs** are larger than the **summer eggs** (Mauchline, 1980; Wittmann, 1981b, 1986a, b; Delgado et al., 1997; Calil & Borzone, 2008). At any given season, egg sizes normally do not or only weakly increase with body size of the parent (Wittmann, 1981b, 1986a, b; San Vicente & Sorbe, 1990; Fenton, 1994; Hanamura, 1999; Hanamura et al., 2009; Biju & Panumpunnayil, 2010; Ramarn et al., 2012; Biju et al., 2013). A stronger correlation, however, is found in a few populations with large winter animals producing large winter eggs (Delgado et al., 1997; Calil & Borzone, 2008). In these latter the reproductive potential of large specimens is clearly diminished in favour of the fitness of the offspring in order to survive the unfavourable season.

Ecological significance of egg size. – When comparing many different species from different latitudes, egg sizes increase strongly, namely with slightly less than the square root of parental body size in epipelagic and coastal species (Wittmann, 1984). Nonetheless, the deep-water species generally produce fewer but larger eggs per brood clutch compared with the epipelagic and coastal species. As in other invertebrates, larger eggs result in larger young. This may be advantageous in the generally trophically poor deep-water environments. Taking parental size into account, exceptionally large eggs are found mainly in bathypelagic species, also in certain hypogean species (e.g., the above-quoted

Spelaeomysis bottazzii) and in a few benthic ones (e.g., the Mysida species *Ischiomysis peterwirtzi*; cf. Wittmann, 2013a).

Steele & Steele (1975) listed six ecological implications of egg size in amphipods, four of which are relevant for Mysida and Lophogastrida according to Wittmann (1984): (1) the **size at release of young** depends only on egg size and does not vary with temperature; (2) the **size at attainment of maturity** is positively correlated with **egg size**, and both increase with increasing latitude (decreasing temperature); (3) the **number of eggs per brood** at a given brood weight decreases with increasing egg size. Nonetheless, egg size and brood size are positively correlated at the interspecific level in mysidaceans: both, together with brood weight, increase with increasing latitude; (4) egg size is positively correlated with **incubation time** at the intraspecific (Wittmann, 1981b) as well as interspecific (Wittmann, 1984) level. By this strategy, any increase of egg size lowers the natality (**birth rate**) at least in **iteroparous mysids**, but supposedly favours the **fitness** of the resulting juveniles.

BROOD SIZE

Mysida. – Mauchline (1980) and Petryashov (1990) listed the **fecundity** in many species of the Mysida. The numbers of eggs produced with each brood clutch vary from 2 to 200, according to individual body size, species, season, and geographical zone. The constant presence of bilaterally identical pairs of ovaries and gonopores suggests a minimum number of two eggs per normal oviposition. In fact, only two eggs per clutch are found in several small-sized tropical taxa, including the above-mentioned *Palaumysis*. Female body length and **egg numbers** are correlated in most species examined in this respect. A linear relation, often varying seasonally, successfully described the data obtained within a given population or species (Murano, 1964b; Murtaugh, 1989; Fenton, 1994; Hanamura, 1999; Okumura, 2003; Feyrer, 2010; Biju et al., 2013). Exponential (log-log) relations have also been applied (San Vicente & Sorbe, 1995, 2003; Delgado et al., 1997) as recommended by Wittmann (1984) based on a mostly linear relation between egg numbers and parental weight in combination with an about third power (log-log) relation between parental length and weight. No statistically significant correlations between parental size and clutch size can be detected if there are only small size differences between the breeding females, such as observed in *Acanthomysis thailandica* from a tropical mangrove estuary in Malaysia (Ramarn et al., 2012). In *Neomysis awatschensis* from a Californian estuary and in *Neomysis mercedis* from Lake Washington, the relation between brood size and parental size varied strongly between seasons and/or years, related to differences in temperature (Heubach, 1969) or **food availability** (Murtaugh, 1989). *Americamysis bahia*, reared by Johns et al. (1981) in the laboratory, showed different survival and fecundity after being fed with different strains of *Artemia*. Feyrer (2010) reported that body size is a stronger determinant of brood size than species-specific differences in four species from coastal marine waters off Vancouver Island (north-eastern Pacific).

A fundamentally different relation between **egg numbers** and **parental size** is obtained when comparing different species from different latitudes: Some small species from the tropics produce only two eggs per clutch; in most species from temperate latitudes the numbers range 10-20, rarely beyond 50. W. M. Tattersall (1951) counted 190 embryos in the marsupium of a female of the boreal *Mysis stenolepis* with 28 mm body size. The wide size range obtained upon simultaneously plotting many species yields a curvilinear relation best described by an **allometric relation**, i.e., by a log-log regression. Such plots showed that numbers of young as well as the egg weight (volume) increase with slightly less than the square root of parental body weight (volume) in epipelagic and coastal species (Mauchline, 1973b; Wittmann, 1984). Accordingly, the portion of body weight (volume) invested into each **brood clutch** does not vary with average size of species and with latitude. This portion averages 10% for body length raised to the third power (as a correlate of volume; Mauchline, 1973b) and 23% for dry weight (Wittmann, 1984). These values are compatible with each other considering that eggs have a much lower water content and a higher lipid content than parental bodies. The order of magnitude of **body mass invested into reproduction** holds for all size classes and also for meso- and bathypelagic species.

Lophogastrida. – Female body size and numbers of eggs are also correlated in lophogastrids, although the results are based on only few species of Gnathophausiidae (Clarke, 1962; Childress & Price, 1983; Wilson & Boehlert, 1993). As a stupendous case, Clarke (1962) found 238 embryos in a female *Gnathophausia ingens* measuring 145 mm length. According to Mauchline (1973b) and Wittmann (1984), however, the **length-fecundity relation** is less obvious in meso- to bathypelagic species that constitute the majority of Gnathophausiidae. In *Eucopia unguiculata* (Eucopiidae), more eggs are produced in the Bay of Biscay (17-23) than in the Mediterranean (8-16), despite the same egg size. In both sea areas, ovigerous females are found throughout the year; there is no marked seasonal cycle in reproduction (J.-P. Casanova, 1977). Nonetheless, also in the Lophogastrida, the periods of propagative activity, egg sizes, and size at attainment of sexual maturity may vary between latitudes, localities, and seasons (Fage, 1941, 1942; Mauchline, 1973b; J.-P. Casanova, 1977; Wittmann, 1984).

Stygiomysida. – Data on **fecundity** are very scarce for the Stygiomysida due to their cryptic, subterranean mode of life (Villalobos, 1951; Pillai & Mariamma, 1964; Nath, 1973; Ortiz et al., 2005; Ariani & Wittmann, 2010). In a brackish well in Apulia, Italy, female *Spelaeomysis bottazzii* measuring 9-11 mm body length carry 8-14 eggs with an average diameter of 0.68 mm (Ariani & Wittmann, 2010). As expected for animals living in nutrient-poor, **subterranean environments**, at a given parental size the mean numbers of young per brood are smaller and egg diameters are larger than in epipelagic and coastal species of Mysidae and Lophogastridae. The *Spelaeomysis* values, however, do not exceed the 95% confidence limits calculated by Wittmann (1984) for these two families. Other bionomical differences from average epigean Mysidae are much stronger: in *Spelaeomysis bottazzii* the incubation period is more than six times as long (>100 days at 20°C) and a resting stage with reduced marsupial plates is adopted at the moult after release of the young (Ariani & Wittmann, 2010).

Incubation

Brood care. – The females show intensive care of their embryos or larvae in the brood pouch. They produce a respiratory current through the marsupium by pumping movements of the oostegites. This may be supported by **ventilation lobes** in a number of Mysida (*Siriella*, *Paramysis*, *Mysis*, etc.). In addition, coxae and endites of the thoracopods, and thoracic sternites, may bear brushes of long, plumose and/or microserrated, setae; these may play some role in ventilating and/or **cleaning the brood** (e.g., *Neomysis integer*; cf. Jancke, 1926; Kinne, 1955). At least a few microserrated setae project into the marsupium in all species of Mysida so far examined in this respect. In several species of *Neomysis*, long, strand-like diverticula with thin walls project from the posterior sternites into the marsupium. B. Casanova & De Jong (1996) reported a strange tool for optimizing the storage of larvae — both with respect to the oostegites and larval interspaces — in the lophogastrid *Gnathophausia zoea* and the mysid *Praunus flexuosus*: the integument of larvae and oostegites is covered by **anchoring structures** such as acute scales or spines; such structures are also found on the eyes of the postnauplioid stage in *Praunus flexuosus*.

Active **manipulation of larvae** with the thoracic endopods, observed by Wittmann (1981a) in females of *Leptomysis lingvura* and by Wortham-Neal & Price (2002) in *Americamysis bahia*, suggest a more intense interaction of females and young than assumed by previous authors. From this perspective, the structures projecting into the marsupium possibly also act as receptors providing information about the spatial distribution and state of the brood. In *Leptomysis lingvura*, the fully developed postnauplioid larvae ready for release are taken out of the frontal opening of the brood pouch, one by one, during the night, by the female with the anterior thoracic endopods; they are then turned around with the endopods and touched with the mouthparts several times (Wittmann, 1981a). Finally, the female jumps backwards with a sudden tail flip and simultaneously releases the larva. The slowly sinking, freshly released larva then starts immediately with moulting to become a free-living juvenile capable of swimming and self-maintenance.

Adoption behaviour. – Some species of Mysida 'supervise' their larvae. Laboratory experiments by Wittmann (1978b) demonstrated that females of three *Leptomysis* species grasp freely sinking premature larvae and introduce some of them into their own brood pouch. This behaviour, termed '**adoption**' (Wittmann, 1978b), is probably also found in nature and probably results from a **larvae replacement behaviour**, which serves to prevent the loss of own young, for example by saving own larvae fallen out from the marsupium. Mauchline & Webster (in Mauchline, 1980) found adoptions in seven species of mysids from British waters, Sato & Murano (1994) in four species from coastal waters of Japan. Wortham-Neal & Price (2002) observed adoptions in field samples of *Americamysis bahia* from the Gulf of Mexico. Johnston & Ritz (2005) showed that *Tenagomysis tasmaniae* adopts its own young in preference to those of a conspecific. In *Leptomysis lingvura*, adoption also works with young from other parents, to a very limited extent also with young from other species (Wittmann, 1978b). The females of *Leptomysis lingvura* and *Leptomysis buergii* adopt in specific patterns according to age and species of young. Embryos and parts of larvae (particularly eyes) are also accepted in a similar pattern, indicating that **chemosensory mechanisms** may be involved. Larvae that are not of an appropriate age to complete larval development in any particular marsupium are rarely adopted. Adoption experiments (A. P. Ariani, unpublished) were also performed in *Diamysis*, showing the same results. In contrast, one female of the Stygiomysida *Spelaeomysis bottazzii*, incubating in the laboratory (Ariani & Wittmann, 2010), resolutely rejected two larvae, one of which was actively motile, coming from a just deceased conspecific female.

In experiments by Wittmann (1978b), the Mediterranean mysid *Leptomysis lingvura* infrequently adopted larvae of its congener *Leptomysis buergii* and consistently rejected larvae from other genera. Mauchline & Webster (in Mauchline, 1980) observed that the north-east Atlantic mysids *Praunus flexuosus* and *Praunus neglectus* adopted each other's larvae and those of *Praunus inermis*, although the latter species did not accept larvae from the other two species. Sato & Murano (1994) observed **intraspecific adoptions** but no interspecific adoptions in a total of four species from three genera in Japan. Johnston & Ritz (2005) obtained the same results for three species from three different genera in Tasmania. The above results suggest that **interspecific adoptions** occur only between certain closely related, congeneric species.

Loss of young. – From field data on eight species of Mysidae (Mysida) from British coasts, Mauchline (1973b) estimated that about 10% (6-26%) of larvae become lost from the brood pouch in a premature condition. Laboratory experiments yielded 11% (5-18%) for *Leptomysis lingvura* from the Gulf of Naples (western Mediterranean; Wittmann, 1981b) and 12-23% for *Mysidopsis californica* from the coast of Matzatlán (Mexico; Ortega-Salas et al., 2008). Such losses were identified as being due to mortality and/or to premature release of living young: Amaratunga & Corey (1975) observed that female *Mysis stenolepis* "were capable of releasing young at any point subsequent to the first embryonic moult". Jepsen (1965) sampled **prematurely released larvae** of *Boreomysis arctica* in the plankton. Berrill (1969a) observed cannibalism by laboratory-kept females of *Mysis diluviana* on their own young. No cannibalism was observed during release of fully developed young in *Mysis stenolepis*, *Leptomysis lingvura*, *Leptomysis truncata sardica*, or *Americamysis bahia* (cf. Amaratunga & Corey, 1975; Wittmann, 1978b; Wortham-Neal & Price, 2002). In three Tasmanian species, laboratory-kept mothers did not feed on their own young before the young moulted after release, but they did feed after this moult (Johnston & Ritz, 2001). Similarly, *Leptomysis lingvura* may eat its own brood under conditions of starvation in the laboratory (Wittmann, 1984). In *Americamysis bahia*, the **survival rate of larvae** increased with increasing age and with decreasing temperature in *in vitro* experiments by Wortham-Neal & Price (2002). In natural populations of many species of Mysidae, a significant part of **marsupial mortality** appears to be due to parasites (see below under 'Parasites').

Duration of incubation. – In Mysida, the **duration of marsupial development** varies mainly with species, season, and temperature: 96 hours in *Mesopodopsis orientalis* from Madras (Nair, 1939); 4 days at 24°C in a laboratory population of *Antromysis juberthiei* taken from subterranean waters of Cuba (Juberthie-Jupeau, 1976); 5-6 days in *Acanthomysis sculpta* from Vancouver Island, in August (Green, 1970); 6-7 days in *Mysidium columbiae* from Jamaica, February to May (Davis, 1966); 12 days in *Praunus flexuosus* from Roscoff, August (Nouvel & Nouvel, 1939); 8-13 days in three Mediterranean species of *Diamysis* at 20°C in ambient water (Ariani & Wittmann, 2000); three months in *Boreomysis arctica* from Norway (Jepsen, 1965); and five months in *Mysis diluviana* from Canada (Berrill, 1969a). In a number of Mysidae, a strong decrease of brood duration with increasing temperature was observed in the laboratory (Murano, 1964a; Wittmann, 1981b; Toda et al., 1984; Johnston et al., 1997; Wortham-Neal & Price, 2002; Fockedey et al., 2006).

Salinity also affects the duration and **success of marsupial development** as shown by Vlasblom & Elgershuizen (1977) and Fockedey et al. (2006) for *Neomysis integer* from brackish waters at the coasts of the Netherlands and Belgium, by McKenney (1996) for *Americamysis bahia* from the Gulf of Mexico, and by Ariani & Wittmann (2000) for *Diamysis mesohalobia* from brackish springs and lagoons at the Adriatic and Ionian coasts of Apulia (eastern Mediterranean). No salinity effects on duration but strong effects on breeding success were reported by Greenwood et al. (1989) for *Mesopodopsis slabberi* from an estuary at Plymouth (north-east Atlantic). At the intraspecific level, the effects of salinity on duration are generally smaller compared to temperature. No significant trend of brood duration was found by Wittmann (1984) in literature on different species of Mysidae breeding at different salinities. According to McLusky & Heard (1971), *Praunus flexuosus* can regulate the marsupial fluid hyper- or hypo-osmotically. Nimmo et al. (1978) observed a delay of brood release under chronic exposure to cadmium in the estuarine mysid *Americamysis bahia*. According to Gentile et al. (1983), increasing mercury concentrations lengthen the brood development of the same species.

A laboratory study by Ariani & Wittmann (1997) on females of the subterranean Stygiomysida *Spelaeomysis bottazzii*, kept in August-November in the dark at 20°C in water taken at the sampling station (with salinity $S = 20$), indicated that embryonic development takes more than 16-22 days, and the total period of marsupial development more than 100-108 days. In contrast to any species of Mysidae so far examined in this respect, the young of any given female *Spelaeomysis bottazzii* developed strongly asynchronously within the same brood pouch.

Breeding cycles. – The **propagative activity** of the various species of the Mysida and Lophogastrida may be continuous or may vary between latitudes, localities, and seasons. Breeding females are found year round in field samples from many populations of Mysidae in temperate to tropical climates (Goodbody, 1965; Heubach, 1969; Brown & Talbot, 1972; Almeida Prado, 1973; Quintero & Zoppi de Roa, 1973; Juberthie-Jupeau, 1976; Wittmann, 1984; Ramarn et al., 2012; Biju et al., 2013). Laboratory experiments by Ariani & Wittmann (2000) on several species of *Diamysis* from temperate waters in Italy showed that the breeding females moult after releasing the mature young and generally lay a new clutch of eggs into the marsupium. Here, the eggs become fertilized upon mating and develop to free-living F1 young. **Continuous breeding** cycles characterize at least certain mysid species in temperate waters. As an alternative, the year-round presence of breeding females may result from the temporal overlap of seasonally alternating, asynchronous generations, as observed by Reynolds & de Graeve (1972) in *Mysis diluviana*, for example. This timing of reproduction was termed '**pseudocontinuous breeding**' by Wittmann (1984).

Most species from cold-temperate and to a lesser extent also from warm-temperate regions show a distinct **minimum of reproduction** or often do not breed at all during winter (e.g., Tattersall & Tattersall, 1951; Murano, 1964a; Heubach, 1969; Borodich & Havlena, 1973; Wittmann, 1978a; Mauchline, 1980; Murtaugh, 1989; Fenton, 1992; Mees et al., 1994). This timing was termed '**warm-season breeding**' by Wittmann (1984). The first brood is released during the spring maximum of biological production, and may in turn reproduce already during summer or during the weaker, qualitatively different (Odum et al., 2004) autumn maximum of production. Generally, there are only two to four generations per year (Mauchline, 1980). However, Borza (2014) inferred as many as 4-6 generations per year from field data on non-indigenous freshwater populations of three highly productive Ponto-Caspian species of Mysidae, which had invaded the Hungarian reach of the Danube River. The females are iteroparous, normally produce a continuous series of brood clutches until autumn, and then disappear from field samples. A **reduction of the oostegites** at the end of the reproductive period is rare in the family Mysidae (see above, 'Oostegites'). Several species show a summer minimum of reproduction between maxima in spring and autumn (Murano, 1964a; Zatkutskiy, 1970), but there are also species with a distinct summer maximum (Hopkins, 1965; Richards & Riley, 1967).

A different situation was observed in the Arctic (Lasenby & Langford, 1972) and assumed also for Antarctic regions (Wittmann, 1996). The animals accumulate storage materials during summer and gain **sexual maturity** in early autumn. In autumn or winter the eggs are deposited in the brood pouch and copulation takes place. Shortly afterwards the adult males may die. The embryos and larvae continuously develop in the brood pouch; after about eight months they are released in spring or early summer, when they can profit from the plankton bloom for fast individual growth. Such animals undergo '**cold-season breeding**' as defined by Wittmann (1984). Under strong Arctic conditions the young need two summers to attain sexual maturity (Larkin, 1948; Geiger, 1969; Fürst, 1972; Lasenby & Langford, 1972; Mauchline, 1980). This process may last even up to four years under extreme oligotrophic regimes (Morgan, 1980).

A normal **interruption of the breeding activity**, not related to seasonal variations, was described by Ariani & Wittmann (2010) for the Stygiomysidae species *Spelaeomysis bottazzii*. Here, the females release their young after a long incubation period passed in deep and trophically poor groundwater, where they live mainly from fat reserves. Each breeding cycle, therefore, requires accumulation of reserves in the shallow, more nutrient-rich groundwater. This may explain why the females adopt a subadult habit at the moult subsequent to the release of young, probably in order to prevent an unsuccessful new breeding cycle under poor trophic conditions. These animals could be considered '**trophically suitable breeders**'.

Adjustment of reproductive parameters

Unlike the Lophogastrida *Eucopia unguiculata*, the Stygiomysida *Spelaeomysis bottazzii*, and the Mysida *Praunus flexuosus* (cf. Nouvel & Nouvel, 1939; J.-P. Casanova, 1977; Ariani & Wittmann, 2010), almost all **iteroparous** Mysida species so far examined in this respect do not show a resting stage after the release of young. Rather, they moult shortly afterwards and deposit a new egg clutch in the marsupium (e.g., Kinne, 1955; Juberthie-Jupeau, 1976; Pezzack & Corey, 1979; Wittmann, 1981b; Ariani & Wittmann, 2000; Johnson et al., 2001; Wortham-Neal & Price, 2002). Nonetheless, females with empty brood pouches are often found in the field; this may be correlated with high population densities (Clutter, 1969). The first brood is aborted by *Siriella armata* in the laboratory, also resulting in females with empty brood pouches (Cuzin-Roudy et al., 1981). The normally non-interrupted **sequence of broods** indicates an effective coupling between female moult cycle, yolk accumulation in the ovaries, and development of the young (Cuzin-Roudy & Tchernigovtzeff, 1985; Okumura, 2003). The eggs of the first brood are produced during 2-3 instars in *Leptomysis lingvura* (cf. Wittmann, 1981b). These combined instars last a total of 1-2 incubation periods (Clutter & Theilacker, 1971; Wittmann, 1978a; Gaudy & Guerin, 1979), suggesting that coupling also plays some role in the first brood clutch of a given female. In aquatic poikilotherms, the duration of development of embryos and of non-feeding larvae is generally highly temperature-dependent. The reproductive biology of mysids, however, is not completely correlated to the temperature regime: for example, the **temperature acceleration of marsupial development** is usually smaller at the intraspecific compared to the interspecific level (but not smaller in *Americamysis bahia*; Wortham-Neal & Price, 2002). Wittmann (1984) showed that the **latitudinal temperature effects** are buffered by coordinated adaptive adjustments of parental weight, brood weight, egg weight, numbers of young, incubation period, and moult instars. These **homeostatic adaptations**, i.e., outcomes of evolutionary trends, make the following reproductive parameters less variable at the interspecific level between latitudes, at least for non-interrupted sequences of broods in iteroparous mysids: (1) the average numbers of **young per female incubatory day**; (2) the portion of **body weight invested into each brood clutch**; (3) the average amount of **yolk invested per egg per day**. Note also that the three parameters are mutually dependent.

DEVELOPMENT AND MOULTING

Marsupial development

Numbers and distinction of marsupial stages

Based on the external appearance of "embryos" in the brood pouch, Kinne (1955) distinguished six **developmental stages** termed by him '*a*' to '*f*' in the mysid *Neomysis integer*. Mauchline (1973b) condensed Kinne's terminology to 'eggs' (*a*), 'eyeless larvae' (*b-e*), and 'eyed larvae' (*f*), whereby the definition of eyeless larvae is inconsistent with hatching from the egg membrane between Kinne's stages *b* and *c*. Based on the events of **egg hatching** and two **larval moults**, Jepsen (1965) distinguished first, second, and third "embryonic" stages in *Boreomysis arctica*. Based on the same events, the biologically and morphologically consistent terminology of Wittmann (1981a) is used here (fig. 54.37), who distinguished an **embryonic stage** (fig. 54.37A-D) between fertilization and hatching from the egg membrane, **nauplioid larvae** (fig. 54.37E-G; named after 2-3 pairs of free naupliar appendages) up to the first moult, and **postnauplioid larvae** (fig. 54.37H) up to the second moult, which leads to the free-living **juvenile stage**. This terminology was adopted by Cuzin-Roudy & Tchernigovtzeff (1985), Johnston et al. (1997), Johnston & Ritz (2001, 2005), Wortham-Neal & Price (2002) and others. Within stages, Wittmann (1981a) distinguished arbitrary phases to indicate important morphological changes that are externally visible and can be recognized easily in formalin- or ethanol-fixed material. Such subdivisions could be practical for physiological and ecological research and aquaculture purposes.

The same number and morphology of main marsupial stages (figs. 54.37, 54.38), i.e., embryonic stage, nauplioid, and postnauplioid larvae, as in the Mysida, are also found in the Stygiomysida (Ariani & Wittmann, 2010) and Lophogastrida (Illig, 1930; Fage, 1941; Mauchline, 1980). It is noteworthy that a pale cornea (fig. 54.38J) is present in the postnauplioid larvae of the anophthalmic, subterranean Stygiomysida species *Spelaeomysis bottazzii*, whereas the cornea is missing in free-living juveniles (fig. 54.38K) and adults (fig. 54.1C) (Ariani & Wittmann, 2010). **Eye reduction** is even stronger in the subterranean *Spelaeomysis longipes*, where the postnauplioids show separate eyestalks without ommatidia and visual pigment. The eyestalks merge successively during the subsequent free-living juvenile stages. In the adult, the eyestalks are fused to a single, transversal **eye plate**, which has rearranged optic ganglia inside (Nath et al., 1972a).

Morphology of marsupial stages

Embryonic stage (fig. 54.37A-D). – **Embryonic development** was studied mainly in species of the family Mysidae (Mysida), namely by Van Beneden (1861), Nusbaum (1887), and Bergh (1893) in *Praunus flexuosus*, by Nair (1939) in *Mesopodopsis orientalis*, by Needham (1937) and Jepsen (1965) in *Boreomysis arctica*, by Scholtz (1984) and Scholtz & Dohle (1996) in *Neomysis integer*, by Manton (1928b) in *Hemimysis lamornae*, and by Wittmann (1981a) in *Leptomysis lingvura*. Data on the embryonic and larval development of mysids are also available in the review given by Johnson et al. (2001) for the marsupial development of the Peracarida.

Nair (1939) observed the maturation of the oocytes, the fecundation, and the first processes of **segmentation** in the mysid *Mesopodopsis orientalis*. The detachment of the first **polar bodies** occurs when the eggs are deposited into the brood pouch. The second polar body is formed after the **penetration** of the spermatozoa (with **polyspermy**). The **amphimixis** occurs in the centre of the egg, as typically in **centrolecithal eggs**. Here, the first **nucleic divisions** take place and the **blastomeres** differentiate and form a vesicle. The latter then spread to a superficial portion of the ectoderm to constitute a **blastoderm**

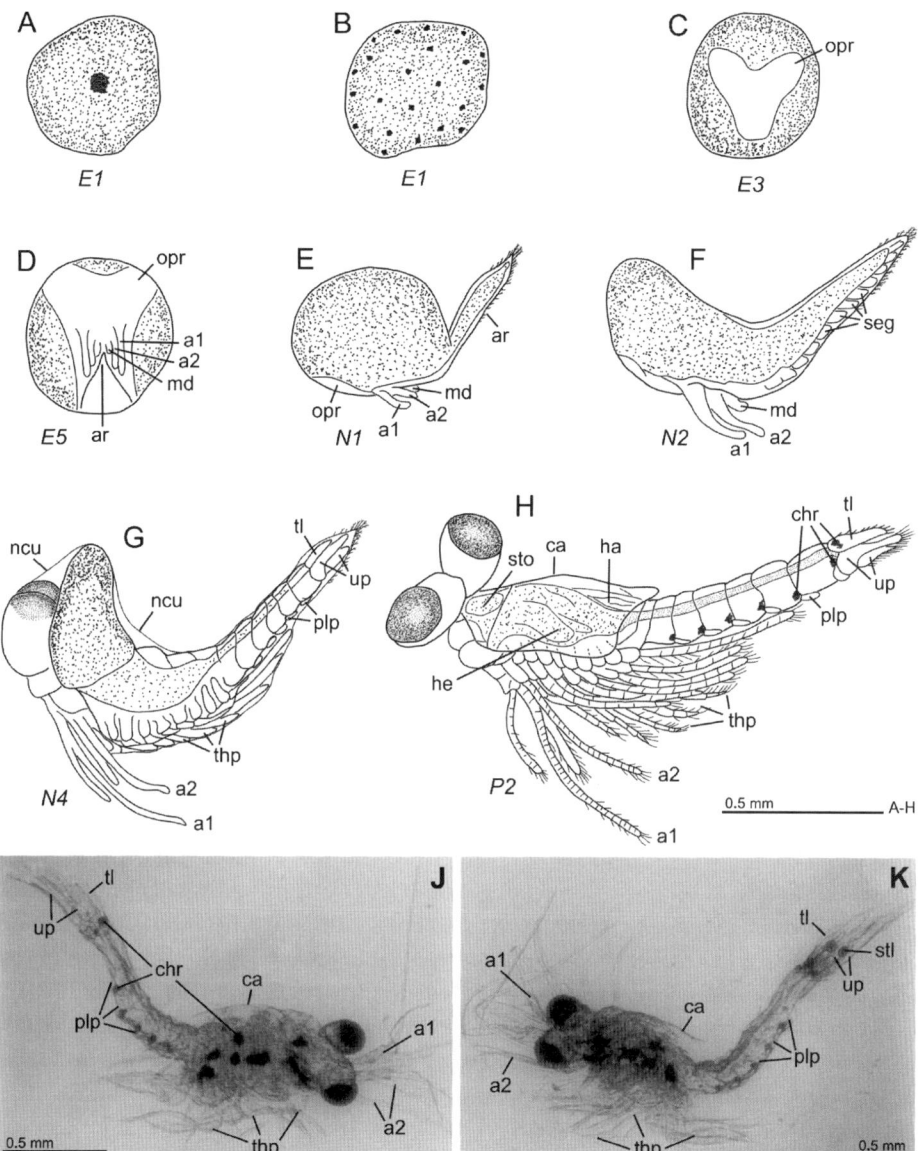

Fig. 54.37. Main stages in the marsupium (A-H) and freshly hatched juveniles (J, K) of the mysid *Leptomysis lingvura* (G. O. Sars, 1868). A-D, phases *E1*, *E3*, and *E5* of embryonic stage; E-G, phases *N1*, *N2*, and *N4* of nauplioid larval stage; H, phase *P2* of postnauplioid larval stage; J, hatchling fixed immediately after release from brood pouch and subsequent second larval ecdysis, note absence of statoliths; K, juvenile fixed 10 h after hatching and subsequent ecdysis, note pair of well-formed, large statoliths. Abbreviation: a1, antennula; a2 antenna; ar, abdominal rudiment; ca, carapace; chr, chromatophores; ha, heart; he, hepatopancreas; md, mandible; ncu, naupliar cuticle; opr, optic rudiment; plp, pleopods; seg, segmentation (metameres); stl, statoliths; sto, stomach; thp, thoracopods; tl, telson; up, uropods. [After Wittmann, 1981a; A-H, modified; J, K, permission granted by Elsevier B.V., Amsterdam.]

disc. The segmentation is described as typically discoidal in *Praunus*. A transversal row of ectodermal **teloblasts** then differentiates, behind of which the endodermal cells meet in enclaves of the **vitellus**, which they absorb. Certain cells, mesodermal teloblasts, migrate downwards from the blastoporic area in order to form a mesoderm, while a genital rudiment is formed at the surface and subsequently sinks down below the disc. The anterior lip of the **blastoporic area** will provide the trunk. Initially, the anterior portion of the embryo consists of two halves (the optic rudiments), which are then united in the sagittal plane. The endodermal cells penetrate downwards, incorporate the vitellus and form the middle intestine, which ultimately becomes sutured with the proctodaeum and the stomodaeum. The hepatic tubes appear to be of mesodermal origin (Nair, 1939). At this time, the **naupliar appendages**, i.e., the antennules, antennae, and mandibles, differentiate. Behind a caudal furrow that transverses the blastoporic area, an anteriorly turned, conical eminence, the **caudal papilla**, is formed. Dohle et al. (2004) doubt for lophogastrids and mysids that this is a "true" caudal papilla. For gene expression and related segmental differentiation in the **germ band**, see the detailed review by these authors (Dohle et al., 2004).

Wittmann (1981a) distinguished six phases in the **embryonic stage**: cleavages and formation of the blastoderm during phase $E1$, formation of the germinal disc ($E2$) and of the optic rudiments ($E3$), separation of the abdominal rudiment ($E4$), formation of naupliar appendages ($E5$), and differentiation and rotation of the abdominal rudiment until hatching ($E6$). The embryonic phases $E1$, $E3$, and $E5$ are illustrated in fig. 54.37A-D.

Nauplioid larva (figs. 54.37E-G, 54.38F-H). – Also **larval development** was studied mainly in species of the family Mysidae (Mysida), in particular by Nair (1939) in *Mesopodopsis orientalis*, by Murano (1964b) in *Neomysis intermedia*, by Jepsen (1965) in *Boreomysis arctica*, by Wittmann (1981a) in several species of mysids, especially in *Leptomysis lingvura*, by Cuzin-Roudy et al. (1981) and Cuzin-Roudy & Tchernigovtzeff (1985) in *Siriella armata*, by Wortham-Neal & Price (2002) in *Americamysis bahia*, and by Fockedey et al. (2006) in *Neomysis integer*; for certain aspects also in species of the Stygiomysida genus *Spelaeomysis* (Nath et al., 1972a; Ariani & Wittmann, 1997, 2010).

A pair of transient glandular, lateral **dorsal organs**, independent from the **germ band**, appear to be related to the hatching from the egg membrane, which releases the nauplioid larva. Below the now uncovered integument, the larva completes its differentiation. The cuticle does not follow the subsequent events of body segmentation and formation of appendages. All somites and their appendages are organized. A pre-antennary somite and a seventh pleomere appear, the latter subsequently becoming fused with the sixth pleomere (Manton, 1928b; Nair, 1939). Wittmann (1981a) distinguished four phases of the **nauplioid stage**: $N1$ is characterized by ramification of the antennula and externally visible signs of body segmentation; $N2$ by a well-segmented thorax and pleon; $N3$ by well-formed thoracopods and uropods, and (if any) also by the development of eye pigment; $N4$ by well-formed maxillae, heart, and pleopods, and by the first heart beats. The nauplioid phases $N1$, $N2$, and $N4$ are illustrated in fig. 54.37E-G.

A comparative study of the nauplioid stage (Wittmann, 1981a) in representatives of the Mysidae genera *Leptomysis*, *Hemimysis*, *Paramysis*, and *Siriella*, showed peculiar

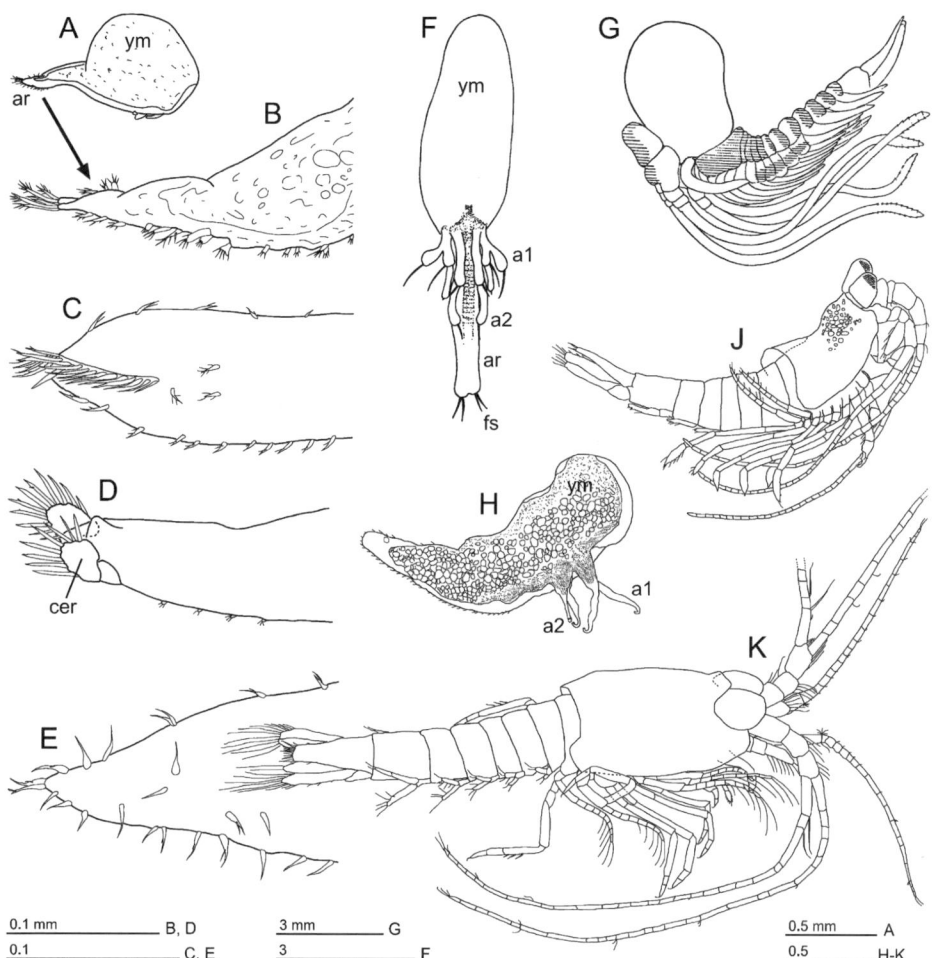

Fig. 54.38. Morphology of larvae and juveniles of the Mysida (A-E), Lophogastrida (F, G), and Stygiomysida (H-K). A, early nauplioid larva of *Leptomysis lingvura* (G. O. Sars, 1868), lateral; B-E, armature of nauplioid abdomen in *Leptomysis lingvura* (B as detail of A), *Siriella gracilipes* H. Nouvel, 1942 (C), *Hemimysis speluncola* Ledoyer, 1963 (D), and *Paramysis helleri* (G. O. Sars, 1877) (E), lateral; F, early nauplioid *Gnathophausia zoea* Willemoës-Suhm, 1873, ventral; G, late nauplioid *Eucopia sculpticauda* Faxon, 1893, lateral; H-K, three successive stages separated by ecdysis in *Spelaeomysis bottazzii* Caroli, 1924: early-mid nauplioid larva (H, lateral), postnauplioid larva (J, obliquely lateral, anterior thoracic exopods omitted), and freshly released juvenile (K, dorsolateral). Abbreviations: a1, antennula; a2, antenna; ar, abdominal rudiment; cer, cercopod; fs, furcal spines; ym, yolk mass. [A-E, after Wittmann, 1981b; F, after Fage, 1941; G, after Illig, 1930; H-K, after Ariani & Wittmann, 2010; A-K, modified.]

characteristics of the armature at the tip of the **nauplioid abdomen** (abdominal rudiment), such as the presence of spines, setae, or **cercopods**. Most species bear setae (fig. 54.38B) or spines (fig. 54.38C, E). A kind of abdominal tip similar to that of *Paramysis helleri* (fig. 54.38E) was observed by Ariani & Wittmann (2000) in *Diamysis mesohalobia*.

Among extant Mysida, cercopods (fig. 54.38D) were so far found in the nauplioid larvae of only five genera, all belonging to the subfamily Mysinae (Manton, 1928b; Green, 1970; Modlin, 1979; Wittmann, 1981a). They possibly represent plesiomorphic characters within the Eumalacostraca, related to the well-formed **caudal furca** (i.e., processes on the telson), which is found throughout (Schram, 1986) in fossil Pygocephalomorpha from the Carboniferous to Permian. The nauplioid abdomen of the lophogastrid *Gnathophausia zoea* bears two pairs of furcal spines (fig. 54.38F) at the apex (Fage, 1941), which suggest a possible homology to cercopods. Such a homology appears less likely for the nauplioids of *Boreomysis arctica* (Boreomysinae, Mysida), which have a pair of "**polyspinal appendices**" (Jepsen, 1965) in a clearly more anterior position on the abdominal rudiment.

The tropical, cavernicolous mysid *Palaumysis philippinensis* (and probably also its congeners) are exceptional in that the nauplioids have very long first and second antennae. Also unusual is the large relative size of eggs and larvae, with the postnauplioids overreaching as much as half the length of the throughout very small adults (<2 mm; Hanamura & Kase, 2002).

Postnauplioid larva (figs. 54.37H, 54.38J). – Upon the first larval **ecdysis** the postnauplioid larva emerges inside the marsupium. This stage is dorsally bent and shows all adult appendages, although these are still incompletely differentiated. Wittmann (1981a) distinguished three phases of the **postnauplioid stage**: The eyes rotate from outer-ventral into anterior position and the carapace is freed during phase *P1*; the stomach becomes well formed and body pigment (if any) becomes increasingly visible during *P2* (fig. 54.37H); the lobes of the hepatopancreas are distinguished (freed) during yolk consumption, and the larvae start to respond to external **stimuli** during *P3*. The **heart beats** are now continuous. Small movements of the gut and appendages, and fluttering of the pleon become visible, interrupted by quiet periods. Nonetheless, artificially released postnauplioids are not capable of swimming unless they moult to the free-living juvenile stage.

Hatching and larval moults

The embryo hatches by stretching the abdominal rudiment, which causes the caudal papilla to point backwards from this time onwards. The stretching stresses and finally ruptures the egg membrane, liberating the nauplioid stage. Later, the **first larval ecdysis** from the nauplioid to the postnauplioid stage is performed, accompanied by body flexing. The old cuticle ruptures above the dorsal yolk mass and it takes considerable time to tear apart its remains. The egg membrane and cuticle of the nauplioid are always shed within the brood pouch, whereas the cuticle of the postnauplioid larva may be shed either within (Davis, 1966), or outside (Wittmann, 1981a; Cuzin-Roudy & Saleuddin, 1985; Johnston & Ritz, 2001) the marsupium, depending on the species. The **second larval ecdysis** is also accompanied by body flexing. The animal emerges dorsally, and the old cuticle falls off in one piece (Wittmann, 1981a). It is always the second **larval ecdysis** that leads to the juvenile stage capable of swimming. The larvae may moult shortly before (Davis, 1966), but normally shortly after release from the brood pouch. In this latter case the freshly released larvae moult while slowly sinking in the water column. In *Leptomysis* *lingvura*, the freshly moulted juveniles need an additional 20-40 min at 16°C to complete their statocysts by precipitation of fluorite during **statolith formation** (Wittmann, 1981a; cf. above, 'Static organs'). Photos of freshly hatched juveniles before and after statolith formation are presented in figs. 54.37J and 54.37K, respectively. The resulting stage is free-living, self-sustaining, and closely resembles the adult, but lacks secondary sexual characteristics.

TIMING OF MARSUPIAL STAGES

Synchronization of marsupial development. – For more than a century we know that the marsupial young of a given female are usually all at about the same stage of development in Mysida and Lophogastrida (for exceptions see above, 'Adoption behaviour'). Johnston et al. (1997) and Johnston & Ritz (2001) observed **synchronous development** and release of broods in three species of Mysidae forming monospecific swarms in coastal waters of Tasmania. Johnston & Ritz (2001) found that development time did not vary between *in vivo* and *in vitro* incubation and concluded that the parent does not influence the synchrony of development. Those authors, however, did not discuss the findings of Cuzin-Roudy & Saleuddin (1985), who observed that the final (= second) **larval moult** in laboratory-kept *Siriella armata* is inhibited when the mother is unable to finish incubation, even when the larvae are artificially released from the brood pouch. According to Cuzin-Roudy & Tchernigovtzeff (1985), unlike the normal release during the night, the artificial release of young during daytime was never followed by the second larval moult in *Siriella armata*. Wittmann (1978b, 1981b) observed in the laboratory that female *Leptomysis lingvura* delayed moulting by 2-5 days (17-42% of moult instars under same culture conditions) following artificial implantation of **premature larvae** into the marsupium. Summarizing the so far published data suggests species-specific differences in the ability to contribute to the synchronization of development by means of limited **moult inhibition**, when one of the counterparts — parent or young — is not ready to finish incubation. The young may contribute by communicating tactile, chemical, or some other kind of signals to the parent. This has never been examined in detail. A hint to chemical or tactile communication could be the adoption of detached eyes or less frequently of other parts of larvae by females of *Leptomysis lingvura* in the laboratory (Wittmann, 1978b). The brood pouch is well equipped with setae and the females certainly perceive the small movements (Davis, 1966; Berrill, 1969a) of the late postnauplioid larvae. These movements merit future examination as potential signals of the approaching **completion of larval development**.

Fenton (1994) and Johnston & Ritz (2001) examined three swarming mysid species and reported that one, namely *Paramesopodopsis rufa*, showed some differentiation according to the marsupial stages among females within swarms. This resulted in partially **synchronous release**, which may assist the initially helpless larvae to escape from cannibalism by older conspecifics in the swarm and from being taken by predators. For many species, a certain degree of 'synchronization' could be expected due to the separation of size- and age-stages in different swarms or in subgroups within swarms. This separation is related to combinations of **social interaction** and environmental stimuli (Clutter, 1969; Mauchline, 1971; Zelickman, 1974; Wittmann, 1977, 1978a; O'Brien, 1988).

Relative duration of marsupial stages. – A close coupling between **moult cycle** and **ovarian cycle** of the parent, and the **development of marsupial young**, was reported by Cuzin-Roudy & Tchernigovtzeff (1985) in laboratory-cultured *Siriella armata*, and by Okumura (2003) in breeding females of *Acanthomysis robusta* collected in the field at sea coasts of Japan. Nonetheless, the relative durations of the marsupial stages in the laboratory show surprisingly strong differences between the various species of Mysidae, and minor, yet significant, temperature-related differences (Wittmann, 1981a; Cuzin-Roudy & Tchernigovtzeff, 1985; Johnston et al., 1997; Wortham-Neal & Price, 2002; Fockedey et al., 2006). The **embryonic stage** occupies mostly an intermediate portion (16-43%) of the incubation period, the **nauplioid stage** is mostly the longest (22-71%), and the **postnauplioid stage** is mostly the shortest (8-36%). Depending on the species, the relative duration of any stage may be shortened or lengthened by a given temperature increase. Comparable results are obtained for the frequencies of the three marsupial stages in field samples: Wittmann (1981b) found means of 29-41, 36-51, and 13-32%, respectively, for eight species from the Adriatic and Tyrrhenian seas (Mediterranean). Very similar results are also obtained for eight species from British coasts (north-eastern Atlantic) by recalculating the data of Mauchline (1973b), who used a different schedule for the distinction of stages.

Moulting and growth

Moulting process. – The process of moulting (fig. 54.39) shows the same pattern in all examined subfamilies of the Mysidae (Mysida; Nouvel, 1958). The cuticle ruptures along a marginal line close to the lateral and frontal margins of the carapace, passing generally below the frontal plate at a certain distance from the tip. A rupture line proceeds also longitudinally and ventrally along each eyestalk. The lateral walls of the thorax show

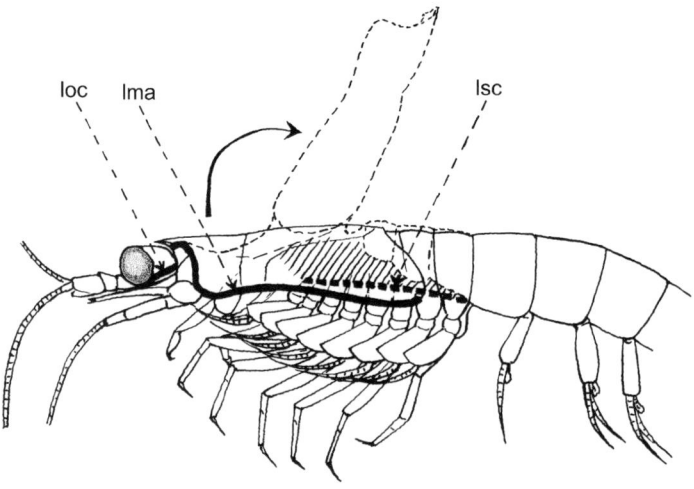

Fig. 54.39. Rupture lines of cuticle upon moulting in Mysida. Abbreviations: lma, marginal line; loc, ocular line; lsc, line of stretched cuticle. [Modified after Nouvel et al., 1999.]

two longitudinal lines where the cuticle becomes considerably stretched, forming a pair of lateral bellows that can more or less lift off from the sclerites, with the borders of the latter resembling a rupture line.

In **antennal aesthetascs** of the mysid *Neomysis integer*, the morphological basis for the sensory function remains intact until ecdysis because the **dendritic connections** with the shaft of the old setae are maintained (Guse, 1980). Such connections are maintained in most arthropod sensilla so far examined, but become lost upon moulting in the antennal aesthetascs of the isopod *Idotea balthica* and in various other sensilla of some spiders and insects (Guse, 1983a).

When moulting approaches, the animals cease feeding, lose transparency, and bend and stretch several times while sinking down to the bottom. Suddenly they liberate themselves from the exuvia through the opening formed by the marginal line of the carapace; here, the outer face of the carapace lifts like a flap and turns backwards around its posterior border, which acts like a hinge. Within a few minutes the animal regains swimming capability.

The morphology of females after the moult at the end of incubation is generally similar to that immediately before moulting. Nonetheless, a **reduction of oostegites** after the reproductive period was reported by Nouvel & Nouvel (1939) for the Mysida species *Praunus flexuosus*. A moulting process leading back to a habit with strongly reduced brood lamellae was observed by Ariani & Wittmann (2010) in the subterranean Stygiomysida *Spelaeomysis bottazzii*. This is probably related to a **breeding strategy** to avoid continuous breeding cycles under critical food conditions (see above, 'Breeding cycles').

Moulting cycle. – For the hormonal control of moulting see above, under 'Endocrine organs'. The duration of the **moulting cycle** generally increases with body size and decreases with increasing temperature. In breeding females, the moult instars are prolonged until the young are released from the brood pouch. The correlations between moulting and the **lunar cycle** or the **tidal amplitude** are poorly explained (Depdolla, 1916; Nair, 1939; Nouvel & Nouvel, 1939; Nouvel, 1940, 1945). According to laboratory observations by Ariani et al. (1982), moulting at the end of incubation in female *Diamysis mesohalobia* generally occurs at dawn.

Experiments by Cuzin-Roudy & Tchernigovtzeff (1985) revealed a **moult cycle** of 12 days at 20°C in reproductive females of the mysid *Siriella armata*. At the end of each cycle, the release of young, moulting, copulation, and egg deposition took place within eight hours. The cycle of the integument appears to be synchronized with the ovarian cycle and with marsupial development. Cuzin-Roudy & Tchernigovtzeff (1985) found that the embryonic phase corresponds to the postmoult period (**postecdysis**) of the parent, the nauplioid phase to the "intermoult period" (**diecdysis**), and the postnauplioid phase to the premoult period (**proecdysis**). A similar timing was observed by Okumura (2003) for *Acanthomysis robusta*. Near the end of the embryonic phase, a row of oocytes becomes visible in each ovary at female moult stages A-B. Yolk accumulation starts at moult stage C during the early nauplioid phase, and continues up to moult stage D2 during the late nauplioid phase.

In juvenile and subadult *Mysis mixta* and *Neomysis integer* taken from the Baltic Sea for laboratory experiments (Gorokhova, 2002), both a decrease of temperature and reduced food availability prolonged the complete moult cycle and the different stages within this cycle. Reduction of food prolonged the relative duration of **diecdysis** and early **proecdysis** stages, at the expense of the stages close to ecdysis. The temperature showed no clear effect on relative stage chronology as long as food supply was high. As expected, the chronology of moulting is particularly important for **epibionts** (Hanamura et al., 2010): The **infestation prevalence** and loads of the tropical, estuarine mysid *Mesopodopsis tenuipes* by the vorticellid *Zoothamnium duplicatum* are lower in females carrying earlier embryos or larvae in the marsupium, *ergo* having a fresher cuticle than those with older young.

In the stygiomysid *Spelaeomysis longipes* from a freshwater well in India, calcium is stored in a pair of rod-like **calcareous bodies** in **midgut caeca** extending forwards from the junction of midgut and hindgut. These bodies appear during the premoult period and attain their maximum length shortly before ecdysis (Nath, 1972).

Growth between moults. – Size increments of the early, free-living juveniles between successive moults are typically in the order of 10-30% in Mysida and Lophogastrida (Murano, 1964a; Clutter & Theilacker, 1971; Amaratunga & Corey, 1975; Childress & Price, 1978). With increasing body size, the **growth factors** of most species decay in a logarithmic manner (Mauchline, 1980), or there may be an initial increase up to a modal value followed by a distinct decay (Matsudaira et al., 1952; Gaudy & Guerin, 1979). Growth slows at low temperatures, but size at attainment of maturity as well as longevity generally increase with decreasing temperature, so that adult sizes and egg sizes generally increase with increasing latitudes. The sizes are often greater in winter than in spring or summer in regions with a distinctly seasonal climate (Mauchline, 1980; Wittmann, 1984; Mees et al., 1994).

'Inter-moult' growth. – Mauchline (1973a) measured 13 species of Mysidae from the north-eastern Atlantic and reported that females with postnauplioids (quoted as "eyed" larvae) have a longer pleon at given carapace length compared to females carrying eggs. He attributed this difference to 'inter-moult' growth by stretching of the inter-segmental joints [note, that Mauchline's "inter-moult" period lasts longer than the diecdysis]. Using regression analysis, he estimated an average increase of body length by about 7%. Fenton (1994) found 2-6% in field populations of three species in Tasmania. Wittmann (1978a) estimated from laboratory-reared specimens of *Leptomysis lingvura* and *Leptomysis buergii* that the 'inter-moult' swelling amounts to about half the total length increase between successive moults, without significant differences according to sex and body size.

Regeneration

As in several groups of animals, particularly invertebrates, crustaceans show some capacity to re-form parts of their body that have been lost or damaged by **injury**, **autotomy**, or otherwise (see review by Charmantier-Daures & Vernet, 2004, in the first volume of the present series). Regeneration following autotomy at a preformed **fracture plane** is common in certain Decapoda, and less frequent in certain Isopoda and

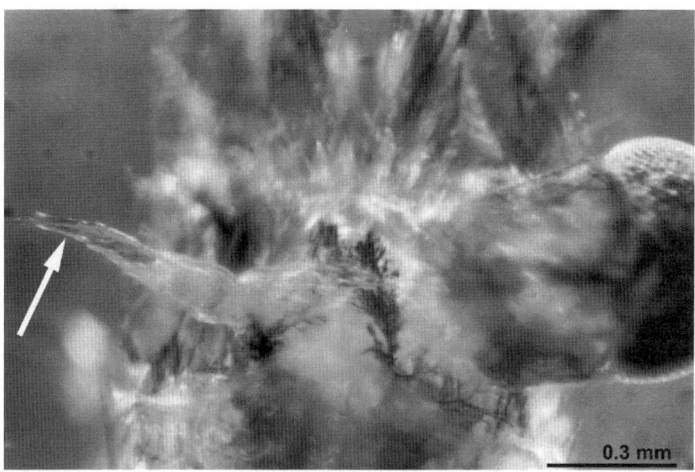

Fig. 54.40. Eye heteromorphosis in the mysid *Diamysis mesohalobia* Ariani & Wittmann, 2000. Basal part of an antenna-like flagellum (arrow) regenerated after excision of left eye in a laboratory-kept female. [A. P. Ariani, unpublished.]

Amphipoda. So far there are no reports of autotomy for Lophogastrida, Stygiomysida, or Mysida.

Băcescu (1940) described the comparatively frequent anomaly "*rotundicauda*" of the telson of *Limnomysis benedeni* in brackish to freshwater coastal lakes of the western Black Sea. **Regeneration** was identified as one of several potential reasons for this. Based on laboratory experiments with *Neomysis integer*, Mees et al. (1995) concluded that irregularly shaped or nearly symmetrically rounded, abnormal telsons are due to regeneration. In this species, abnormal telsons, along with **intersexuality** (see above, 'Intersexuality'), are comparatively frequent and widespread in estuaries from northern to south-western Europe. By comparison, both phenomena seem to be rare or absent, respectively, in the partly co-occurring mysids *Schistomysis kervillei* and *Gastrosaccus spinifer*.

In a classical finding, Herbst (1896) obtained regeneration of an antenna after extirpating an eyestalk together with the optical ganglion in a marine decapod. An analogous experiment was conducted on the mysid *Diamysis mesohalobia* from a brackish spring in southern Italy. Also in this case there was evidence of heteromorphic regeneration, which gave rise to the basal part of an antennal flagellum, partly segmented and equipped with small setae (fig. 54.40; A. P. Ariani, unpublished). Replacement of eyes or appendages by appendages of another somite is known in various crustaceans (Charmantier-Daures & Vernet, 2004), particularly decapods (Nevin & Malecha, 1991). Morphogenetic signals for the replacement of antennal structures by pediform structures are well known in insects (Angelini & Kaufmann, 2005); data on eyestalks *versus* legs are still lacking. From a phylogenetic point of view, the heteromorphic replacement of eyestalks seems to support the as yet unsettled hypothesis (cited in Gruner & Scholtz, 2004) that the malacostracan eyestalks originate from an antenna-like appendage. This, in turn, is connected with the also unresolved question (Dohle et al., 2004; Gruner & Scholtz, 2004) of a putative **preantennary somite**, often also quoted as "**ocular somite**".

Life cycle

Attainment of sexual maturity. – The 'giant' deep-sea lophogastrid *Gnathophausia ingens* needs 10-11 post-larval instars to attain maturity (Childress & Price, 1978). In general, the Mysida need 10-13 (range 8-21) instars between hatching of the free-living juvenile and **attainment of sexual maturity** (Ishikawa & Oshima, 1951; Matsudaira et al., 1952; Gaudy & Guerin, 1979; Berill & Lasenby, 1983; Toda et al., 1984). When comparing different epipelagic and coastal species, the size at attainment of sexual maturity is positively correlated with egg size and shows additional variations between latitudes, localities, and seasons (Blegvad, 1922; Nouvel & Nouvel, 1939; Labat, 1957; Mauchline, 1973b; Wittmann, 1984). Low temperatures and **starvation** may slow down or suppress the advancement towards **sexual maturity** (Matsudaira et al., 1952; Gaudy & Guerin, 1979; Morgan, 1980; Toda et al., 1984). In the estuarine mysid *Americamysis bahia* from the Gulf of Mexico (McKenney, 1996), **time to maturation** and to **release of first brood** decrease, and percentage of reproductively active females increases with increasing temperature (range 19-31°C) and salinity ($S = 3$-31), whereby both environmental factors interact.

Life span. – In temperate climates, the **life span** of coastal mysids generally varies in the range of 2-18 months, rarely up to two years (San Vicente & Sorbe, 1993, 1995). *Neomysis intermedia* from warm-temperate (6-30°C) freshwater lakes of Japan shows life spans of only 1.5, 2, 2, or 5 months in summer, spring, autumn, or winter, respectively (Murano, 1964a, 1999b). It is 5, 7, or 9 months, respectively, for three **overlapping generations** observed per year in the mysid *Schistomysis spiritus* from waters off Arcachon (north-eastern Atlantic; San Vicente & Sorbe, 1995, 2003). When there is more than one generation per year in waters with strong seasonality, the summer generations may live only 2-3 months (Matsudaira, 1952), while the **overwintering generation** (released in late summer to autumn) may live 2-3 times longer. In Subarctic to Arctic waters the normal life span is 1-4 years, depending on **trophic conditions** and average size of the species (Geiger, 1969; Hakala, 1978). Life spans of that order are also found in populations artificially introduced in subalpine lakes (Morgan, 1980, 1981). Ward (1984) estimated generation times of 2 or 4 years for two Subantarctic populations of *Antarctomysis maxima*, respectively. For Antarctic populations, Siegel & Mühlenhardt-Siegel (1988) gave minimum estimates of 3-4 years for the life span of the intermediate-sized *Mysidetes posthon*, versus 5-6 years for the much larger *Antarctomysis ohlinii* and *Antarctomysis maxima*.

Calil & Borzone (2008) suggest a direct relation between egg volume and generation longevity in the mysid *Metamysidopsis neritica* from sandy beaches in southern Brazil. This continuously breeding species shows three main generations, namely summer, fall, and winter generations, with the largest eggs produced in winter. In agreement with this, the **overwintering generation** of *Mesopodopsis slabberi* in a coastal lagoon of the Ebro delta (north-western Mediterranean) shows a longer **life expectancy** and produces larger eggs compared to the generation from spring and summer (Delgado et al., 1997).

Mauchline (1972) compiled literature data on euphausiids and mysidaceans living in different bathymetric zones and concluded that bathypelagic species probably live 3-7 times as long as epipelagic species. Chikugo et al. (2013) estimated life spans of three years for pelagic populations of the lophogastrid *Eucopia australis* and the mysid *Boreomysis californica* sampled in ⩽1000 m depth in oceanic waters off northern Japan. Wilson & Boehlert (1993) estimated a two-year life span for the lophogastrid *Gnathophausia longispina* off north-east Hawaii, but they conceded that this duration could increase with increasing latitudes and also that older specimens possibly live at depths beyond that reached by their sampling gears. According to Childress & Price (1978), a life span of eight years is plausible for the 'giant' lophogastrid *Gnathophausia ingens*.

Laboratory investigations by Ariani (1982) showed that specimens of the subterranean Stygiomysida species *Spelaeomysis bottazzii* may survive for over 16 months at 20°C in water from the sampling well. There, the only food source is a thin film of green algae, diatoms, and Cyanobacteria partially covering soft calcareous (Quaternary calcarenites) rocks (from the walls and the bottom of the well).

ECOLOGY AND ETHOLOGY

Habitat and distribution

Pelagic environments

Lophogastrida. – The lophogastrids are exclusively marine and mostly pelagic. During daytime, the adults of most species in the genera *Lophogaster* and *Paralophogaster* stay on the bottom or close to the bottom, generally in 200-700 m depth, while the juveniles show a planktonic mode of life. J.-P. Casanova (1993) reported this kind of **stratified distribution** for six species from New Caledonian waters. The remaining species of the same genera plus one *Ceratolepis* are pelagic and inhabit more superficial layers (about 200 m): *Lophogaster spinosus*, *Lophogaster schmidti*, *Ceratolepis hamata*, and four small species of *Paralophogaster*, living in the Red Sea and/or the northern Indian Ocean. All remaining Lophogastrida are bathypelagic: the genera *Chalaraspidum*, *Gnathophausia*, and *Eucopia* are found mostly below 2000 m depth and are restricted to narrow ranges of hydrological conditions, particularly regarding temperature and oxygen content (Fage, 1942). The supposed rarity of certain species may be due to sparse exploration of their habitat. In addition, Childress et al. (1989) observed a *Gnathophausia*, later described as *Gnathophausia childressi*, as being strictly tied to the turbid layer above the bottom: this layer was characterized by the constant presence of suspended matter. This species is closely related to *Gnathophausia affinis*, which also occupies the same type of bentho-pelagic habitat.

Mysida. – Some **holopelagic** species, such as certain *Siriella*, only rarely approach the coast. This genus contains most of the (few) **epipelagic** species with transoceanic, in part world-wide distribution. The greatest numbers of **mesopelagic** species of Mysidae are in the subfamilies Erythropinae and Boreomysinae. Fosså & Brattegard (1990) found the distribution of 18 mysid species in Norwegian fjords clearly separated in four distinct depth zones within the range of 32-1260 m.

Certain mysids are **bathypelagic** (e.g., Cartes & Sorbe, 1995; Hargraeves & Murano, 1996; Hargraeves, 1999; Frutos & Sorbe, 2013). Those inhabiting the continental slope may be divided into three main faunistic groups occurring in different depth zones. The **zones of faunistic change** vary between latitudes and are positioned around 500 and 1000 m depth in temperate zones (Lagardère, 1977b; Elizalde et al., 1991) and less deep (300-400 and 800 m) in boreal waters (Mauchline, 1986; Fosså & Brattegard, 1990) of the North Atlantic. Cartes & Sorbe (1995) noted marked seasonal variations of mysid abundance on the upper and middle slope, but only very small ones on the lower slope (>1200 m) of the Catalan Sea (western Mediterranean). Similarly, Sorbe & Elizalde (2013) found marked seasonal variations in the abundance of suprabenthic mysids on the middle slope (400 m) off Arcachon (Bay of Biscay, north-eastern Atlantic).

Even fewer Mysida are known to be markedly **abyssal**. The abyssal plane is preferred by the family Petalophthalmidae, as indicated by the amazing diversity of the genera *Hansenomysis* and *Bacescomysis* (Birstein & Tchindonova, 1970; Băcescu, 1971b; Lagardère, 1983; Murano & Krygier, 1985). Five species of the genus *Boreomysis* (Mysidae: Boreomysinae) were reported in macroplankton and micronekton samples taken at depths between 1900 m and 5430 m in the north-eastern Atlantic (Hargreaves & Murano, 1996).

COASTAL ENVIRONMENTS

Vertical distribution. – Most marine mysids are **neritic** or **littoral**. Many species are found in intertidal rocky pools or soft-bottom pools (e.g., *Praunus flexuosus*, certain *Leptomysis* and *Paramysis*). Many neritic species stay a few or up to tens of meters above the bottom for most of the day. Depth is certainly not the sole factor influencing their distribution. For example, during the day in waters off Arcachon, *Schistomysis spiritus* lives mostly at 50-100 m above the bottom, the upper 50 cm of which is always sandy-muddy (San Vicente & Sorbe, 1995). In contrast, a study of night surface plankton supported by sonar bathymetry in the Tyrrhenian Sea (western Mediterranean; Ariani & Spagnuolo, 1976) showed — for nine species of Siriellinae and Gastrosaccinae — distribution patterns according to depth but not related to sandy or rocky types of substratum. In many species, the **vertical distribution** varies with the time of day and/or photic brightness in the context of migratory activity (cf. above, 'Migration'). In a Baltic population of *Neomysis integer*, body size increases with depth and decreases with *in situ* daytime light. According to Ogonowski et al. (2013b) size variations with depth are better related to light intensity than to temperature. From this they concluded that the size variations observed in *Neomysis integer* are a response to predation rather than to size-related thermal preferences.

Substrate relations. – In general, the coastal marine and brackish-water Mysida are benthopelagic or benthic during daytime. These species show substrate relations to different degrees, visible by many biological features such as body colour, migration, feeding, social behaviour, etc. (e.g., Clutter, 1969; Wittmann, 1977, 1978a; Fosså, 1986; Wooldridge & Webb, 1988; Wooldridge, 1989; Macquart-Moulin & Ribera Maycas, 1995). The bottom type is important for many species that dwell on or above substrates such as mud, sand, gravel, and stone, or on certain types of vegetation. Most Mediterranean species of *Diamysis* are strictly benthic or bentho-planktonic, living in marine to freshwater environments: during the day they occupy vegetation stands, where they show marked preferences (Wittmann & Ariani, 2012a) for certain **vegetable substrata** such as the sea grasses *Posidonia* in *Diamysis bacescui*, *Zostera* in *Diamysis bahirensis*, and *Cymodocea* in *Diamysis cymodoceae*. The semi-hypogean *Diamysis camassai* prefers calcareous rock coated by microalgae, in this respect resembling the Stygiomysida species *Spelaeomysis bottazzii* (see below, 'Feeding').

Many taxa of Gastrosaccinae (Mysida) are typically **psammophilic** and can bury themselves down to 5 cm deep into the sediment, particularly into the sand of the intertidal zone upon falling tide. The burrowing Gastrosaccinae are active swimmers for feeding and propagative activities during the night, and bury themselves into the sand during the day; at falling tide they may bury themselves or, alternatively, may migrate offshore. The intertidal zonation, migratory, and **burrowing behaviour** of Gastrosaccinae are strongly influenced by tidal and diurnal variations of hydrodynamics, salinity, food availability, and risk of predation (Băcescu, 1940; Wooldridge, 1989; Webb & Wooldridge, 1990; Nonomura et al., 2007).

Salinity relations. – Salinity is among the most important ecological factors governing distribution. As in most aquatic invertebrates, the effects of salinity on osmoregulation, reproduction, distribution, and survival typically show strong variations with temperature (Vlasblom & Elgershuizen, 1977; McKenney, 1996; Greenwood, 2007; Hanamura et al., 2009; Paul et al., 2013). Many littoral Mysida are **stenohalobious** (stenohaline), never found in estuaries or in seas with low salinity. By contrast, there are also more or less **euryhalobious** (euryhaline) species (Stammer, 1936), whose populations are found from the open sea to estuaries and distinctly brackish lagoons. Certain populations attain enormous densities in 'marine' waters influenced by the influx of fresh water, as exemplified by several Black Sea populations of *Paramysis* and *Mesopodopsis* (Băcescu, 1940; Gomoiu, 1978). Most brackish-water species penetrate into the sea but only few of them into freshwater reaches. In Catalan populations of the essentially marine *Gastrosaccus sanctus*, all age stages can migrate up to the mixing zone between infiltrating seawater and subterranean fresh water (Delamare-Deboutteville, 1955). Within the family Mysidae, most species of the subfamily Rhopalophthalminae inhabit brackish waters, particularly estuaries of tropical to subtropical zones. Nonetheless, most brackish-water and almost all freshwater mysids belong to the subfamily Mysinae. An almost paradoxical case is that of the Mediterranean *Diamysis bahirensis* (Mysinae), long considered extremely euryhaline, with records attributed to populations from metahaline to fresh waters. Recently, however, this taxon was split into several species with more restricted ecological ranges (Wittmann & Ariani, 1998, 2012a, b; Ariani & Wittmann, 2000, 2002). A very wide salinity range of S = 2.6-65 was recorded *in situ* for the (sub)tropical *Mesopodopsis africana* from estuaries in South Africa (Carrasco & Perissinotto, 2011a).

Other environmental factors. – Food availability as a primary factor of distribution is expressly or (in)directly treated in various (sub)sections of the present contribution. Other important ecological factors are oxygen content, particularly in brackish-water species, along with pH and water movement. The latter determines the re-suspension of food particles — an important factor for the establishment and maintenance of dense mysid populations (Clutter, 1967). High silt levels in the water, however, may negatively affect feeding rate and health and may increase mortality in laboratory-kept *Mesopodopsis africana* (cf. Carrasco et al., 2007).

Home site preference and homing behaviour. – Certain Mysidae show some degree of permanence at their home site, suggesting a clear preference for a distinct biotope (Clutter, 1969; Wittmann, 1977; McFarland & Kotchian, 1982; Hahn & Itzkowitz, 1986; Twining et al., 2000). The main investigations in this respect were performed on *Mysidium gracile*, whose swarms usually live in association with damselfish or *Diadema* urchins, but were also rarely found in other biotopes such as bare sand or *Thalassia* sea grass (Hahn & Itzkowitz, 1986). After nocturnal dispersal, marked specimens of *Leptomysis lingvura* showed a daily return rate of 0-93%, mainly related to the number of swarms at suitable microhabitats in the surroundings (Wittmann, 1977, 1978a). Similarly, 77% of the radio-labelled specimens from one swarm of the coral reef mysid *Mysidium gracile* returned to the same microhabitat in a single experiment by Twining et al. (2000). As shown by displacement tests (Hahn & Itzkowitz, 1986), the **home site permanence** may vary widely, but certain mysids nonetheless have definite habitat preferences as well as clear homing abilities.

GROUNDWATER AND CONTINENTAL ENVIRONMENTS

Groundwater. – Both Stygiomysida genera so far known, *Spelaeomysis* and *Stygiomysis*, are found in fresh to brackish water, rarely marine, mostly in groundwater habitats, almost always near the coast (Caroli, 1924, 1937; Fage, 1924; Villalobos, 1951; Gordon, 1960; Pillai & Mariamma, 1964; Băcescu & Orghidan, 1971; Bowman, 1973, 1976; Pesce, 1976a, b; Pesce et al., 1978; Ariani, 1981c, 1982; Bowman et al., 1984). Most species from Central America and the Caribbean were recorded from freshwater wells or from freshwater strata in **anchihaline** habitats (Bowman, 1973, 1976; Bowman et al., 1984; Wagner, 1992; García-Garza et al., 1996; Kallmeyer & Carpenter, 1996). Often no detailed data on salinity are available, particularly at the sampling depths in habitats with stratified salinity. At least four species — *Spelaeomysis cardisomae*, *Spelaeomysis cochinensis*, *Stygiomysis hydruntina*, and *Stygiomysis ibarrae* — are known only from brackish to marine waters (Bowman, 1973; Pesce, 1985; Panampunnayil & Viswakumar, 1991; Ortiz et al., 1996; Inguscio, 1998). *Spelaeomysis bottazzii* from coastal **groundwater habitats** in south-eastern

Italy is very rarely found below S = 2 (Ruffo, 1957; Pesce et al., 1978; Ariani & Wittmann, 2010). The estimate by Porter et al. (2008) that 12 out of the 16 currently known Stygiomysida species may have invaded "inland" waters (S = 0-3) reflects mainly the oligohaline range (S ⩾ 0.5) corresponding to coastal rather than inland groundwater. We agree with Ruffo (1957) that the two genera are essentially marine ("thalassoid") **Tethyan faunistic elements** that invaded the groundwater through the marine-brackish interface (see also below, 'Historical biogeography').

When studied in detail, and under experimental conditions, the animals may show preference for higher salinity than expected from sampling in shallow groundwater alone. This may reflect the fact that the animals also often have access to the generally more saline **deep groundwater** through underground connecting paths. A three-year study on a natural population of *Spelaeomysis bottazzii* in a superficial brackish-water well in south-eastern Italy yielded a salinity range of S = 2.3-7.1 and seasonal temperature variations of 9.8-21.6°C (Ariani & Wittmann, 2010). Females taken from the same well, acclimatized in the laboratory over one month at S = 7.0-7.5, and 16-18°C, consistently chose higher salinities (S = 9-10; Cesaro et al., 1984) and temperature (21-22°C; Ariani et al., 1984) during subsequent preference experiments in the laboratory. In line with the hydrogeological situation of the sampling locality (Ariani, 1982), the chosen values correspond to those prevailing in deep groundwater (Cotecchia, 1977; Cotecchia et al., 1978). This is where the animals probably breed and develop, unlike the superficial groundwater where they dwell mainly for feeding (Ariani & Wittmann, 2010).

Most species of Stygiomysida are **cavernicolous** or **phreaticolous**, showing **reduced eyes** of which the cornea is reduced to a small number of ommatidia or completely missing (Caroli, 1924, 1937; Fage, 1925; Villalobos, 1951; Gordon, 1960; Pillai & Mariamma, 1963; Băcescu & Orghidan, 1971; Ingle, 1972; Bowman, 1976; Bowman et al., 1984). Well-developed or vestigial eyestalks are present throughout. This pattern is also found in the comparatively small number of cavernicolous Mysidae (Mysida), mostly belonging to the subfamilies Mysinae and Heteromysinae (Creaser, 1936; Stammer, 1936; W. M. Tattersall, 1951; Băcescu & Orghidan, 1971; Ledoyer, 1989). Remarkably, certain species of *Hemimysis* constitute the dominant faunistic element in dark **submarine caves** of the Mediterranean and the north-eastern Atlantic (Ledoyer, 1989; Breton et al., 1995). These species do not show particular adaptations to this habitat and, notably, have well-developed eyes. Accordingly, they are not completely dependent on this habitat, but leave their caves during the night to feed and then return in the morning. Unlike the non-cavernicolous *Leptomysis lingvura*, the cavernicolous *Hemimysis speluncola* shows a very marked **negative phototaxis** (Macquart-Moulin, 1979; Bourdillon et al., 1980; Macquart-Moulin & Passelaigue, 1982; Passelaigue & Bourdillon, 1986; Passelaigue, 1989). Unlike certain non-cavernicolous species, the photophobia of *Hemimysis* swarms on the bottom of submarine caves does not appear to be related to reproduction.

Fresh water. – Freshwater species are dominated by the genus *Mysis* in boreal to Subarctic waters of the circum-arctic region, by *Taphromysis* in the Mid-Atlantic drainage of America, *Diamysis* in the Mediterranean, *Paramysis* in the Ponto-Caspian, and *Neomysis* in inland waters of Japan. Several species penetrate more than 1000 km into river systems. Only few **potamophilous** species are so far known from Africa and South America.

Locomotion, orientation, and taxis

In the Lophogastrida, swimming is mainly effected by the beats of the pleopods, which are well developed in both sexes. At least in the lophogastrid *Gnathophausia ingens*, also the oar-like beats of the thoracic exopods help propelling the body forward (Hessler, 1985). In Mysida, the pleopods are rudimentary in the females and often also in males. Both sexes swim by the rotating movements of the thoracic exopods. The legs work in a synergistic fashion, also producing the **respiratory current** together with **filtration currents** (Schabes & Hamner, 1992).

Swimming by thoracic exopods appears to have developed secondarily. All three orders considered here normally maintain a horizontal, often slightly inclined, orientation of the body when swimming. A number of Mysidae, particularly species of *Praunus* and *Limnomysis*, often hover with slightly inclined vertical orientation of the body axis. In the Mysida and Stygiomysida, the thoracic endopods are to some extent used for crawling and/or holding onto the substrate. During daytime, most littoral Mysidae do not swim actively; they remain in contact with the substrate and may hover among algae or gravel. Some species bury themselves shallowly in soft sediment. During the night, all littoral species show stronger **swimming activity** and many of them tend to rise up from the bottom. Foxon (1940) argued that this behaviour reflects negative **geotaxis**, which turns to positive geotaxis during light hours. Some species show daytime activity and may continue to swim in bright sunlight.

The **locomotive orientation** of the Mysida is conditioned by three types of reflexes: responses to gravity and inertia mediated by the statocysts, dorsal light reaction (in certain species) mediated by the eyes, and finally a general position reflex of still unknown nature. Bainbridge & Waterman (1957, 1958) studied the locomotive orientation in polarized light. Statistically, *Mysidium gracile* adjusts its longitudinal body axis perpendicular to the **plane of polarization**. This reaction is more distinct in turbid water (Waterman, 1960). The orientation of certain species may also reflect variations in **hydrostatic pressure** (Rice, 1961; Knight-Jones & Morgan, 1966). At the same time, there are reports of an escape reflex, occurring when the animal becomes attacked, when the water shows sudden agitation, or when the illumination suddenly changes. Under these conditions, the pleon is withdrawn under the thorax and then immediately straightened; this induces powerful backward jumps by tail flipping, which may even project the animal out of the water.

Migration

Seasonal migration. – The seasonal **migratory rhythms** vary between species and are mostly synchronized with the period of gonad development and sexual activity — both of which are strongly affected by temperature and probably also by the photoperiod.

The Mysida normally do not show large-scale **horizontal migrations**, but seasonal shifts in distribution often occur at an intermediate scale: According to Băcescu (1940), *Mesopodopsis slabberi* is quite common during the cold season in 'marine' waters along the Black Sea coast of Romania. Here the densities become very low in summer, whereas in that same season densities strongly increase in the oligo- to mesohaline waters along the shores near the mouth of the Danube River and in brackish ponds and lagoons. Along these Black Sea coasts, *Paramysis pontica* approaches closer to the shore in winter, while several fluvio-lacustrine species show a wider dispersion with more upstream limits in the rivers. In the Bay of Biscay, *Schistomysis spiritus* lives close to the beach in summer but more distant from the shore at maximum depths of 30-90 m during the rest of the year (San Vicente & Sorbe, 1995). W. M. Tattersall (1939) observed seasonal shifts in the occurrence of 19 species off Plymouth (north-eastern Atlantic) and attributed these to horizontal and/or vertical movements. A number of species are to some extent passively shifted by currents to localities where they may have little chance of long-term survival (Künne, 1937, 1939).

Diurnal migration. – Some coastal species perform diurnal inshore and offshore migrations (Debus et al., 1992; Macquart-Moulin & Ribera Maycas, 1995), often related to water movement and food availability (Webb et al., 1988; Wooldridge, 1989; Webb & Wooldridge, 1990). Passive or active, diurnal horizontal shifts appear to be a common feature in populations living in estuaries, fjords, and in the surf zone and other coastal habitats (Välipakka, 1992; Beyst et al., 2001; Jumars, 2007). A number of Mediterranean species disperse mainly horizontally in layers close to the sea bottom at night (Wittmann, 1977, 1982). *Hemimysis speluncola* performs nocturnal inward and outward migrations from submarine caves, mainly for outside feeding during the night and for inside hiding during the day (Macquart-Moulin & Passelaigue, 1982; Passelaigue & Bourdillon, 1986; Passelaigue, 1989; Riera et al., 1991; Rastorgueff et al., 2011).

Detailed studies of **diurnal vertical migration** were conducted by Russel (1925, 1931), Fage (1932, 1933), Beeton (1958, 1960), Herman (1963), Macquart-Moulin (1973b, 1975), Toda et al. (1983b), Jumars (2007), and Ogonowski et al. (2013a). Many marine to freshwater Mysida remain at or close to the bottom during the day, and migrate upwards mainly for feeding during the night (Grossnickle, 1979; Kouassi et al., 2006). In most cases the nocturnal migrations result in a wider dispersion within the water column (Toda et al., 1983b; Apel, 1992; Jumars, 2007). These vertical migrations, most often of endogenous origin, are synchronized by temporal variations in brightness (Macquart-Moulin, 1973a). In *Mysis relicta* and *Mysis diluviana* this periodicity is influenced by varying conditions of brightness but also of temperature and possibly even pH. The effects of temperature variations on this **phototropism** are sufficient to explain the diurnal and seasonal vertical migrations of juveniles of various marine coastal Mysidae (Fage, 1932). Geotaxis and the effects of temperature on geotaxis may also play a role (Foxon, 1940; Bourdillon & Castelbon, 1983). According to Ogonowski et al. (2013a), the Baltic species *Mysis salemaai* shows two subsets of the population: one group shifts upwards in the pelagic zone and the other remains near the bottom at night. The two migratory strategies are consistently coupled with diet differences, with pelagic mysids having a more uniform and carnivorous diet. Sequencing of mitochondrial genes revealed differences between geographical locations but not between the subsets predisposed for different migratory activity (Ogonowski et al., 2013a).

Waterman et al. (1939) studied the **nocturnal vertical migrations** of several bathypelagic lophogastrids (*Eucopia*) and mysids (*Boreomysis microps*) from waters off the north-western Atlantic coast. The amplitude of daily shifts is on the order of 400 m and may reflect variations in light intensity, even though the intensity is low at these depths (200-1200 m). In the north-eastern Atlantic, the lophogastrids *Gnathophausia* and *Eucopia* show much more restricted nocturnal vertical migrations (Hargreaves, 1985).

Tidal migration. – The **tidal migrations** of *Gastrosaccus psammodytes* at high-energy beaches in South Africa may have endogenous components in addition to the obvious external ones (McLachlan et al., 1979). In order to minimize seaward flushing by tidal currents in a South African estuary, *Gastrosaccus brevifissura* employs several strategies, including utilization of flood-currents, to maximize up-estuary transport (Schlacher & Wooldridge, 1994). Several littoral species, such as *Praunus flexuosus* and *Schistomysis spiritus*, show a nearshore drift with the tides during very dark nights (Elmhirst, 1931, 1932).

Ontogenetic migration. – **Ontogenetic migrations** were demonstrated for a number of pelagic and benthic mysidaceans. A frequent pattern is that juveniles migrate to surface layers at night while adults remain close to the bottom (Almeida Prado, 1973). As an inverse example, adult *Mysis relicta* tend to live in deeper areas than juveniles during the day in two freshwater lakes in north-east Germany, while the vertical distribution pattern is apparently inverted at night (Waterstraat et al., 2005). According to Biju et al. (2013), the numbers of breeding females of *Siriella gracilis* from tropical offshore waters in the northern Indian Ocean are not correlated with temperature in the range of 26-31°C, but significantly increase with salinity in the range of $S = 30$-37. This is probably due to the migration to deeper, more saline water layers for selection of the **optimal salinity** for **embryonic development**. The females of the lophogastrid *Gnathophausia ingens* and the mysid *Mysis stenolepis* breed in deeper waters and migrate to more superficial strata prior to releasing the young (Amaratunga & Corey, 1975; Childress & Price, 1978). An inshore migration of adults in spring coincided with the onset of reproduction of the overwintering generation of *Americamysis bigelowi* in a New Jersey estuary (north-western Atlantic; Allen, 1984).

Social aggregation

Types of aggregation. – Considerable attention has been focused on the distributional, behavioural, and physiological aspects of swarming in the Mysida (Clutter, 1967, 1969; Berrill, 1969b; Macquart-Moulin, 1970, 1971, 1973a, b; Mauchline, 1971; Zelickman, 1974; Wittmann, 1977; O'Brien, 1988; Twining et al., 2000). **Definitions of aggregation** in mysids by Clutter (1967), Mauchline (1971), and Zelickman (1974) were mainly related to **social interaction**. O'Brien (1988) classified **grouping** according to the mechanisms stimulating their formation and maintenance. Definitions by Wittmann (1977) were primarily geometric and, therefore, more practical for comparative field observations on different species. He considered essentially: inter-individual distance excluding social interaction (**dispersion**); no or reduced regularity in spatial arrangement (**aggregation** *sensu stricto*); regularity in spatial arrangement (**swarms**); swimming in the same direction (**schools**) [apart from mere effects of water movement]; presence inside a swarm of structures differing in abundance of distinguishable individual types (**sub-swarms**); presence of non-dominant species in an association (**guests**).

Structure of aggregations. – The internal arrangement of mysid and euphausiid swarms has in common that the individuals avoid positions directly above and below their neighbours (O'Brien, 1989). Different age groups are often organized in **separate swarms** or **sub-swarms** during the day (Macquart-Moulin, 1970; Wittmann, 1977; O'Brien, 1988; Ohtsuka et al., 1995). Juvenile sub-swarms of Mediterranean *Leptomysis* typically stay closer to the substrate where they are less visible for (bentho)pelagic **predators** and at lower risk to be consumed by larger conspecifics. Here, however, they may be exposed to an increased risk of being attacked by benthic predators, for example by epibenthic gobies from below (Wittmann, 1978a). Modlin (1990) observed an inverse stratification in the mangrove mysid, *Mysidium columbiae*, with the juveniles near surface and the breeding females near bottom. Based on multifrequency acoustic echo-sounding on three coastal species, particularly *Neomysis rayii*, Kaltenberg & Benoit-Bird (2013) concluded that intra-patch clustering of mixed-size and mixed-species aggregations may greatly affect the interactions among mysids and their predators. Preferential grouping of sexually mature individuals is often observed and may represent breeding aggregations (Clutter, 1969; Mauchline, 1971). Daytime swarming may also help to find mating partners, as suggested by residual dense aggregations of the Mediterranean *Leptomysis lingvura*, in which mating normally starts after sundown, while most of the remaining population is already dispersed over the sea bottom (Wittmann, 1982).

Cohesiveness of aggregations. – As first noted by Steven (1961), mysid schools perform cohesive and precise locomotion comparable to that of fish. According to Zelickman (1974) these patterns are controlled by the combined effects of external stimuli and collective (inter-individual) responses. Monospecificity and separation of age groups are controlled by **distributional separation** such as bathymetric zonation, substrate-preference, and activity periods. They are also controlled by **active avoidance** of individuals of different size and/or preference for same size. Clutter (1969) assumes that mysids respond mainly to the visual image of their neighbours, but tactile and olfactory stimuli may also be important. According to Wittmann (1977) several swarming species are unable to identify others exactly like themselves; there may be some capability for a general mysid-identification. **Cohesiveness of swarms** is controlled by preference for the same habitat and probably by an often non-specific **gregariousness**.

According to Wittmann's (1977) investigations, mysids seem to have evolved two different strategies of optimizing swarm stability and cohesiveness. One (shown by *Mesopodopsis slabberi*) is to reduce **substrate relations** and to intensify **social interaction**. Such species can form motile, polarized, cohesive groups. The second method (shown by *Leptomysis lingvura*) is to maintain a relatively low level of collective responses and to intensify habitat-relations such that the population occurs only at a few distinct places within its area. Such species form non-motile swarms (fig. 54.41A, C) or open near-bottom aggregations. The motility of swarms may change with season and sexual maturity (Ohtsuka et al., 1995). This second method allows the species to profit from the protection afforded by the substrate, for example, by means of **camouflage** and/or by hiding among vegetation or on the sediment upon predator attack (Wittmann, 1977, 1978a). As a striking example of substrate relations, the **homing behaviour** of the demersal tropical mysid *Mysidium gracile* facilitates swarm reformation each morning after nocturnal dispersal (Twining et al., 2000; see also above, 'Home site preference and homing behaviour'). In general agreement with the above concept, the relative inter-individual distances and other indices of internal swarm structure, as far as examined, vary between substrate-specialized mysids, non-substrate-specialized mysids, and the again more pelagic euphausiids (O'Brien, 1989).

In most mysid species, the groupings tend to disintegrate and the individuals disperse over the substrate or in the water column during the night (Macquart-Moulin, 1970, 1973b; Wittmann, 1977, 1978a, 1982; Twining et al., 2000). In laboratory experiments by Macquart-Moulin (1973b), the test specimens of *Leptomysis lingvura* dispersed when luminosity dropped below 1 lux, and reformed swarms above about 30 lux. Nonetheless, certain species aggregate throughout the night (O'Brien, 1988; Patzner, 2004). In *Mysidium columbiae*, the adults become solitary at night, whereas the juveniles continue to form compact swarms (Modlin, 1990).

Benefits and disadvantages of aggregation. – Based on published results, the **benefits of gregariousness** for mysids are facilitation of position maintenance, increased reproductive potential, trapping of prey, protection from predators, and energy saving (Clutter, 1969; Mauchline, 1971; Wittmann, 1977; O'Brien, 1988; Modlin, 1990; Ohtsuka et al., 1995; Ritz et al., 2011). Ritz (2000) showed that cohesive mysid schools consume less energy for swimming compared to un-cohesive small groups. By measuring oxygen uptake, Ritz et al. (2001) showed that large mysid swarms consume less energy *pro capite* for escape responses compared to small swarms and again to solitary specimens. Swarms of species of *Gastrosaccus*, *Anisomysis*, and *Tenagomysis* aggregate for feeding (Wooldridge, 1989; Ohtsuka et al., 1995). Ritz & Metillo (1998) found, however, no significant effect of swarming on success of food capture in *Paramesopodopsis rufa* from coastal waters of Tasmania. Apparent **disadvantages of aggregation** are enhanced competition for food, facilitation of cannibalism, and vulnerability to large predators that could break up swarms (Emery, 1968; Hahn & Itzkowitz, 1986; Ohtsuka et al., 1995; Johnston & Ritz, 2001).

Heterospecific aggregations. – Mysids are often found in mixed swarms or schools formed by 2-5 species (Macquart-Moulin, 1970; Wittmann, 1977; Ohtsuka et al., 1995). The swarms stay mostly (not always) at the habitat usually preferred by the numerically dominant species. The NE-Atlantic-Mediterranean *Leptomysis heterophila* got its specific name from its regular habit to co-occur with other mysid species (Wittmann, 1986a). Such associations may be **mutualistic** because potential predators may be confused by differences in **anti-predatory behaviour** within the same association. However, such associations may be one-sided if the predator is directed to the other mysid species. Several mysid species tend to prey on larvae and early juveniles of (their own and) other swarming species (Johnston & Ritz, 2001). Mysid swarms are often associated with sessile benthic invertebrates (fig. 54.41D; see below, 'Symbiotic associations').

In shallow tropical backreef habitats, McFarland & Kotchian (1982) observed two species of *Mysidium* to form **mixed schools** with larvae of the fish *Haemulon flavolineatum*. These associations break down at night, when both partners disperse for feeding, and restructure at the same reef each morning. This kind of association has some mutualistic elements, but appears ambivalent for the mysids, as they may fall prey to the fish larvae as these grow.

Grooming

Behavioural and SEM-observations by Acosta & Poirrier (1992) showed that the mysid *Americamysis bahia* uses specialized setae on the mandibular palps and on the thoracic endopods 2-8 to clean the body from epizoites and particulate matter. The second thoracic endopod is the most active in **grooming** the antennae, setae of the marsupium, and other thoracic appendages. The eighth thoracic endopods clean the outer face of the marsupium. The telson and uropods are also cleaned. Prevention of fouling may be particularly important for the females during the long period when they breed their young and do not moult (see above, 'Incubation'). Newly emerged young are poorly equipped with **cleaning setae**, suggesting that the higher moult frequency is the main mechanism to counter fouling in juveniles.

Sets of specialized setae on the terminal segment of the mandibular palp are present in most species of the family Mysidae. Mauchline (1980) figured such setae for *Schistomysis ornata* and interpreted these as components for cleaning the cutting and grinding surfaces of the mandibles as well as for removing food from the posterior mouthparts and to push it into the mouth. Combining the observations by Mauchline (1980) and Acosta & Poirrier (1992) points to an integrative role of the mandibular palp for feeding and grooming.

Food and feeding

Feeding mechanisms. – The **feeding mechanisms** and digestion were studied in species of Lophogastrida (Manton, 1928a), Mysida (Depdolla, 1923; Cannon & Manton, 1927a; Băcescu, 1940; Tattersall & Tattersall, 1951; Foulds & Mann, 1978), and Stygiomysida (Nath & Pillai, 1972a; Ariani, 1982). Feeding is generally based on three independent, often coexisting procedures:

The first procedure is **fragmentation** of large to medium-sized, living or dead particles that are grasped with the thoracic endopods and then held to the mouth with the aid of the anterior endopods and the mandibular palp. The food is cut into small fragments by the incisor processes of the mandibles. The fragments are reduced to a very fine pulp by the molar processes of the mandibles and the internal armature of the stomach. This mode of feeding was observed in the laboratory in three Mediterranean species of *Diamysis* when fed thin flakes (Tetramin®) composed largely of dried algae (Ariani, 1966, 1979, 1981b; Ariani et al., 1982, 1999; Ariani & Wittmann, 2000).

The second procedure, observed in Lophogastrida and Mysida, is the **filtration** of fine particles. The **filtration current** is provided by rotating movements of the thoracic exopods and by vibrations of the maxillae (Manton, 1928a). Food enters the **feeding basket** formed by thoracic endopods and maxillae from the front (Schabes & Hamner, 1992). The particles accumulate in a 'feeding tank' anteriorly limited by the paragnaths. From here they are carried to the mouth or swept away as they arrive at the long setae of the basal endites of the maxillae and of the first pair of thoracopods. Lucas (1936) calculated that the mysid *Neomysis integer* can consume up to 6 million diatoms per hour. This feeding mechanism is often supported by strong movements of the animal, agitating water against the substrate. Certain species are exclusively limivorous and strongly perform this procedure. However, increased silt levels in the water may hinder the food-collecting ability in laboratory-kept *Mesopodopsis africana* (cf. Carrasco et al., 2007).

The third procedure is **rasping** the coat of autotrophic microorganisms from the substrate surface with the strong mandibles. In the essentially phreaticolous but locally also cavernicolous Stygiomysida species *Spelaeomysis bottazzii* (cf. Ariani, 1982; Ariani & Wittmann, 2010), this mode of feeding was observed in the field at a lighted well and in the laboratory under both light and dark conditions. The ingested microorganisms were diatoms, green algae, and Cyanobacteria; the substrate consisted of soft calcareous rock (Quaternary calcarenite). This is generally in line with the geological characteristics of the geographical area, where the species is found in caves or wells (Ariani, 1981b, 1982). When living under lighted conditions, animals accumulate fat reserves as **subcuticular fat bodies**. This is visible in a change in body colour from white to pale up to marked yellow. Both field and laboratory observations showed that *Spelaeomysis bottazzii* produces cylindrical faecal pellets consisting mostly of amorphous calcareous grains, rarely with additional 'enigmatic' chains of well-formed calcite crystals (Ariani, 1982). These crystals might result from unknown processes of dissolving and re-crystallization of calcium carbonate anywhere in the digestive system. During an observation time of two weeks in the laboratory, *Spelaeomysis bottazzii* produced almost equal volumes of faecal pellets per mm of body length in the presence of only fragments of bare porous rock, compared to fragments covered by autotrophic micro-organisms. This suggests that the species can survive during the long incubation period in deep groundwater (Ariani & Wittmann, 1997, 2010) by using the organic matter infiltrated into the porous rock or contained in a fossil state.

Digestion. – **Digestion** takes about 2-4 hours in various Mysidae (Benko, 1962; Fockedey et al., 2005). By contrast, it takes about 12 hours in the cavernicolous Stygiomysida *Spelaeomysis longipes* (cf. Nath & Pillai, 1972a). In laboratory specimens of the mysid *Neomysis integer*, **egestion rates** as a measure of **ingestion** generally increased with increasing temperature and salinity, but were suppressed by the combination of high temperature and high salinity (Roast et al., 2000). In the laboratory, the faecal pellets of this species were generally enriched in nitrogen, probably due to bacterial growth on the pellets and on the peritrophic membrane, as well as to the disintegration of intestinal cells (Fockedey et al., 2005).

Certain species, such as *Mysis stenolepis*, can digest the cellulose of vegetal debris (Foulds & Mann, 1978) by means of endogenous cellulases (Friesen et al., 1986a). Niiyama et al. (2012) found **cellulase activity** in all six mysid species tested from a mangrove estuary in Malaysia. In three coexisting mysids from the coasts of Tasmania, Metillo & Ritz (2001, 2003) found differential activity of **laminarinase** as a measure of herbivorousness on algae. They also reported differential responses to feeding stimulants and suppressants, both findings pointing to different partitioning of the feeding niche.

Nature of food. – The **nature of the food** was studied in species of Mysida (Blegvad, 1915; Depdolla, 1923; Băcescu, 1940; Tattersall & Tattersall, 1951; Mauchline, 1968; Bowers & Grossnickle, 1978; Grossnickle, 1979; Siegfried & Kopache, 1980; Wittmann & Ariani, 2000; Hanamura et al., 2012; Hanselmann et al., 2013) and Stygiomysida (Nath & Pillai, 1972b; Ariani, 1982; Ariani & Wittmann, 2010). Most species are **omnivorous**, chasing actively after planktonic larvae and small crustaceans, but also ingesting fragments of algae and vegetal debris (Bowers & Grossnickle, 1978; Fenton, 1996; Thiel, 1996; Fockedey & Mees, 2005; Cibinetto et al., 2006; Jumars, 2007; Lehtiniemi & Nordström, 2008; Lehtiniemi et al., 2009).

Analysis of stomach contents and food choice experiments suggest that the mysid *Limnomysis benedeni* is essentially **herbivorous** to **detritivorous** and probably shows no major **predatory impact** on zooplankton in the field (Wittmann & Ariani, 2000; Gergs et al., 2008; Hanselmann et al., 2013). From stable isotope analysis of an invasive population of *Limnomysis benedeni* in a gravel pit lake in the catchment of the Rhine River, Fink & Harrod (2013) concluded that pelagic food sources prevail at least in eutrophic and turbid lakes with low benthic primary production. In the laboratory it selects leaf litter according to factors related to carbon content and polyphenol content; fungal colonization and conditioning by fungi also play a role (Aßmann et al., 2009). Some mysids feed mainly on zooplankton (Hansson et al., 1989; Cartes & Sorbe, 1998; Chigbu, 2004; Borcherding et al., 2006). Certain mysids prey upon eggs and larvae of fish when these are offered in the laboratory (Tornainen & Lehtiniemi, 2008). Cannibalism is often reported from mysids in captivity.

Food search and selection. – Crouau (1978a, 1983, 1987, 1989) observed **food search** and **feeding behaviour** in the cavernicolous tropical mysid *Antromysis juberthiei*. He described a great variety of chemosensitive and mechanosensitive setae on antennulae, antennae, and mouthparts, which could be important for detecting, chasing, and grasping food. According to Metillo & Ritz (1994, 2003), food resource partitioning of three co-occurring species of Mysidae from coastal marine waters of Tasmania is characterized by differences in gut volume and internal armature of the cardiac region, and by differential chemosensory feeding behaviour.

Mysids typically have a high capability of adapting their **selection of food sources** and feeding behaviour to local environmental conditions and food availability (Johannsson et al., 2001; Lehtiniemi & Nordström, 2008; Vilas et al., 2008; Fink et al., 2012;

Hanamura et al., 2012; Marty et al., 2012). A good example is the (sub)tropical, euryhaline *Mesopodopsis africana*, which modifies its diet rapidly under natural conditions. In certain parts of the same South African estuary, it utilizes mostly particulate organic matter, in another part mainly macroalgae, and again elsewhere copepods (Carrasco & Perissinotto, 2010a, 2011b). Nonetheless, in this estuary it is also an important grazer of microalgae and its energetic requirements may be met by a microalgal diet alone (Carrasco & Perissinotto, 2010b). In *Mysis diluviana* from Lake Ontario, its diurnal vertical migration activity and the resulting nocturnal pelagic versus benthic feeding vary with the season, mainly according to the availability of diatoms (Johannsson et al., 2001). The high **plasticity of food selection** may be a major obstacle in transferring laboratory results to the field. Fink et al. (2012) observed selective feeding of *Limnomysis benedeni* on two *Daphnia* species in the laboratory and in outdoor mesocosm experiments, but Wittmann & Ariani (2000) and Hanselmann et al. (2013) found no correspondence in stomach contents of field animals. According to field studies by Borcherding et al. (2006) on *Hemimysis anomala*, the proportion of zooplankton consumed is higher during the nocturnal stay in more superficial water layers compared to daytime dwelling on the bottom. Cartes & Sorbe (1998) found strong depth patterns in three marine deep-water species of mysids: the relative amount of phytodetritus versus pelagic food in the stomachs increased with depth. **Zooplankton clearance rates** of freshwater *Mysis* in the laboratory did not vary with time of day, light intensity or temperature (Cooper & Goldman, 1982), suggesting that the differences observed in nature to some degree reflect different prey density at different water depths during vertical migration. Light has a negative effect on predation rates in laboratory specimens of the marine littoral mysid *Gastrosaccus roscoffensis*, which in nature tends to **bury itself** a few centimetres into the sediment during the day (Escánez et al., 2012). In a freshwater population of *Hemimysis anomala*, Borcherding et al. (2006) reported that small specimens feed relatively more on phytoplankton, large ones more on zooplankton. In line with this, a stable nitrogen isotope study by Branstrator et al. (2000) provided evidence for increasing **zoophagy** with increasing body size and maturity in freshwater populations of *Mysis diluviana*. By contrast, Fockedey & Mees (2005) found no qualitative differences in diet according to sex or developmental stage in the omnivorous estuarine mysid *Neomysis integer*.

Adaption to food scarcity. – The hypogean Stygiomysida *Spelaeomysis longipes* counters the **food scarcity** in subterranean environments by feeding on decaying vegetable matter, as observed in a well with a supply of plant material (Nath & Pillai, 1972b). A different strategy is shown by the hypogean *Spelaeomysis bottazzii* and the semi-hypogean mysid *Diamysis camassai*, which exhibit **micro-phytophagy** in or near the margins of the photic zone during part of their life cycle. Both species show a strong mandibular apparatus capable of scratching microorganisms from hard substrate. This appears to be an important prerequisite for life in deeper, dark groundwater, to which the females may escape from predation by epigean animals when incubating their young (Ariani & Wittmann, 2002, 2010).

Trophic interactions

According to Chigbu (2004), the mysid *Neomysis mercedis* feeds selectively on *Daphnia* and can **control the abundance** of this cladoceran in Lake Washington. Populations of several cladocerans were diminished by intentional or non-intentional introductions of *Mysis relicta* and *Hemimysis anomala* in European lakes and impoundment basins (Langeland, 1981; Fürst et al., 1984; Koksvik et al., 1991; Langeland et al., 1991; Ketelaars et al., 1999). The **frequency of predation** by two *Mysis* species on the cladoceran *Cercopagis pengoi* in the laboratory was consistently higher in juveniles than in adults, and in *Mysis mixta* higher than in *Mysis relicta* (cf. Gorokhova & Lehtiniemi, 2007). Both species adjust their feeding behaviour under the risk of predation by fish (Lehtiniemi & Lindén, 2006).

The predatory mysid *Rhopalophthalmus terranatalis* controls the density of the mysid *Mesopodopsis wooldridgei* in a South African estuary by feeding on the juveniles of the latter species (Wooldridge & Webb, 1988). Three mysids are involved in **trophic interactions** in the Guadalquivir Estuary (south-western Spain), where the more carnivorous *Rhopalophthalmus tartessicus* and the more omnivorous *Neomysis integer* compete for feeding on juveniles of the more micro-herbivorous *Mesopodopsis slabberi* and on copepods (Vilas et al., 2008). A **predator-prey relationship** between various species of mysids is also suggested by laboratory data on *Praunus flexuosus* as a predator of small specimens of *Neomysis integer* from a North Sea estuary (Winkler & Greve, 2004).

Seasonally increased **predation pressure** of grey whales on the dominant mysid species along the coast of British Columbia (north-eastern Pacific) favours the propagation of subdominant species. This is an example for elevated species diversity being supported by **top-down control** (Feyrer & Duffus, 2011).

Symbiotic associations

Diversity of symbiotic relationships (fig. 54.41). – Many Mysidae (Mysida), particularly species belonging to the Heteromysinae genera *Heteromysis* (fig. 54.41B) and *Ischiomysis* (fig. 54.41D), show **associations with benthic invertebrates** such as sponges, madreporarian corals, sea anemones, sabellids, pagurids, and ophiuroids (Bonnier & Pérez, 1902; W. M. Tattersall, 1922; Clarke, 1955; Pillai, 1968; Bǎcescu, 1970, 1976; Bǎcescu & Bruce, 1980; Ortiz & Gomez, 1988; Vannini et al., 1993; Wirtz, 1997, 2009; Williams & McDermott, 2004; Wittmann, 2008, 2013a). *Heteromysis harpax* is usually found as a family group formed by a pair of adults together with juveniles in shells inhabited by *Dardanus* hermit crabs (Vannini et al., 1993). These mysids probably identify suitable shells by visual and chemical cues and avoid other species of hermit crabs. Among species of the subfamily Mysinae, *Idiomysis tsurnamali* is associated with jellyfish (Bǎcescu, 1973a; Chadwick et al., 2008; Niggl & Wild, 2009), *Idiomysis inermis* with sea anemones (Greenwood & Hadley, 1982), and *Anisomysis levi* with gorgonians (Bǎcescu, 1973b). Swarms of several Mediterranean species of *Leptomysis* (Leptomysinae) show rather loose, transient associations with the sea anemones *Anemonia viridis* (fig. 54.41C) and *Aiptasia mutabilis* (Wittmann, 1977, 1978a, 1982, 1986b; Patzner & Debelius, 1984; Patzner, 2004). Swarms

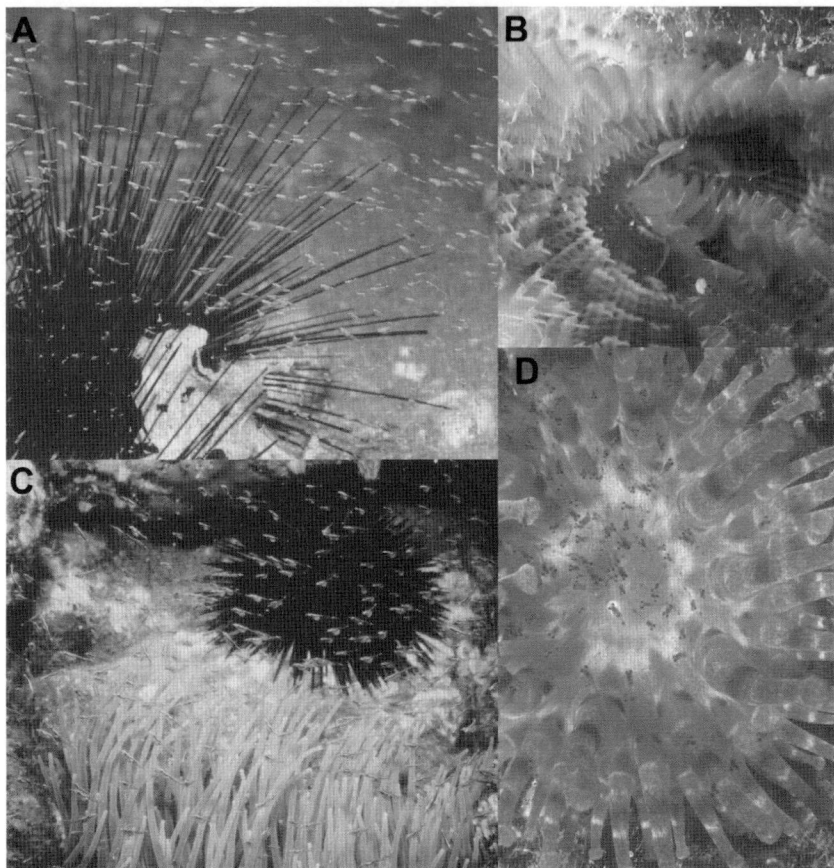

Fig. 54.41. Associations between Mysidae (Mysida) and sessile benthic invertebrates. A, swarm of *Leptomysis* sp. A at the long-spined sea urchin *Diadema antillarum* Philippi, 1845, from Madeira, mysids facing water current; B, adult female *Heteromysis* sp. inside open fan of the sabellid worm *Branchiomma nigromaculatum* (Baird, 1865) from Cape Verde Islands; C, swarm constituted mainly by breeding females of *Leptomysis lingvura marioni* Gourret, 1888, above the snakelocks anemone *Anemonia viridis* (Forskål, 1775) from the Gulf of Naples; D, dense aggregation of undetermined Mysidae above oral disk of the club-tipped anemone *Telmatactis cricoides* (Duchassaing, 1850) from Trindade, a small island in the tropical south-west Atlantic. [A, after Wirtz, 1995; B, photo Peter Wirtz; C, after Wittmann, 1978a; D, after Wittmann, 2013a (photo Lisandro de Almeida).]

of *Leptomysis*, *Mysidium*, and other genera may show facultative associations with sea urchins (fig. 54.41A; Randall et al., 1964; Wirtz, 1995; Wittmann & Wirtz, 1998).

Diurnal periodicity of symbionts. – In southern France, Patzner (2004) observed *Leptomysis lingvura* to remain at the anemones over night. In shallow water of the northern Adriatic Sea, Wittmann (1977, 1978a) observed that swarms of this and several other species of *Leptomysis* disintegrate during the night and re-gather in the morning over various types of substrates, including substrates other than anemones (see above, 'Habitat and distribution'). Inverse periodicity may also be shown, e.g., by a so far un-described

Heteromysis species from the Cape Verde Islands, which is closely associated with the sabellid annelid *Branchiomma nigromaculatum* only during the night (fig. 54.41B; Wirtz, 2009).

Antipredator behaviour by symbiosis. – Some direct observations on mysids underline the role of **symbiosis** as **antipredator adaptation**. According to Băcescu (1973a), *Idiomysis tsurnamali* hovers in swarms above the jellyfish *Cassiopea* and hides between the tentacles when potential predators approach. *Heteromysis actiniae* swims closely around the tentacles of the anemone *Bartholomea annulata* and may rest on the basis of the tentacles, where nematocysts are less dense (Clarke, 1955). As far as observed, other Mysidae associated with cnidarians avoid physical contact with their host. Upon approach of predators from the side or from above, swarms of several species of *Leptomysis* swim closer to whatever type of substrate is available, including sea anemones, but do not seek shelter between tentacles; the mysids are killed upon forced physical contact with the nematocysts (Wittmann, 1978a). *Mysidium gracile* and a few other species of Mysidae seek shelter between the spines of sea urchins when threatened (Randall et al., 1964). In coral reefs of Florida, *Mysidium gracile* shows a **facultative association** with nesting pomacentrid fish. Upon approach of predators, the mysids actively crowd into the nest cave defended by the pomacentrid (Emery, 1968).

Commensalism. – A **commensal relationship** between mysids and their hosts is often assumed, but only few direct observations are available. *Heteromysis harpax* feeds on suspended particles and plankton; apparently it does not clean its hermit crab host and does not feed on the host's faeces (Vannini et al., 1994). By contrast, *Heteromysis actiniae* does feed on the faeces expelled by its sea anemone host (Clarke, 1955). Niggl et al. (2010) showed by stable element labelling experiments that the symbiotic mysid *Idiomysis tsurnamali* feeds on organic matter other than faeces released by its jellyfish host *Cassiopea*. It clearly consumes organic matter derived from mucus released by the jellyfish.

Parasites

Several Lophogastrida and Mysida were found to be affected by **epibionts** or even by true external or internal **parasites**. Their integument may be colonized by algae and protists (Hoenigman, 1960). Several species of Peritrichia and Apostomida (phylum Ciliophora = Ciliata) are of major importance among the **epizoites** (i.e., not 'epizooties') (Hanamura & Nagasaki, 1996). The ciliates appear to be capable of re-attaching rapidly on the fresh cuticle after moulting of the host (Ohtsuka et al., 2006). Prevalence of the vorticellid *Zoothamnium duplicatum* as an epibiont on the cuticle of the mysid *Mesopodopsis orientalis* is negatively correlated with salinity in a mangrove estuary in Malaysia (Hanamura et al., 2012). True **ectoparasites** are mainly represented by crustaceans, namely at least three species of nicothoid copepods, and a number of epicarideans (Isopoda, currently in the suborder Cymothoida, families Asconiscidae and Dajidae). The latter are mainly present in the marsupium, occasionally also on the carapace or the pleon (Nouvel, 1951). Nicothoid copepods and epicaridean isopods show marked effects on the population dynamics of several mysids by infesting the marsupium (Daly & Damkaer, 1986; Ohtsuka et al., 2006). The leech *Mysidobdella borealis* sucks on the body fluid of its mysid hosts (Utevsky & Sorbe, 2012). According to field and laboratory studies by Allen & Allen (1981), recruitment and survival of this Boreoarctic leech are determined at the southern range of its distribution by water temperature and by the seasonal onshore-offshore migrations of its main host, the mysid *Neomysis americana*. The externally established Ellobiopsidae (phylum Myzozoa) implant a fixation organ into the tissue of the host in order to **suppress moulting** and to modify the development of secondary sexual characteristics (Coutière, 1911; Fage, 1936; Nouvel, 1941, 1954; Nouvel & Hoenigman, 1955; Vader, 1973). This may induce **castrating** and/or **feminizing effects** on males (Ohtsuka et al., 2003, 2006). Reports of **endoparasitism** regard: microsporidians (Mercier & Poisson, 1926); metacercariae of the trematode *Bunocotyle cingulata* (cf. Popova & Nikitina, 1972); larvae of the cestode *Amphilina foliacea*, whose adults infest sturgeons; first larvae of the nematode *Anisakis simplex*, whose second larva infests marine fish and whose final stage infests marine mammals (Makings, 1981); larvae of the acanthocephalan *Echinorhynchus leidyi*, which infest freshwater fish as definitive hosts (Prychitko & Nero, 1983).

ECOLOGICAL AND ECONOMIC IMPORTANCE

Contribution to biodiversity

Global diversity. – Based on a thorough evaluation of potential synonymies in 2012, a total of 1195 described extant species in 182 genera were confirmed for the orders Lophogastrida, Stygiomysida, and Mysida (see below, 'Classification'). This is 1.8% of the total of 66 914 extant crustacean species reported by Ahyong et al. (2011), who somewhat overestimated the numbers of described mysidaceans by indicating 1247 species. More conservative estimates of **global species numbers** were provided by Schram (2013), who indicated 1141 mysidaceans [data on orders pooled by us] and only 49 658 crustaceans.

Wittmann (1999) documented a more than linear but less than exponential increase of numbers described since 1860, and extrapolated a minimum number of around 4000 living species of mysidaceans. In rough accordance with this, Appeltans et al. (2012) estimated 1180 accepted plus 2090-4120 so far unknown species of marine Mysida, yielding a figure of 3270-5300 living marine species (not including continental species). The trend of descriptions is not homogeneous and currently mainly reflects benthic and benthopelagic species, whereas first descriptions of pelagic species are slowing down (Wittmann, 1999). The fraction of pelagic species among total numbers known peaked at 42% in 1910 and decreased to about 20% in 2012. Most species are strictly marine. Porter et al. (2008) estimated that "inland" species ($S = 0$-3) represent only 6.7% of "mysid" (i.e., Mysida plus Stygiomysida) diversity and are concentrated in the Palaearctic and Neotropical regions. As indicated below ('Ecological biogeography'), only about 5% live at least partly in fresh water (S = 0-0.5), with the **hotspot of diversity** in the Ponto-Caspian region. Numbers of described mysidacean species show a distinct peak at 30-40°N, with a sharp decline towards polar regions (Wittmann, 1999). The strong **hemispherical asymmetry** may reflect a major bias in the degree of research effort. In fact, a globally more symmetrical distribution pattern is obtained by expressing diversity as the number of species per 1000 individuals sampled with epi- to hyperbenthic nets in shelf areas. However, some asymmetry may have natural causes in as far as there are more extended shelf areas in the **Northern Hemisphere**.

Regional and local diversity. – Many species of **Lophogastrida** are meso- to bathypelagic with a widespread, often **transoceanic** distribution. Price et al. (2009) listed nine species for the Gulf of Mexico, eight of which show a panoceanic distribution; only one is restricted to the western Atlantic. Wittmann & Riera (2012) listed twelve species from waters off eastern Atlantic islands: ten of these are **panoceanic**, one is **endemic** to the Atlantic Ocean, and another endemic to the eastern Atlantic. There are no lophogastrids in the Arctic ocean and only a few species in the Antarctic. San Vicente (2010b) listed only panoceanic species, namely two Gnathophausiidae and one Eucopiidae, for waters south of 60°S. Unsurprisingly, there is a strong **latitudinal gradient** in the Southern Ocean, as suggested by a total of eight species, including two Lophogastridae, in the Subantarctic region, which is only fuzzily defined by the Subtropical Convergence as its northern limit.

Compared with Lophogastrida, a much higher **diversity** and degree of **endemism** are found in **Mysida**, mainly due to a larger fraction of benthopelagic and benthic species. San Vicente (2010b) reported 34 species from Antarctic and 52 from Subantarctic waters. Petryashov (2004) indicated 28 species for the Euro-Asiatic sub-basin of the Arctic Basin and adjacent seas. Price & Heard (2009) listed 52 species for the Gulf of Mexico, 13 being endemic. Striking hotspots of endemism are found in continental and semi-enclosed 'seas': Petryashov & Daneliya (2006) documented 21 Mysidae for the Caspian Sea, all belonging to the flock of Ponto-Caspian endemics. Skolka (2005) listed 20 species from coastal waters of the western Black Sea, mainly Ponto-Caspian endemics but also several species derived from Mediterranean immigrants. San Vicente (2010a) indicated 37 endemics (39%) among a total of 95 species in marine waters of the Mediterranean Sea. High levels of **molecular diversity** within and between populations of *Mesopodopsis slabberi* in the Mediterranean and north-eastern Atlantic (Remerie et al., 2006) suggest a high additional (cryptic) diversity in the Mediterranean.

Most **Stygiomysida** show a **subterranean mode of life** in an- to (mixo)euhaline waters including anchihaline groundwater, caves, and wells of subtropical to tropical zones. All seven species of *Stygiomysis* and seven out of the nine species of *Spelaeomysis* currently known are **stenoendemic** to karstic coastal waters within very restricted areas. Eleven out of these 16 species make an important contribution to the amazing subterranean biodiversity of Mexico and the Caribbean. Only *Spelaeomysis cardisomae* shows a wider distribution range, being found in small pools at the bottom of crab burrows in the Caribbean as well as along the coast of Peru (Bowman, 1973).

Environmental change and biodiversity. – Anthropogenic range expansion of certain biota is among the most important drivers of **environmental change**. Potentially detrimental effects of anthropogenic species introductions are discussed below ('Impact of bioinvaders'). According to Aladin et al. (2006), the unique fauna, including 20 endemic Mysidae, of the Caspian Sea is endangered by **negative impacts** from river flow control, poaching, water level fluctuations, pollution, exotic species, and climate change. Compared with the state of knowledge in 1930, Wittmann (2001) counted increased species numbers in the peripheral islands and submarine banks of the Gulf of Naples, probably due to intensified faunistic research. By contrast, there was a strong decrease in species numbers in the inner parts of the gulf, probably due to human impact on freshwater input, disappearance of brackish waters and fresh waters by drainage and canalization, and by pollution leading to eutrophication of coastal waters and impoverishment of the benthos. Chevaldonné & Lejeusne (2003) documented the range expansion of *Hemimysis margalefi* and the corresponding regression of the cold-water congeneric *Hemimysis speluncola* in marine caves of the Mediterranean and attributed this to **regional warming**. These processes appear to be part of a general trend of 'tropicalization' of the Mediterranean marine flora and fauna. This is linked to **climate change** and **invasion of warm-water biota** due to anthropogenic introduction and via immigration across the Suez Canal and the Strait of Gibraltar (Bianchi & Morri, 2000; Bianchi, 2007). In a wider context, Coll et al. (2010) observed a general **decreasing trend in biodiversity** from west to east and north to south in the Mediterranean Sea, linked with habitat loss and degradation, pollution, climate change, over-exploitation, and invasive species.

A **counter-balance of potential benefits and damage** suggests, that negative effects of human-driven environmental change appear to strongly prevail over positive effects. Due to current global threats to biodiversity, there is reasonable concern that current adverse trends involving human impact could already wipe out a substantial portion of the estimated minimum of 4000 mysidacean species before they have been described by science (Wittmann, 1999).

Impact on ecosystems

Role as environmental engineers. – In marine ecosystems, diurnally migrating mysids represent important **trophic linkages** between pelagic and benthic food webs by being predators and prey in both habitats (Jumars, 2007). The role of mysidaceans in transforming microplankton, detritus, and debris into animal matter is far from being negligible (Thiel, 1996; Lehtiniemi & Nordström, 2008). Mysid grazing activity can significantly affect the survival and settlement of kelp zoospores (VanMeter & Edwards, 2013). The roles of the invasive mysid *Limnomysis benedeni* in leaf litter decomposition (Aßmann et al., 2009) and in biosynthesis of polyunsaturated fatty acids (Fink, 2013) in freshwater ecosystems are discussed in the next paragraph below. In the Gulf of Finland, the least saline part of the Baltic Sea, two pelagic *Mysis* species increase the total chlorophyll-a concentration of phytoplankton in summer and the biomass of small-sized phytoplankton by means of their feeding activity and **excretion of growth-limiting nutrients** (Lindén

& Kuosa, 2004). The **bioturbation activity** of *Mysis relicta* can oxygenize the sediment surface by breaking the diffusive boundary layer; this improves benthic oxygen conditions and promotes colonization of oxygen-poor bottoms by benthic animals (Lindström & Sandberg-Kilpi, 2008).

Impact of bioinvaders. – Among the few recently published examples for **ecological benefits** of non-indigenous species, the invasive mysid *Limnomysis benedeni* may contribute to higher **secondary production** and species diversity by its role in **leaf litter decomposition** in fresh waters (Aßmann et al., 2009). Based on the **biosynthesis of polyunsaturated fatty acids** by this mysid, Fink (2013) concluded a potential upgrading of the aquatic food web for native predator species in a recently invaded gravel pit lake, temporarily connected to the Rhine River, Germany.

A major paradigm change in aquatic ecology took place with culminating evidence of **adverse impacts** from the anthropogenic **expansion of non-indigenous organisms**, including mysids: The introduction of *Mysis relicta* in Norwegian lakes caused decreased zooplankton density, particularly of cladocerans, and reduced growth of certain fish (Langeland, 1981; Koksvik et al., 1991; Langeland et al., 1991). Daphniids were diminished in a storage reservoir in the Netherlands due to the invasion of *Hemimysis anomala* (cf. Ketelaars et al., 1999). The ecosystems in lakes and coastal lagoons were significantly altered by deliberate as well as non-deliberate introductions of invertebrates, including mysids (Lasenby et al., 1986; Olenin & Leppäkoski, 1999; Ojaveer et al., 2002; Bernauer & Jansen, 2006; Walsh et al., 2012). **Biomagnification** of toxicants (PCB, Hg) may be enhanced by insertion of an additional trophic level into food webs (Rasmussen et al., 1990; Wittmann, 2005; Southward Hogan et al., 2007; Wittmann et al., 2010).

As an important component of prevention projects, Ricciardi & Rasmussen (1998) underlined the necessity of identification and risk assessment of potential **biological invaders**. Accordingly, Ricciardi et al. (2012) called for developing predictive models on the ecological impact of *Hemimysis anomala*, which is considered a major hazard to the zooplankton communities of the Laurentian Great Lakes. Ovčarenko et al. (2006) tested the tolerance of *Paramysis lacustris* and *Limnomysis benedeni* to sudden salinity changes and concluded that a salinity of $S \geqslant 30$ could be used as an appropriate biocide in **ballast water treatment** to prevent potential transfer of Ponto-Caspian mysids.

Importance in fisheries

Role as food organisms. – In marine to freshwater environments, numerous Lophogastrida and Mysida are found in the stomachs of many species of decapods (Lagardère, 1972, 1977a, b; Siegfried, 1982; Cartes & Abello, 1992; Cartes, 1993a-c) and fish (Sorbe, 1981; Mauchline, 1982; Astthorsson, 1984; Mauchline & Gordon, 1984a, b; Thiel, 1996; Dürr & González, 2002; Wittmann et al., 2004; Specziár, 2005; Specziár & Rezsu, 2009; Carrasco et al., 2012). In certain cases they may play an important role in predator diets. As a striking example, the mysid *Neomysis awatschensis* provides 88% of the winter-spring food mass for Saffron cod in a brackish pool in Kamchatka (Danilin et al., 2012). Abundance of the mysid *Orientomysis mitsukurii* shows significant effects on the growth of juvenile

Japanese flounder *Paralichthys olivaceus* in the wild (Tomiyama et al., 2013). The mysid *Limnomysis benedeni* may be important for fish nutrition because it can biosynthesize essential polyunsaturated fatty acids from dietary precursors (Fink, 2013). Mysidaceans are also important **food organisms** for marine birds, seals, and whales (Feyrer & Duffus, 2011). Upon mass occurrence near the mouth of the Danube River in coastal waters of the Black Sea, the mysid *Mesopodopsis slabberi* was formerly used by the local population to feed domestic animals (Gomoiu, 1978). Considerable amounts of certain marine mysids, such as *Neomysis*, *Schistomysis*, and *Praunus*, are sold as frozen food for aquarium fish. In India, Malaysia, and Japan, mysids are used as food for cultured fish, for water birds in small farms, for pigs, as well as for human nutrition; in this last respect, an important role is played by the dense populations of *Nanomysis* in the Java lagoons (Schuster, 1952; Nouvel, 1957). On certain coasts, mysids are used as a kind of roe. Artisanal fishermen in Java collect *Gastrosaccus yuyu* and cook it into a fried rice-flour crisp biscuit, known locally as "peyek" (Bamber & Morton, 2012). *Archaeomysis vulgaris* is used in Japan as dried food called "Ami" and as bait for fish (Matsudaira et al., 1952). *Neomysis intermedia* and *Neomysis japonica* are used in that country for "tsuku-dani", a popular food boiled in soy sauce (Murano, 1999b).

Mysids in fisheries management. – Several species of Mysinae (Mysida: Mysidae) were successfully implanted ("acclimatized") into impoundment reservoirs and various other artificial and natural water bodies of eastern and northern Europe, North America, northern Asia (Lake Aral), and Japan in order to enrich the food basis for commercially important fish (Zhuravel, 1950, 1969; Karpevich & Bokova, 1963; Murano, 1963, 1966; Sparron et al., 1964; Gasiunas, 1965, 1968; Linn & Frantz, 1965; Dediu, 1966; Daribaev, 1967; Stringer, 1967; Tjutenkov et al., 1967; Sergeeva & Ckhonelidze, 1968; Gosho, 1975; Fürst, 1981; Langeland, 1981; Bowles et al., 1991; Arbačiauskas, 2002; Aladin et al., 2003, 2004; Arbačiauskas et al., 2010). The appearance of implanted species in the stomachs of many species of fish was for several decades emphasized as a confirmation that the intentional introductions were benefitting fisheries. However, a retrospective evaluation by Arbačiauskas et al. (2010) did not support a significant enhancement of fish production after **species introductions** in Lithuanian lakes. Fürst et al. (1984) recommended introducing *Mysis relicta* into Swedish lakes as a measure to reduce parasitism by tapeworms on benthic char and whitefish. Up to the 1980s there was also the expectation that the detrimental effects of constructing impoundment reservoirs on aquatic food chains and biodiversity could be partially compensated by introducing non-indigenous species (Fürst, 1981).

Very soon, however, the expectations of generally positive effects of species introductions turned out to have been unrealistic. A major paradigm change in fisheries management took place in the 1980/90s. The accumulated data pointed to unexpected effects of deliberately introducing fish food organisms on fish growth, and more generally at the ecosystem level (see above, 'Impact of bioinvaders').

PHYLOGENY AND BIOGEOGRAPHY

Fossil record

Macrofossils. – The earliest **fossil records** of mysidaceans (or closely related taxa) are preserved as moulds in marine sedimentary rock from the Carboniferous to Jurassic. Most **Palaeozoic fossils** belong to the extinct Carboniferous to Permian Pygocephalomorpha (a species-rich order with five families), which were mostly listed as Mysidacea (compilation in Tattersall & Tattersall, 1951; Taylor et al., 2001), but also as a separate order within the Eumalacostraca in general (Schram, 1984). Schram (1986) based a separate family, the Peachocarididae (definition given below, 'Classification'), on *Peachocaris strongi* and *Peachocaris acanthouraea* from the Carboniferous of North America, and assigned his new family to the Lophogastrida.

Among **Mesozoic fossils**, true Lophogastrida with striking similarity to extant forms were *Lophogaster voultensis* (family Lophogastridae) and *Eucopia precursor* (Eucopiidae) from the Middle Jurassic of France (Secretan & Riou, 1986). The assignment of the Triassic genus *Schimperella* to the Eucopiidae also appears to be well founded, namely *Schimperella beneckei* and *Schimperella kessleri* from deposits of France (Bill, 1914), and *Schimperella acanthocercus* from Guizhou, China (Taylor et al., 2001). The previously supposed lophogastrids *Dollocaris ingens* and *Kilianicaris lerichei* from the Upper Jurassic of France are now identified as Thylacocephala (Secretan, 1985). Among Mysida, clear records are given for the Middle Jurassic *Siriella antiqua* and *Siriella carinata* (cf. Secretan & Riou, 1986). According to Schram (1986), *Elder unguiculata* and *Francocaris grimmi* from the Jurassic of Bavaria are too poorly known for a reliable assignment to the Mysida.

In contrast to microfossils, **Tertiary macrofossils** are almost unknown. The family status as Mysidae appears clear, but not so the generic status of Oligocene fossils of *Mysidopsis oligocenicus*, due to a suboptimal state of preservation (De Angeli & Rossi, 2006).

Microfossils. – In most living Mysida, hard mineralized parts predisposed for fossilization are represented mainly by fluorite or carbonate (vaterite) statoliths. So far, fluorite statoliths were recorded not as fossils but only as subfossil remains from North American seashores (Enbysk & Linger, 1966). Voicu (1974, 1981) first identified calcareous (calcite) **microfossils** (previously considered foraminiferans or calcareous algae) from Miocene deposits of the **Paratethys** as representing mysid statoliths (fig. 54.27C). At that time the Paratethys was a large brackish basin extending across Eurasia from the Vienna Basin to Lake Aral. Likewise, statoliths discovered by Fuchs (1979) in the Vienna Basin were assigned to the Recent genus *Paramysis*, but later transferred by Maissuradze & Popescu (1987) to the fossil (Miocene) genus *Sarmysis*. The basic morphology of these microfossils resembles that of Recent **calcareous statoliths**. Common peculiar features are the arch of pores enabling establishment of a 'statolith formula' (Voicu, 1974) and the hilum (Wittmann et al., 1993). The discovery of vaterite (fig. 54.27F; a hexagonal metastable polymorph of crystalline calcium carbonate) as a normal mineral component of extant calcareous statoliths (Ariani et al., 1981, 1983) revealed that the most stable crystal phase of calcite (fig. 54.27G) present in the fossil material is of secondary nature, namely derived from phase transformation of the metastable vaterite to the stable calcite (for an overview of 'mysid statolithology' see Ariani, 2004).

No fossil record is known for the Stygiomysida, which have no statoliths. Living specimens are not completely devoid of mineral structures, as *Spelaeomysis longipes* shows spherulite calcareous bodies for calcium storage in caeca of the midgut (Nath, 1972).

Phylogeny

Position within the Malacostraca. – Due to striking morphological coincidence (see above, 'Outline of the orders'), Boas (1883) accommodated the Lophogastrida and Mysida as separate suborders within the order Mysidacea, distinct from the Euphausiacea. This system remained widely accepted until the early 2000s, even though Siewing (1953), Watling (1983), Schram (1984), Dahl (1992), Martin & Davis (2001), Kobusch (1999), and others already had concluded from **phylogenetic reconstructions** based on morphological data on Recent and fossil taxa that the Mysida and Lophogastrida are to be placed separately either within, or outside the **Peracarida**, respectively. Watling (1983, 1999) discarded the Peracarida and placed the two orders separately within the Eucarida. Schram (1984) placed the Lophogastrida within the Peracarida, but the Mysida more distantly as a **sister group** of the Peracarida. Based on **foregut morphology**, Kobusch (1999) argued for a **paraphyly**, De Jong-Moreau & J.-P. Casanova (2001) for a **monophyly of the Mysidacea** within the Peracarida. A potential monophyly of the Lophogastrida and Mysida as most basal **sister taxa** within the Peracarida was also supported by the **parsimonious analysis** of 92 morphological characters by Poore (2005).

Based on the structure of the **arterial system**, particularly the descending artery, Mayrat et al. (2006) argued against the validity of the Peracarida and proposed a modified revival of the Podophthalmata, i.e., Malacostraca with stalked eyes and at least a **caridoid facies**, extended by the Cumacea. Wirkner & Richter (2007) acknowledged the homology of the descending artery and a ventral vessel, but emphasized these characters as part of the plesiomorphic caridoid facies, and thus not useful for uniting taxa into more encompassing groups. From distinct ostia patterns of the heart and other characters of the circulatory system, Wirkner & Richter (2007, 2010) and Wirkner (2009) supported the arguments brought forward by De Jong-Moreau & J.-P. Casanova (2001) for the **unity of the Mysidacea** within the Peracarida.

Jenner et al. (2009) performed, among other methods, a parsimony analysis based on 177 morphological characters. This yielded a closer relationship of the Lophogastrida and Mysida, with both taxa together forming a sister group of the remaining Peracarida. This is in line with the classical comparative morphology of these taxa, but may also reflect the persistence of shared plesiomorphic characters in the two (sub)orders.

Jarman et al. (2000) concluded from **28S rDNA** data in favour of a **polyphyly of the Mysidacea**, with the Mysida closer to the Euphausiacea, and rather distant from the

Lophogastrida. Martin & Davis (2001) critically evaluated the morphological and genetic literature available at that time and concluded that the Lophogastrida and Mysida are to be installed as separate orders, but both within Peracarida. Based on **18S rDNA** phylogenetic analysis, Spears et al. (2005) obtained the Lophogastrida as part of a monophyletic Peracarida **clade**, whereas the Mysida were allied with some uncertainty to the Stomatopoda, Syncarida, and Eucarida — in any case at great distance from the Lophogastrida. Similarly, based on 18S and 28S rDNA sequences, in the analysis of Babbitt & Patel (2005) the Mysida emerged close to the Euphausiacea and Anaspidacea (Syncarida), but distant from the Peracarida (Lophogastrida not examined). Meland & Willassen (2007) added a great number of taxa to the dataset of Spears et al. (2005) and concluded to a **disunity of the Mysidacea**. In that scheme, the Mysida form a monophyletic group with low resolution allied to the Euphausiacea and Stomatopoda (unlike Spears et al., 2005, this clade not including Syncarida and Decapoda). The Lophogastrida and Stygiomysida, in contrast, are located at distant positions from each other within the Peracarida, with the Stygiomysida therein consistently close to the Mictacea. Jenner et al. (2009) compared different **methods of phylogenetic analysis** based on four nuclear ribosomal and mitochondrial loci, a morphological dataset, and combinations of these. From the various methods and their variants — supporting strikingly different trees — they concluded "that existing molecular and morphological evidence is unable to resolve a well-supported eumalacostracan phylogeny". Wills et al. (2009) extended the study of Jenner et al. (2009) with palaeontological data and obtained very poor coincidence between morphological and stratigraphic signals. A major obstacle to robust phylogenetic reconstruction can be an **early initial radiation**. In fact, this is indicated by the fossil record, insofar as the Eumalacostraca might have appeared in the Devonian, and their major branches, including the Lophogastrida, were already well formed in the Carboniferous.

Evolution within the Mysida and Lophogastrida. – The uncanny similarity of certain Jurassic Lophogastrida and Mysida with Recent genera (see above, 'Fossil record') also points to a comparatively **early radiation**, in this case within these orders. Unlike most large crustacean orders, only few **genetic analyses** are available to highlight the relationships within the mysidacean orders discussed here. Based on foregut morphology and **16S mitochondrial rRNA** sequences, J.-P. Casanova et al. (1998) showed that Eucopiidae are well defined and derived from **common ancestors** with the Gnathophausiidae (quoted as Lophogastridae). Based solely on **foregut morphology**, De Jong-Moreau & J.-P. Casanova (2001) proposed a basal state of *Gnathophausia* versus *Lophogaster* and *Eucopia*. In contrast, a cladogram by Meland & Willassen (2007: fig. 2b), derived from nuclear 18S rDNA sequences, hints at Eucopiidae and Gnathophausiidae being potential sister groups, with common roots with/in the Lophogastridae. Nonetheless, additional data are required for reliable conclusions in this context.

Remerie et al. (2004) presented a **parsimonious tree** based on **18S rRNA** sequence data of 25 species of Mysidae. This approach supported the monophyly of subfamilies, morphologically and taxonomically already distinguished by Hansen (1910), namely the Siriellinae and the Gastrosaccinae, whereas part of the Mysinae (as Mysini-A clade) un-

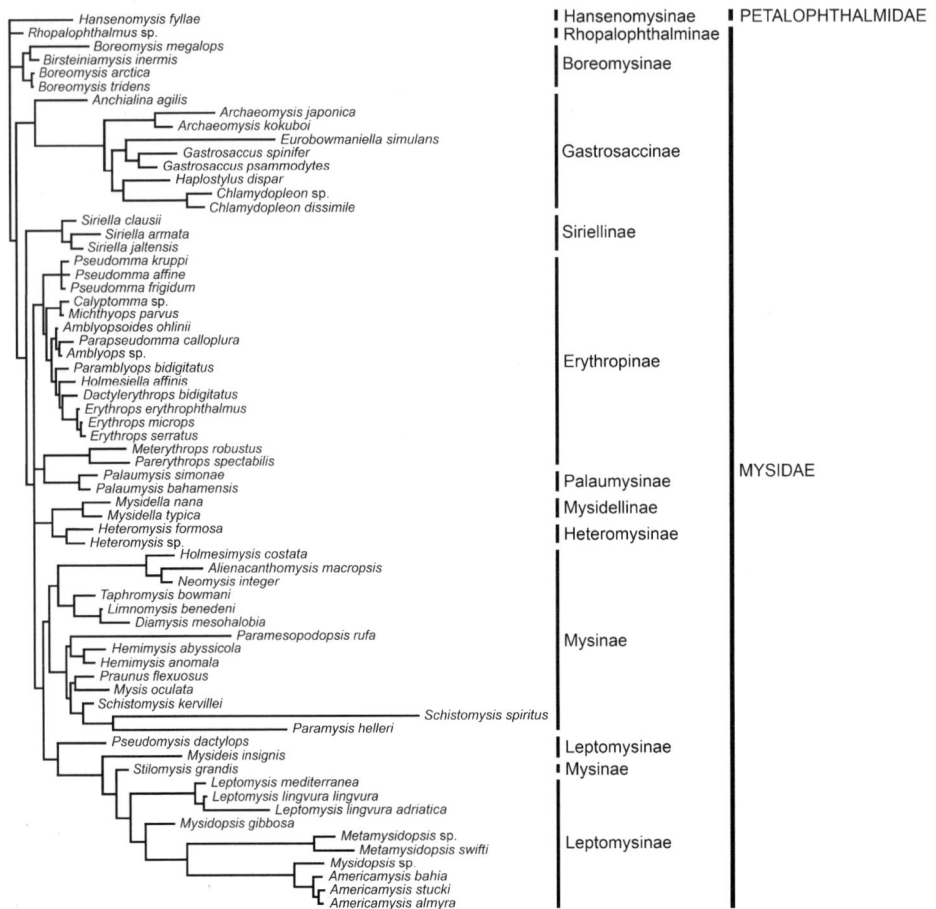

Fig. 54.42. Molecular phylogeny of the Mysida estimated from a Bayesian analysis of nuclear 18S rDNA. The tree represents all families and almost all subfamilies, except Petalophthalminae, acknowledged in this order. For authorities of species names see 'Appendix', for genus names see 'Classification'. [Modified after Meland & Willassen, 2007.]

expectedly appeared as a sister group of the Siriellinae. Based on 67 (sub)species (fig. 54.42), Meland & Willassen (2007) obtained indication of monophyly for additional subfamilies of Hansen (1910), namely the Boreomysinae, Gastrosaccinae, and Mysidellinae, whereas the Mysinae appeared again as a heterogeneous assemblage but with a different pattern compared to that obtained by Remerie et al. (2004). Meland & Willassen (2007) diminished the numbers of poly-/paraphyletic taxa by raising the former Mysinae tribes Erythropini, Mancomysini (now Palaumysinae), Heteromysini, and Leptomysini to subfamily level. Nonetheless, some problems remained in the now upgraded Erythropinae and Leptomysinae as a task for future research. Mitochondrial and nuclear DNA sequences of no less than 77 species of Mysidae, but covering a smaller number of subfamilies compared to Meland & Willassen (2007), were presented by Porter (2005). Her parsimonious tree showed more or less similar clades but in some case fixed at different branches. It also included a greater number of unexpected single taxa into tribes and subfamilies.

Morphological and biomineralogical analysis. – Evidence from the morphology of extant animals, **macrofossils** and **microfossils**, as well as from the biomineralogy of static bodies (Ariani et al., 1993), enables tracing a possible **phylogeny of the static organ**.

Among mysidaceans without a distinct static organ, the Lophogastrida are undoubtedly the most primitive forms, as shown by their fossil record dating back to the Carboniferous and the Triassic. Most of their recent species show an oceanic life habit. Together with the family Petalophthalmidae among Mysida, and with the Stygiomysida, they have conserved ancestral (plesiomorphic) characters of the Peracarida (if considered a natural group), particularly seven pairs of oostegites, biramous male pleopods, and the absence of statocysts in the tail fan.

As mentioned above, due to uncanny morphological coincidences, Secretan & Riou (1986) assigned Middle Jurassic fossils to the Recent Lophogastrida genera *Eucopia* and *Lophogaster* and to the Recent Mysida genus *Siriella*. They concluded that there was little morphological change between the Jurassic forms and their extant congenerics. For the Mysida, however, they were unable to establish the potential presence or absence of a statocyst in the endopods of the uropods, which is among the most important **autapomorphic features** of the family Mysidae within this order. Similarly, the (badly preserved) body remains of an Oligocene species are strongly reminiscent of Recent mysids, so they were tentatively ascribed as belonging to the living genus *Mysidopsis* (De Angeli & Rossi, 2006).

The comparatively large (with respect to body size) statoliths contained in the statocysts can shed some light on the evolution within the family Mysidae (Ariani et al., 1993). As discussed by Schlacher et al. (1992), the subfamilies Boreomysinae (pelagic) and Rhopalophthalminae (benthopelagic) have conserved certain **plesiomorphic features** such as subdivided exopods of uropods and biramous, multi-segmented male pleopods or, as in Boreomysinae, seven pairs of oostegites. In both subfamilies, the statoliths are exclusively non-crystalline (organic). In contrast, the remaining subfamilies (Siriellinae, Gastrosaccinae, Mysinae, Mysidellinae) show mineralized statoliths and are characterized by two or three pairs of brood lamellae, reduced pleopods in the females, and mostly modified pleopods in males. The apparent correlation with plesiomorphic characteristics suggests that **organic statoliths** may have preceded mineral ones, as also shown by ontogenetic evidence from the post-moult development of calcareous statoliths (Ariani et al., 1982). In contrast to carbonate statoliths, **fluorite statoliths** are found in most recent species, but are still unknown in the fossil record. The fluorite-bearing species constitute the bulk of the extant, benthopelagic to benthic forms inhabiting the coastal and littoral zones of all oceans. The late appearance of **carbonate statoliths** as fossil remains (Lower Miocene according to Voicu, 1981) suggests a post-Cretaceous origin of at least some genera in the subfamily Mysinae.

Both the scarcity of fossil mysid statoliths outside **Paratethyan sediments** and biogeographical considerations suggest that carbonate statoliths developed in fresh to brackish waters of the Ponto-Caspian region during the Miocene. Selective pressure may have been at work here, inducing precipitation of a mineral with lower specific gravity than fluorite, and with a lower demand for fluorine, often poorly available in fresh water (Ariani et al., 1983). Thus, the precipitation of calcium carbonate as the crystal phase of vaterite is found in species of *Diamysis*, *Katamysis*, *Limnomysis*, *Schistomysis*, and particularly *Paramysis*, most of which live or have their relatives in the Ponto-Caspian region. Nonetheless, mysids with **vaterite statoliths** belonging to the genus *Antromysis* as well as to certain recently discovered taxa (K. J. Wittmann, unpublished) live in the Caribbean-Amazonian region, where an analogous evolutionary process may have occurred.

Biogeography

Historical biogeography. – The problematic distinction between the models of **ecological biogeography** and those of **historical biogeography** (Parenti & Ebach, 2009) is exemplified by the distribution patterns of certain taxa of 'Mysidacea'. For example, the distribution of the subterranean genus *Stygiomysis* (family Stygiomysidae) is restricted to the Mediterranean and the Caribbean plus surrounding areas, thus well marking the western to central range of the **Tethys Sea** (Ruffo, 1957; Pesce, 1985; Pesce & Iliffe, 2002). During the Mesozoic this sea extended from the predecessor of the Caribbean eastward to what is now the Indian Ocean. Only one additional genus in this order, namely *Spelaeomysis* (family Lepidomysidae), shows a more circumtropical distribution, but is missing in the central and western Pacific. The 16 known species of Stygiomysida are characterized by mostly separate eyestalks (fused in *Spelaeomysis longipes*) with only few ommatidia or no visual elements at all. Most species live in fresh water to (mixo)euhaline subterranean waters, with only *Spelaeomysis cardisomae* inhabiting small pools at the bottom of crab burrows (Bowman, 1973), and *Spelaeomysis cochinensis* inhabiting prawn culture fields (Panampunnayil & Viswakumar, 1991). The species of *Spelaeomysis* typically prefer dark to dimly lit environments; only *Spelaeomysis cochinensis* is fully exposed to daylight. Remarkably, *Spelaeomysis bottazzii* from a brackish well in Apulia (south-eastern Italy) dwells around the margins of the **photic zone** for feeding, but seeks shelter in the deep groundwater during the very long incubation of the young (Ariani & Wittmann, 2010).

Among Mysida from temperate regions, the mainly freshwater to brackish genus *Paramysis* offers a good example for a centre of origin (Parenti & Ebach, 2009) as hypothesized according to classical historical biogeography. This is the **Ponto-Caspian** region (a successor of the **Paratethys**) where 15 out of 23 species live today. The few remaining species, some having a marine habit, are considered to be of the same origin from both geographical and ecological points of view. This is also supported by genetic data of Audzijonyte et al. (2008a). In fact, with only one exception, i.e., the fluorite-bearing *Paramysis arenosa*, they share with the Ponto-Caspian species the carbonate (vaterite) nature of their statoliths (Wittmann & Ariani, 2011). Similarly, all Mediterranean species of *Diamysis* have vaterite statoliths (Ariani et al., 1981; Wittmann & Ariani, 1998, 2012a, b; Ariani & Wittmann, 2000, 2002), unlike the great majority of Mediterranean mysids with fluorite statoliths. In light of the post-Miocene geo-historical events (Hsü et al., 1977; Hsü, 1978) the *Diamysis* species also appear to have a **Paratethyan origin**; in contrast, most other Mediterranean mysids (like marine animals of this region in general) are undoubtedly of **Atlantic origin**.

So far there is no clear evidence that Red Sea mysidaceans contributed to the Mediterranean fauna prior to the opening of the Suez Canal. The lophogastrid *Lophogaster affinis*, first described from the Red Sea by Colosi (1930), was until just recently cited in faunal lists for the Mediterranean (Nikoforos, 2002; San Vicente, 2010a), but the Mediterranean populations actually belong to *Lophogaster subglaber*, known only from the East Atlantic and Mediterranean (Wittmann & Riera, 2012). Conversely, previous erroneous reports of the East-Atlantic-Mediterranean endemic *Lophogaster typicus* from the Red Sea (Băcescu, 1985) and Indian Ocean (Pillai, 1973) have been attributed (Wittmann et al., 2004) to the Red Sea endemic *Lophogaster erythraeus*.

Many populations from cold-temperate to Subarctic freshwater lakes of the Northern Hemisphere are considered **glacial relicts** and were previously lumped together as *Mysis relicta* (cf. Thienemann, 1925, 1928a, b; Banner, 1948; Holmquist, 1959, 1962). Based on molecular and morphological data, Audzijonyte & Väinölä (2005, 2006) split *Mysis relicta* into four species, and demonstrated that the relict populations belong to *Mysis relicta* and *Mysis salemaai* in northern Europe, to *Mysis diluviana* in North America. Four additional species of this genus described by G. O. Sars (1895, 1907) are steno-endemic of the Caspian Sea. Their ancestors may have invaded this lake from the north, along drainage systems redirected to the south by **glacier barrages** in the **Pleistocene** (Holmquist, 1959).

Ecological biogeography. – The **Lophogastrida** appear to be generally **stenoecious**, at least when compared with the Mysida. Unlike the remaining two mysidacean orders, the Lophogastrida are exclusively marine. In addition, they are more stenothermic as indicated by their absence in the Arctic Ocean and by modest species numbers in Antarctic and Subantarctic waters (Petryashov, 2007, 2009; San Vicente, 2010b), even though most are deep-water species that are less exposed to temperature variations. Compared to the mainly bathypelagic and in part **panoceanic** Gnathophausiidae and Eucopiidae, the more species-rich Lophogastridae show a more (near-)coastal, benthopelagic distribution, and a somewhat higher **degree of endemism**. According to Hargreaves (1989), the large-scale horizontal distribution of four bathypelagic species of Gnathophausiidae in the eastern Atlantic is correlated with hydrographical boundaries defined by temperature and oxygen content.

The **Mysida** range from tropical to Arctic zones. The numbers of both benthic and benthopelagic species exceed the pelagic ones. The few panoceanic species show a mostly **circumtropical distribution**, as exemplified by the epipelagic *Anchialina typica* (Gastrosaccinae) and *Siriella thompsonii* (Siriellinae). Most pelagic forms are mesopelagic representatives of the Erythropinae. This subfamily also contains **bathypelagic** and **archibenthic** species, mostly belonging to the species-rich genus *Pseudomma*. Adaptations to deep-water environments are evident in the long thoracic endopods of many *Pseudomma* spp. and in the fused eyestalks without ommatidia in all of them (Meland, 2004). Most mysids are strictly marine, nonetheless about 5% live at least partly in freshwater environments (S = 0-0.5). Most **freshwater species** belong to the subfamily Mysinae. According to

Porter et al. (2008), the "inland" Mysida are mainly euryhaline estuarine species, autochthonous Ponto-Caspian endemics, or glacial relicts.

Areal biogeography. – Petryashov (2005, 2007, 2009) provided extensive information on **areal biogeography** and **bathymetric distribution** of the **Lophogastrida** in the Arctic, North Atlantic, North Pacific, and the (Sub-)Antarctic. Several bathypelagic species have a panoceanic distribution (Brandt et al., 2012), but all of them are missing in the Arctic Ocean (e.g., *Eucopia australis*, *Gnathophausia zoea*, *Gnathophausia gigas*) and most of them also in the Southern Ocean (e.g., *Eucopia grimaldii*, *Eucopia sculpticauda*, *Chalaraspidum alatum*) (Brandt et al., 1998; Petryashov, 2007, 2009; Wittmann & Riera, 2012). Species richness in the Indo-Pacific realms is generally well above that in the Atlantic, particularly regarding *Lophogaster* and *Paralophogaster*. A few examples for regional vicariance exist, such as the NE-Atlantic-Mediterranean *Lophogaster typicus* versus the West Atlantic *Lophogaster longirostris*.

All species-rich subfamilies of the Mysidae (**Mysida**) are found around the globe, again with some exceptions for the Arctic Ocean. However, the frequency of congeneric species in the Indo-Pacific and the Atlantic realms may differ considerably. For example, besides the **cosmopolitan** *Siriella thompsonii*, 48 species of *Siriella* are reported from the Indo-Pacific, and only 13 from the Atlantic (Murano & Fukuoka, 2008). From distributional and rDNA-based **molecular clock data**, Meland & Willassen (2004) explained the differences between ancestral Atlantic and Pacific lineages in the genus *Pseudomma* by the presence of this genus in the **Tethys Sea** already in the Oligocene, followed by subsequent allopatric divergence due to isolation of the two oceans in the Miocene. Petryashov (2005, 2007, 2009) provided ample data on the areal distribution of mysidaceans in temperate to Arctic zones, Daneliya & Petryashov (2011) and Daneliya et al. (2012) for mysids of the **Ponto-Caspian** and additional Eurasian waters. Using similarity indices based on species presence, they delineated biogeographical subdivisions for the Arctic, North Atlantic, Ponto-Caspian, northern Pacific, Subantarctic, and Antarctic. Notably, Petryashov (2007) distinguished a Notantarctic realm for the pelagic species of Lophogastrida plus Mysida versus separate Antarctic and Notal realms for the more numerous benthopelagic species of Mysida (fig. 54.43). In both cases the realms were further subdivided into provinces and regions.

The Arctic and the Antarctic mysid faunas differ greatly, although a **bipolar distribution** pattern was reported for three deep-water species (Petryashov, 2007). The warm-temperate to tropical faunas of benthopelagic and benthic mysids show great differences at the species level between the east and west coasts of both the Atlantic and Pacific oceans, less strongly also of the Indian Ocean. Several benthopelagic and benthic genera (e.g., *Acanthomysis* and *Heteromysis*) are found around the globe, but most show **endemism** at all levels: from oceans (*Indomysis*, *Kainommatomysis*), small sea basins (*Calyptomma*), down to certain submarine caves (*Bermudamysis*) or even a single freshwater cave (*Troglomysis*).

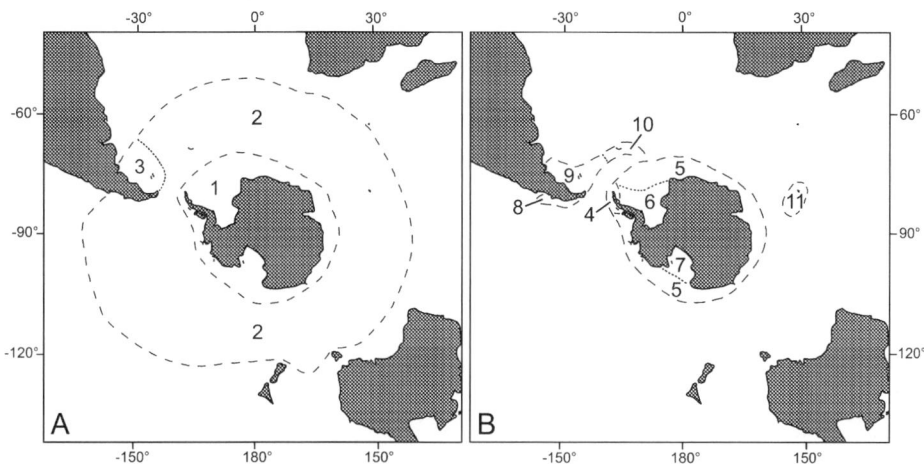

Fig. 54.43. Biogeographical division of pelagic (A) and benthopelagic (B) mysidacean (Lophogastrida plus Mysida) faunas in the Antarctic and Subantarctic. A, Notantarctic realm (1-3) comprises Circumantarctic (1) and Notal (2-3) provinces, including Patagonian region (3); B, Antarctic realm (4-7) comprises West Antarctic (4) and East Antarctic (5) provinces, the latter includes the Weddell Sea region (6) and Ross Sea region (7); Notal realm (8-11) comprises South Chilean (8), Patagonian (9, 10), and Kerguelen (11) provinces; South Georgia region (10) pertains to the Patagonian province (9, 10). Dashed lines delimit provinces, dotted (almost continuous) lines regions within provinces. [Modified after Petryashov, 2007, using his terminology.]

Anthropogenic range extension. – Recent, sometimes ample, enlargements of the previously known distribution range of certain species of the subfamily Mysinae are mainly due to **anthropogenic dispersal** mechanisms. As already discussed above ('Importance for fisheries'), several species were intentionally introduced into fresh waters of Europe, northern Asia, and North America in order to enrich the food basis for commercially important fish. The freshwater mysids involved were species of *Neomysis* implanted into lakes of Japan, **glacial relict** species of *Mysis* into waters of northern Europe and North America, and to a major extent the **Ponto-Caspian** endemics *Paramysis*, *Limnomysis*, and *Hemimysis* into waters ranging from central Europe to eastern Asia (Aral Sea). From the sites of introduction they often spread elsewhere on their own means or as a consequence of human activity. Some older authors emphasized potential benefits of species introductions for biodiversity: introductions, including those of mysids, may increase species numbers and favour recovery from the losses caused by impoundment of water bodies (Fürst, 1981) or from the heavy natural losses at the last **glaciation** (Leppäkoski, 1984). Today, the prevailing concern is that native species will be crowded out by introduced ones. Scenarios presented by Sala et al. (2000) point to **biotic exchange** as the greatest **threat to freshwater biodiversity** for the 21st century.

Non-intentional introductions in freshwater bodies were mainly related to fish stocking, aquaculture, aquaristics, and navigation; the latter involved transport in ballast water, bilge water, and cooling water filters (Gollasch, 1996; Reinhold & Tittizer, 1998; Wittmann et al., 1999; Minchin & Rosenthal, 2002; Grigorovich et al., 2003; Nehring, 2005; Ovčarenko et al., 2006; Habermehl, 2008; Wittmann & Ariani, 2009; Borza et al., 2011; Arbačiauskas et al., 2011, 2012). The most important factor for **transcontinental dispersion** appears to be the connection of previously separate drainage systems by newly constructed waterways (fig. 54.44; Tittizer et al., 2000; Leppäkoski et al., 2002; Wittmann, 2007; Wittmann & Ariani, 2009). The importance of **connected waterways** for the dispersion of Ponto-Caspian invertebrates across continental Europe is underlined by the popularized term 'invasion highways' (Bij de Vaate et al., 2002; Slynko et al., 2002; Ketelaars, 2004; Pienimäki & Leppäkoski, 2004; Pöckl et al., 2011).

A stupendous case is that of *Limnomysis benedeni* (fig. 54.44A), which went up the Danube River from the Black Sea (Wittmann, 1995, 2007) in at least three independent waves (Audzijonyte et al., 2009). From here it colonized waters in vast areas of continental Europe; remarkably it reached the Mediterranean coast by dispersing along rivers and artificial waterways of France (Wittmann & Ariani, 2000, 2009). Within less than two decades the Ponto-Caspian freshwater to brackish species *Hemimysis anomala* (fig. 54.44B) spread mainly along waterways to most of continental Europe, England, Ireland, and even to the North American Great Lakes (Geissen, 1997; Ricciardi & Rasmussen, 1998; Schleuter et al., 1998; Wittmann & Ariani, 2000; Holdich et al., 2006; Ricciardi, 2007; Audzijonyte et al., 2008b; Minchin & Holmes, 2008; Stubbington et al., 2008; Borza et al., 2011; Borza & Boda, 2013). The essentially brackish-water species *Diamysis lagunaris* reached the Atlantic coasts of Spain and Portugal, possibly by **anthropogenic transfer** from Mediterranean lagoons (Cunha et al., 2000), although an indigenous status of the Atlantic populations is not excluded (Wittmann & Ariani, 2012a).

The Mysida also contributed to the world-wide anthropogenic dispersal of non-indigenous marine invertebrates. This is commonly attributed to transport in **ballast water** (Carlton, 1985; Carlton & Geller, 1993), whereas other dispersal mechanisms such as escape from aquaria, aquaculture, and trade with living aquatic plants or animals appear unlikely but cannot be excluded for mysids. Similar to the findings in fresh water, most currently known marine cases pertain to the subfamily Mysinae. Probably ballast water mediated, **transoceanic transfers** of Mysinae are exemplified by *Praunus flexuosus* from Europe to the U.S. east coast, *Neomysis americana* from the U.S. east coast to Argentina and Europe, *Neomysis japonica* from Japan to Australia, and *Acanthomysis aspera* from Japan to the U.S. west coast (Wigley, 1963; Hoffmeyer, 1990; Carlton & Geller, 1993; Ruiz et al., 2000; Wittmann et al., 2012).

In contrast to fish, decapods, opisthobranchs, and numerous other invertebrates (Ariani & Serra, 1969; Por, 1978; Dulčić et al., 2004; Yokes & Rudman, 2004; Koukouras et al., 2010), there is few evidence for a '**Lessepsian migration**' of Lophogastrida and Mysida from the Red Sea via the Suez Canal to the Mediterranean or *vice versa*. The mysid *Kainommatomysis foxi* is probably endemic to the Red Sea (Băcescu, 1973c; Almeida Prado-Por, 1980) but is also found at three stations along the Suez Canal (Por & Ferber, 1972), whereby the northern-most station is the type locality at the Mediterranean coast (W. M. Tattersall, 1927). No additional records of *Kainommatomysis* have so far been reported for the Mediterranean. *Pyroleptomysis rubra* is endemic to the Mediterranean, and so far only one specimen was recorded in the Red Sea (Wittmann, 1985).

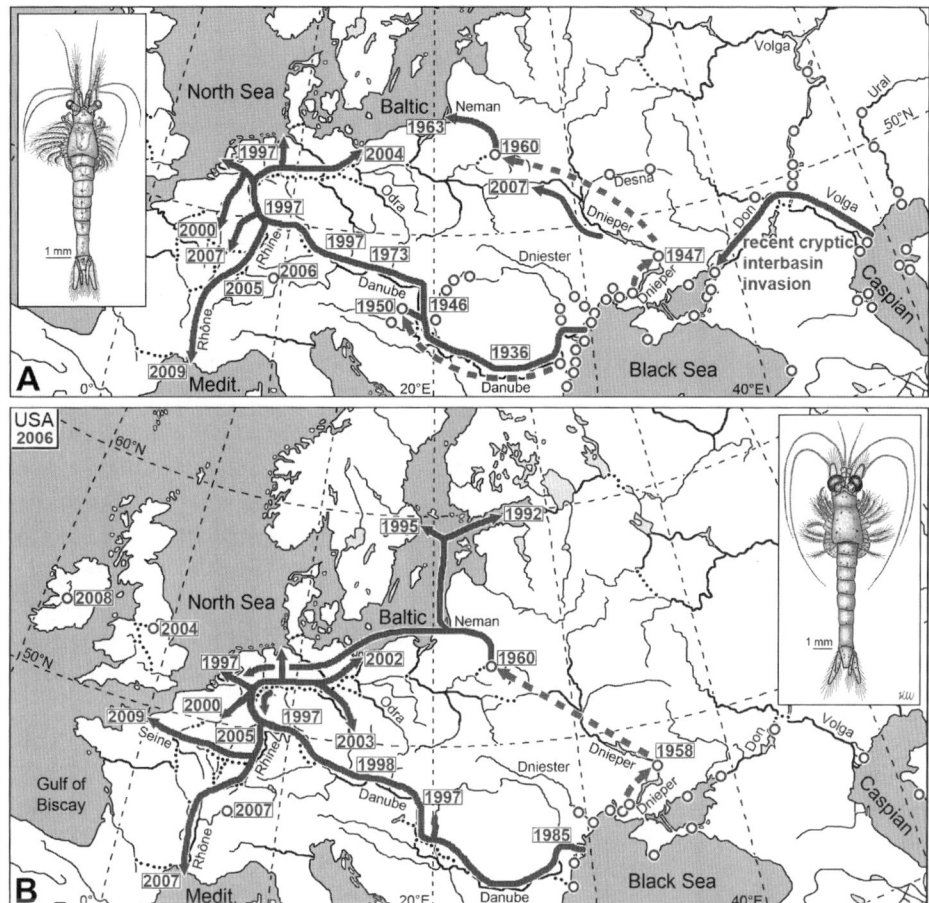

Fig. 54.44. Expansion pathways of the mysids *Limnomysis benedeni* Czerniavsky, 1882 (A), and *Hemimysis anomala* G. O. Sars, 1907 (B), from the Ponto-Caspian to western Europe and beyond. Continuous heavy arrows indicate spread along waterways; dashed heavy arrows indicate deliberate transplantations. Small dotted lines are artificial navigation canals. Years stand for first records or for transplantations, respectively. [Original; data updated from Wittmann & Ariani, 2009.]

SYSTEMATICS

Guidelines to classification

Although **phylogenetic trees** for the three extant orders traditionally assigned to the Mysidacea are far from being consolidated (see above, 'Phylogeny'), for the time being the scheme of Meland & Willassen (2007) is accepted. Already much earlier, Tchindonova (1981) had concluded from morphological data that (in her terminology) Lophogastrina, Stygiomysina, Petalophthalmina, and Mysina represent different entities, altogether ranked by her at suborder level within the Mysidacea. Meland & Willassen (2007) estimated from genetic data that the Mysidacea actually represent a **polyphyletic assemblage** of three orders, the Lophogastrida, Stygiomysida, and Mysida. Within the order Mysida, they elevated most former tribes to subfamily level. They did not comment on the **classification** by Zimmer (1909), who already a century earlier had defined — with only one exception — the same families and subfamilies at the same rank. The main difference was that he placed them within the "suborder Mysidacea" (of course not considering taxa based on animals discovered after 1909). The recent re-establishment of subfamily ranks liberates space to re-establish previously neglected tribes, well based by previous authors on morphological evidence, and to establish eight new tribes within these subfamilies.

Genetic evidence can give supplementary as well as complementary information for estimation of phylogenies and thus also for classification of morphologically similar taxa. The trees published by Remerie et al. (2004), Porter (2005), and Meland & Willassen (2007) (fig. 54.42) for taxa of Mysidae are based on various mitochondrial and nuclear DNA (RNA) sequences. They show some congruent patterns but also strong inconsistencies between and especially within subfamilies, possibly due to different taxa coverage, gene sampling, and statistical treatment. Particularly taxon sampling (numbers and diversity of taxa) can strongly affect **phylogenetic reconstruction** (Lecointre et al., 1993). In conclusion, available genetic data can provide only limited information in the present context. The limitations are not surprising considering that phylogenetic reconstruction is hampered by the great age, i.e., early radiation, of mysid morphotypes, as suggested by the stupefying coincidence between Jurassic fossils (Secretan & Riou, 1986) and living genera.

Based on the above considerations, the classical scheme developed by Hansen (1910), continued and refined by Illig (1930), Nouvel (1943), W. M. Tattersall (1951), O. S. Tattersall (1955), Tchindonova (1981), Băcescu & Iliffe (1986), Murano (1986, 1999b), Nouvel et al. (1999), and many others, is essentially adopted here, with certain higher taxa ranked according to Zimmer (1909) and Meland & Willassen (2007). Norman (1892) based the classification within the family Mysidae mainly on the structure of male pleopods. This character has been in continuous use as the most important **diagnostic tool**. This and a number of additional morphological features (including the mineral composition of static bodies, if present) were used in the present contribution to give *ex novo* more precise and consistent diagnoses of the higher taxa. The present approach was based on physical examination of 73 genera and on published diagnoses available for 186 currently acknowledged genera (including fossils) pertaining to the three orders discussed here. Previously forgotten and dismissed taxa above genus level were carefully reconsidered. The characters of these and proposed new taxa were cross-checked among each other for consistency within the ranges of uncertainty inherent to **phylogenetic analyses** (see above, 'Phylogeny'). Summing up, the (re)insertion of tribes as an additional level of classification of the Mysida is certainly at its beginning and could profit from improvement and consolidation. Such a higher level of **taxonomic resolution** is intended to help prompt all branches of biology to take a closer view at the Mysida, for example regarding morphology, biogeography, biodiversity, ecology, physiology, biomineralogy, and biochemistry. Finally, the keys given below could be helpful for a short, synthetic look at the structure of the morphological relationships between the various higher taxa.

Classification

The first author's taxonomic database has been in continuous development since 1981. It was updated by a thorough evaluation of potential synonymies and by crosschecking with the base of Mees & Meland (2012). This resulted in many modifications of both data holdings. At the cut-off date 31 December 2012, the total numbers of acknowledged taxa were 182 genera with 1195 extant species described for the three orders pooled together. The Lophogastrida contributed with seven genera containing 52 species, the Mysida with 173 genera and 1127 species, and the Stygiomysida with two genera and 16 species. An additional 13 fossil species were ascribed to a total of four extant and three exclusively fossil genera (details below). Additions after the cut-off date were one species to the extant Lophogastrida and three genera and five species (minus one put into synonymy) to the extant Mysida. The additions and the one withdrawal were integrated in the lists of taxa, below.

For each order, diagnostic characters are given below down to the tribe level. The respective taxa inventories are listed down to generic level together with species numbers in parentheses. The sign † indicates **fossil taxa**. The families, subfamilies, and tribes are listed approximately in order of decreasing morphological affinity:

Order LOPHOGASTRIDA Boas, 1883. Carboniferous-Recent (fig. 54.2).

Definition. – Eumalacostraca with well developed, stalked eyes; carapace projects freely over several posterior thoracomeres; external gills on the coxa of thoracopods 2-7 (8) present at least in extant taxa, gills covered entirely or in part by the lateral extensions of the branchiostegal carapace; very large (fig. 54.16A, B) foliaceous thoracic epipod 1 penetrates into carapace cavity; thoracic exopod 1 reduced to a blade or rod in extant taxa, exopods 2-8 normally well developed, natatory (exopod 8 missing in *Ceratolepis*); thoracic endopods vary according to family; females with 7 pairs of oostegites; at least extant taxa with gonopores on the coxae of thoracopods 6 (females) or 8 (males), the simple slot-like male orifices are not elevated or are on top of a small elevation, no lobes around orifice and no tubular penes present; all pleopods biramous, well developed in both sexes, not or weakly modified in males; tail fan strong, endopods of uropods without statocyst; developmental stages as given below for the Mysida.

Remarks. – Generally large species. Three extant and one fossil family acknowledged. The four families can be distinguished by the numbers of thoracic endopods specialized as maxillipeds (gnathopods).

†PEACHOCARIDIDAE Schram, 1986. Carboniferous.

Definition (modified from Schram, 1986). – Integument weakly calcified; thoracic endopod 1 not specialized, all thoracopods alike, well-developed peduncle at the base of the exopods; pleomeres 1-5 with well-developed pleural plates, ultimate pleomere (as far as known) not divided by a transversal furrow; flat flap-like pleopods; endopod of uropods undivided and without statocyst.

One fossil genus (type). – †*Peachocaris* Schram, 1976 (2 species).

LOPHOGASTRIDAE G. O. Sars, 1870. Jurassic-Recent (fig. 54.2B, C).

Definition. – Thoracic endopods 1, 2 developed as maxillipeds; thoracopods 3-8 normal, usually almost uniform (however, exopod 8 missing in *Ceratolepis hamata*); all pleomeres with well-developed pleural plates; ultimate pleomere more or less distinctly divided by a transversal furrow; both rami of the uropods undivided.

Type genus. – *Lophogaster* M. Sars, 1856.

Inventory. – Five genera with a total of 34 extant species included: *Ceratolepis* G. O. Sars, 1884 (1 species), *Chalaraspidum* Willemoës-Suhm, 1874 (1), *Lophogaster* (21 extant and one fossil species), *Paralophogaster* M. Sars, 1870 (10), *Pseudochalaraspidum* Birstein & Tchindonova, 1962 (1).

GNATHOPHAUSIIDAE Udrescu, 1984 (figs. 54.1A, 54.2A).

Definition. – Integument strongly calcified; rostrum elongate, triquetrous, denticulate; thoracic exopod 1 very small or missing, exopods 2-8 well developed and multi-segmented; thoracic endopod 1 developed as maxilliped; thoracopods 2-8 nearly uniform, their endopods pediform; all pleomeres with well-projecting, bilobate pleural plates; ultimate pleomere more or less distinctly divided by a transversal furrow; exopod of uropods divided by a subapical suture, endopod undivided; telson large, dorsally with two long keels, telson constricted near the base, its apex crescent-shaped.

Only the type genus acknowledged. – *Gnathophausia* Willemoës-Suhm, 1873 (11 species).

EUCOPIIDAE G. O. Sars, 1885. Triassic-Recent (figs. 54.2D, 54.19A).

Definition. – Integument comparatively soft; antennal scale with short apical segment; thoracic endopods 1-4 relatively small and specialized as maxillipeds (gnathopods), endopods 2-4 short but strong and subchelate (fig. 54.19A, B), endopods 5-7 very long, thin and also subchelate, only endopod 8 with normal, pediform structure; pleomeres without projecting pleural plates; exopod of uropods divided by a subapical suture, endopod undivided; telson without apical cleft.

Type genus. – *Eucopia* Dana, 1852 (8 extant and one fossil species). No additional recent genera known.

One fossil genus. – †*Schimperella* Bill, 1914 (3 species).

Order STYGIOMYSIDA Tchindonova, 1981 (fig. 54.3).

Definition. – Eumalacostraca with eyestalks generally separate (fused only in *Spelaeomysis longipes*), ranging from vestigial to well developed and moveable, cornea missing or reduced to a small number of ommatidia; sizes of the respiratory carapace and of the foliaceous thoracic epipod 1 vary according to family; gills absent; first pair of thoracopods with the exopod reduced to a leaf-like blade, exopods 2-8 well developed, natatory; anterior 3-4 pairs of thoracic endopods modified as maxillipeds, remaining endopods pediform (among a total of 8 pairs); 4-7 pairs of oostegites; gonopores on the coxae of thoracopods 6 (females) or 8 (males), male gonopores flanked by an anterior setose and a posterior bare lobe, no penes present; pleopods biramous, but somewhat rudimentary in both sexes, sympod and endopod unsegmented, but exopod with a small number of segments; tail fan strong, sympod of uropod with spinose inner lobe, statocysts absent; developmental stages as given below for the Mysida.

Remarks. – Two families with one genus each, total of 16 species. Most species stygobiotic (cavernicolous or phreatobious). No fossils known.

LEPIDOMYSIDAE Clarke, 1961 (figs. 54.1C, 54.3B-D, 54.19C, D).

Definition. – Antennal scale well developed; the large carapace projects freely backward over several posterior thoracomeres, cervical sulcus absent; comparatively large (fig. 54.17A) thoracic epipod 1 penetrates into the well-formed carapace cavity; maxillula with palp; thoracic endopods 1-3 modified as maxillipeds, endopods 4-8 pediform; oostegites on thoracopods 2-8; pleopods with 3-4 segmented exopod; sympod of uropod with small, spinose inner lobe, exopod unsegmented or divided by a distinct or indistinct suture, endopod unsegmented; telson linguiform, its margins with spines and distally also plumose setae.

Type genus. – *Lepidomysis* Clarke, 1961a (junior synonym of *Spelaeomysis*).

Only one genus acknowledged. – *Spelaeomysis* Caroli, 1924 (9 species).

STYGIOMYSIDAE Caroli, 1937 (figs. 54.1B, 54.3A).

Definition. – Antennal scale vestigial; carapace short, dorsally completely fused with the anterior thoracomeres — this makes the body appear vermiform; maxillula without palp; only a small (fig. 54.17B) thoracic epipod 1 penetrates into the small carapace cavity; thoracic endopods 1-4 modified as maxillipeds, endopods 5-8 pediform; oostegites on thoracopods

4-7; pleopods with 2-4 segmented exopod; sympod of uropod with large inner lobe with its apical spines extending beyond the endopod, exopod and endopod unsegmented; telson about lozenged to subrectangular, terminally with spines.

Only type genus known. – *Stygiomysis* Caroli, 1937 (7 species).

Order MYSIDA Boas, 1883. Jurassic-Recent (figs. 54.4, 54.5).

Definition. – Eumalacostraca with eyes mostly well developed, stalked, less frequently reduced or modified in various ways; carapace always well developed, projecting freely over several posterior thoracomeres (except in *Palaumysis philippinensis*), its cavity lined with respiratory tissue; no gills developed; mostly intermediate-sized (fig. 54.15M), foliaceous thoracic epipod 1 penetrates into carapace cavity; thoracic exopods 1-8 usually well developed, natatory (exopods 1, 2 reduced or missing in certain taxa); thoracic endopods 1, 2 specialized as maxillipeds, endopods 4-7 pediform, endopods 3, 8 pediform or modified; 1-7 pairs of functional oostegites; gonopores on the coxae of thoracopods 6 (females) or 8 (males), tubular penes present (except in Rhopalophthalminae), penes with (sub)terminal orifice and one or more terminal lobes; pleopods 2-5 (mostly also 1) reduced to unsegmented plates or rods in females (but biramous in *Archaeomysis*); partly also reduced in males; 1-2 pairs of pleopods with particular modifications in males of most taxa; tail fan well-formed, sympod of uropods without endite, statocysts present only in the family Mysidae; embryonic and two larval stages (nauplioids, postnauplioids) develop in the brood pouch; all post-marsupial stages free-living, self-sustaining; freshly hatched juveniles resemble miniature adults but lack secondary sexual characteristics.

Nomenclatorial note. – There is a current trend (e.g., Anderson, 2010) to quote Haworth (1825) as the author of the Mysida. However, Haworth (1825) established the Mysidae only at family level. Article 35.1. of the code of nomenclature (ICZN, 1999) applies only up to superfamily level. Boas (1883) defined the Mysida for the first time by establishing this taxon at suborder level. His authorship was in continuous use until just recently and, therefore, is recommended to be applied further.

PETALOPHTHALMIDAE Czerniavsky, 1882 (figs. 54.4A, 54.5A).

Definition. – Eyes modified in various ways, less frequently normal; maxillula without palp; thoracopod 1, often also thoracopod 2, without natatory exopod; merus of thoracic endopod 2 with a large lobe-like expansion (= endite); females with seven pairs of oostegites; tubular penes present; pleomeres without projecting pleural plates; female pleopods uniramous or biramous; male pleopods always biramous, all their exopods and most endopods multi-segmented; exopods of uropods with or without subterminal suture, endopods not subdivided, without statocyst.

Inventory. – Two subfamilies with a total of 6 genera and 38 species. Most species bathypelagic and/or archibenthic. No fossils known.

PETALOPHTHALMINAE Czerniavsky, 1882 (figs. 54.4A, 54.12B).

Definition. – Eyes normal or modified, visual elements present or absent; males with basally moderately thickened, inner antennular flagellum; mandibular palp long, powerful, and prehensile; only first thoracopod without exopod; female pleopods uniramous or biramous; exopods of uropods with subterminal suture.

Type genus. – *Petalophthalmus* Willemoës-Suhm, 1874.

Inventory. – Three genera with a total of 8 species: *Parapetalophthalmus* Murano & Bravo, 1998 (1 species), *Petalophthalmus* (5), *Pseudopetalophthalmus* Bravo & Murano, 1997 (2).

HANSENOMYSINAE new subfamily (figs. 54.5A, 54.12A, 54.25M).

Definition. – Eyes fused in a single plate, mostly without or with few visual elements; males with modified, strongly thickened, inner antennular flagellum; mandibular palp normal, not prehensile; first and second thoracopods without exopods; pleomeres without projecting pleural plates (but with spiniform extensions in *Ceratomysis*); female pleopods uniramous, reduced to (1-3)-segmented rods; exopods of uropods with or without (*Bacescomysis*) subterminal suture.

Type genus. – *Hansenomysis* Stebbing, 1893, by present designation.

Inventory. – Three genera with a total of 30 species: *Bacescomysis* Murano & Krygier, 1985 (7 species), *Ceratomysis* Faxon, 1893 (5), *Hansenomysis* (18).

MYSIDAE Haworth, 1825. Jurassic-Recent (figs. 54.4B-M, 54.5B-M).

Definition. – All eight pairs of thoracopods normally with natatory exopod (first pair reduced to laminae in *Mysimenzies*); merus of thoracic endopod 2 without large lobe-like expansion; well-developed tubular penes except in *Rhopalophthalmus*; endopods of uropods normally undivided (subdivided in *Rhopalophthalmus*), always with statocyst. Statoliths non-mineralized (organic) or composed of fluorite (CaF_2) or calcium carbonate ($CaCO_3$), the latter in the crystal phases of vaterite or calcite in the extant and fossil forms, respectively.

Inventory. – Ten extant subfamilies are acknowledged here according to Meland & Willassen (2007), except for the *nomen nudum* 'Mancomysinae', which was replaced by Palaumysinae Wittmann, 2013b. The Mysidae comprise 170 genera with a total of 1093 living species; six extinct species are assigned to one exclusively fossil plus three Recent genera.

BOREOMYSINAE Holt & Tattersall, 1905 (figs. 54.4B, 54.5B).

Definition. – Smooth outer margin of the antennal scale ending in a non-articulate spine; penes well developed, tubular; seven pairs of oostegites; thoracopods normal; pleomeres without projecting pleural plates in both sexes; females with pleopods reduced to unsegmented rods; male pleopods biramous, endopod 1 undivided, remaining endopods and all exopods multi-segmented, without apparent modifications except for some elongation of exopods 2, 3; exopod of uropods divided by a proximal, incomplete suture; statocyst containing a small, non-mineralized statolith; telson with apical cleft.

Type genus. – *Boreomysis* M. Sars, 1869.

Two genera included. – *Birsteiniamysis* Tchindonova, 1981 (2 species), *Boreomysis* (36).

RHOPALOPHTHALMINAE Hansen, 1910 (figs. 54.4C, 54.5C).

Definition. – Antennal scale well developed, with smooth outer margin ending in a non-articulate spine; thoracic endopods 1-7 as normal in Mysidae, endopod 8 small or minute, reduced to 1-3 segments showing distinct sexual dimorphism; three pairs of oostegites; no penes developed, bilobate male gonopores located on the inner terminal margin of the strongly enlarged coxae of thoracopod 8; male pleomere 1 with well-projecting, rounded pleural plates; remaining pleomeres of males and all pleomeres of females without projecting pleurae; female pleopods reduced to unsegmented, setose rods; male pleopods biramous, both rami multi-segmented with exception of the unsegmented, plate-like, first endopod; male pleopod 2 with modified, elongate exopod; both rami of the uropods subdivided by a transverse suture, exopod setose all around and without spines; statocyst with almost spherical, non-mineralized statolith; telson entire, linguiform, its margins distally armed with spines.

Only the type genus known. – *Rhopalophthalmus* Illig, 1906 (25 species).

SIRIELLINAE Czerniavsky, 1882 (figs. 54.4D, 54.5D, 54.26A).

Definition. – Eyes normal; antennal scale well developed, with smooth outer margin ending in a non-articulate spine; labrum mostly with long frontal, spiniform process; thoracic endopods 3-8 with unsegmented or rarely with two-segmented propodus, peculiar brush of setae around dactylus; penes well developed, tubular; three pairs of oostegites; pleomeres without projecting pleural plates; female pleopods uniramous, unsegmented, representing small rods (plates); male pleopods vary according to tribe; endopod of uropods undivided, exopod with spines on outer margin of the basal segment, short terminal segment separated by a mostly oblique suture, this segment setose all around; statocyst with flattened, fluorite statolith; telson linguiform with spines on lateral margins.

Inventory. – Two tribes are distinguished according to the structure of the male pleopods. They comprise a total of 3 genera with 85 extant plus 2 fossil species.

Siriellini Czerniavsky, 1882 (figs. 54.4D, 54.5D, 54.26A).

Definition. – Male pleopods biramous, multi-segmented (endopod of the first pair rudimentary), basal segment of endopods with spirally coiled or curved, less frequently straight exites (= pseudobranchial lobes); modified setae may be present (or not) on exopod or endopod or both of any among pleopods 2-4.

Type genus. – *Siriella* Dana, 1850.

Two genera included. – *Hemisiriella* Hansen, 1910 (4 species), *Siriella* (80 extant plus 2 fossil species).

Metasiriellini Murano, 1986.

Definition. – Male pleopods 1-3, 5 rudimentary, uniramous, and unsegmented as in females; male pleopod 4 with two multi-segmented rami, basal segment of endopod with straight, rod-like exite (= pseudobranchial lobe).

Only the type genus known. – *Metasiriella* Murano, 1986 (1 species).

GASTROSACCINAE Norman, 1892 (figs. 54.4E, F, 54.5E, F).

Definition. – Eyes normal; antennal scale well developed, smooth outer margin ending in a non-articulate spine; two pairs of oostegites; first pleomere with pleural plates normally in females only (in *Archaeomysis* in both sexes); pleopods of both sexes vary according to tribe; both rami of the uropods undivided, exopod with spines on outer margin in most taxa; statocyst with (generally small) fluorite statolith. Telson with spines on lateral margins, its apical cleft armed with spine-like laminae.

Nomenclatorial note. – The precedence of the family group taxon Pontomysidae Czerniavsky, 1882 (as "Divisio" = tribus), over the junior synonym Gastrosaccinae Norman, 1892, is inverted here according to Article 23.9.1 of the code of nomenclature (ICZN, 1999). The senior synonym was last used by Czerniavsky (1887). The junior synonym has been in continuous use since Norman (1892), by far exceeding the requirements of Article 23.9.1.2.

Inventory. – Three tribes are distinguished based on to the structure of pleural plates and pleopods. They comprise a total of 10 genera with 96 species.

Archaeomysini Czerniavsky, 1882.

Definition. – First pleomere (in part also the second) of females with large pleural plates covering and supporting the marsupium from behind, in part also from the side; such plates also present in males, but smaller; certain taxa with plates on additional pleomeres, whereby plate size decreases from pleura 2 to 5; pleopods biramous in both sexes, exopod of pleopod 3 elongated in males only.

Only the type genus known. – *Archaeomysis* Czerniavsky, 1882 (6 species).

Gastrosaccini Norman, 1892 (figs. 54.4F, 54.5F).

Definition. – Only first pleomere of females with large pleural plates covering and supporting the marsupium from behind, in part also from the side, no such plates in males; female pleopod 1 biramous, pleopods 2-5 uniramous, unsegmented; male pleopods biramous or reduced in various ways, exopod 3 (exceptionally exopod 4) particularly long and modified (mostly styliform or with complex structure in certain genera).

Type genus. – *Gastrosaccus* Norman, 1868.

Six genera included. – *Chlamydopleon* Ortmann, 1893 (3 species), *Coifmanniella* Heard & Price, 2006 (4), *Eurobowmaniella* Murano, 1995 (2), *Gastrosaccus* (24), *Haplostylus* Kossmann, 1880 (25), *Iiella* Băcescu, 1968 (9).

Anchialinini new replacement name (figs. 54.4E, 54.5E).

Formation of the replacement name. – According to article 39 of the code of nomenclature, a family group name is invalid if based on a junior homonym (ICZN, 1999). The family group name Anchialidae Czerniavsky, 1882, is based on the mysid genus *Anchialus* Krøyer, 1861, a junior homonym of the coleopteran genera *Anchialus* Gistel, 1834, and *Anchialus* Thomson, 1859, and is therefore to be replaced by a family group name based on the valid replacement name *Anchialina* Norman & Scott, 1906, for the former type genus.

Definition. – Only first pleomere of females with small pleural plates covering a small part of the marsupium, no such plates in males; female pleopods uniramous, unsegmented; male pleopod 1, in part also 4, 5, rudimentary and resembling those of females, remaining pleopods biramous, exopod 3 modified in various, sometimes complex ways.

Revised type genus. – *Anchialina* Norman & Scott, 1906 (replacement name for *Anchialus* Krøyer, 1861), by present designation.

Three genera included. – *Anchialina* (17 species), *Paranchialina* Hansen, 1910 (2), *Pseudanchialina* Hansen, 1910 (4).

ERYTHROPINAE Hansen, 1910 (figs. 54.4G, 54.5G).

Definition. – The eyes show a great diversity from well-developed to reduced, plate-like, according to taxon; antennal scale normally well-developed (rudimentary in *Arachnomysis*, *Caesaromysis*, *Chunomysis*, *Gymnerythrops*) with mostly smooth outer margin ending in an articulate or a non-articulate spine; thoracic endopods 3-8 with the carpus separated from the propodus in most genera by an oblique articulation, less frequently with transverse articulation; females with 2-3 (4) pairs of oostegites; pleomeres without projecting pleural plates in both sexes; pleopods 1-5 in females are uniramous and unsegmented, endopod of pleopod 1 in males uniramous, exopod 1 and both rami of pleopods 2-5 vary in males according to tribe, exites of endopods never curved or coiled; both rami of the uropods undivided; statocyst with fluorite statolith; telson entire or rarely terminally indented or cleft.

Nomenclatorial note. – The senior synonym Protomysidellinae Czerniavsky, 1882, was no longer used as a valid taxon after Czerniavsky (1887) and is invalid because of being not based on an available generic name (Article 11.7.1.1 of ICZN, 1999).

Inventory. – Eight tribes are distinguished according to the structure of eyes, antennular peduncle, antennal scale, antennal gland, labrum, articulation between carpus and propodus of the posterior thoracic endopods, male pleopods, and telson. Altogether, the eight tribes comprise 54 genera with a total of 243 species.

Erythropini Hansen, 1910 (fig. 54.5G).

Definition. – Eyes mostly with, or in a few taxa without visual elements, eyestalks well separated from each other; antennal scale well developed, antennal gland normal or hypertrophic; labrum normal; thoracic endopods 3-8 with mostly oblique (rarely transverse) articulation between carpus and propodus; 2-3 pairs of oostegites; endopod of male pleopod 1 reduced to a single segment, exopod 1 rarely entire but mostly multi-segmented, male pleopods 2-5 well developed, biramous; in several genera modified setae may be present on any of male pleopods 3-5, mostly on endopods; telson entire, lateral margins generally smooth (never serrate), in some taxa with spines laterally on terminal portions, in certain taxa with a pair of medio-apical setae.

Type genus. – *Erythrops* G. O. Sars, 1869.

Thirty-five genera included. – *Aberomysis* Băcescu & Iliffe, 1986 (1 species), *Amathimysis* Brattegard, 1969 (9), *Atlanterythrops* Nouvel & Lagardère, 1976 (1), *Australerythrops* W. M. Tattersall, 1928 (2), *Dactylamblyops* Holt & Tattersall, 1906 (15), *Dactylerythrops* Holt & Tattersall, 1905 (5), *Echinomysides* Murano, 1977 (1), *Echinomysis* Illig, 1905 (3), *Erythrops* (17), *Euchaetomera* G. O. Sars, 1883 (9), *Euchaetomeropsis* W. M. Tattersall, 1909 (2), *Gibbamblyops* Murano & Krygier, 1985 (1), *Gibberythrops* Illig, 1930 (4), *Heteroerythrops* O. S. Tattersall, 1955 (3),

Holmesiella Ortmann, 1908 (3), *Hyperamblyops* Birstein & Tchindonova, 1958 (5), *Hypererythrops* Holt & Tattersall, 1905 (8), *Illigiella* Murano, 1981 (1), *Indoerythrops* Panampunnayil, 1998 (1), *Katerythrops* Holt & Tattersall, 1905 (4), *Liuimysis* Wang, 1998 (1), *Longithorax* Illig, 1906 (6), *Meierythrops* Murano, 1981 (2), *Metamblyops* W. M. Tattersall, 1907 (3), *Meterythrops* S. I. Smith, 1879 (7), *Nakazawaia* Murano, 1981 (2), *Nipponerythrops* Murano, 1977 (1), *Parerythrops* G. O. Sars, 1869 (6), *Pleurerythrops* Ii, 1964 (5), *Pseudamblyops* Ii, 1964 (1), *Pseuderythrops* Coifmann, 1936 (3), *Pteromysis* Ii, 1964 (1), *Shenimysis* Wang, 1998 (1), *Synerythrops* Hansen, 1910 (3), *Teraterythrops* Ii, 1964 (2).

Remark. – The morphological heterogeneity together with genetic data of Meland & Willassen (2007) suggest that this tribe is polyphyletic. More data on a greater diversity of genera are wanted.

Arachnomysini Holt & Tattersall, 1905 (figs. 54.7F, 54.12C).

Revised definition. – Eyes well developed, separate and stalked; carapace anteriorly without or with series of spines; antennal scale missing or vestigial; labrum normal; thoracic endopods 3-8 with the carpus separated from the propodus by a transverse or a slightly oblique articulation; endopod of male pleopod 1 reduced to a single segment, exopod multi-segmented, male pleopods 2-5 well developed, biramous; marsupium with two pairs of oostegites; telson subtriangular, without or with small medio-apical indentation, lateral margins smooth.

Type genus. – *Arachnomysis* Chun, 1887.

Four genera included. – *Arachnomysis* (2 species), *Caesaromysis* Ortmann, 1893 (1), *Chunomysis* Holt & Tattersall, 1905 (1), *Gymnerythrops* Hansen, 1910 (3 species, males unknown).

Inusitatomysini new tribe.

Definition. – Eyes with well-developed cornea on separate eyestalks; antennal scale well developed, its outer margin proximally smooth, centrally and distally serrate, ending in a non-articulate spine; antennal gland and labrum normal; posterior thoracic endopods with some sub-segments of the carpopropodus separated by oblique articulations from each other; three pairs of well-developed oostegites plus a pair of vestigial ones on thoracic endopod 5; all pleopods uniramous, represented by endopods in both sexes, these endopods reduced to unsegmented plates, with the only exception that endopod 4 is elongate, multi-segmented, and apically furnished with modified setae in males; uropods setose all around, endopod with one spine on outer margin below statocyst; telson large, its lateral margins with spines, but not serrate, strong apical cleft furnished with a pair of plumose setae emerging from its bottom, margins of cleft lined by spine-like laminae.

Only the type genus known. – *Inusitatomysis* Ii, 1940 (1 species), by present designation.

Remarks. – Based on the reduction of most male pleopods, this aberrant genus was placed by W. M. Tattersall (1951) and Ii (1964) in the Mysini (now Mysinae). In the present study it is shifted as a separate tribe to the Erythropinae due to the modification of the endopod — unlike the exopod as in Mysinae — of the fourth male pleopod. The structure of the antennal scale and the eyes closely resemble those in *Erythrops serratus* and *Erythrops abyssorum*. The reduced oostegite on thoracic endopod 5 is reminiscent of the small but well-developed brood plate found in the same position in *Thalassomysis*.

Thalassomysini Nouvel, 1942b.

Definition. – Eyes with well-developed separate eyestalks, cornea reduced to a varying extent; antennal scale well developed, setose all around; labrum strongly asymmetrical; thoracopods 3-8 with the carpus separated from the propodus by a transverse articulation; four pairs of oostegites; uropods setose all around, without spines; telson entire, elongate, its lateral margins with spines, but not serrate.

Only the type genus known. – *Thalassomysis* W. M. Tattersall, 1939 (2 species, males unknown).

Remark. – The placement of *Thalassomysis* is difficult as long as the males remain unknown. W. M. Tattersall (1939) placed it with query in the Erythropini, Nouvel (1942b) into a special subfamily, the Thalassomysinae, Pillai (1965) in the Leptomysini, Murano & Krygier (1985) and Nouvel et al. (1999) back to the Thalassomysinae, and Meland & Willassen (2007) again back to the Erythropinae. Here, it is also placed in the Erythropinae, but in a separate tribe due its aberrant morphology.

Amblyopsini Tchindonova, 1981 (figs. 54.4G, 54.18D).

Revised definition. – Eyes separate, eyestalks reduced to immotile plates, visual elements rudimentary or absent; antennal scale well developed, its smooth outer margin ending in a non-articulate spine, antennal gland mostly hypertrophic; labrum normal; thoracopods 3-8 with the carpus separated from the propodus by an oblique articulation; 2-3 pairs of oostegites; endopod of male pleopod 1 reduced to a single segment, exopod multi-segmented, male pleopods 2-5 well developed, biramous; a number of species shows (sub)apical, modified setae in some of male pleopods 2-4, most frequently on exopod 4; telson entire, lateral margins smooth (not serrate), only terminally or all along with spines, two (sub)apical setae present in some species.

Type genus. – *Amblyops* G. O. Sars, 1872.

Six genera included. – *Amblyops* (22 species), *Amblyopsoides* O. S. Tattersall, 1955 (4), *Eoamblyops* Murano, 2013 (1), *Paramblyops* Holt & Tattersall, 1905 (8), *Scolamblyops* Murano, 1974 (1), *Teratamblyops* Murano, 2001 (3).

Mysimenziesini Tchindonova, 1981.

Revised definition. – Eyes rudimentary, without visual pigment, the rudimentary eyestalks resemble horns and are connected by a median bridge; basal segment of antennular peduncle medio-dorsally with sensory fossette or with setose elevation; antennal scale well developed, with smooth outer margin ending in a non-articulate spine; labrum normal; thoracic endopods 3-8 with oblique or transverse articulation between carpus and propodus; three pairs of oostegites; male pleopod 4 biramous with both rami multi-segmented, modified setae at tip of endopod 4, remaining male pleopods uniramous or biramous; telson truncate (in part terminally slightly indented), its lateral margins distinctly serrate.

Type genus. – *Mysimenzies* Băcescu, 1971a.

Two genera included. – *Marumomysis* Murano, 1999a (2 species), *Mysimenzies* (2).

Remark. – The first detection of males and of additional taxa (Murano, 1999a; San Vicente, 2007; Fukuoka, 2009) confirmed previous assumptions (Băcescu, 1971a; Nouvel et al., 1999; Meland & Willassen, 2007) that *Mysimenzies* is to be placed in the Erythropinae. The recent findings confirmed the aberrant morphology of these mysids, leading here to revalidation of the Mysimenziesinae at tribe rank.

Pseudommini new tribe (figs. 54.20C, 54.21J, 54.25P).

Definition. – Both eyes fused to a common plate with anterior, median cleft, visual elements missing, posterior parts of eyeplate ventrally fused with basal parts of antennulae and antennae; antennal scale well developed, its smooth outer margin ending in a non-articulate spine, antennal gland hypertrophic; labrum normal; thoracopods 3-8 with the carpus separated from the propodus by an oblique articulation; two pairs of oostegites; endopod of male pleopod 1 reduced to a single element, exopod 1 multi-segmented, male pleopods 2-5 well developed, biramous: modified setae, if present, mainly on fourth endopod; telson entire, linguiform to nearly lozenged, its lateral margins entirely smooth or only proximally smooth with spines along terminal portions.

Type genus. – *Pseudomma* G. O. Sars, 1870, by present designation.

Three genera included. – *Neoamblyops* Fukuoka, 2009 (1 species), *Parapseudomma* Nouvel & Lagardère, 1976 (1), *Pseudomma* (45). The species *Parapseudomma calloplura* and *Pseudomma oculospinum* are included here based on eye morphology (for different views see Murano, 1974; Meland, 2004; Meland & Willassen, 2007).

Calyptommini W. M. Tattersall, 1909.

Definition. – Both eyes fused to a common lamina without cleft, visual elements rudimentary or absent; antennal scale well developed, its smooth outer margin ending in a non-articulate spine; antennal gland hypertrophic; labrum normal; pleopods of both sexes reduced to unsegmented, rod-like endopods that increase in length caudally, the ultimate pleopods 4, 5 most elongate in males; telson entire, lateral margins not serrated but mostly smooth, terminal portions with spines.

Type genus. – *Calyptomma* W. M. Tattersall, 1909.

Two genera included. – *Calyptomma* (1 species), *Michthyops* W. M. Tattersall, 1911 (3).

LEPTOMYSINAE Czerniavsky, 1882 (figs. 54.4H, 54.5H).

Definition. – Eyes mostly normal, a few taxa with modified cornea; antennal scale setose all around and in most taxa with small terminal segment; thoracic endopods 3-8 with carpopropodus subdivided by transverse articulations into more than two segments; females with three, rarely two, pairs of oostegites; pleomeres without projecting pleural plates in both sexes; female pleopods reduced to small, unsegmented endopods; male pleopod 1 biramous with unsegmented endopod and multi-segmented exopod (exopod missing in *Pyroleptomysis*); male pleopods 2-5 biramous, both rami multi-segmented, exopod of pleopod 4 modified and/or with modified setae; both rami of the uropods undivided, exopods without spines, setose all around; statocyst with fluorite statolith.

Inventory. – Three tribes are distinguished based on the structure of maxilla, thoracic endopod 1, and telson. Altogether, they comprise 30 genera with 164 extant plus one fossil species.

Mysidopsini new tribe (figs. 54.15L, 54.18F, 54.22N).

Definition. – Antennal scale setose all around, mostly with small terminal segment; exopod of maxilla narrow or absent; first thoracic endopod only (5-6)-segmented, its ischium fused with the merus; coxa may show a small conical endite, no endites on basis or the fused merischium (fig. 54.15L versus 54.15M); females with three pairs of well-formed oostegites (the first one very small); telson mostly entire, or with small apical indentation, its margins with spines but without setae.

Type genus. – *Mysidopsis* G. O. Sars, 1864, by present designation.

Five genera included. – *Americamysis* Price & Stuck, 1994 (6 species), *Brasilomysis* Băcescu, 1968 (2), *Cubanomysis* Băcescu, 1968 (2), *Metamysidopsis* W. M. Tattersall, 1951 (8), *Mysidopsis* (48 extant plus one fossil species).

Leptomysini Czerniavsky, 1882 (figs. 54.4H, 54.5H, 54.21N, 54.36E-F, 54.37).

Definition. – Antennal scale setose all around, with small terminal segment; exopod of maxilla laterally well expanded, endopod normal; first thoracic endopod seven-segmented, i.e., ischium and merus not fused, endites may be present on coxa and basis, often also on ischium; females with three pairs of well-formed oostegites (fig. 54.19K-H); telson entire or with minute cleft, telson lined by spines along most of its margins, never by laminae or apical setae (plumose setae may be present on ventral face but not on margins).

Type genus. – *Leptomysis* G. O. Sars, 1869.

Six genera included. – *Antichthomysis* Fenton, 1991 (1 species), *Leptomysis* (9), *Megalopsis* Panampunnayil, 1987 (1), *Paraleptomysis* Liu & Wang, 1983 (5), *Pyroleptomysis* Wittmann, 1985 (2), *Notomysis* Wittmann, 1986 (1).

Afromysini new tribe (fig. 54.25F).

Definition. – Antennal scale setose all around, without or in most taxa with small terminal segment; exopod of maxilla laterally well expanded, distal segment of endopod often terminally widened and with spines or teeth on terminal margin; first thoracic endopod seven-segmented, i.e., ischium and merus not fused, endites present on coxa and basis, often also on ischium and merus; females mostly with three pairs of well-developed oostegites, rarely with only two; telson apically with cleft or less frequently narrowly truncate, lateral margins lined by spines; apical cleft, if present, mostly with two plumose setae emerging from or close to its bottom, margins of cleft smooth or lined by spine-like laminae.

Type genus. – *Afromysis* Zimmer, 1916, by present designation.

Nineteen genera included. – *Afromysis* (7 species), *Australomysis* W. M. Tattersall, 1927 (5), *Bathymysis* W. M. Tattersall, 1907 (2), *Ceratodoxomysis* Murano, 2003 (1), *Dioptromysis* Zimmer, 1915 (5), *Doxomysis* Hansen, 1912 (17), *Hyperiimysis* H. Nouvel, 1966 (1), *Iimysis* H. Nouvel, 1966 (3), *Mysideis* G. O. Sars, 1869 (2), *Neobathymysis* Bravo & Murano, 1996 (2), *Neodoxomysis* Murano, 1999 (3), *Nouvelia* Băcescu & Vasilescu, 1973 (3), *Prionomysis* W. M. Tattersall, 1922 (4), *Promysis* Dana, 1850 (2), *Pseudobranchiomysis* Carcedo, Fiori & Hoffmeyer, 2013 (1), *Pseudomysis* G. O. Sars, 1879 (2), *Pseudoxomysis* H. Nouvel, 1973 (3), *Rostromysis* Panampunnayil, 1987 (1), *Tenagomysis* G. M. Thomson, 1900 (15).

MYSINAE Haworth, 1825 (figs. 54.1E, 54.4J, 54.5J, 54.36A, B).

Definition. – Eyes well developed, eyestalks always separate, only few taxa with modified or reduced cornea; antennal scale variable; thoracic endopods 3-8 with carpopropodus subdivided by transverse articulations, or exceptionally not divided; females with 2-3 pairs of oostegites; pleomeres without projecting pleural plates; female pleopods reduced to simple, unsegmented rods; male pleopods 1 and 2 (mostly also 5) rudimentary as in females; male pleopod 3 biramous and/or reduced to different degrees, male pleopod 4 normally biramous, with elongate exopod and modified setae; both rami of the uropods undivided, exopods without spines, setose all around; statocyst with statolith composed of fluorite or less frequently of vaterite (calcite in fossil taxa; cf. above, 'Microfossils').

Inventory. – Four tribes are distinguished based on the structure of antennal scale, oostegites, male pleopods, and telson. Altogether, they comprise 53 extant genera with 297 species plus one fossil genus with 3 species.

Mysini Haworth, 1825 (figs. 54.1E, 54.5J).

Definition. – Antennal scale mostly with short apical segment, scale setose all around or with (over part of its length) bare outer margin ending in an articulate or nonarticulate spine; carpopropodus of thoracic endopod 6 with 3-10 (in *Arthromysis* up to 26) segments; two pairs of well-developed oostegites, rudimentary oostegite on thoracic endopod 6; all pleopods of females and pleopods 1, 2 of males rudimentary; male pleopod 3 generally biramous (uniramous in *Kainommatomysis*), with unsegmented endopod and sub-segmented exopod, male pleopod 4 biramous, exopod elongate, with modified setae, pleopod 5 rudimentary or less frequently biramous; statoliths composed of fluorite, less frequently of vaterite; telson variable, mostly with apical cleft, this cleft (if present) mostly armed with spine-like laminae and in several taxa also with a pair of medio-apical plumose setae.

Type genus. – *Mysis* Latreille, 1802.

Seventeen genera included. – *Antarctomysis* Coutière, 1906 (3 species), *Arthromysis* Colosi, 1924 (1), *Caspiomysis* G. O. Sars, 1907 (1), *Hemimysis* G. O. Sars, 1869

(9), *Hyperstilomysis* Fukuoka, Bravo & Murano, 2005 (1), *Kainommatomysis* W. M. Tattersall, 1927 (3), *Katamysis* G. O. Sars, 1893 (1), *Mesopodopsis* Czerniavsky, 1882 (8), *Mysis* (15), *Nanomysis* W. M. Tattersall, 1921 (3), *Paramysis* Czerniavsky, 1882 (8 subgenera, 23 species), *Parastilomysis* Ii, 1936 (4), *Praunus* Leach, 1814 (3), †*Sarmysis* Maissuradze & Popescu, 1987 (3 fossil species), *Schistomysis* Norman, 1892 (6), *Stilomysis* Norman, 1892 (4), *Tasmanomysis* Fenton, 1985 (1).

Remarks. – *Kainommatomysis* is assigned here with some reservation — see discussion in W. M. Tattersall (1927). The allocation of this genus together with the remaining morphological heterogeneity of this tribe require further investigation.

Diamysini new tribe (figs. 54.4J, 54.12D-F, 54.18J-O, 54.21O, P, 54.35H, J, 54.44A).

Definition. – Antennal scale setose all around, generally with short apical segment (rarely missing); carpopropodus of thoracic endopod 6 with 2-3 segments; two pairs of well-developed oostegites, rudimentary oostegite on thoracic endopod 6; pleopods rudimentary in both sexes, except male pleopods 3, 4; male pleopod 3 reduced to small endopod fused with larger 2-segmented sympod, less frequently rudimentary as in female; pleopod 4 with large, 2-segmented sympod, its exopod frequently styliform, ending in 1 (2-3) large, modified setae; endopod of uropod with 1 (0-2) spines on ventral face near statocyst, statoliths composed of vaterite, less frequently of fluorite; telson shorter than ultimate pleonite, mostly with short apical cleft or at least apically truncate, spines (if present) on lateral margins arranged in continuous series, not arranged in groups of large spines with smaller spines in between; each lateral margin ending in a comparatively large spine, cleft (if distinct) lined with spine-like laminae or with small spines.

Type genus. – *Diamysis* Czerniavsky, 1882, by present designation.

Nine genera included. – *Antromysis* Creaser, 1936 (6 species), *Diamysis* (14), *Gangemysis* Derzhavin, 1924 (1), *Indomysis* W. M. Tattersall, 1914 (2), *Limnomysis* Czerniavsky, 1882 (1), *Parvimysis* Brattegard, 1969 (3), *Surinamysis* Bowman, 1977 (3), *Taphromysis* Banner, 1953 (3), *Troglomysis* Stammer, 1933 (1).

Anisomysini new tribe (fig. 54.19G).

Definition. – Antennal scale setose all around (exceptionally with blunt spine on outer margin) and with short apical segment; carpopropodus of thoracic endopod 6 with 1-3 segments; two pairs of well-developed oostegites, rudimentary oostegite on thoracic endopod 6; pleopods rudimentary in both sexes, except male pleopod 4 and to a minor extent also pleopod 3; third male pleopod uniramous, unsegmented, mostly rudimentary as in females or reduced to small endopod fused with sympod; exopod of fourth male pleopod (sub)apically with 2 (1) large modified setae; uropods without spines, statoliths composed of fluorite; telson shorter than ultimate pleonite, terminally rounded or with apical cleft, lateral margins bare or furnished with spines; spines (if present) on lateral margins arranged in continuous series, not arranged in groups of large spines with smaller spines in between; telson always devoid of setae or laminae.

Type genus. – *Anisomysis* Hansen, 1910, by present designation.

Seven genera included. – *Anisomysis* (2 subgenera; 55 species), *Carnegieomysis* W. M. Tattersall, 1943 (4 species), *Halemysis* Bǎcescu & Udrescu, 1984 (1), *Idiomysis* W. M. Tattersall, 1922 (5), *Javanisomysis* Bǎcescu, 1992 (1), *Mysidium* Dana, 1852 (6), *Paramesopodopsis* Fenton, 1985 (1).

Neomysini new tribe.

Definition. – Antennal scale setose all around, with short apical segment; carpopropodus of thoracic endopod 6 with 4-20 segments; two pairs of well-developed oostegites, anterior pair occasionally with a posterior lobe, sixth thoracic endopod with additional rudimentary oostegite; pleopods rudimentary in both sexes, except male pleopod 4; exopod of male pleopod 4 typically large, (sub)apically with mostly 2 large modified setae; the rudimentary pleopod 5 may be longer in males than in females; endopod of uropods normally with closely set spines on inner margin near statocyst, with the spines increasing in length distally (rarely without spines or spines loosely set all along inner margin), statoliths composed of fluorite; telson prolonged, linguiform to subtriangular, subequal to longer than ultimate pleonite, lateral margins partly or entirely armed with spines, spines in most species arranged in groups of large spines with small spines in between.

Type genus. – *Neomysis* Czerniavsky, 1882, by present designation.

Twenty-one genera included. – *Acanthomysis* Czerniavsky, 1882 (17 species), *Alienacanthomysis* Holmquist, 1981 (1), *Boreoacanthomysis* Fukuoka & Murano, 2004 (1), *Columbiaemysis* Holmquist, 1982 (1), *Disacanthomysis* Holmquist, 1981 (1), *Exacanthomysis* Holmquist, 1981 (3), *Hemiacanthomysis* Fukuoka & Murano, 2002 (1), *Hippacanthomysis* Murano & Chess, 1987 (1), *Holmesimysis* Holmquist, 1979 (5), *Hyperacanthomysis* Fukuoka & Murano, 2000 (2), *Lycomysis* Hansen, 1910 (2), *Mesacanthomysis* Nouvel, 1967 (1), *Neomysis* (17), *Nipponomysis* Takahashi & Murano, 1986 (19), *Notacanthomysis* Fukuoka & Murano, 2000 (2), *Orientomysis* Derzhavin, 1913 (22), *Pacifacanthomysis* Holmquist, 1981 (1), *Paracanthomysis* Ii, 1936 (4), *Proneomysis* W. M. Tattersall, 1933 (1), *Telacanthomysis* Fukuoka & Murano, 2001 (1), *Xenacanthomysis* Holmquist, 1980 (1).

PALAUMYSINAE Wittmann, 2013b (figs. 54.4K, 54.5K, 54.7G, 54.21U).

Definition. – Eyes well developed; inner antennular flagellum very short, reduced to only 2-5 segments; antennal scale missing or vestigial; labrum normal; thoracopods 3-8 with the carpus separated from the propodus by a transverse articulation; two pairs of small oostegites; penes of moderate size; all pleopods reduced to uniramous, unsegmented rods in both sexes, pleopod 4 with modified apical spine (seta) in males only; both rami of the uropods undivided, endopod with statocyst composed of fluorite; telson (sub)triangular, without or with small medio-apical indentation, its lateral margins bare and each ending in a spine.

Type genus. – *Palaumysis* Băcescu & Iliffe, 1986.

Two genera included. – *Gironomysis* Ortiz, García-Debrás & Pérez, 2000 (1 species), *Palaumysis* (4).

HETEROMYSINAE Norman, 1892 (figs. 54.1D, 54.4M, 54.5M).

Diagnosis. – Eyes well developed, rarely weakly reduced; appendix masculina varies from well developed to very small or reduced to a brush of setae; antennal scale setose all around; labrum, maxillula, and thoracic endopod 1 normal; digitus mobilis present generally on both or rarely on only one mandible; thoracic endopod 3 pediform or more frequently subchelate, prehensile; 1-3 pairs of oostegites; pleomeres without projecting pleural plates; pleopods normally reduced to unsegmented endopods in both sexes (pleopod 3 sub-divided in males of *Harmelinella*); penes tubular, in most genera strongly developed; both rami of the uropods undivided; statocyst with fluorite statolith; telson variable.

Inventory. – Three tribes are distinguished according to the structure of thoracic endopod 3 and male pleopod 3. Altogether, they comprise 14 genera with a total of 124 species.

Heteromysini Norman, 1892 (figs. 54.1D, 54.4M, 54.5M, 54.18K, H, 54.20G, F, 54.21R, S, 54.35F, G, 54.36C).

Diagnosis. – Appendix masculina very small or reduced to a brush of setae; thoracic endopod 3 swollen, prehensile, carpopropodus two-segmented, rarely fused; thoracic endopods 4-8 developed as pereiopods, normal; 2-3 pairs of oostegites; penes large (in length and/or in diameter) or rarely small, mostly tubular, mostly with distinct terminal lobes; males of certain species with some of the pleopods bearing small (flagellate) spines or modified setae; telson with apical cleft armed with spine-like laminae, lateral margins smooth or distally with spines.

Type genus. – *Heteromysis* S. I. Smith, 1873.

Five genera included. – *Heteromysis* (4 subgenera; 83 species), *Heteromysoides* Băcescu, 1968 (10 species), *Ischiomysis* Wittmann, 2013a (2), *Platymysis* Brattegard, 1980 (1), *Retromysis* Wittmann, 2004 (1).

Harmelinellini new tribe.

Definition. – Appendix masculina well developed, rounded; thoracic endopods 3-8 normal, endopod 3 pediform, not prehensile and not conspicuously swollen, its carpopropodus two-segmented; penes large, tubular, stiff, not changeable in size and form; all pleopods of the females and pleopods 1, 2, 4, 5 of males rudimentary, without modified setae or spines; pleopod 3 of males uniramous, elongate, sub-segmented, apically modified; telson with apical cleft armed with spine-like laminae, lateral margins distally with spines.

Only the type genus known. – *Harmelinella* Ledoyer, 1989, by present designation (1 species).

Mysidetini Holt & Tattersall, 1906 (figs. 54.20E, 54.22M).

Diagnosis. – Appendix masculina varies from well developed to very small or reduced; thoracic endopods 3-8 normal, endopod 3 pediform, not prehensile and not conspicuously swollen, its carpopropodus with two or more segments; 1-3 pairs of oostegites; penes mostly long, tubular, stiff or changeable in size and form; pleopods rudimentary and non-dimorphic, without modified setae or spines; telson entire or with apical cleft, lateral margins distally with spines.

Discussion. – This family group taxon was defined as the subfamily Mysidetinae by Holt & Tattersall (1906). Hansen (1910) did not accept this subfamily and retained the genus *Mysidetes* in the tribe Leptomysini (Mysinae). Zimmer (1914) followed the proposal of Hansen (1910), but with some reservation based on the reduction of the male pleopods. Finally, W. M. Tattersall (1923) himself withdrew the Mysidetinae. However, this does not diminish the nomenclatorial availability of the taxon (see Article 11 in ICZN, 1999). Here, the taxon is revived at tribus level due to the structure of thoracic endopod 3, penes, male pleopods, and telson.

Type genus. – *Mysidetes* Holt & Tattersall, 1906.

Eight genera included. – *Bermudamysis* Băcescu & Iliffe, 1986 (1 species), *Burrimysis* Jaume & Garcia, 1993 (1), *Deltamysis* Bowman & Orsi, 1992 (1), *Kochimysis* Panampunnayil & Biju, 2007 (1), *Mysidetes* (16), *Mysifaun* Wittmann, 1996 (1), *Platyops* Băcescu & Iliffe, 1986 (1), *Pseudomysidetes* W. M. Tattersall, 1936 (4).

MYSIDELLINAE Czerniavsky, 1882 (figs. 54.4L, 54.5L, 54.21T).

Diagnosis. – Antennal scale setose all around; labrum with strongly asymmetrical posterior lobes; mandibles with large cutting edge without teeth; maxillula with inwards bent lobes; thoracic endopod 1 with strongly expanded propodus; thoracic endopod 3 normal, not prehensile; three pairs of oostegites; penes tubular, strongly developed; pleomeres without projecting pleural plates; pleopods rudimentary in both sexes; uropods without spines, both rami of the uropods undivided; statocyst with fluorite statolith; telson with apical cleft.

Only the type genus known. – *Mysidella* G. O. Sars, 1872 (16 species).

Keys to the families, subfamilies, and tribes of the Lophogastrida, Stygiomysida, and Mysida

Note. – Habitus figures are given in figs. 54.1-54.5 as examples for all extant taxa at order, family, and subfamily level.

Eumalacostraca with caridoid habitus; eyes generally stalked; carapace well developed, forming ventrally open, lateral chambers, with its posterior portions freely projecting over some of the ultimate thoracomeres in most taxa; eight pairs of generally biramous thoracopods; oostegites present on (mostly the ultimate) 1-7 pairs of thoracopods; gonopores on the sixth thoracomere in females, and on the coxae of the eighth thoracopods in males; well-developed pleon, capable of tail flipping; strong tail fan formed by the telson and a pair of biramous uropods 1

1a. Branchiae emerge from 5-7 pairs of thoracopods; females with (so far known) seven pairs of oostegites; pleopods well developed, biramous in both sexes; uropods without statocyst (†Pygocephalomorpha, Lophogastrida) 2

1b. No branchiae present; instead, the inner wall of the lateral carapace chambers is lined with respiratory tissue; female pleopods reduced to some extent in most taxa; 1-7 pairs of oostegites; uropods with or without statocyst (Stygiomysida, Mysida) 6

2a. Carapace broad, with branchiostegal development of the pleura in order to accommodate the branchiae; no tubular penes known; telson with well-developed, lobe- or spine-like apical and furcal processes .... Order †Pygocephalomorpha Beurlen, 1930. Carboniferous-Permian.

2b. Ramified branchiae entirely or only partly covered by the carapace in lateral view; male gonopores on the coxa of each thoracopod 8 in extant forms, the orifices are not elevated or are on top of a small elevation, no lobes around orifice and no tubular penes present; pleopods not modified or only weakly modified in males
Order Lophogastrida Boas, 1883. Carboniferous-Recent 3

3a. Pleomeres without projecting pleural plates, ultimate pleomere not divided by a transversal furrow; thoracic endopods 1-4 developed as maxillipeds (gnathopods)
.... Eucopiidae G. O. Sars, 1885 (figs. 54.2D, 54.19A). Triassic-Recent.

3b. Pleomeres with well-developed pleural plates; at least thoracic endopods 3-7 are pediform
.... 4

4a. All thoracic endopods pediform, not specialized as maxillipeds; ultimate pleomere (as far as known) not divided by a transversal furrow
.... †Peachocarididae Schram, 1986. Carboniferous.

4b. At least thoracic endopod 1 developed as maxilliped; ultimate pleomere more or less distinctly divided by a transversal furrow 5

5a. Only thoracic endopod 1 developed as maxilliped
.... Gnathophausiidae Udrescu, 1984 (figs. 54.1A, 54.2A). Recent.

5b. Thoracic endopods 1, 2 developed as maxillipeds
.... Lophogastridae G. O. Sars, 1870 (fig. 54.2B, C). Jurassic-Recent.

6a. Eyes normal or less frequently reduced; well-developed tubular penes (except in Rhopalophthalminae); pleopods 2-5 (mostly also 1) reduced to unsegmented plates or rods in females (but well developed, biramous in Archaeomysini), partly also reduced in males; sympod of uropods without inner lobe, endopod with or without statocyst (Mysida) 8

6b. Cornea reduced to varying extent; gonopores flanked by two lobes, no tubular penes developed; all pleopods biramous but rudimentary in both sexes, sympod and endopod unsegmented, but exopod with a small number of segments; sympod of uropod with spinose inner lobe, statocysts absent
Order Stygiomysida Tchindonova, 1981. Recent 7

7a. Antennal scale vestigial; carapace short, completely fused with the anterior thoracomeres, giving the body a vermiform shape; maxillula without palp; thoracic endopods 1-4 modified as maxillipeds, endopods 5-8 pediform, females with oostegites on thoracopods 4-7 only; sympod of uropod with large inner lobe whose apical spines extend beyond the endopod
.... Stygiomysidae Caroli, 1937 (figs. 54.1B, 54.3A).

7b. Antennal scale well developed; carapace large, projecting freely over several posterior thoracomeres and giving the body a shrimp-like appearance; maxillula with palp; thoracic endopods 1-3 modified as maxillipeds, endopods 4-8 pediform, females with oostegites on thoracopods 2-8; sympod of uropod with small, spinose inner lobe
.... Lepidomysidae Clarke, 1961 (figs. 54.1C, 54.3B-D, 54.19C, D).

8. Carapace projects freely over several posterior thoracomeres (except in *Palaumysis philippinensis*); 1-2 pairs of pleopods with particular modifications in males of most taxa
Order Mysida Boas, 1883. Jurassic-Recent 9

9a. Endopods of uropods always with statocyst; merus of thoracic endopod 2 without large lobe-like expansion; all thoracopods normally with natatory exopod; maxillula mostly with palp; females with 1-7 pairs of oostegites... 11

9b. Endopods of uropods without statocyst; merus of thoracic endopod 2 with a large lobe-like expansion; thoracopod 1, often also thoracopod 2, without natatory exopod; maxillula without palp; females with 7 pairs of oostegites
Petalophthalmidae Czerniavsky, 1882. Recent 10

10a. Mandibular palp strongly enlarged, prehensile; only first thoracopod without natatory exopod
.... Petalophthalminae Czerniavsky, 1882 (figs. 54.4A, 54.12B).

10b. Mandibular palp normal, not prehensile; first as well as second thoracopod without natatory exopod .... Hansenomysinae, new subfamily (figs. 54.5A, 54.12A).

11. Well-developed tubular penes except in *Rhopalophthalmus*; endopods of uropods normally undivided (subdivided in *Rhopalophthalmus*)
Mysidae Haworth, 1825. Jurassic-Recent 12

12a. Seven pairs of oostegites; exopod of uropods divided by a proximal, incomplete suture, endopod not subdivided; smooth outer margin of the antennal scale ending in a non-articulate spine; male pleopods biramous throughout; telson with apical cleft
.... Boreomysinae Holt & Tattersall, 1905 (figs. 54.4B, 54.5B).

12b. Females with less than 5, mostly 2-3, pairs of oostegites; exopod of uropods not subdivided or with a subterminal, but never a proximal suture....................................... 13

13a. Both rami of the uropods not subdivided (7 subfamilies)............................. 16

13b. Exopod of uropods divided by a subapical suture; telson elongate, linguiform; antennal scale well developed, with smooth outer margin ending in a non-articulate spine (Rhopalophthalminae, Siriellinae) 14

14a. Both rami of the uropods divided by a suture at about 25-30% length from tip; thoracic endopod 8 small or minute; no tubular penes developed; exopod of uropods without spines, setose all around; male pleopods biramous, exites of endopods not strongly curved or coiled
.... Rhopalophthalminae Hansen, 1910 (figs. 54.4C, 54.5C).

14b. Endopods of uropods not subdivided; thoracic endopods 3-8 well developed, with peculiar brush of setae around dactylus; penes well developed; exopod of uropods with spines on outer margin of the basal segment
Siriellinae Czerniavsky, 1882 15

15a. Male pleopods 1-5 biramous, multi-segmented except in endopod 1, basal segment of endopods with spirally coiled or curved, less frequently straight exites (= pseudobranchial lobes)
.... Siriellini Czerniavsky, 1882 (figs. 54.4D, 54.5D, 54.26A).

15b. Male pleopods 1-3, 5 rudimentary, uniramous and unsegmented as in females; male pleopod 4 with two multi-segmented rami, basal segment of endopod with straight, rod-like exite
.... Metasiriellini Murano, 1986.

16a. First pleomere without projecting pleural plates in both sexes; exopod of male pleopod 3 not conspicuously elongate, terminally not or only weakly modified (6 subfamilies)............19
16b. First pleomere of female with pleural plates supporting the marsupium; male pleopod 3 with elongate, strongly modified exopod; antennal scale with smooth outer margin ending in a non-articulate spine; telson with spines on lateral margins, its apical cleft armed with spine-like laminae
Gastrosaccinae Norman, 1892 17
17a. First pleomere with small pleural plates covering a small part of the marsupium, no such plates in males; female pleopods uniramous, unsegmented; male pleopod 1, in part also 4, 5, rudimentary, resembling those in females, remaining male pleopods biramous, exopod 3 modified in diverse, in part complex ways
.... Anchialinini, new replacement name (figs. 54.4E, 54.5E).
17b. First pleomere with large pleural plates covering and supporting the marsupium from behind, in part also from the side; female pleopod 1 biramous (Archaeomysini, Gastrosaccini).....18
18a. First pleomere without pleural plates in males; female pleopods 2-5 uniramous, unsegmented; male pleopods biramous or reduced in various ways, exopod 3 (exceptionally exopod 4) particularly long and modified in diverse, in part complex ways
....Gastrosaccini Norman, 1892 (figs. 54.4F, 54.5F).
18b. First pleomere with pleural plates also present in males, but smaller than in females; certain taxa with plates on additional pleomeres, whereby plate size decreases from pleura 2 to 5; pleopods biramous in both sexes, exopod of pleopod 3 elongated in males only
.... Archaeomysini Czerniavsky, 1882.
19a. All pleopods uniramous, representing endopods in both sexes; endopods 4, 5 not elongate in males (Heteromysinae, Mysidellinae, Palaumysinae)...................................35
19b. All pleopods uniramous in both sexes; endopods of male pleopods 4, 5 elongate (Erythropinae: Calyptommini)..28
19c. All male pleopods or at least male pleopod 4 biramous, all female pleopods uniramous (Mysinae, Leptomysinae, Erythropinae in part)...20
20a. Exopod of male pleopod 4 not conspicuously elongate, mostly without modified setae; oblique or less frequently transverse articulation between carpus and propodus of thoracic endopods 3-8 (Erythropinae in part) 28
20b. Exopod of male pleopod 4 elongate, with modified setae (never on endopod); always with transverse articulation between carpus and propodus of thoracic endopods 3-8 (Mysinae, Leptomysinae) 21
21a. Male pleopods 1, 2 always uniramous, rudimentary as in females; male pleopod 4 and in certain taxa also pleopod 3, exceptionally also pleopod 5, biramous; antennal scale variable; females with two pairs of well-developed oostegites, rarely with three
Mysinae Haworth, 1825 24
21b. Male pleopods 2-5 always biramous, mostly multi-segmented; antennal scale setose all around; females with three pairs of well-developed oostegites, rarely with only two
Leptomysinae Czerniavsky, 1882..22
22a. Exopod of maxilla narrow; first thoracic endopod (5-6)-segmented, its ischium fused with the merus; telson mostly entire, or with small apical indentation, its margins with spines but without setae............................. Mysidopsini, new tribe (figs. 54.15L, 54.18F, 54.22N).
22b. Exopod of maxilla laterally well expanded; first thoracic endopod seven-segmented, its ischium and merus not fused; telson variable (Leptomysini, Afromysini)..........................23
23a. Telson entire or with minute cleft, its margins with spines but without laminae or apical setae; distal segment of endopod of maxilla not conspicuously widened
.... Leptomysini Czerniavsky, 1882 (figs. 54.4H, 54.5H, 54.21N, 54.36E-F).

23b. Telson apically with distinct cleft or less frequently narrowly truncate; apical cleft, if present, mostly with two plumose setae emerging from or close to its bottom, margins of cleft smooth or lined by spine-like laminae; distal segment of endopod of maxilla often terminally widened and with spines or teeth on terminal margin .... Afromysini, new tribe (fig. 54.25F).
24a. Male pleopod 3 uniramous 25
24b. Male pleopod 3 biramous
.... Mysini (without genus *Kainommatomysis*) (figs. 54.1E, 54.4J, 54.5J, 54.36A, B).
25a. Cornea bipartite, posterior part with large, backward-oriented ommatidium bearing a conical lens .... Mysini (genus *Kainommatomysis*).
25b. Cornea normal, not bipartite (Diamysini, Anisomysini, Neomysini) 26
26a. Carpopropodus of thoracic endopod 6 with 4-20 segments; telson entire, linguiform to elongate subtriangular, its lateral margins with spines organized in groups of large spines with smaller ones in between, less frequently in nearly continuous series; male pleopod 3 reduced to small plate as in females .... Neomysini, new tribe (fig. 54.19G).
26b. Carpopropodus of thoracic endopod 6 with only 1-3 segments; telson shorter, with variable shape, its lateral margins bare or with spines arranged in a nearly continuous series of size, i.e., not arranged in groups of different size (Diamysini, Anisomysini) 27
27a. Male pleopod 3 reduced to small endopod fused with larger 2-segmented sympod, less frequently rudimentary as in female; endopod of uropod with mostly one (0-2) spine on ventral face near statocyst; telson mostly with short apical cleft or at least apically truncate; lateral margins ending in a distinctly larger spine compared to the remaining ones, cleft (if distinct) lined with spine-like laminae or with small spines
.... Diamysini, new tribe (figs. 54.4J, 54.18J, M, O, 54.21O, P, 54.35H, J).
27b. Male pleopod 3 normally rudimentary as in females, rarely reduced to small endopod fused with sympod; uropods without spines; telson terminally rounded or with apical cleft, lateral margins bare or furnished with spines of mostly subequal size; telson always devoid of laminae
.... Anisomysini, new tribe.
28. Antennal scale with its outer margin ending in a spine, less frequently setose all around or vestigial; telson entire or rarely terminally indented or cleft
Erythropinae Hansen, 1910 29
29a. Eyestalks separate; antennal gland normal or hypertrophic (5 tribes) 31
29b. Eyestalks fused to a common plate or connected by a median bridge; antennal gland hypertrophic (Calyptommini, Pseudommini, Mysimenziesini) 30
30a. Both eyes fused to a transverse lamina without cleft, visual elements rudimentary or absent; basal segment of antennular peduncle normal; lateral margins of telson not serrate but smooth, terminally with spines .... Calyptommini W. M. Tattersall, 1909.
30b. Both eyes fused to a common plate with anterior, median cleft, visual elements missing; basal segment of antennular peduncle normal; lateral margins of telson not serrate but smooth, terminally with or without spines ..... Pseudommini, new tribe (figs. 54.20C, 54.21J, 54.25P).
30c. Rudimentary, horn-like eyestalks connected by a median bridge; basal segment of antennular peduncle medio-dorsally with sensory fossette or with setose elevation; lateral margins of telson distinctly serrate .... Mysimenziesini Tchindonova, 1981.
31a. Antennal scale absent or vestigial; eyes well developed; labrum normal; 2 pairs of oostegites; thoracic endopods 3-8 with the carpus separated from the propodus by a transverse or a slightly oblique articulation .... Arachnomysini Holt & Tattersall, 1905 (figs. 54.7F, 54.12C).
31b. Antennal scale well developed; eyes normal or reduced to varying extent 32
32a. Eyestalks reduced to a pair of immotile plates, cornea rudimentary or absent; thoracic endopods 3-8 with transverse articulation between carpus and propodus; labrum normal
.... Amblyopsini Tchindonova, 1981 (figs. 54.4G, 54.18D).
32b. Eyes mostly normal or reduced to varying degrees, not plate-like 33

33a. Labrum strongly asymmetrical; four pairs of oostegites; endopods 3-8 with carpus separated from propodus by transverse articulation .... Thalassomysini Nouvel, 1942b.
33b. Labrum normal; 2-3 pairs of well-formed oostegites; thoracic endopods 3-8 with oblique or less frequently with transverse articulation between carpus and propodus................. 34
34a. All pleopods uniramous in both sexes; male pleopod 4 represented as elongate, multi-segmented endopod; remaining pleopods unsegmented, rudimentary
.... Inusitatomysini, new tribe.
34b. All pleopods biramous in males, whereas uniramous, rudimentary in females
.... Erythropini Hansen, 1910 (fig. 54.5G).
35a. Antennal scale missing or vestigial; inner antennular flagellum very short, reduced to only 2-5 segments; penes of moderate size; labrum normal; telson (sub)triangular, without or with small medio-apical indentation
.... Palaumysinae Wittmann, 2013b (figs. 54.4K, 54.5K, 54.7G, 54.21U).
35b. Antennal scale well developed, setose all around; inner antennular flagellum long, multi-segmented as usual; penes strongly developed in most taxa, rarely small (Mysidellinae, Heteromysinae) 36
36a. Labrum with asymmetrical posterior lobes; mandibles with large cutting edge without teeth; maxillula with inwards bent lobes; thoracic endopod 1 with strongly expanded propodus; thoracic endopod 3 normal, not prehensile; telson with apical cleft
.... Mysidellinae Czerniavsky, 1882 (figs. 54.4L, 54.5L, 54.21T).
36b. Labrum, mandibles, maxillula, and thoracic endopod 1 normal; thoracic endopod 3 normal or prehensile; structure of telson varies between tribes
Heteromysinae Norman, 1892 37
37a. Thoracic endopod 3 developed as gnathopod, swollen, prehensile, its carpopropodus two-segmented, rarely fused; males of certain taxa with some of the pleopods bearing small (flagellate) spines or modified setae; telson with apical cleft
.... Heteromysini Norman, 1892 (figs. 54.1D, 54.4M, 54.5M, 54.18K, H, 54.21R, S).
37b. Thoracic endopod 3 pediform, not prehensile and not conspicuously swollen, its carpopropodus with two or more segments 38
38a. Carpopropodus of thoracic endopod 3 with two or more segments; all pleopods reduced to unsegmented plates, without modified setae or spines in both sexes; telson entire or with apical cleft.......................... Mysidetini Holt & Tattersall, 1906 (figs. 54.20E, 54.22M).
38b. Carpopropodus of thoracic endopod 3 with two segments; most pleopods of males rudimentary as in females, without modified setae or spines; by contrast, male pleopod 3 elongate, sub-segmented, apically modified; telson with apical cleft .... Harmelinellini, new tribe.

ACKNOWLEDGEMENTS

The authors are greatly indebted to Jean-Paul Casanova (France) for consent to translate and integrate his contributions to the treatise of Nouvel et al. (1999) into the present work. Consent to reproduce previously published figures was generously given (besides institutional copyright holders) by the authors Lisandro de Almeida (Brazil), Jean-Paul Casanova (figures other than from 1999), Mikhail Daneliya (Helsinki), Laetitia De Jong-Moreau (Marseille), Marc Espeel (Ghent), Yukio Hanamura (Yokohama), Salvatore Inguscio (Nardò), Kenneth Meland (Bergen), Masaaki Murano (Tokyo), Victor Petryashov (St. Petersburg), César Vilas-Fernández (Cádiz), Christian Wirkner (Rostock), and Peter Wirtz (Funchal). Originals of published drawings (fig. 54.31A, C) were kindly provided by C. Wirkner. Original photos were generously supplied by L. de Almeida, José Antonio González (Las Palmas), S. Inguscio, and P. Wirtz. Sincere thanks to the editors J. C. von Vaupel Klein (Bilthoven) and Frederick R. Schram (Langley, WA) for their invitation to write this chapter, for proposing improvements, and for their great patience with our repeated emendations.

APPENDIX

Taxa at species and genus level in the text and figures of this chapter are listed with authorities and dates of their description. Data on synonymy or homonymy are indicated wherever necessary. For names of genera not given below see above, 'Classification'

Acanthomysis aspera Ii, 1964
Acanthomysis longicornis (H. Milne Edwards, 1837)
Acanthomysis robusta Murano, 1984
Acanthomysis sculpta (W. M. Tattersall, 1933)
Acanthomysis thailandica Murano, 1986
Aiptasia mutabilis (Gravenhorst, 1831)
Amblyopsoides crozetii (Willemoës-Suhm in G. O. Sars, 1884)
Amblyopsoides ohlinii (W. M. Tattersall, 1951)
Americamysis almyra (Bowman, 1964)
Americamysis bahia (Molenock, 1969)
Americamysis bigelowi (W. M. Tattersall, 1926)
Amphilina foliacea (Rudolphi, 1819)
Anchialina agilis (G. O. Sars, 1877)
Anchialina truncata (G. O. Sars, 1883)
Anchialina typica (Krøyer, 1861)
Anchialus Krøyer, 1861, junior homonym of the coleopteran genera *Anchialus* Gistel, 1834, and *Anchialus* Thomson, 1859; replaced by the valid *Anchialina* Norman & Scott, 1906
Anemonia viridis (Pennant, 1777)
Anisakis simplex (Rudolphi, 1809)
Anisomysis levi Băcescu, 1973
Antarctomysis maxima Holt & Tattersall, 1906
Antarctomysis ohlinii Hansen, 1908
Antromysis cubanica Băcescu & Orghidan, 1971
Antromysis juberthiei Băcescu & Orghidan, 1977
Archaeomysis articulata Hanamura, 1997
Archaeomysis grebnitzkii Czerniavsky, 1882
Archaeomysis vulgaris (Nakazawa, 1910)
Artemia Leach, 1819

Bacescomysis abyssalis (Lagardère, 1983)
Bartholomea annulata (Lesueur, 1817)
Birsteiniamysis inermis (Willemoës-Suhm, 1874)
Boreomysis arctica (Krøyer, 1861)
Boreomysis californica Ortmann, 1894
Boreomysis megalops G. O. Sars, 1872
Boreomysis microps G. O. Sars, 1883
Boreomysis obtusata G. O. Sars, 1884
Boreomysis scyphops G. O. Sars, 1879, junior synonym of *Birsteiniamysis inermis*
Branchiomma nigromaculatum (Baird, 1865)
Bunocotyle cingulata Odhner, 1928

Caesaromysis hispida Ortmann, 1893
Cancer Linnaeus, 1758
Cancer bipes Fabricius, 1780, outdated original combination of *Nebalia bipes*
Cancer flexuosus O. F. Müller, 1776, outdated original combination of *Praunus flexuosus*
Cancer oculatus Fabricius, 1780, outdated original combination of *Mysis oculata*
Cancer pedatus Fabricius, 1780, suppressed name (ICZN, 1959) (euphausiid?)
Cassiopea Péron & Lesueur, 1810
Ceratolepis hamata G. O. Sars, 1883
Ceratomysis spinosa Faxon, 1893
Cercopagis pengoi (Ostroumov, 1891)
Chalaraspidum alatum (Willemoës-Suhm, 1876)
Chlamydopleon aculeatum Ortmann, 1893
Chunomysis diadema Holt & Tattersall, 1905
Coifmanniella johnsoni (W. M. Tattersall, 1937)
Cymodocea K. D. König, 1806

Dactylerythrops dactylops Holt & Tattersall, 1905
Daphnia O. F. Müller, 1785
Dardanus Paulson, 1875
Diadema Gray, 1825
Diadema antillarum Philippi, 1845
Diamysis bacescui Wittmann & Ariani, 1998
Diamysis bahirensis (G. O. Sars, 1877)
Diamysis camassai Ariani & Wittmann, 2002
Diamysis cymodoceae Wittmann & Ariani, 2012
Diamysis fluviatilis Wittmann & Ariani, 2012
Diamysis lagunaris Ariani & Wittmann, 2000
Diamysis mesohalobia Ariani & Wittmann, 2000
Dioptromysis paucispinosa Brattegard, 1969
Dioptromysis perspicillata Zimmer, 1915
Dollocaris ingens van Straelen, 1923

Echinorhynchus leidyi Van Cleve, 1924
Elder unguiculata Münster, 1839
Erythrops abyssorum (G. O. Sars, 1869)
Erythrops serratus (G. O. Sars, 1863)
Euchaetomera zurstrasseni (Illig, 1906)
Eucopia australis Dana, 1852
Eucopia crassicornis Casanova, 1997
Eucopia grimaldii H. Nouvel, 1942
Eucopia major Hansen, 1910
Eucopia precursor Secretan & Riou, 1986
Eucopia sculpticauda Faxon, 1893
Eucopia unguiculata (Willemoës-Suhm, 1875)

Francocaris grimmi Broili, 1917

Gastrosaccus brevifissura O. S. Tattersall, 1952
Gastrosaccus psammodytes O. S. Tattersall, 1958
Gastrosaccus roscoffensis Băcescu, 1970
Gastrosaccus sanctus (Van Beneden, 1861)
Gastrosaccus spinifer (Goës, 1864)

Gastrosaccus widhalmi (Czerniavsky, 1882)
Gastrosaccus yuyu Bamber & Morton, 2012
Gnathophausia affinis G. O. Sars, 1884
Gnathophausia childressi Casanova, 1996
Gnathophausia gigas Willemoës-Suhm, 1873
Gnathophausia ingens (Dohrn, 1870)
Gnathophausia longispina G. O. Sars, 1883
Gnathophausia zoea Willemoës-Suhm, 1873

Haemulon flavolineatum (Desmarest, 1823)
Hansenomysis atlantica Lagardère, 1983
Hansenomysis abyssalis Lagardère, 1983
Hansenomysis falklandica O. S. Tattersall, 1955
Hansenomysis nouveli Lagardère, 1983
Hansenomysis pseudofyllae Lagardère, 1983
Haplostylus lobatus (H. Nouvel, 1951)
Haplostylus magnilobatus (Băcescu & Schiecke, 1974)
Hemimysis anomala G. O. Sars, 1907
Hemimysis lamornae (Couch, 1856)
Hemimysis lamornae mediterranea Băcescu, 1937
Hemimysis margalefi Alcaraz, Riera & Gili, 1986
Hemimysis speluncola Ledoyer, 1963
Heteromysis actiniae Clarke, 1955
Heteromysis arianii Wittmann, 2000
Heteromysis armoricana H. Nouvel, 1940
Heteromysis dardani Wittmann, 2008
Heteromysis harpax (Hilgendorf, 1879)
Heteromysis harpaxoides Băcescu & Bruce, 1980
Heteromysis wirtzi Wittmann, 2008
Heteromysoides cotti (Calman, 1932)

Idiomysis inermis W. M. Tattersall, 1922
Idiomysis tsurnamali Băcescu, 1973
Idotea balthica (Pallas, 1772)
Ischiomysis peterwirtzi Wittmann, 2013
Ischiomysis telmatactiphila Wittmann, 2013

Kainommatomysis foxi W. M. Tattersall, 1927
Kainommatomysis schieckei Băcescu, 1973
Kilianicaris lerichei van Straelen, 1923

Lepidomysis Clarke, 1961, junior synonym of *Spelaeomysis* Caroli, 1924
Lepidophthalmus servatus Fage, 1924, invalid original combination for *Spelaeomysis servata*; genus name is a junior homonym of the decapod genus *Lepidophthalmus* Holmes, 1904
Lepidops Zimmer, 1927, invalid replacement name for *Lepidophthalmus* Fage, 1924, in turn a junior synonym of *Spelaeomysis* Caroli, 1924; junior homonym of the decapod genus *Lepidops* Miers, 1878
Leptomysis buergii Băcescu, 1966
Leptomysis gracilis (G. O. Sars, 1864)
Leptomysis heterophila Wittmann, 1986
Leptomysis lingvura (G. O. Sars, 1868)
Leptomysis lingvura marioni Gourret, 1888

Leptomysis mediterranea G. O. Sars, 1877
Leptomysis posidoniae Wittmann, 1986
Leptomysis truncata sardica G. O. Sars, 1877
Limnomysis benedeni Czerniavsky, 1882
Longithorax fuscus Hansen, 1908
Lophogaster affinis Colosi, 1930
Lophogaster erythraeus Colosi, 1930
Lophogaster longirostris Faxon, 1896
Lophogaster pacificus Băcescu, 1985
Lophogaster schmidti Fage, 1940
Lophogaster spinosus Ortmann, 1907
Lophogaster subglaber Hansen, 1927
Lophogaster typicus M. Sars, 1857
Lophogaster voultensis Secretan & Riou, 1986

Megalophthalmus Fabricianus Leach, 1830, suppressed name (ICZN, 1959) (an euphausiid?)
Mesopodopsis aegyptia Wittmann, 1992
Mesopodopsis africana O. S. Tattersall, 1952
Mesopodopsis orientalis (W. M. Tattersall, 1908)
Mesopodopsis slabberi (Van Beneden, 1861)
Mesopodopsis tenuipes Hanamura et al., 2008
Mesopodopsis wooldridgei Wittmann, 1992
Metamysidopsis elongata (Holmes, 1900)
Metamysidopsis neritica Bond-Buckup & Tavares, 1992
Mysidella biscayensis Lagardère & Nouvel, 1980
Mysidella typhlops G. O. Sars, 1872
Mysidella typica G. O. Sars, 1872
Mysidetes intermedia O. S. Tattersall, 1955
Mysidetes posthon Holt & Tattersall, 1906
Mysidium columbiae (Zimmer, 1915)
Mysidium gracile (Dana, 1852)
Mysidium integrum W. M. Tattersall, 1951
Mysidobdella borealis (Johansson, 1898)
Mysidopsis californica W. M. Tattersall, 1932
Mysidopsis gibbosa G. O. Sars, 1864
Mysidopsis oligocenica De Angeli & Rossi, 2006 [originally described as *Mysidopsis oligocenicus*; disagreement in gender with the generic name corrected here]
Mysidopsis robustispina Brattegard, 1969
Mysifaun erigens Wittmann, 1996
Mysis diluviana Audzijonyte & Väinölä, 2005
Mysis Fabricii Leach, 1815, junior synonym of *Mysis oculata*
Mysis mixta Lilljeborg, 1852
Mysis oculata (Fabricius, 1780)
Mysis relicta Lovén, 1862
Mysis salemaai Audzijonyte & Väinölä, 2005
Mysis stenolepis S. I. Smith, 1873

Nebalia bipes (Fabricius, 1780)
Nebaliopsis typica G. O. Sars, 1887
Neomysis americana (S. I. Smith, 1873)
Neomysis awatschensis (Brandt, 1851)

Neomysis integer (Leach, 1814)
Neomysis intermedia (Czerniavsky, 1882)
Neomysis japonica Nakazawa, 1910
Neomysis mercedis Holmes, 1896
Neomysis rayii (Murdoch, 1885)

Orientomysis mitsukurii (Nakazawa, 1910)

Palaumysis philippinensis Hanamura & Kase, 2002
Palaumysis pilifera Hanamura & Kase, 2003
Palaumysis simonae Băcescu & Iliffe, 1986
Paraleptomysis dimorpha Wittmann, 1986
Paralichthys olivaceus Temminck & Schlegel, 1846
Paralophogaster glaber Hansen, 1910
Paralophogaster foresti Băcescu, 1981
Paramesopodopsis rufa Fenton, 1985
Paramysis arenosa (G. O. Sars, 1877)
Paramysis bakuensis G. O. Sars, 1895
Paramysis helleri (G. O. Sars, 1877)
Paramysis inflata (G. O. Sars, 1907)
Paramysis lacustris (Czerniavsky, 1882)
Paramysis nouveli Labat, 1953
Paramysis pontica Băcescu, 1940
Paramysis portzicensis H. Nouvel, 1950
Paramysis ullskyi Czerniavsky, 1882
Paranchialina angusta (G. O. Sars, 1883)
Parapseudomma calloplura (Holt & Tattersall, 1905)
Peachocaris acanthouraea Schram, 1984
Peachocaris strongi (Brooks, 1962)
Petalophthalmus armiger Willemoës-Suhm, 1875
Posidonia K. D. König, 1805
Praunus flexuosus (O. F. Müller, 1776)
Praunus inermis (Rathke, 1843)
Praunus neglectus (G. O. Sars, 1869)
Pseudomma affine G. O. Sars, 1870
Pseudomma armatum Hansen, 1913
Pseudomma oculospinum W. M. Tattersall, 1951
Pyroleptomysis rubra Wittmann, 1985

Rhopalophthalmus egregius Hansen, 1910
Rhopalophthalmus longicauda O. S. Tattersall, 1957
Rhopalophthalmus tartessicus Vilas-Fernández, Drake & Sorbe, 2008
Rhopalophthalmus tattersallae Pillai, 1961
Rhopalophthalmus terranatalis O. S. Tattersall, 1957

Schimperella acanthocercus Taylor, Schram & Yan-Bin, 2001
Schimperella beneckei Bill, 1914
Schimperella kessleri Bill, 1914
Schistomysis assimilis (G. O. Sars, 1877)
Schistomysis kervillei (G. O. Sars, 1885)
Schistomysis ornata (G. O. Sars, 1864)
Schistomysis spiritus (Norman, 1860)

Sinelobus stanfordi (Richardson, 1901)
Siriella adriatica Hoenigman, 1960, junior synonym of *Siriella gracilipes*
Siriella antiqua Secretan & Riou, 1986
Siriella armata (H. Milne Edwards, 1837)
Siriella carinata Secretan & Riou, 1986
Siriella castellabatensis Ariani & Spagnuolo, 1976
Siriella clausii G. O. Sars, 1877
Siriella gracilipes H. Nouvel, 1942
Siriella gracilis Dana, 1852
Siriella thompsonii (H. Milne Edwards, 1837)
Spelaeomysis bottazzii Caroli, 1924
Spelaeomysis cardisomae Bowman, 1973
Spelaeomysis cochinensis Panampunnayil & Viswakumar, 1991
Spelaeomysis longipes (Pillai & Mariamma, 1964)
Spelaeomysis nuniezi Băcescu & Orghidan, 1971
Spelaeomysis olivae Bowman, 1973
Spelaeomysis quinterensis (Villalobos, 1951)
Spelaeomysis servata (Fage, 1924) [Caroli (1924) clearly indicated female gender upon first description of the genus by spelling "la *Spelaeomysis*", where in Italian "la" is the female, singular, definite article. Clarke (1961a) used the correct ending of the species name by recombining *Lepidophthalmus servatus* Fage, 1924, to the binomen *Lepidomysis servata* (Fage, 1924). Ingle (1972) did not acknowledge the genus *Lepidomysis* Clarke, 1961a, and proposed the new combination *Spelaeomysis servatus* (Fage, 1924), however, without taking account to the previously defined gender of *Spelaeomysis*. Disagreement in gender of species name is corrected by present authors (Article 34.2 of ICZN, 1999)]
Stygiomysis aemete Wagner, 1992
Stygiomysis cokei Kallmeyer & Carpenter, 1996
Stygiomysis holthuisi (Gordon, 1958)
Stygiomysis hydruntina Caroli, 1937
Stygiomysis ibarrae Ortiz, Lalana & Perez, 1996
Stygiomysis major Bowman, 1976
Surinamysis merista (Bowman, 1980)

Telmatactis cricoides (Duchassaing, 1850)
Tenagomysis tasmaniae Fenton, 1991
Thalassia Banks ex K. D. König, 1805
Troglomysis vjetrenicensis Stammer, 1933

Zoothamnium duplicatum Kahl, 1933
Zostera Linnaeus, 1753

BIBLIOGRAPHY

ACOSTA, C. & M. POIRRIER, 1992. Grooming behaviour and associated structures of the mysid *Mysidopsis bahia*. — Journal of Crustacean Biology, **12**: 383-391.
AGASSIZ, L., 1843. Nomenclator zoologicus. Fasciculus IV. Continens Crustacea et Vermes i.e. Entozoa, Turbelloria et Annulata. Nomina systematica generum Crustaceorum, tam viventium quam fossilium, secundum ordinem alphabeticum disposita, adjectis auctoribus, libris in quibus reperiuntur, anno editionis, etymologia et familiis, ad quas pertinent. — Recognovit H. Burmeister, pp. i-vii, 1-28. (Jent et Gassmann, Soloduri.)

AHYONG, S. T., J. K. LOWRY, M. ALONSO, R. N. BAMBER, G. A. BOXSHALL, P. CASTRO, S. GERKEN, G. S. KARAMAN, J. W. GOY, D. S. JONES ET AL., 2011. Subphylum Crustacea Brünnich, 1772. — In: Z.-Q. ZHANG (ed.), Animal biodiversity: an outline of higher-level classification and survey of taxonomic richness. — Zootaxa, **3148**: 165-191.

ALADIN, N., I. PLOTNIKOV, A. BOLSHOV & A. PICHUGIN, 2006. Biodiversity of the Caspian Sea. — In: Caspian Sea Biodiversity Project under umbrella of Caspian Sea Environment Program, pp. 1-85. http://www.zin.ru/projects/caspdiv/biodiversity_report.html [16 Nov. 2010]

ALADIN, N. V., I. S. PLOTNIKOV & A. A. FILIPPOV, 2003. Opportunistic settlers in the Aral Sea. — In: First International Meeting "The invasion of the Caspian Sea by the comb jelly *Mnemiopsis* — problems, perspectives, need for action", Baku, Azerbaijan, 24-26 April 2002, attachment, **9**: 3 pp. http://www.caspianenvironment.org/mnemiopsis/mnem_attach9.htm [28 Nov. 2005]

ALADIN, N., I. PLOTNIKOV & R. LETOLLE, 2004. Hydrobiology of the Aral Sea. — In: J. C. J. NIHOUL, P. O. ZAVYALOV & PH. P. MICKLIN (eds.), Dying and dead seas. NATO ARW/ASI Series, pp. 125-158. (Kluwer Academic Publishers, Dordrecht.)

ALEXANDROWICZ, J. S., 1955. Innervation of the heart of *Praunus flexuosus* (Mysidacea). — Journal of the Marine Biological Association of the United Kingdom, **34**: 47-53, 1 pl.

ALLEN, D. M., 1984. Population dynamics of the mysid shrimp *Mysidopsis bigelowi* W. M. Tattersall in a temperate estuary. — Journal of Crustacean Biology, **4** (1): 25-34.

ALLEN, D. M. & W. B. ALLEN, 1981. Seasonal dynamics of a leech-mysid shrimp interaction in a temperate salt marsh. — Biological Bulletin, Woods Hole, **160** (1): 1-10.

ALMEIDA PRADO, M. S. DE, 1973. Distribution of Mysidacea (Crustacea) in the Cananeia region. — Boletim de Zoologia e Biologia Marinha, (N. S.) **30**: 395-417.

ALMEIDA PRADO-POR, M. S., 1980. Mysidacea from the Gulf of Elat (Gulf of 'Aqaba). — Israel Journal of Zoology, **29**: 188-191.

AMARATUNGA, T. & S. COREY, 1975. Life history of *Mysis stenolepis* Smith (Crustacea, Mysidacea). — Canadian Journal of Zoology, **53**: 942-952.

ANDERSON, G., 2010. Mysida classification, January 20, 2010. — http://peracarida.usm.edu/MysidaTaxa.pdf, pp. 1-24. (Instant Web Publishing, U.S.A.). [21 Jan. 2010]

ANGELINI, D. R. & T. C. KAUFMANN, 2005. Comparative developmental genetics and the evolution of arthropod body plans. — Annual Review in Genetics, **39**: 95-119.

APEL, M., 1992. Spatial distribution and seasonal occurrence of Mysidacea in the Jade estuary (North Sea, Germany), with some comments on diurnal migrations. — In: J. KÖHN, M. B. JONES & A. MOFFAT (eds.), Taxonomy, biology and ecology of (Baltic) mysids (Mysidacea, Crustacea), pp. 98-108. (Rostock University Press, Rostock.)

APPELTANS, W., S. T. AHYONG, G. ANDERSON, M. V. ANGEL, T. ARTOIS, N. BAILLY, R. BAMBER, A. BARBER, I. BARTSCH, A. BERTA ET AL., 2012. The magnitude of global marine species diversity. — Current Biology, **22** (23): 2189-2202.

ARBAČIAUSKAS, K., 2002. Ponto-Caspian amphipods and mysids in the inland waters of Lithuania: history of introduction, current distribution and relations with native malacostracans. — In: E. LEPPÄKOSKI, S. GOLLASCH & S. OLENIN (eds.), Invasive aquatic species of Europe — distribution, impacts and management, pp. 104-115. (Kluwer Academic Publishers, Dordrecht.)

ARBAČIAUSKAS, K., V. RAKAUSKAS & T. VIRBICKAS, 2010. Initial and long-term consequences of attempts to improve fish-food resources in Lithuanian waters by introducing alien peracaridan species: a retrospective overview. — Journal of Applied Ichthyology, **26** (Suppl. 2): 28-37.

ARBAČIAUSKAS, K., G. VIŠINSKIENĖ & S. SMILGEVIČIENĖ, 2012. Non-indigenous macroinvertebrate species in Lithuanian fresh waters. Part 2: Macroinvertebrate assemblage deviation from naturalness in lotic systems and the consequent potential impacts on ecological quality assessment. — Knowledge and Management of Aquatic Ecosystems, **402** (13): 13p1-13p18. www.kmae-journal.org [31 Jan. 2012]

ARBAČIAUSKAS, K., G. VIŠINSKIENĖ, S. SMILGEVIČIENĖ & V. RAKAUSKAS, 2011. Nonindigenous macroinvertebrate species in Lithuanian fresh waters. Part 1: Distributions, dispersal and future. — Knowledge and Management of Aquatic Ecosystems, **402** (12): 12p1-12p18. www.kmae-journal.org [31 Jan. 2012]
ARIANI, A. P., 1966. Su una forma di *Diamysis bahirensis* (G.O. Sars) rinvenuta in territorio pugliese. — Bollettino di Zoologia, **33**: 227-229.
— —, 1979. Contribution à l'étude écotaxonomique et biogéographique des *Diamysis* d'eau saumâtre de la Méditerrannée. — Rapports et Procès verbaux de la Commission internationale pour l'Exploration scientifique de la Mer Méditerranée, **25-26** (3): 159-160.
— —, 1981a. Systématique du genre *Diamysis* et paléogéographie de la Méditerranée. — In: Journées d'Études sur la Systématique évolutive et la Biogéographie en Méditerranée, C.I.E.S.M., **1980**, pp. 121-130.
— —, 1981b. Expériences d'hybridation entre populations méditerranéennes du genre *Diamysis*. — Rapports et Procès Verbaux de la Commission Internationale de la Mer Méditerrannée, **27** (4): 177-180.
— —, 1981c. *Spelaeomysis bottazzii* Caroli (Crustacea, Mysidacea) nella falda freatica del litorale Brindisino. — Annuario dell'Istituto e Museo di Zoologia dell'Università di Napoli, **23**: 157-166. [1979-1980]
— —, 1982. Osservazioni e ricerche su *Typhlocaris salentina* (Crustacea, Decapoda) e *Spelaeomysis bottazzii* (Crustacea, Mysidacea). Approccio idrogeologico e biologico sperimentale allo studio del popolamento acquatico ipogeo della Puglia. — Annuario dell'Istituto e Museo di Zoologia dell'Università di Napoli, **25**: 201-326.
— —, 2004. Statolitologia dei Crostacei Misidacei: tra biologia e scienze della terra. — Atti della Accademia Pontaniana, (N. S.) **52**: 349-377. [English summary.] [2003]
ARIANI, A. P., G. BALASSONE, G. MIRONE & K. J. WITTMANN, 1999. Experimentally induced mineral phase change and morphological aberrations in $CaCO_3$ statoliths of Mysidacea. — In: F. R. SCHRAM & J. C. VON VAUPEL KLEIN (eds.), Crustaceans and the biodiversity crisis. Proceedings of the Fourth International Crustacean Congress, Amsterdam, The Netherlands, July 20-24, 1998, **1**: 859-870. (Brill, Leiden.)
ARIANI, A. P., G. CESARO & A. BALBONI, 1984. Ricerche sulle preferenze termiche di *Spelaeomysis bottazzii* (Mysidacea, Lepidomysidae). — Bollettino di Zoologia, **51** (Suppl.): 4.
ARIANI, A. P., F. MARMO, G. BALSAMO, G. CESARO & N. MARESCA, 1982. Prime osservazioni sullo sviluppo degli statoliti di crostacei misidacei. — Annuario dell'Istituto e Museo di Zoologia dell'Università di Napoli, **25**: 327-341.
ARIANI, A. P., F. MARMO, G. BALSAMO & E. FRANCO, 1981. Vaterite in the statoliths of a mysid crustacean (*Diamysis bahirensis*). — Annuario dell'Istituto e Museo di Zoologia dell'Università di Napoli, **24**: 69-78.
ARIANI, A. P., F. MARMO, G. BALSAMO, E. FRANCO & K. J. WITTMANN, 1983. The mineral composition of statoliths in relation to taxonomy and ecology in mysids. — Rapports de la Commission Internationale pour l'Exploration Scientifique de la Mer Méditerranée, **28** (6): 333-336.
ARIANI, A. P. & V. SERRA, 1969. Sulla presenza del *Portunus pelagicus* (L.) in acque italiane, con osservazioni sulla morfologia della specie (Crustacea, Decapoda). — Archivio Botanico e Biogeografico Italiano, **54** (4ª Serie) – **14** (4): 187-206, pls. 1-3.
ARIANI, A. P. & G. SPAGNUOLO, 1976. Ricerche sulla misidofauna del Parco di Santa Maria di Castellabate (Salerno) con descrizione di una nuova specie di *Siriella*. — Bollettino della Società dei Naturalisti in Napoli, **84**: 441-481.
ARIANI, A. P. & K. J. WITTMANN, 1997. Alcuni aspetti della biologia della riproduzione in *Spelaeomysis bottazzii* Caroli (Mysidacea, Lepidomysidae). — In: Il carsismo dell'area Mediterranea. — Thalassia Salentina, **23** (Suppl.): 193-200.

— — & — —, 2000. Interbreeding versus morphological and ecological differentiation in Mediterranean *Diamysis* (Crustacea, Mysidacea), with description of four new taxa. — Hydrobiologia, **441**: 185-236.

— — & — —, 2002. The transition from an epigean to a hypogean mode of life: morphological and bionomical characteristics of *Diamysis camassai* sp. nov. (Mysidacea, Mysidae) from brackish-water dolinas in Apulia, SE-Italy. — Crustaceana, **74** (11): 1241-1265.

— — & — —, 2010. Feeding, reproduction, and development of the subterranean peracarid shrimp *Spelaeomysis bottazzii* (Lepidomysidae) from a brackish well in Apulia (southeastern Italy). — Journal of Crustacean Biology, **30** (3): 384-392.

ARIANI, A. P., K. J. WITTMANN & E. FRANCO, 1993. A comparative study of static bodies in mysid crustaceans: evolutionary implications of crystallographic characteristics. — Biological Bulletin, Woods Hole, **185** (3): 393-404.

AßMANN, C., E. VON ELERT & R. GERGS, 2009. Impact of leaf litter and its fungal colonisation on the diet of *Limnomysis benedeni* (Crustacea: Mysida). — Hydrobiologia, **636** (1): 439-447.

ASTTHORSSON, O. S., 1984. The distribution and biology of mysids in Icelandic subarctic waters as demonstrated by analysis of cod stomach contents. — Sarsia, **69** (2): 197-116.

ATTRAMADAL, Y. G., 1981. On a non-existent ventral filtration current in *Hemimysis lamornae* (Couch) and *Praunus flexuosus* (Müller) (Crustacea: Mysidacea). — Sarsia, **66** (4): 283-286.

ATTRAMADAL, Y. G., J. M. FOSSÅ & H. L. NILSSON, 1985. Changes in behaviour and eye-morphology of *Boreomysis megalops* G.O. Sars (Crustacea: Mysidacea) following exposure to short periods of artificial and natural daylight. — Journal of Experimental Marine Biology and Ecology, **85** (3): 135-148.

AUDZIJONYTE, A., M. E. DANELIYA, N. MUGUE & R. VÄINÖLÄ, 2008 (cf. a). Phylogeny of *Paramysis* (Crustacea: Mysida) and the origin of Ponto-Caspian endemic diversity: resolving power from nuclear protein-coding genes. — Molecular Phylogenetics and Evolution, **46**: 738-759.

AUDZIJONYTE, A. & R. VÄINÖLÄ, 2005. Diversity and distributions of circumpolar fresh- and brackish-water *Mysis* (Crustacea: Mysida): descriptions of *M. relicta* Lovén, 1862, *M. salemaai* n. sp., *M. segerstralei* n. sp. and *M. diluviana* n. sp., based on molecular and morphological characters. — Hydrobiologia, **544**: 89-141.

— — & — —, 2006. Phylogeographic analyses of a circumarctic coastal and a boreal lacustrine mysid crustacean, and evidence of fast post-glacial mtDNA rates. — Molecular Ecology, **15**: 3287-3301.

AUDZIJONYTE, A., K. J. WITTMANN, I. OVCARENKO & R. VÄINÖLÄ, 2009. Invasion phylogeography of the Ponto-Caspian crustacean *Limnomysis benedeni* dispersing across Europe. — Diversity and Distributions, **15**: 346-255.

AUDZIJONYTE, A., K. J. WITTMANN & R. VÄINÖLÄ, 2008 (cf. b). Tracing recent invasions of the Ponto-Caspian mysid shrimp *Hemimysis anomala* across Europe and to North America with mitochondrial DNA. — Diversity and Distributions, **14** (2): 179-186.

BABBITT, C. C. & N. H. PATEL, 2005. Relationships within the Pancrustacea: examining the influence of additional malacostracan 18S and 28S rDNA. — In: S. KOENNEMAN & R. A. JENNER (eds.), Crustacea and arthropod relationships. — Crustacean Issues, **16**: 275-294. (CRC Press, Taylor & Francis, Boca Raton, FL.)

BĂCESCU, M., 1940. Les Mysidacés des eaux Roumaines (étude taxonomique, morphologique, biogéographique et biologique). — Annales Scientifiques de l'Université de Jassy, **26**: 453-804, pls. I-IV.

— —, 1954. Crustacea. Mysidacea. — In: Fauna Republicii Populare Romîne, **4** (3): 1-126. (Editura Academiei Republicii Populare Romîne, Bucureşti.)

— —, 1968a. Heteromysini nouveaux des eaux cubaines: trois espèces nouvelles de *Heteromysis* et *Heteromysoides spongicola* n.g. n.sp. — Revue Roumaine de Biologie, (Zoologie) **13** (4): 221-237.

— —, 1968b. Contributions to the knowledge of the Gastrosaccinae psammobionte of the tropical America, with the description of a new genus (*Bowmaniella*, n.g.) and three new species of its frame. — In: The Centennary Grigore Antipa, 1867-1967. — Travaux du Muséum d'Histoire Naturelle « Grigore Antipa », **8**: 355-373.

— —, 1970. New spongicolous *Heteromysis* of the Caribbean Sea (*H. gomezi* n.sp. and *H. mariani* n.sp.). — Revue Roumaine de Biologie, (Zoologie) **15** (1): 11-16.

— —, 1971a. *Mysimenzies hadalis* g.n. sp.n., a benthic mysid of the Peru Trench, found during cruise XI/1965 of R/V Anton Bruun (U.S.A.). — Revue Roumaine de Biologie, (Zoologie) **16** (1): 3-8.

— —, 1971b. Contributions to the mysid Crustacea from the Peru-Chile Trench (Pacific Ocean). — In: Scientific Results of the Southeast Pacific Expedition. — Anton Bruun Report, **7**: 3-24.

— —, 1973a. A new case of commensalism in the Red Sea: the mysid *Idiomysis tsurnamali* n.sp. with the Coelenterata *Megalactis* and *Cassiopea*. — Revue Roumaine de Biologie, (Zoologie) **18** (1): 3-7.

— —, 1973b. *Anisomysis levi* n.sp. from the Red Sea and the dichotomic key of the species belonging to the genus, with description of a new taxon, *Paranisomysis* n.sg. — Revue Roumaine de Biologie, (Zoologie) **18** (3): 173-180.

— —, 1973c. Contribution à la connaissance des Mysidés benthiques de la mer Rouge. — Rapports de la Commission Internationale pour l'Exploration Scientifique de la Mer Méditerranée (CIESM), **21**: 643-646.

— —, 1976. Contribution à la connaissance des Mysidacés (Crustacés) de la côte Lybienne, avec la description de deux nouvelles espèces, *Neoheteromysis mülleri* n.sp. n.sp. et *Heteromysis lybiana* n.sp. — Revue Roumaine de Biologie, (Biologie Animale) **21** (2): 85-91.

— —, 1981. Crustacés: Mysidacea. — In: J. FOREST (ed.), Resultats des Campagnes MUSORSTOM. I. Philippines (18-28 mars 1976), **1**. — Mémoires ORSTOM, **91**: 261-276.

— —, 1985. Crustacés Mysidacés [MUSORSTOM II]. — In: Résultats des Campagnes MUSORSTOM II. — Mémoires du Muséum National d'Histoire Naturelle, (A, Zoologie) **133**: 355-366.

BĂCESCU, M. & A. J. BRUCE, 1980. New contributions to the knowledge of the representatives of genus *Heteromysis* s.l. from the Australian coral reefs. — Travaux du Muséum d'Histoire Naturelle « Grigore Antipa », **21** (1): 63-72.

BĂCESCU, M. & T. M. ILIFFE, 1986. Contribution to the knowledge of Mysidacea from western Pacific: *Aberomysis muranoi* n.gen., n.sp. and *Palaumysis simonae* n.gen., n.sp. from marine caves on Palau, Micronesia. — Travaux du Muséum d'Histoire Naturelle « Grigore Antipa », **28**: 25-35.

BĂCESCU, M. & T. ORGHIDAN, 1971. *Antromysis cubanica* n.sp. et *Spelaeomysis nuniezi* n.sp., *Mysis cavernicoles* nouvelles de Cuba. — Revue Roumaine de Biologie, (Zoologie) **16** (4): 225-231.

BAINBRIDGE, R. & T. H. WATERMAN, 1957. Polarized light and the orientation of two marine Crustacea. — The Journal of Experimental Biology, **34**: 342-364.

— — & — —, 1958. Turbidity and the polarized light orientation of the crustacean *Mysidium*. — The Journal of Experimental Biology, **35** (3): 487-493.

BAMBER, R. N. & B. MORTON, 2012. A new and commercial species of *Gastrosaccus* Norman, 1868 (Peracarida: Mysida: Mysidae) from Java, Indonesia. — Zootaxa, **3546**: 43-52.

BANNER, A. H., 1948. Part II. Mysidacea, from tribe Mysini through subfamily Mysidellinae. — In: A taxonomic study of the Mysidacea and Euphausiacea (Crustacea) of the northeastern Pacific. — Transactions of the Royal Canadian Institute, **27**: 65-125.

BEETON, A. M., 1958. The vertical migration of *Mysis relicta* in lakes Huron and Michigan. — Dissertation Abstracts, **19** (3): 601.

— —, 1959. Photoreception in the opossum shrimp *Mysis relicta* Lovén. — Biological Bulletin, Woods Hole, **116** (2): 204-216.

— —, 1960. The vertical migration of *Mysis relicta* in lakes Huron and Michigan. — Journal of the Fisheries Research Board of Canada, **17** (4): 517-539.
BELMAN, B. W. & J. J. CHILDRESS, 1976. Circulatory adaptations to the oxygen minimum layer in the bathypelagic mysid *Gnathophausia ingens*. — Biological Bulletin, Woods Hole, **150**: 15-37.
BENKO, K. I., 1962. Some results on quantitative study of food of mysids. — Trudy Karadagskoi Biologicheskoi Stantsii, **18**: 37-43. [In Russian.]
BENZID, D., L. DE JONG, C. LEJEUSNE, P. CHEVALDONNÉ & X. MOREAU, 2006. Serotonin expression in the optic lobes of cavernicolous crustaceans during the light-dark transition phase: role of the lamina ganglionaris. — Journal of Experimental Marine Biology and Ecology, **335**: 74-81.
BERGH, R. S., 1893. Beiträge zur Embryologie der Crustaceen. 1. Zur Bildungsgeschichte des Keimstreifens von *Mysis*. — Zoologische Jahrbücher, (Abtheilung für Anatomie und Ontogenie der Thiere) **6**: 491-528, pls. 26-29.
BERNAUER, D. & W. JANSEN, 2006. Recent invasions of alien macroinvertebrates and loss of native species in the upper Rhine River, Germany. — Aquatic Invasions, **1** (2): 55-71.
BERRILL, M., 1969a. The embryonic behavior of the mysid shrimp, *Mysis relicta*. — Canadian Journal of Zoology, **47**: 1217-1221.
— —, 1969b. The schooling and antipredatory behavior of mysid shrimp. — Dissertation, Princeton University. Dissertation Abstracts International, **29** (8-B): 3130.
BERRILL, M. & D. C. LASENBY, 1983. Life cycles of the freshwater mysid shrimp *Mysis relicta* reared at two temperatures. — Transactions of the American Fisheries Society, **112** (4): 551-553.
BETHE, A., 1895. Die Otocyste von *Mysis*. Bau, Innervierung, Entwicklung und physiologische Bedeutung. — Zoologische Jahrbücher, (Abteilung für Anatomie und Ontogenie der Tiere) **8**: 544-564.
BEYST, B., D. BUYSSE, A. DEWICKE & J. MEES, 2001. Surf zone hyperbenthos of Belgian sandy beaches: seasonal patterns. — Estuarine, Coastal and Shelf Science, **53**: 877-895.
BIANCHI, C. N., 2007. Biodiversity issues for the forthcoming tropical Mediterranean Sea. — In: Biodiversity in enclosed seas. — Hydrobiologia, **580**: 7-21.
BIANCHI, C. N. & C. MORRI, 2000. Biodiversity of the Mediterranean Sea: situation, problems and prospects for future research. — Marine Pollution Bulletin, **40** (5): 367-376.
BIJ DE VAATE, A., G. JAŻDŻEWSKI, H. A. M. KETELAARS, S. GOLLASCH & G. VAN DER VELDE, 2002. Geographical patterns in range extension of Ponto-Caspian macroinvertebrate species in Europe. — Canadian Journal of Fisheries and Aquatic Sciences, **59**: 1159-1174.
BIJU, A., U. K. HONEY, K. K. KUSUM & L. JAGADEESAN, 2013. Population structure and biological characteristics of *Siriella gracilis* (Mysida) from the Exclusive Economic Zone of India. — Crustaceana, **86** (5): 515-536.
BIJU, A. & S. U. PANUMPUNNAYIL, 2010. Seasonality, reproductive biology and ecology of *Mesopodopsis zeylanica* (Crustacea: Mysida) from a tropical estuary (Cochin backwater) in India. — Plankton & Benthos Research, **5** (2): 49-55.
BILL, PH. C., 1914. Über Crustaceen aus dem Voltziensandstein des Elsasses. — Mitteilungen der Geologischen Landesanstalt von Elsaß-Lothringen, **8**: 289-338, pls. X-XVI.
BIRSTEIN, J. A. & YU. G. TCHINDONOVA, 1970. New mysids (Crustacea, Mysidacea) from the Kurile-Kamtchatka Trench. — Trudy Instituta Okeanologii, **86**: 277-291. [In Russian.]
BLEGVAD, H., 1915. Food and conditions of nourishment among the communities of invertebrate animals found on or in the sea bottom in Danish water. — Report of the Danish Biological Station to the Board of Agriculture, **22**: 41-78.
— —, 1922. On the biology of some Danish gammarids and mysids (*Gammarus locusta*, *Mysis flexuosa*, *Mysis neglecta*, *Mysis inermis*). — Report of the Danish Biological Station to the Board of Agriculture, **28**: 1-103.

BOAS, J. E. V., 1883. Studien über die Verwandtschaftsbeziehungen der Malakostraken. — Morphologisches Jahrbuch, **8**: 485-579, pls. XXI-XXIV.

BONNIER, J. & C. PÉREZ, 1902. Sur un crustacé commensal des pagures, *Gnathomysis gerlachei*, nov. sp., type d'une famille, nouvelle de schizopodes. — Comptes Rendus de l'Académie des Sciences, **134**: 117-119.

BORCHERDING, J., S. MURAWSKI & H. ARNDT, 2006. Population ecology, vertical migration and feeding of the Ponto-Caspian invader *Hemimysis anomala* in a gravel-pit lake connected to the River Rhine. — Freshwater Biology, **51**: 2376-2387.

BORODICH, N. D. & F. K. HAVLENA, 1973. The biology of mysids acclimatized in the reservoirs of the Volga River. — Hydrobiologia (Bucharest), **42** (4): 527-539.

BORZA, P., 2014. Life history of invasive Ponto-Caspian mysids (Crustacea: Mysida): a comparative study. — Limnologica — Ecology and Management of Inland Waters, **44**: 9-17.

BORZA, P. & P. BODA, 2013. Range expansion of Ponto-Caspian mysids (Mysida, Mysidae) in the River Tisza: first record of *Paramysis lacustris* Czerniavsky, 1882) for Hungary. — Crustaceana, **86** (11): 1316-1327.

BORZA, P., A. CZIROK, C. DEÁK, M. FICSÓR, V. HORVAI, Z. HORVÁTH, P. JUHÁSZ, K. KOVÁCS, T. SZABÓ & C. F. VAD, 2011. Invasive mysids (Crustacea: Malacostraca: Mysida) in Hungary: distributions and dispersal mechanisms. — North-Western Journal of Zoology, **7** (2): 222-228.

BOURDILLON, A. & C. CASTELBON, 1983. Influence des variations de température sur la géotaxie de deux espèces de mysidacés. — Journal of Experimental Marine Biology and Ecology, **71** (2): 105-117.

BOURDILLON, A., C. CASTELBON & C. MACQUART-MOULIN, 1980. Ecophysiologie comparée des mysidacés *Hemimysis speluncola* Ledoyer (cavernicole) et *Leptomysis lingvura* G.O. Sars (noncavernicole). L'orientation à la lumière: tests de longe durée — étude expérimentale des mouvements nycthéméraux. — Journal of Experimental Marine Biology and Ecology, **43** (1): 61-86.

BOWERS, J. A. & N. E. GROSSNICKLE, 1978. The herbivorous habits of *Mysis relicta* in Lake Michigan. — Limnology and Oceanography, **23** (4): 767-776.

BOWLES, E. C., B. E. RIEMAN, G. R. MAUSER & D. H. BENNET, 1991. Effects of introductions of *Mysis relicta* on fisheries in northern Idaho. — American Fisheries Society Symposium, **9**: 65-74.

BOWMAN, T. E., 1973. Two new American species of *Spelaeomysis* (Crustacea: Mysidacea) from a Mexican cave and land crab burrows. — In: R. W. MITCHELL & J. R. REDDELL (eds.), Studies on the cavernicole fauna of Mexico and adjacent regions. — Association for Mexican Cave Studies Bulletin, **5**: 13-20.

— —, 1976. *Stygiomysis major*, a new troglobitic mysid from Jamaica, and extension of the range of *S. holthuisi* to Puerto Rico (Crustacea: Mysidacea: Stygiomysidae). — International Journal of Speleology, **8**: 365-373.

— —, 1977. A review of the genus *Antromysis* (Crustacea: Mysidacea), including the new species from Jamaica and Oaxaca, Mexico, and a redescription and new records for *A. cenotensis*. — In: J. R. REDDELL (ed.), Studies on the caves and cave fauna of the Yucatan Peninsula. — Association of Mexican Cave Studies Bulletin, **6**: 27-38.

— —, 1980. *Antromysis* (*Surinamysis*) *merista*, a new freshwater mysid from Venezuela (Crustacea: Mysidacea). — Proceedings of the Biological Society of Washington, **93** (1): 208-215.

BOWMAN, T. E., T. M. ILIFFE & J. YAGER, 1984. New records of the troglobitic mysid genus *Stygiomysis*: *S. clarkei*, new species, from the Caicos Islands, and *S. holthuisi* (Gordon) from Grand Bahama Island (Crustacea: Mysidacea). — Proceedings of the Biological Society of Washington, **97** (3): 637-644.

BRANDT, A., M. BŁAŻEWICZ-PASZKOWYCZ, R. BAMBER. U. MÜHLENHARDT-SIEGEL, M. MALYUTINA, S. KAISER, C. DE BROYER & C. HAVERMANS, 2012. Are there widespread peracarid species in the deep sea (Crustacea: Malacostraca)? — Polish Polar Research, **33** (2): 139-162.
BRANDT, A., U. MÜHLENHARDT-SIEGEL & V. SIEGEL, 1998. An account of the Mysidacea (Crustacea, Malacostraca) of the Southern Ocean. — Antarctic Science, **10** (1): 3-11.
BRANSTRATOR, D. K., G. CABANA, A. MAZUMDER & J. B. RASMUSSEN, 2000. Measuring life-history omnivory in the opossum shrimp, *Mysis relicta*, with stable nitrogen isotopes. — Limnology and Oceanography, **45** (2): 463-467.
BRATTEGARD, T., 1973. Mysidacea from shallow water on the Caribbean coast of Colombia. — Sarsia, **54**: 1-65.
BRETON, G., A. GIRARD & J.-P. LAGARDÈRE, 1995. Espèces animales benthiques des bassins du port du Havre (Normandie, France) rares, peu connues ou nouvelles pour la région. — Bulletin Trimestriel de la Société Géologique de Normandie et des Amis du Muséum du Havre, **82** (3): 7-28.
BROOKS, H. K., 1962. The Paleozoic Eumalacostraca of North America. — Bulletins of American Paleontology, **44** (202): 163-338.
BROWN, A. C. & M. S. TALBOT, 1972. The ecology of the sandy beaches of the Cape Peninsula, South Africa, Part 3: a study of *Gastrosaccus psammodytes* Tattersall (Crustacea: Mysidacea). — Transactions of the Royal Society of South Africa, **40** (4): 309-333.
BURMEISTER, H., 1837. Handbuch der Naturgeschichte. Zum Gebrauch bei Vorlesungen entworfen. Zweyte Abtheilung. Zoologie, **2**. — Pp. i-xii, 369-858. (Verlag von Theodor Christoph Friedrich Enslin, Berlin.)
CALDWELL, R. L., 1991. Variation in reproductive behavior in stomatopod Crustacea. — In: R. T. BAUER & J. W. MARTIN (eds.), Crustacean sexual biology, pp. 67-90. (Columbia University Press, New York, NY.)
CALIL, P. & C. A. BORZONE, 2008. Population structure and reproductive biology of *Metamysidopsis neritica* (Crustacea: Mysidacea) in a sand beach in south Brazil. — Revista Brasileira de Zoologia, **25** (3): 403-412.
CALMAN, W. T., 1904. On the classification of Malacostraca. — Annals and Magazine of natural History, (7) **13**: 144-158.
— —, 1909. Crustacea. — In: R. LANKESTER (ed.), A treatise on zoology, **7** (3): i-viii, 1-346. (Adam and Charles Black, London.)
— —, 1932. A cave dwelling crustacean of the family Mysidae from the island of Lanzarote. — Annals and Magazine of natural History, (10) **10** (55): 127-131.
CANNON, H. G. & S. M. MANTON, 1927a. On the feeding mechanism of a mysid crustacean, *Hemimysis lamornae*. — Transactions of the Royal Society of Edinburgh, **55** (1) (10): 219-253, pls. I-IV.
— — & — —, 1927b. Notes on the segmental excretory organs of Crustacea. — Journal of the Linnean Society, (Zoology) **36**: 439-456.
CARLTON, J. T., 1985. Transoceanic and interoceanic dispersal of coastal marine organisms: the biology of ballast water. — Oceanography and Marine Biology — An Annual Review, **23**: 313-371.
CARLTON, J. T. & J. B. GELLER, 1993. Ecological roulette: the global transport of nonindigenous marine organisms. — Science, New York, **261**: 78-82.
CAROLI, E., 1924. Su di un misidaceo cavernicolo (*Spelaeomysis bottazzii* n.g., n.sp.) di Terra d'Otranto. — Rendiconti dell'Accademia Nazionale dei Lincei, (Classe di Scienze Fisiche, Matematiche e Naturali) **33** (2): 512-513.
— —, 1937. *Stygiomysis hydruntina* n.g., n.sp., Misidaceo cavernicolo di Terra d'Otranto rappresentante di una nuova famiglia. Nota preliminare. — Bollettino di Zoologia, **8**: 219-227.

CARRASCO, N. K. & R. PERISSINOTTO, 2010a. Spatial and temporal variations in the diet of the mysid *Mesopodopsis africana* in the St Lucia Estuary (South Africa). — Marine Ecology Progress Series, **417**: 127-138.

— — & — —, 2010b. In situ feeding rates and grazing impact of *Mesopodopsis africana* O. Tattersall in the St Lucia Estuary, South Africa. — Journal of Experimental Marine Biology and Ecology, **396**: 61-68.

— — & — —, 2011a. Temperature and salinity tolerance of *Mesopodopsis africana* O. Tattersall in the freshwater-deprived St Lucia Estuary, South Africa. — Journal of Experimental Marine Biology and Ecology, **399**: 93-100.

— — & — —, 2011b. The comparative diet of the dominant zooplankton species in the St Lucia Estuary, South Africa. — Journal of Plankton Research, **33**: 479-490.

CARRASCO, N. K., R. PERISSINOTTO & N. A. F. MIRANDA, 2007. Effects of silt loading on the feeding and mortality of the mysid *Mesopodopsis africana* in the St Lucia Estuary, South Africa. — Journal of Experimental Marine Biology and Ecology, **352**: 152-164.

CARRASCO, N. K., R. PERISSINOTTO & H. A. NEL, 2012. Diet of selected fish species in the freshwater-deprived St Lucia Estuary, South Africa, assessed using stable isotopes. — Marine Biology Research, **8**: 701-714.

CARTES, J. E., 1993a. Day-night feeding by decapod crustaceans in a deep-water bottom community in the western Mediterranean. — Journal of the Marine Biological Association of the United Kingdom, **73**: 795-811.

— —, 1993b. Diets of two deep-sea decapods: *Nematocarcinus exilis* (Caridea: Nematocarcinidae) and *Munida tenuimana* (Anomura: Galatheidae) on the western Mediterranean slope. — Ophelia, **37** (3): 213-229.

— —, 1993c. Feeding habits of pasiphaeid shrimps close to the bottom on the western Mediterranean slope. — Marine Biology, Berlin, **117**: 459-468.

CARTES, J. E. & P. ABELLO, 1992. Comparative feeding habits of polychelid lobsters in the western Mediterranean deep-sea communities. — Marine Ecology Progress Series, **84**: 139-150.

CARTES, J. E. & J. C. SORBE, 1995. Deep-water mysids of the Catalan Sea: species composition, bathymetric and near-bottom distribution. — Journal of the Marine Biological Association of the United Kingdom, **75** (1): 187-197.

— — & — —, 1998. Aspects of population structure and feeding ecology of the deep-water mysid *Boreomysis arctica*, a dominant species in western Mediterranean slope assemblages. — Journal of Plankton Research, **20** (12): 2273-2290.

CASANOVA, B., 1987. Relations entre les tergites thoraciques et la carapace dans la série des Mysidacés, Euphausiacés et Décapodes Dendrobranchiata (Crustacés). — Comptes Rendus de l'Académie des Sciences, **305** (3): 655-660.

— —, 1991. Origine protocéphalique antennaire de la carapace chez les Leptostracés, Mysidacés et Eucarides (Crustacés). — Comptes Rendus de l'Académie des Sciences, **312** (3): 461-468.

CASANOVA, B. & L. DE JONG, 1996. External and internal morphological aspect of embryos in Mysidacea. — In: 2nd European Crustacean Conference, Liège, Belgium. Abstracts, p. 130.

CASANOVA, B., L. DE JONG & X. MOREAU, 2002. Carapace and mandibles ontogeny in the Dendrobranchiata (Decapoda), Euphausiacea, and Mysidacea (Crustacea): a phylogenetic interest. — Canadian Journal of Zoology, **80**: 296-306.

CASANOVA, J.-P., 1977. La faune pélagique profonde (zooplancton et micronecton) de la province atlanto-méditerranéenne. Aspects taxonomique, biologique et zoogéographique. — Pp. i-ix, 1-500. (Thèse, Université de Provence, Aix-Marseille I.)

— —, 1993. Crustacea Mysidacea: les Mysidacés Lophogastrida et Mysida (Petalophthalmidae) de la région neo-calédonienne. — In: A. CROSNIER (ed.), Résultats des Campagnes MUSORSTOM, **10** (3). — Mémoires du Muséum National d'Histoire Naturelle, (A, Zoologie) **156**: 33-53.

— —, 1996a. Crustacea Mysidacea: les Lophogastridés d'Indonésie, de Nouvelle Calédonie et des îles Wallis et Futuna. — In: A. CROSNIER (ed.), Résultats des Campagnes MUSORSTOM, **15** (4). — Mémoires du Muséum National d'Histoire Naturelle, (A, Zoologie) **168**: 125-146.

— —, 1996b. *Gnathophausia childressi*, new species, a mysid from deep near-bottom waters off California, with remarks on the mouthparts of the genus *Gnathophausia*. — Journal of Crustacean Biology, **16** (1): 192-200.

— —, 1997. Les mysidacés Lophogastrida (Crustacea) du canal de Mozambique (côte de Madagascar). — Zoosystema, **19** (1): 91-109.

CASANOVA, J.-P., L. DE JONG & E. FAURE, 1998. Interrelationships of the two families constituting the Lophogastrida (Crustacea: Mysidacea) inferred from morphological and molecular data. — Marine Biology, Berlin, **132** (1): 59-65.

CECCALDI, H. J., 2006. The digestive tract: anatomy, physiology, and biochemistry. — In: J. FOREST & J. C. VON VAUPEL KLEIN (eds.), The Crustacea. Revised and updated from the Traité de Zoologie, **2**: 85-203. (Brill, Leiden.)

CESARO, G., A. P. ARIANI, A. BALBONI & V. SAGGIOMO, 1984. Preferenze di salinità in *Spelaeomysis bottazzii* (Mysidacea, Lepidomysidae). — Bollettino di Zoologia, **51** (Suppl.): 27.

CHADWICK, N. E., Z. ĎURIŠ & I. HORKÁ, 2008. 15. Biodiversity and behaviour of shrimps and fishes symbiotic with sea anemones in the Gulf of Aqaba, northern Red Sea. — In: F. D. POR (ed.), Aqaba — Eilat, the improbable gulf. Environment, biodiversity and preservation, pp. 209-223, pls. I-II + erratum. (The Hebrew University Magnes Press — Natural Science, Jerusalem.)

CHARMANTIER-DAURES, M. & G. CHARMANTIER, 2006. The endocrine organs. — In: J. FOREST & J. C. VON VAUPEL KLEIN (eds.), The Crustacea. Revised and updated from the Traité de Zoologie, **2** (12): 309-352. (Brill, Leiden.)

CHARMANTIER-DAURES, M. & G. VERNET, 2004. Moulting, autotomy, and regeneration. — In: J. FOREST & J. C. VON VAUPEL KLEIN (eds.), The Crustacea. Revised and updated from the Traité de Zoologie, **1** (5): 161-255. (Brill, Leiden.)

CHEVALDONNÉ, P. & CH. LEJEUSNE, 2003. Regional warming-induced species shift in north-west Mediterranean marine caves. — Ecology Letters, **6**: 371-379.

CHIGBU, P., 2004. Assessment of the potential impact of the mysid shrimp, *Neomysis mercedis*, on *Daphnia*. — Journal of Plankton Research, **26** (3): 295-306.

CHIKUGO, K., A. YAMAGUCHI, K. MATSUNO, R. SAITO & I. IMAI, 2013. Life history and production of pelagic mysids and decapods in the Oyashio region, Japan. — Crustaceana, **86** (4): 449-474.

CHILDRESS, J. J., 1971. Respiratory adaptations to the oxygen minimum layer in the bathypelagic mysid *Gnathophausia ingens*. — Biological Bulletin, Woods Hole, **141** (1): 109-121.

CHILDRESS, J. J., D. L. GLUCK, R. S. CARNEY & M. M. GOWING, 1989. Benthopelagic biomass production and oxygen consumption in a deep-sea benthic boundary layer dominated by gelatinous organisms. — Limnology and Oceanography, **34** (5): 913-930.

CHILDRESS, J. J. & M. H. PRICE, 1978. Growth rate of the bathypelagic crustacean *Gnathophausia ingens* (Mysidacea: Lophogastridae). 1. Dimensional growth and population structure. — Marine Biology, Berlin, **50**: 47-62.

— — & — —, 1983. Growth rate of the bathpelagic crustacean *Gnathophausia ingens* (Mysidacea: Lophogastridae). 2. Accumulation of material and energy. — Marine Biology, Berlin, **76** (2): 165-177.

CHUN, C., 1896. Atlantis. Biologische Studien über pelagische Organismen. V. Über pelagische Tiefseeschizopoden. VI. Leuchtorgan und Facettenauge. — Bibliotheca Zoologica, **7** (19): 137-262, pls. VIII-XX.

CIBINETTO, T., G. CASTALDELLI, S. MANTOVANI & E. A. FANO, 2006. Valutazione della nicchia trofica potenziale di *Palaemonetes antennarius* (Crustacea: Palaemonidae) e di *Diamysis mesohalobia* (Crustacea: Mysidacea). — In: C. COMOGLIO, E. COMINO & F. BONA (eds.), XV Congresso della Societá Italiana di Ecologia — Torino 2005. Articoli in extenso presentati al XV Congresso, pp. 1-6. (Università degli Studi di Torino, Torino, Italia.) http://www.xvcongresso.societaitalianaecologia.org/articles/Cibinetto.pdf [25 Feb. 2009]

CLARKE, W. D., 1955. A new species of the genus *Heteromysis* (Crustacea, Mysidacea) from the Bahama Islands, commensal with a sea-anemone. — American Museum Novitates, **1716**: 1-13.

— —, 1961a. Proposal of a new name, *Lepidomysis*, for the preoccupied mysidacean generic name *Lepidops* Zimmer, 1927. — Crustaceana, **2** (3): 251-252.

— —, 1961b. A giant specimen of *Gnathophausia ingens* (Dohrn, 1870) (Mysidacea) and remarks on the asymmetry of the paragnaths in the suborder Lophogastrida. — Crustaceana, **2** (4): 313-324.

— —, 1962. The genus *Gnathophausia* (Mysidacea, Crustacea), its systematics and distribution in the Pacific Ocean. — Pp. 1-251. (Doctoral Thesis, University of California, San Diego.)

CLAUS, C., 1876. Untersuchungen zur Erforschung der genealogischen Grundlage des Crustaceen-Systems. Ein Beitrag zur Descendenzlehre. — Pp. i-viii, 1-114, pls. I-XIX. (C. Gerold's Sohn, Wien.)

— —, 1884. Zur Kenntnis der Kreislauforgane der Schizopoden und Decapoden. — Arbeiten des Zoologischen Institutes zu Wien, **5** (3): 271-318, pls. I-IX.

CLUTTER, R. I., 1967. Zonation of nearshore mysids. — Ecology, **48** (2): 200-208.

— —, 1969. The microdistribution and social behaviour of some pelagic mysid shrimps. — Journal of Experimental Marine Biology and Ecology, **3**: 125-155.

CLUTTER, R. I. & G. H. THEILACKER, 1971. Ecological efficiency of a pelagic mysid shrimp; estimates from growth, energy budget and mortality studies. — Fisheries Bulletin, U.S., **69**: 93-115.

COHEN, M. J., 1955. The function of receptors in the statocyst of the lobster *Homarus americanus*. — The Journal of Physiology, **130** (1): 9-34.

COLL, M., C. PIRODDI, J. STEENBEEK, K. KASCHNER, F. BEN RAIS LASRAM, J. AGUZZI, E. BALLESTEROS, C. N. BIANCHI, J. CORBERA, T. DAILIANI ET AL., 2010. Biodiversity of the Mediterranean Sea: estimates, patterns & threats. — PLoS ONE, **5** (8): e11842: 334 pp. www.plosone.org [30 Mar. 2010]

COLOSI, G., 1930. Lofogastridi nuovi. — Bollettino di Zoologia, **1** (4): 119-125.

COOPER, S. D. & C. R. GOLDMAN, 1982. Environmental factors affecting predation rates of *Mysis relicta*. — Canadian Journal of Fisheries and Aquatic Sciences, **39** (1): 203-208.

COSTA, O. G., 1847. Ordine II. Crostacei Stomatopodi. Caratteri gener. Gen. *Misis, Phyllosoma*. — In: O. G. COSTA & A. COSTA (eds.) (1838-71), Fauna del Regno di Napoli, **2** (29): 1-8, pls. XI, fig. 5. (Tipografia Azzolino e Compagno, Naples.)

COTECCHIA, V., 1977. Studi e ricerche sulle acque sotterranee e sull'intrusione marina in Puglia (Peninsola Salentina). — Quaderni dell'Istituto di Ricerca sulle Acque, **20**: 1-462.

COTECCHIA, V., T. TADOLINI & L. TULIPANO, 1978. Groundwater temperature in the Murgia karst aquifer (Puglia — Southern Italy). — In: International Symposium on Karst Hydrology, **2**: 18 pp. (Budapest, Hungary.)

COUTIÈRE, H., 1911. Les Ellobiopsidae des crevettes bathypélagiques. — Bulletin Scientifique de la France et de la Belgique, **45**: 186-206.

COVI, J. A., E. S. CHANG & D. L. MYKLES, 2012. Neuropeptide signaling mechanisms in crustacean and insect molting glands. — Invertebrate Reproduction and Development, **56**: 33-49.

CREASER, E. P., 1936. Crustaceans from Yucatan. — Carnegie Institute of Washington Publications, **457**: 117-132.

CROUAU, Y., 1978a. Organes sensoriels d'un Mysidacé souterrain anophthalme, *Antromysis juberthiei*: étude ultrastructurale des aesthetascs. — Bulletin du Muséum National d'Histoire Naturelle, (3) (Zoologie) **352**: 165-175.

— —, 1978b. Ultrastructure des phanères spinués mécanorécepteurs d'un Crustacé Mysidacé souterrain anophthalme. — Comptes Rendus de l'Académie des Sciences, (D) **287**: 1215-1218.

— —, 1979. Ultrastructure d'un mechanorecepteur externe de type scolopidial d'*Antromysis juberthiei*, mysidacé souterrain anophtalme. — Bulletin de la Société des Sciences Naturelles de la Tunisie, **14**: 35-42.

— —, 1981. Cytology of various antennual setae in a troglobitic Mysidacea (Crustacea). — Zoomorphology, **98** (2): 121-134.

— —, 1982. Primary stages in the sensory mechanism of the setulate sensilla, external mechanoreceptors of a cavernicolous Mysidacea. — Biologie Cellulaire, **44**: 45-56.

— —, 1983. Etude du comportement de recherche de la nourriture chez le crustacé mysidacé cavernicole *Antromysis juberthiei*. — Mémoires de Biospéologie, **10**: 385-393.

— —, 1987. Morphologie des soies des appendices buccaux et des pattes d'un mysidacé cavernicole. — Crustaceana, **52** (3): 287-297.

— —, 1989. Feeding mechanisms of the Mysidacea. — In: B. E. FELGENHAUER, L. WATLING & A. B. THISTLE (eds.), Functional morphology of feeding and grooming in Crustacea. — Crustacean Issues, **6**: 153-171. (A. A. Balkema, Rotterdam.)

CUNHA, M. R. DA, M. H. MOREIRA & J. C. SORBE, 2000. *Diamysis bahirensis*: a mysid species new to the Portuguese fauna and first record from the west European coast. — In: J. C. VON VAUPEL KLEIN & F. R. SCHRAM (eds.), The biodiversity crisis and Crustacea. — Crustacean Issues, **12**: 139-152. (A. A. Balkema, Rotterdam.)

CUZIN-ROUDY, J., J. BERREUR-BONNENFANT & M. C. FRIED-MONTAUFIER, 1981. Chronology of post-embryonic development in *Siriella armata* (M. Edw.) (Crustacea: Mysidacea) reared in the laboratory: growth and sexual differentiation. — International Journal of Invertebrate Reproduction, **4**: 193-208.

CUZIN-ROUDY, J. & A. S. M. SALEUDDIN, 1985. A study of the neurosecretory centres of the eyestalk in *Siriella armata* M. Edw. (Crustacea: Mysidacea): their involvement in molting and reproduction. — Canadian Journal of Zoology, **63** (12): 2783-2788.

CUZIN-ROUDY, J. & C. TCHERNIGOVTZEFF, 1985. Chronology of the female molt cycle in *Siriella armata* M. Edw. (Crustacea: Mysidacea) based on marsupial development. — Journal of Crustacean Biology, **5**: 1-14.

CZERNIAVSKY, V., 1882. Monographia Mysidarum inprimis Imperii Rossici. Fasc. 1, 2. — Trudy Sankt-Peterburgskogo Obshchestva Estestvoispytatelei, **12**: 1-170, **13**: 1-85, pls. I-IV. (St.-Peterburg.)

— —, 1887. Monographia Mysidarum inprimis Imperii Rossici. Fasc. 3. — Trudy Sankt-Peterburgskogo Obshchestva Estestvoispytatelei, **18**: i-viii, 1-102, pls. V-XXXII. (St.-Peterburg.)

DAHL, E., 1992. Aspects of malacostracan evolution. — Acta Zoologica, **73** (5): 339-346.

DAHL, E. & C. VON MECKLENBURG, 1969. The sensory papilla X-organ in *Boreomysis arctica* (Krøyer) (Crustacea Malacostraca Mysidacea). — Zeitschrift für Zellforschung und Mikroskopische Anatomie, **101**: 88-97.

DALY, K. L. & D. M. DAMKAER, 1986. Population dynamics and distribution of *Neomysis mercedis* and *Alienacanthomysis macropsis* (Crustacea: Mysidacea) in relation to the parasitic copepod *Hansenulus trebax* in the Columbia River estuary. — Journal of Crustacean Biology, **6** (4): 840-857.

DANA, J. D., 1852. Crustacea. Part I. — In: United States Exploring Expedition during the years 1838-1842 under the command of Charles Wilkes, U.S.N., **13**: i-viii, 1-685. (C. Sherman, Philadelphia.)

DANELIYA, M. E., 2012. Description of *Heteromysis* (*Olivemysis*) *ningaloo* new species and interesting records of *H.* (*Gnathomysis*) *harpaxoides* Băcescu and Bruce (Crustacea: Mysida: Mysidae) from Australian coral reefs. — Records of the Western Australian Museum, **27**: 135-147.

DANELIYA, M. E., A. AUDZIJONYTE & R. VÄINÖLÄ, 2007. Diversity within the Ponto-Caspian *Paramysis baeri* Czerniavsky sensu lato revisited: *P. bakuensis* G.O. Sars restored (Crustacea: Mysida: Mysidae). — Zootaxa, **1632**: 21-36.

DANELIYA, M. E. & V. V. PETRYASHOV, 2011. Biogeographic zonation of the Black Sea and Caspian Sea basin based on mysid fauna (Crustacea: Mysidacea). — Russian Journal of Marine Biology, **37** (2): 85-97.

DANELIYA, M. E., V. V. PETRYASHOV & R. VÄINÖLÄ, 2012. Continental mysid crustaceans of northern Eurasia. — In: N. M. KOROVCHINSKY, S. M. ZHDANOVA & A. V. KRYLOV (eds.), Modern problems of the continental crustacean research. Proceedings of the lectures and presentations of the international school-conference. I. D. Papanin's Institute of Biology of Continental Waters RAS, Borok, November 5-9, 2012, pp. 21-30. (Kostromskoi Pechatnyi Dom, Kostroma.)

DANIEL, R. J., 1928. The abdominal muscular systems of *Praunus flexuosus* (Müller). — Lancashire Sea-Fisheries Laboratory Report for 1927, **36**: 5-41.

— —, 1933. Comparative study of the abdominal musculature in Malacostraca. Part III. The abdominal muscular systems of *Lophogaster typicus* M. Sars and *Gnathophausia zoea* Suhm, and their relationships with the musculatures of other Malacostraca. — Lancashire Sea-Fisheries Laboratory Report for 1932, **41**: 71-133.

DANILIN, D. D., P. N. PANFILOVA, L. L. BUDNIKOVA, V. V. PETRYASHOV, T. N. TRAVINA & A. V. BOGDANOV, 2012. Feeding of Navaga *Eleginus gracilis* in brackish water pool (Lake Nerpichie, eastern Kamchatka) in winter-spring season. — In: Conservation of biodiversity of Kamchatka and coastal waters. Materials of XIII international scientific conference Petropavlovsk-Kamchatsky, November 14-15, 2012, pp. 81-84. (Izdatelstvo "Kamchatpress", Petropavlovsk-Kamchatsky.) [In Russian.]

DARIBAEV, A. K., 1967. Essai d'acclimatation des Mysides et des *Calanipeda* dans la zone méridionale de la Mer d'Aral. — Gidrobiologiceskij Zhurnal, **3** (4): 69-70. [In Russian.]

DAVIS, C. C., 1966. A study of the hatching process in aquatic invertebrates. XXII. Multiple membrane shedding in *Mysidium columbiae* (Zimmer) (Crustacea: Mysidacea). — Bulletin of Marine Science, **16** (1): 124-131.

DE ANGELI, A. & A. ROSSI, 2006. Crostacei ologocenici di Peralolo (Vicenza — Italia settentrionale), con la descrizione di una nuova specie di Mysida e di Isopoda. — Società Veneziana di Scienze Naturali — Lavori, **31**: 85-93.

DE JONG, L., 1996. Functional morphology of the foregut of *Lophogaster typicus* and *L. spinosus* (Crustacea, Mysidacea, Lophogastrida). — Cahiers de Biologie Marine, **37**: 341-347.

DE JONG, L. & B. CASANOVA, 1997. Comparative morphology of the foregut of three *Eucopia* species (Crustacea, Mysidacea, Lophogastrida). — Journal of Natural History, London, **31**: 389-402.

DE JONG, L. & J.-P. CASANOVA, 1997. Comparative morphology of the foregut of four *Gnathophausia* species (Crustacea, Mysidacea, Lophogastrida). Relationships with some related taxa. — Journal of Natural History, London, **31**: 1029-1040.

DE JONG, L., X. MOREAU, R.-M. BARTHÉLÉMY & J.-P. CASANOVA, 2002. Relevant role of the labrum associated with the mandibles in the *Lophogaster typicus* digestive function. — Journal of the Marine Biological Association of the United Kingdom, **82**: 219-227.

DE JONG-MOREAU, L., M. BRUNET, J.-P. CASANOVA & J. MAZZA, 2000. Comparative structure and ultrastructure of the midgut and hepatopancreas of five species of Mysidacea (Crustacea): functional implications. — Canadian Journal of Zoology, **81**: 235-241.

DE JONG-MOREAU, L., B. CASANOVA & J.-P. CASANOVA, 2001. Detailed comparative morphology of the peri-oral structures of the Mysidacea and Euphausiacea (Crustacea): an indication for the food preference. — Journal of the Marine Biological Association of the United Kingdom, **81**: 235-241.

DE JONG-MOREAU, L. & J.-P. CASANOVA, 2001. The foreguts of the primitive families of the Mysida (Crustacea, Peracarida): a transitional link between those of the Lophogastrida (Crustacea, Mysidacea) and the most evolved Mysida. — Acta Zoologica, Stockholm, **82**: 137-147.

DEBAISIEUX, P., 1947. Statocystes de *Praunus* et leurs poils sensoriels. — Annales de la Sociètè Royale Zoologique de Belgique, **78**: 26-31.

— —, 1949a. Les poils sensoriels d'Arthropodes et l'histologie nerveuse. I. *Praunus flexuosus* Müller et *Crangon crangon* L. — La Cellule, **52** (3): 311-360.

— —, 1949b. Équipement sensoriel et croissance chez les Arthropodes. — Académie Royale de Belgique, (Bulletin de la Classe des Sciences) **35**: 1164-1177.

— —, 1954. Histogenèse des muscles et charpentes chez les Crustacés. — La Cellule, **56** (3): 265-305.

DEBUS, L., T. MEHNER & R. THIEL, 1992. Spatial and diel patterns of migration for *Neomysis integer*. — In: J. M. KÖHN, M. B. JONES & A. MOFFAT (eds.), Taxonomy, biology and ecology of (Baltic) mysids (Mysidacea, Crustacea), pp. 79-82. (Rostock University Press, Rostock.)

DEDIU, I. I., 1966. Répartition et caractéristique écologique des Mysides des bassins des rivières Dniestr et Pruth. — Revue Roumaine de Biologie, (Zoologie) **11** (3): 233-239.

DEGNER, E., 1912a. Über Bau und Funktion der Krusterchromatophoren. Eine histologisch-biologische Untersuchung. — Zeitschrift für Wissenschaftliche Zoologie, **102**: 1-78, pls. I-III.

— —, 1912b. Weitere Beiträge zur Kenntnis der Crustaceen-Chromatophoren. — Zeitschrift für Wissenschaftliche Zoologie, **102**: 701-710.

DELAGE, Y., 1883. Circulation et respiration chez les Crustacés Schizopodes. — Archives de Zoologie Expérimentale et Générale, (2) **1**: 105-130, pl. X.

— —, 1887. Sur une fonction nouvelle des otocystes comme organes d'orientation locomotrice. — Archives de Zoologie Expérimentale et Générale, **5**: 1-26.

DELAMARE-DEBOUTTEVILLE, C., 1955. Eaux souterraines littorales de la côte catalane française (mise au point faunistique). — Vie et Milieu, **5** (3): 408-450.

DELGADO, L., G. GUERAO & C. RIBERA, 1997. Biology of the mysid *Mesopodopsis slabberi* (Van Beneden, 1861) (Crustacea, Mysidacea) in a coastal lagoon of the Ebro delta (NW Mediterranean). — Hydrobiologia, **357**: 27-35.

DEPDOLLA, P., 1916. Biologische Notizen über *Praunus flexuosus* (Müller). — Zoologischer Anzeiger, **47**: 43-47.

— —, 1923. Nahrung und Nahrungserwerb bei *Praunus flexuosus* (Müller). — Biologisches Zentralblatt, **43** (5): 534-546.

DOHLE, W., M. GERBERDING, A. HEJNOL & G. SCHOLTZ, 2004. Cell lineage, segment differentiation, and gene expression in crustaceans. — In: G. SCHOLTZ (ed.), Evolutionary developmental biology of Crustacea. — Crustacean Issues, **15**: 95-133. (A. A. Balkema, Rotterdam.)

DRACH, P. & C. TCHERNIGOVTZEFF, 1967. Sur la méthode de détermination des stades d'intermue et son application générale aux Crustacés. — Vie et Milieu, **18**: 595-610.

DULČIĆ, J., L. LIPEJ, A. PALLAORO & A. SOLDO, 2004. The spreading of Lessepsian fish migrants into the Adriatic Sea: a review. — Rapports de la Commission Internationale pour l'Exploration Scientifique de la Mer Méditerranée, **37**: 349.

DÜRR, J. & J. A. GONZÁLEZ, 2002. Feeding habits of *Beryx splendens* and *Beryx decadactylus* (Berycidae) off the Canary Islands. — Fisheries Research, **54**: 363-374.

ELIZALDE, M., J.-C. DAUVIN & J. C. SORBE, 1991. Les Mysidacés suprabenthiques de la marge sud du canyon du Cap-Ferret (golfe de Gascogne): répartition bathymétrique et activité natatoire. — Annales de'l Institut Océanographique, Monaco, **67** (2): 129-144.

ELMHIRST, R., 1931. Studies in the Scottish marine fauna. The Crustacea of the sandy and muddy areas of the tidal zone. — Proceedings of the Royal Society of Edinburgh, **51** (2): 169-175.

— —, 1932. Quantitative studies between tide marks. — Glasgow Naturalist, **10**: 56-62.

ELOFSSON, R. & E. DAHL, 1970. The optic neuropiles and chiasmata of Crustacea. — Zeitschrift für Zellforschung und Mikroskopische Anatomie, **107**: 343-360.

ELOFSSON, R. & M. HAGBERG, 1986. Evolutionary aspects on the construction of the first optic neuropil (lamina) in Crustacea. — Zoomorphology, **106** (3): 174-178.

ELOFSSON, R. & E. HALLBERG, 1977. Compound-eyes of some deep-sea fjord mysid crustaceans. — Acta Zoologica, **58**: 169-177.

EMERY, A. R., 1968. Preliminary observations on coral reef plankton. — Limnology and Oceanography, **13** (2): 293-303.

ENBYSK, B. J. & F. E. LINGER, 1966. Mysid statoliths in shelf sediments off northwest North America. — Journal of Sedimentary Petrology, **36**: 839-840.

ESCÁNEZ, E., R. RIERA, L. MÁRQUEZ, A. SKALLI, B. C. FELIPE, I. GARCÍA-HERRERO, D. REIS, C. RODRÍGUEZ & E. ALMANSA, 2012. A general survey of the feasibility of culturing the mysid *Gastrosaccus roscoffensis* (Peracarida, Mysida): growth, survival, predatory skills, and lipid composition. — Ciencias Marinas, **38** (3): 475-490.

ESPEEL, M., 1985. Fine structure of the statocyst sensilla of the mysid shrimp *Neomysis integer* (Leach, 1814) (Crustacea, Mysidacea). — Journal of Morphology, **186** (2): 149-165.

— —, 1986. Morphogenesis during moulting of the setae in the statocyst sensilla of the mysid shrimp *Neomysis integer* (Leach, 1814) (Crustacea, Mysidacea). — Journal of Morphology, **187** (1): 61-68.

— —, 1987. On the fine structure of the statolith and the caudal statocyst gland of the mysid shrimp *Neomysis integer* (Leach, 1814) (Crustacea, Mysidacea). — Academiae Analecta. Koninklijke Vlaamse Academie voor Wetenschappen, Letteren en Schone Kunsten van België, **49** (1): 93-109.

FABRICIUS, O., 1780. Fauna Groenlandica. — Pp. i-xvi, 1-452, 1 fig.-pl. (Ioannis Gottlob Rothe, Hafniae et Lipsiae.)

FAGE, L., 1924. Sur un type nouveau de Mysidacé des eaux souterraines de l'île de Zanzibar. — Comptes Rendus de l'Académie des Sciences, **178**: 2127-2129.

— —, 1925. *Lepidophthalmus servatus* Fage. Type nouveau de Mysidacé des eaux souterraines de Zanzibar. — Biospeologica **LI**. — Archives de Zoologie Expérimentale et Générale, **63**: 525-532.

— —, 1932. La migration verticale saisonnière des Mysidacés. — Comptes Rendus de l'Académie des Sciences, **194**: 313-315.

— —, 1933. Pêches planctoniques à la lumière effectuées à Banyuls-sur-Mer et à Concameaux. III. Crustacés. — Archives de Zoologie expérimentale et générale, **76** (3): 105-248.

— —, 1936. Sur un Ellobiopsidé nouveau *Amallocystis fasciatus* g. et sp. nov. parasite des Mysidacés bathypélagiques. — Archives de Zoologie Expérimentale et Générale, **78** (3): 145-154.

— —, 1940. Sur le déterminisme des caractères sexuels secondaires des Lophogastrides (Crustacés-Mysidacés). — Comptes Rendus Hebdomadaires des Séances de l'Académie des Sciences, **211** (16): 335-337.

— —, 1941. Mysidacea — Lophogastrida. I. — Dana Reports, **4** (19): 1-52.

— —, 1942. Mysidacea — Lophogastrida. II. — Dana Reports, **4** (23): 1-67.

FAIN-MAUREL, M. A., J. F. REGER & P. CASSIER, 1975. Le gamète mâle des Schizopodes et ses analogies avec celui des autres Péracarides. 1. Le spermatozoïde. — Journal of Ultrastructure Research, **51** (2): 269-280.

FAXON, W., 1895. The stalk-eyed Crustacea. — In: Reports on an exploration off the west coasts of Mexico, Central and South America, and off the Galapagos Islands, in charge of Alexander Agassiz, by the U.S. Fish Commission Steamer Albatross during 1891. — Memoirs of the Museum of Comparative Zoölogy at Harvard College, **18**: 1-292, 67 pls.
FELDMAN, T., M. YAKOVLEVA, M. LINDSTRÖM, K. DONNER & M. OSTROVSKY, 2010. Eye adaption to different light environments in two populations of *Mysis relicta*: a comparative study of carotenoids and retinoids. — Journal of Crustacean Biology, **30** (4): 636-642.
FENTON, G. E., 1992. Population dynamics of *Tenagomysis tasmaniae* Fenton, *Anisomysis mixta australis* (Zimmer) and *Paramesopodopsis rufa* Fenton from south-eastern Tasmania (Crustacea, Mysidacea). — Hydrobiologia, **246** (3): 173-193.
— —, 1994. Breeding biology of *Tenagomysis tasmaniae* Fenton, *Anisomysis mixta australis* (Zimmer) and *Paramesopodopsis rufa* Fenton from south-eastern Tasmania (Crustacea, Mysidacea). — Hydrobiologia, **287**: 259-276.
— —, 1996. Diet and predators of *Tenagomysis tasmaniae* Fenton, *Anisomysis mixta australis* (Zimmer) and *Paramedopodopsis rufa* Fenton from south-eastern Tasmania (Crustacea: Mysidacea). — Hydrobiologia, **323**: 31-44.
FEYRER, L. J., 2010. Differences in embryo production between sympatric species of mysids (family Mysidae) in the shallow coastal waters off Vancouver Island, BC. — Marine Biology, Berlin, **157**: 2461-2465.
FEYRER, L. J. & D. A. DUFFUS, 2011. Predatory disturbance and prey species diversity: the case of gray whale (*Eschrichtius robustus*) foraging on a multispecies mysid (family Mysidae) community. — Hydrobiologia, **678**: 37-47.
FINK, P., 2013. Invasion of quality: high amounts of essential fatty acids in the invasive Ponto-Caspian mysid *Limnomysis benedeni*. — Journal of Plankton Research, **35** (4): 907-913.
FINK, P. & C. HARROD, 2013. Carbon and nitrogen stable isotopes reveal the use of pelagic resources by the invasive Ponto-Caspian mysid *Limnomysis benedeni*. — Isotopes in Environmental and Health Studies, **49** (3): 312-317.
FINK, P., A. KOTTSIEPER, M. HEYNEN & J. BORCHERDING, 2012. Selective zooplanktivory of an invasive Ponto-Caspian mysid and possible consequences for the zooplankton community structure of invaded habitats. — Aquatic Sciences, **74**: 191-202.
FOCKEDEY, N., A. GHEKIERE, S. BRUWIERE, C. R. JANSSEN & M. VINCX, 2006. Effect of salinity and temperature on the intra-marsupial development of the brackish water mysid *Neomysis integer* (Crustacea: Mysidacea). — Marine Biology, Berlin, **148** (6): 1339-1356.
FOCKEDEY, N. & J. MEES, 2005. Feeding of the hyperbenthic mysid *Neomysis integer* in the maximum turbidity zone of the Elbe, Westerschelde and Gironde estuaries. — Journal of Marine Systems, **22**: 207-228.
FOCKEDEY, N., J. MEES & M. VINX, 2005. Addendum 2: some experimental observations on gut passage time, egestion rate and faecal pellet production of brackish water mysid *Neomysis integer* (Mysidacea: Crustacea) feeding on different food items. — In: N. FOCKEDEY (ed.), Diet and growth of *Neomysis integer* (Leach, 1814) (Crustacea, Mysidacea), pp. 215-225. (University of Ghent, Ghent, Belgium.)
FOREST, J., 2004. The Crustacea: definition, primitive forms, and classification. — In: J. FOREST & J. C. VON VAUPEL KLEIN (eds.), The Crustacea. Revised and updated from the Traité de Zoologie, **1** (1): 3-12. (Brill, Leiden.)
FOSSÅ, J. H., 1986. Aquarium observations on vertical zonation and bottom relationships of some deep-living hyperbenthic mysids (Crustacea: Mysidacea). — Ophelia, **25** (2): 107-117.
FOSSÅ, J. H. & T. BRATTEGARD, 1990. Bathymetric distribution of Mysidacea in fjords of western Norway. — Marine Ecology Progress Series, **67** (1): 7-18.
FOULDS, J. B. & K. H. MANN, 1978. Cellulose digestion in *Mysis stenolepis* and its ecological implications. — Limnology and Oceanography, **23** (4): 760-766.

Fox, H. M., 1952. Anal and oral intake of water by Crustacea. — The Journal of Experimental Biology, **29**: 583-599.

Foxon, G. E. H., 1940. The reactions of certain mysids to stimulation by light and gravity. — Journal of the Marine Biological Association of the United Kingdom, **29** (1): 89-97.

Frank, T. M., E. A. Widder, M. I. Latz & J. F. Case, 1984. Dietary maintenance of bioluminescence in a deep-sea mysid. — Journal of Experimental Biology, **109**: 385-389.

Friesen, J. A., K. H. Mann & J. A. Novitsky, 1986 (cf. a). *Mysis* digests cellulose in the absence of a gut microflora. — Canadian Journal of Zoology, **64** (2): 442-446.

Friesen, J. A., K. H. Mann & J. H. M. Willison, 1986 (cf. b). Gross anatomy and fine structure of the gut of the marine mysid shrimp *Mysis stenolepis* Smith. — Canadian Journal of Zoology, **64** (2): 431-441.

Frutos, I. & J. C. Sorbe, 2013. Bathyal suprabenthic assemblages from the southern margin of the Capbreton Canyon ("Kostarrenkala" area), SE Bay of Biscay. — Deep-Sea Research Part II, (Topical Studies in Oceanography) **2013**: 19 pp. http://dx.doi.org/10.1016/j.dsr2.2013.09.010i [early online, 10 Nov. 2013]

Fuchs, R., 1979. Das Vorkommen von Statolithen fossiler Mysiden (Crustacea) im obersten Sarmatien (O-Miozän) der Zentralen Paratethys. — Beiträge zur Paläontologie von Österreich, **6**: 61-69.

Fukuoka, K., 2009. Deep-sea mysidaceans (Crustacea: Lophogastrida and Mysida) from the northwestern North Pacific off Japan, with descriptions of six new species. — In: T. Fujita (ed.), Deep-sea fauna and pollutants off Pacific coast of northern Japan. — National Museum of Nature and Science Monographs, **39**: 405-446.

Fürst, M., 1972. Livscykler, tillväxt och reproduktion hos *Mysis relicta* Lovén. — Information från Sötvattenslaboratoriet Drottningholm, **11**: 1-41.

— —, 1981. Results of introduction of new fish food organisms into Swedish lakes. — Reports of the Institute of Freshwater Research, Drottningholm, **59**: 33-47.

Fürst, M., J. Hammar, C. Hill, U. Boström & B. Kinsten, 1984. Effekter av introduktion av *Mysis relicta* i reglerade sjöar i Sverige. — Information från Sötvattenslaboratoriet Drottningholm, **1984** (1): 1-84.

Gabe, M., 1953. Sur l'existence, chez quelques Crustacés Malacostracés, d'un organe comparable à la glande de la mue des Insectes. — Comptes Rendus de l'Académie des Sciences, **237** (18): 1111-1113.

— —, 1956. Histologie comparée de la glande de mue (organe Y) des crustacés malacostracés. — Annales des Sciences Naturelles, (Zoologie et Biologie Animale) **18**: 145-152.

Gadzikiewicz, W., 1905. Über den feineren Bau des Herzens bei Malakostraken. — Jenaische Zeitschrift für Naturwissenschaft, **39**: 203-234.

García-Garza, M. E., G. A. Rodriguez-Almarez & T. E. Bowman, 1996. *Spelaeomysis villalobosi*, a new species of mysidacean from northeastern Mexico (Crustacea: Mysidacea). — Proceedings of the Biological Society of Washington, **109** (1): 97-102.

Gasiunas, I., 1965. On the results of the acclimatization of food invertebrates of the Caspian complex in Lithuanian waterbodies. — Zoologicheskij Zhurnal, **44**: 340-343. [In Russian with English summary.]

— —, 1968. A mysid's *Hemimysis anomala* Sars acclimatization in the water reservoir of the Kaunas HEPS. — Trudy Akademii Nauk Litovskoj SSR, **3**: 71-73. [In Russian with English summary.]

Gaudy, R. & J. P. Guérin, 1979. Ecophysiologie comparée des mysidacés *Hemimysis speluncola* Ledoyer (cavernicole) et *Leptomysis lingvura* G.O. Sars (non cavernicole). Action de la temperature sur la croissance en élevage. — Journal of Experimental Marine Biology and Ecology, **38**: 101-119.

Geiger, S. R., 1969. Distribution and development of mysids (Crustacea, Mysidacea) from the Arctic Ocean and confluent seas. — Bulletin of the Southern California Academy of Sciences, **68** (2): 103-111.

GEISSEN, H.-P., 1997. Nachweis von *Limnomysis benedeni* Czerniavski (Crustacea: Mysidacea) im Mittelrhein. — Lauterbornia, **31**: 125-127.
GELDERD, C., 1909. Research on the digestive system of the Schizopoda. Anatomy, histology and physiology. — La Cellule, **25** (1): 6-70.
GENTILE, J. H., S. M. GENTILE, G. HOFFMANN, J. F. HELTSHE & N. HAIRSTON, 1983. The effects of a chronic mercury exposure on survival, reproduction and population dynamics of *Mysidopsis bahia*. — Environmental Toxicology and Chemistry, **2**: 61-68.
GERGS, R., A. J. HANSELMANN, I. EISELE & K.-O. ROTHHAUPT, 2008. Autecology of *Limnomysis benedeni* Czerniavsky, 1882 (Crustacea: Mysida) in Lake Constance, southwestern Germany. — Limnologica — Ecology and Management of Inland Waters, **38**: 139-146.
GOLLASCH, S., 1996. Untersuchungen des Arteintrages durch den internationalen Schiffsverkehr unter besonderer Berücksichtigung nichtheimischer Arten. — Pp. 1-215, Anhang Tabs. 1-13. (Dissertation, University of Hamburg, Verlag Dr. Kovač, Hamburg.)
GOMOIU, M.-T., 1978. Quantitative data concerning the distribution and ecology of the *Mesopodopsis slabberi* (Van Beneden) at the Danube River mouth area. — Cercetări Marine, **11**: 103-112.
GOODBODY, I., 1965. Continuous breeding in populations of two tropical crustaceans, *Mysidium columbiae* (Zimmer) and *Emerita portoricensis* Schmidt. — Ecology, **46**: 195-197.
GORDON, I., 1960. On a *Stygiomysis* from the West Indies, with a note on *Spelaeogriphus* (Crustacea, Peracarida). Bulletin of the British Museum (Natural History), (Zoology) **6** (5): 285-324.
GOROKHOVA, E., 2002. Moult cycle and its chronology in *Mysis mixta* and *Neomysis integer* (Crustacea, Mysidacea): implications for growth assessment. — Journal of Experimental Marine Biology and Ecology, **278**: 179-194.
GOROKHOVA, E. & M. LEHTINIEMI, 2007. A combined approach to understand trophic interactions between *Cercopagis pengoi* (Cladocera: Onychopoda) and mysids in the Gulf of Finland. — Limnology and Oceanography, **52** (2): 685-695.
GOSHO, M. E., 1975. The introduction of *Mysis relicta* into freshwater lakes (a literature survey). — Circular No. **75-2**: 1-66. (Fisheries Research Institute, University of Washington, Seattle, WA.)
GRABE, S. A., 1989. Some aspects of the biology of *Rhopalophthalmus tattersallae* Pillai, 1961 (Crustacea, Mysidacea) and extension of range into the Khor al Sabiya, Kuwait (Arabian Gulf). — Proceedings of the Biological Society of Washington, **102** (3): 726-731.
GREEN, J. M., 1970. Observations on the behaviour and larval development of *Acanthomysis sculpta* (Tattersall) (Mysidacea). — Canadian Journal of Zoology, **48** (2): 289-292.
GREENWOOD, J. G. & D. J. HADLEY, 1982. A redescription of the mysid *Idiomysis inermis* Tattersall, 1922 (Mysidacea) to include the previously unknown female. — Crustaceana, **42** (2): 174-178.
GREENWOOD, J. G., M. B. JONES & J. GREENWOOD, 1989. Salinity effects on brood maturation of the mysid crustacean *Mesopodopsis slabberi*. — Journal of the Marine Biological Association of the United Kingdom, **69** (3): 683-694.
GREENWOOD, M. F. D., 2007. Nekton community change along estuarine salinity gradients: can salinity zones be defined? — Estuaries and Coasts, **30**: 537-542.
GRIGOROVICH, I. A., I. R. COLAUTTI, E. L. MILLS, K. HOLECK, A. G. BALLERT & H. J. MACISAAC, 2003. Ballast-mediated animal introductions in the Laurentian Great Lakes: retrospective and prospective analyses. — Canadian Journal of Fisheries and Aquatic Sciences, **60**: 740-756.
GROBBEN, C., 1881. Die Antennendrüse der Crustaceen. — Arbeiten aus dem Zoologischen Institut der Universität Wien und der Zoologischen Station in Triest, **3**: 93-110.
GROSSNICKLE, N. E., 1979. Nocturnal feeding patterns of *Mysis relicta* in Lake Michigan, based on gut content fluorescence. — Limnology and Oceanography, **24** (4): 777-780.

GRUNER, H.-E. & G. SCHOLTZ, 2004. Segmentation, tagmata, and appendages. — In: J. FOREST & J. C. VON VAUPEL KLEIN (eds.), The Crustacea. Revised and updated from the Traité de Zoologie, **1** (2): 13-57. (Brill, Leiden.)

GUSE, G. W., 1980. Development of antennal sensilla during moulting in *Neomysis integer* (Leach) (Crustacea, Mysidacea). — Protoplasma, **105** (1-2): 53-67.

— —, 1983a. Ultrastructure, development, and moulting of the aesthetascs of *Neomysis integer* and *Idotea balthica* (Crustacea, Malacostraca). — Zoomorphology, **103** (2): 121-133.

— —, 1983b. Die Sensillen des Lobus masculinus von Mysidaceen (Crustacea): Ihre Ultrastruktur und mögliche Bedeutung als Pheromonrezeptoren. — Verhandlungen der Deutschen Zoologischen Gesellschaft, **76**: 189.

HABERMEHL, G., 2008. Nachweise von *Hemimysis anomala* Sars und *Limnomysis benedeni* Czerniavsky (Crustacea: Mysida) in Mittelfranken/Bayern, ein Beitrag zur Verbreitung dieser Arten. — Lauterbornia, **62**: 33-39.

HAFFER, K., 1965. Zur Morphologie der Malacostraca: Der Kaumagen der Mysidacea im Vergleich zu dem verschiedener Peracarida und Eucarida. — Helgoländer Wissenschaftliche Meeresuntersuchungen, **12** (1-2): 156-206.

HAHN, P. & M. ITZKOWITZ, 1986. Site preference and homing behavior in mysid shrimp, *Mysidium gracile* (Dana). — Crustaceana, **51** (2): 215-219.

HAKALA, I., 1978. Distribution, population dynamics and production of *Mysis relicta* (Lovén) in southern Finland. — Annales Zoologici Fennici, **15**: 243-258.

HALLBERG, E., 1977. The fine structure of the compound eyes of mysids (Crustacea: Mysidacea). — Cell & Tissue Research, **184**: 45-65.

HALLBERG, E., M. ANDERSSON & D. E. NILSSON, 1980. Responses of the screening pigments in the compound eye of *Neomysis integer* (Crustacea: Mysidacea). — The Journal of Experimental Zoology, **212** (3): 397-402.

HALLBERG, E. & J. CHAIGNEAU, 2004. The non-visual sense organs. — In: J. FOREST & J. C. VON VAUPEL KLEIN (eds.), The Crustacea. Revised and updated from the Traité de Zoologie, **1** (7): 301-380. (Brill, Leiden.)

HALLBERG, E., K. U. I. JOHANSSON & R. ELOFSSON, 1992. The aesthetasc concept: structural variations of putative olfactory receptor cell complexes in Crustacea. — Microscopy Research and Technique, **22**: 325-335.

HANAMURA, Y., 1999. Seasonal abundance and life cycle of *Archaeomysis articulata* (Crustacea: Mysidacea) on a sandy beach of western Hokkaido, Japan. — Journal of Natural History, London, **33**: 1811-1830.

HANAMURA, Y., G. FERNANDEZ-LEBORANS, R. SIOW, A. MAN & P.-E. CHEE, 2010. Prevalence and seasonality of *Zoothamnium duplicatum* (Protozoa: Ciliophora) epibiont on an estuarine mysid (Crustacea: Mysida) in tropical mangrove brackish water. — Plankton & Benthos Research, **5** (1): 39-43.

HANAMURA, Y. & T. KASE, 2002. Marine cave mysids of the genus *Palaumysis* (Crustacea: Mysidacea), with a description of a new species from the Philippines. — Journal of Natural History, London, **36**: 253-263.

— — & — —, 2003. *Palaumysis pilifera*, a new species of cave-dwelling mysid (Crustacea: Mysidacea) from Okinawa, southwestern Japan, with an additional note on *P. simonae* Băcescu & Iliffe, 1986. — Hydrobiologia, **497**: 145-152.

HANAMURA, Y., M. MURANO & A. MAN, 2011. Review of eastern Asian species of the mysid genus *Rhopalophthalmus* Illig, 1906 (Crustacea, Mysida) with descriptions of three new species. — Zootaxa, **2788**: 1-37.

HANAMURA, Y. & K. NAGASAKI, 1996. Occurrence of the sandy beach mysids *Archaeomysis* spp. (Mysidacea) infested by epibiontic peritrich ciliates (Protozoa). — Crustacean Research, **25**: 25-33.

HANAMURA, Y., R. SIOW, P.-E. CHEE & F. M. KASSIM, 2009. Seasonality and biological characteristics of the shallow-water mysid *Mesopodopsis orientalis* (Crustacea: Mysida) on a tropical sandy beach, Malaysia. — Plankton & Benthos Research, **4** (2): 53-61.
HANAMURA, Y., K. TANAKA, A. MAN & F. M. KASSIM, 2012. Ecological characteristics of hyperbenthic crustaceans in mangrove estuaries on the north-west coast of peninsular Malaysia: an overview. — In: K. TANAKA, S. MORIOKA & S. WATANABE (eds.), Sustainable stock management and development of aquaculture technology suitable for southeast Asia. JIRCAS Working Report, **75**: 25-34. (Japan International Research Center for Agricultural Sciences, Tsukuba, Ibaraki, Japan.)
HANSELMANN, A. J., B. HODAPP & K.-O. ROTHHAUPT, 2013. Nutritional ecology of the invasive freshwater mysid *Limnomysis benedeni* — field data and laboratory experiments on food choice and juvenile growth. — Hydrobiologia, **705** (1): 75-86.
HANSEN, H. J., 1908. Crustacea Malacostraca (I). — The Danish Ingolf Expedition, **3** (2): 1-120, pls. I-V. (Bianco Luno, Copenhagen.)
— —, 1910. The Schizopoda of the Siboga Expedition. — Siboga Expeditie, Monographie, **37**: 1-123. (Leyden.)
— —, 1925. Studies on Arthropoda II. — Pp. 1-176. (Gyldendalska Boghandel, Copenhagen.)
HANSSON, S., U. LARSSON & S. JOHANSSON, 1989. Selective predation by herring and mysids, and zooplankton community structure in a Baltic coastal area. — In: S. HANSSON (ed.), Biotic interactions in fish and mysid communities, studies in two Baltic coastal areas, pp. 1-24. (Doctoral Thesis, Department of Zoology, University of Stockholm, Stockholm.)
HANSTRÖM, B., 1947. The brain, the sense organs, and the incretory organs of the head in the Crustacea Malacostraca. — Lunds Universitets Årsskrift, (N. F.) (Avd. 2) **43** (9): 1-45.
— —, 1948. The brain, the sense organs, and the incretory organs of the head in the Crustacea Malacostraca. — Bulletin Biologique de la France et de la Belgique, (Suppl.) **33**: 98-126.
HARGREAVES, P. M., 1985. The distribution of Mysidacea in the open ocean and near-bottom over slope regions in the northern North-east Atlantic Ocean during 1979. — Journal of Plankton Research, **7** (2): 241-261.
— —, 1989. The vertical and horizontal distribution of four species of the genus *Gnathophausia* (Crustacea: Mysidacea) in the eastern North Atlantic Ocean. — Journal of Plankton Research, **11** (4): 687-702.
— —, 1999. The vertical distribution of micronektonic decapod and mysid crustaceans across the Goban Spur of the Porcupine Seabight. — Sarsia, **84**: 1-18.
HARGREAVES, P. M. & M. MURANO, 1996. Mysids of the genus *Boreomysis* from abyssopelagic regions of the north-eastern Atlantic. — Journal of the Marine Biological Association of the United Kingdom, **76**: 665-674.
HARZSCH, S., D. SANDEMAN & J. CHAIGNEAU, 2012. Morphology and development of the central nervous system. — In: J. FOREST & J. C. VON VAUPEL KLEIN (eds.), The Crustacea. Revised and updated from the Traité de Zoologie, **3** (15): 9-237. (Brill, Leiden.)
HAWORTH, A. H., 1825. XXIX. A new binary arrangement of the Macrurous Crustacea. — The Philosophical Magazine and Journal, London, **65** (323): 183-184.
HENTSCHEL, E. & G. WAGNER, 1976. Zoologisches Wörterbuch (2nd ed.). — Pp. 1-672. (UTB, Gustav Fischer Verlag, Stuttgart.)
HERBST, C., 1896. Über die Regeneration von antennenähnlichen Organen an Stelle von Augen. I. — Archiv für Entwicklungsmechanik, **2** (4): 544-558.
HERMAN, S. S., 1963. Vertical migration of the opossum shrimp, *Neomysis americana* Smith. — Limnology and Oceanography, **8** (2): 228-238.
HESSLER, R. R., 1985. Swimming in Crustacea. — Transactions of the Royal Society of Edinburgh, **76** (2-3): 115-122.
HEUBACH, W., 1969. *Neomysis awatschensis* in the Sacrament-San Joaquin river estuary. — Limnology and Oceanography, **14**: 533-546.

HOENIGMAN, J., 1960. Faits nouveaux concernant les Mysidacés (Crustacea) et leurs épibiontes dans l'Adriatique. — Rapports de la Commission Internationale pour l'Exploration Scientifique de la Mer Méditerranée, **15**: 339-343.

HOFFMEYER, M. S., 1990. The occurrence of *Neomysis americana* in two new localities of the South American coast (Mysidacea). — Crustaceana, **58** (2): 186-192.

HOGSTAD, A. M., 1969. The X organ and the sinus gland of *Mysis relicta* Lovén (Crustacea: Mysidacea). — Nyt Magazin för Zoologi, **17**: 105-109.

HOLDICH, D., S. GALLAGHER, L. RIPPON, P. HARDING & R. STUBBINGTON, 2006. The invasive Ponto-Caspian mysid, *Hemimysis anomala*, reaches the UK. — Aquatic Invasions, **1** (1): 4-6.

HOLMQUIST, C., 1957. On aberrant specimens of *Praunus flexuosus* and some other opossum shrimps. — Acta Borealia, **13A**: 1-29, tabs. I-III.

— —, 1958. Proposed use of the Plenary Powers to validate a neotype for the nominal species "Cancer oculatus" Fabricius (O.), 1780, to designate the species so named to be the type species of the genus "Mysis" Latreille, [1802-1803] (Class Crustacea, Order Mysidacea) and matters incidental thereto. — The Bulletin of Zoological Nomenclature, **16** (2): 51-61.

— —, 1959. Problems on marin-glacial relicts on account of investigations on the genus *Mysis*. — Pp. 1-270. (Berlingska Boktryckeriet, Lund.)

— —, 1962. The relict concept — is it a merely zoogeographical conception? — Oikos, **13** (2): 262-292.

HOLT, E. W. L. & W. M. TATTERSALL, 1905. Schizopoda from the north-east Atlantic slope. — Scientific Investigations. Fisheries Branch, Department of Agriculture for Ireland, Dublin — Annual Report, **2**: 99-152.

— — & — —, 1906. Schizopodous Crustacea from the north-east Atlantic slope. — Report on the sea and inland fisheries of Ireland. Part II. Scientific Investigations, Appendix **IV** (1905) (2): 99-152; pls. XV-XXV. (Fisheries Research Board of Ireland, Dublin.)

HOPKINS, T. L., 1965. Mysid shrimp abundance in surface waters of Indian River Inlet, Delaware. — Chesapeake Science, **6** (2): 86-91.

HOSFELD, B. & H. K. SCHMINKE, 1997. Discovery of segmental extranephridial podocytes in Harpacticoida (Copepoda) and Bathynellacea (Syncarida). — Journal of Crustacean Biology, **17** (1): 13-20.

HOUGH, A. R., N. J. BANNISTER & E. NAYLOR, 1992. Intersexuality in the mysid *Neomysis integer*. — Journal of Zoology, London, **226** (4): 585-588.

HSÜ, K. J., 1978. When the Black Sea was drained. — Scientific American, **238** (5): 53-63.

HSÜ, K. J., L. MONTANDERT, D. BERNOULLI, M. B. CITA, A. ERICKSON, R. E. GARRISON, R. B. KIDD, F. MÈLIERÉS, C. MÜLLER & R. WRIGHT, 1977. History of the Mediterranean salinity crisis. — Nature, London, **267**: 399-403.

ICZN (International Commission on Zoological Nomenclature), 1959. Opinion 578. Use of the plenary power to validate a neotype for the nominal species *Cancer oculatus* O. Fabricius, 1780, and to designate that species as the type-species of the nominal genus *Mysis* Latreille, [1802-1803] (class Crustacea, order Mysidacea). — The Bulletin of Zoological Nomenclature, **17**: 143-145.

— —, 1999. International Code of Zoological Nomenclature (4th ed.). — Pp. 1-306. (International Trust for Zoological Nomenclature, London; also available online at: http://www.iczn.org/iczn/index.jsp)

II, N., 1964. Fauna Japonica. Mysidae (Crustacea). — Pp. i-x, 1-610. (Biogeographical Society of Japan, Tokyo.)

ILLIG, G., 1912. *Echinomysis chuni*, eine neue pelagisch lebende Mysidee. — Zoologica, **67**: 129-138, pls. XV-XVIII.

— —, 1930. Die Schizopoden der Deutschen Tiefsee-Expedition. — In: C. CHUN (ed.), Wissenschaftliche Ergebnisse der Deutschen Tiefsee-Expedition auf dem Dampfer "Valdivia" 1898-1899, **22** (6): 397-620.

INGLE, R. W., 1972. A redescription of *Spelaeomysis servatus* (Fage) comb. nov. (Mysidacea: Lepidomysidae) from the material collected on Aldabra Atoll, with a key to species of Lepidomysidae. — Bulletin of the British Museum (Natural History), (Zoology) **22** (7): 197-210.

INGUSCIO, S., 1998. Misidacei stigiobionti di Puglia. — Pp. 1-95. (ideemultimediali, Nardò, Italia.)

ISHIKAWA, M. & Y. OHSHIMA, 1951. On the life history of a mysid crustacean, *Neomysis japonica* Nakazawa. — Bulletin of the Japanese Society of Scientific Fisheries, **16**: 461-472.

JANCKE, O., 1926. Über die Brutpflege einiger Malakostraken. — Archiv für Hydrobiologie, **17** (4): 678-698.

JARMAN, S. N., S. NICOL, N. G. ELLIOT & A. MCMINN, 2000. 28S rDNA evolution in the Eumalacostraca and the phylogenetic position of krill. — Molecular Phylogenetics and Evolution, **17** (1): 26-36.

JENNER, R. A., C. N. DHUBHGHAILL, M. P. FERLA & M. A. WILLS, 2009. Eumalacostracan phylogeny and total evidence: limitations of the usual suspects. — BMC Evolutionary Biology, **2009**: 9/21 (20 pp.). (BioMed Central.) DOI:10.1186/1471-2148-9-21. http://www.biomedcentral.com/1471-2148/9/21 [5 Dec. 2012]

JEPSEN, J., 1965. Marsupial development of *Boreomysis arctica* (Kröyer, 1861). — Sarsia, **20**: 1-8.

JOHANNSSON, O. E., M. F. LEGGETT, L. G. RUDSTAM, M. R. SERVOS, M. A. MOHAMMADIAN, G. GAL, R. M. DERMOTT & R. H. HESSLEIN, 2001. Diet of *Mysis relicta* in Lake Ontario as revealed by stable isotope and gut content analysis. — Canadian Journal of Fisheries and Aquatic Sciences, **58** (10): 1975-1986.

JOHANSSON, K. U. I., L. GEFORS, R. WALLÈN & E. HALLBERG, 1996. Structure and distribution patterns of aesthetascs and male-specific sensilla in *Lophogaster typicus* (Mysidacea). — Journal of Crustacean Biology, **16** (1): 45-53.

JOHANSSON, K. U. I. & E. HALLBERG, 1992. Male-specific structures in the olfactory system of mysids (Mysidacea; Crustacea). — Cell & Tissue Research, **268**: 359-368.

JOHNS, D. M., W. J. BERRY & W. WALTON, 1981. International study on *Artemia*. 16. Survival, growth and reproductive potential of the mysid, *Mysidopsis bahia* Molenock fed various geographical strains of the brine shrimp *Artemia*. — Journal of Experimental Marine Biology and Ecology, **53** (2-3): 209-219.

JOHNSON, W. S., M. STEVENS & L. WATLING, 2001. Reproduction and development of marine peracaridans. — Advances in Marine Biology, **39**: 105-260.

JOHNSTON, N. M. & D. A. RITZ, 2001. Synchronous development and release of broods by the swarming mysids *Anisomysis mixta australis* (Zimmer), *Paramesopodopsis rufa* Fenton and *Tenagomysis tasmaniae* Fenton (Mysidacea: Crustacea). — Marine Ecology Progress Series, **223**: 225-233.

— — & — —, 2005. Kin recognition and adoption in mysids (Mysidacea: Crustacea). — Journal of the Marine Biological Association of the United Kingdom, **85** (6): 1441-1447.

JOHNSTON, N. M., D. A. RITZ & G. E. FENTON, 1997. Larval development in Tasmanian coastal mysids: *Anisomysis mixta australis* (Zimmer), *Paramesopodopsis rufa* Fenton and *Tenagomysis tasmaniae* Fenton (Peracarida: Mysidacea). — Marine Biology, Berlin, **130**: 93-99.

JUBERTHIE-JUPEAU, L., 1976. Données sur la reproduction et le cycle vital d'un mysidacé souterrain tropical, *Antromysis juberthiei* Bacesco et Orghidan. — Comptes Rendus Hebdomadaires des Séances de l'Académie des Sciences, (D, Sciences Naturelles) **282** (13): 1321-1323.

JUBERTHIE-JUPEAU, L. & Y. CROUAU, 1977. Ultrastructure des aesthetascs d'un Mysidacé souterrain anophthalme. — Comptes Rendus de l'Académie des Sciences, (D) **284**: 2257-2259.

JUCHAULT, P., 1963. Sur la glande androgène d'un certain nombre de Péracarides (Cumacés, Mysidacés, Tanaidacés). — Comptes Rendus Hebdomadaires des Séances et Mémoires de la Société de Biologie, **157**: 613-615.

JUMARS, P. A., 2007. Habitat coupling by mid-latitude, subtidal, marine mysids: import-subsidised omnivores. — Oceanography and Marine Biology — An Annual Review, **45**: 89-138.

KALLMEYER, D. E. & J. H. CARPENTER, 1996. *Stygiomysis cokei*, new species, a troglobitic mysid from Quintana Roo, Mexico (Mysidacea: Stygiomysidae). — Journal of Crustacean Biology, **16** (2): 418-427.

KALTENBERG, A. M. & K. J. BENOIT-BIRD, 2013. Intra-patch clustering in mysid swarms revealed through multifrequency acoustics. — ICES Journal of Marine Science, **70** (4): 883-891.

KARPEVICH, A. F. & E. N. BOKOVA, 1963. Acclimatation des poissons et des invertébrés aquatiques effectuée en URSS en 1960-1961. — Voprosij Ikhtiologii, **3** (2): 366-395. [In Russian.]

KASAOKA, I. D., 1974. The male genital system in two species of mysid Crustacea. — Journal of Morphology, **143** (3): 259-284.

KAURI, T. & E. DAHL, 1975. Fine structure of the organ of Bellonci (SPX) in *Boreomysis arctica* (Kroyer) (Crustacea, Mysidacea). — Zoologica Scripta, **4** (1): 41-47.

KEEBLE, F. & F. W. GAMBLE, 1904. The colour-physiology of higher Crustacea. — Philosophical Transactions of the Royal Society, (B, Biological Sciences) **196**: 295-388.

KETELAARS, H. A. M., 2004. Range extensions of Ponto-Caspian aquatic invertebrates in continental Europe. — In: H. DUMONT, A. T. SHIGANOVA & U. NIERMANN (eds.), Aquatic invasions in the Black, Caspian, and Mediterranean seas. The ctenophores *Mnemiopsis leidyi* and *Beroe* in the Ponto-Caspian and other aquatic invasions. NATO Science Series, (IV, Earth and Environmental Sciences) **35** (2): 209-236.

KETELAARS, H. A. M., F. E. LAMBREGTS-VAN DE CLUNDERT, C. J. CARPENTIER, A. J. WAGENVOORT & W. HOOGENBOEZEM, 1999. Ecological effects of the mass occurrence of the Ponto-Caspian invader, *Hemimysis anomala* G.O. Sars, 1907 (Crustacea: Mysidacea), in a freshwater storage reservoir in the Netherlands, with notes on its autecology and new records. — Hydrobiologia, **394**: 233-248.

KIKUCHI, S. & M. MATSUMASA, 1993. Two ultrastructurally distinct types of transporting tissues, the branchiostegal and the gill epithelia, in an estuarine tanaid, *Sinelobus stanfordi* (Crustacea, Peracarida). — Zoomorphology, **113**: 253-260.

KINNE, O., 1955. *Neomysis vulgaris* Thompson, eine autökologisch-biologische Studie. — Biologisches Zentralblatt, **74**: 160-202.

KLEPAL, W. & R. T. KASTNER, 1980. Morphology and differentiation of non-sensory cuticular structures in Mysidacea, Cumacea and Tanaidacea (Crustacea, Peracarida). — Zoologica Scripta, **9** (4): 271-281.

KNIGHT-JONES, E. W. & E. MORGAN, 1966. Responses of marine animals to changes in hydrostatic pressure. — Oceanography and Marine Biology — An Annual Review, **4**: 267-299.

KOBUSCH, W., 1998. The foregut of the Mysida (Crustacea, Peracarida) and its phylogenetic relevance. — Philosophical Transactions of the Royal Society, (B, Biological Sciences) **353**: 559-581.

— —, 1999. The phylogeny of Peracarida (Crustacea, Malacostraca): morphological investigations of the peracaridan foreguts, their phylogenetic implications, and an analysis of peracaridan characters. — Vol. **1**: 1-277; Vol. **2**: figs. 1-76. (Dissertation University of Bielefeld; Cuvillier Verlag, Göttingen.)

KOKSVIK, J. I., H. REINERTSEN & A. LANGELAND, 1991. Changes in plankton biomass and species composition in Lake Jonsvatn, Norway, following the establishment of *Mysis relicta*. — American Fisheries Society Symposium, **9**:115-125.

KOUASSI, E., M. PAGANO, L. SAINT-JEAN & J. C. SORBE, 2006. Diel vertical migrations and feeding behavior of the mysid *Rhopalophthalmus africana* (Crustacea: Mysidacea) in a tropical lagoon (Ebrié, Côte d'Ivoire). — Estuarine, Coastal and Shelf Science, **67**: 355-368.

KOUKOURAS, A., M.-S. KITSOS, TH. TZOMOS & A. TSELEPIDES, 2010. Evolution of the entrance rate and of the spatio-temporal distribution of Lessepsian Crustacea Decapoda in the Mediterranean Sea. — Crustaceana, **83** (12): 1409-1430.

KULAKOVSKII, E. E., 1969. Neurosecretory system of *Mysis oculata* (Fabricius) Crustacea Malacostraca. — Doklady Akademii Nauk SSSR, **192** (1): 226-228. [In Russian.]

— —, 1971. Neurohormonal control for chromatophores in mysids. — Doklady Akademii Nauk SSSR, **196** (1): 234-236. [In Russian.]

KÜNNE, C., 1937. Über als «Fremdlinge» zu bezeichnende Grossplanktonten in der Ostsee. — Rapports Conseil International pour l'Exploration de la Mer, **102** (2): 1-7.

— —, 1939. Beiträge zur Kenntnis der Mysideenfauna der südlichen Nordsee. — Zoologische Jahrbücher, (Abteilung für Systematik, Ökologie und Geographie der Tiere) **72** (5-6): 329-358.

LABAT, R., 1953. *Paramysis nouveli* n. sp. et *Paramysis bacescoi* n. sp. deux espèces de Mysidacés confondues, jusqu'à présent, avec *Paramysis helleri* (G.O. Sars, 1877). — Bulletin de l'Institut Océanographique, Monaco, **1034**: 1-24.

— —, 1954. Observations sur l'accouplement et la ponte de *Paramysis nouveli* (Crustacé, Mysidacé). — Bulletin de la Société d'Histoire Naturelle de Toulouse, **89** (3-4): 406-409.

— —, 1957. Observations sur le cycle annuel et la sexualité de *Paramysis nouveli* et *Paramysis bacescoi* (Crustacés Mysidacés) dans la région de Roscoff. — Archives de Zoologie Expérimentale et Générale, **94** (3): 162-173.

— —, 1961. L'appareil génital mâle de *Praunus flexuosus* (Crustacé Mysidacé). — Bulletin de la Société d'Histoire Naturelle de Toulouse, **96** (1-2): 60-66.

— —, 1962. La spermatogenèse chez *Praunus flexuosus* [Crustacé Mysidacé]. — Bulletin de la Société d'Histoire Naturelle de Toulouse, **97** (1-2): 51-60.

LACHAISE, F., A. LE ROUX, M. HUBERT & R. LAFONT, 1993. The molting gland of crustaceans: localization, activity and endocrine control (a review). — Journal of Crustacean Biology, **13** (2): 198-234.

LAGARDÈRE, J.-P., 1972. Recherches sur l'alimentation des crevettes de la pente continentale marocaine. — Téthys, **3** (3): 655-675.

— —, 1977a. Recherches sur le régime alimentaire et le comportement prédateur des Décapodes benthiques de la pente continentale de l'Atlantique nord-oriental (Golfe de Gascogne et Maroc). — In: B. F. KEEGAN, P. O'CEIDIGH & P. J. S. BOADEN (eds.), Biology of benthic organisms. Proceedings of the Eleventh European Marine Biology Symposium, pp. 397-408. (Pergamon Press, Oxford.)

— —, 1977b. Recherches sur la distribution verticale et sur l'alimentation des Crustacés Décapodes benthiques de la pente continentale du golfe de Gascogne — Analyse des groupements carcinologiques. — Bulletin du Centre d'Études et de Recherches Scientifiques, Biarritz, **11** (4): 367-440.

— —, 1983. Les Mysidacés de la plaine abyssale du golfe de Gascogne. I. Familles des Lophogastridae, Eucopiidae et Petalophthalmidae. — Bulletin du Muséum National d'Histoire Naturelle, (4) (A, Zoologie, Biologie, Écologie Animale) **5** (3): 809-843.

LAGARDÈRE, J.-P. & H. NOUVEL, 1980a. Les Mysidacés du talus continental du golfe de Gascogne. II. Familles des Lophogastridae, Eucopiidae et Mysidae (tribu des Erythropini exceptée). — Bulletin du Muséum National d'Histoire Naturelle, (4) (A, Zoologie, Biologie, Écologie Animale) **2** (2): 375-412.

— — & — —, 1980b. Les Mysidacés du talus continental du golfe de Gascogne. II. Familles des Lophogastridae, Eucopiidae et Mysidae (tribu des Erythropini exceptée) (suite et fin). — Bulletin du Muséum National d'Histoire Naturelle, (4) (A, Zoologie, Biologie, Écologie Animale) **2** (3): 845-887.

LAND, M. F., 2004. Eyes and vision. — In: J. FOREST & J. C. VON VAUPEL KLEIN (eds.), The Crustacea. Revised and updated from the Traité de Zoologie, **1** (6): 257-299. (Brill, Leiden.)

LANGELAND, A., 1981. Decreased zooplankton density in two Norwegian lakes caused by predation of recently introduced *Mysis relicta*. — Verhandlungen der Internationalen Vereinigung für Limnologie, **21** (2): 926-937.

LANGELAND, A., J. I. KOKSVIK & J. NYDAL, 1991. Impact of the introduction of *Mysis relicta* on the zooplankton and fish populations in a Norwegian lake. — American Fisheries Society Symposium, **9**: 98-114.

LARKIN, P. A., 1948. *Pontoporeia* and *Mysis* in Athabasca, Great Bear and Great Slave Lakes. — Bulletin of the Fisheries Research Board of Canada, **88**: 1-33.

LASENBY, D. C. & R. R. LANGFORD, 1972. Growth, life history and respiration of *Mysis relicta* in an Arctic and temperate lake. — Journal of the Fisheries Research Board of Canada, **29**: 1701-1708.

LASENBY, D. C., T. G. NORTHCOTE & M. FÜRST, 1986. Theory, practice and effects of *Mysis relicta* introductions to North American and Scandinavian Lakes. — Canadian Journal of Fisheries and Aquatic Sciences, **43**: 1277-1284.

LATREILLE, P. A., 1802. Histoire naturelle, génerale et particulière des Crustacés et des Insectes. — Vol. **3**: i-xii, 1-468. (Imprimerie de F. Dufart, Paris.)

— —, 1817. Les Crustacés, les Arachnides et les Insectes. — In: G. CUVIER (ed.), Le règne animal, distribué d'après son organisation pour servir de base à l'histoire naturelle des animaux et d'introduction à l'anatomie comparée, **3**: i-xxix, 1-653. (Librairie Deterville, Paris.)

— —, 1825. Familles naturelles du règne animal, exposées succinctement et dans un ordre analytique, avec l'indication de leurs genres. — Pp. 1-570. (Libraire de J.-B. Baillière, Paris.)

LEACH, W. E., 1815. A tabular review of the external characters of four classes of animals, which Linné arranged under Insecta; with the distribution of the genera composing three of these classes into orders, &c. and descriptions of several new genera and species. — The Transactions of the Linnean Society of London, **11** (31): 306-400.

— —, 1830. On the genus *Megalophthalmus*, a new and very interesting genus, completely proving the theory of Jules-Caesar Savigny to be correct. — Transactions of the Plymouth Institution, **1**: 176-178.

LECOINTRE, G., H. PHILIPPE, H. L. V. LE & H. LE GUYADER, 1993. Species sampling has a major impact on phylogenetic inference. — Molecular Phylogenetics and Evolution, **2**: 205-224.

LEDOYER, M., 1989. Les Mysidacés (Crustacea) des grottes sous-marines obscures de Méditerranée nord-occidentale et du proche Atlantique (Portugal et Madère). — Marine Nature, **2** (1): 39-62.

LEHTINIEMI, M., M. KILJUNEN & R. I. JONES, 2009. Winter food utilisation by sympatric mysids in the Baltic Sea, studied by combined gut content and stable isotope analyses. — Marine Biology, Berlin, **156** (4): 619-628.

LEHTINIEMI, M. & E. LINDÉN, 2006. *Cercopagis pengoi* and *Mysis* spp. alter their feeding rate and prey selection under predation risk of herring (*Clupea harengus membras*). — Marine Biology, Berlin, **149**: 845-854.

LEHTINIEMI, M. & H. NORDSTRÖM, 2008. Feeding differences among common littoral mysids, *Neomysis integer*, *Praunus flexuosus* and *P. inermis*. — In: U. M. AZEITEIRO, I. JENKINSON & M. J. PEREIRA (eds.), Plankton studies. — Hydrobiologia, **614** (1): 309-320.

LEPPÄKOSKI, E., 1984. Introduced species in the Baltic Sea and its coastal ecosystems. — Ophelia, (Suppl.) **3**: 123-135.

LEPPÄKOSKI, E., S. GOLLASCH, P. GRUSZKA, H. OJAVEER, S. OLENIN & V. PANOV, 2002. The Baltic — a sea of invaders. — Canadian Journal of Fisheries and Aquatic Sciences, **59** (7): 1175-1188.

LINDÉN, E. & H. KUOSA, 2004. Effect of grazing and excretion by pelagic mysids (*Mysis* ssp.) on the size structure and biomass of the phytoplankton community. — In: H. KAUTSKY & P. SNOEIJS (eds.), Biology of the Baltic Sea. — Hydrobiologia, **514** (1-3): 73-78.

LINDSTRÖM, M., 1992. Spectral sensitivity and light tolerance of mysid species in the Baltic area. — In: J. KÖHN, M. B. JONES & A. MOFFAT (eds.), Taxonomy, biology and ecology of (Baltic) mysids (Mysidacea, Crustacea), pp. 120-126. (Rostock University Press, Rostock.)
— —, 2000. Eye function of Mysidacea (Crustacea) in the northern Baltic Sea. — Journal of Experimental Marine Biology and Ecology, **246**: 85-101.
LINDSTRÖM, M. & V. B. MEYER-ROCHOW, 1987. Near infra-red sensitivity of the eye of the crustacean *Mysis relicta*. — Biochemical and Biophysical Research Communications, **147** (2): 747-752.
LINDSTRÖM, M. & H. L. NILSSON, 1988. Eye function of *Mysis relicta* Lovén (Crustacea) from two photic environments. Spectral sensitivity and light tolerance. — Journal of Experimental Marine Biology and Ecology, **120**: 23-37.
LINDSTRÖM, M. & E. SANDBERG-KILPI, 2008. Breaking the boundary — the key to bottom recovery? The role of mysid crustaceans in oxygenizing bottom sediments. — Journal of Experimental Marine Biology and Ecology, **354** (2): 161-168.
LINN, J. D. & T. C. FRANTZ, 1965. Introduction of the opossum shrimp (*Mysis relicta* Lovén) into California and Nevada. — California Fish and Game, **51** (1): 48-51.
LOWRY, J. K., 1986. The callynophore, a eucaridan/peracaridan sensory organ prevalent among the Amphipoda (Crustacea). — Zoologica Scripta, **15** (4): 333-349.
LUCAS, C. E., 1936. On certain inter-relations between phytoplankton and zooplankton under experimental conditions. — Journal du Conseil. Conseil Permanent International pour l'Exploration de la Mer, **11** (3): 343-362.
MACQUART-MOULIN, C., 1970. Le comportement d'essaim chez les Mysidacés. Observation et analyse de quelques essaims littoraux. — Rapports de la Commission Internationale pour l'Exploration Scientifique de la Mer Méditerranée, **20** (3): 439-441.
— —, 1971. Modifications des réactions photocinétiques des péracarides de l'hyponeuston nocturne en function de l'importance de l'éclairement. — Téthys, **3** (4): 897-920.
— —, 1973a. L'activité natatoire rythmique chez les péracarides bentho-planctoniques. Déterminisme endogène des rythmes nycthéméraux. — Téthys, **5** (1): 209-231.
— —, 1973b. Le comportement d'essaim chez les Mysidacés. Influence de l'intensité lumineuse sur la formation, le maintien et la dissociation des essaims de *Leptomysis lingvura*. — Rapports de la Commission Internationale pour l'Exploration Scientifique de la Mer Méditerranée, **21** (8): 449-501.
— —, 1975. Les Pécarides benthiques dans le plancton nocturne, Arnphipodes, Cumacés, Isopodes, Mysidacés. Analyse des comportements migratoires dans le golfe de Marseille. Recherches expérimentales sur l'origine des migrations et le contrôle de la distribution des espèces. — Pp. 1-376. (Thèse, Université Aix-Marseille II, Arch. orig. CNRS n° 10864.)
— —, 1979. Écophysiologie comparée des Mysidacés *Hemimysis speluncola* Ledoyer (cavernicole) et *Leptomysis lingvura* G.O. Sars (non-cavernicole). L'orientation à la lumière: tests ponctuels. — Journal of Experimental Marine Biology and Ecology, **38** (3): 287-299.
MACQUART-MOULIN, C. & F. PASSELAIGUE, 1982. Mouvements nycthéméraux d'*Hemimysis speluncola* Ledoyer, espèce cavernicole, et de *Leptomysis lingvura* G.O. Sars, espèce non cavernicole (Crustacea, Mysidacea). — Tethys, **10** (3): 221-228.
MACQUART-MOULIN, C. & E. RIBERA MAYCAS, 1995. Inshore and offshore diel migrations in European benthopelagic mysids, genera *Gastrosaccus*, *Anchialina* and *Haplostylus* (Crustacea, Mysidacea). — Journal of Plankton Research, **17** (3): 531-555.
MAISSURADZE, L. S. & G. POPESCU, 1987. Carpatho-Caucasian comparative study of Sarmatian mysids. — Dări de seamă ale şedinţelor — Institutul de Geologie şi Geofizică. 3. Paleontologie, **72-73**/3: 75-80.
MAKINGS, P., 1981. *Mesopodopsis slabberi* (Mysidacea) at Millport, W. Scotland, with the parasitic nematode *Anisakis simplex*. — Crustaceana, **41** (3): 310-312.

MANTON, S. M., 1928a. On some points in the anatomy and habits of the lophogastrid Crustacea. — Transactions of the Royal Society of Edinburgh, **56** (1): 103-119, pls. I-III.

— —, 1928b. On the embryology of a mysid crustacean, *Hemimysis lamornae*. — Philosophical Transactions of the Royal Society, (B, Biological Sciences) **216**: 363-463, pls. 21-25.

MARTIN, J. W. & G. E. DAVIS, 2001. An updated classification of the Recent Crustacea. — Natural History Museum of Los Angeles County, Science Series, **39**: 1-124.

MARTY, J., J. IVES, Y. DE LAFONTAINE, S. DESPATIE, M. A. KOOPS & M. POWER, 2012. Evaluation of carbon pathways supporting the diet of invasive *Hemimysis anomala* in a large river. — Journal of Great Lakes Research, **38** (Suppl. 2): 45-51.

MATSUDAIRA, C., T. KARIYA & T. TSUDA, 1952. The study of the biology of a mysid, *Gastrosaccus vulgaris* Nakazawa. — Tohoku Journal of Agricultural Research, **3** (1): 155-174.

MAUCHLINE, J., 1968. The biology of *Erythrops serrata* and *E. elegans* (Crustacea, Mysidacea). — Journal of the Marine Biological Association of the United Kingdom, **48** (2): 455-464.

— —, 1971. Seasonal occurrence of mysids (Crustacea) and evidence of social behaviour. — Journal of the Marine Biological Association of the United Kingdom, **51**: 809-825.

— —, 1972. The biology of bathypelagic organisms, especially Crustacea. — Deep-Sea Research, **19** (11): 753-780.

— —, 1973a. Intermoult growth of species of Mysidacea (Crustacea). — Journal of the Marine Biological Association of the United Kingdom, **53**: 569-572.

— —, 1973b. The broods of British Mysidacea (Crustacea). — Journal of the Marine Biological Association of the United Kingdom, **53**: 801-817.

— —, 1977. The integumental sensilla and glands of pelagic Crustacea. — Journal of the Marine Biological Association of the United Kingdom, **57**: 973-994.

— —, 1980. The biology of mysids and euphausiids. Part one. The biology of Mysids. — In: J. H. S. BLAXTER, F. S. RUSSEL & C. M. YONGE (eds.), Advances in Marine Biology, **18**: 3-369. (Academic Press, London.)

— —, 1982. The predation of mysids by fish of Rockall Trough, northeastern Atlantic Ocean. — Hydrobiologia, **93**: 85-99.

— —, 1986. The biology of the deep-sea species of Mysidacea (Crustacea) of the Rockall Trough. — Journal of the Marine Biological Association of the United Kingdom, **66**: 803-824.

MAUCHLINE, J. & J. D. M. GORDON, 1984a. Feeding and bathymetric distribution of the gadoid and morid fish of the Rockall Trough. — Journal of the Marine Biological Association of the United Kingdom, **64**: 657-665.

— — & — —, 1984b. Diets and bathymetric distributions of the macrourid fish of the Rockall Trough, northeastern Atlantic Ocean. — Marine Biology, Berlin, **81**: 107-121.

MAUCHLINE, J. & M. MURANO, 1977. World list of the Mysidacea, Crustacea. — Journal of the Tokyo University of Fisheries, **64** (1): 39-88.

MAYRAT, A., 1955. Mise en évidence de tendons chez les Crustacés dans le muscle attracteur du sinciput de *Praunus flexuosus* O.F. Müller. — Bulletin de la Société Zoologique de France, **80** (2-3): 81-85.

— —, 1956a. Oeil, centres optiques et glandes endocrines de *Praunus flexuosus* (O.F. Müller). — Archives de Zoologie Expérimentale et Génerale, **93** (4): 319-366.

— —, 1956b. Le système artériel de *Praunus flexuosus* et le prétendu cœur frontal des Malacostracés. — Bulletin — Station Océanographique de Salammbô, **53**: 44-49.

MAYRAT, A., B. R. MCMAHON & K. TANAKA, 2006. The circulatory system. — In: J. FOREST & J. C. VON VAUPEL KLEIN (eds.), The Crustacea. Revised and updated from the Traité de Zoologie, **2**: 3-84. (Brill, Leiden.)

MCFARLAND, W. N. & N. M. KOTCHIAN, 1982. Interaction between schools of fish and mysids. — Behavioral Ecology and Sociobiology, **11** (2): 71-76.

MCKENNEY, C. L., JR., 1996. The combined effects of salinity and temperature on various aspects of the reproductive biology of the estuarine mysid, *Mysidopsis bahia*. — Invertebrate Reproduction and Development, **29** (1): 9-18.
MCLACHLAN, A., T. WOOLDRIDGE & G. VAN DER HORST, 1979. Tidal movements of the macrofauna on an exposed sandy beach in South Africa. — Journal of Zoology, London, **187**: 433-442.
MCLUSKY, D. S. & V. E. J. HEARD, 1971. Some effects of salinity on the mysid *Praunus flexuosus*. — Journal of the Marine Biological Association of the United Kingdom, **51** (3): 709-715.
MEES, J., Z. ABDULKERIM & O. HAMERLYNCK, 1994. Life history, growth and production of *Neomysis integer* in the Westerschelde estuary (SW Netherlands). — Marine Ecology Progress Series, **109** (1): 43-57.
MEES, J., N. FOCKEDEY, A. DEWICKE, C. R. JANSSEN & J. C. SORBE, 1995. Aberrant individuals of *Neomysis integer* and other Mysidacea: intersexuality and variable telson morphology. — Netherlands Journal of Aquatic Ecology, **29** (2): 161-166.
MEES, J. & K. MELAND, 2012. World List of Lophogastrida, Stygiomysida and Mysida. — http://www.marinespecies.org/mysidacea [30 Jan. 2013]
MELAND, K., 2004. Species diversity and phylogeny of the deep-sea genus *Pseudomma* (Crustacea: Mysida). — Zootaxa, **649**: 1-30.
MELAND, K. & E. WILLASSEN, 2004. Molecular phylogeny and biogeography of the genus *Pseudomma* (Peracarida: Mysida). — Journal of Crustacean Biology, **24** (4): 541-557.
— — & — —, 2007. The disunity of "Mysidacea" (Crustacea). — Molecular Phylogenetics and Evolution, **44** (3): 1083-1104.
MERCIER, L. & R. POISSON, 1926. Microsporidies parasites de *Mysis* (Crust. Schizopodes). — Comptes Rendus de l'Académie des Sciences, **182**: 1576-1578.
METILLO, E. B. & D. A. RITZ, 1994. Comparative foregut functional morphology of three co-occuring mysids (Crustacea: Mysidacea) from south-eastern Tasmania. — Journal of the Marine Biological Association of the United Kingdom, **74**: 323-336.
— — & — —, 2001. Laminarinase activity in three co-occurring mysids (Crustacea: Mysidacea) from southeastern Tasmania, Australia. — The Philippine Scientist, **38**: 82-101.
— — & — —, 2003. Differential chemosensory feeding behaviour by three co-occurring mysids (Crustacea, Mysidacea) from southeastern Tasmania. — Comparative Biochemistry and Physiology, (A) **134**: 399-408.
MEUSY, J.-J., 1963. Description de la glande androgène chez deux Crustacés Péracarides: *Paramysis nouveli* Labat (Mysidacé) et *Eocuma dollfusi* Calman (Cumacé). — Comptes Rendus de l'Académie des Sciences, **256**: 5425-5428.
MEYER-ROCHOW, V. B. & L. JUBERTHIE-JUPEAU, 1987. An electron microscope study of the eye of the cave mysid *Heteromysoides cotti* from the island of Lanzarote (Canary Islands). — Stygologia, **3** (1): 24-34.
MILNE EDWARDS, H., 1837. Histoire naturelle des Crustacés, comprenant l'anatomie, la physiologie et la classification de ces animaux. — In: Collection des suites à Buffon formant avec les oeuvres de cet auteur un cours complet d'histoire naturelle, **2**: 1-531. (Libraire Encyclopédique de Roret, Paris.)
MINCHIN, D. & J. M. C. HOLMES, 2008. The Ponto-Caspian mysid, *Hemimysis anomala* G.O. Sars 1907 (Crustacea), arrives in Ireland. — Aquatic Invasions, **3** (2): 257-259.
MINCHIN, D. & H. ROSENTHAL, 2002. Exotics for stocking and aquaculture, making correct decisions. — In: E. LEPPÄKOSKI, S. GOLLASCH & S. OLENIN (eds.), Invasive aquatic species of Europe — distribution, impacts and management, pp. 206-216. (Kluwer Academic Publishers, Dordrecht.)
MODLIN, R. F., 1979. Development of *Mysis stenolepis* (Crustacea: Mysidacea). — The American Midland Naturalist, **101** (1): 250-254.

— —, 1990. Observations on the aggregative behavior of *Mysidium columbiae*, the mangrove mysid. — Pubblicazioni della Stazione Zoologica di Napoli, (I, Marine Ecology) **11** (3): 263-275.

MOELLER, J. F. & J. F. CASE, 1995. Temporal adaptations in visual systems of deep-sea crustaceans. — Marine Biology, Berlin, **123**: 47-54.

MOLLOY, F. M. E., 1958. The comparative anatomy and histophysiology of the alimentary canal in certain Mysidacea. — Pp. 1-149, figs. 1-59. (Ph. D. Thesis, University of London, London.)

MONOD, T. & J. FOREST, 2012. A history of crustacean classification. — In: J. FOREST & J. C. VON VAUPEL KLEIN (eds.), The Crustacea. Revised and updated from the Traité de Zoologie, **3** (19): 403-444. (Brill, Leiden.)

MORGAN, M. D., 1980. Life history characteristics of two introduced populations of *Mysis relicta*. — Ecology, **61** (3): 551-561.

— —, 1981. Abundance, life history, and growth of introduced populations of the opossum shrimp (*Mysis relicta*) in subalpine California lakes. — Canadian Journal of Fisheries and Aquatic Sciences, **38** (8): 989-993.

MÜLLER, O. F., 1776. Zoologiae Danicae prodromus, seu animalium Daniae et Norvegiae indigenarum: characteres, nomina, et synonyma imprimis popularium. — Pp. i-xxxii, 1-282. (Typis Hallageriis, Havniae.)

MURANO, M., 1963. Fisheries biology of a marine relict mysid *Neomysis intermedia* Czerniawsky. I. Role of the mysid on the production of fish in lakes. — Suisan-zoushoku (The Aquiculture), **11** (3): 149-158.

— —, 1964a. Fisheries biology of a marine relict mysid *Neomysis intermedia* Czerniawsky. III. Lifecycle, with special reference to the reproduction of the mysid. — Suisan-zoushoku (The Aquiculture), **12** (1): 19-30.

— —, 1964b. Fisheries biology of a marine relict mysid *Neomysis intermedia* Czerniawsky. IV. Lifecycle, with special reference to growth. — Suisan-zoushoku (The Aquiculture), **12** (2): 109-117.

— —, 1966. Fisheries biology of a marine relict mysid *Neomysis intermedia* Czerniawsky. VI. The transplantation of the mysid. — Suisan-zoushoku (The Aquiculture), **14** (2): 79-84.

— —, 1974. *Scolamblyops japonicus* gen. nov., sp. nov. (Mysidacea) from Suruga Bay, Japan. — Crustaceana, **26** (3): 225-228.

— —, 1986. Description of *Metasiriella kitaroi* n.gen. n.sp. (Mysidacea) with revision of the subfamily Siriellinae. — Crustaceana, **51** (3): 235-240.

— —, 1988. Mysidacea from Thailand with descriptions of two new species. — Crustaceana, **55** (3): 293-305.

— —, 1999a. *Marumomysis hakuhoae* new genus, new species, from the Sulu Sea (Crustacea: Mysidacea: Mysidae: Erythropini). — Plankton Biology & Ecology, **46** (2): 148-152.

— —, 1999b. Mysidacea. — In: D. BOLTOVSKOY (ed.), South Atlantic zooplankton, **2**: 1099-1140. (Backhuys Publishers, Leiden.)

MURANO, M. & K. FUKUOKA, 2008. A systematic study of the genus *Siriella* (Crustacea: Mysida) from the Pacific and Indian oceans, with descriptions of fifteen new species. — National Museum of Nature and Science Monographs, **36**: 1-173.

MURANO, M. & E. E. KRYGIER, 1985. Bathypelagic mysids from the northeastern Pacific. — Journal of Crustacean Biology, **5** (4): 686-706.

MURTAUGH, P. A., 1989. Fecundity of *Neomysis mercedis* Holmes in Lake Washington (Mysidacea). — Crustaceana, **57** (2): 194-200.

MYERS, A. A. & S. DE GRAVE, 2000. Endemism: origins and implications. — Vie et Milieu, **50**: 195-203.

NAIR, K. B., 1939. The reproduction, oogenesis and development of *Mesopodopsis orientalis* Tatt. — Proceedings of the Indian Academy of Sciences, **9** (B): 175-223.

NATH, C. N., 1972. On the storage of calcium in *Spelaeomysis*, a subterranean mysid. — Crustaceana, (Suppl.) **3**: 351-353.

— —, 1973. Breeding and fecundity in a subterranean mysid, *Lepidomysis longipes* (Pillai and Mariamma). — International Journal of Speleology, **5** (3-4): 319-323.

— —, 1974. Studies on the abdominal musculature of the subterranean mysid *Lepidomysis longipes* (Pillai and Mariamma). — International Journal of Speleology, **6** (2): 173-180.

NATH, C. N. & N. K. PILLAI, 1972a. On the alimentary canal of *Spelaeomysis longipes* (Crustacea: Mysidacea). — Journal of the Zoological Society of India, **23** (2): 95-108. [1971]

— — & — —, 1972b. On the food and feeding habits of *Lepidomysis longipes* (Pillai & Mariamma) (Crustacea Mysidacea). — International Journal of Speleology, **4**: 45-50.

— — & — —, 1976. The alimentary system of the littoral mysid *Gastrosaccus simulans* (Van Beneden). — Journal of the Marine Biological Association of India, **15** (2): 577-586. [1973]

NATH, C. N., D. M. THAMPY & N. K. PILLAI, 1972a. Optic regression in a subterranean mysid (Crustacea, Mysidacea). — International Journal of Speleology, **4**: 51-54.

— —, — — & — —, 1972b. A note on the androgenic gland of a subterranean mysidacean, *Spelaeomysis longipes* (Pillai and Mariamma). — Crustaceana, (Suppl.) **3**: 354-356.

NEEDHAM, A. B., 1937. Some points in the development of *Neomysis vulgaris*. — Quarterly Journal of Microscopical Science, **79**: 559-588.

NEHRING, S., 2005. International shipping — a risk for aquatic biodiversity in Germany. — In: W. NENTWIG, S. BACHER, M. J. W. COCK, H. DIETZ, A. GIGON & R. WITTENBERG (eds.), Biological invasions — from ecology to control. — Neobiota, **6**: 125-143.

NEIL, D. M., 1975a. Statocyst control of eyestalk movements in mysid shrimps. — Fortschritte der Zoologie, **23** (1): 98-109.

— —, 1975b. The mechanism of statocyst operation in the mysid shrimp *Praunus flexuosus*. — The Journal of Experimental Biology, **62**: 685-700.

NEVIN, P. A. & S. R. MALECHA, 1991. The occurrence of a heteromorph antennule in a cultured freshwater prawn, *Macrobrachium rosenbergii* (De Man) (Decapoda, Caridea). — Crustaceana, **60** (1): 105-107.

NIGGL, W., M. S. NAUMANN, U. STRUCK, R. MANASRAH & C. WILD, 2010. Organic matter release by the benthic upside-down jellyfish *Cassiopea* sp. fuels pelagic food webs in coral reefs. — Journal of Experimental Marine Biology and Ecology, **384**: 99-106.

NIGGL, W. & C. WILD, 2009. Spatial distribution of the upside-down jellyfish *Cassiopea* sp. within fringing coral reef environments of the northern Red Sea: implications for its life cycle. — Helgoland Marine Research, online: 7 pp. (Springer Verlag and Alfred Wegener Institut (AWI), Bremerhaven). DOI: 10.1007/s10152-009-0181-8 [23 June 2010]

NIIYAMA, T., Y. HANAMURA, K. TANAKA & H. TOYOHARA, 2012. Occurrence of cellulase activities in mangrove estuarine mysids and *Acetes* shrimps. — In: K. TANAKA, S. MORIOKA & S. WATANABE (eds.), Sustainable stock management and development of aquaculture technology suitable for Southeast Asia. JIRCAS Working Report, **75**: 35-39. (Japan International Research Center for Agricultural Sciences, Tsukuba, Ibaraki, Japan.)

NIKOFOROS, G., 2002. Fauna del Mediterraneo. — Pp. 1-366. (Giunti, Firenze.)

NILSSON, D.-E. & R. F. MODLIN, 1994. A mysid shrimp carrying a pair of binoculars. — The Journal of Experimental Biology, **189**: 213-236.

NIMMO, D. R., R. A. RIGBY, L. H. BAHNER & J. M. SHEPPARD, 1978. The acute and chronic effects of cadmium on the estuarine mysid *Mysidopsis bahia*. — Bulletin of Environmental Contamination and Toxicology, **19**: 80-85.

NOËL, P. Y. & C. CHASSARD-BOUCHAUD, 2004. Chromatophores and pigmentation. — In: J. FOREST & J. C. VON VAUPEL KLEIN (eds.), The Crustacea. Revised and updated from the Traité de Zoologie, **1** (4): 145-160. (Brill, Leiden.)

NONOMURA, T., Y. HAYAKAWA, Y. SUDA & J. OHTOMI, 2007. Habitat zonation of the sand-burrowing mysids (*Archaeomysis vulgaris*, *Archaeomysis japonica* and *Iiella ohshimai*), and diel and tidal distribution of dominant *Archaeomysis vulgaris*, in an intermediate sandy beach at Fukiagehama, Kagoshima Prefecture, southern Japan. — Plankton & Benthos Research, **2** (1): 38-48.

NORMAN, A. M., 1892. On British Mysidae, a family of Crustacea Schizopoda. — Annals and Magazine of Natural History, (6) **10**: 143-166, 242-263, pls. IX, X.

NOUVEL, H., 1937. Observation de l'accouplement chez une espèce de Mysis: *Praunus flexuosus*. — Comptes Rendus Hebdomadaires des Séances de l'Académie des Sciences, **205**: 1184-1186.

— —, 1940. Observations sur la sexualité d'un Mysidacé, *Heteromysis armoricana* n.sp. — Bulletin de l'Institut Océanographique, Monaco, **789**: 1-11.

— —, 1941. Sur les Ellobiopsidés des Mysidacés provenant des campagnes du Prince de Monaco. — Bulletin de l'Institut Océanographique, Monaco, **809**: 1-8.

— —, 1942a. Sur la sexualité des Mysidacés du genre *Eucopia* (Caractères sexuels secondaires, taille et maturité sexuelle, anomalies et action possible d'un Epicaride). — Bulletin de l'Institut Océanographique, Monaco, **820**: 1-10.

— —, 1942b. Diagnoses préliminaires de Mysidacés nouveaux provenant des campagnes du Prince Albert 1er de Monaco. — Bulletin de l'Institut Océanographique, Monaco, **831**: 1-12.

— —, 1943. Mysidacés provenant des campagnes du Prince Albert 1er de Monaco. — In: J. RICHARD (ed.), Résultats des Campagnes Scientifiques Accomplies sur son Yacht par Albert Ier, **105**: 1-128, pls. I-V.

— —, 1945. Les relations entre la périodicité lunaire, les marées et la mue des crustacés. — Bulletin de l'Institut Océanographique, Monaco, **878**: 1-4.

— —, 1951. *Gastrosaccus normani* G. O. Sars 1877 et *Gastrosaccus lobatus* n.sp. (Crust. Mysid.) avec précision de l'hôte de *Prodajus lobiancoi* (Crust. Isop. Epicar.). — Bulletin de l'Institut Océanographique, Monaco, **993**: 1-12.

— —, 1954. Un Ellobiopsidae nouveau (*Amallocystis boschmai* n.sp.) parasite d'un mysidacé en Méditerranée (Note préliminaire). — Vie et Milieu, **4** (1): 57-58.

— —, 1957. Mysidacés provenant de deux échantillons de « Djembret » de Java. — Zoologische Mededelingen, Leiden, **25** (22): 315-331.

— —, 1958. L'exuviation chez les Mysidae (Crustacés, Mysidacés). — Bulletin de la Société Zoologique de France, **82** (5-6): 395-400.

NOUVEL, H., J.-P. CASANOVA & J.-P. LAGARDÈRE, 1999. Ordre des Mysidacés (Mysidacea Boas, 1883). — In: J. FOREST (ed.), Traité de Zoologie, **VII**, Crustacés (IIIA, Crustacés Péracarides). Mémoires de l'Institut Océanographique, Monaco, **19**: 39-86.

NOUVEL, H. & J. HOENIGMAN, 1955. *Amallocystis boschmai* Nouvel, 1954, Ellobiopsidé parasite du Mysidacé *Leptomysis gracilis* G.O. Sars. — In: Résultats de Campagnes du Pr. Lacaze-Duthiers. Vie et Milieu, (Suppl.) **2**: 7-19.

NOUVEL, H. & J.-P. LAGARDÈRE, 1976. Les Mysidacés du talus continental du golfe de Gascogne. I. Tribu des Erythropini (genre *Erythrops* excepté). — Bulletin du Muséum National d'Histoire Naturelle, (3) (Zoologie, 414) **291**: 1243-1324.

NOUVEL, H. & L. NOUVEL, 1939. Observations sur la biologie d'une Mysis: *Praunus flexuosus* (Müller, 1788). — Bulletin du Musée Océanographique, Monaco, **761**: 1-10.

NUNN, A. D. & I. G. COWX, 2012. Diel and seasonal variations in the population dynamics of *Hemimysis anomala*, a non-indigenous mysid: implications for surveillance and management. — Aquatic Invasions, **7** (3): 357-365.

NUSBAUM, J., 1887. L'embryologie de *Mysis chameleo* (Thompson). — Archives de Zoologie Expérimentale et Générale, (2) **5**: 123-202, pls. V-XII.

O'BRIEN, D. P., 1988. Direct observations of clustering (schooling and swarming) behaviour in mysids (Crustacea: Malacostraca). — Marine Ecology Progress Series, **42** (3): 235-246.

— —, 1989. Analysis of the internal arrangement of individuals within crustacean aggregations (Euphausiacea, Mysidacea). — Journal of Experimental Marine Biology and Ecology, **128** (1): 1-30.

ODUM, E. P., G. W. BARRETT & R. BREWER, 2004. Fundamentals of ecology (5th ed.). — Pp. 1-624. (Thomson/Brooks Cole, Australia.)

OGONOWSKI, M., J. DUBERG, S. HANSSON & E. GOROKHOVA, 2013 (cf. a). Behavioral, ecological and genetic differentiation in an open environment — a study of a mysid population in the Baltic Sea. — PLoS ONE, **8** (3): e57210: 1-11. www.plosone.org [23 Oct. 2013]

OGONOWSKI, M., S. HANSSON & J. DUBERG, 2013 (cf. b). Status and vertical size-distributions of a pelagic mysid community in the northern Baltic proper. — Boreal Environment Research, **18**: 1-18.

OHTSUKA, S., Y. HANAMURA, S. HARADA & M. SHIMOMURA, 2006. Recent advances in studies of parasites on mysid crustaceans. — Bulletin of Plankton Society of Japan, **53** (1): 37-44.

OHTSUKA, S., T. HORIGUCHI, Y. HANAMURA, K. NAGASAWA & T. SUZAKI, 2003. Intersex in the mysid *Siriella japonica izuensis* Ii: the possibility it is caused by infestation with parasites. — Plankton Biology & Ecology, **50** (2): 65-70.

OHTSUKA, S., H. INAGAKI, T. ONBE, K. GUSHIMA & Y. H. YOON, 1995. Direct observations of groups of mysids in shallow coastal waters of western Japan and southern Korea. — Marine Ecology Progress Series, **123**: 33-44.

OJAVEER, H., E. LEPPÄKOSKI, S. OLENIN & A. RICCIARDI, 2002. Ecological impacts of alien species in the Baltic Sea and in the Great Lakes: an inter-ecosystem comparison. — In: E. LEPPÄKOSKI, S. OLENIN & S. GOLLASCH (eds.), Invasive aquatic species of Europe — distributions, impacts, and management, pp. 412-425. (Kluwer Scientific Publishers, Dordrecht.)

OKUMURA, T., 2003. Relationship of ovarian and marsupial development to the female molt cycle in *Acanthomysis robusta* (Crustacea: Mysida). — Fisheries Science, **69** (5): 995-1000.

OLENIN, S. & E. LEPPÄKOSKI, 1999. Non-native animals in the Baltic Sea: alteration of benthic habitats in coastal inlets and lagoons. — In: E. M. BLOMQVIST, E. BONSDORFF & K. ESSINK (eds.), Geochemical features of enclosed and semi-enclosed marine systems. — Hydrobiologia, **393**: 233-234.

ORTEGA-SALAS, A. A., A. NÚÑEZ-PASTÉN & H. A. CAMACHO M., 2008. Fecundity of the crustacean *Mysidopsis californica* (Mysida, Mysidae) under semi-controlled conditions. — International Journal of Tropical Biology, **56** (2): 535-539.

ORTIZ, M., A. GARCÍA-DEBRÁS, A. PÉREZ & R. LALANA, 2005. Estudio del marsupio y las larvas de *Spelaeomysis nuniezi* Bacescu y Orghidan, 1971 (Crustacea, Mysidacea). — Revista Biología, **19** (1/2): 83-84.

ORTIZ, M. & O. GOMEZ, 1988. Nueva localidad para un misidáceo (Crustacea), asociado a la anémona *Bartholomea annulata* (Coelenterata). — Revista de Investigaciones Marinas, **9** (2): 111-113.

ORTIZ, M., R. LALANA & A. PEREZ, 1996. El primer registro del genero *Stygiomysis* (Crustacea, Mysidacea) en la Isla de Cuba y description de una especie nueva. — Revista de Investigaciones Marinas, **17** (2-3): 107-115.

OSHEL, P. E. & D. H. STEELE, 1988. SEM morphology of the foreguts of gammaridean amphipods compared to *Anaspides tasmaniae* (Anaspidacea: Anaspidae), *Gnathophausia ingens* (Mysidacea: Lophogastridae), and *Idotea balthica* (Isopoda: Idoteidae). — Crustaceana, (Suppl.) **13**: 209-219.

OVČARENKO, I., A. AUDZIJONYTE & Z. A. GASIUNAITE, 2006. Tolerance of *Paramysis lacustris* and *Limnomysis benedeni* (Crustacea, Mysida) to sudden salinity changes: implications for ballast water treatment. — Oceanologia, **48** (S): 231-242.

PANAMPUNNAYIL, S. U. & A. BIJU, 2006. Four new species of the genus *Rhopalophthalmus* (Mysidacea: Crustacea) from the northwest coast of India. — Journal of Natural History, London, **4** (23-24): 1389-1406.

PANAMPUNNAYIL, S. U. & M. VISWAKUMAR, 1991. *Spelaeomysis cochinensis*, a new mysid (Crustacea: Mysidacea) from a prawn culture field in Cochin, India. — Hydrobiologia, **209** (1): 71-78.

PARENTI, L. R. & M. C. EBACH, 2009. Comparative biogeography: discovering and classifying biogeographical patterns of a dynamic earth. — Pp. i-xiii, 1-295. (University of California Press, Berkeley, CA.)

PARKER, G. H., 1891. The compound eyes in crustaceans. — Bulletin of the Museum of Comparative Zoology, **21** (2): 45-140, pls. I-X.

PASSELAIGUE, F., 1989. Les migrations journalières du mysidacé marin cavernicole *Hemimysis speluncola*. Comparaison avec les migrations verticales du plancton. — Pp. 1-209. (Thèse de Doctorat d'État, Université d'Aix-Marseille II.)

PASSELAIGUE, F. & A. BOURDILLON, 1986. Distribution and circadian migrations of the cavernicolous mysid *Hemimysis speluncola* Ledoyer. — Stygologia, **2** (1-2): 112-118.

PATZNER, R. A., 2004. Associations with sea anemones in the Mediterranean Sea: a review. — Ophelia, **58** (1): 1-11.

PATZNER, R. A. & H. DEBELIUS, 1984. Partnerschaft im Meer. — Pp. 1-120. (Engelbert Pfriem, Wuppertal.)

PAUL, S., M. KRKOSEK, P. K. PROBERT & G. P. CLOSS, 2013. Osmoregulation and survival of two mysid species of *Tenagomysis* in southern estuaries of New Zealand. — Marine and Freshwater Research, **64** (4): 340-347.

PESCE, G. L., 1976a. A new locality for *Spelaeomysis bottazzii* with redescription of the species (Crustacea, Mysidacea). — Bollettino del Museo Civico di Storia Naturale, Verona, **2**: 345-354.

— —, 1976b. Stato attuale delle conoscenze sui Misidacei cavernicoli e freatici (Crustacea). — Notiziario del Circolo Speleologico Romano, **1**: 47-57.

— —, 1985. The groundwater fauna of Italy: a synthesis. — Stygologia, **1** (2): 129-159.

PESCE, G. L., G. FUSACCHIA, D. MAGGI & P. TETÈ, 1978. Ricerche faunistiche in acque freatiche del Salento. — Thalassia Salentina, **8**: 1-51.

PESCE, G. L. & T. M. ILIFFE, 2002. New records of cave-dwelling mysids from the Bahamas and Mexico with description of *Palaumysis bahamensis* n. sp. (Crustacea: Mysidacea). — Journal of Natural History, London, **36**: 265-278.

PETRYASHOV, V. V., 1990. Reproduction and fecundity of mysids (Crustacea, Mysidacea) of Arctic Ocean and north-west Pacific. — USSR Academy of Sciences. Proceedings of the Zoological Institute, Leningrad [= St. Petersburg], **218**: 140-160. [In Russian with English summary.]

— —, 2004. Mysids (Crustacea, Mysidacea) of the Euroasiatic subbasin of the Arctic Basin and the adjacent seas: the Barents, Kara and Laptev seas. — In: Fauna i ecosistemi morja Laptevih i Sopredelnih glubokovodnih utschastkov Arktitscheskogo Bassejna. Tschast I. Explorations of the Fauna of the Seas, **54** (62): 124-145.

— —, 2005. Biogeographical division of the North Pacific sublittoral and upper bathyal zones by the fauna of Mysidacea and Anomura (Crustacea). — Russian Journal of Marine Biology, **31** (4): 233-250.

— —, 2007. Biogeographical division of Antarctic and Subantarctic by mysid (Crustacea: Mysidacea) fauna. — Russian Journal of Marine Biology, **33** (1): 1-16.

— —, 2009. The biogeographical division of the Arctic and North Atlantic by the mysid (Crustacea: Mysidacea) fauna. — Russian Journal of Marine Biology, **35** (2): 97-116.

PETRYASHOV, V. V. & M. DANELIYA, 2006. Check-list for Caspian Sea mysids (opossum shrimps). — In: Caspian Sea Biodiversity Project: 5 pp. http://www.zin.ru/projects/caspdiv/caspian_mysidacea.html [6 Apr. 2006]

PEZZACK, D. S. & S. COREY, 1979. The life history and distribution of *Neomysis americana* (Smith): Crustacea, Mysidacea. — Canadian Journal of Zoology, **57** (4): 785-793.

PIENIMÄKI, M. & E. LEPPÄKOSKI, 2004. Invasion pressure on the Finnish Lake District: invasion corridors and barriers. — Biological Invasions, **6** (3): 331-346.

PILLAI, N. K., 1965. A review of the work of shallow-water Mysidacea of the Indian waters. — In: Proceedings of the Symposium on Crustacea, held at Ernakulam from January 12 to 15, 1965. Part V: 1681-1728. (Marine Biological Association of India, Mandapam Camp.)
— —, 1968. *Heteromysis zeylanica* Tattersall (Crustacea: Mysidacea), an associate of madreporarian corals in south Indian waters. — Journal of the Bombay Natural History Society, **65** (1): 45-57.
— —, 1973. Mysidacea of the Indian Ocean. — I.O.B.C. Handbook, **4**: 1-125.
PILLAI, N. K. & T. MARIAMMA, 1963. On the discovery of the primitive mysidacean family Lepidomysidae in India. — Current Science, **32** (5): 219-220.
— — & — —, 1964. On a new lepidomysid from India. — Crustaceana, **7** (2): 113-124.
PÖCKL, M., M. GRABOWSKI, J. GRABOWSKA, K. BACELA-SPYCHALSKA & K. J. WITTMANN, 2011. Large European rivers as biological invasion highways. — In: H. HABERSACK, B. SCHOBER & D. WALLING (eds.), International Conference on the Status and Future of the World's Large Rivers, 11-14 April 2011, Vienna; Conference Abstract Book, p. 215 [abstract booklet + CD]. (Tribun EU, Vienna.)
POORE, G. C. B., 2005. Peracarida: monophyly, relationships and evolutionary success. — Nauplius, **13** (1): 1-27.
POPOVA, T. I. & E. N. NIKITINA, 1972. The finding of *Bunocotyle cingulata* Odhner, 1928, in *Mysis microphthalma* in the Caspian Sea. — Vestnik Moskovskogo Universiteta Biologiya, Pochvevodstvo, **27** (3): 102-104. [In Russian.]
POR, F. D., 1972. Hydrobiological notes on the high-salinity waters of the Sinai Peninsula. — Marine Biology, Berlin, **14**: 111-119.
— —, 1978. Lessepsian migration. The influx of Red Sea biota into the Mediterranean by way of the Suez Canal. — Pp. 1-228. (Springer-Verlag, Berlin.)
POR, F. D. & I. FERBER, 1972. The Hebrew University – Smithsonian Institution collections from the Suez Canal (1967 - 1972). — In: Contributions to the knowledge of Suez Canal migration. Israel Journal of Zoology, **21** (3-4): 149-166.
PORTER, M. L., 2005. Crustacean phylogenetic systematics and opsin evolution. — Pp. i-xi, 1-189. (Ph. D. Dissertation, Brigham Young University.) http://patriot.lib.byu.edu/ETD/image/etd859.pdf [10 Nov. 2008]
PORTER, M. L., K. MELAND & W. PRICE, 2008. Global diversity of mysids (Crustacea-Mysida) in freshwater. — Hydrobiologia, **595**: 213-218.
PRICE, W. W. & R. W. HEARD, 2009. Mysida (Crustacea) of the Gulf of Mexico. — In: D. L. FELDER & D. K. CAMP (eds.), Gulf of Mexico — origins, waters, and biota. Biodiversity, **1**: 929-938. (Texas A&M Press, College Station, TX.)
— — & — —, 2011. Two new species of *Heteromysis* (*Olivemysis*) (Mysida, Mysidae, Heteromysinae) from the tropical northwest Atlantic with diagnostics on the subgenus *Olivemysis* Băcescu, 1968. — Zootaxa, **2823**: 32-46.
PRICE, W. W., R. W. HEARD, P. AAS & K. MELAND, 2009. Lophogastrida (Crustacea) of the Gulf of Mexico. — In: D. L. FELDER & D. K. CAMP (eds.), Gulf of Mexico — origins, waters, and biota. Biodiversity, **1**: 923-927. (Texas A&M Press, College Station, TX.)
PRYCHITKO, S. B. & R. W. NERO, 1983. Occurrence of the acanthocephalan *Echinorhynchus leidyi* (van Cleave, 1924) in *Mysis relicta*. — Canadian Journal of Zoology, **61** (2): 460-462.
QUINTERO, R. C. & E. ZOPPI DE ROA, 1973. Notas bioecologicas sobre *Metamysidopsis insularis* Brattegard (Crustacea – Mysidacea) en una laguna littoral de Venezuela. — Acta Biologica Venezuelica, **8** (2): 245-278.
RAMARN, T., V.-C. CHONG & Y. HANAMURA, 2012. Population structure and reproduction of the mysid shrimp *Acanthomysis thailandica* (Crustacea: Mysidae) in a tropical mangrove estuary, Malaysia. — Zoological Studies, **51** (6): 768-782.
RANDALL, J. E., R. E. SCHROEDER & W. A. STARCK, 1964. Notes on the biology of the echinoid *Diadema antillarum*. — Caribbean Journal of Science, **4** (2-3): 421-433.

RASMUSSEN, J. B., D. J. ROWAN, D. R. S. LEAN & J. H. CAREY, 1990. Food chain structure in Ontario lakes determines PCB levels in lake trout (*Salvelinus namaycush*) and other pelagic fish. — Canadian Journal of Fisheries and Aquatic Sciences, **47**: 2030-2038.

RASTORGUEFF, P.-A., M. HARMELIN-VIVIEN, P. RICHARD & P. CHEVALDONNÉ, 2011. Feeding strategies and resource partitioning mitigate the effects of oligotrophy for marine cave mysids. — Marine Ecology Progress Series, **440**: 163-176.

REINHOLD, M. & T. TITTIZER, 1998. *Limnomysis benedeni* Czerniavsky, 1882 (Crustacea: Mysidacea), ein weiteres pontokaspisches Neozoon im Main-Donau-Kanal. — Lauterbornia, **33**: 37-40.

REMERIE, T., T. BOURGOIS, D. PEELAERS, A. VIERSTRAETE, J. VANFLETEREN & A. VANREUSEL, 2006. Phylogeographic patterns of the mysid *Mesopodopsis slabberi* (Crustacea, Mysida) in western Europe: evidence for high molecular diversity and cryptic speciation. — Marine Biology, Berlin, **149**: 465-481.

REMERIE, T., B. BULCKAEN, J. CALDERON, T. DEPREZ, J. MEES, J. VANFLETEREN, A. VANREUSEL, A. VIERSTRAETE, M. VINCX, K. J. WITTMANN & T. WOOLDRIDGE, 2004. Phylogenetic relationships within the Mysidae (Crustacea, Peracarida, Mysida) based on nuclear 18S ribosomal RNA sequence. — Molecular Phylogenetics and Evolution, **32**: 770-777.

RETZIUS, G., 1909. Die Spermien der Crustaceen. — Biologische Untersuchungen, (Neue Folge) **14** (1): 4-54. (Gustav Fischer, Jena.)

REYNOLDS, J. B. & G. M. DE GRAEVE, 1972. Seasonal population characteristics of the opossum shrimp, *Mysis relicta*, in southeastern Lake Michigan, 1970-71. — In: Proceedings of the 15th Conference on Great Lakes Research, pp. 117-131. (International Association of Great Lakes Research.)

RICCIARDI, A., 2007. Forecasting the impacts of *Hemimysis anomala*: the newest invader discovered in the Great Lakes. — Aquatic Invaders, **18** (1): 1, 4-7.

RICCIARDI, A., S. AVLIJAS & J. MARTY, 2012. Forecasting the ecological impacts of the *Hemimysis anomala* invasion in North America: lessons from other freshwater mysid introductions. — In: Special issue on mysids of the Great Lakes. Journal of Great Lakes Research, **38** (Suppl. 2): 7-13.

RICCIARDI, A. & J. B. RASMUSSEN, 1998. Predicting the identity and impact of future biological invaders: a priority for aquatic resource management. — Canadian Journal of Fisheries and Aquatic Sciences, **55**: 1759-1765.

RICE, A. L., 1961. The responses of certain mysids to changes in hydrostatic pressure. — The Journal of Experimental Biology, **38** (2): 391-401.

RICHARDS, S. W. & G. A. RILEY, 1967. The benthic epifauna of Long Island Sound. — Bulletin of the Bingham Oceanographic Collection, **19** (2): 89-135.

RICHTER, S., 2003. The mouthparts of two lophogastrids, *Chalaraspidum alatum* and *Pseudochalaraspidum hanseni* (Lophogastrida, Peracarida, Malacostraca), including some remarks on the monophyly of the Lophogastrida. — Journal of Natural History, London, **37**: 2773-2786.

RICHTER, S. & G. SCHOLTZ, 2001. Phylogenetic analysis of the Malacostraca (Crustacea). — Journal of Zoological Systematics and Evolutionary Research, **39**: 113-136.

RIERA, T., M. ZABALA & J. PEÑUELAS, 1991. Mysids from a submarine cave emerge each night to feed. — Scientia Marina, **55** (4): 605-609.

RITZ, D. A., 2000. Is social aggregation in aquatic crustaceans a strategy to conserve energy? — Canadian Journal of Fisheries and Aquatic Sciences, **57** (Suppl.): 59-67.

RITZ, D. A., E. G. FOSTER & K. M. SWADLING, 2001. Benefits of swarming: mysids in larger swarms save energy. — Journal of the Marine Biological Association of the United Kingdom, **81**: 543-544.

RITZ, D. A., A. J. HOBDAY, J. C. MONTGOMERY & A. J. W. WARD, 2011. Social aggregation in the pelagic zone with special reference to fish and invertebrates. — Advances in Marine Biology, **60**: 163-230.
RITZ, D. A. & E. B. METILLO, 1998. Costs and benefits of swarming behaviour in mysids: does orientation and position in the swarm matter? — Journal of the Marine Biological Association of the United Kingdom, **78**: 1011-1014.
ROAST, S. D., J. WIDDOWS & M. B. JONES, 2000. Egestion rates of the estuarine mysid *Neomysis integer* (Peracarida: Mysidacea) in relation to a variable environment. — Journal of Experimental Marine Biology and Ecology, **245**: 69-81.
RUFFO, S., 1957. Le attuali conoscenze sulla fauna cavernicola della regione Pugliese. — Memorie di Biogeografia Adriatica, **3**: 1-143.
RUIZ, G. M., P. W. FOFONOFF, J. T. CARLTON, M. J. WONHAM & A. H. HINES, 2000. Invasion of coastal marine communities in North America: apparent patterns, processes, and biases. — Annual Review of Ecology and Systematics, **31**: 481-531.
RUSSEL, F. S., 1925. The vertical distribution of marine macroplankton and observation of diurnal changes. — Journal of the Marine Biological Association of the United Kingdom, **13**: 769-809.
— —, 1931. The vertical distribution of marine macroplankton. XI. Further observations on diurnal changes. — Journal of the Marine Biological Association of the United Kingdom, **17** (3): 767-784.
SALA, O. E., F. S. CHAPIN, J. J. ARMESTO, E. BERLOW, J. BLOOMFIELD, R. DIRZO, E. HUBER-SANWALD, L. F. HUENNEKE, R. B. JACKSON, A. KINZIG ET AL., 2000. Global biodiversity scenarios for the year 2100. — Science, New York, **287** (5459): 1770-1774.
SAN VICENTE, C., 2007. A new species of *Marumomysis* (Mysidacea: Mysidae: Erythropini) from the benthos of the Bellingshausen Sea (Southern Ocean). — Scientia Marina, **71** (4): 683-690.
— —, 2010a. Mysidaceans. — In: M. COLL, C. PIRODDI, J. STEENBEEK, K. KASCHNER, F. BEN RAIS LASRAM, J. AGUZZI, E. BALLESTEROS, C. N. BIANCHI, J. CORBERA, T. DAILIANIS ET AL. (eds.), Biodiversity of the Mediterranean Sea: estimates, patterns & threats. — PLoS ONE, **5** (8): e11842: 254-275. www.plosone.org [30 Mar. 2011]
— —, 2010b. Chapter 1. Species diversity of Antarctic mysids (Crustacea: Lophogastrida and Mysida). — In: T. J. MULDER (ed.), Antarctica: global, environmental and economic issues, pp. 1-80. (Nova Science Publishers, Inc., New York, NY.)
SAN VICENTE, C. & J. C. SORBE, 1990. Biología del Misidáceo suprabentónico *Schistomysis kervillei* (Sars, 1885) en la plataforma continental Aquitana (suroeste de Francia). — Bentos, **6**: 245-267.
— — & — —, 1993. Biologie du Mysidacé suprabenthique *Schistomysis parkeri* Norman, 1892 dans la zone sud du golfe de Gascogne (Plage D'Hendaye). — Crustaceana, **65** (2): 222-252.
— — & — —, 1995. Biology of the suprabenthic mysid *Schistomysis spiritus* (Norman, 1860) in the southeastern part of the Bay of Biscay. — Scientia Marina, **59** (Suppl. 1): 71-86.
— — & — —, 2003. Biology of the suprabenthic mysid *Schistomysis assimilis* (Sars, 1877) on Creixell beach, Tarragona (northwestern Mediterranean). — Boletín Instituto Español de Oceanografía, **19** (1-4): 391-406.
SARS, G. O., 1867. Histoire naturelle des Crustacés d'eau douce de Norvège. I. Les Malacostracés — Pp. i-iiii, 1-146, pls. I-X. (Christiania.)
— —, 1870. Carcinologiske Bidrag til Norges Fauna. I. Monographi over de ved Norges Kyster forekommende Mysider. — Pt. **1**: 1-64, pls. I-V. (K. Norske Viidenskab., Trondhjem, Christiania.)
— —, 1877. Nye Bidrag til Kundskaben om Middelhavets Invertebratfauna. I. Middelhavets Mysider. — Archiv før Mathematik og Naturvidenskaberne, **2**: 10-119, pls. 1-36.
— —, 1879. Carcinologiske Bidrag til Norges Fauna I. Monographi over de ved Norges Kyster forekommende de Mysider. — Pt. **3**: i-v, 1-131, pls. IX-XLII. (A.W. Brøgger, Christiania.)

— —, 1885a. Report on the Schizopoda collected by HMS Challenger during the years 1873-76. — In: G. S. NARTES (ed.), Report on the Scientific Results of the Voyage of H.M.S. Challenger during the years 1873-76, **13** (37): 1-228, pls. I-XXXVIII. (Longmans & Co., London.)

— —, 1885b. Crustacea I. — In: Den Norske Nordhavs-Expedition, 1876-1878, XIV, Zoologi, **6** (1): 1-280, pls. 1-21. (Grøndahl & Son, Christiania.)

— —, 1893. Crustacea Caspia. Contribution to the knowledge of the carcinological fauna of the Caspian Sea. Part. I — Mysidae. — Bulletin de l'Académie Impériale des Sciences de St.-Pétersbourg, (N. S. IV) **36** (1): 51-74, pls. I-VIII.

— —, 1895. Crustacea Caspia. Account of the Mysidae in the collection of Dr. O. Grimm. — Bulletin de l'Académie Impériale des Sciences de St.-Pétersbourg, (5) **3** (5): 433-458, pls. I-VIII.

— —, 1907. Mysidae. — In: Report of the Caspian Expedition 1904 [English part of bilingual edition], **1**: 278-313 [in one volume] or **1**: 36-71 [in separate bindings], pls. I-XII. (St.-Peterburg.)

SARS, M., 1857. Om 3 nye norske Krebsdyr. — In: Forhandlinger ved de Skandinaviske naturforskeres syvende møde i Christiania den 12-18 Juli 1856, pp. 160-175. (C. C. Werner & Co., Christiania.)

SATO, H. & M. MURANO, 1994. Adoption of larvae escaped from the marsupium in four mysid species. — Umi. La mer (Bulletin de la Société Franco-Japonaise d'Océanographie), **32**: 71-74.

SCHABES, M. & W. HAMNER, 1992. Mysid locomotion and feeding: kinematics and waterflow patterns of *Antarctomysis* sp., *Acanthomysis sculpta*, and *Neomysis rayii*. — Journal of Crustacean Biology, **12** (1): 1-10.

SCHLACHER, T. A., K. J. WITTMANN & A. P. ARIANI, 1992. Comparative morphology and actuopalaeontology of mysid statoliths (Crustacea, Mysidacea). — Zoomorphology, **112**: 67-79.

SCHLACHER, T. A. & T. H. WOOLDRIDGE, 1994. Tidal influence on distribution and behaviour of the estuarine opossum shrimp *Gastrosaccus brevifissura*. — In: K. D. DYER & R. J. ORTH (eds.), Changes in fluxes in estuaries: implications form science to management. International Symposium Series, pp. 307-312. (Olsen & Olsen, Fredesborg, Denmark.)

SCHLEUTER, A., H.-P. GEISSEN & K. J. WITTMANN, 1998. *Hemimysis anomala* G. O. SARS 1907 (Crustacea: Mysidacea), eine euryhaline pontokaspische Schwebgarnele in Rhein und Neckar. Erstnachweis für Deutschland. — Lauterbornia, **32**: 67-71.

SCHOLTZ, G., 1984. Untersuchungen zur Bildung und Differenzierung des postnauplialen Keimstreifs von *Neomysis integer* Leach (Crustacea, Malacostraca, Peracarida). — Zoologische Jahrbücher, (Abteilung für Anatomie) **112** (3): 295-349.

SCHOLTZ, G. & W. DOHLE, 1996. Cell lineage and cell fate in crustacean embryos — a comparative approach. — International Journal of Developmental Biology, **40**: 211-220.

SCHÖNE, H., 1954. Statozystenfunktion und statische Lageorientierung bei dekapoden Krebsen. — Zeitschrift für Vergleichende Physiologie, **36**: 241-260.

SCHRAM, F. R., 1984. Relationships within eumalacostracan Crustacea. — Transactions of the San Diego Society of Natural History, **20** (16): 301-312.

— —, 1986. Crustacea. — Pp. i-xiv, 1-606. (Oxford University Press, New York & Oxford.)

— —, 2013. Comments on crustacean biodiversity and disparity of body plans. — In: L. WATLING & M. THIEL (eds.), The natural history of Crustacea, **1**, Functional morphology & diversity, pp. 1-33. (Oxford University Press, U.S.A.)

SCHUSTER, W. H., 1952. Fish-culture in brackish-water ponds of Java. — Indo-Pacific Fisheries Council, Special Publications, **1**: 1-143.

SECRETAN, S., 1985. Conchyliocarida, a class of fossil crustaceans: relationships to Malacostraca and postulated behavior. — Transactions of the Royal Society of Edinburgh, **76**: 381-389.

SECRETAN, S. & B. RIOU, 1986. Les Mysidacés (Crustacea, Peracarida) du Callovien de la Voulte-sur-Rhône. — Annales de Paléontologie, (Vertébrés-Invertébrés) **72** (4): 295-323.

SERGEEVA, Z. N. & O. L. CKHONELIDZE, 1968. Reconstitution de la faune des petits bassins de Géorgie. — Gidrobiologiceskij Zhurnal, **4** (4): 41-46.
SIEGEL, V. & U. MÜHLENHARDT-SIEGEL, 1988. On the occurrence and biology of some Antarctic Mysidacea (Crustacea). — Polar Biology, **8**: 181-190.
SIEGFRIED, C. A., 1982. Trophic relations of *Crangon franciscorum* Stimpson and *Palaemon macrodactylus* Rathbun: predation on the opossum shrimp, *Neomysis mercedis* Holmes. — Hydrobiologia, **89** (2): 129-139.
SIEGFRIED, C. A. & M. E. KOPACHE, 1980. Feeding of *Neomysis mercedis* (Holmes). — Biological Bulletin, Woods Hole, **159** (1): 193-205.
SIEWING, R., 1953. Morphologische Untersuchungen an Tanaidaceen und Lophogastriden. — Zeitschrift für Wissenschaftliche Zoologie, **157** (3/4): 333-426.
— —, 1956. Untersuchungen zur Morphologie der Malacostraca (Crustacea). — Zoologische Jahrbücher, (Abteilung für Anatomie und Ontogenie der Tiere) **75**: 39-176.
SKOLKA, M., 2005. Biodiversitatea Marii Negre. — Pp. 1-134. (University of Constanta, Romania.) http://www.univ-vidius.ro/faculties/Nat_Science/main/Cercetare%20biodiversitate.htm [20 Mar. 2010]
SLABBER, M., 1778. Natuurkundige verlustigingen, 15, Tweede waarneeming van een steur-garnaal met trompetswyze oogen. — Pp. 136-139, pl. 15 figs. 3-4. (J. Bosch, Haarlem.)
SLYNKO, Y. V., L. G. KORNEVA, I. K. RIVIER, V. P. PAPCHENKOV, G. H. SCHERBINA, M. I. ORLOVA & T. W. THERRIAULT, 2002. The Caspian-Volga-Baltic invasion corridor. — In: E. LEPPÄKOSKI, S. GOLLASCH & S. OLENIN (eds.), Invasive aquatic species of Europe — distribution, impacts and management, pp. 399-411. (Kluwer Academic Publishers, Dordrecht.)
SNODGRASS, R. E., 1952. Comparative studies of the jaws of mandibulate arthropods. — Smithsonian miscellaneous Collections, **116** (1): 1-85.
SORBE, J. C., 1981. Rôle du benthos dans le régime alimentaire des poissons démersaux du secteur Sud-Gascogne. — Kieler Meeresforschungen, (Sonderheft) **5**: 479-489.
SORBE, J. C. & M. ELIZALDE, 2013. Temporal changes in the structure of a slope suprabenthic community from the Bay of Biscay (NE Atlantic Ocean). — Deep-Sea Research Part II, (Topical Studies in Oceanography) **2013**: 13 pp. http://dx.doi.org/10.1016/j.dsr2.2013.09.041i [early online, 10 Nov. 2013]
SOUTHWARD HOGAN, L. S., E. MARSCHALL, C. FOLT & R. A. STEIN, 2007. How non-native species in Lake Erie influence trophic transfer of mercury and lead to top predators. — In: Special issue on mysids of the Great Lakes. Journal of Great Lakes Research, **33** (1): 46-61.
SPARRON, R. A. H., P. A. LARKIN & R. A. RUTHERGLEN, 1964. Successful introduction of *Mysis relicta* Loven into Kootenay Lake, British Columbia. — Journal of the Fisheries Research Board of Canada, **21** (5): 1325-1327.
SPEARS, T., R. W. DEBRY, L. G. ABELE & K. CHODYLA, 2005. Peracarid monophyly and interordinal phylogeny inferred from nuclear small-subunit ribosomal DNA sequences (Crustacea: Malacostraca: Peracarida). — Proceedings of the Biological Society of Washington, **118** (1): 117-157.
SPECZIÁR, A., 2005. First year ontogenetic diet patterns in two coexisting *Sander* species, *S. lucioperca* and *S. volgensis* in Lake Balaton. — Hydrobiologia, **549**: 115-130.
SPECZIÁR, A. & E. T. REZSU, 2009. Feeding guilds and food resource partitioning in a lake fish assemblage: an ontogenetic approach. — Journal of Fish Biology, **75**: 247-267.
STAMMER, H.-J., 1936. Ein neuer Höhlenschizopode, *Troglomysis vjetrenicensis* n.g. n.sp. Zugleich eine Übersicht der bisher aus dem Brack- und Süßwasser bekannten Schizopoden, ihrer geographischen Verbreitung und ihrer ökologischen Einteilung – sowie eine Zusammenstellung der blinden Schizopoden. — Zoologische Jahrbücher, (Abteilung für Systematik) **68**: 53-104.
STEELE, D. H. & V. J. STEELE, 1975. Egg size and duration of embryonic development in Crustacea. — Internationale Revue der Gesamten Hydrobiologie, **60** (5): 711-715.

STEVEN, D. M., 1961. Shoaling behaviour in a mysid. — Nature, London, **192**: 280-281.
STORCH, V., 1989. Scanning and transmission electron microscopic observations on the stomach of three mysid species (Crustacea). — Journal of Morphology, **200** (1): 17-27.
STRAUSFELD, N. J. & D. R. NÄSSEL, 1981. Neuroarchitecture of brain regions that subserve the compound eyes of Crustacea and insects. — In: H. AUTRUM (ed.), Handbook of sensory physiology, **7** (6B), Vision in invertebrates, B, Invertebrate visual centers and behaviour, **I** (1): 1-132. (Springer Verlag, Berlin.)
STRINGER, G. E., 1967. Introduction of *Mysis relicta* Lovén into Kalamalka and Pinaus Lakes, British Columbia. — Journal of the Fisheries Research Board of Canada, **24** (2): 463-465.
STUBBINGTON, R., C. TERRELL-NIELD & P. HARDING, 2008. The first occurrence of the Ponto-Caspian invader, *Hemimysis anomala* G. O. Sars, 1907 (Mysidacea) in the U.K. — Crustaceana, **81** (1): 43-55.
TATTERSALL, O. S., 1952. Report on a small collection of Mysidacea from estuarine waters of South Africa. — Transactions of the Royal Society of South Africa, **33** (2): 153-187.
— —, 1955. Mysidacea. — Discovery Reports, **28**: 1-190.
— —, 1957. Report on a small collection of Mysidacea from the Sierra Leone estuary together with a survey of the genus *Rhopalophthalmus* Illig and a description of a new species of *Tenagomysis* from Lagos, Nigeria. — Proceedings of the Zoological Society of London, **129**: 81-128.
— —, 1961. Report on some Mysidacea from the deeper waters of the Ross Sea. — Proceedings of the Zoological Society of London, **137** (4): 553-571.
TATTERSALL, W. M., 1922. Indian Mysidacea. — Records of the Indian Museum, **24** (4): 445-504.
— —, 1923. Crustacea. Pt. VII. Mysidacea. — In: British Antarctic (Terra Nova) Expedition, 1910. Natural History Report, (Zoology) **3** (10): 273-304. (British Museum (Natural History), London.)
— —, 1925. Mysidacea and Euphausiacea of Marine Survey, South Africa. — Fisheries and Marine Biological Survey, South Africa. Reports, **4** (1924) (Special Reports) V: 1-12, pls. I, II.
— —, 1927. XI. Report on the Crustacea Mysidacea. — In: Zoological results of the Cambridge expedition to the Suez Canal, 1924. Transactions of the Zoological Society of London, **22**: 185-198.
— —, 1939. The Euphausiacea and Mysidacea of the John Murray Expedition to the Indian Ocean. — In: John Murray Expedition 1933-1934. Scientific Reports, **5**: 203-246. (British Museum (Natural History), London.)
— —, 1951. A review of the Mysidacea of the United States National Museum. — Bulletin of the United States National Museum, **201**: 1-292.
TATTERSALL, W. M. & O. S. TATTERSALL, 1951. The British Mysidacea. — Ray Society Monograph, **136**: 1-460. (Ray Society, London.)
TAYLOR, R. S., F. R. SCHRAM & S. YAN-BIN, 2001. A new Upper Middle Triassic shrimp (Crustacea: Lophogastrida) from Guizhou China, with discussion regarding other fossil mysidaceans. — Journal of Palaeontology, **75** (2): 310-318.
TCHINDONOVA, YU. G., 1981. New data on the systematic position of some deep-sea mysids (Mysidacea, Crustacea) and their distribution in the world ccean. — In: Proceedings of the XIV Pacific Science Congress (Khabarovsk, August 1979), Section Marine Biology. Biology of the Pacific Ocean Depths, issue **1**: 24-33. (Acad. Sci. USSR, Far East Science Center, Inst. mar. Biol., Vladivostok.) [In Russian.]
THIEL, R., 1996. The impact of fish predation on the zooplankton community in a southern Baltic bay. — Limnologica — Ecology and Management of Inland Waters, **26** (2): 123-137.
THIENEMANN, A., 1925. *Mysis relicta*. Fünfte Mitteilung der Untersuchungen über die Beziehungen zwischen dem Sauerstoffgehalt des Wassers und der Zusammensetzung der Fauna norddeutscher Seen. — Zeitschrift für Morphologie und Ökologie der Tiere, **3**: 389-440.

— —, 1928a. *Mysis relicta* im sauerstoffarmen Tiefenwasser der Ostsee und das Problem der Atmung im Salzwasser und Süsswasser. — Zoologische Jahrbücher, (Abteilung für Allgemeine Zoologie und Physiologie der Tiere) **45**: 371-384.

— —, 1928b. Die Reliktenkrebse *Mysis relicta, Pontoporeia affinis, Pallasea quadrispinosa* und die von ihnen bewohnten norddeutschen Seen. — Archiv für Hydrobiologie, **19**: 521-582.

TITTIZER, T., F. SCHÖLL, M. BANNING, A. HAYBACH & M. SCHLEUTER, 2000. Aquatische Neozoen im Makrozoobenthos der Binnenwasserstraßen Deutschlands. — Lauterbornia, **39**: 1-72.

TJUTENKOV, S. K., N. B. VOROBEVAN & A. M. SAMONOV, 1967. Modifications du benthos et du régime alimentaire des Poissons du Lac Balkhash en rapport avec l'acclimatation des Mysidaceae. — Gidrobiologiceskij Zhurnal, **3** (3): 48-54. [In Russian.]

TODA, H., S. NISHIZAWA, M. TAKAHASHI & S. ICHIMURA, 1983 (cf. a). Temperature control on the post-embryonic growth of *Neomysis intermedia* Czerniawsky in a hypereutrophic temperate lake. — Journal of Plankton Research, **5** (3): 377-392.

TODA, H., M. TAKAHASHI & S. ICHIMURA, 1983 (cf. b). Diel vertical movement of *Neomysis intermedia* Czerniawsky (Crustacea, Mysidacea) population in a shallow eutrophic lake. — Japanese Journal of Limnology, **44** (4): 277-282.

— —, — — & — —, 1984. The effect of temperature on the post-embryonic growth of *Neomysis intermedia* Czerniawsky (Crustacea, Mysidacea) under laboratory conditions. — Journal of Plankton Research, **6** (4): 647-662.

TOMIYAMA, T., S. UEHARA & Y. KURITA, 2013. Feeding relationships among fishes in shallow sandy areas in relation to stocking of Japanese flounder. — Marine Ecology Progress Series, **479**: 163-175.

TORNAINEN, J. & M. LEHTINIEMI, 2008. Potential predation pressure of littoral mysids on herring (*Clupea harengus membras* L.) eggs and yolk-sac larvae. — Journal of Experimental Marine Biology and Ecology, **367**: 247-252.

TWINING, B. S., J. J. GILBERT & N. S. FISHER, 2000. Evidence of homing behavior in the coral reef mysid *Mysidium gracile*. — Limnology and Oceanography, **45** (8): 1845-1849.

UTEVSKY, S. YU. & J. SORBE, 2012. First record of the boreal-arctic marine leech *Mysidobdella borealis* (Hirudinida, Piscicolidae) from the southern Bay of Biscay. — Vestnik Zoologii, **46** (2): e35-e38.

VADER, W., 1973. A bibliography of the Ellobiopsidae, 1959-1971, with a list of *Thalassomyces* species and their hosts. — Sarsia, **52**: 175-180.

VÄLIPAKKA, P., 1992. Distribution of mysid shrimps (Mysidacea) in the Bay of Mecklenburg (western Baltic Sea). — In: J. KÖHN, M. B. JONES & A. MOFFAT (eds.), Taxonomy, biology and ecology of (Baltic) mysids (Mysidacea, Crustacea), pp. 61-72. (Rostock University Press, Rostock.)

VAN BENEDEN, P.-J., 1861. Recherches sur les Crustacés du littoral de Belgique. — Mémoires de l'Académie Royale des Sciences, des Lettres et des Beaux-Arts de Belgique, **33**: 1-174, pls. I-XXI.

VANMETER, K. & M. S. EDWARDS, 2013. The effects of mysid grazing on kelp zoospore survival and settlement. — Journal of Phycology, **49** (5): 896-901.

VANNINI, M., G. INNOCENTI & R. K. RUWA, 1993. Family group structure in mysids, commensals of hermit crabs (Crustacea). — Tropical Zoology, **6**: 189-205.

VANNINI, M., R. K. RUWA & G. INNOCENTI, 1994. Notes on the behaviour of *Heteromysis harpax*, a commensal mysid living in hermit crab shells. — Ethology Ecology & Evolution, (Special Issue) **3**: 137-142.

VILAS, C., P. DRAKE & N. FOCKEDEY, 2008. Feeding preferences of estuarine mysids *Neomysis integer* and *Rhopalophthalmus tartessicus* in a temperate estuary (Guadalquivir estuary, SW Spain). — Estuarine, Coastal and Shelf Science, **77** (3): 345-356.

VILAS-FERNÁNDEZ, C., P. DRAKE & J. C. SORBE, 2008. *Rhopalophthalmus tartessicus* sp. nov. (Crustacea: Mysidacea), a new mysid species from the Guadalquivir estuary (SW Spain). — Organisms Diversity and Evolution, **7** (4): 292.e1-292.e13.
VILLALOBOS, A., 1951. Un nuevo Misidaceo de las Grutas de Quintero en el estado de Tamaulipas. — Anales del Instituto de Biología Universidad Nacional Autónoma de México, **22**: 191-218.
VLASBLOM, A. G. & J. H. B. W. ELGERSHUIZEN, 1977. Survival and oxygen consumption of *Praunus flexuosus* and *Neomysis integer*, and embryonic development of the latter species, in different temperature and chlorinity combinations. — Netherlands Journal of Sea Research, **11**: 305-315.
VOGT, W., 1932. Über die Morphologie und Histologie der Antennendrüse und der thoracalen Arthrocytenorgane der Mysideen. — Zeitschrift für Morphologie und Ökologie der Tiere, **24**: 288-318.
— —, 1933. Über die Antennendrüse von *Mysis relicta*. — Zoologische Jahrbücher, (Abteilung für Anatomie und Ontogenie der Tiere) **56**: 373-386.
— —, 1935a. Die Entwicklung der Antennendrüse der Mysideen. — Zeitschrift für Morphologie und Ökologie der Tiere, **29** (4): 481-506.
— —, 1935b. Über ein Seitenorgan der Mysideen. — Zeitschrift für Morphologie und Ökologie der Tiere, **29** (4): 507-510.
VOICU, GH., 1974. Identification des mysidés fossiles dans les dépots du Miocene Supérieur de la Paratethys Centrale et Orientale et leur importance paléontologique, stratigraphique et paléogéographique. — Geologicky Zbornik — Geologica Carpathica, **25** (2): 231-239.
— —, 1981. Upper Miocene and Recent mysid statoliths in Central and Eastern Paratethys. — Micropaleontology, **27** (3): 227-247.
WÄGELE, J. W., 1994. Review of methodological problems of "computer cladistics" exemplified with a case study on isopod phylogeny (Crustacea: Isopoda). — Journal of Zoological Systematics and Evolutionary Research, **32** (2): 81-107.
WAGNER, H. P., 1992. *Stygiomysis aemete* n.sp., a new subterranean mysid (Crustacea, Mysidacea, Stygiomysidae) from the Dominican Republic, Hispaniola. — Bijdragen tot de Dierkunde, **62** (2): 71-79.
WALSH, M. G., B. T. BOSCARINO, J. MARTY & O. E. JOHANNSSON, 2012. *Mysis diluviana* and *Hemimysis anomala*: Reviewing the roles of a native and invasive mysid in the Laurentian Great Lakes region. — Journal of Great Lakes Research, **38** (Suppl. 2): 1-6.
WARD, P., 1984. Aspects of the biology of *Antarctomysis maxima* (Crustacea: Mysidacea). — Polar Biology, **3** (2): 85-92.
WATERMAN, T. H., 1960. Interaction of polarized light and turbidity in the orientation of *Daphnia* and *Mysidium*. — Zeitschrift für Vergleichende Physiologie, **43** (2): 149-172.
WATERMAN, T. H., R. F. NUNNEMACHER, F. A. CHACE, JR. & G. L. CLARKE, 1939. Diurnal vertical migrations of deep-water plankton. — Biological Bulletin, Woods Hole, **76** (2): 256-279.
WATERSTRAAT, A., M. KRAPPE, PH. RIEL & M. RUMPF, 2005. Habitat shifts of *Mysis relicta* (Decapoda, Mysidacea) in the lakes Breiter and Schmaler Luzin (NE Germany). — Crustaceana, **78** (6): 685-699.
WATLING, L., 1981. An alternative phylogeny of peracarid crustaceans. — Journal of Crustacean Biology, **1** (2): 201-210.
— —, 1983. Peracaridan disunity and its bearing on eumalacostracan phylogeny with a redefinition of eumalacostracan superorders. — In: F. R. SCHRAM (ed.), Crustacean phylogeny. Crustacean Issues, **1**: 213-228. (A. A. Balkema, Rotterdam.)
— —, 1999. Towards understanding the relationship of the peracaridan orders: the necessity of determining exact homologies. — In: F. R. SCHRAM & J. C. VON VAUPEL KLEIN (eds.), Crustaceans and the biodiversity crisis. Proceedings of the Fourth International Crustacean Congress, Amsterdam, The Netherlands, July 20-24, 1998, **1**: 73-89. (Brill, Leiden.)

WEBB, P., R. PERISSINOTTO & T. H. WOOLDRIDGE, 1988. Diet and feeding of *Gastrosaccus psammodytes* (Crustacea, Mysidacea) with special reference to the surf diatom *Anaulus birostratus*. — Marine Ecology Progress Series, **45**: 255-261.

WEBB, P. & T. H. WOOLDRIDGE, 1990. Diel horizontal migration of *Mesopodopsis slabberi* (Crustacea: Mysidacea) in Algoa Bay, southern Africa. — Marine Ecology Progress Series, **62**: 73-77.

WIGLEY, R. L., 1963. Occurrence of *Praunus flexuosus* (O.F. Müller) (Mysidacea) in New England waters. — Crustaceana, **6**: 158.

WILLIAMS, J. D. & J. J. MCDERMOTT, 2004. Hermit crab biocoenoses: a worldwide review of the diversity and natural history of hermit crab associates. — Journal of Experimental Marine Biology and Ecology, **305**: 1-128.

WILLS, M. A., R. A. JENNER & C. NÍ DHUBHGHAILL, 2009. Eumalacostracan evolution: conflict between three sources of data. — Arthropod Systematics & Phylogeny, **67** (1): 71-90.

WILSON, C. D. & G. W. BOEHLERT, 1993. Population biology of *Gnathophausia longispina* (Mysidacea: Lophogastrida) from a central North Pacific seamount. — Marine Biology, Berlin, **115**: 537-543.

WINKLER, G. & W. GREVE, 2004. Trophodynamics of two interacting species of estuarine mysids, *Praunus flexuosus* and *Neomysis integer*, and their predation on the calanoid copepod *Eurytemora affinis*. — Journal of Experimental Marine Biology and Ecology, **308**: 127-146.

WIRKNER, C. S., 2009. The circulatory system in Malacostraca — evaluating character evolution on the basis of differing phylogenetic hypotheses. — Arthropod Systematics & Phylogeny, **67** (1): 57-70.

WIRKNER, C. S. & S. RICHTER, 2007. The circulatory system of Mysidacea revisited — implications for the phylogenetic position of Mysida and Lophogastrida (Malacostraca, Crustacea). — Journal of Morphology, **268**: 311-328.

— — & — —, 2010. Evolutionary morphology of the circulatory system in Peracarida (Malacostraca; Crustacea). — Cladistics, **26**: 143-167.

— — & — —, 2013. Circulatory system and respiration. — In: L. WATLING & M. THIEL (eds.), The natural history of Crustacea, **1**, Functional morphology & diversity, pp. 376-412. (Oxford University Press, U.S.A.)

WIRTZ, P., 1995. Unterwasserführer Madeira, Kanaren / Azoren. Niedere Tiere. — Pp. 1-247. (Delius Klasing. Edition Naglschmid, Stuttgart.)

— —, 1997. Crustacean symbionts of the sea anemone *Telmatactis cricoides* at Madeira and the Canary Islands. — Journal of Zoology, London, **242**: 799-811.

— —, 2009. Thirteen new records of marine invertebrates and two of fishes from Cape Verde Islands. — Arquipélago, (Life and Marine Sciences) **26**: 51-56.

WITTMANN, K. J., 1977. Modification of association and swarming in North Adriatic Mysidacea in relation to habitat and interacting species. — In: B. F. KEEGAN, P. O'CEIDIGH & P. J. S. BOADEN (eds.), Biology of benthic organisms. Proceedings of the Eleventh European Marine Biology Symposium, pp. 493-502. (Pergamon Press, Oxford.)

— —, 1978a. Biotop- und Standortbindung mediterraner Mysidacea. — Pp. 1-211. (Dissertation, University of Vienna.)

— —, 1978b. Adoption, replacement and identification of young in marine Mysidacea (Crustacea). — Journal of Experimental Marine Biology and Ecology, **32**: 259-274.

— —, 1981a. Comparative biology and morphology of marsupial development in *Leptomysis* and other Mediterranean Mysidacea (Crustacea). — Journal of Experimental Marine Biology and Ecology, **52** (2-3): 243-270.

— —, 1981b. On the breeding biology and physiology of marsupial development in Mediterranean *Leptomysis* (Mysidacea: Crustacea), with special reference to the effects of temperature and egg size. — Journal of Experimental Marine Biology and Ecology, **53** (2-3): 261-279.

— —, 1982. Untersuchungen zur Sexualbiologie einer mediterranen Mysidacee (Crustacea), *Leptomysis lingvura* G. O. Sars. — Zoologischer Anzeiger, **209** (5-6): 362-375.

— —, 1984. Ecophysiology of marsupial development and reproduction in Mysidacea (Crustacea). — Oceanography and Marine Biology — An Annual Review, **22**: 393-428.

— —, 1985. Freilanduntersuchungen zur Lebensweise von *Pyroleptomysis rubra*, einer neuen bentho-pelagischen Mysidacee aus dem Mittelmeer und dem Roten Meer. — Crustaceana, **48** (2): 153-166.

— —, 1986a. Saisonale und morphogeographische Differenzierung bei *Leptomysis lingvura* und zwei verwandten Spezies (Crustacea, Mysidacea). — Annalen des Naturhistorischen Museums in Wien, **87B**: 265-294, pls. 1, 2.

— —, 1986b. Untersuchungen zur Lebensweise und Systematik von *Leptomysis truncata* und zwei verwandten Formen (Crustacea, Mysidacea). — Annalen des Naturhistorischen Museums in Wien, **87B**: 295-323, pl. 1.

— —, 1986c. A revision of the genus *Paraleptomysis* Liu & Wang (Crustacea: Mysidacea). — Sarsia, **71**: 147-160.

— —, 1992a. Morphogeographic variations in the genus *Mesopodopsis* Czerniavsky with descriptions of three new species (Crustacea, Mysidacea). — Hydrobiologia, **241**: 71-89.

— —, 1992b. Cyclomorphosis in temperate zone Mysidacea: evidence and possible adaptive and taxonomical significance. — In: J. KÖHN, M. B. JONES & A. MOFFAT (eds.), Taxonomy, biology and ecology of (Baltic) mysids (Mysidacea, Crustacea), pp. 25-32. (Rostock University Press, Rostock.)

— —, 1995. Zur Einwanderung potamophiler Malacostraca in die obere Donau: *Limnomysis benedeni* (Mysidacea), *Corophium curvispinum* (Amphipoda) und *Atyaephyra desmaresti* (Decapoda). — Lauterbornia, **20**: 77-85.

— —, 1996. Morphological and reproductive adaptations in Antarctic meso- to bathypelagic Mysidacea, with description of *Mysifaun erigens* n.g. n.sp. — In: F. UIBLEIN, J. OTT & M. STACHOWISCH (eds.), Deep-sea and extreme shallow-water habitats: affinities and adaptations. — Biosystematics and Ecology Series, **11**: 221-231.

— —, 1999. Global biodiversity in Mysidacea, with notes on the effects of human impact. — In: F. R. SCHRAM & J. C. VON VAUPEL KLEIN (eds.), Crustaceans and the biodiversity crisis, **1**: 511-525. (Brill, Leiden.)

— —, 2000. *Heteromysis arianii* sp.n., a new benthic mysid (Crustacea, Mysidacea) from coralloid habitats in the Gulf of Naples (Mediterranean Sea). — Annalen des Naturhistorischen Museums in Wien, **102B**: 279-290.

— —, 2001. Centennial changes in the near-shore mysid fauna of the Gulf of Naples (Mediterranean Sea), with description of *Heteromysis riedli* sp. n. (Crustacea, Mysidacea). — Pubblicazioni della Stazione Zoologica di Napoli, (I, Marine Ecology) **22** (1-21): 85-109.

— —, 2005. Anthropogenic dispersal and environmental impact of invasive species of Mysidae (Mysidacea: Crustacea) on a world-wide scale. — In: EGU General Assembly 2005, Vienna. Geophysical Research Abstracts, **7**: 10881. [CD-ROM.]

— —, 2007. Continued massive invasion of Mysidae in the Rhine and Danube river systems, with first records of the order Mysidacea (Crustacea: Malacostraca: Peracarida) for Switzerland. — Revue Suisse de Zoologie, **114** (1): 65-86.

— —, 2008. Two new species of Heteromysini (Mysida, Mysidae) from the Island of Madeira (N.E. Atlantic), with notes on sea anemone and hermit crab commensalisms in the genus *Heteromysis* S. I. Smith, 1873. — Crustaceana, **81** (3): 351-374.

— —, 2009. Revalidation of *Chlamydopleon aculeatum* Ortmann, 1893, and its consequences for the taxonomy of Gastrosaccinae (Crustacea: Mysida: Mysidae) endemic to coastal waters of America. — Zootaxa, **2115**: 21-33.

— —, 2013a. Mysids associated with sea anemones from the tropical Atlantic: descriptions of *Ischiomysis* new genus, and two new species in this taxon (Mysida: Mysidae: Heteromysinae). — Crustaceana, **86** (4): 487-506.

— —, 2013b. Comparative morphology of the external male genitalia in Lophogastrida, Stygiomysida, and Mysida (Crustacea, Eumalacostraca). — Zoomorphology, **132** (4): 389-401.

WITTMANN, K. J. & A. P. ARIANI, 1996. Some aspects of fluorite and vaterite precipitation in marine environments. — Pubblicazioni della Stazione Zoologica di Napoli, (I, Marine Ecology) **17** (1): 213-219.

— — & — —, 1998. *Diamysis bacescui* n.sp., a new benthopelagic mysid (Crustacea: Peracarida) from Mediterranean seagrass meadows: description and comments on statolith composition. — Travaux du Muséum National d'Histoire Naturelle « Grigore Antipa », **40**: 35-49.

— — & — —, 2000. *Limnomysis benedeni*: Mysidacé ponto-caspien nouveau pour les eaux douces de France (Crustacea, Mysidacea). — Vie et Milieu, **50** (2): 117-122.

— — & — —, 2009. Reappraisal and range extension of non-indigenous Mysidae (Crustacea, Mysida) in continental and coastal waters of eastern France. — Biological Invasions, **11** (2): 401-407.

— — & — —, 2011. An adjusted concept for a problematic taxon, *Paramysis festae* Colosi, 1921, with notes on morphology, biomineralogy, and biogeography of the genus *Paramysis* Czerniavsky, 1882 (Mysida: Mysidae). — Crustaceana, **84** (7): 849-868.

— — & — —, 2012a. *Diamysis cymodoceae* sp. nov. from the Mediterranean, Marmora, and Black Sea basins, with notes on geographical distribution and ecology of the genus (Mysida, Mysidae). — Crustaceana, **85** (3): 301-332.

— — & — —, 2012b. The species complex of *Diamysis* Czerniavsky, 1882, in fresh waters of the Adriatic basin (NE Mediterranean), with descriptions of *D. lacustris* Băcescu, 1940, new rank, and *D. fluviatilis* sp. nov. (Mysida, Mysidae). — Crustaceana, **85** (14): 1745-1779.

WITTMANN, K. J., A. P. ARIANI & A. STANZIONE, 1990. Implicazioni tassonomiche ed ecologiche di alcune caratteristiche biometriche degli statoliti dei Misidacei. — Oebalia, (Suppl.) **16-2**: 805-807.

WITTMANN, K. J., G. GRABER & C. GUNDACKER, 2010. Anthropogenic modification of food chains may show marked impact on levels of mercury (Hg) in fish food. — In: 32nd Annual Meeting of the Austrian Society for Hygiene, Microbiology and Preventive Medicine, 17-20 May 2010, (Abstract) **S9-5**: 52-53.

WITTMANN, K. J., F. HERNÁNDEZ, J. DÜRR, E. TEJERA, J. A. GONZÁLEZ & S. JIMÉNEZ, 2004. The epi- to bathypelagic Mysidacea (Peracarida) off the Selvagens, Canary, and Cape Verde Islands (NE Atlantic), with first description of the male of *Longithorax alicei* H. Nouvel, 1942. — Crustaceana, **76** (10): 1257-1280.

WITTMANN, K. J., L. MORO & R. RIERA, 2011. Sobre la distribución de *Gastrosaccus roscoffensis* (Crustacea: Mysida) en el Atlántico nororiental y primer registro para las Islas Canarias. — Revista de la Academia Canaria de Ciencias, **XXII** (4): 91-101. [2010]

WITTMANN, K. J. & R. RIERA, 2012. Check-list of lophogastrids (Crustacea, Peracarida) from the Canary Islands. — Revista de la Academia Canaria de Ciencias, **XXIV** (3): 63-80.

WITTMANN, K. J., T. A. SCHLACHER & A. P. ARIANI, 1993. Structure of Recent and fossil mysid statoliths (Crustacea, Mysidacea). — Journal of Morphology, **215**: 31-49.

WITTMANN, K. J., J. THEISS & M. BANNING, 1999. Die Drift von Mysidaceen und Dekapoden und ihre Bedeutung für die Ausbreitung von Neozoen im Main-Donau-System. — Lauterbornia, **35**: 53-66.

WITTMANN, K. J., T. J. VANAGT, M. A. FAASSE & J. MEES, 2012. A new transoceanic invasion? First records of *Neomysis americana* (Crustacea: Mysidae) in the East Atlantic. — The Open Marine Biology Journal, **6**: 62-66.

WITTMANN, K. J. & P. WIRTZ, 1998. A first inventory of the mysid fauna (Crustacea: Mysidacea) in coastal waters of the Madeira and Canary archipelagos. — Boletim do Museu Municipal do Funchal, (Suppl.) **5**: 511-533.
WOLLNER, E., 1924. Zur Kenntnis des Baues und der Muskulatur des Vorderkopfs und seiner Anhänge von *Nebalia* und den Schizopoden. — Meddelanden från Göteborgs Musei, (Zoologiska Avdelning) **25**: 1-23.
WOOLDRIDGE, T. H., 1989. The spatial and temporal distribution of mysid shrimps and phytoplankton accumulations in a high energy surfzone. — Vie et Milieu, **39** (3/4): 127-133.
WOOLDRIDGE, T. H. & P. WEBB, 1988. Predator-prey interactions between two species of estuarine mysid shrimps. — Marine Ecology Progress Series, **50**: 21-28.
WORTHAM-NEAL, J. L. & W. W. PRICE, 2002. Marsupial developmental stages in *Americamysis bahia* (Mysida: Mysidae). — Journal of Crustacean Biology, **22** (1): 98-112.
YAMASHITA, Y., T. OKUMURA & H. YAMADA, 2001. Intersexuality in *Acanthomysis mitsukurii* (Mysidacea) in Sendai Bay, northeastern Japan. — Plankton Biology & Ecology, **48** (2): 128-132.
YOKES, B. & W. B. RUDMAN, 2004. Lessepsian opisthobranchs from southwestern coast of Turkey; five new records for Mediterranean. — Rapports de la Commission Internationale pour l'Exploration Scientifique de la Mer Méditerranée, **37**: 557.
ZATKUTSKIY, V. P., 1970. Some biological features of mysids in the hyponeuston of the Black Sea and Sea of Azov. — Gidrobiologiceskij Zhurnal, **6** (6): 17-22.
ZELICKMAN, E. A., 1974. Group orientation in *Neomysis mirabilis* (Mysidacea: Crustacea). — Marine Biology, Berlin, **24** (3): 251-258.
ZHARKOVA, I. S., 1970. Réduction des organes de la vue chez les Mysidacés abyssaux. — Zoologicheskij Zhurnal, **49** (5): 685-693.
ZHURAVEL, P. A., 1950. K probleme obogashchenija kormnosti vodohranilishch yugo-vostoka Ukrainy. — Zoologicheskij Zhurnal, **29** (2): 128-139. [In Russian.]
— —, 1969. O rasshchirenii arealov nekotorih limanno-kaspijskih bezbozvonochnih. — Gidrobiologiceskij Zhurnal, **5** (3): 1152-1162 (76-80). [In Russian.]
ZIEMANN, H. & C.-J. SCHULZ, 2011. Methods for biological assessment of salt-loaded running waters. Fundamentals, current positions and perspectives. — Limnologica — Ecology and Management of Inland Waters, **41**: 90-95.
ZIMMER, C., 1909. Die nordischen Schizopoden. — In: K. BRANDT & C. APSTEIN (eds.), Nordisches Plankton, **6**: 1-178. (Lipsius und Tischler, Kiel and Leipzig.)
— —, 1914. Die Schizopoden der deutschen Südpolar-Expedition 1901-1903. — In: E. VON DRYGALSKI (ed.), Deutsche Südpolar-Expedition 1901-1903, **XV** (Zoologie, 7): 377-445, pls. XXIII-XXVI. (Georg Reimer, Berlin.)
— —, 1915. Schizopoden des Hamburger Naturhistorischen (Zoologischen) Museums. — Mitteilungen aus dem Naturhistorischen Museum in Hamburg, **32** (2): 159-182.
— —, 1927. Mysidacea. — In: W. KÜKENTHAL (ed.), Handbuch der Zoologie, **3** (1): 607-650. (W. de Gruyter, Berlin and Leipzig.)
— —, 1932. Beobachtungen an lebenden Mysidaceen und Cumaceen. — Sitzungsberichte der Gesellschaft Naturforschender Freunde zu Berlin, **18**: 326-347.

Colour figures of vol. 4B

Fig. 27B.1. Marine fauna, including a fight between an octopus and a spiny lobster. Note the shrimp at the upper part of the picture. Faun House, Pompei (before 79 AD). Soprintendenza Speciale per i Beni Archeologici di Napoli e Pompei. [Reproduced with permission.]

Fig. 27B.2. Oil lamp decorated with a crab, presumably *Pachygrapsus marmoratus*; early 2nd century AD. Potter mark on the back side. Inscription: LHOSCRI. Terracotta, beige paste, brown varnish; 10.9 × 7.9 × 3.1 cm. Musée de l'Arles Antique, France, inv. FAN.91.00.2044, © M. Lacanaud. [Reproduced with permission.]

Fig. 27B.3. Sculpture of the zodiacal Cancer sign; 12th century. Apse of the Saint-Austremoine Abbey, Issoire, France. [Photograph G. Charmantier.]

Fig. 27B.5. Pieter Claesz, *Still Life with Glass of White Wine and Crab*, 1644. Oil on wood. Musée des Beaux-Arts de Strasbourg. Photograph E. Bacher. [Reproduced with permission.]

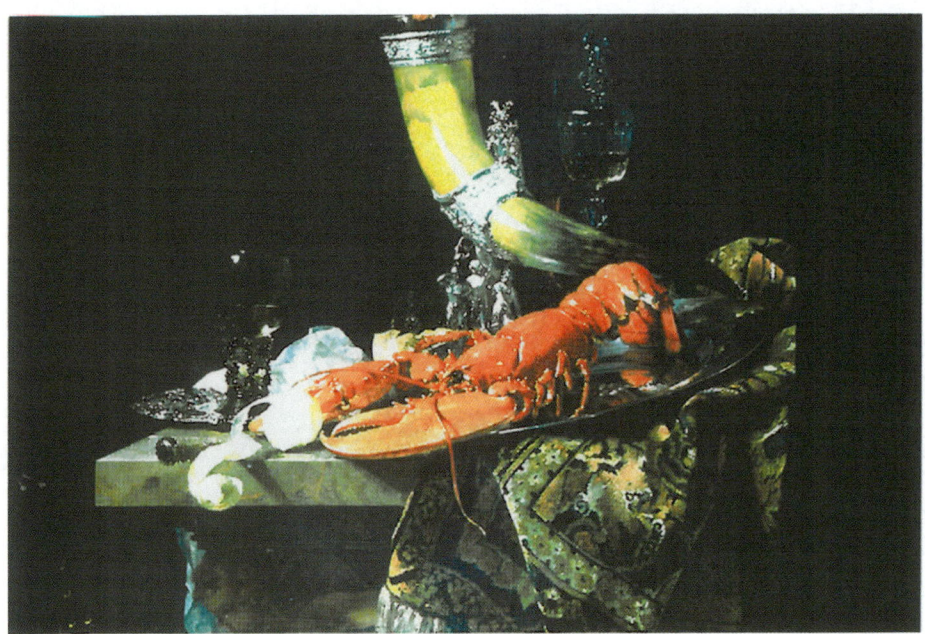

Fig. 27B.6. Willem Kalf, *Still Life with the Drinking Horn of the Saint Sebastian Archers' Guild, Lobster and Glasses*, c. 1653. Oil on canvas. National Gallery, London. [Reproduced with permission.]

Fig. 27B.7. Attributed to Abraham Janssens, *Inconstancy*, c. 1617. Oil on canvas. Statens Museum for Kunst, Copenhagen. [Reproduced with permission.]

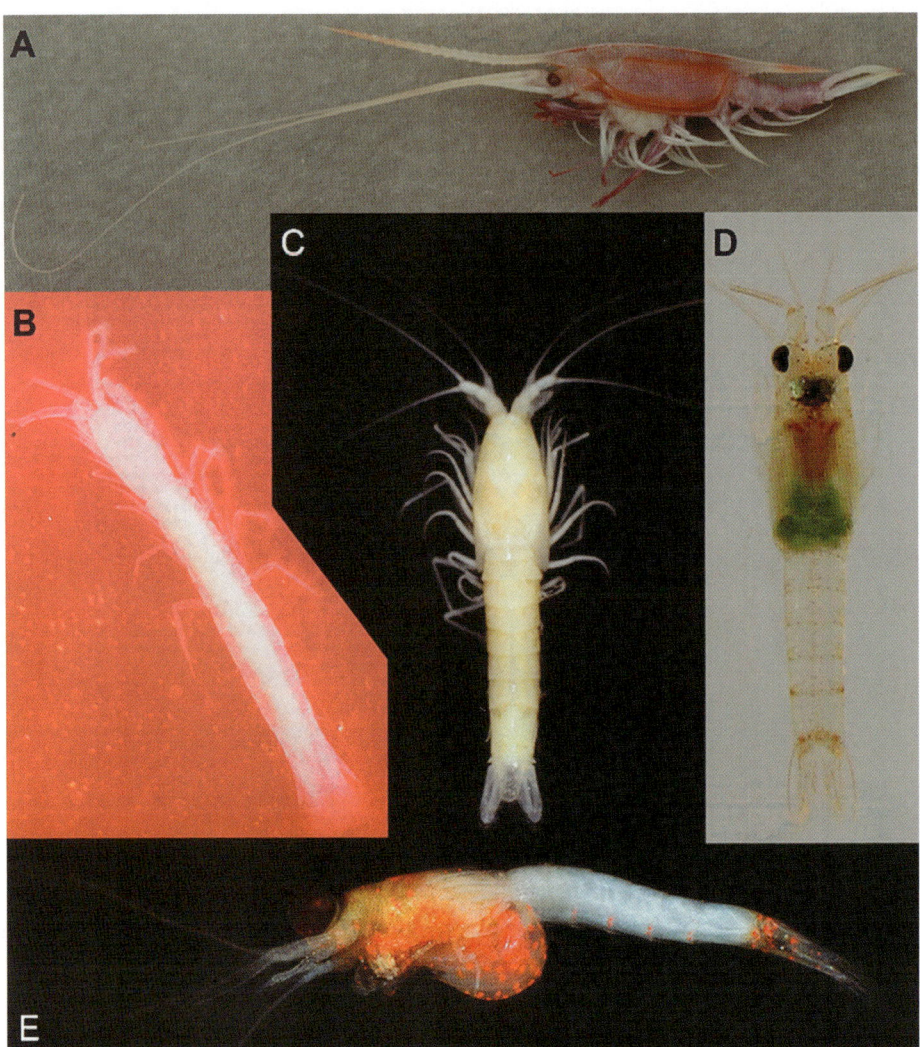

Fig. 54.1. Diversity of Lophogastrida (A), Stygiomysida (B, C), and Mysida (D, E). A, *Gnathophausia zoea* Willemoës-Suhm, 1873; B, *Stygiomysis hydruntina* Caroli, 1937; C, *Spelaeomysis bottazzii* Caroli, 1924; D, *Heteromysis wirtzi* Wittmann, 2008; E, *Hemimysis lamornae mediterranea* Băcescu, 1937. [A, photo José Antonio González; B, after Inguscio, 1998; C, E, photo Antonio P. Ariani; D, after Wittmann, 2008 (photo Peter Wirtz).]

Fig. 54.41. Associations between Mysidae (Mysida) and sessile benthic invertebrates. A, swarm of *Leptomysis* sp. A at the long-spined sea urchin *Diadema antillarum* Philippi, 1845, from Madeira, mysids facing water current; B, adult female *Heteromysis* sp. inside open fan of the sabellid worm *Branchiomma nigromaculatum* (Baird, 1865) from Cape Verde Islands; C, swarm constituted mainly by breeding females of *Leptomysis lingvura marioni* Gourret, 1888, above the snakelocks anemone *Anemonia viridis* (Forskål, 1775) from the Gulf of Naples; D, dense aggregation of undetermined Mysidae above oral disk of the club-tipped anemone *Telmatactis cricoides* (Duchassaing, 1850) from Trindade, a small island in the tropical south-west Atlantic. [A, after Wirtz, 1995; B, photo Peter Wirtz; C, after Wittmann, 1978a; D, after Wittmann, 2013a (photo Lisandro de Almeida).]

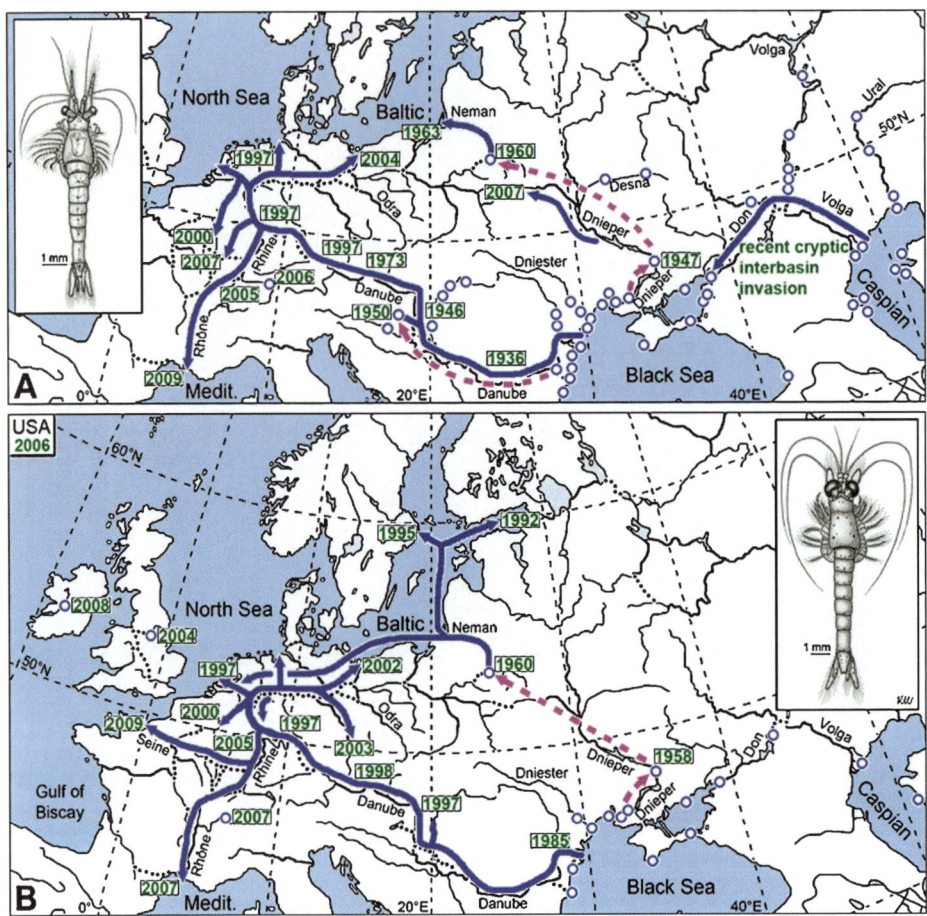

Fig. 54.44. Expansion pathways of the mysids *Limnomysis benedeni* Czerniavsky, 1882 (A), and *Hemimysis anomala* G. O. Sars, 1907 (B), from the Ponto-Caspian to western Europe and beyond. Continuous heavy arrows indicate spread along waterways; dashed heavy arrows indicate deliberate transplantations. Small dotted lines are artificial navigation canals. Years stand for first records or for transplantations, respectively. [Original; data updated from Wittmann & Ariani, 2009.]

LIST OF CONTRIBUTORS

Addresses of contributors and their (former) affiliations

Editors

DR. J. C. VON VAUPEL KLEIN
Bosuillaan 311
NL-3722 XM BILTHOVEN
Netherlands
jcvvk@xs4all.nl

> *former affiliation*
> Division of Systematic Zoology, Leiden University, Leiden, Netherlands

DR. M. CHARMANTIER-DAURES
mgcharmantier@yahoo.fr

> *former affiliation*
> Equipe Adaptation Ecophysiologique et Ontogenèse
> UMR 5119 Ecosym
> Université Montpellier 2
> Cc 092, Place Eugène Bataillon
> F-34095 MONTPELLIER Cedex 05
> France

PROF. DR. F. R. SCHRAM
Post Box 1567
LANGLEY, WA 98260
U.S.A.
fschram@u.washington.edu

> *and* Professor emeritus, University of Amsterdam, Amsterdam, Netherlands;
> *and* Research Associate, Burke Museum of Natural History and Cultures, University of Washington, Seattle, WA, U.S.A.

Chapter 26

DR. P. NOËL
pierre@flgb.net or *pierre.noel@mnhn.fr*

> *former affiliation*
> Département Milieux et Peuplements Aquatiques
> Muséum national d'Histoire naturelle
> F-75005 PARIS
> France

PROF. DR. TH. MONOD (†)

> *former affiliation*
> Muséum national d'Histoire naturelle
> PARIS
> France

DR. L. LAUBIER (†)

> *former affiliation*
> Institut Océanographique, Paris

195 rue Saint-Jacques
F-75005 PARIS
France

Chapter 27A

DR. H.-M. CAUCHIE
CRP-Gabriel Lippmann
EVA
41, rue du Brill
L-4422 BELVAUX
Luxembourg
cauchie@lippmann.lu

PROF. DR. TH. MONOD (†)
[*see above*]

DR.L. LAUBIER (†)
[*see above*]

Chapter 27B

PROF. DR. G. CHARMANTIER
Professor emeritus
Equipe AEO (Adaptation Ecophysiologique et Ontogenèse)
UMR 5119
Université Montpellier 2
Cc 092, Place Eugène Bataillon
F-34095 MONTPELLIER Cedex 05
France
guy.charmantier@univ-montp2.fr

Chapter 54

PROF. DR. K. J. WITTMANN
Medizinische Universität Wien
Institut für Umwelthygiene, Abteilung für Ökotoxikologie
Kinderspitalgasse 15
A-1090 VIENNA
Austria
karl.wittmann@meduniwien.ac.at

PROF. DR. A. P. ARIANI
Via S.G.M. Pignatelli, 38
I-80134 NAPLES
Italy
antonio.ariani@gmail.com

    former affiliation
    Università di Napoli Federico II
    Dipartimento di Biologia
    NAPLES
    Italy

Dr. J.-P. LAGARDÈRE
49, chemin de la Tuilerie
F-40110 MORCENX
France
jplagard@free.fr

> *former affiliation*
> CNRS-IFREMER
> CREMA L'Houmeau
> L'HOUMEAU
> France

TAXONOMIC INDEX

Aberomysis, 334
Abludomelita obtusata, 64, 73
Abylopsis tetragona, 40, 80
Abyssorchomene abyssorum, 6, 64, 74
Acanthephyra, 62
Acanthocaris, 23, 24
Acanthomolgus, 41, 50
Acanthomysis, 324, 340
Acanthomysis aspera, 326, 347
Acanthomysis longicornis, 211, 213, 347
Acanthomysis robusta, 290, 291, 347
Acanthomysis sculpta, 281, 347
Acanthomysis thailandica, 277, 347
Acanthonotozomatidae, 40
Acanthopleura granulata, 50, 82
Acanthoserolis schythei, 62, 64, 77
Acanthosquilla digueti, 51, 64
Acari, 35
Acarnus ternatus, 39, 80
Acartia, 62
Acasta, 11, 38, 41, 42
Achteinus, 63
Acinetidae, 62
Acmaea, 50
Acontiophorus, 38
Acontiophorus scutatus, 38, 64
Acromitus, 40
Acropora, 44, 45, 47
Acroporidae, 45
Acropyga acutiventris, 35, 82
Acrothoracica, 11, 44, 52
Actaea superciliaris, 45, 64
Actinia equina, 43, 81
Actiniaria, 43
Actinodendron, 44
Actinodinium, 62
Actinoloba, 43
Actinoloba dianthus, 43, 81
Adamsia palliata, 43, 81
Adeleana chapmani, 31, 64
Aechmea paniculigera, 32
Aega spongiophila, 38, 64
Aegidae, 9, 38
Aegiochus spongiophila, 64
Aegla, 27
Aegla cavernicola, 31
Aeromonas, 125
Afrogitanopsis paguri, 64, 70
Afromysini, 338, 344
Afromysini [in key], 345
Afromysis, 338
Agelas, 38

Aglaophenia, 39
Aglaophenia cupressina, 39, 80
Aiptasia mutabilis, 309, 347
Akentrogonida, 52
Alachosquilla digueti, 64
Albunea, 17
Albunea paretii, 17, 64
Albunea symmista, 17, 64
Albunea symmysta, 64
Alcyonacea, 40, 41
Alcyonium palmatum, 42, 81
Alcyonohippolyte commensalis, 64, 71
Alepas, 40
Alexandrium, 125
Alicella gigantea, 5, 64
Alienacanthomysis, 340
Allanaspides, 55
Allantogynus, 57
Allogalathea, 57
Allogaussia recondita, 43, 64
Allograea, 25
Alloschizidium cottarellii, 35, 64, 79
Alona affinis, 33, 64
Alpheidae, 27, 31, 36, 39, 45, 50, 58
Alpheopsis, 31
Alpheus, 21, 39, 45
Alpheus armatus, 44, 64
Alpheus dentipes, 15
Alpheus djiboutensis, 19, 64
Alpheus frontalis, 16, 65
Alpheus laevis, 65
Alpheus lottini, 15, 65
Alpheus malleodigitus, 15, 65
Alpheus pachychirus, 16, 65
Alpheus saxidomus, 12, 65
Alpheus simus, 15, 65
Alteromonas, 125
Alvinella, 25
Alvinocarididae, 5, 25
Alvinocaridides, 25
Alvinocaris, 25
Amalocystis, 62
Amalopenaeus, 24
Amaroucium, 16
Amathimysis, 334
Ambassia, 57
Amblyops, 229, 336
Amblyopsini, 336
Amblyopsini [in key], 345
Amblyopsoides, 336
Amblyopsoides crozetii, 200, 347
Amblyopsoides ohlinii, 347
Amboine, 51
Ambolana, 30
Americamysis, 337

412   TAXONOMIC INDEX

Americamysis almyra, 347
Americamysis bahia, 277, 279–281, 283, 286, 294, 305, 347
Americamysis bigelowi, 302, 347
Americorophium spinicorne, 10, 65
Ameson, 61, 62
Amigdoscalpellum hispidum, 65, 77
Ampelisca, 53
Ampelisca cristata, 16, 65
Ampeliscoidea, 8
Amphibalanus amphitrite, 13, 38, 53, 65
Amphibalanus improvisus, 54
Amphibia, 83
Amphilina foliacea, 312, 347
Amphilochidae, 40, 57
Amphilochus, 57
Amphilodridae, 8
Amphineura, 49
Amphipoda, 4–8, 10, 12, 18–20, 24, 25, 27, 28, 31, 33–35, 38, 40, 43, 50, 55, 57–59, 62, 209, 267, 268, 276, 293
Amphitrite, 14, 51
Amphiura, 57
Amphiurophilus, 56
Ampithoe, 21
Ampithoe humeralis, 65
Ampithoe rubricata, 16, 20, 65
Anachlorocurtis commensalis, 46, 65
Anadiaptomus, 30
Anamixidae, 8
Anamixis, 72
Anamixis linsleyi, 38, 65
Anamixis pacifica, 38, 65
Anapagurus, 43
Anaspidacea, 19
Anchialidae, 333
Anchialina, 229, 334, 347
Anchialina agilis, 347
Anchialina truncata, 200, 203, 219, 347
Anchialina typica, 323, 347
Anchialinini, 333
Anchialinini [in key], 344
Anchialus, 334, 347
Anchicaligus nautili, 49, 65
Anchimolgus, 44
Anchistioides, 39
Anchistus, 50
Ancylomenes holthuisi, 40, 44, 60, 65, 75
Ancylomenes magnificus, 45, 60, 65, 76
Ancylomenes pedersoni, 60, 65
Anelasma squalicola, 59, 65
Anemonia sulcata, 43, 81
Anemonia viridis, 309, 310, 347, 405
Angeliera, 30
Animalia, 64, 80, 130

Anisakis simplex, 312, 347
Anisomolgus, 41
Anisomysini, 339, 345
Anisomysini [in key], 345
Anisomysis, 304, 339
Anisomysis levi, 309, 347
Ankylocythere, 30, 53
Annelida, 7, 13, 61, 82
Annina kumari, 18, 65
Anomalocera patersoni, 22, 36, 65
Anomopsyllidae, 51
Anomopsyllus pranizoides, 51, 65
Anomura, 5, 6, 13, 24, 31, 34, 35, 51, 54, 55, 57, 166
Anophrys, 62
Anoplodelphys, 58
Anoplophrya, 62
Anostraca, 6, 26, 32, 119
Antarctomysis, 338
Antarctomysis maxima, 294, 347
Antarctomysis ohlinii, 294, 347
Antarcturus, 24
Antecaridina, 31
Antecaridina lauensis, 21, 65
Antheacheres duebenii, 43, 65
Antheacheridae, 43
Anthessius, 49
Anthessius alatus, 49, 65
Anthessius amicalis, 49, 65
Anthessius discipedatus, 49, 65
Anthessius solidus, 49, 65
Anthopleura elegantissima, 43, 81
Anthoptilum murrayi, 41, 81
Anthosomatidae, 58
Anthozoa, 41, 81
Anthuridae, 20, 30
Anthuridea, 13, 20, 27
Antias uniramea, 57, 65
Antichthomysis, 211, 338
Antillesia, 55
Antipathes, 46
Antrolana, 30
Antromysis, 30, 322, 339
Antromysis anophelinae, 56, 65
Antromysis cubanica, 244, 347
Antromysis juberthiei, 251–253, 280, 307, 347
Antromysis peckorum, 65
Antromysis pectorum, 56, 65
Anura, 60
Aora typica, 13, 16, 55, 65
Apanthura corsica, 20, 65
Aphanodomus, 51
Aphanomyces astaci, 124, 130
Aphroditiformia, 50
Apocorophium acutum, 13, 65
Apodomyzon brevicorne, 38, 65
Apodomyzon longicorne, 38, 65
Apohyale prevostii, 65

Aporobopyrus, 55
Apostomida, 311
Apseudes, 6
Arachnida, 35, 82
Arachnomysini, 335
Arachnomysini [in key], 345
Arachnomysis, 207, 217, 334, 335
Aratus, 28
Archaea, 24
Archaeomysini, 333, 342, 344
Archaeomysini [in key], 344
Archaeomysis, 210, 233, 331, 333
Archaeomysis articulata, 273, 347
Archaeomysis grebnitzkii, 267, 347
Archaeomysis vulgaris, 316, 347
Arcidae, 49
Arcoscalpellum botellinae, 38, 65
Arcturidae, 24
Areiopontonza, 58
Arenicola, 14
Aretopsis amabilis, 55, 65
Argulidae, 124
Argulus, 124
Aristeus coruscans, 65
Aristeus virilis, 56, 65
Aristias, 57
Aristias neglectus, 43, 48, 65, 81
Armadillidium, 149, 168
Armadillidium vulgare, 122, 129
Armases roberti, 65, 77
Armatobalanus, 45
Armatobalanus durhami, 65
Aroui, 57
Artemia, 6, 27, 32, 119, 277, 347
Artemia salina, 65, 119, 129
Arthrochordeumium, 56
Arthromysis, 338
Arthropoda, 33, 82, 130
Artotrogidae, 38
Artotrogus, 49
Artystone, 27
Ascidioxymus, 58
Ascomyzon, 56
Asconiscidae, 311
Ascothoracida, 42, 46, 57
Ascothorax, 57
Asellidae, 30
Asellota, 6, 24, 27, 30, 53
Asellus, 30
Asellus aquaticus, 33, 65
Asotana, 27
Aspidoconcha, 53
Aspidomolgus, 43
Aspidophryxus, 53
Astacidea, 5, 31, 36, 166, 185

Astacilla, 53
Astacilla longicornis, 39, 65
Astacilloechus, 53
Astacopsis gouldi, 5, 66
Astacus astacus, 162, 185
Astacus leptodactylus, 124, 129
Aster, 22
Asterias, 57
Astericola, 56
Asterocheres, 38, 56
Asterocheres parvus, 38, 66
Asterocheridae, 40, 43, 44
Asteropontides, 43
Asteropontius, 43, 44
Asthenognathus, 14, 51
Astralione, 55
Astroxynus, 56
Astyra zenkevitchi, 66
Astyra zenkevithchi, 6, 66
Astyroides carinatus, 6, 66
Athanas, 50, 52, 58
Athanas amazone, 19
Athelges, 55
Atlanterythrops, 334
Attheyella (Chappuisiella) inopinata, 66
Attheyella (Chappuisiella) ruttneri, 66
Attheyella inopinata, 33, 66
Attheyella pilosa, 30, 66
Attheyella ruttneri, 32, 66
Atthyella pilosa, 54, 66
Atthyella trispinosa, 54, 66
Atya, 27, 28
Atyidae, 21, 27, 31
Aurelia, 40
Austinograea, 25
Australerythrops, 334
Australomysis, 338
Australophialus, 11, 52
Austromegabalanus psittacus, 5, 66
Austrominius, 11, 52
Austroniphargus bryophilus, 33, 66
Austroongerthona, 56
Austroongerthona picta, 66
Austropotamobius pallipes, 162, 185
Austrorgathona, 27
Autonoe longipes, 16, 66, 72
Avdeevia, 49
Avicennia, 22
Axiidae, 24
Axiidea, 18, 24
Axius, 55
Baccalaureus, 42
Baccalaureus argalicornis, 47, 66
Baccalaureus japonicus, 46, 66
Baccalaureus maldivensis, 47
Baccalaureus maldivensis maldivensis, 66
Bacescomysis, 247, 253, 296, 331, 332
Bacescomysis abyssalis, 217, 347

Bacteria, 130
Bagatus minutus, 66
Balaenophilus unisetus, 61, 66
Balaenoptera, 52, 61
Balaenoptera borealis, 118, 131
Balaenoptera musculus, 118, 131
Balaenoptera physalus, 118, 131
Balanidae, 44
Balanodytes, 11, 45
Balanomorpha, 11, 44, 59
Balanus, 10, 11, 13, 36, 41, 52, 61, 62, 124
Balanus concavus pacificus, 57, 66
Balanus crenatus, 48, 52, 66
Balanus improvisus, 59
Balanus tintinnabulum, 52
Balssia, 42
Barbouria, 31
Barbouria cubensis, 22, 28, 66
Bartholomea, 44
Bartholomea annulata, 43, 81, 311, 347
Basipodella harpacticola, 51, 66
Bathyactis, 43
Bathyceradocus stephenseni, 6, 66
Bathycuma, 24
Bathymysis, 338
Bathynectes, 56
Bathynella natans, 30, 66
Bathynellacea, 19, 26, 28, 30
Bathynellidae, 30
Bathynomus, 24, 53
Bathynomus giganteus, 5, 66
Bathyporeia robertsoni, 18, 66
Bathyporeia sarsi, 18, 66
Bathyschraderia magnifica, 6, 66
Belonidae, 58
Benthesicymus, 24
Benthesicymus bartletti, 6, 66
Benthoctopus, 49
Benthonectes, 24
Bermudamysis, 324, 341
Berndtia, 11, 45
Betaeus ensenadensis, 55, 66
Birgus latro, 35, 66
Birsteiniamysis, 332
Birsteiniamysis inermis, 201, 202, 217, 221, 238, 239, 241, 244, 347
Bittium, 13
Bivalvia, 9–11, 15, 24
Blastodinium, 62
Boeckosimus normani, 43, 66
Bogidiella, 31
Bogidiellidae, 31
Bogidielloidea, 8, 31
Bolocera, 43
Bomolochidae, 58

Bonnierilla, 58
Bopyridae, 53, 55, 56
Bopyrissa diogeni, 66, 76
Bopyrissa fraissei, 66, 76
Bopyroidea, 53
Bopyrus, 53
Boreoacanthomysis, 340
Boreomysinae, 200, 202, 217, 229, 233, 235, 250, 253, 270, 272, 288, 296, 320, 321, 332
Boreomysinae [in key], 343
Boreomysis, 24, 207, 225, 228, 240, 245, 247, 253, 296, 332
Boreomysis arctica, 241, 262, 280, 281, 284, 286, 288, 347
Boreomysis californica, 295, 347
Boreomysis megalops, 235, 245, 347
Boreomysis microps, 209, 302, 347
Boreomysis obtusata, 200, 347
Boreomysis scyphops, 239, 241, 347
Boreonymphon robustum, 51, 82
Boscia, 45
Boscia anglia, 66
Boscia anglicum, 45
Botachus, 58
Botellina pinnata, 38, 66
Bothroponera tesserinoda, 35, 82
Brachiopoda, 81
Brachycarpus biunguiculatus, 60, 66
Brachyura, 5, 6, 14, 17, 20, 23–25, 27, 29, 31, 32, 34, 35, 45, 46, 51, 52, 57, 62, 63, 120, 121, 166, 185
Bradophila, 50
Bradypontius, 44
Braga, 27
Branchiomma nigromaculatum, 310, 311, 347, 405
Branchiopoda, 4, 5, 25, 26, 32, 36, 166
Branchiura, 4, 36, 59, 124
Branta leucopsis, 128, 131
Brasilomysis, 337
Brementia, 58
Brescianiana, 49
Bresiliidae, 25
Briarella, 49
Brincoxelia abyssalis, 66
Brissopsis, 57
Bromeliaceae, 28, 32, 33
Brugelia, 24
Brychiopontius, 57
Bryocamptus, 30, 33
Bryocamptus pygmaeus, 33, 66
Bryocyclops anninae, 33, 66
Bryocyclops bogoriensis, 32, 33, 66
Bryocyclops chappuisi, 33, 66
Bryozoa, 9, 11, 48, 61, 81
Buccinidae, 50
Buccinum undatum, 13, 82
Bullana, 50
Bunocotyle cingulata, 312, 347

TAXONOMIC INDEX   415

Burrimysis, 341
Bursa, 50
Byblis, 24
Bythograea, 25
Bythograea galapagensis, 25
Bythograea intermedia, 25
Bythograea laubieri, 25
Bythograea microps, 25, 66
Bythograea thermydron, 25, 66
Bythograea vrijenhoeki, 25
Bythograeidae, 5, 25
Cabirops, 53
Caecidotea, 30
Caecosphaeroma, 31, 53
Caesaromysis, 199, 217, 242, 334, 335
Caesaromysis hispida, 218, 348
Calanoida, 30, 49
Calanus hyperboreus, 22, 66
Calappa flammea, 56, 66
Calcinus tubularis, 10, 66
Caligidae, 49, 58, 62, 123
Caligoida, 49
Caligus, 62
Calliactis armillata, 43, 81
Calliactis parasitica, 43, 81
Callianassa, 19, 55
Callianassa californiensis, 55, 67
Callianassidae, 12
Calliasmata pholidota, 21, 67
Callinectes, 22, 28, 55, 63
Callinectes ornatus, 55, 67
Callinectes sapidus, 55, 67, 124, 129
Calliobothrium, 62
Calliopiella michaelseni, 50, 67
Calliopiidae, 31
Callistocypris zlotini, 35, 67
Calvocheres, 56
Calyptomma, 247, 324, 337
Calyptommini, 337, 344, 345
Calyptommini [in key], 345
Cambaridae, 27, 30
Cambaroides, 126
Cambarus, 30, 31, 53
Campecopea, 13, 34, 35
Campecopea hirsuta, 13, 67
Camphyra, 56
Cancellus, 14
Cancer, 143, 144, 152, 159, 160, 166, 172, 173, 183, 191, 348, 400
Cancer bipes, 191, 192, 348
Cancer flexuosus, 191, 348
Cancer irroratus, 176, 178, 179, 185
Cancer magister, 125, 129
Cancer oculatus, 191–193, 348
Cancer pagurus, 146–150, 152–154, 161, 162, 164, 166, 168, 175–178, 185

Cancer pedatus, 191–193, 348
Cancerillopsis, 56
Cancricepon, 56
Cancrincola, 55
Cancrion, 56
Candona, 30
Candona compressa, 33, 67
Candona pratensis, 33, 67
Candonopsis anisitsi, 32, 67
Cantabroniscus primitivus, 31, 67
Cantellius, 45
Canthocamptus pilosa, 66, 67
Canthocamptus trispinosus, 66, 67
Canthocamptus vejdovskyi, 67
Caphyra alcyoniophila, 42, 67
Caprella, 57
Caprella acanthifera, 13, 20, 67
Caprella erethizon, 20, 67
Caprella fretensis, 20, 67
Caprella linearis, 20, 67
Caprella tuberculata, 20, 67
Caprellidae, 40, 57, 62
Caprellidea, 8, 27, 62
Carchesium, 62
Carcinonemertes, 61, 63
Carcinus, 52, 56, 62, 63, 144
Carcinus maenas, 55, 61, 67, 150, 160, 168, 173, 185
Cardiidae, 49
Cardiophilus baeri, 50, 67
Cardisoma, 19, 28, 36, 55, 56
Cardisoma hirtipes, 64, 67
Cardium, 14, 16, 49
Caridea, 5, 6, 10, 12, 16, 20, 24, 31, 36, 45, 46, 56, 60, 118, 166
Caridina, 31
Caridina troglodytes, 31, 67
Carnegieomysis, 242, 339
Carpias minutus, 4, 66, 67
Caspiomysis, 203, 338
Cassidinidea quadricarinata, 9, 67, 69
Cassiopea, 40, 311, 348
Cassiopea xamachana, 40, 80
Catinia plana, 47, 67
Caudella, 54
Cecidomyzon, 40
Cecrinis, 61
Cellepora samboangensis, 81
Cellepora senegambiensis, 48, 81
Celleporaria oculata, 81
Cephalocarida, 4, 26, 36
Cephaloidophora, 62
Cephalolobus, 61
Cephalopoda, 49
Cephalorhyncha, 63
Cephalorhynchus hectori, 61, 83
Cephalothrix galatheae, 63, 67, 81

Ceramium, 62
Cerapus abditus, 67
Cerapus brasiliensis, 16, 67
Cerapus tubularis, 16, 67
Cerastocheres, 49
Ceratodoxomysis, 338
Ceratolana papuae, 12, 67
Ceratolepis, 257, 258, 295, 329
Ceratolepis hamata, 198, 214, 217, 227, 236, 295, 329, 348
Ceratomysis, 225, 255, 258, 331, 332
Ceratomysis spinosa, 244, 348
Cerberusa caeca, 31, 67
Cercopagis pengoi, 309, 348
Ceriodaphnia cornuta, 29, 67
Cestoda, 61, 62
Cetacea, 8, 61
Cetorhinus maximus, 59, 82
Chaenostoma boscii, 67, 70
Chaetopterus, 14, 51
Chalaraspidum, 190, 257, 258, 295, 329
Chalaraspidum alatum, 324, 348
Chaoborus, 118
Charybdis, 56
Charybdis longicollis, 124, 129
Chauliodoniscus reyssi, 4, 67
Chauliolobion, 57
Cheiriphotis megacheles, 21, 67
Cheiroplatea, 39
Chelicorophium robustum, 124, 129
Chelina, 39
Chelonibia, 60
Chelonibia manati, 60, 67
Chelonibia patula, 51, 67
Chelorchestia costaricana, 21, 67
Chelura, 12
Chelura terebrans, 12, 53, 67
Cheluridae, 12
Chilodochona, 62
Chiromantes haematocheir, 67, 77
Chirostylidae, 23
Chirostyloidea, 24
Chiton, 50
Chiton (Chiton) tuberculatus, 82
Chiton tuberculatus, 50, 82
Chlamidae, 49
Chlamydopleon, 333
Chlamydopleon aculeatum, 235, 348
Chlamydotheca, 159
Chlamys, 49
Chlorodiella nigra, 45, 67
Chlorotocella gracilis, 40, 67
Cholidya polypi, 49, 67
Cholidyella, 49
Cholidyinae, 49

Cholomyzon, 44
Cholydiella polypi, 67
Chondracanthidae, 58
Choniomyzon panuliri, 54, 67
Choniosphaera, 55
Choniostomatidae, 53, 55
Chordata, 82, 131
Chorocaris, 25
Chromista, 62, 130
Chrysaora, 40
Chrysaora quinquecirrha, 80
Chthamalidae, 52
Chthamalophilus delagei, 52, 67
Chthamalus, 10
Chthamalus stellatus, 13, 52, 67
Chunomysis, 217, 334, 335
Chunomysis diadema, 206, 348
Chydorus ovalis, 33, 67
Chydorus sphaericus, 33, 68
Chytriodinium, 62
Ciliata, 61, 62, 311
Cilicaea splendida, 9, 68
Ciliophora, 62, 311
Cirolana, 53, 61
Cirolana lineata, 57, 68
Cirolana narica, 68
Cirolana venusticauda, 9, 68
Cirolanidae, 9, 12, 17, 27, 30, 38
Cirolanides, 30
Cirolanoidea, 9
Cirripathes, 46
Cirripedia, 5, 6, 10, 25, 26, 38, 40, 44, 46, 49, 51–53, 57–60, 119, 166
Cithadius, 53
Cladocera, 5, 10, 25, 26, 29, 32, 33
Clausia, 58
Clausidiidae, 49, 50
Clausidium, 55
Clausiidae, 44, 50
Cleantis, 15
Cleantis phryganea, 15, 68, 80
Cleantis prismatica, 13, 15, 68, 80
Cleantis tubicola, 13, 68
Cleonardo longipes, 6, 68
Cleonardo longirostris, 68
Clibanarius, 43
Clibanarius erythropus, 54, 68
Cliona, 39
Cloridopsis dubia, 122, 129
Clymene, 50
Clymenella, 14, 51
Clypeoniscus, 53
Clytia, 39
Clytia hemisphaerica, 62, 63, 80
Clytia johnstoni, 62, 63, 80
Cnidaria, 80
Coccospora, 62
Cochlodelphys, 58

TAXONOMIC INDEX 417

Codium arabicum, 38
Codoba, 56
Coelenterata, 80
Coenobita, 6, 14, 28, 35
Coifmanniella, 333
Coifmanniella johnsoni, 235, 348
Colidotea rostrata, 57, 68
Colocasia, 32
Colomastigidae, 8, 57
Colomastix, 57
Colomastix pusilla, 16, 38, 68
Columbiaemysis, 340
Comanthina, 57
Comanthus, 57
Conasellus, 30
Conchoderma, 52, 61
Conchoderma auritum, 52, 59, 68
Conchoderma virgatum, 52, 55, 59, 60, 68
Conchodytes, 50, 58
Conchodytes tridacnae, 58, 68
Conchoecetes, 14
Concholepas, 11, 52
Conchostraca, 26
Conchyliurus, 49
Condylactis, 43
Conopea, 41
Conopeum commensale, 48, 81
Conopora, 40
Contracaecum, 61
Conus, 9, 14, 50
Copepoda, 4, 5, 7, 10, 19, 25, 26, 29, 30, 32–34, 36, 38–44, 46, 49, 50, 52, 54–58, 61, 62, 119, 123
Corallanidae, 9, 27
Coralliocaris, 45
Corallovexia, 44
Corallovexiidae, 44
Coronula, 61
Coronula diadema, 52, 68
Corophiidea, 8
Corophioidea, 8
Corophium, 13, 16, 21, 125
Corophium shoemakeri, 16, 68
Corophium uenoi, 68
Corophium volutator, 18, 68
Corycaeus, 62
Corycella, 62
Corynactis viridis, 43, 81
Coryphaena hippurus, 59, 82
Cothurnia, 62
Cotylomolgus, 50
Cotylomyzon, 51
Crangon, 62
Crangonidae, 6, 17, 18
Crangonyctidae, 31

Crangonyctoidea, 8, 31
Crangonyx, 31
Craniata, 82
Crassicorophium bonellii, 13, 68
Creaseria, 31
Creaseriella, 30
Creusia, 11, 45
Crinuma, 32
Cruregens, 27, 30
Crustacea, 3, 4, 6, 7, 10, 11, 13, 15, 17, 19, 20, 22–26, 28, 29, 33, 36, 38, 40, 47, 49–51, 56, 60–64, 80, 127, 129, 130, 189, 191
Cryphiops, 31
Crypthelia, 40
Cryptocandona, 30
Cryptocentrus cryptocentrus, 19, 82
Cryptochirus, 45
Cryptocope, 24
Cryptolepas rachianecti, 61, 68
Cryptoniscoidea, 52
Cryptophallus magnus, 47
Cryptophialidae, 11
Cryptophialus, 11, 45, 52
Cryptopleura, 38
Cryptopontius capitalis, 38, 68
Cryptopontius minor, 38, 68
Ctenopontonia cyphastreophila, 45, 68
Cubanomysis, 337
Cubaris granulatus, 35, 68
Cucullaea labiata, 49, 82
Cumacea, 5, 6, 25, 27, 36, 53, 318
Cumaoechus, 53
Cumopsis goodsir, 27, 68
Curtandra, 33
Cyamidae, 8, 61
Cyamus, 61
Cyamus rhytinae, 60, 68
Cyanagraea, 25
Cyanea, 40
Cyanophyta, 16
Cyathura, 27, 30, 53
Cyclactinia, 47
Cyclocypris laevis, 33, 68
Cyclopoida, 30, 49
Cyclops quadricornis, 29, 68
Cylindroleberis, 15
Cylindrolepas, 60
Cymadusa, 16
Cymatocarpus, 62
Cymo andreossyi, 45, 68
Cymo melanodactylus, 45, 68
Cymodoce, 9, 11
Cymodoce bifida, 9, 68
Cymodocea, 20, 297, 348
Cymodocea japonica, 12
Cymodopsis gorgoniae, 42, 68
Cymothoida, 9, 27, 30, 36, 53, 59, 311
Cymothoidae, 27, 59

Cynops pyrrhogaster, 83
Cypridina, 59
Cypridina parasitica, 59, 68
Cypridina squamosa, 59, 68
Cypridinopsis, 30
Cypris balnearia, 29, 68
Cyproniscus cypridinae, 68
Cyproniscus cyprinidinae, 52, 68
Cyrtomaia, 5
Cyrtophium, 13
Cystomyzon, 40
Cystoseira, 21
Dactylamblyops, 334
Dactylerythrops, 334
Dactylerythrops dactylops, 244, 348
Dactylocythere, 30
Dactylometra quinquecirrata, 40, 80
Dactylopusia neglecta, 53, 68
Dahlella caldariensis, 25, 68
Dajidae, 311
Danalia, 52, 56
Danalia hapalocarcini, 56, 68
Danalia ypsilon, 55, 68
Daphnia, 62, 308, 309, 348
Daphnia ambigua, 29, 68
Dardanus, 39, 54, 309, 348
Dardanus arrosor, 43, 55, 68
Dardanus sanguinolentus, 55, 68
Dardanus tinctor, 68, 74
Darwinula, 30
Darwinula malayica, 29, 35, 68
Darwinula zimmeri, 33, 68
Dasia, 58
Dasycaris, 46
Decapoda, 4–6, 10, 16, 17, 19, 23–27, 35, 40, 46, 51, 55, 57, 62, 118, 122, 124, 165, 166, 185, 209, 219, 237, 250, 268, 292, 319
Deltamysis, 341
Dendraster excentricus, 57
Dendrobranchiata, 5, 185
Dendrogaster, 57
Dendrophyllia, 44
Dendrosomides paguri, 62, 80
Dentalium, 13, 14
Deoterthridae, 52
Deoterthron, 51
Desmidiaceae, 62
Desmocarididae, 27
Dexaminidae, 8, 57
Dexaminoidea, 8
Dhalella caldariensis, 5, 68
Diacyclops, 30
Diacyclops bisetosus, 33, 69
Diacyclops nanus, 69
Diacyclops pulchellus, 33, 69

Diadema, 298, 348
Diadema antillarum, 310, 348, 405
Diamysini, 339, 345
Diamysini [in key], 345
Diamysis, 204, 210, 211, 213, 227, 247, 250, 280, 281, 297, 299, 305, 322, 339
Diamysis bacescui, 211, 297, 348
Diamysis bahirensis, 206, 237, 297, 348
Diamysis camassai, 213, 229, 231, 297, 308, 348
Diamysis cymodoceae, 201, 204, 211, 213, 235, 297, 348
Diamysis fluviatilis, 219, 348
Diamysis lagunaris, 197, 213, 219, 244, 326, 348
Diamysis mesohalobia, 213, 219, 229, 247, 249, 281, 287, 291, 293, 348
Diaponticus, 43
Diarthrodes feldmanni, 38, 69
Diastylis, 24
Diastylis goodsiri, 5, 69
Dichelesthiidae, 58
Dichelina phormosomae, 56, 69
Didacna baeri, 50, 82
Dienictylus pyrrhogaster, 60, 83
Dies quadricarinatus, 69
Dikerogammarus bispinosus, 124, 129
Dikerogammarus villosus, 124, 129
Dinophyceae, 61, 62
Diodon, 52
Diodon hystrix, 59, 82
Diogenella, 57
Diogenes, 43
Diogenes ovatus, 47
Diogenes pugilator, 48, 54, 69, 74
Diogenidion, 57
Dioptromysis, 242, 338
Dioptromysis paucispinosa, 242, 348
Dioptromysis perspicillata, 244, 348
Diphyllobothrium, 126
Diplaneis, 80
Diplodontidae, 49
Diploneis smithii, 62, 80
Diploria, 44
Diplosoma, 59
Diptera, 64, 118, 130
Disacanthomysis, 340
Disciadidae, 39
Discias exul, 39, 69
Discias mvitae, 69
Discoporella umbellata, 48, 81
Dissodactylus, 57
Distichopora, 45
Dofleinia, 44
Dogielinotidae, 17
Doliolida, 58
Dollocaris ingens, 317, 348
Domecia glabra, 45, 69
Domecia hispida, 45, 69
Donsiella limnoriae, 53, 69

Doridicola, 41, 43, 46, 49, 50, 56, 57
Doropygus, 58
Doryphallophora, 52
Dosima fascicularis, 22, 40, 51, 52, 69, 72
Dotilla, 19
Dotilla fenestrata, 17, 69
Dotilla myctiroides, 17, 69
Doxomysis, 211, 338
Dracunculus medinensis, 126, 130
Drepanorchis, 56
Dromiacea, 39
Dromiidae, 14
Dugong, 60
Dulichia, 57
Duplorbis, 53
Duplorbis calathurae, 53, 69
Durvillea, 21
Dynamene bidentata, 13
Dynamene ramuscula, 9, 69, 70
Dynamenella, 9
Dynamenella australis, 50, 69
Dynamenella moorei, 50, 69
Dynamenella octoloba, 9, 69
Dynamenella perforata, 50, 69
Dynamenopsis, 9
Dynamopsis dianae, 50, 69
Dysidea, 38
Dysidea fragilis, 38, 80
Ebalia, 62
Echinaster, 57
Echinobothrium, 62
Echinocardium, 57
Echinodermata, 57
Echinogammarus trichiatus, 124, 129
Echinomysides, 334
Echinomysis, 199, 334
Echinopora, 44
Echinorhynchus leidyi, 312, 348
Echiura, 47, 81
Echiurophilus, 47
Ectocarpus, 62
Ectocyclops medius, 32, 69
Ectocyclops phaleratus, 32, 69
Eisothistos macrurus, 13, 69
Eisothistos pumilus, 13, 69
Elaphoidella, 30
Elaphoidella bromeliaecola, 32, 33, 69
Elaphoidella cornuta, 33, 69
Elaphoidella elegans, 32, 69
Elaphoidella malayica, 33, 69
Elaphoidella sewelli, 32, 69
Elaphoidella thienemanni, 33, 69
Elasmopus, 57
Elasmopus calliactis, 43, 69
Elatostemma, 33

Elder unguiculata, 317, 348
Electra pilosa, 61
Ellobiopsidae, 62, 312
Ellobiopsis, 62
Elpidium bromeliarum, 32, 69
Emerita, 17
Emerita analoga, 17
Enalcyonium parvum, 42, 69
Enalcyonium rubicundum, 42, 69
Enalcyonium setigerum, 42, 69
Engaeus, 55, 62
Enterognathus, 57
Entocythere, 30, 53
Entocytheridae, 30
Entomolepis adriae, 69
Entomopsyllus adriae, 38, 69
Entoniscidae, 56
Eoamblyops, 336
Eochionelasmus ohtai, 25, 69
Epactophanes, 33
Epactophanes muscicola, 33, 69
Epactophanes richardi, 32, 33, 69
Ephelota, 62
Ephydridae, 119
Epicaridea, 52, 53, 55, 56, 61
Epilobocera, 28, 32
Epimeria, 57
Epimolgus, 49
Epinephelus lanceolatus, 82, 83
Episesarma mederi, 19, 69, 77
Epistylis, 61, 62
Epizoanthus paguriphilus, 47
Epizoanthus paguropsidis, 47
Ergasilidae, 49, 58, 123
Ericthonius, 125
Ericthonius brasiliensis, 16, 69
Ericthonius pugnax, 16, 69
Ericthonius punctatus, 16, 21, 67, 69, 76
Eriocheir, 22, 26, 126
Eriocheir sinensis, 123, 124, 130
Eriphia, 145, 161
Erugosquilla massavensis, 119, 124, 130
Eryonidae, 24
Erythropinae, 200, 202, 204, 211, 217, 224, 225, 233, 235, 253, 267, 272, 296, 320, 323, 334–336, 344
Erythropinae [in key], 345
Erythropini, 320, 334, 336
Erythropini [in key], 346
Erythrops, 24, 242, 334
Erythrops abyssorum, 335, 348
Erythrops serratus, 335, 348
Ethusina abyssicola, 6, 69
Ethusina alba, 6, 69
Etmopterus spinax, 59, 82
Eucarida, 166, 185, 190, 318, 319
Euchaetomera, 241, 334
Euchaetomera zurstrasseni, 244, 348

Euchaetomeropsis, 334
Eucopia, 232, 238, 240, 245, 258, 267–269, 295, 302, 319, 321, 330
Eucopia australis, 194, 198, 214, 217, 224, 230, 236, 257, 295, 324, 348
Eucopia crassicornis, 229, 263, 348
Eucopia grimaldii, 324, 348
Eucopia major, 244, 348
Eucopia precursor, 317, 348
Eucopia sculpticauda, 205, 210–212, 239, 245, 257, 258, 260, 287, 324, 348
Eucopia unguiculata, 232, 243, 260, 264, 271, 278, 283, 348
Eucopidae, 194
Eucopiidae, 194, 198, 201, 203–205, 209, 210, 214, 223, 224, 227, 233, 235, 236, 278, 313, 317, 319, 323, 330
Eucopiidae [in key], 342
Eucyclops, 30
Eucypris latissima, 5, 69
Eucypris neumanni, 5, 70
Eudorella, 24
Eumalacostraca, 166, 185, 190, 191, 195, 288, 317, 319, 329–331, 342
Eumedon, 57
Eumunida, 40
Eunicicolidae, 50
Euonyx, 57
Euphausia, 62
Euphausia pacifica, 118, 130
Euphausia superba, 118, 130
Euphausiacea, 5, 26, 36, 62, 118, 193, 194, 219, 242, 318, 319
Euplax bosci, 17, 70
Euplectella, 39
Euplectella aspergillum, 38, 80
Eupolyodontes, 51
Eurobowmaniella, 333
Eurydice, 17
Eurydice affinis, 17, 70
Eurydice pulchra, 17, 70
Euryrhynchidae, 27
Eurysilemum, 50
Eurystheus maculatus, 13, 55, 70
Eurythenes gryllus, 5, 70
Eusiroidea, 8, 31
Eusirus, 62
Eusirus fragilis, 6, 70
Eusphaeroma, 12
Eusphaeroma ovatum, 70
Eutetrarhynchidae, 62
Euthylacus, 52
Exacanthomysis, 340
Excirolana, 17
Excirolana natalensis, 17, 70

Excirolana orientalis, 17, 70
Exocetus, 52
Exosphaeroma, 50, 53
Exosphaeroma ramusculum, 70
Fabia, 57
Faucheria, 30
Favia, 44
Fecampia, 61
Fecampia balanicola, 62, 81
Fecampia spiralis, 62, 81
Flabellicola, 51
Flabellifera, 9, 27
Flabelligera, 51
Foettingeria, 62
Folliculina, 62
Folliculinopsis, 62
Fortimesus gigas, 6, 70
Francocaris grimmi, 317, 348
Fungi, 130
Fungia, 45
Fusinus, 49, 50
Gaillardiellus superciliaris, 64, 70
Galapsiellus leleuporum, 4, 21, 70
Galathea, 42, 57
Galatheoidea, 6, 24, 25, 31, 40, 54, 55, 57, 63
Gallocaris, 31
Gammaridae, 17
Gammaridea, 7, 8, 27, 28, 31, 40, 53
Gammaroidea, 7, 31
Gammaropsis, 57
Gammaropsis anaculata, 70
Gammaropsis maculata, 13, 70
Gammaropsis nitida, 55, 70, 76
Gandalfus, 25
Gangemysis, 339
Gasterocheres, 41
Gasterosteus aculeatus, 118, 131
Gastrodelphyidae, 51
Gastropoda, 10, 11, 24, 28, 49, 50
Gastrosaccinae, 200–204, 207, 210, 217, 233, 235, 253, 270, 272, 296, 297, 319–321, 323, 333
Gastrosaccinae [in key], 344
Gastrosaccini, 333, 344
Gastrosaccini [in key], 344
Gastrosaccus, 62, 204, 304, 333
Gastrosaccus brevifissura, 302, 348
Gastrosaccus psammodytes, 302, 348
Gastrosaccus roscoffensis, 206, 222, 308, 348
Gastrosaccus sanctus, 297, 348
Gastrosaccus spinifer, 203, 269, 272, 293, 348
Gastrosaccus widhalmi, 231, 349
Gastrosaccus yuyu, 316, 349
Gebia, 19
Gebiidea, 18
Gecarcinidae, 18, 31, 35
Gecarcinus, 6, 28, 36
Gecarcoidea, 6
Gecarcoidea natalis, 36, 64, 70

TAXONOMIC INDEX

Gelastocaris, 39
Gemmosaccus, 54
Geograpsus, 35
Geosesarma noduliferum, 70
Geryon, 55
Geryonidae, 23, 24, 63
Gevgeliella, 31
Gibbamblyops, 334
Gibberythrops, 334
Gibbosphaeroma, 9
Gigantocypris agassizi, 5, 70
Gironomysis, 340
Gitanopsis aff. *pusilla*, 40
Gitanopsis paguri, 55, 70
Gitanopsis pusilla, 70
Glyphocrangonidae, 24
Glyptelasma carinatum, 52, 70
Glyptelasma gracilius, 52, 70
Gnathia maxillaris, 13, 70
Gnathiidea, 27, 59
Gnathomysis gerlachei, 55, 70
Gnathophausia, 62, 190, 203, 205, 207, 229, 232, 240, 258, 263, 269, 295, 302, 319, 330
Gnathophausia affinis, 295, 349
Gnathophausia childressi, 219, 220, 295, 349
Gnathophausia gigas, 324, 349
Gnathophausia ingens, 5, 70, 198, 201, 205, 214, 236, 243, 257, 260, 263, 267, 278, 294, 295, 300, 302, 349
Gnathophausia longispina, 214, 217, 220, 224, 239, 262, 295, 349
Gnathophausia zoea, 192, 198, 207, 209, 224, 237, 279, 287, 288, 324, 349, 404
Gnathophausiidae, 198, 201, 203–205, 209, 210, 214, 217, 221, 223–225, 233, 235, 236, 245, 267, 278, 313, 319, 323, 330
Gnathophausiidae [in key], 342
Gnathophyllidae, 58
Gnathostomulida, 7
Gnomoniscus, 53
Gnorimosphaeroma oregonensis, 12, 70
Gnorimosphaeroma ovatum, 12, 70
Goidelia, 47
Gomphocythere angresta, 70
Gomphocythere angulata, 5, 70
Gomphopodaria, 57
Goniopsis, 10, 55
Gonodactylaceus randalli, 125, 130
Gonophysema gullmarensis, 58, 70
Gorgonacea, 40–42
Gorgonolaureus, 42
Graeteriella, 30
Gramineae, 13
Graneledone, 49
Grapsidae, 27, 31, 32, 35

Grapsus, 10, 19, 28, 52
Gregarina, 61, 62
Guinotia, 32
Gymnangium, 39
Gymnerythrops, 207, 334, 335
Gymnodinioides, 61, 62
Gymnodinium catenatum, 125, 130
Gymnothorax, 59
Hadrothoe crosnieri, 56, 70
Hadzia, 31
Hadziidae, 8, 28, 31
Haematodinium, 62
Haemulon flavolineatum, 305, 349
Halacari, 38
Halacarsantia uniramea, 65, 70
Halemysis, 339
Halice macronyx, 6, 70, 78
Halice subquarta, 6, 70
Halichondria, 16, 38, 39
Haliclona, 38
Halicometra, 62
Halicyclops caridophilus, 55, 70
Halimeda, 21
Haliporoides triarthrus, 56, 70, 71
Halirages, 24
Halisarca, 39
Halocaridina rubra, 21, 70
Halosbaena, 27, 30
Hammatimyzon, 40
Hamopontonia, 44
Hansenomysinae, 214, 225, 331
Hansenomysinae [in key], 343
Hansenomysis, 214, 247, 253, 296, 332
Hansenomysis abyssalis, 206, 229, 235, 349
Hansenomysis atlantica, 222, 223, 349
Hansenomysis falklandica, 202, 349
Hansenomysis nouveli, 206, 229, 235, 349
Hansenomysis pseudofyllae, 218, 223, 349
Hapalocarcinidae, 45, 56
Hapalocarcinus marsupialis, 45, 56, 70
Haplomesus, 6
Haplomesus gigas, 70
Haploniscus, 24
Haploniscus unicornis, 4, 70
Haploops tubicola, 16, 70
Haplostylus, 204, 333
Haplostylus lobatus, 200, 349
Haplostylus magnilobatus, 249, 349
Haptolana, 30
Harmelinella, 233, 235, 340, 341
Harmelinellini, 341
Harmelinellini [in key], 346
Harpacticoida, 19, 30, 49, 51
Harpacticus pulex, 60, 61, 70
Harpilius consobrinus, 70, 75
Harpilius lutescens, 45, 70, 76
Harpinia, 24
Harpinia excavata, 6, 70

Harpinia spaercki, 70
Harpiniopsis spaercki, 6, 70
Harrietella simulans, 53, 70
Harrovia elegans, 45, 70
Haustoriidae, 17, 57
Hedyphanella, 50
Heike monogatori, 128
Heikeopsis japonica, 128, 130
Heliogabalus, 47
Heliopora, 45
Hematodinium, 61
Hemiacanthomysis, 340
Hemichordata, 51
Hemicyclops, 54
Hemicyclops acanthosquillae, 52, 71
Hemicyclops perinsignis, 38, 71
Hemidoxyphium, 58
Hemigrapsus oregonensis, 125, 130
Hemilaophonte janinae, 55, 71
Hemimysis, 210, 286, 299, 325, 338
Hemimysis anomala, 124, 130, 273, 308, 309, 315, 326, 327, 349, 406
Hemimysis lamornae, 221, 260, 262, 284, 349
Hemimysis lamornae mediterranea, 192, 349, 404
Hemimysis margalefi, 314, 349
Hemimysis speluncola, 217, 241, 251, 287, 299, 301, 314, 349
Hemioniscus balani, 52
Hemisiriella, 333
Hemithiris psittacea, 81
Hemiurus, 62
Hepatus, 43
Hepatus epheliticus, 56, 71
Hermodice, 50
Herpotanais kirkegaardi, 6, 71
Herpyllobiidae, 50, 53
Herpyllobius, 50
Hersiliodes, 50
Heteralepas, 55
Heteralepas quadrata, 52, 71
Heterocarpus alphonsi, 71
Heterocarpus dorsalis, 24, 71
Heteroerythrops, 334
Heteromysinae, 200, 202, 204, 211, 214, 223, 227, 233, 235, 270, 272, 299, 309, 340, 344, 346
Heteromysinae [in key], 346
Heteromysini, 320, 340
Heteromysini [in key], 346
Heteromysis, 30, 55, 213, 214, 231, 235, 253, 270, 309–311, 324, 341, 405
Heteromysis (*Gnathomysis*) *gerlachei*, 70, 71
Heteromysis (*Olivemysis*) *actiniae*, 71
Heteromysis actiniae, 43, 311, 349
Heteromysis arianii, 222, 223, 349
Heteromysis armoricana, 232, 273, 274, 349

Heteromysis dardani, 229, 349
Heteromysis harpax, 309, 311, 349
Heteromysis harpaxoides, 201, 349
Heteromysis wirtzi, 192, 203, 229, 235, 271, 349, 404
Heteromysoides, 341
Heteromysoides cotti, 21, 71, 244, 245, 349
Heterosaccus, 56
Heterotanais oerstedii, 15, 71
Hexabathynella halophila, 26, 71
Hexacreusia durhami, 44, 65, 71
Hexapoda, 34, 82, 130
Hexelasma, 24
Himerometra, 57
Hippa, 17
Hippa pacifica, 17, 71
Hippa testudinaria, 17, 71
Hippacanthomysis, 340
Hippidea, 17, 36
Hippolysmata grabhami, 71
Hippolyte, 58
Hippolyte commensalis, 42, 71
Hippolytidae, 31, 39, 40, 60
Hippopus, 49
Hiroa, 45
Hirondellea gigas, 5, 71
Hoekia, 45
Holmesiella, 335
Holmesimysis, 340
Homarus, 54, 55, 62, 169, 183
Homarus americanus, 4, 5, 71, 125, 130
Homarus gammarus, 144–153, 155, 161, 164, 166, 168, 176, 177, 179, 185
Homodactylus, 42
Homolidae, 23, 24
Homolodromiidae, 23, 24
Homostichanthus, 43
Hoplocarida, 25, 166, 185
Hoplopleon medusarum, 40, 71
Hourstonius pusilla, 70, 71
Huenia proteus, 21, 71
Hyachelia tortugae, 60, 71
Hyale grandicornis, 50, 71
Hyale grimaldii, 60, 71
Hyale milloti, 33
Hyale nilssoni, 50
Hyale perieri, 50, 71
Hyalella, 31
Hyalella aff. *curvispina*, 5
Hyalella curvispina, 71
Hyalellidae, 31
Hydractinia echinata, 40
Hydrodamalis gigas, 60, 83
Hydroida, 39
Hydroides, 13
Hydrozoa, 9, 62, 80
Hymeniacidon, 38
Hymenopenaeus triarthrus, 71

TAXONOMIC INDEX

Hyperacanthomysis, 340
Hyperamblyops, 335
Hypererythrops, 335
Hyperiidea, 8, 10, 26, 27, 40
Hyperiimysis, 227, 338
Hyperoedesipus plumosus, 29, 71
Hyperstilomysis, 339
Hypocamptus, 33
Hypocamptus brehmi, 33, 71
Hypoconcha, 14
Hyssuridae, 20
Iais, 53
Ibla cumingi, 52, 71
Idiomysis, 339
Idiomysis inermis, 309, 349
Idiomysis tsurnamali, 309, 311, 349
Idotea, 53
Idotea balthica, 53, 71, 291, 349
Idotea chelipes, 53, 71
Idotea neglecta, 20, 71
Idotea pelagica, 53, 71
Idotea viridis, 20, 71
Idunella, 57
Iiella, 333
Iimysis, 338
Illigiella, 335
Ilyoplax, 19
Ilyoplax ceratophorus, 18, 71
Ilyoplax delsmani, 18, 19, 71
Ilyoplax elegans, 19, 71
Indoclausia, 44
Indoerythrops, 335
Indomolgus, 43
Indomysis, 324, 339
Ingolfiella, 31
Ingolfiella fontinalis, 31, 71
Ingolfiellidea, 8, 31
Inodosporus, 62
Insecta, 35, 64, 82
Intersunaristes dardani, 71
Inusitatomysini, 335
Inusitatomysini [in key], 346
Inusitatomysis, 335
Iphimedia eblanae, 71
Ircinia, 38
Irus mitis, 82
Isaacsicalanus, 25
Isaea, 55
Isaeidae, 57
Ischiomysis, 231, 309, 341
Ischiomysis peterwirtzi, 277, 349
Ischiomysis telmatactiphila, 229, 232, 349
Ischnochitonika, 49
Ischnopontonia lophos, 45, 71
Ischyroceridae, 57

Ischyrocerus, 57
Ischyromene australis, 69, 71
Isidicola, 41
Isocyamus, 61
Isonebula, 27
Isopoda, 4–8, 15, 18–20, 24–27, 29, 30, 33–35, 38, 50, 52, 53, 55, 57, 59, 62, 122, 166, 185, 209, 267, 268, 292, 311
Jaera, 27, 34, 71
Jaera (*Jaera*) *nordmanni nordica insulana*, 71
Jaera hopeana, 27, 53, 71
Jaera nordica insulana, 27, 71
Jaera nordmanni guernei, 27, 71
Janaria mirabilis, 48, 81
Janthina, 50
Jassa, 57, 125
Jassa falcata, 16, 21, 71
Javanisomysis, 339
Jocastes, 45
Johannella, 30
Kainommatomysis, 242, 324, 326, 338, 339, 345
Kainommatomysis foxi, 326, 349
Kainommatomysis schieckei, 242, 349
Katamysis, 27, 203, 322, 339
Katamysis warpachowskyi, 27, 72
Katerythrops, 335
Kelleria, 57
Kilianicaris lerichei, 317, 349
Kochimysis, 341
Kochlorine, 11, 45
Kochlorinopsis, 11, 48
Koleolepas, 49
Kombia, 44
Koonunga cursor, 32, 72
Kronborgia amphipodicola, 62, 81
Kronborgia caridicola, 62, 81
Kronborgia spiralis, 81
Kupellonura serritelson, 20, 72
[*L.*] *anserifera*, 128
Lafystiidae, 8
Lafystius sturionis, 59
Lagenophrys, 62
Lamellibrachia, 51
Lamellibranchia, 49, 50
Lamiaceae, 13
Laminaria, 21
Lamippe, 41
Lamippe bouligandi, 41, 72
Lamippella faurei, 42, 72
Lamippidae, 41
Lamippina aciculifera, 42, 72
Lamippula parva, 69, 72
Langitanais willemoesi, 15, 72
Laophonte, 53
Laphystiopsidae, 8, 57
Laphystiopsis, 57
Laticorophium baconi, 16, 21, 72
Latreutes anoplonyx, 40, 72

424   TAXONOMIC INDEX

Latreutes mucronatus, 40, 72
Laura gerardiae, 46, 72
Lecanurius, 57
Lecithomyzon, 55
Leiochone, 50
Leiocomes, 56
Lembos longipes, 72
Lepadomorpha, 59
Lepas, 22, 36, 40, 50, 61
Lepas anatifera, 52, 72, 124, 130
Lepas anserifera, 128, 130
Lepas australis, 52, 72
Lepas fascicularis, 72
Lepas hilli, 52, 72
Lepechinella ultraabyssalis, 6, 72
Lepechinella wolffi, 6
Lepechinella wolffi wolffi, 72
Lepidasthenia digueti, 51, 82
Lepidomysidae, 194, 196, 199, 201, 203, 206, 207, 214, 216, 221, 226, 227, 229, 235, 236, 257, 267, 322, 330
Lepidomysidae [in key], 343
Lepidomysis, 194, 330, 349, 352
Lepidomysis servata, 352
Lepidonotus, 50
Lepidophthalmidae, 194
Lepidophthalmus, 349
Lepidophthalmus servatus, 194, 349, 352
Lepidopidae, 194
Lepidops, 30, 194, 349
Lepidopsidae, 194
Lepisosteus, 59
Leptasterias, 57
Leptastraea, 45
Leptocheirus pilosus, 16, 72
Leptochiton cancellatus, 82
Leptodora kindti, 5, 72
Leptomithrax longipes, 56, 72, 75
Leptomysinae, 200, 202, 210, 211, 225, 233, 235, 242, 263, 272, 309, 320, 337, 344
Leptomysinae [in key], 344
Leptomysini, 320, 336, 337, 341, 344
Leptomysini [in key], 344
Leptomysis, 213, 279, 286, 296, 303, 309–311, 337, 338, 405
Leptomysis buergii, 213, 279, 280, 292, 349
Leptomysis gracilis, 201, 203, 211, 349
Leptomysis heterophila, 235, 304, 349
Leptomysis lingvura, 241, 265, 273–275, 279, 280, 283–287, 289, 292, 298, 299, 303, 304, 310, 349
Leptomysis lingvura marioni, 231, 310, 349, 405
Leptomysis mediterranea, 213, 350
Leptomysis posidoniae, 213, 350
Leptomysis truncata sardica, 280, 350

Leptostraca, 5, 26, 209
Lernaea, 62
Lernaea cyprinacea, 60, 72
Lernaea esocina, 60, 72
Lernaea ranae, 60, 72
Lernaeenicus, 62
Lernaeidae, 58
Lernaeocera, 62, 63
Lernaeocera branchialis, 63, 72
Lernaeoida, 49
Lernaeolophidae, 58
Lernaeopodidae, 58
Lernaeosaccus, 56
Leucothoe incisa, 13, 72
Leucothoe spinicarpa, 13, 72
Leucothoidae, 8
Leucothoidea, 8
Leucothoides, 72
Leucothoides pacifica, 72
Lichina pygmaea, 13
Lichomolgidae, 39–41, 43, 44, 49, 50
Lichomolgidium, 58
Lichomolgus, 40, 44, 49
Lichomolgus hippopi, 49, 72
Lichomolgus longicauda, 49, 72
Lichomolgus tridacnae, 49, 72
Lichothuria, 57
Ligia, 10, 34, 62
Ligia latissima, 34
Ligia perkinsi, 34, 72
Ligia philoscoides, 34, 72
Ligia platycephala, 34, 72
Ligia simoni, 34, 72
Ligur uveae, 22, 72
Liljeborgia caeca, 6, 72
Liljeborgiidae, 8, 57
Liljeborgioidea, 8, 31
Limnomysis, 27, 300, 322, 325, 339
Limnomysis benedeni, 27, 72, 235, 273, 293, 307, 308, 314–316, 326, 327, 350, 406
Limnoria, 11, 12, 53, 62, 63
Limnoria lignorum, 53, 72
Limnoria tripunctata, 53, 72
Limnoriidae, 9, 12
Limnosbaena, 30
Limulus, 51, 164
Linaresia, 41
Linaresia mammillifera, 41, 42, 72
Lineus longissimus, 48
Liocarcinus, 52, 56, 62
Lirceus, 30
Liriopsidae, 56
Liriopsis, 52
Lironeca, 27
Lissocarcinus, 57
Lissocephala powelli, 64, 82
Listinogaster, 56
Listriella, 57

TAXONOMIC INDEX

Lithocarpus, 39
Lithodidae, 24
Lithodomus, 15
Lithoglyptes, 11, 45
Lithoglyptidae, 11
Lithophaga, 15
Lithophaga lithophaga, 15
Lithotrya, 11
Litopenaeus vannamei, 121, 130
Litoscalpellum giganteum, 72, 77
Liuimysis, 335
Loimia, 14, 51
Loligo, 49
Longithorax, 204, 207, 335
Longithorax fuscus, 203, 213, 350
Lophogaster, 194, 204, 205, 209, 210, 232, 257–259, 267, 295, 319, 321, 324, 329
Lophogaster affinis, 323, 350
Lophogaster erythraeus, 323, 350
Lophogaster longirostris, 324, 350
Lophogaster pacificus, 205, 350
Lophogaster schmidti, 295, 350
Lophogaster spinosus, 295, 350
Lophogaster subglaber, 323, 350
Lophogaster typicus, 194, 198, 204, 205, 209, 214, 217, 236, 237, 252, 259, 260, 262, 268, 269, 271, 323, 324, 350
Lophogaster voultensis, 317, 350
Lophogastrida, 190–192, 194–198, 203–205, 207, 209, 211, 212, 214, 219, 220, 224, 225, 227, 229–232, 235–237, 239–241, 243, 244, 251, 252, 255, 257–260, 262, 264, 267–272, 276–278, 281, 283, 284, 287, 289, 292, 293, 295, 300, 305, 306, 311–313, 317–319, 321, 323–326, 328, 329, 342, 404
Lophogastrida [in key], 342
Lophogastridae, 194, 198, 201, 203–205, 209, 210, 214, 217, 236, 257, 258, 278, 313, 317, 319, 323, 329
Lophogastridae [in key], 342
Lophogastrina, 327
Lottia, 50
Loxothylacus, 56
Lucifer, 26
Lucioperca, 59
Lybia, 43, 56
Lychnorhiza malayensis, 40, 80
Lycomysis, 340
Lysianassidae, 57
Lysianassoidea, 7
Lysiosquilla, 19
Lysiosquilla maculata, 72
Lysiosquillina maculata, 19, 72
Lysmata, 39
Lysmata amboinensis, 60, 72

Lysmata californica, 60, 72
Lysmata grabhami, 60, 71, 72
Lysmata seticaudata, 60, 72
Macandrevia cranium, 48, 81
Macrobrachium, 22, 27, 28, 31, 56, 121
Macrobrachium americanum, 21, 73
Macrobrachium carcinus, 5, 73
Macrobrachium grandimanus, 21, 73
Macrocheira, 24
Macrocheira kaempferi, 5, 73
Macrochiron, 39, 56, 58
Macrocoeloma, 56
Macrocylindrus, 24
Macrocystis, 12, 21
Macrophthalmidae, 14
Macrophthalmus, 19, 22
Macrophthalmus convexus, 17, 73
Macrophthalmus dilatatus, 18, 73
Macrophthalmus grandidieri, 17, 73
Macrophthalmus japonicus, 18, 73
Macropipus, 52, 62
Macropodia, 56
Macrostylis galatheae, 6, 73
Mactra, 49
Magniezia, 30
Magnippe, 41
Maja brachydactyla, 55, 73
Maja squinado, 55, 73
Majidae, 23, 39, 45, 46, 61
Majoidea, 24
Makaira, 52
Malacolepas conchicola, 49, 73
Malacostraca, 4, 25, 32, 34, 50, 165, 166, 185, 190, 197, 223, 264, 318
Maldanidae, 50
Mammalia, 83
Mancasellus, 30
Mancomysinae, 332
Mancomysini, 320
Maraenobiotus, 33
Maraenobiotus vejdovskyi, 33, 73
Marsupenaeus japonicus, 124, 130
Marumomysis, 255, 336
Mastigias papua, 40, 80
Maxillopoda, 119, 166
Mebis holothuriae, 57
Meckelia, 31
Megabalanus, 124
Megabalanus decorus, 56, 73
Megabalanus stultus, 44, 73
Megabalanus tintinnabulum, 51, 73
Megalasma, 24
Megalasma striatum, 52, 73
Megaleledone, 49
Megalophthalmus Fabricianus, 193, 350 [= *Mysis oculata*]
Megalopsis, 338
Meganyctiphanes, 62

Meganyctiphanes norvegica, 22, 73, 118, 130
Megathura, 50
Megatrema anglicum, 66, 73
Megatrema madreporarum, 38
Meierythrops, 335
Melicertus hathor, 124, 130
Melita obtusata, 13, 55, 57, 73
Melitoidea, 8, 31
Menaethius monoceros, 21, 73
Menigrates, 57
Menippe, 51
Menippe mercenaria, 56, 73
Meomicola, 56
Mephidippoidea, 8
Mesacanthomysis, 340
Mesnilia, 50
Mesocypris, 33
Mesocypris terrestris, 35, 73
Mesoglicola delagei, 43, 73
Mesomysis intermedia, 73
Mesopodopsis, 210, 214, 297, 339
Mesopodopsis aegyptia, 197, 350
Mesopodopsis africana, 298, 306, 308, 350
Mesopodopsis orientalis, 264, 265, 273, 280, 284, 286, 311, 350
Mesopodopsis slabberi, 191, 213, 255, 257, 281, 294, 301, 303, 309, 313, 316, 350
Mesopodopsis tenuipes, 292, 350
Mesopodopsis wooldridgei, 309, 350
Mesopontonia, 42
Metabetaeus lohena, 21, 73
Metabetaeus minutus, 22, 73
Metacarcinus magister, 125, 129, 130
Metaceradocoides vitjazi, 6, 73
Metacrinus, 57
Metacypris bromeliarum, 32, 73
Metacypris laesslei, 32, 73
Metamblyops, 335
Metamysidopsis, 223, 337
Metamysidopsis elongata, 273, 350
Metamysidopsis neritica, 294, 350
Metaniphargus, 31
Metapenaeopsis commensalis, 36, 73
Metapenaeus monoceros, 124, 130
Metaplax, 19
Metaplax elegans, 71, 73
Metasiriella, 333
Metasiriellini, 333
Metasiriellini [in key], 343
Metavargula, 51, 52
Metaxymolgus, 41, 43
Metazoa, 62
Meterythrops, 335
Methypocoelis ceratophora, 18
Metis holothuriae, 73

Metopa, 13
Metopa alderi, 39, 73
Metopa borealis, 39, 73
Metopa glacialis, 50, 73
Metopa groenlandica, 50, 73
Metopa solsbergi, 43, 73
Metopaulias, 28, 32, 62
Metopaulias depressus, 32, 73
Metopograpsus, 35
Metridia longa, 22, 73
Metridium senile, 81
Mexiconiscus laevis, 31, 73
Mexilana, 30
Michthyops, 247, 337
Micrallecto, 49
Micraspides, 55
Microcephalus, 61
Microcerberidae, 4, 19
Microcharon, 30
Microdeutopus gryllotalpa, 16, 73
Microdiaptomus, 30
Microfolliculina, 62
Microlistra, 31
Microparasellidae, 4, 30
Microparasellus, 30
Microphallus, 61
Microphrys, 56
Micropontius, 56
Microprotopus maculatus, 16, 73
Microsporidia, 61, 62
Mictyris, 19
Mictyris longicarpus, 18, 73
Millepora, 44
Miniacina miniacea, 38
Minyaspis aurivillii, 73, 74
Mirocaris, 25
Mitella, 10
Mitella mitella, 52, 73
Mithraculus cinctimanus, 43, 73
Mithrax, 56
Mithrax (Mithraculus) commensalis, 73
Modiola, 63
Modiolaria discors, 50, 82
Modiolicola, 49
Mola, 52
Mola mola, 59, 82
Mollusca, 9, 23, 49, 50, 63, 82
Monocheres, 44
Monocorophium acherusicum, 16, 20, 73
Monocorophium insidiosum, 16, 21, 73
Monodella, 30
Monolistra, 31, 53
Monoplacophora, 49
Monstrilla, 49
Monstrillidae, 51
Monstrilloida, 49
Montastraea, 44
Montipora, 44

Moraria, 30, 33
Moraria arboricola, 33, 73
Moraria monticola, 33, 74
Moraria sphagnicola, 33, 74
Morlockia ondinae, 21, 74
Moschites, 49
Mullus surmuletus, 63, 82
Munidopsidae, 25
Munidopsis, 6, 23–25
Munidopsis polymorpha, 21, 31, 74
Munnopsis, 24
Munnopsurus, 24
Murex, 50
Musculus discors, 82
Mychophilus, 58
Myicola, 49
Myicolidae, 49
Myocheres, 49
Myoschiston, 62
Myriapoda, 35
Myriocladus, 57
Mysida, 25, 27, 30, 43, 53, 55, 62, 190–197, 200–204, 206, 207, 209–212, 214, 217–219, 221, 222, 225, 227–232, 235–244, 246, 247, 249–252, 254, 255, 257–260, 262–274, 276, 277, 279–281, 283, 284, 286–294, 296, 297, 299–302, 305–307, 309–313, 315–326, 328–331, 342, 343, 404, 405
Mysida [in key], 342
Mysidacea, 5, 27, 189–191, 193–195, 317–319, 322, 327, 328
Mysidae, 193, 194, 200, 202, 203, 206, 207, 209–211, 213, 214, 217–219, 221–223, 227, 228, 233, 235–237, 241, 246–250, 253–255, 257, 263, 270, 272–274, 278, 280–282, 284, 286, 289, 290, 292, 296–301, 305–307, 309–311, 313, 314, 316, 317, 319–321, 324, 328, 331, 332, 405
Mysidae [in key], 343
Mysidea, 193
Mysideis, 338
Mysidella, 219, 231, 341
Mysidella biscayensis, 235, 350
Mysidella typhlops, 201, 350
Mysidella typica, 203, 350
Mysidellinae, 200, 202, 217, 233, 235, 272, 320, 321, 341, 344, 346
Mysidellinae [in key], 346
Mysidetes, 231, 275, 341
Mysidetes intermedia, 236, 350
Mysidetes posthon, 294, 350
Mysidetinae, 341
Mysidetini, 341
Mysidetini [in key], 346
Mysidium, 305, 310, 339

Mysidium columbiae, 281, 303, 304, 350
Mysidium gracile, 298, 300, 304, 311, 350
Mysidium integrum, 249, 350
Mysidobdella borealis, 312, 350
Mysidopsini, 337
Mysidopsini [in key], 344
Mysidopsis, 229, 242, 321, 337
Mysidopsis californica, 273, 280, 350
Mysidopsis gibbosa, 223, 350
Mysidopsis oligocenica, 350
Mysidopsis oligocenicus, 317, 350
Mysidopsis robustispina, 236, 350
Mysifaun, 341
Mysifaun erigens, 231, 232, 275, 350
Mysiformida, 193
Mysimenzies, 225, 253, 332, 336
Mysimenziesinae, 336
Mysimenziesini, 336, 345
Mysimenziesini [in key], 345
Mysina, 327
Mysinae, 200, 202–204, 210, 211, 214, 217, 225, 233, 235, 242, 248, 250, 251, 263, 272, 288, 297, 299, 309, 316, 319–321, 323, 325, 326, 335, 338, 341, 344
Mysinae [in key], 344
Mysini, 319, 335, 338, 345
Mysini [in key], 345
Mysis, 27, 192, 193, 207, 240, 245, 268, 279, 299, 308, 309, 314, 325, 338, 339
Mysis diluviana, 243, 280–282, 301, 308, 323, 350
Mysis Fabricii, 193, 350
Mysis mixta, 235, 291, 309, 350
Mysis oculata, 192, 193, 266, 269, 348, 350
Mysis relicta, 238, 239, 243, 264, 265, 268, 275, 301, 302, 309, 315, 316, 323, 350
Mysis salemaai, 243, 301, 323, 350
Mysis stenolepis, 254, 258, 259, 278, 280, 302, 307, 350
Mystacocarida, 4, 19, 26, 36
Mytilicola, 49, 123, 124
Mytilicola intestinalis, 124, 130
Mytilicola orientalis, 124, 130
Mytilicolidae, 49
Mytilidae, 15, 49
Mytilus edulis, 63, 82
Mytilus edulis platensis, 82
Mytilus platensis, 63, 82
Myxicola, 50
Myxomolgus, 50
Myzomolgus stupendus, 47, 74
Myzopontius australis, 38, 74
Myzotheridion, 49
Myzozoa, 312
Nakazawaia, 335
Namakosiramia, 57
Nanaspis, 57
Nannalecto, 49
Nannoniscus, 24, 53

Nanomysis, 316, 339
Nasomolgus, 50
Natatolana narica, 61, 68, 74
Nautilicaris, 25
Nautilograpsus, 63
Nautilus, 49, 50, 148, 149, 152, 178
Navanax, 50
Nebalia, 62
Nebalia bipes, 192, 348, 350
Nebaliopsis typica, 5, 74, 211, 350
Nectonema agile, 63, 81
Nematoda, 7, 38, 61, 63, 130
Nematoscelis megalops, 22, 74
Nemertea, 61, 63, 81
Neoaiptasia commensali, 43, 81
Neoamblyops, 337
Neobathymysis, 338
Neocyamus, 61
Neodoxomysis, 338
Neoglyphea inopinata, 18, 53, 74
Neohela monstrosa, 16, 74
Neohyssura spinicauda, 20, 74
Neolepas zevinae, 25, 74
Neomysini, 339, 345
Neomysini [in key], 345
Neomysis, 27, 210, 242, 250, 279, 300, 316, 325, 340
Neomysis americana, 312, 326, 350
Neomysis awatschensis, 267, 277, 315, 350
Neomysis integer, 231, 243, 246, 253, 254, 272, 275, 279, 281, 283, 284, 286, 291, 293, 296, 306, 308, 309, 351
Neomysis intermedia, 201, 273, 286, 294, 316, 351
Neomysis japonica, 316, 326, 351
Neomysis mercedis, 277, 309, 351
Neomysis rayii, 221, 303, 351
Neoniphargidae, 31
Neopetrolisthes maculatus, 74, 76
Neopetrolisthes oshimai, 43, 74
Neophreatoicus, 31
Neopontonides, 42
Neotanais hadalis, 6, 74
Neotrypaea californiensis, 67, 74
Nephropidea, 5, 23
Nephrops, 23
Nephrops norvegicus, 18, 74
Nephropsis, 23, 24
Neptunus, 55
Neptunus pelagicus, 55, 74
Nereicola, 50
Nereicolidae, 50
Nereidae, 9, 50
Nerocila, 27
Nesotanais lacustris, 26, 74

Nichollsia, 31
Nicothoe, 54
Nicothoe astaci, 54, 74
Nicothoe tumulosa, 53, 74
Niphargidae, 31
Niphargoidea, 8, 31
Niphargus, 31
Nippochelura, 12
Nippochelura brevicaudata, 12, 74
Nipponasellus, 30
Nipponerythrops, 335
Nipponomysis, 340
Nitokra divaricata, 54, 74
Nitokra medusaea, 40, 74
Nitophyllum, 38
Nobia, 45
Nogagella siphonophoriae, 40, 74
Nordmanni nordica insulana, 71
Nosema, 61
Notacanthomysis, 340
Notodelphyidae, 58
Notodelphyoida, 49
Notodelphys, 58
Notomysis, 211, 338
Notostraca, 26, 32
Nouvelia, 338
Nucella lapillus, 50, 82
Nudibranchia, 50
Nyctiphanes couchi, 22, 74
Obelia geniculata, 62, 80
Obelia longissima, 62, 80
Obione, 22
Octocorallia, 40–42
Octolasmis, 41, 46, 50, 53, 55
Octolasmis aymonini, 53, 74
Octolasmis bathynomi, 53, 74
Octolasmis muelleri, 55, 74
Octolasmis orthogonia, 52, 74
Octolasmis warwicki, 51, 59, 74
Octopicola, 49
Octopus, 49
Octosporea, 62
Ocypode, 17, 19, 28, 29
Ocypode ceratophthalmus, 17, 18, 74
Ocypode cordimanus, 17, 18, 74
Ocypodidae, 18
Odontomolgus, 44
Oedicerotidae, 17
Oedicerotoidea, 8
Oedomyzon, 40
Oestrella, 50
Onceroxenus, 52
Onchocerca volvulus, 126, 130
Oneirophanta, 57
Oniscidae, 34
Oniscidea, 6, 31, 34
Oniscoidea, 35
Onisimus, 43

TAXONOMIC INDEX 429

Onychocaris, 39
Onychopygos, 56
Opaepele, 25
Opecarcinus crescentus, 74, 77
Opecoeloides, 61
Ophelicola, 51
Opheliidae, 51
Ophiocten, 57
Ophioika, 56
Ophionotus, 57
Ophioseides, 58
Ophiura, 57
Ophryotrocha geryonicola, 63, 82
Ophyoderma, 62
Opisthobranchia, 49
Opisthopus, 50, 57
Opisthoteuthis, 49
Oplophoroidea, 10
Oplophorus, 62
Orchestia platensis, 5
Orchomene commensalis, 13, 74
Orchomenella abyssorum, 74
Orchomenella commensalis, 74
Orconectes, 30, 31, 53, 54
Orconectes limosus, 124, 130
Orientomysis, 340
Orientomysis mitsukurii, 272, 315, 351
Ormieresia, 61, 62
Orstomella, 44
Orthagoriscicola muricatus, 52, 74
Oscarella, 38
Oscarella malabonensis, 16
Oscillatoria, 16
Ostracoda, 4, 5, 7, 10, 11, 19, 20, 25, 29, 30, 32, 33, 35, 52, 53, 59, 166
Ostracotheres, 52
Ostreicola, 49
Ostreidae, 9, 49
Ostrincola, 49
Oxynaspis, 46
Oxynaspis aurivillii, 46, 74
Oxypora, 45
Pachycondyla sulcata, 82
Pachycondyla sulcata Mayr var. *sulcatotesserinoda*, 82
Pachygrapsus, 10, 28
Pachygrapsus marmoratus, 142, 157, 158, 167, 185, 399
Pachylasma, 41
Pacifacanthomysis, 340
Pacifastacus leniusculus, 124, 130
Pagurapseudes bouryi, 74
Paguristes, 39
Paguroidea, 6, 10, 13, 24, 39, 40, 43, 47, 54, 55, 62
Pagurolepas conchicola, 50, 74

Paguropsis typica, 43, 74
Pagurotanais bouryi, 13, 74
Pagurus, 39, 125
Pagurus alcocki, 48, 74
Pagurus bernhardus, 13, 43, 54, 74
Pagurus brevipes, 55, 74
Pagurus prideaux, 43, 74
Pagurus varians, 48, 74
Palaemon, 28, 62, 167
Palaemon adspersus, 60, 74
Palaemon debilis, 21, 75
Palaemon elegans, 60, 63, 75
Palaemon serratus, 63
Palaemonella, 58
Palaemonella burnsi, 21, 75
Palaemonetes, 27, 31, 167
Palaemonias, 31
Palaemonidae, 27, 28, 31, 60, 121
Palaemoninae, 60
Palaumysinae, 200, 202, 207, 233, 235, 272, 320, 332, 340, 344
Palaumysinae [in key], 346
Palaumysis, 207, 217, 235, 263, 276, 277, 340
Palaumysis philippinensis, 201, 207, 288, 331, 343, 351
Palaumysis pilifera, 235, 351
Palaumysis simonae, 203, 351
Palinura, 5, 166, 185
Palinuridae, 55
Palinuroidea, 40
Palinurus elephas, 167, 185
Pallenopsis fluminensis, 82
Palythoa, 46, 47
Palythoa senegambiensis, 47
Pandalidae, 24, 40, 46
Pandalus montagui, 63, 75
Pandora glacialis, 50, 82
Panopea, 49
Panopeus herbstii, 56, 75
Panoplea eblanae, 40
Panulirus homarus, 54, 75
Panulirus interruptus, 55, 75
Panulirus japonicus, 11, 75
Panulirus laevicauda, 125, 130
Panulirus ornatus, 11, 75
Panulirus penicillatus, 141, 142, 185
Panulirus versicolor, 11, 75
Parabathynellidae, 30
Parabogidiella, 31
Paracalanus, 62
Paracalliactis, 43
Paracanthomysis, 340
Parachristianella, 61
Paracilicaea, 9
Paracilicaea mossambica, 9, 75
Paracilicaea teretron, 9, 75
Paracineta homari, 62, 80
Paracirolana platysoma, 9, 75

Paracis squamata, 42, 75, 81
Paracleistostoma depressum, 19, 75
Paracyclops, 30
Paracymothoa, 27
Paradella dianae, 69, 75
Paradynamenella psammophila, 9, 75
Paragnathia formica, 18, 75
Paragonimus, 126
Paraidya, 54
Parajassa pelagica, 16, 75
Paralepas quadrata, 71, 75
Paraleptamphopus, 31
Paraleptomysis, 242, 338
Paraleptomysis dimorpha, 244, 351
Paralichthys olivaceus, 316, 351
Paralithodes camtschaticus, 5, 75
Paralophogaster, 210, 232, 257, 263, 276, 295, 324, 330
Paralophogaster foresti, 244, 247, 351
Paralophogaster glaber, 201, 351
Paramacrochiron ennorense, 40, 75
Paramacrochiron japonicum, 40, 75
Paramacrochiron rhizostomae, 40, 75
Paramacrochiron sewelli, 40, 75
Paramblyops, 336
Paramesopodopsis, 339
Paramesopodopsis rufa, 289, 304, 351
Paramithrax longipes, 75
Paramolgus, 41, 43, 46
Paramuricea clavata, 41, 42, 81
Paramysis, 203, 210, 213, 225, 231, 250, 268–270, 279, 286, 296, 297, 299, 317, 322, 325, 339
Paramysis (Mesomysis) intermedia, 27
Paramysis arenosa, 322, 351
Paramysis bakuensis, 271, 351
Paramysis helleri, 287, 351
Paramysis inflata, 203, 351
Paramysis intermedia, 73, 75
Paramysis lacustris, 248, 249, 315, 351
Paramysis nouveli, 267, 273, 274, 351
Paramysis pontica, 227, 231, 271, 301, 351
Paramysis portzicensis, 210, 351
Paramysis ullskyi, 203, 351
Paranaeitis, 49
Paranchialina, 334
Paranchialina angusta, 235, 351
Paranchistus, 50
Paranephrops, 62
Paranicothoe, 53, 56
Paranthessius, 43, 49
Paranthuridae, 20
Paraonidae, 50
Parapagurus, 6, 14, 43
Parapagurus pictus, 75
Parapetalophthalmus, 211, 331

Parapleustes, 57
Parapleustes commensalis, 55, 75
Parapontonza, 58
Parapseudomma, 247, 337
Parapseudomma calloplura, 233, 235, 337, 351
Parargissa galatheae, 5, 75
Parascothorax, 57
Parasicyonis, 44
Parasphaeroma granosum, 9, 75
Parastacidae, 27, 55
Parastacoides, 55
Parastenasellus, 30
Parastenocaris, 30
Parasterope, 52
Parastilomysis, 339
Parathelges, 55
Parathemisto, 62
Paratya, 31, 56
Paratymolus, 45
Paratypton, 45
Paratypton siebenrocki, 45, 47, 75
Pardaliscoidea, 8
Pardaliscoides longicaudatus, 6, 75
Pareledone, 49
Parerythrops, 24, 335
Parhippolyte uveae, 72, 75
Parhyale fascigera, 21, 75
Parhyale hawaiensis, 50, 75
Pariambus, 57
Pariambus typicus, 40, 75
Parione, 55
Parionella, 55
Parisia, 31
Parthenopidae, 45
Parvimysis, 339
Pasiphaea, 62
Pasiphaea princeps, 5, 75
Pasiphaeoidea, 10
Patella, 14, 50
Patella vulgata, 50, 82
Pavona, 44, 45
Pavona gigantea, 46, 81
Peachocarididae, 225, 317, 329
Peachocarididae [in key], 342
Peachocaris, 329
Peachocaris acanthouraea, 317, 351
Peachocaris strongi, 317, 351
Pecten, 49
Pectinaria, 14, 51
Pectinidae, 49
Pectinophilus ornatus, 124, 130
Pedunculata, 46, 48, 53
Pelobates cultripes, 60, 83
Peltogaster, 52, 54
Peltomyzon, 44
Penaeidae, 17, 18, 22–24, 61
Penaeoidea, 6, 10, 24, 36, 166
Penaeus, 28

Penaeus kerathurus, 167, 185
Penaeus monodon, 5, 75
Peniculus, 62
Peniculus fistula, 63, 75
Pennatula, 41
Pennatulacea, 40, 41
Pennatulicola, 41
Pennella, 52, 61
Pennella balaenoptera, 5, 61, 75
Pennella varians, 49, 75
Pentastomida, 4
Peracarida, 53, 166, 185, 190, 194, 195, 231, 284, 318, 319, 321
Peramphithoe humeralis, 12, 65, 75
Percnon, 55
Percnon gibbesi, 124, 130
Perezia, 52, 62
Perforatus perforatus, 13, 75
Pericharax, 38
Periclimenaeus, 39, 58
Periclimenaeus hecate, 59, 75
Periclimenella petitthouarsii, 45, 75
Periclimenes, 37, 39, 40, 42, 46, 58
Periclimenes amethysteus, 60, 75
Periclimenes brevicarpalis, 44, 75
Periclimenes consobrinus, 45, 75
Periclimenes holthuisi, 75
Periclimenes imperator, 50, 76
Periclimenes lutescens, 76
Periclimenes madreporae, 45, 76
Periclimenes magnificus, 76
Periclimenes ornatus, 44, 76
Periclimenes pholeter, 21, 76
Periclimenes rathbunae, 44, 76
Periclimenes yucatanicus, 60, 76
Peritrichia, 311
Petalophthalmidae, 194, 200, 202, 203, 206, 214, 218, 222, 225, 228, 229, 235, 236, 247, 253, 255, 257, 258, 296, 321, 331
Petalophthalmidae [in key], 343
Petalophthalmina, 327
Petalophthalminae, 211, 320, 331
Petalophthalminae [in key], 343
Petalophthalmus, 211, 219, 255, 331
Petalophthalmus armiger, 200, 218, 223, 236, 244, 257, 351
Petrarca bathyactidis, 43, 76
Petricolaria pholadiformis, 11
Petrolisthes, 28, 57
Petrolisthes maculatus, 43, 76
Petrosia, 38
Phallusiella, 50
Phellia, 43
Philichthyidae, 58
Philorthagoriscus serratus, 52, 76

Philoscia, 149, 168
Philostomella, 27
Pholeterides, 58
Pholetischus, 55
Phoretophrya, 61, 62
Phoxichilidium fluminense, 51, 82
Phoxinus phoxinus, 118, 131
Phoxocephalidae, 17
Phoxocephaloidea, 7
Phreatoicidae, 26, 27, 31
Phreatoicopsis terricola, 35, 62, 76
Phreatoicus, 31
Phreatomerus, 31
Phreatoniphargus, 31
Phycolimnoria, 12
Phyllocarida, 36
Phyllodiaptomus, 119
Phyllodicolidae, 50
Phyllognathopus camptoides, 34, 76
Phyllognathopus coecus, 32, 33, 76, 79
Phymodius, 56
Phymodius ungulatus, 45, 76
Physophora hydrostatica, 40, 80
Pilsbryscalpellum subalatum, 76, 77
Pinnaxodes, 51, 57
Pinnidae, 49, 50
Pinnixa, 14, 47, 50, 51, 57
Pinnixa franciscana, 47, 55, 76
Pinnixa lunzi, 47, 76
Pinnixa schmitti, 55, 76
Pinnotheres, 14, 50, 51
Pinnotheridae, 14, 50, 124
Pinnotherion, 56
Pionodesmotes phormosomae, 56, 76
Pisa, 56
Pisces, 82
Plagusia, 10, 28
Planes minutus, 40, 60, 76
Plantae, 80, 130
Platorchestia platensis, 5, 76
Platyarthrus acropyga, 35, 76
Platyarthrus caudatus, 35, 76
Platyarthrus hoffmannseggii, 35, 76
Platycaris, 45
Platycyamus, 61
Platygyra, 44
Platyhelminthes, 81
Platylepas, 60
Platylepas hexastylos ichthyophila, 59, 76
Platymaia, 24
Platymysis, 341
Platyops, 341
Platypontonia, 50
Platysympodidae, 24
Plecostidae, 8
Plectonema, 16
Pleonexes brevirostris, 76
Plesionika ensis, 56, 76

TAXONOMIC INDEX

Plesiopenaeus, 24
Plesiopenaeus armatus, 6, 76
Plesiopenaeus coruscans, 24, 65, 76
Pleurerythrops, 335
Pleurocrypta, 55
Pleustidae, 57
Poaceae, 13
Pocillopora, 44, 45
Pocillopora damicornis, 44, 81
Pocilloporidae, 45
Podascon, 53
Podoceridae, 57
Podoceropsis nitida, 76
Podocerus, 125
Podocerus brasiliensis, 21, 76
Poecilasma, 55
Poecilostomatoida, 47
Pollicipes, 10
Pollicipes elegans, 119, 130
Pollicipes pollicipes, 73, 76, 119, 130
Pollicipes polymerus, 119, 130
Polyascus gregaria, 26, 76, 77
Polychaeta, 9, 13, 23, 63
Polycheles, 24
Polychelidae, 23
Polycheria antarctica, 16, 76
Polycheria osborni, 16, 76
Polycirrus, 51
Polydectus cupulifer, 43, 76
Polydora, 50
Polyneura, 38
Polynoidae, 50
Polyonyx, 51
Polyplacophora, 11, 49
Polyspira, 62
Pomatoceros, 13, 61
Pontella mediterranea, 36, 76
Pontogeloides, 17
Pontomysidae, 333
Pontonia, 58
Pontonides, 46
Pontoniinae, 36, 37, 39, 45, 46, 50, 58, 60
Pontoniopsis, 58
Pontoporeioidea, 7
Porcellana platycheles, 62, 76
Porcellanidae, 43, 55, 57
Porcellanopagurus, 14
Porcellidium, 54
Porcellidium echinophilum, 56, 76
Porcellio scaber, 143, 159, 185
Porifera, 9, 15, 38, 80
Porites, 11, 44, 45
Porospora, 61, 62
Porpita, 40
Portunidae, 10, 45

Portunion, 56
Portunion maenadis, 61
Portunus, 56
Portunus pelagicus, 74, 76, 124, 130
Posidonia, 13, 20, 297, 351
Potamidae, 27, 31, 34
Potamocypris philotherma, 5, 29, 76
Potamon, 126, 128
Potamon fluviatile, 29, 76
Potamonautes, 126
Potiicoara brasiliensis, 30, 76
Pourtalesia, 57
Praunus, 213, 228, 229, 240, 242, 269, 286, 300, 316, 339
Praunus flexuosus, 191, 203, 209, 212, 219, 231, 237, 238, 247, 250, 252, 254, 260, 262, 264–268, 273, 274, 279–281, 283, 284, 291, 296, 302, 309, 326, 348, 351
Praunus inermis, 264, 266, 280, 351
Praunus neglectus, 280, 351
Priapon, 56
Primno, 40
Princaxelia abyssalis, 6, 66, 76
Prionomysis, 211, 338
Procambarus, 28, 30, 31, 53
Procambarus clarkii, 124, 130
Procampilaspidae, 24
Procarididea, 21
Procaris ascensionis, 21, 76
Processidae, 17
Prochristianella, 61
Promysis, 338
Proneomysis, 340
Prosobranchia, 49
Prostoma, 63
Protista, 62, 80
Protomysidellinae, 334
Protoraphis, 62
Psammocora, 44, 45
Pseudamblyops, 335
Pseudanchialina, 334
Pseudanthessiidae, 44, 50
Pseudanthessius, 50, 56
Pseudanthessius latus, 47, 76
Pseudanthessius nemertophilus, 48, 76
Pseuderythrops, 335
Pseudione, 55
Pseudione diogeni, 53, 76
Pseudione fraissei, 53, 76
Pseudionella, 55
Pseudoasellus, 30
Pseudobranchiomysis, 338
Pseudocaligus, 62
Pseudocarcinus gigas, 5, 24, 77
Pseudochalaraspidum, 330
Pseudochiron, 57
Pseudocrangonyx, 31
Pseudocryptochirus, 45

TAXONOMIC INDEX 433

Pseudocryptochirus crescentus, 46, 77
Pseudohapalocarcinus ransoni, 45, 77
Pseudohimanthidium, 62
Pseudomacrochiron stocki, 40, 77
Pseudomma, 247, 323, 324, 337
Pseudomma affine, 244, 351
Pseudomma armatum, 232, 351
Pseudomma oculospinum, 337, 351
Pseudommini, 336, 345
Pseudommini [in key], 345
Pseudomolgus, 46
Pseudomonas, 125
Pseudomyicola, 49
Pseudomysidetes, 227, 231, 341
Pseudomysis, 24, 338
Pseudoniphargus, 31
Pseudopagurus granulimanus, 48, 77
Pseudopetalophthalmus, 211, 331
Pseudophilomedes, 52
Pseudothelphusa, 28, 32
Pseudothelphusidae, 27, 31, 32
Pseudotiron longicaudatus, 6, 77, 78
Pseudoxomysis, 338
Psilomyzon pauciseta, 38, 77
Pteriidae, 49
Pteromysis, 335
Pteropontius, 44
Pteropontius cristatus, 38, 77
Ptychascus, 22, 56
Pycnogonida, 82
Pygocephalomorpha, 288, 317, 342
Pygocephalomorpha [in key], 342
Pygodelphys, 58
Pylocheles, 39
Pylocheles (Pylocheles) mortensenii, 14
Pylocheles (Xylocheles) macrops, 14
Pylocheles mortensenii, 77
Pylochelidae, 12, 14, 16, 24
Pylopagurus, 14
Pyrgoma, 11, 45
Pyrgoma stockesi, 38, 77
Pyrgomatidae, 45
Pyrgopsella, 45
Pyrodinium bahamense, 125, 130
Pyroleptomysis, 337, 338
Pyroleptomysis rubra, 326, 351
Pyrosomatida, 58
Pyrotheca, 62
Quadrella cyrenae, 46, 77
Quadrella maculosa, 77
Racilius, 45
Racilius compressus, 45, 77
Radianthus, 43, 44
Rana clamitans, 60, 83
Rapipontonia galene, 39, 77

Rectisura brachycephala, 6, 77
Rectisura herculea, 6, 77
Rectisura tenuispinis, 6, 77
Rectisura vitjazi, 6, 77
Remipedia, 4, 21
Reptantia, 5, 23, 36, 166
Retromysis, 341
Rhabdophrya, 62
Rhabdopus, 50
Rhachotropis, 62
Rhizocephala, 22, 26, 52, 53, 56, 61
Rhizorhina ampeliscae, 53, 77
Rhizostoma, 40
Rhodactis, 43
Rhodinicola, 50
Rhodophyceae, 38
Rhodophyllis, 38
Rhodymenia, 38
Rhopalophthalminae, 200, 202, 210, 211, 231, 233, 235, 237, 250, 270, 272, 297, 321, 331, 332, 343
Rhopalophthalminae [in key], 342, 343
Rhopalophthalmus, 210, 227, 231, 332, 343
Rhopalophthalmus egregius, 200, 235, 351
Rhopalophthalmus longicauda, 236, 351
Rhopalophthalmus tartessicus, 202, 309, 351
Rhopalophthalmus tattersallae, 273, 351
Rhopalophthalmus terranatalis, 232, 249, 309, 351
Rhynchomolgus, 44
Rhynchopus, 50
Ridgewayia fosshageni, 43, 77
Riggia, 27
Rimicaris, 25
Rimicaris exoculata, 25, 77, 118, 130
Rissoa, 13
Rochinia, 55
Rodriguezia mensabak, 77, 79
Rostromysis, 338
Rotundicauda, 293
Ruffohyale milloti, 77
Rynchonella psittacea, 48, 81
Sabella, 50
Sabellaria, 13, 15
Sabellastarte, 50
Sabellida, 51
Sabelliphilidae, 51
Sabelliphilus, 50
Saccolepis, 50
Sacculina, 22
Sacculina carcini, 56, 61, 77
Sacculina gregaria, 77
Sacodiscus ovalis, 54, 77
Sadella, 58
Sagittocythere, 30
Sagmariasus verreauxi, 5, 77
Saharolana, 30
Salentinella, 31
Salentinellidae, 8, 31

TAXONOMIC INDEX

Salicornia, 22
Salmacina, 13
Salmacina dysteri, 13
Salmincola, 62
Salpida, 58
Sameioneis, 62
Sanguinolariidae, 49
Sapphirina, 58
Sarmysis, 317, 339
Sarsiella, 52
Sarsiflustra abyssicola, 48, 81
Sarsilenium, 50
Savignium, 45
Saxidomus, 49
Scalopidia spinosipes, 55, 77
Scalpellum, 41, 51
Scalpellum aduncum, 51, 77
Scalpellum cancellatum, 48, 77
Scalpellum giganteum, 52, 77
Scalpellum hispidum, 48, 77
Scalpellum stearnsii, 52, 77
Scalpellum subalatum, 48, 77
Scambicornus, 57
Scaphopoda, 14, 49
Schimperella, 317, 330
Schimperella acanthocercus, 317, 351
Schimperella beneckei, 317, 351
Schimperella kessleri, 317, 351
Schistomysis, 273, 316, 322, 339
Schistomysis assimilis, 249, 351
Schistomysis kervillei, 293, 351
Schistomysis ornata, 305, 351
Schistomysis spiritus, 201, 294, 296, 301, 302, 351
Schizodinium, 62
Schizophrys, 45
Schizopoda, 193, 194
Schizopodes, 193
Scleractinia, 9, 45
Scleroplax granulata, 55, 77
Scolamblyops, 336
Scopimera bitympana, 18, 77
Scopimera globosa, 18, 77
Scorpaena scrofa, 59, 83
Sculeolaria chuni, 40, 80
Scutocyamus, 61
Scylla, 19, 22, 55
Scyllaridea, 36
Scyllarus, 152
Scyphozoa, 80
Sebidae, 8, 31
Seborgia, 31
Segonzacia, 25
Segonzacia mesatlantica, 25
Semibalanus balanoides, 13, 77
Sepia, 49

Septosaccus, 52, 54
Seriatipora, 45
Serolidae, 53
Serolis, 53
Serolis scythei, 77
Seroloniscus, 53
Serpula, 50
Serpulidicola, 50
Serpulidicolidae, 50
Serpuliphilus, 50
Serranus lanceolatus, 59, 83
Sesarma, 19, 22, 26–28, 31, 35, 55, 56
Sesarma bidentatum, 32, 77
Sesarma bromeliarum, 32, 77
Sesarma haematocheir, 18, 77
Sesarma jarvisi, 32, 77
Sesarma nodulifera, 32, 77
Sesarma obtusifrons, 64, 77
Sesarma taeniolata, 19, 77
Sesarma verleyi, 32, 77
Sesarmaxenos, 22, 26, 56
Sessilia, 5, 25, 52, 53, 58, 60, 61
Sestropontius bullifer, 38, 77
Sewellochiron fidens, 40, 77
Shenimysis, 335
Shinkalcaris, 25
Siboglinidae, 51
Sicyiodelphys, 58
Sideropora, 45
Simulium naevi, 126, 130
Simulium naevis, 130
Simulium naevus, 130
Sinelobus stanfordi, 21, 26, 77, 204, 352
Sinopotamon, 126
Siphonaria, 14
Siphonoecetes, 13, 16
Sipunculida, 47
Sipunculus nudus, 47
Siriella, 62, 224, 228, 242, 268, 269, 279, 286, 296, 321, 324, 333
Siriella adriatica, 201, 352
Siriella antiqua, 317, 352
Siriella armata, 217, 239–241, 260, 269, 283, 286, 289–291, 352
Siriella carinata, 317, 352
Siriella castellabatensis, 219, 352
Siriella clausii, 235, 246, 352
Siriella gracilipes, 201, 246, 287, 352
Siriella gracilis, 236, 302, 352
Siriella thompsonii, 200, 202, 323, 324, 352
Siriellinae, 200, 202, 204, 210, 217, 233, 235, 251, 263, 270, 272, 296, 319–321, 323, 332, 343
Siriellinae [in key], 343
Siriellini, 333
Siriellini [in key], 343
Skogsbergia squamosa, 68, 78
Skotobaena, 30, 35
Smilium, 41

Solaster, 57
Spartina, 22
Specirolana, 30
Spelaeogammarus, 31
Spelaeogriphacea, 26, 30, 36
Spelaeogriphus lepidops, 30, 78
Spelaeomysis, 27, 30, 194, 198, 199, 210, 217, 232, 245, 258, 270, 278, 286, 298, 313, 322, 330, 349, 352
Spelaeomysis bottazzii, 192, 194, 199, 216, 230, 233, 236, 245, 276–278, 280–284, 287, 291, 295, 297–299, 306, 308, 322, 352, 404
Spelaeomysis cardisomae, 199, 209, 216, 227, 236, 245, 298, 313, 322, 352
Spelaeomysis cochinensis, 216, 244, 298, 322, 352
Spelaeomysis longipes, 207, 216, 217, 226, 237, 245, 258, 259, 266, 267, 284, 292, 306, 308, 318, 322, 330, 352
Spelaeomysis nuniezi, 245, 352
Spelaeomysis olivae, 226, 227, 244, 352
Spelaeomysis quinterensis, 257, 352
Spelaeomysis servata, 349, 352
Spelaeomysis servatus, 352
Spelaeoniscidae, 35
Speleonectes, 21
Speleonectes lucayensis, 21, 78
Speocyclops, 30
Sphaeroma, 23, 53
Sphaeroma pentodon, 11, 12, 78
Sphaeroma peruvianum, 12, 78
Sphaeroma polydemum, 9, 78
Sphaeroma retrolaeve, 12, 78
Sphaeroma serratum, 27, 53, 78
Sphaeroma sieboldii, 9, 12, 78
Sphaeroma terebrans, 9, 12, 22, 78
Sphaeroma venustissimum, 11, 78
Sphaeroma walkeri, 53, 78
Sphaeromatidae, 34
Sphaeromatidea, 9, 27, 30, 53
Sphaeromatoidea, 9
Sphaeromicola, 53
Sphaeromicola dudichi, 53
Sphaeromides, 30, 53
Sphaeromonopsis amathitis, 78
Sphaeromopsis amathitis, 17, 78
Sphaeronella, 53
Sphaeronellopsis, 52
Sphagnum, 33
Sphyrapus, 24
Sphyrna zygaena, 59, 83
Spiophanicolidae, 51
Spirobranchus, 50, 61
Spirochona, 62
Spirographis, 50
Spirophrya, 62

Spirula, 50
Splanchnotrophus, 49
Spondylidae, 49
Spongicola venustus, 16, 39, 78
Sponginticola, 38
Sponginticolidae, 38
Spongiocnizon, 38
Spongiocnizontidae, 38
Squilla mantis, 19, 119, 130, 162, 185
Staurosoma parasiticum, 43, 78
Stegocephaloidea, 8
Stegopontonia, 58
Stellicola, 56
Stellicomes, 56
Stenasellidae, 30
Stenasellus, 30
Stenetriidae, 30
Stenogramma, 38
Stenopodidae, 39, 58, 60
Stenopodidea, 60
Stenopus hispidus, 60, 78
Stenopus scutellatus, 60, 78
Stenopus spinosus, 60, 78
Stenothoe valida, 21, 78
Stenothoidae, 8, 39, 57
Stenothoides, 57
Stenula rubrovittata, 13, 55, 78
Stephanolepas, 60
Stichodactyla helianthus, 81
Stichopathes, 46
Stichopus, 57
Stigeoclonium australense, 62, 80
Stilomysis, 339
Stockia, 44
Stoichactis, 43, 44
Stoichactis helianthus, 43, 81
Stomatolepas, 60
Stomatopoda, 19, 26, 51, 122, 166, 185
Storthyngura, 6
Strongylocentrotus purpuratus, 57
Stygiasellus, 30
Stygiocaris, 31
Stygiomysida, 190–192, 195–197, 199, 201, 203, 204, 206, 207, 209, 211, 214, 216, 217, 219, 225, 226, 229–232, 235–237, 240, 244, 255, 257–260, 263, 264, 266, 272, 276, 278, 280, 281, 283, 284, 286, 287, 291, 293, 295, 297–300, 305–308, 312, 313, 318, 319, 321, 322, 328–330, 342, 404
Stygiomysida [in key], 342
Stygiomysidae, 194, 196, 199, 201, 203, 206, 207, 216, 223, 226, 227, 229, 235, 236, 257, 282, 322, 330
Stygiomysidae [in key], 343
Stygiomysina, 327
Stygiomysis, 27, 30, 198, 199, 217, 219, 232, 258, 264, 270, 298, 313, 322, 331
Stygiomysis aemete, 207, 216, 217, 226, 227, 352

Stygiomysis cokei, 199, 352
Stygiomysis holthuisi, 209, 226, 227, 236, 352
Stygiomysis hydruntina, 192, 194, 216, 232, 233, 264, 298, 352, 404
Stygiomysis ibarrae, 298, 352
Stygiomysis major, 257, 258, 352
Stygobromus, 31
Stygocarididae, 19
Stygonectes, 31
Stygotantulus stocki, 4, 78
Stylaster, 40
Stylasterina, 40
Stylophora, 44
Stynocoides longicaudatus, 78
Suberites, 38
Suberites carnosus, 16
Suberites domuncula, 16, 38, 80
Sulcatotesserinoda, 82
Sulculeolaria chuni, 80
Sunaristes, 54
Sunaristes dardani, 54
Sunaristes inaequalis, 54, 78
Sunaristes paguri, 54, 78
Surinamysis, 214, 270, 339
Surinamysis merista, 271, 352
Sympagurus pictus, 43, 75, 78
Synagoga, 57
Synagoga mira, 46, 78
Synalpheus, 39, 45, 58
Synapticola, 57
Synaptiphilus, 57
Synasellus, 30
Syncarida, 19, 29, 30, 36, 55
Syncyamus, 61
Syndelphys, 58
Syndinium, 62
Synerythrops, 335
Synophrya, 61, 62
Synopioidea, 8
Synopioides secundus, 6, 78
Synstellicola, 56
Synurella, 31
Syrrhoidae, 57
Syscenus infelix, 62, 78
Tachaea caridophaga, 56, 78
Tachaea picta, 56, 66, 78
Tachidium brevicornis, 78
Tachidius brevicornis, 33, 78
Tachypleus gigas, 51, 82
Talitridae, 5, 6, 8, 17, 35
Talitroidea, 8, 31
Talitroides alluaudi, 35, 78
Talitroides hortulanus, 35, 78
Talitroides pacificus, 35, 78
Talitroides topitotum, 78

Talitrus, 35
Talitrus (Talitroides) pacificus, 33
Talitrus (Talitroides) topitotum, 33
Talitrus alluaudi, 78
Talitrus saltator, 18, 78
Talorchestia, 17
Tanaidacea, 6–8, 26, 36, 38
Tanais, 21
Tanais cavolinii, 15, 78
Tanais dulongii, 15, 78
Tantulocarida, 4, 51, 52
Taphromysis, 27, 299, 339
Tardigrada, 63
Tasmanomysis, 339
Telacanthomysis, 340
Telestacicola, 39
Telmactis decora, 43
Telmatactis cricoides, 310, 352, 405
Telmessus acutidens, 125, 130
Telotha, 27
Temnaspis, 50
Temnaspis bathynomi, 74, 78
Temnaspis tridens, 55, 78
Temnocephala, 62
Temnologus, 46
Tenagomysis, 211, 304, 338
Tenagomysis tasmaniae, 279, 352
Teratamblyops, 336
Teraterythrops, 335
Terebellides stroemii, 51, 82
Terebratulina caputserpentis, 48, 81
Terebratulina retusa, 81
Teredicola, 49
Teredo, 9, 49
Teredoika, 49
Terrestricypris arborea, 35, 78
Terrestricythere, 35
Terrestricythere ivanovae, 35, 78
Terrestricythere pratenesis, 35, 78
Tethysbaena, 30
Tetraclita, 9, 10, 45
Tetraclitella costata, 11, 78
Tetraclitidae, 9
Tetralia, 45
Tetramorium brevicorne, 35, 82
Tetrarhynchidae, 62
Teximeckelia, 31
Thalamita pilumnoides, 45, 78
Thalamitoides quadridens, 45, 78
Thalassia, 298, 352
Thalassina, 19, 55
Thalassina anomala, 19, 78, 123, 130
Thalassinidea, 18
Thalassomyces, 62
Thalassomysinae, 336
Thalassomysini, 335
Thalassomysini [in key], 346
Thalassomysis, 219, 229, 335, 336

TAXONOMIC INDEX

Thalestris rhodymeniae, 38, 79
Thaliacea, 58
Thammolgus, 46
Thaumastocaris, 39
Thaumastocheles, 23, 24
Thecacineta cypridinae, 62, 80
Thecostraca, 4
Thelohania, 61, 62
Theosbaena, 30
Thermobathynella adami, 5, 29, 79
Thermocyclops, 34
Thermosbaena, 30
Thermosbaena mirabilis, 5, 29, 79
Thermosbaenacea, 19, 26, 28–30, 36
Thermosphaeroma, 29
Thor amboinensis, 44, 79
Thoracica, 5, 11, 38, 44, 57
Thylacocephala, 317
Thysanoessa inermis, 22, 79
Thysanoessa longicaudata, 22, 79
Thysanopoda cornuta, 5, 79
Thysanostoma thysanura, 40, 80
Tigriopus, 5, 21
Tima bairdii, 39, 80
Tisbe cucumariae, 57, 79
Tisbe elongata, 54, 79
Tisbe furcata, 53, 79
Tisbe holothuriae, 57, 79
Tisbe japonica, 57, 79
Tivela, 49
Tmethypocoelis ceratophora, 71, 79
Tokophrya, 62
Tonna, 43
Trandosia jenkinsae, 5, 79
Trapezia, 45
Trematoda, 61, 62
Triactis producta, 43, 81
Triangulus, 54
Trichechus, 60
Trichechus manatus, 60, 83
Trichodactylidae, 31, 32
Trichodactylus, 32
Trichodactylus mensabak, 31, 79
Trichoniscidae, 31
Tridacna, 49, 58
Trifur tortuosus, 63, 79
Triops, 5
Triops longicaudatus, 123, 130
Tripartisoma, 49
Trischizostoma, 24
Trischizostoma nicaeense, 59, 79
Tritaeta gibbosa, 38, 57, 79
Tritodymania, 14
Trizocheles, 16, 39
Trochicola, 49

Trochilioides, 62
Troglocambarus, 31
Troglocarcinus, 45, 56
Troglocaris, 31
Troglocirolana, 30
Troglocubanus, 28, 31
Trogloleleupia, 31
Troglomysis, 30, 324, 339
Troglomysis vjetrenicensis, 241, 352
Tropichelura, 12
Tropichelura insulae, 12, 79
Tropocyclops prasinus, 32, 79
Tropodiaptomus, 30
Trypetesa, 11
Trypetesa lateralis, 11, 79
Trypetesidae, 11
Tubicinella, 61
Tubularia, 39
Tubularia indivisa, 39, 80
Tuleariocaris, 58
Tulumella, 30
Tuphacheres micropus, 38, 79
Turbellaria, 7, 61, 62, 81
Turbinaria, 45
Turbo, 49
Turniella, 61
Tursiops truncatus, 61, 83
Tychidion, 51
Tylocarcinus, 45
Tylos, 17
Tylos punctatus, 22, 79
Typanodinium, 62
Typhlatya, 28, 31
Typhlatya galapagensis, 21, 79
Typhlatya rogersi, 21, 79
Typhlocarididae, 31
Typhlocaris, 31
Typhlocirolana, 27, 30
Typhlogammaridae, 31
Typhlogammarus, 31
Typhlopatsa, 31
Typhlopseudothelphusa juberthiei, 31, 79
Typhlopseudothelphusa mitchelli, 31, 79
Typhlopseudothelphusa mocinoi, 31, 79
Typhloschizidium cottarellii, 79
Typhlotanais, 24
Typhlotricholigioides aquaticus, 31, 79
Typton, 39
Uca, 19, 22, 28
Uca (Australuca) bellator, 79
Uca (Australuca) signata, 19, 79
Uca (Gelasimus) vocans, 79
Uca (Tubuca) arcuata, 79
Uca (Tubuca) urvillei, 79
Uca arcuata, 18
Uca consobrinus, 19, 79
Uca formosensis, 18, 79
Uca lactea, 18, 19, 79

Uca marionis, 18, 79
Uca marionis vocans, 79
Uca signatus, 19, 79
Uca tangeri, 19, 79
Uca urvillei, 19, 79
Uca vocans, 18
Udonella, 62
Ulophysema, 57
Umbricella, 21
Unciala, 13
Uncinocythere, 30, 53
Upogebia, 19, 55, 63
Upogebiidae, 12, 39
Uradiophora, 61
Urechis caupo, 47, 81
Urocaridella, 60
Uroptychus, 23, 24, 42
Urothoe, 17, 57
Vahinius, 46
Valvifera, 13, 15, 27
Vanhoeffenura bicornis, 6, 79
Vanhoeffenura chelata, 6, 79
Vargula parasitica, 68, 79
Varuna, 22
Varunidae, 14
Vectoriella, 50
Velella, 40
Veleronia, 42
Veneridae, 49
Venerupis mitis, 49, 82
Ventriculina, 47
Ventriculinidae, 47, 49
Vermilia, 13
Vermiliopsis, 13
Vermiliopsis infundibulum, 13, 79
Verongia, 38
Verruca, 24, 42

Verruca stroemia, 48, 79
Vertebrata, 131
Verum cancellatum, 77, 79
Vestimentifera, 24
Vibrio, 125
Vibrio cholerae, 126, 130
Viguierella caeca, 32, 33, 79
Vir, 45
Weltneria, 11, 45
Weltnerium aduncum, 77, 80
Weltnerium nymphocola, 51, 80
Willemoesia, 24
Xanthidae, 45
Xantho, 56
Xarifia, 44
Xarifiidae, 44
Xenacanthomysis, 340
Xenobalanus globicipitis, 61, 80
Xenocarcinus, 46
Xenocoeloma, 51
Xiphias, 52, 59
Xiphias gladius, 59, 83
Xiphocarididae, 27
Xiphosura, 82
Xylocheles macrops, 80
Xylopagurus, 14
Zebrida, 57
Zebrida adamsii, 57, 80
Zenobia prismatica, 13, 53, 80
Zenobiana, 13, 15
Zenobiana phryganea, 80
Zingiber macradenium, 32, 80
Zoantharia, 9, 42
Zoothamnium, 62
Zoothamnium duplicatum, 292, 311, 352
Zostera, 13, 20, 297, 352
Zygomolgus, 58

SUBJECT INDEX

ω-3 polyunsaturated fatty acids, 119
16S mitochondrial rRNA, 319
18S rDNA, 319, 320
18S rRNA, 319
28S rDNA, 318, 319
Abbreviated development, 27
Abdominal artery, 261
Abdominal rudiment, 285–288
Abnormality, 272
Abyssal, 6–8, 18, 23, 41, 51, 57, 296
Abyssal fauna, 23
Abyssal forms, 7
Abyssal mud flats, 23
Abyssal plane, 296
Acanthocephalan, 312
Accessible water, 33
Accessory cell, 242
Accessory sex organs, 269
Accumulation, 177, 282, 283, 291
Acetazolamide, 250
Acorn barnacles, 13, 22, 52, 53
Acrothoracicans, 11, 45
Active avoidance, 303
Adaptations, 7, 18, 19, 26, 28, 29, 34, 35, 199, 299, 323
Adoption, 279, 280, 289
Adriatic, 281, 290
Adriatic Sea, 310
Adult females, 40, 197, 209, 270, 272
Adult males, 201, 209, 269, 270, 272, 282
Adult size, 4
Adulthood, 201, 272
Adults, 10, 21, 26, 38, 62, 191, 201, 243, 245, 272, 284, 288, 295, 302, 304, 309, 312
Adverse impacts, 315
Aerial humidity, 34
Aerial respiration, 34
Aesthetasc, 252, 253
Aesthetascs, 251, 253, 269
Africa, 5, 31, 119, 126, 149, 156, 169, 194, 300
Aggregation, 303, 310, 405
Agonistic behaviour, 273
Agriculture, 174, 176
Alcyonarian, 42
Aldabra, 22
Algae, 9, 18, 20, 21, 61, 118, 295, 300, 306, 307, 311, 317
Algal determinant, 9
Algeria, 29
Alimentary belt, 254, 257
Allergenic protein, 120
Allergic reactions, 120

Allometric relation, 278
Alpheid, 16, 19, 55
Altarpiece, 160
Alternative hypothesis, 190
Altitude, 26, 32
Alvinocaridid, 25
Amazonian, 322
Ambitus, 247
America, 32, 122, 159, 299
American, 28, 32, 53, 54, 123, 124, 196, 317, 326
Americas, 7, 35
Ampeliscid, 62
Amphibious species, 6
Amphimixis, 284
Amphipod, 10, 13, 16, 20, 21, 39, 40, 48, 53, 60, 62
Amphipods, 4, 13, 16, 17, 20, 21, 24, 35, 43, 53, 55, 57, 59–61, 125, 253, 272, 277
Amplification, 251
Anal drinking, 260
Ancestors, 26, 250, 323
Ancestral somites, 209
Anchihaline, 21, 298, 313
Anchihaline cave, 21
Anchihaline stations, 22
Anchoring structures, 279
Andalucia, 19
Andes, 5
Androgenic glands, 267
Anemones, 310
Animal, 3, 12, 15, 17, 61, 139, 140, 145, 157–159, 161, 164, 167, 169, 171, 181, 199, 242, 288, 291, 300, 306
Animal Kingdom, 3, 4, 164
Animal matter, 314
Animals, 9, 10, 12, 16–19, 23, 50, 64, 80, 117–119, 127, 140, 141, 156, 159–164, 166–169, 171, 172, 176, 183, 199, 243, 250, 268, 274, 276, 278, 282, 291, 292, 299, 306, 308, 315, 316, 321, 323, 326, 328
Annam, 45
Annelid, 13, 311
Annelids, 13, 56
Annual production, 118
Anomaly, 293
Anomuran, 21, 55, 168, 172, 176
Anophthalmic, 241, 251–253, 284
Anophthalmy, 20
Antarctic, 7, 22, 118, 275, 282, 294, 313, 323–325
Antennae, 35, 39, 47, 167, 178, 179, 197, 214, 216, 218, 237, 251, 263, 286, 288, 305, 307, 336
Antennal aesthetascs, 291
Antennal arteries, 261
Antennal gland, 214, 219, 268, 334–337, 345
Antennal glands, 267

Antennal peduncle, 214
Antennal scale, 197, 214, 217, 219, 270, 330, 332–341, 343–346
Antennal sensilla, 253
Antennal somite, 203, 204, 207
Antennal sympod, 214
Antennulae, 35, 197, 251, 270, 307, 336
Antennular flagellum, 252, 253, 269, 331, 340, 346
Antennular peduncle, 213, 214, 253, 270, 334, 336, 345
Antennular somite, 204
Antennules, 17, 214, 286
Anterior aorta, 260, 262
Anterior intestine, 255
Anterior projections, 203
Anthropogenic dispersal, 325, 326
Anthropogenic range expansion, 314
Anthropogenic transfer, 326
Anti-predatory behaviour, 304
Antiboreal regions, 8
Antilles, 27, 28, 30, 36, 43
Antioxidant levels, 119
Antipatharian coral, 46
Antipatharians, 11
Antiperistalsis, 260
Antipredator adaptation, 311
Ants, 35
Anus, 210, 237, 254, 259
Aorta, 238, 260–263
Apolysis, 268
Apomorphic characters, 195
Appendages, 7, 18, 64, 158, 191, 194, 195, 199, 201, 209, 211, 217, 225, 231, 233, 236, 237, 255, 258, 261, 270, 286, 288, 293
Appendix masculina, 214, 219, 252, 253, 270, 272, 340, 341
Apulia, 194, 278, 281, 322
Aquaculture, 120, 121, 123, 124, 126, 284, 325, 326
Aquaristics, 325
Aquatic crustaceans, 5
Aquatic environments, 6
Arabian Gulf, 273 [see also Persian Gulf]
Arcachon, 17, 185, 201, 294, 296
Archaeobacteria, 24
Archibenthic, 323, 331
Arctic, 7, 22, 192, 282, 294, 299, 313, 324
Arctic Ocean, 313, 323, 324
Arctic zones, 201, 323, 324
Areal biogeography, 324
Arginine kinase, 120
Arid coasts, 21
Aristotle, 127, 140, 157, 173
Art, 127, 128, 139–142, 147–150, 152–154, 156–167, 169, 170, 172, 174, 175, 179, 180, 183, 184

Arterial system, 262, 318
Arteries, 260–263
Arterioles, 261
Arthrocytes, 267
Arthropod, 49, 291
Arthropods, 4
Articulation of pleopod, 217, 224, 233, 334–336, 340, 344–346
Articulation of the leg, 223
Artificial implantation, 289
Ascension, 31
Ascension Island, 21
Ascothoracid cirripede, 43
Asia, 35, 121, 126, 149, 169, 316, 325
Asian, 8
Asiatic, 313
Assam, 31
Associates, 47, 53, 57, 58, 163
Association, 19, 24, 36, 38, 40, 47, 56, 60, 118, 156, 170, 173, 183, 298, 303–305
Associations, 15, 36, 38, 40, 43, 44, 49–51, 57, 59, 174, 304, 305, 309, 310, 405
Associations with benthic invertebrates, 309
Assyria, 139, 141
Asteroids, 56, 57
Asymmetric mandibles, 217
Asymmetrical lobes, 217
Asymmetrical paragnaths, 219
Athrocytes, 267, 268
Atlantic, 5, 17, 22, 27, 43, 50, 166, 191, 194, 231, 243, 270, 272, 273, 280, 281, 290, 292, 294, 296, 299, 301, 302, 304, 310, 313, 323, 324, 326, 405
Atlantic origin, 323
Attachment, 47
Attainment of sexual maturity, 201, 278, 294
Auditory functions, 250
Australia, 13, 15, 27, 31, 32, 44, 152, 211, 326
Autapomorphic features, 321
Autotomy, 292, 293
Autotrophic bacteria, 118
Autotrophic microorganisms, 306
Axons, 260
Azores, 27, 34
Bacterial mats, 24, 25
Bahama Island, 21, 55
Ballast water, 125, 325, 326
Ballast water treatment, 315
Baltic, 191, 243, 272, 291, 296, 301, 314
Bamboo, 14, 23
Banquets, 175, 177, 179, 180
Baobab, 14
Barnacle, 9, 11, 13, 54, 59, 119, 128
Barnacle geese, 128
Barnacles, 11, 23, 44, 48, 51, 119, 124, 128, 276
Basal plate, 190, 223, 225, 227
Basaltic lavas, 23, 25

SUBJECT INDEX 441

Base, 141, 144, 161, 197, 219, 223, 241, 247, 253, 259, 261, 262, 264, 267, 329, 330
Basis, 35, 117, 183, 201, 213, 214, 217, 219, 221, 223, 225, 227, 229, 232, 233, 242, 247, 258, 263, 271, 291, 311, 316, 325, 337, 338
Bathymetric distribution, 324
Bathypelagic, 213, 241, 295, 296, 302, 313, 323, 331
Bathypelagic species, 199, 201, 276, 278, 295, 323, 324
Bavaria, 317
Bay of Biscay, 51, 273, 278, 296, 301
Bayesian analysis, 320
Behaviour, 12, 35, 167, 181, 196, 245, 279, 289, 300
Belgian, 164
Belgium, 141, 184, 281
Benefits of gregariousness, 304
Benthic, 7, 8, 10, 17, 18, 24, 44, 405
Benthic [predator], 277, 297, 302, 303, 305, 307, 308, 310, 312–316, 321, 323, 324
Bentho-planktonic, 297
Benthopelagic, 194, 213, 295, 297, 312, 313, 321, 323–325
Biochemistry, 328
Biocide, 315
Biocoenoses, 22, 64
Biocoenosis, 10
Biodiversity, 10, 184, 313, 314, 316, 325, 328
Biodiversity loss, 124
Biofouling, 124, 125
Biogeography, 30, 250, 328
Biological determinant, 9
Biological invaders, 315
Biological invasion, 124
Biological production, 282
Biological-control agent, 123
Bioluminescent secretion, 267
Biomagnification, 315
Biomineralogy, 321, 328
Biosphere, 64
Biosynthesis of polyunsaturated fatty acids, 314, 315
Biosynthetic routes, 125
Biotic exchange, 325
Biotope, 22, 23, 32, 298
Biotopes, 9, 18, 21, 24, 32, 34, 64, 298
Bioturbation activity, 315
Bipolar distribution, 324
Biramous [appendage], 191, 193, 197, 233, 321, 329–338, 342–346
Bird, 32, 127, 128, 142, 177, 303
Birds, 36, 60, 127, 128, 140, 172, 316
Birth rate, 277
Bismarck Archipelago, 29

Bisphenol-A, 272
Bivalve, 14
Bivalve molluscs, 9
Bivalves, 14, 124
Black Sea, 50, 250, 293, 297, 301, 313, 316, 326
Bladder, 268
Blastoderm disc, 286
Blastomeres, 284
Blastoporic area, 286
Blister-like cells, 259
Blood sinus, 239, 240
Blue alga, 16
Boar, 177
Bodegones, 162, 163
Body cavity, 261
Body colour, 213, 214, 297, 306
Body flexing, 288
Body flipping, 191
Body lengths, 201
Body mass invested into reproduction, 278
Body position, 250
Body segmentation, 195, 286
Body size, 19, 23, 196, 201, 258, 263, 276–278, 291, 292, 296, 308, 321
Body wall, 211, 213, 255
Body weight invested into each brood clutch, 283
Bohemia, 33
Bonaire, 31
Book, 121, 122, 127, 128, 144, 159, 160, 172, 173
Books of Hours, 160, 168, 173
Boreal, 201, 276, 278, 299
Boreal waters, 296
Boreoarctic, 312
Borneo, 31, 55
Bottom, 23, 210, 255, 264, 291, 295–297, 299–304, 308, 313, 322, 335, 338, 345
Brachyuran, 51
Brachyurans, 18, 21, 45, 56, 167
Brackish [waters], 7, 8, 22, 26, 27
Brackish lagoons, 297
Brackish waters, 7, 8, 22, 243, 250, 281, 297, 314, 322
Brain, 237, 240, 241, 261, 262, 268
Branchiopods, 123, 165
Branchiostegal carapace, 204, 329
Branchiostegal epithelia, 204, 207
Brazil, 30, 31, 169, 294, 346
Brazilian, 163
Breakdown of food, 217
Breathing physiology, 19
Breeding aggregations, 303
Breeding cycle, 282
Breeding females, 269, 277, 281, 282, 290, 291, 302, 303, 310, 405
Breeding season, 273
Breeding strategy, 291
Brine, 6, 120
Brine shrimp, 3, 27

British coasts, 280, 290
British Columbia, 309
British Isles, 163
British waters, 279
Bronze, 159, 213
Brood clutch, 275–278, 283
Brood duration, 281
Brood lamellae, 229, 291, 321
Brood pouch, 191, 193, 197, 229, 231, 272–276, 279–285, 288, 289, 291, 331
Brood protection, 201
Brood size, 277
Brood weight, 277, 283
Buccal region, 53
Burgundy, 174, 175
Burrow, 17, 18
Burrower, 17, 52
Burrowing activities, 123
Burrowing barnacles, 11, 44, 45
Burrowing behaviour, 297
Bury itself, 308
Butterflies, 140
Butterfly, 140
Cadmium, 281
Caeca, 259, 318
Calcareous bodies, 292, 318
Calcareous statoliths, 317, 318, 321
Calcite, 248–250, 306, 317, 318, 332, 338
Calcium, 259, 292, 318
Calcium carbonate, 122, 248, 306, 317, 322, 332
California, 16, 20, 22, 43, 47, 51, 55, 57
Californian, 50, 277
Callynophore, 219, 253
Cameo, 143, 170
Camouflage, 304
Canada, 119, 281
Canary Islands, 21, 30, 31, 245
Cannibalism, 280, 289, 304, 307
Cape Town, 30
Cape Verde Islands, 310, 311, 405
Caprellid, 20
Carapace, 5, 39, 55, 56, 61, 170, 190, 193, 196–199, 203–207, 209, 211–214, 229, 238, 255, 261–264, 285, 288, 290–292, 311, 329–331, 335, 342, 343
Carapace cavity, 190, 204, 207, 225, 264, 329–331
Carapace duplicature, 204, 263
Carapace formation, 203, 207
Carbon, 307
Carbon transfer, 118
Carbonate [statolith], 248, 317, 322
Carbonate statoliths, 248, 321
Carboniferous, 11, 225, 288, 317, 319, 321, 329, 342
Cardiac arteries, 263

Cardiac chamber, 254, 257, 258
Cardiac filter, 258
Cardiac filters, 257
Cardiac ridge, 255
Cardiac tooth, 257
Cardiac valve, 255
Cardial position, 211
Cardio-oesophageal valves, 255
Cardio-pyloric valves, 257
Caribbean, 242, 298, 313, 322
Caridean, 27, 40, 52, 56, 142, 167
Caridoid facies, 197, 318
Carnivorous, 7, 8, 17, 217, 302, 309
Carnivory, 118
Carpopropodus, 197, 225, 227, 270, 335, 337–341, 345, 346
Carpus, 196, 223–225, 227, 229, 271, 334–336, 340, 344–346
Caspian Sea, 26, 50, 313, 314, 323
Castrating, 312
Catalan, 297
Catalan Sea, 296
Catchment areas, 26
Catiniid, 47
Cattle farming, 176
Caudal furca, 288
Caudal papilla, 286, 288
Cave-dwelling, 207, 263
Cavernicolous, 20, 27, 30, 31, 35, 53, 237, 241, 245, 267, 288, 299, 306, 307, 330
Cavernicolous freshwater Crustacea, 29
Caves, 26, 29, 31, 276, 299, 306, 313, 314
Cellulase activity, 307
Cellulases, 307
Cellulose, 307
Central America, 159, 298
Central cavity, 248
Centre of origin, 322
Centrolecithal eggs, 284
Cephalic arteries, 261
Cephalic artery, 261
Cephalic mass, 238
Cephalic region, 190, 207, 218
Cephalic somites, 197, 203
Cephalic tergites, 203
Cephalothorax, 142, 156, 167, 169, 179, 190, 197, 199, 203, 205, 207, 209, 237, 238, 268
Ceramic, 145, 159, 162, 172
Cercopods, 287, 288
Cerebral ganglia, 237
Cervical position, 211
Cervical sulcus, 190, 204, 213, 330
Cestode, 312
Cestodes, 126
Cetaceans, 52, 61
Ceylon, 44
Char, 316
Chelicerate, 129, 164

SUBJECT INDEX

Chelurids, 12
Chemical digestion, 259
Chemoreceptors, 251
Chemosensory abilities, 245
Chemosensory mechanisms, 280
Chemotactic substances, 273
Chiasma, 241
Chiasmata, 240
China, 128, 129, 317
Chinese mitten crab, 123
Chitin, 122, 123, 255
Chitosan, 122, 123
Chlorophyll-a, 314
Cholera, 126
Cholesterol levels, 119
Choniostomatid, 52, 54
Chorion, 275
Christian era, 156
Christmas Island, 36, 64
Chromatophores, 213, 247, 269, 285
Ciliates, 311
Circadian rhythms, 243
Circular muscle, 254
Circulatory system, 213, 260, 318
Circumtropical distribution, 322, 323
Cirolanid, 18, 61
Cirripede, 3, 40, 48, 60, 150, 168
Cirripedes, 4, 10, 11, 24, 25, 36, 38, 41, 49, 51–53, 55, 60, 61, 169
Clade, 319
Clades, 321
Cladoceran, 309
Cladocerans, 118, 123, 309, 315
Cladogram, 319
Clams, 123
Classical antiquity, 127
Classification, 4, 7–9, 157, 195, 312, 317, 320, 328, 347
Classification history, 191
Clausidiid, 47
Claw, 170, 223, 224, 227
Clean, 305, 311
Cleaning, 55, 59, 60, 305
Cleaning lobes, 229
Cleaning setae, 227, 305
Cleaning shrimps, 60
Cleaning the brood, 279
Climate change, 314
Closing apparatus, 231, 232
Clutch size, 277
Cnidarians, 8, 36, 56, 311
Coast, 12, 17, 43, 127, 237, 280, 296, 298, 301, 302, 309, 313, 326
Coastal habitats, 251, 301
Coastal lagoons, 22, 315

Coastal waters, 8, 191, 211, 272, 279, 289, 304, 313, 314, 316
Cockles, 123
Coconuts, 14, 23
Cod, 176, 315
Cohesiveness of swarms, 303
Cold-season breeding, 282
Colombia, 34
Colonization, 6, 7, 26, 307, 315
Colour, 24, 45, 128, 140, 161, 162, 169, 171, 175, 177–180, 183, 213, 214, 247, 255, 397
Colour patterns, 213
Commensal, 10, 35, 37, 47, 50, 55, 58, 61, 63
Commensal relationship, 311
Commensalism, 13, 14, 36, 44, 57
Commensals, 7, 8, 13, 27, 30, 35, 50, 53, 55, 56, 58, 61, 124
Commercial fishing, 118, 123
Commissures, 240
Common ancestors, 319
Community, 19, 23
Comores, 33
Compartmented biotopes, 27
Competing hypotheses, 189, 190
Competition, 34, 175, 304
Completion of larval development, 289
Compound eyes, 241, 242
Compound organ, 211, 213
Cone, 242
Congeneric species, 229, 280, 324
Congo, 34
Connected waterways, 326
Connective tissue, 241
Connectives, 238
Consortes, 36
Conspecific, 279, 280
Constructions, 10, 15
Contact with the sea, 28, 43
Continental slope, 296
Continental waters, 29
Continents, 28
Continuous breeding, 281, 291
Control the abundance, 309
Convergent evolution, 19
Copepod, 5, 21, 32, 36, 38, 40, 47–49, 52, 53, 58, 60, 63, 124
Copepods, 4, 7, 17, 21, 25, 30, 32, 33, 36, 38–41, 43, 44, 46–58, 61, 117, 118, 123, 124, 126, 276, 308, 309, 311, 312
Copulation, 15, 274, 275, 282, 291
Coral reef, 298
Coral reefs, 10, 11, 44, 45, 311
Corals, 8, 11, 14, 40, 44, 45
Core, 247, 248
Cornea, 190, 196, 213, 241, 242, 245, 247, 284, 299, 330, 335, 337, 338, 342, 345
Corneagenous cells, 242
Corpus centrale, 240

SUBJECT INDEX

Cosmopolitan, 8, 324
Costa Rica, 12
Counter-balance of potential benefits and damage, 314
Coxa, 214, 217, 221, 223, 229, 231–233, 267, 271, 329, 337, 338, 342
Crab, 3, 25, 35, 36, 40, 42, 43, 46, 53, 54, 56, 60–62, 120, 122, 123, 125–129, 141–154, 156, 158, 160–164, 167, 168, 170–173, 175–178, 181, 311, 313, 322, 399, 401
Crabs, 5, 10, 14, 15, 18, 19, 21–23, 25–28, 31, 32, 36, 39, 42–45, 47, 50, 51, 55, 60, 64, 119–129, 139, 141, 142, 144, 146, 148–151, 153, 154, 156, 159, 161, 162, 164–173, 181, 183
Crayfish, 5, 26, 27, 30, 53–55, 120, 122–124, 126, 127, 139, 142–149, 152–154, 156, 159, 161, 162, 165–168, 170–173, 179, 181, 183
Crayfish plague, 124
Cretaceous, 23, 25, 321
Crinoids, 57, 58
Crustacean, 4, 11, 21, 28, 32, 43, 61, 64, 118, 120, 122, 124, 125, 141–156, 158, 162, 165, 169–173, 180, 184, 194, 213, 240, 260, 267, 272, 312, 319
Crustaceans, 3–7, 10, 12, 15, 18, 20–29, 33, 34, 38, 43, 44, 46, 51–56, 58, 60–62, 64, 117–129, 139–143, 156–177, 181, 183, 211, 240, 241, 260, 269, 275, 292, 293, 307, 311, 312
Crustaceans in art, 139–141, 162, 183
Cryptoniscid, 52
Crystalline cells, 242
Crystalline cone, 242
Crystalline tract, 242
Crystallographic characteristics, 250
Cuba, 13, 22, 29–31, 280
Culture, 119, 129, 166, 169, 176, 289
Cumaceans, 24, 196, 276
Cures, 121
'Curiosity rooms', 163
Cuticle, 169, 196, 211–213, 229, 241, 242, 248, 255, 259, 263, 268, 285, 286, 288, 290–292, 311
Cyclomorphosis, 201
Cysts, 264, 265
Cytoplasm, 242
Czech Republic, 30
Dactylus, 197, 223, 224, 227, 229, 270, 332, 343
Damselfish, 298
Danube, 27
Danube River, 282, 301, 316, 326
Daphniids, 315
Dark and light adaptation, 243
Dark habitats, 11
Dark regeneration, 243
Dark-adaptation, 243

Dark-adapted eye, 243
Daylight, 181, 245, 322
Decapod, 54, 119, 122, 128, 165, 181, 194, 269, 293, 349
Decapods, 19, 22, 24, 26, 35, 36, 44, 55, 119, 120, 125, 126, 129, 139, 141, 154, 156, 161, 165, 167, 169, 183, 194, 242, 243, 250, 253, 255, 259, 276, 293, 315, 326
Deep benthos, 23
Deep groundwater, 299, 306, 322
Deep sea, 6, 23, 24, 250
Deer, 151, 176, 177
Defecation, 259
Deferent canals, 264, 265, 267
Definitions of aggregation, 302
Degree of endemism, 323
Dendritic connections, 291
Dendritic segment, 251
Dendritic segments, 251, 253
Denmark, 33
Depth, 5, 23, 243, 295, 296, 308
Descending artery, 261, 262, 318
Described species, 4, 30, 31
Deserts, 32, 35
Desiccation, 9, 27, 32
Detection of predators, 253
Determinant relating to shelter, 9
Detritivores, 117
Detritivorous, 18, 307
Detritivory, 118
Deutocerebrum, 237, 240
Development of marsupial young, 290
Development of the young, 283
Developmental stages, 283, 329, 330
Devonian, 319
Diagnoses, 328
Diagnostic tool, 328
Diameters of fertilized eggs, 276
Diapausing eggs, 32
Diaphragm, 260, 261
Diatoms, 15, 19, 62, 242, 295, 306, 308
Didactism, 168
Diecdysis, 195, 291, 292
Diet, 19, 118, 168, 176, 268, 301, 302, 308
Diets, 3, 315
Differential mortality, 273
Differentiation, 175, 286, 289
Digesting food, 217
Digestion, 255, 259, 260, 305, 306
Digestion of cellulose, 12
Digestive cycle, 259
Digestive system, 254, 255, 306
Digestive tract, 32, 47, 63, 254
Digestive tube, 255, 261, 264
Digitus mobilis, 219, 340
Dinoflagellates, 125
Diphyllobothriasis, 126
Diplopod, 34

SUBJECT INDEX 445

Direct development, 26
Directional swimming, 250
Disadvantages of aggregation, 304
Disease, 126
Diseases, 122, 126, 129
Dispersion, 301, 303, 326
Distribution, 4, 28, 29, 194, 204, 211, 213, 273, 279, 296–298, 301, 310, 312, 313, 322–325
Distributional separation, 303
Disunity of the Mysidacea, 319
Diurnal horizontal shifts, 301
Diurnal pigment migrations, 243
Diurnal variations, 297
Diurnal vertical migration, 301, 308
Diurnally migrating [species], 241, 314
Diversification, 8, 25, 26
Diversity, 3, 7, 23, 44, 64, 140, 192, 196, 204, 225, 241, 244, 251, 270, 296, 309, 312, 313, 315, 328, 334, 335, 404
Dogs, 122
Dolphin, 61, 171
Dorsal caecum, 254, 258
Dorsal diverticulum, 254, 258
Dorsal organs, 286
Dracunculiasis, 126
Dugong, 60
Duration of marsupial development, 280
Dutch, 128, 139, 151, 162–164, 166, 168, 169, 174, 175, 177, 179–181, 183
Dwarfing, 207, 263, 264
Early initial radiation, 319
Early radiation, 319, 328
East Africa, 5
Ebro delta, 295
Ecdysis, 248, 253, 285, 287, 288, 291, 292
Echinoderms, 15, 36, 56–58
Echinoids, 56, 58
Ecological adaptations, 10
Ecological benefits, 315
Ecological biogeography, 312, 322
Ecological categories, 9
Ecological classification, 27
Ecological efficiency, 118
Ecological factors, 297, 298
Ecological implications, 277
Ecological ranges, 298
Ecological role, 117
Ecological specializations, 7
Ecology, 250, 315, 328
Economic importance, 117
Ecosystem, 316
Ecosystems, 117, 118, 124, 125, 314, 315
Ectoderm, 284
Ectodermal origin, 255, 258
Ectoparasites, 51, 124, 311

Ecuador, 5
Eelgrass beds, 10, 11, 20
Efferent canals, 264
Egestion rates, 306
Egg clutch, 273, 274, 283
Egg deposition, 291
Egg diameters, 276, 278
Egg hatching, 284
Egg membrane, 284, 286, 288
Egg numbers, 277
Egg sac, 275
Egg size, 24, 196, 207, 264, 276–278, 294
Egg volume, 294
Egg weight, 278, 283
Eggs, 7, 17, 19, 20, 22, 27, 32, 36, 61–64, 126, 213, 214, 230, 264, 269, 275–278, 281–284, 288, 292, 294, 295, 307
Egypt, 128, 139, 141, 142
Electric signal, 251
Ellobiopsid parasites, 272
Embryonic [stage], 191, 195, 259, 280, 284, 286, 331
Embryonic development, 20, 209, 275, 281, 284, 302
Embryonic phase, 291
Embryonic stage, 284–286, 290
Embryos, 210, 269, 272, 278, 279, 282, 283, 292
End sac, 214, 219, 268
Endemic, 25, 26, 313, 314, 323, 326
Endemic forms, 7
Endemics, 24, 313, 324, 325
Endemism, 196, 313, 324
Endite, 214, 217, 221, 223, 225, 331, 337
Endocrine control, 268
Endocrine glands, 240
Endodermal cells, 286
Endodermal origin, 255
Endoecism, 52
Endoparasitism, 312
Endopsammon, 19
Endosternite, 237
Endosymbiosis, 24
Energy, 24, 118, 120, 302, 304
England, 160, 172, 273, 326
English Channel, 13, 20, 38, 45, 48
Environmental change, 314
Environmental stimuli, 289
Enzymatic digestion, 259
Eocene, 23
Epibenthic lifestyle, 242
Epibiont, 311
Epibionts, 169, 292, 311
Epibiotic community, 118
Epicaridean, 312
Epicarideans, 311
Epigean species, 28, 237
Epimeron, 203, 236
Epipelagic, 201, 242, 251, 276, 278, 294–296, 323

Epiphyta, 55, 61
Epiphytes, 124
Epipod, 190, 223, 224, 227, 262, 263, 329–331
Epirhabdome, 242
Epithelium, 260, 265
Epizoites, 305, 311
Epizoonts, 55, 60–62
Equator, 7
Equilibrium organ, 250
Erectile penes, 275
Erectile tissue, 231
Erection, 275
Escape movements, 191
Estuaries, 22, 26–28, 293, 297, 298, 301
Ethanol, 250, 284
Ethiopia, 29
Etiology, 126
Etymology, 192, 196
Euphausiacean, 191
Euphausiids, 118, 295, 304
Eurasia, 317
Europe, 21, 31, 35, 39, 40, 122, 124, 149, 158, 169, 181, 248, 273, 293, 316, 323, 325–327, 406
European, 8, 43, 54, 118, 124, 141, 158, 174, 184, 309
Eurybathic, 24
Euryhaline, 15, 297, 308, 324
Euryhalinity, 7, 20
Euryhalobious, 297
Eurythermy, 20
Eutrophic, 118, 307
Evaporation, 21, 32, 35
Evolution, 20, 28, 177, 255, 321
Evolutionary history, 28
Evolutionary trends, 283
Excretion, 267
Excretion of growth-limiting nutrients, 314
Exite, 223, 235, 333, 343
Exites, 197, 233, 333, 334, 343
Exopod of the maxilla, 263
Exoskeleton, 203, 255
Exoskeletons, 122, 164
Exotic species, 314
Expansion of non-indigenous organisms, 315
Experiments, 243, 250, 279–281, 291, 293, 304, 311
Extant taxa, 225, 329, 342
Extensor, 223
Extensors, 237
Exuvia, 291
Eye modifications, 199, 242, 244
Eye plate, 218, 245, 247, 284, 336
Eye reduction, 198, 245, 284
Eye reductions, 199, 245
Eyed larvae, 283

Eyeless larvae, 283
Eyes, 7, 17, 24, 123, 177, 181, 190, 193, 194, 207, 241, 243, 245, 247, 250, 261, 267, 279, 288, 289, 293, 299, 300, 331–338, 340, 342, 345
Eyestalk, 54, 213, 239–241, 245, 247, 268, 269, 290, 293
Eyestalk movements, 242, 250
Eyestalk papilla, 241, 244
Eyestalk papillae, 241, 247
Eyestalks, 178, 179, 181, 190, 196, 237, 240, 245, 247, 255, 268, 284, 293, 299, 322, 323, 330, 334–336, 338, 345
Facet, 242
Facultative association, 311
Faecal pellets, 306
Faeces, 55, 126, 311
Fairy shrimp, 119
Faith, 166
Family group, 309, 333, 334, 341
Far East, 128, 159
Fat bodies, 213
Fat reserves, 282, 306
Fauna, 10, 20–22, 32, 33, 44, 60, 139–143, 156, 157, 163, 166, 167, 172, 314, 323, 398
Faunas, 22, 25, 32, 324, 325
Faunistic element, 299
Faunistic richness, 23
Fecundity, 277, 278
Feeding, 18, 19, 55, 118, 123, 183, 191, 217, 219, 262, 276, 283, 291, 297–299, 301, 304–309, 314, 322
Feeding basket, 306
Feeding behaviour, 307, 309
Feeding mechanisms, 305
Female genital apparatus, 264, 265
Female genital orifice, 264
Feminized males, 272
Feminizing effects, 272, 312
Fenestra paracornealis, 196, 219, 244, 247
Fern, 35, 202, 203, 270, 346, 351
Fertilization, 264, 275, 284
Fibrillary cells, 259
Fiddler crabs, 19
Field studies, 308
Fiji, 22, 31, 36, 125
Filters, 27, 258, 259, 325
Filtration, 126, 196, 258, 306
Filtration belt, 254, 257
Filtration channel, 254, 258
Filtration current, 306
Filtration currents, 190, 300
Finland, 243, 275
First larval ecdysis, 288
Fish, 7, 8, 19, 27, 49, 52, 58–60, 117–119, 123, 126, 140–142, 146, 147, 150, 152, 153, 156, 158, 162–164, 167, 169–172, 175, 176, 183, 303, 305, 307, 309, 311, 312, 315, 316, 325, 326

SUBJECT INDEX

Fish lice, 124
Fish markets, 140, 176
Fish stocking, 325
Fishery, 119
Fitness, 276, 277
Fitness for reproduction, 275
Fitness for survival, 275
Fjords, 8, 301
Flagella, 24, 214, 251, 253
Flagellum, 190, 214, 217, 223, 225, 253, 267, 293
Flamingos, 32
Flanders, 163, 173, 174, 177
Flemish, 128, 139, 162, 163, 166, 168, 170, 174, 175, 181, 183
Flexor, 223
Flexors, 237
Flora, 314
Florida, 12, 184, 311
Flounder, 316
Fluoride, 250
Fluorine, 250, 322
Fluorite, 248–250, 289, 317, 321, 322, 332–334, 337–341
Fluorite precipitation, 250
Fluorite statoliths, 248, 250, 251, 317, 321, 322
Fly, 4, 36, 64
Foliaceous epipod, 225
Food, 17, 20, 117, 118, 120, 122, 123, 129, 140, 157, 163, 167, 171, 172, 175–180, 183, 191, 196, 253, 254, 257, 259, 268, 291, 292, 295, 304–308, 315, 316, 325
Food availability, 277, 292, 297, 298, 301, 307
Food capture, 304
Food chains, 316
Food choice experiments, 307
Food organisms, 316
Food particles, 255, 298
Food scarcity, 308
Food search, 307
Food supply, 292
Food web, 315
Food webs, 314, 315
Foraminiferans, 317
Foregut, 195, 254, 255, 257, 258
Foregut morphology, 318, 319
Form vision, 245
Formalin, 284
Fossil [statolith], 248, 249, 288, 306, 317–319, 321, 329, 330, 332, 333, 337–339
Fossil records, 317
Fossilization, 248, 317
Fossils, 4, 11, 317, 321, 328, 330, 331
Fouling, 20, 125, 305
Fracture plane, 292
Fragmentation, 305

France, 6, 17, 20, 31, 60, 119, 128, 141–144, 151, 156, 158–160, 163–165, 167, 168, 172, 173, 175, 201, 310, 317, 326, 346, 399, 400
Frequency of predation, 309
Fresh water, 7, 8, 21, 22, 26, 27, 32, 34, 117, 118, 120–122, 124–126, 156, 162, 165, 170, 172, 243, 259, 275, 282, 292–294, 297, 298, 301, 302, 308, 312–315, 322–326
Fresh waters, 6–8, 20, 22, 26, 27, 29, 34, 59, 121, 159, 297, 314, 315, 325
Freshwater [habitat], 19, 26–28, 35, 55
Freshwater population, 273, 308
Freshwater species, 28, 299, 323
Fringes, 196, 204, 207, 211, 213
Frog, 151, 162, 167
Frontal cornea, 242
Frontal organ, 240, 268
Frontal plate, 240, 247, 268, 270, 290
Fruit, 146–149, 151, 153, 156, 162, 176, 177–179
Funafuti atoll, 22
Functional ovaries, 272
Fungi, 307
Funnel, 35, 254, 257, 258
Funnel region, 207, 211, 251, 257, 282, 292, 312, 322, 324, 325
Furcal spines, 287, 288
Galapagos, 21, 31
Gallery, 18, 19, 51, 149, 150, 152–154, 165, 180, 184, 402
Galls, 38, 40, 41, 43, 45, 56
Gammarids, 124, 273
Ganglia, 210, 238–240, 284
Ganglion, 238, 240, 247, 260, 293
Ganglion chain, 238
Gas exchange, 263
Gastric mill, 254, 257, 258, 260
Gastroliths, 122
Gastropod, 170
Gastropod shells, 9, 11, 13, 50, 168
Gecarcinids, 34
Gene expression, 286
Generations, 282, 294
Genetic analyses, 195, 319
Genetic evidence, 190, 328
Genital papillae, 272
'Genre painting', 156
Genre painting, 161–164, 175–177
Genre scenes, 164, 166
Geographical distribution, 28, 29
Geographical distributions, 26
Geographical locations, 302
Geographical zone, 277
Geological evolution, 26
Geomorphological structures, 23
Geomorphology, 9
Geotaxis, 300, 301
Germ band, 286
German, 161, 177, 180

Germany, 123, 153, 165, 302, 315
Germinal disc, 286
Giant penes, 275
Gill chamber, 54, 60, 63
Gill chambers, 53, 56
Gills, 25, 30, 35, 49, 53, 54, 62, 123, 190, 191, 204, 205, 207, 227, 229, 230, 263, 269, 271, 329–331
Glacial relict, 325
Glacial relicts, 323, 324
Glaciation, 325
Glaciations, 26
Glacier barrages, 323
Glands, 16, 211, 223, 254, 255, 261, 267, 269
Glandular units, 217
Global geology, 28
Global species numbers, 312
Glomerulus, 240
Glycoproteins, 250
Glypheid, 18
Gnathiid, 18
Gnathopods, 35, 190, 194, 223, 227, 329, 330, 342
Gobies, 303
Gobiid, 19
Golden Age, 162, 173, 175, 183
Gonad development, 300
Gonads, 213, 269
Gonopore, 224, 231, 232
Gonopores, 191, 275, 277, 329–332, 342
Goose barnacles, 119, 128
Gorgonian, 41
Gorgonians, 11, 309
Gothic period, 160, 167, 168, 173
Grapsids, 34
Gravity, 300, 322
Grazer, 308
Grazing activity, 118, 314
Great Lakes, 326
Greek, 127, 156, 157, 161, 168, 170, 172, 193, 194
Greek art, 156, 157
Greenland, 191
Gregariousness, 303
Grinding lamellae, 219
Grinding surface, 219
Grooming, 227, 231, 305
Groundwater, 194, 282, 299, 308, 313
Groundwater habitats, 199, 298
Grouping, 302, 303
Growth factors, 292
Growth rates, 124, 201
Guadalquivir Estuary, 309
Guatemala, 31
Guests [in an association], 303
Guinea worm, 126
Guizhou, 317

Gulf of Aqaba, 242
Gulf of Finland, 314
Gulf of Mexico, 279, 281, 294, 313
Gulf of Naples, 213, 280, 310, 314, 405
Gulf of Sinai, 242
Guyots, 23
Habitat, 5, 9, 10, 17, 20–22, 26, 33, 39, 47, 64, 169, 213, 243, 245, 273, 295, 299, 303, 304, 310, 314
Habitat preferences, 298
Habitats, 4–7, 10, 20, 23, 25–28, 30, 32, 34, 44
Habitus, 198–200, 202, 207, 342
Hadal zone, 23
Haemocoel, 261
Haemolymph channels, 204, 263
Hatching, 36, 284–286, 294
Hawaii, 21, 34, 43, 295
Head, 17, 58, 122, 127, 163, 167, 170, 197, 261, 273
Hearing, 251
Heart, 122, 238, 260–266, 285, 286, 318
Heart beats, 286, 288
Hemi-tergites, 203
Hemispherical asymmetry, 313
Hepatic caeca, 259
Hepatopancreas, 238, 254, 255, 259, 261, 265, 266, 268, 285, 288
Herbivorous, 17, 217, 307, 309
Herculaneum, 142, 156
Hermaphroditism, 58, 264
Hermatypic corals, 44
Hermit crabs, 14, 20, 53, 54, 125, 149, 166, 309
Herring, 174
Heteromorphic regeneration, 293
Heteromorphic replacement, 293
Hexactinellid, 16
Hg, 254, 315
Hila, 248
Hilum, 248, 249, 317
Hindgut, 254, 255, 259, 292
Hippids, 17
Historical biogeography, 299, 322
Historical depictions, 161
History, 29, 127, 129, 140, 163, 166, 169, 173, 174, 183, 184, 190
Holarctic temperate regions, 7
Holopelagic, 199, 207, 296
Holothurians, 57, 58
Holothuroids, 57, 58
Homarid lobster, 142, 145, 153–156, 162–166, 168, 172, 179, 182
Home site, 298, 304
Home site permanence, 298
Homeostatic adaptations, 283
Homing abilities, 298
Homing behaviour, 304
Homology, 194, 288, 318
Homonymy, 194, 347

Horizontal burrowers, 18
Horizontal migration, 251
Horizontal migrations, 301
Hormonal control, 269, 291
Horseshoe crabs, 51
Hosts, 37, 38, 42, 44, 46, 49, 64, 126, 311, 312
Hot smoker vents, 5
Hot water, 5
Hotspot of diversity, 313
Human, 124–129, 140, 162, 163, 167, 170, 171, 173, 174, 177, 183, 314, 316, 325
Human consumption, 119, 120
Human health, 125
Human impact, 314
Humanist, 166
Humans, 117, 119, 120, 126, 140, 161, 167, 181, 250
Hungarian, 282
Hydrodynamic stimuli, 251
Hydrogen sulfide, 24
Hydrographical boundaries, 323
Hydromedusae, 39
Hydrostatic pressure, 300
Hydrothermal vent sites, 24
Hydrothermal vents, 5, 118
Hydrozoans, 8, 15, 16, 39
Hyper- or hypo-osmotically, 281
Hyperparasites, 52, 53
Hypersaline waters, 21, 27
Hypodermis, 213
Hypogean species, 30, 31, 213, 276
Hypotheses, 29
Hyssurid, 10
Illumination change, 241
Immigration, 23, 314
Immunohistochemical methods, 241
Impressionism, 168
Impressionists, 164
Incubation, 207, 269, 283, 289, 291, 305, 322
Incubation period, 278, 282, 283, 290, 306
Incubation time, 277
India, 31, 40, 54, 124, 211, 292, 316
Indian Ocean, 295, 302, 322–324
Indiana, 30
Indicators, 22, 178
Indo-Pacific, 33, 35, 324
Indonesia, 32, 35
Indonesian islands, 201
Infestation prevalence, 292
Infralittoral zone, 17
Infrared light, 243
Ingestion, 255, 306
Inhibitor, 250
Injury, 292
Inner lobe, 235, 237, 330, 331, 342, 343

Innervation, 260
Inquilines, 7, 8, 13
Inquilinism, 13, 36, 44
Insect, 118
Insects, 4, 34, 127, 140, 156, 162, 183, 240, 269, 291, 293
Instars, 38, 283, 289, 291, 294
Intake of water, 260
Integument, 53, 212, 247, 251, 255, 279, 286, 291, 311, 329, 330
Integumental sensilla, 211
Inter-individual distance, 303
Interaction, 279
Intermediate hosts, 125, 126
Intermediate intestine, 255
Intermoult period, 195, 291
Internal organs, 213, 237, 238
Internal skeleton, 237
Interruption of the breeding activity, 282
Intersexes, 264, 272
Intersexuality, 264, 272, 293
Interspecific adoptions, 280
Interstitial [habitat], 19
Interstitial environment, 26
Intertidal rocky pools, 296
Intertidal zonation, 19
Intestine, 49, 62, 238, 258, 259, 261, 265, 266, 286
Intestine contents, 213
Intraspecific adoptions, 280
Introductions, 309, 315, 316, 325
Introgression, 26
Invaders, 124
Invasion highways, 326
Invasion of warm-water biota, 314
Invasive population, 307
Invasive species, 124, 314
Invertebrate, 167
Invertebrates, 7, 24, 25, 36, 56, 140, 156, 276, 292, 297, 305, 310, 315, 326, 405
Ion transport, 204
Ionian, 281
Ions, 122, 250
Ireland, 326
Ischium, 223, 224, 229, 337, 338, 344
Islam, 159
Island populations, 29
Isopod, 12, 16, 20, 22, 25, 34, 39, 42, 61–63, 143, 154, 159, 173, 291
Isopods, 4, 6, 9, 10, 12, 13, 17, 19, 21, 23, 24, 35, 36, 52, 53, 56, 57, 59, 149, 165, 168, 171, 253, 276, 312
Israel, 31
Italian, 161, 162, 171, 175, 352
Italy, 13, 30, 31, 58, 127, 143, 144, 156, 163, 194, 278, 281, 293, 299, 322
Iteroparous, 282, 283
Iteroparous mysids, 277, 283
Jamaica, 31, 32, 62, 281

Jameos del Agua, Lanzarote, 21
Japan, 40, 44, 46, 49, 60, 125, 128, 272, 273, 279, 280, 290, 294, 295, 300, 316, 325, 326
Java, 33, 316
Jellyfish, 40, 156, 309, 311
Jewel, 143, 170
Jewelry, 158
Jumping behaviour, 35
Junior homonym, 194, 333, 334, 347, 349
Jurassic, 23, 317, 319, 321, 328, 329, 331, 332, 342, 343
Juvenile stage, 284, 288
Juveniles, 10, 12, 15, 21, 32, 39, 54, 203, 209, 243, 245, 269, 270, 277, 284, 285, 287, 289, 292, 295, 301–305, 309, 331
Kalahari, 32
Kamchatka, 315
Karstic, 30, 313
Kelp, 12, 314
Kentucky, 30
Kenya, 17, 36, 39
'Kitchen scenes', 176
Knee of thoracic endopod, 223, 225
Krill, 118, 119
Labium, 214, 217, 219, 221, 222
Laboratory culture, 273
Labrum, 214, 216, 217, 219–222, 259, 262, 332, 334–337, 340, 341, 345, 346
Lacinia mobilis, 219, 221
Lacuna, 259
Lacunae, 261, 262
Lagoon, 10, 237, 294
Lake Aral, 316, 317
Lake Baikal, 7, 26, 30
Lake Constance, 273
Lake Tanganyika, 26
Lake Washington, 277, 309
Lamella ventralis, 254
Lamina, 225, 239–241, 254, 337, 345
Lamina ganglionaris, 240
Laminarinase, 307
Land masses, 29
Landscapes, 161, 163, 164
Lanzarote, 31, 245
Large minimum egg size, 276
Larva, 51, 279, 286–288, 312
Larvae, 19, 25, 26, 28, 34, 59, 61, 63, 118, 119, 126, 207, 269, 279, 280, 282, 283, 287–289, 292, 305, 307, 312
Larvae replacement behaviour, 279
Larval development, 34, 203, 219, 280, 284, 286
Larval ecdysis, 288
Larval growth, 203
Larval instars, 26, 294
Larval moult, 269, 289

Larval moults, 284
Larval stages, 7, 27, 28, 191, 195, 276, 331
Last Supper, 144, 161, 171
Lateral arteries, 261, 263
Lateral cornea, 242
Lateral organ, 269
Lateralia, 254
Latitude, 276–278
Latitudinal gradient, 313
Latitudinal temperature effects, 283
Laurentian Great Lakes, 315
Leaf litter decomposition, 314, 315
Leech, 312
Length-fecundity relation, 278
Lens, 81, 242, 345
Lepadomorphs, 11
Leptostracan, 5, 25, 192
Leptostracans, 193, 194
Lessepsian migration, 326
Libya, 31, 119
Life expectancy, 295
Life span, 201, 294, 295
Light, 24, 177, 178, 180, 213, 241, 243, 245, 251, 257, 296, 300, 306, 308, 321, 322
Light intensities, 243, 245
Light intensity, 243, 296, 302, 308
Light penetration, 247
Light-adaptation, 243
Light-adapted eyes, 243
Lignicolous [species], 12, 22
Limivorous, 306
Limpet, 50
Limpets, 50
Lineage, 27
Lineages, 27, 324
Lipoprotein metabolism, 259
Literal sense, 166
Literature, 4, 129, 195, 196, 267, 281, 295, 319
Lithuanian lakes, 316
Littoral, 6–8, 15, 18–20, 22, 23, 26, 28, 29, 34, 213, 296, 297, 300, 302, 308, 321
Littoral zone, 10, 19
Living fossils, 25
Living resource, 119
Lobsters, 53–55, 119, 125, 127, 129, 139, 143, 153–155, 164–166, 168, 169, 172, 173, 176, 179, 181, 183
Lobula, 239, 240
Lobus masculinus, 270
Locomotion, 199, 207, 303
Locomotive orientation, 300
Locomotor reflexes, 250
Locomotory appendages, 251
Lodges, 11, 12, 15, 45
Longevity, 201, 273, 292, 294
Longitudinal muscle, 254

Lophogastrid, 194, 211, 217, 227, 229, 238, 240, 243, 245, 247, 257, 258, 260, 264, 268, 279, 288, 294, 295, 300, 302, 323
Lophogastrids, 194, 195, 233, 237, 240, 273, 278, 286, 295, 302, 313, 317
Low-salinity environment, 26
Loyalty Islands, 22
Luminescence, 24, 268
Luminescent capacity, 268
Luminescent organs, 267
Luminescent prey, 268
Luminescent secretion, 223
Luminosity, 304
Lunar cycle, 291
Macro-evolutionary breaks, 20
Macroalgae, 308
Macrofossils, 317, 321
Macroplankton, 296
Macruran, 167, 168
Madagascar, 8, 11, 17, 31, 33, 36, 38, 46, 50, 52, 56
Madeira, 8, 310, 405
Madras, 280
Madreporarian corals, 309
Magdalenian, 141
Majid, 20, 42, 55, 149, 172, 176
Majid crab, 142, 146, 155, 156, 165, 166
Malacostracan, 189, 196, 293
Malagasy, 20
Malaysia, 18, 277, 307, 311, 316
Maldives Islands, 44
Male genital apparatus, 264, 266, 267
Male genital orifice, 265
Male pleopod, 224, 233, 235, 270, 332–340, 343–346
Male pleopods, 197, 233, 235, 272, 321, 328, 331–339, 341, 343, 344
Male-specific sensilla, 269
Mammals, 127, 140, 141, 162, 164, 167, 176, 312
Manatee, 60
Mandibles, 190, 204, 214, 217, 219, 220, 222, 255, 286, 305, 306, 341, 346
Mandibular palp, 219, 305, 331, 343
Mandibular somite, 219
Mangrove, 12, 19, 23, 277, 303, 307, 311
Mangroves, 12, 18, 22, 23, 34
Manipulation of larvae, 279
Mantis shrimp, 19, 51
Mantis shrimps, 127
Mantle, 49, 247, 248
Marine environment, 7, 28, 50
Marine species, 26, 28, 156, 312
Marine species entering estuaries, 28
Marine transgression, 23
Market scenes, 175–177

Marquesas Islands, 46
Marsupia, 53
Marsupial development, 245, 281, 284, 291
Marsupial mortality, 280
Marsupial stages, 284, 289, 290, 331
Marsupium, 191, 196, 201, 209, 210, 213, 229, 264, 265, 270, 272, 273, 278–281, 283, 285, 288, 289, 292, 305, 311, 312, 333–335, 344
Masculinized adult females, 272
Masculinized females, 272
Mastication, 219, 257, 259
Masticatory structure, 219
Mating, 15, 61, 231, 233, 270, 273–275, 281, 303
Mating partners, 253, 303
Maturation, 284
Maui Island, Hawaii, 21
Mauritania, 13, 21
Mauritius, 38
Maxilla, 205, 214, 216, 222, 223, 263, 268, 337, 338, 344, 345
Maxillae, 217, 255, 261, 267, 286, 306
Maxillary glands, 267
Maxillary palp, 223
Maxilliped, 196, 223, 225, 330, 342
Maxillipedes, 35, 194
Maxillipeds, 167, 178, 190, 196, 219, 222, 223, 227, 329–331, 342, 343
Maxillopodans, 165
Maxillula, 205, 209, 214, 216, 221, 222, 263, 330, 331, 340, 341, 343, 346
Maxillules, 217
ME-MI X-organ, 241, 268
Mechanical digestion, 259
Mechanical signal, 251
Mechanoreceptors, 251
Median dorsal tooth, 257
Medicinal animals, 122
Medicine, 122
Medieval period, 139, 159, 161, 167, 172
Mediolittoral zone, 17
Mediterranean, 17, 20, 27, 119, 124, 127, 156, 165, 167, 194, 237, 243, 250, 278, 280, 281, 290, 295–297, 299, 301, 303–305, 309, 313, 314, 322–324, 326
Medulla, 239, 240
Medulla externa, 240
Medulla externa – medulla interna X-organ, 268
Medulla externa X-organ, 239
Medulla interna, 239, 240
Medulla terminalis, 239–241
Medulla terminalis X-organ, 239, 268
Megalography, 161, 163
Meiobenthos, 7
Meiofauna, 19
Melanin pigment, 242
Mercury, 127, 281
Merovingians, 158

Merus, 196, 223, 224, 229, 331, 332, 337, 338, 343, 344
Mesenteron, 255
Mesocosm experiments, 308
Mesoderm, 286
Mesodermal teloblasts, 286
Mesohaline, 301
Mesopelagic, 245, 296, 313, 323
Mesozoic, 7, 29, 322
Mesozoic fossils, 317
Metacercariae, 312
Metahaline, 297
Methods of phylogenetic analysis, 319
Mexico, 29–31, 119, 143, 159, 280, 313
Micro-phytophagy, 308
Microalgae, 242, 297, 308
Microfossils, 317, 321, 338
Microhabitat, 298
Microhabitats, 298
Micronekton, 296
Microorganisms, 117, 306, 308
Microphagy, 20
Microsporidians, 312
Mid-Atlantic Ridge, 25, 118
Middle East, 159
Middle Jurassic, 317, 321
Midgut, 254, 255, 258, 259, 261, 292, 318
Midgut caeca, 255, 259, 292
Midgut gland, 255
Migratory activity, 296, 302
Migratory rhythms, 300
Mineral, 25, 317
Mineral [statolith], 248, 249, 318, 321, 328
Mineral content, 123
Mineral precipitation, 250
Mineralization, 247, 249, 250
Mineralized statoliths, 321
Miniature adults, 276, 331
Minimum of reproduction, 282
Minnow, 118
Miocene, 27, 248, 249, 317, 321, 322, 324
Mitochondrial genes, 302
Mitochondrial loci, 319
Mixed schools, 305
Mode of life, 198, 295
Models, 28, 315, 322
Modified genital structures, 270
Modified male pleopods, 274
Modified pleopods, 272, 321
Moist, 6, 33
Molar structures, 219
Molecular characters, 245
Molecular clock data, 324
Molecular diversity, 313
Mollusc, 14, 40, 141, 172

Molluscs, 11, 14, 36, 45, 49, 50, 56, 123, 156, 162, 171, 172
Moluccas, 22, 39
Monkey, 172
Monophyly, 195, 318–320
Monophyly of the Mysidacea, 318
Monospecificity, 303
Monotony, 23
Moral, 166, 174
Morphological analysis, 195
Morphological concept, 193
Morphological evidence, 190, 319, 328
Morphological relationships, 328
Morphological specializations, 7
Morphological types, 9
Morphotypes, 328
Mortality, 124, 280, 298
Mosquitoes, 123
Mosses, 32, 33
Moult, 17, 231, 247–250, 255, 273, 275, 278, 280–284, 288, 289, 291, 292, 305, 321
Moult cycle, 195, 283, 290–292
Moult increments, 201
Moult inhibition, 289
Moulting, 181, 182, 248, 250, 253, 259, 268, 269, 279, 289–292, 311
Moulting cycle, 291
Moulting hormone, 269
Mount Kenya, 5
Mount Tabwemasana, 6
Mouth, 22, 27, 59, 60, 217, 220, 238, 255, 261, 301, 305, 306, 316
Mouthparts, 195, 203, 214, 216, 217, 221, 222, 237, 238, 240, 251, 252, 259, 279, 305, 307
Mozambique, 229, 263
MT X-organ, 241, 268
Mucopolysaccharides, 250
Mucus, 265, 311
Mucus canals, 58, 59
Mud flats, 22
Mud shrimp, 18
Muddy beaches, 18
Muscicolous [species], 33–35
Muscle bundles, 255
Muscle fibres, 260
Muscles, 221, 237, 238, 255, 260
Musculature, 209, 219, 221, 223, 225, 237, 261
Mussel, 124
Mussels, 15, 123–125
Mutualism, 36
Mutualistic, 304, 305
Myoarterial formation, 260, 261
Myosin light chain, 120
Mysid, 124, 191–195, 231, 237, 238, 240, 241, 243, 245, 247, 248, 250–253, 258, 264, 269, 272, 273, 275, 279–281, 283–285, 288, 289, 291–296, 298, 302–309, 311, 312, 314–318, 321, 324, 326, 328, 333

Mysidacean, 21, 194, 195, 198, 199, 313, 314, 319, 323, 325
Mysidaceans, 24, 195, 196, 223, 277, 295, 302, 312, 314, 316, 317, 321, 323, 324
Mysids, 55, 56, 193, 195, 196, 201, 207, 213, 240, 243, 245, 250, 251, 253, 260, 264, 267, 273, 275, 279, 280, 283, 284, 286, 293, 294, 296–298, 301–305, 307–312, 314–316, 321–327, 336, 405, 406
Mystacocarid, 17
Mystacocarids, 4
Mythology, 127, 172, 181
Namibia, 32
Naples, 20, 127, 142
Natatory exopods, 225
Native species, 124, 325
Natural populations, 273, 280
Naturalist, 166
Nature of the food, 307
Naupliar appendages, 284, 286
Naupliar eye, 240
Nauplioid abdomen, 287, 288
Nauplioid larvae, 199, 272, 284, 288
Nauplioid stage, 286, 288, 290
Nauplioids, 196, 288, 331
Nebaliacean, 211
Negative impacts, 314
Negative phototaxis, 299
Nekton, 25
Nematocysts, 311
Nematode, 126, 312
Nematodes, 126
Nemertean, 48
Neoteny, 20
Neotropical, 312
Nephrostome, 267
Neritic, 7, 8, 18, 251, 296
Neritic biotopes, 23
Neritic zone, 17
Nerve fibre, 242
Nerve tract, 238
Nervous chain, 261
Nervous elements, 260
Nervous system, 213, 237, 239, 240, 260, 268, 269
Netherlands, 151, 163, 173–176, 281, 315
Neural chain, 238–240
Neurohaemal functions, 241, 269
Neurohaemal organs, 268
Neurohormones, 268
Neuropils, 239–241
Neurosecretory [organ], 239, 241, 268, 269
Neurosecretory cells, 239, 269
Neurotoxins, 125
Neurotransmitter, 241
New Britain, 29

New Caledonia, 34
New Guinea, 12
New Ireland, 31
New Jersey, 302
New World, 161
New Zealand, 30, 31, 61, 124
Niche, 34, 307
Niches, 7, 9, 64
Nicothoid, 311, 312
Nitrogen, 306, 308
Nocturnal dispersal, 298, 304
Nocturnal migrations, 301
Nocturnal vertical migrations, 302
Nomenclatural code, 193
Non-crystalline [statolith], 250, 321
Non-indigenous species, 125, 315, 316
Non-migrating [species], 241
Non-rocky substrates, 10
Non-sensory cuticular structures, 211
Nonylphenol, 272
North America, 30, 31, 119, 164, 316, 317, 323, 325
North Atlantic, 18, 41, 118, 201, 296, 324
North Sea, 27, 168, 309
Northern Hemisphere, 7, 8, 173, 313, 323
Norway, 50, 54, 214, 281
Norway lobster, 18
Norwegian fjords, 296
Nosy-Bé, 36, 52
Notal realms, 324
Notantarctic realm, 324, 325
Nuclear DNA sequences, 320
Nucleic divisions, 284
Nucleus, 248, 264, 266
Number of eggs per brood, 277
Number of ommatidia, 245, 299, 330
Numbers of ommatidia, 243
Numbers of oostegites, 229
Numbers of species, 4, 10
Nurseries, 22
Nursery, 15
Nutrient-rich, 282
Nutrition, 316
Nutritional value, 119
Oceanic ridges, 23
Oceanic waters, 22, 295
Oceanographic expeditions, 194
Octopus, 156, 157, 167, 398
Ocular somite, 197, 294
Ocypodids, 34
Oesophagus, 255, 257
Offspring, 276
Ohio, 60
Oil globules, 213
Oil lamp, 142, 156, 158, 167, 399
Olfactory lobes, 240
Olfactory stimuli, 303
Oligocene, 317, 321, 324

Oligohaline, 299
Oligotrophic, 275, 282
Ommatidia, 241–243, 245, 247, 284, 322, 323
Ommatidium, 242, 345
Omnivorous, 217, 307–309
Onchocerciasis, 126
Ontogenetic migrations, 302
Oocyte, 264
Oocytes, 264, 284, 291
Oomycete, 124
Oostegites, 35, 191, 194, 209, 213, 229–231, 269, 272, 275, 279, 282, 321, 329–346
Ophiurans, 58
Ophiuroids, 56, 309
Ophthalmic artery, 261
Opisthobranchs, 326
Optic neuropils, 240
Optimal salinity, 302
Organ of Bellonci, 239, 241, 244, 247, 269
Organic, 10, 15, 122
Organic [statolith], 248–250, 306, 308, 311, 321, 332
Organic statoliths, 250, 321
Orientation, 250, 300
Orthogenetic, 20
Osmoregulation, 297
Osmoregulatory function, 204
Ostia, 260, 263, 318
Ostracode, 15, 51
Ostracodes, 4, 24, 30, 33–35, 52, 53, 159, 165
Ostrich shrimps, 17
Otocyst, 250
Otter, 176
Ovarian bridge, 265
Ovarian cycle, 290, 291
Ovarian tubes, 238, 264, 265, 273
Ovary, 264, 265, 291
Over-exploitation, 314
Overlapping generations, 294
Overwintering generation, 201, 294, 302
Overwintering males, 273
Oviduct, 238, 264, 265, 275
Oviducts, 264, 265, 275
Oviposition, 264, 275, 277
Oxydation, 24
Oxygen, 21
Oxygen [uptake], 229, 263, 295, 298, 304, 315, 323
Oyster, 150
Oysters, 11, 12, 123, 175, 176
Pacific, 5, 8, 12, 25, 37, 50, 58, 118, 121, 125, 277, 309, 322, 324
Pagurid, 10, 14, 43
Pagurids, 11, 14, 48–50, 55, 309
Painting, 128, 139–142, 144, 145, 156, 159, 161–164, 167–173, 175–181

Pair of adults, 309
Palaearctic, 312
Palaeolithic times, 140
Palaeontological data, 319
Palaeozoic fossils, 317
Paleoclimatological conditions, 26
Palp, 35, 217, 330, 331, 343
Panama, 43
Panchronic forms, 25
Panoceanic, 313, 323
Panoceanic distribution, 313, 324
Papilla, 223, 247
Paracentral lobe, 240
Paradactylary setae, 197, 227, 229
Paragnaths, 214, 217, 219–222, 306
Paragonimiasis, 126
Paraphyletic taxa, 320
Paraphyly, 318
Parasite, 27, 52, 56, 63, 124, 126
Parasites, 7, 8, 11, 25, 27, 36, 38, 45, 49–53, 56–59, 61, 64, 123–125, 272, 280, 311
Parasitism, 7, 36, 44, 57–60, 64, 272, 316
Paratethyan fossils, 248
Paratethyan origin, 250, 322
Paratethyan sediments, 321
Paratethys, 248, 249, 317, 322
Parental body size, 276
Parental size, 276–278
Parental weight, 277, 283
Parolfactory lobes, 240
Pars centralis, 219, 221
Pars incisiva, 217, 221
Pars molaris, 219, 221
Parsimonious analysis, 318
Parsimonious tree, 319, 321
Parsimony analysis, 318
Partes molares, 219
Parthenogenesis, 7
Pathogenic bacteria, 125
PCB, 315
Pea crabs, 124
Peacock, 176, 177
Pediform thoracic endopod, 225, 227, 293, 330, 331, 340–343, 346
Peduncle, 214, 253, 329
Pelagic, 7, 8, 10, 17, 24–26, 61, 62, 117, 118, 199, 211, 295, 301–304, 307, 308, 312, 314, 321, 323–325
Pelagic fauna, 25
Penaeid, 120, 121, 125, 142, 154, 165, 167, 194
Penes, 191, 231, 269, 272, 273, 275, 329–332, 340–343, 346
Penetration, 20, 27, 34, 284
Penguins, 118
Penis, 229, 231, 232, 265, 266, 274
Peracarids, 4, 275
Perception of vibrations, 251

Pereiopods, 156, 167–169, 173, 179, 181, 191, 196, 341
Peri-oesophageal connectives, 238
Peri-oral structures, 217
Pericardium, 261, 263, 264
Peristaltic movements, 255
Peritrophic membrane, 258, 307
Permanent burrow, 18
Permian, 288, 317, 342
Persian, 173, 179
Persian Gulf, 273
Peru, 119, 313
Pests, 123
Petrels, 118
Petricolous [species], 11, 12
pH, 25, 298, 301
Pharmacopeia, 122
Philippines, 16, 22, 125
Phoresia, 36
Photic brightness, 296
Photic zone, 308, 322
Photoperiod, 300
Photophobia, 299
Photoreception signal, 241
Photoreceptor cells, 242
Photoreceptor damage, 245
Photosensitive function, 241
Phototropism, 301
Phreaticolous, 299, 306
Phyletic lineages, 245
Phyllosomas, 26, 40
Phylogenetic analyses, 328
Phylogenetic reconstruction, 195, 319, 328
Phylogenetic reconstructions, 195, 318
Phylogenetic trees, 327
Phylogeny, 30, 190, 319, 320, 327, 328
Phylogeny of the static organ, 321
Physicochemical parameters, 23
Physiological factors, 27
Phytodetritus, 308
Phytophagous, 217
Phytoplankton, 117, 118, 308, 314
Phytothelmes, 28, 32, 33, 62
Pigment cells, 213, 243
Pigment migration, 243
Pigmentation, 7, 213, 270
Pigs, 316
Pinnotherids, 57
Plaice, 176
Plane of polarization, 300
Plankton, 22, 25, 280, 282, 296, 311
Planktonic, 118, 123, 126, 295
Planktonic larvae, 307
Plant, 13, 23, 32, 211, 308
Plaster castings, 19

Plasticity of food selection, 308
Plate tectonics, 26, 28, 29
Pleistocene, 323
Pleomere, 169, 200, 201, 210, 229, 231, 240, 260, 263, 270, 286, 329, 330, 332–334, 342, 344
Pleon, 167, 169, 179, 197, 199, 209–211, 226, 227, 233, 238, 240, 259, 261, 286, 288, 292, 300, 312, 342
Pleonites, 209, 210, 261
Pleopods, 34, 35, 191, 197, 209, 224, 226, 231, 233, 235, 262, 263, 269, 272, 274, 285, 286, 300, 321, 329–335, 337–344, 346
Plesiomorphic, 190, 318, 321
Plesiomorphic characteristics, 321
Plesiomorphic characters, 288, 318
Plesiomorphic features, 321
Pleura, 209, 333, 342, 344
Pleural plate, 200, 231
Pleural plates, 209, 210, 229, 270, 329–334, 337, 338, 340–342, 344
Pleurites, 190, 201, 203, 219
Plymouth, 50, 281, 301
Podo-pericardial sinuses, 261, 263
Podobranchiae, 229
Podocytes, 267
Poikilotherms, 283
Polar bodies, 284
Polar body, 284
Polar regions, 35, 313
Polarization sensitivity, 245
Polarized light, 300
Polluted, 272
Pollution, 314
Polychaetes, 13, 14, 50, 51
Polyclad turbellarian, 47
Polynesia, 34
Polynoid, 51
Polyphenol, 307
Polyphyletic assemblage, 327
Polyphyly of the Mysidacea, 318
Polysaccharide, 122
Polyspermy, 284
Polyspinal appendices, 288
Polyunsaturated fatty acids, 316
Pomacentrid, 311
Pompei, 142, 156, 157, 181, 184, 398
Pons protocerebralis, 240
Ponto-Caspian, 27, 124, 203, 282, 299, 313, 315, 322, 324–327, 406
Pontomediterranean, 270
Pontoniid, 10
Population densities, 283
Populations, 17, 18, 24, 119, 123, 124, 243, 272, 276, 281, 282, 292, 294, 295, 297, 298, 301, 308, 309, 313, 316, 323, 326
Porcellanid, 51
Pores, 195–197, 211, 212, 217, 248, 249, 255, 317
Porpoise, 176

Portraits, 161, 162, 164, 175
Portugal, 119, 326
Portuguese, 163
Post-cerebral masses, 238
Post-oesophageal commissure, 238
Postage stamps, 129, 141
Postecdysis, 291
Posterior intestine, 255
Postmoult period, 291
Postnauplioid larvae, 245, 265, 276, 279, 284, 289
Postnauplioid stage, 209, 279, 288, 290
Postnauplioids, 196, 247, 284, 288, 292, 331
Potamids, 34
Potamophilous, 300
Praecoxa, 214, 221, 223, 229, 267
Praeischium, 224, 229
Prague, 30
Prawn culture, 322
Prawns, 22, 27, 63, 120
Pre-antennary somite, 286, 294
Pre-epipods, 229
Precipitation of a mineral, 322
Precopula, 273
Predation, 59, 123, 296, 297, 308, 309
Predation pressure, 309
Predator-prey relationship, 309
Predators, 34, 289, 303, 304, 311, 314
Predatory impact, 307
Predatory mode of life, 242
Preference experiments, 299
Prehensile structure, 225
Prehistoric times, 167
Premature larvae, 279, 289
Prematurely released larvae, 280
Premoult period, 291, 292
Prey, 118, 242, 251, 304, 305, 307, 308, 314
Primary production, 118, 307
Priority rules, 193
Processus incisivus, 217, 221
Processus incisivus accessorius, 219, 221
Processus masculinus, 270
Processus molaris, 217, 219, 221
Proctodaeum, 255, 286
Production of digestive enzymes, 260
Proecdysis, 291, 292
Propagative activity, 278, 281
Propodal formula, 197
Propodus, 54, 197, 223–225, 227, 229, 332, 334–336, 340, 341, 344–346
Protein, 119, 120
Protists, 311
Protocerebrum, 237, 240
Psammophilic, 297
Pseudo-tracheae, 34
Pseudobranchial lobes, 197, 233, 263, 272, 333, 343

Pseudocontinuous breeding, 282
Pseudoexopodite, 221
Pseudorostrum, 204
Pteridine, 243
Pyloric chamber, 254, 255, 257, 258
Pyloric filter, 258
Pyloric filters, 257
Pyloric region, 258
Pyloric ridge, 255, 257
Pyloro-intestinal valve, 257
Quaternary calcarenite, 306
Quaternary calcarenites, 295
Réunion, 33, 185
Radiation, 20
Rain forests, 34
Random dispersal, 28
Ras Muhamad, Sinai, 21
Rasping, 306
rDNA, 195, 324
Realism, 129, 162, 168
Recent, 7, 18, 34, 118
Recent [statolith], 195, 248, 249, 273, 317–319, 321, 325, 328–332, 336, 342, 343
Receptacula seminis, 264
Receptaculum seminis, 275
Reception, 165, 193, 251
Receptive females, 253, 275
Receptors, 253, 279
Recognition of conspecifics, 253
Rectum, 255, 259
Red Sea, 127, 242, 295, 323, 326
Red worm disease, 124
Red worms, 124
Reduced carapace, 194, 199, 207
Reduced eyes, 31, 241, 251, 299
Reduction of [the] oostegites, 231, 282, 291
Reductions, 7, 124
Reflux of food, 255
Refracting superposition type, 242
Regeneration, 292, 293
Regional warming, 314
Regression, 29, 278, 292, 314
Relationships, 37, 167, 195, 319
Release of broods, 289
Release of first brood, 294
Release of [the] young, 231, 278, 282, 283, 289, 291, 302
Relict distribution, 29
Relict forms, 26
Religious, 139, 158–161, 166, 168–171, 174
Religious meaning, 166
Remaining in fresh water, 28
Remedies, 121, 122
Renaissance, 128, 139, 140, 161, 166, 168, 170, 191, 195
Rennell Island, 26
Replacement [name], 194, 293, 333, 334, 344, 349

SUBJECT INDEX 457

Representation of crustaceans in art, 128
Reproduction, 7, 34, 268, 278, 282, 297, 299, 302
Reproductive behaviour, 275
Reproductive biology, 283
Reproductive males, 253
Reproductive parameters, 264, 283
Reproductive period, 231, 264, 282, 291
Reproductive potential, 276, 304
Reptiles, 127, 140, 162
Resorptive cells, 259
Respiration, 34, 262, 263
Respiratory carapace, 207, 262–264, 330
Respiratory current, 190, 263, 279, 300
Respiratory tissue, 190, 204, 207, 263, 331, 342
Resting eggs, 7, 32
Resting stage, 264, 278, 283
Resurrection, 145, 171
Retina, 242
Retinoids, 243
Retinol, 243
Retinula [cell], 242
Retinula cells, 242
Retroverted palp, 214, 221
Reversal of precedence, 193
Reversal of sex, 264
Rhabdome, 242, 243
Rhabdomeres, 242
Rhabdomes, 242, 243
Rhine River, 273, 307, 315
Rhopography, 161, 163
Ribosomal [loci], 319
Rice crops, 123
Rice fields, 123
River bank erosion, 123
River blindness, 126
River Trent, 273
Rivers, 26, 28, 119, 301, 326
Rocks exposed to waves, 10
Roman, 7, 8, 20, 127, 139, 156–158, 161, 162, 165, 167, 168, 170, 172, 175, 181, 183, 194
Romanesque period, 168
Romania, 27, 301
Roscoff, 38, 281
Rostrum, 17, 198, 203, 206, 214, 330
Rudimentary flagellum, 214
Rudimentary gill, 229
Rupture line, 290, 291
Sabellid, 310, 311, 405
Sabellids, 309
Sacculinid, 3
Sacculus, 219, 267, 268
Saharian wadies, 32
Salinities, 6, 281, 299
Salinity, 6, 21, 22, 27, 28, 32, 196, 281, 294, 297–299, 302, 306, 311, 315

Salinity classification, 196
Salinity crisis, 250
Salmon, 176
Salmonid industry, 123
Salps, 8
Salt marshes, 22
Sampling locality, 299
Sand tube, 15
Sandy beaches, 17, 18, 34, 273, 294
Saprophagous, 217
Sarcoplasmic calcium-binding protein, 120
Saxitoxin, 125
Scallops, 123, 124
Scandinavia, 243
Scaphognathite, 53
Scaphopods, 14
Scenario, 20, 29
Schools, 180, 303, 304
Sciaphilic habitats, 11
Scientific drawing, 169
Scissor-like structure, 225
Scleractinian, 10
Scleractinians, 44, 45
Sclerites, 291
Scorpion, 169
Scotland, 38
Screening pigment cells, 242
Scutella paracaudalia, 210
Scutellum paracaudale, 197, 237
Sea, 3, 9, 21, 22, 26–29, 34, 36, 45, 57, 119, 124, 129, 152, 156, 163, 169–172, 176, 243, 250, 278, 290, 291, 294, 297, 301, 303, 310, 313, 314, 322, 324, 325, 405
Sea anemone, 36, 43, 44, 311
Sea anemones, 43, 309, 311
Sea cucumbers, 57
Sea fan, 42
Sea food, 120, 152, 164, 168, 175, 176
Sea grass, 16, 20, 298
Sea grasses, 9, 20, 297
Sea level, 5, 29
Sea lice, 123
Sea urchins, 57, 58, 310, 311
Seal, 176
Seals, 118, 316
Seamounts, 23
Search for food, 253
Season, 176, 201, 276, 277, 280, 301, 304, 308
Seasonal cycle, 278
Seasonal shifts, 301
Seasonality, 294
Seasons, 143, 273, 277, 278, 281, 294
Seaweeds, 11, 16, 170
Second larval ecdysis, 285, 288
Secondary producers, 117
Secondary production, 315
Secretory activity, 267
Secretory canal, 267

458 SUBJECT INDEX

Secretory glands, 15
Sediment, 9, 17, 18, 24, 164, 173, 201, 297, 300, 304, 308, 315
Sedimentary determinant, 9
Sedimentation, 22
Segmental border, 209, 224
Segmental glands, 267
Segmentation, 158, 284–286
Selection of food sources, 307
Selective pressures, 19
Seminal vesicle, 231, 232, 238, 265, 266
Seminal vesicles, 265
Semiterrestrial crustaceans, 34
Sense organs, 211, 240, 255, 275
Sensilla, 251, 253, 291
Sensillum, 251, 253
Sensitivity to vibrations, 250
Sensory cushion, 247, 248
Sensory fossette, 255, 336, 345
Sensory function, 241, 291
Sensory papilla, 247
Sensory papilla X-organ, 241
Sensory setae, 214, 246–252, 270
Separate swarms, 303
Separation of fluids and fine particles, 259
Sequence of broods, 283
Serotonin, 241
Serpulids, 10
'Served tables', 177
Sessile, 4, 10, 124, 305, 310, 405
Sessile animals, 124
Sessile fauna, 23
Sex determination, 272
Sex organs, 269
Sex ratio, 273
Sex ratios, 273
Sex verification, 272
Sexes, 191, 197, 209, 210, 233, 253, 269, 270, 272, 273, 300, 329, 330, 332–335, 337, 339–342, 344, 346
Sexual activity, 300
Sexual differences, 270
Sexual dimorphism, 214, 269, 271, 332
Sexual dimorphisms, 269
Sexual maturity, 22, 269, 270, 282, 294, 304
Shelf, 313
Shellfish, 123, 177
Shells, 9, 10, 13, 14, 28, 45, 48–50, 52, 53, 123, 127, 142, 147, 153, 154, 172, 309
Shelter, 9–11, 13, 14, 17, 20, 311, 322
Shield, 9, 203
Shipworms, 9
Shrimp, 15, 16, 20–22, 25, 27, 36, 39, 42, 44, 45, 50, 52, 55, 56, 58, 60–63, 118–123, 125, 139, 142, 143, 146–149, 153, 154, 156, 157, 162, 164, 165, 167, 183 190, 343, 398

Shrimps, 5, 10, 17, 19, 21, 24, 25, 27, 28, 31, 40, 44, 45, 58, 118–120, 124, 142, 143, 147–152, 154–156, 163–166, 181, 183
Sinus gland, 239, 240, 268
Sipunculidan, 47
Sirenian, 60
Sister group, 318, 320
Sister taxa, 318
Size, 4, 7, 12, 24, 118, 197, 201, 209, 210, 214, 219, 225, 231, 233, 240, 247, 248, 257, 259, 271, 273, 277, 278, 288, 289, 294, 296, 303, 333, 340, 341, 344–346
Size at attainment of maturity, 277, 292
Size at release of young, 277
Size increments, 292
Size variations, 201, 296
Snake, 145, 151, 162
Snapping shrimp, 12, 19
Social behaviour, 297
Social interaction, 289, 302, 303
Soft-bottom pools, 296
Solitary, 304
Solomon Islands, 26
South Africa, 7, 40, 50, 298, 302
South America, 27, 34, 119, 300
South-East Asia, 32
South-East Asia (Indonesia), 32
Southern Hemisphere, 8, 35
Southern Ocean, 118, 313, 324
Spain, 19, 119, 163, 174, 181, 309, 326
Spanish, 162, 174, 175, 181
Spatial arrangement, 303
Specializations, 7
Species introductions, 314, 316, 325
Species of brackish lagoons, 28
Species-specific differences, 277, 289
Spectral sensitivity, 243
Sperm competition, 273
Sperm transfer, 233, 274, 275
Spermatogenesis, 265
Spermatogonia, 265
Spermatophore, 265, 274
Spermatozoa, 265, 267, 272, 274, 284
Sphaeromatidae, 9, 12, 13, 17, 27, 31, 38, 50
Sphere, 9
Spider crab, 5
Spider-like morphological types, 23
Spiders, 291
Spindle, 9
Spine row, 219, 221
Spinose, 24, 55, 191, 217, 219, 235, 330, 342, 343
Spiny lobster, 11, 142, 143, 145, 146, 152–154, 156, 157, 169, 170, 398
Spiny lobsters, 53, 55, 127, 128, 141–143, 152, 156, 165–167
Sponges, 8, 9, 13, 15, 16, 36, 38, 39, 56, 309
Spread along waterways, 327, 406

SUBJECT INDEX 459

SPX-organ, 241, 269
Squids, 118
St.-Helena, 26
Stabilization, 250
Stalked eyes, 197, 318, 329
Starvation, 280, 294
Static bodies, 197, 250, 253, 321, 328
Statocyst, 237, 246–248, 250, 252, 253, 321, 329, 331–335, 337–343, 345
Statocyst slit, 247, 248
Statocysts, 191, 194, 247, 250, 251, 289, 300, 321, 330, 331, 342
Statolith, 237, 246–250, 332–334, 337, 338, 340, 341
Statolith formation, 250, 289
Statolith formula, 197, 248, 317
Steller's sea cow, 60
Stem cells, 259
Steno-endemic, 323
Stenobathic, 24
Stenoecious, 323
Stenoecy, 7
Stenoendemic, 313
Stenohaline, 297
Stenohalobious, 297
Stenopodidean, 16
Stenothermy, 7
Sternal artery, 261
Sternal lamellae, 209, 227, 237
Sternal projections, 209, 269, 270
Sternites, 209, 223, 229, 231, 232, 259, 271, 279
Stickleback, 118
'Still life', 156
Still life, 161, 163, 164, 168, 175, 177, 178, 180
Still life paintings, 166, 168, 175
Stimuli [external], 288, 303
Stimulus, 251
Stomach, 122, 194, 213, 238, 255, 257–262, 264–266, 268, 285, 288, 305
Stomach contents, 168, 307, 308
Stomachs, 242, 308, 315, 316
Stomatopod, 124, 125, 162
Stomatopods, 16, 52, 119, 165, 172, 194, 207
Stomodaeum, 255, 286
Storage cells, 267
Storage materials, 282
Strait of Gibraltar, 314
Stratified distribution, 295
Stratigraphic signals, 319
Sturgeon, 176
Sturgeons, 312
Stygiomysid, 245, 259, 292
Stygiomysids, 195
Stygophilic, 199
Sub-oesophageal ganglion, 269

Sub-swarms, 303
Subadult males, 270
Subalpine lakes, 294
Subantarctic, 294, 313, 323–325
Subarctic, 275, 294, 299, 323
Subbasal suture, 235
Subchela, 225, 227
Subcuticular fat bodies, 306
Subdivided cornea, 242
Submarine caves, 11, 251, 299, 301, 324
Subrostral processes, 203
Substrate, 10, 18, 19, 23, 24, 196, 213, 300, 303, 304, 306, 308, 311
Substrate relations, 297, 303, 304
Substrates, 16, 18, 25, 28, 297, 310
Substratum, 7, 8, 10, 13, 15, 45, 46, 296
Subterranean, 8, 26, 29, 30, 190, 198, 199, 245, 251, 252, 259, 281, 284, 291, 295, 297, 313, 322
Subterranean environments, 278, 308
Subterranean lake, 21
Subterranean mode of life, 264, 278, 313
Subterranean species, 28
Subterranean waters, 7, 194, 280, 322
Subtropical, 276, 297, 313
Subtropical Convergence, 313
Subungulary setae, 197, 227
Success of marsupial development, 281
Suez Canal, 314, 323, 326
Sulfo-oxydant bacteria, 24
Sulphated mucopolysaccharides, 250
Summer eggs, 276
Summer generations, 294
Superficial layers, 295
Superficial strata, 302
Suppress moulting, 312
Suprabenthic, 296
Supralittoral pools, 21
Supralittoral zone, 17
Survival, 277, 297, 301, 312, 314
Survival rate of larvae, 280
Suspended matter, 295
Sustainable management, 119
Swan, 176, 177, 247
Swarm, 289, 298, 303, 304, 310, 405
Swarming, 289, 302–304
Swarming species, 274, 303, 305
Swarms, 36, 118, 289, 298, 299, 303–305, 309–311
Swedish lakes, 316
Swimming, 8, 10, 36, 44, 170, 190, 237, 262, 275, 279, 288, 291, 300, 303, 304
Swimming activity, 237, 300
Swimming crabs, 10, 22
Swiss, 171
Symbiosis, 13, 36, 60, 64, 311
Symbolic sense, 166

Symbolism, 139–141, 156, 158, 161, 163, 167, 168, 170, 172, 177, 181
Symphyses, 203
Symphysis, 203, 204
Sympod, 214, 221, 223, 225, 227, 233, 235, 237, 247, 252, 330, 331, 339, 342, 343, 345
Synchronization of development, 289
Synchronous development, 289
Synchronous release, 289
Systematics, 139, 141, 165, 166, 190, 195, 196
Tactile [abilities], 245, 251, 289, 303
Tagmata, 195, 197
Tail fan, 191, 197, 235, 246, 321, 329–331, 342
Tail flipping, 197, 209, 300, 342
Talitrids, 18, 34, 35
Tanaidacean, 13, 17, 21
Tanaidaceans, 15, 24, 52, 60
Tanaids, 9
Tantulocarid, 52
Tapestries, 160, 163, 169
Tapeworms, 126, 316
Tardigrade, 63
Tasmania, 211, 280, 289, 292, 304, 307
Taxonomic, 28, 121, 196, 329
Taxonomic resolution, 328
Taxonomic value, 211, 213, 248
Teloblasts, 286
Telson, 179, 182, 191, 196, 197, 210, 211, 213, 235–237, 247, 259, 261, 270, 285, 288, 293, 305, 330–346
Telson cleft, 210
Temperate, 8, 20, 32, 35
Temperate [latitudes], 201, 276, 278, 281, 282, 294, 296, 322–324
Temperate climates, 275, 276, 294
Temperature, 24, 25, 201, 276, 277, 280, 281, 283, 290–292, 294–297, 299–302, 306, 308, 312, 323
Temperature acceleration of marsupial development, 283
Temperature regime, 283
Temporarily drying ponds, 32
Temporary hole, 18
Temporary waters, 32
Tendons, 237
Tergite, 199, 203, 207, 209, 229
Terminal claw, 197, 225
Terminology, 195, 252, 283, 284, 325, 327
Terrestrial organisms, 22
Terrestrial species, 28, 35, 140
Tertiary, 29, 317
Tertiary macrofossils, 317
Testes, 264, 265
Testicular tissue, 272
Testis, 264, 266

Tethyan distribution, 30
Tethyan faunistic elements, 299
Tethys Sea, 322, 324
Tetrodotoxins, 125
Texas, 29
Thermal chimneys, 5
Thigmotaxis, 20
Thoracic appendages, 21, 190, 191, 204, 205, 223, 229, 261, 263, 264, 267, 269, 305
Thoracic endopods, 190, 196–198, 203, 227, 230, 251, 252, 263, 270, 271, 279, 300, 305, 306, 323, 329–332, 334–338, 341–346
Thoracic exopods, 190, 263, 268, 269, 287, 300, 306, 331
Thoracic ganglia, 238, 240
Thoracic lacuna, 261
Thoracic somites, 190, 197, 203, 204, 207, 209, 261
Thoracomere, 199, 209, 229, 240, 260, 267, 342
Thoracopods, 190, 191, 193, 194, 196, 197, 223–229, 231, 232, 238, 262, 263, 267, 271, 275, 279, 285, 286, 306, 329–332, 336, 340, 342, 343
Thorax, 197, 204, 209, 228, 238, 240, 259–261, 263, 264, 286, 290, 300
Threat to freshwater biodiversity, 325
Tidal amplitude, 291
Tidal migrations, 302
Tide, 9, 10, 17, 18, 297
Time of day, 296, 308
Time to maturation, 294
Tokelau, 35
Tokelau Archipelago, 22
Tolerance, 243, 315
Tonga, 25
Top-down control, 309
Toxic food items, 125
Toxic substance, 250
Toxicants, 315
Traditional hypothesis, 190
Transcontinental dispersion, 326
Transduction, 251
Transitions, 27, 28
Transmission, 126, 251
Transoceanic [distribution], 296, 313
Transoceanic transfers, 326
Transport of fluids, 259
Transversal folds, 210
Transversal groove, 209
Transverse furrow, 204
Transverse suture, 235, 237, 332
Tree trunks, in Europe, 33
Trees, 23, 32, 35, 128, 319, 328
Trematode, 126, 312
Trematodiasis, 126
Trends, 7, 163, 165, 314
Triassic, 317, 321, 330, 342
Tributaries, 26

SUBJECT INDEX

Tritocerebrum, 237
Troglobitic [forms], 7, 29, 31
Trompe l'oeil, 168
Trophic conditions, 282, 294
Trophic interactions, 309
Trophic levels, 117, 118
Trophic linkages, 314
Trophic networks, 117
Trophically suitable breeders, 282
Tropical, 6, 8, 10, 12, 20, 22, 32, 34, 43, 50, 201, 207, 251, 252, 276, 277, 288, 292, 297, 298, 302, 304, 305, 307, 308, 310, 313, 323, 324, 405
Tropical climates, 281
Tropical regions, 8
Tropics, 18, 278
Tropomyosin, 120
Troughs, 23
Truncated prism, 9
Trunk, 33, 199, 210, 212, 214, 217, 219, 221, 261, 263, 286
Tubes, 8, 10, 13–16, 50, 51, 264, 265, 286
Tubular heart, 260
Tuléar, 8, 17, 20, 50
Tunicates, 8, 13, 58
Turbid layer, 295
Turbulences, 251
Turkey, 127, 177
Turtles, 60, 176
Type genus, 193, 329–341
Tyrrhenian, 290
Tyrrhenian Sea, 296
U.S., 326
U.S.A., 8, 31, 60, 119
Uniform environment, 26
Uniramous [appendage], 197, 233, 235, 331–336, 338–341, 343–346
Unity of the Mysidacea, 318
Unity or disunity of the Mysidacea, 189, 195
Upper Cretaceous, 29
Upper Jurassic, 317
Uptake of ions, 250
Urchins, 57, 298
Uropodal sympods, 194
Uropods, 167, 179, 191, 209, 235–237, 247, 252, 261, 270, 285, 286, 305, 321, 329–343, 345
Urosome, 35
Vagile, 10, 49, 125
Valve, 232, 254, 257, 259, 260
Valves, 231, 257, 258, 260, 261
Valvula dorsalis oesophagi, 255
Vancouver Island, 277, 281
Vanuatu, 6, 125
Variable salinities, 22
Vas deferens, 231, 266

Vascular system, 260
Vaterite, 248–250, 317, 318, 322, 332, 338, 339
Vaterite precipitation, 250
Vaterite statoliths, 248, 250, 322
Vegetable matter, 308
Vegetable substrata, 297
Vegetal debris, 19, 23, 307
Vegetation, 22, 33, 297, 304
Venice System, 196
Ventilation, 229, 231
Ventilation lobes, 231, 279
Ventilation movements, 263
Ventilation plates, 231
Vermiform body, 198, 207
Vertebrates, 127, 140, 141, 156, 164, 183, 272
Vertical distribution, 296, 302
Vertical migrations, 301
Vesicle, 241, 247, 267, 284
Vestimentiferans, 25
Vibrations, 251, 306
Vicariance, 324
Vicariance or fragmentation, 29
Vienna, 146, 150, 244, 247, 249
Vienna [basin], 317
Viscera, 125, 261
Visual abilities, 245
Visual field, 242, 250
Visual pigment, 243, 284, 336
Visual pigments, 243
Visual sensibility, 243
Vitellogenesis, 269
Vitellus, 286
Vorticellid, 292, 311
Vulnerability, 304
Warm [water], 5, 22, 27
Warm-season breeding, 282
Wastewater stabilization ponds, 123
Wastewater treatment, 123
Water flea, 3
Water movement, 298, 301, 303
Water treatment agent, 123
Wave breaking zone, 17
Wave lengths, 243
Ways of life, 3
West Asian, 8
Westphalia, 177
Whales, 118, 309, 316
Whitefish, 316
Winter eggs, 276
Wood boring, 11
Woodlice, 6, 22, 34, 35, 122, 164, 171
World-wide distribution, 296
Xenia, 143, 156, 157, 162, 175
Xeno-oestrogens, 272
Xenobiotics, 272
Y-organ, 269
Yolk, 213, 214, 264, 283, 288, 291
Yolk invested per egg per day, 283

Yolk mass, 287, 288
Young, 11, 38, 63, 118, 123, 152, 171, 172, 177, 181, 191, 231, 265, 269, 276, 278–283, 289, 291, 292, 305, 308, 322
Young per female incubatory day, 283
Yugoslavia, 30
Zaïre, 29
Zanzibar, 30, 194
Zoantharians, 46, 47
Zodiac, 143, 144, 172, 173, 181, 183
Zodiacal constellation, 127
Zodiacal sign, 159, 161, 172, 173
Zones of faunistic change, 296
Zoophagy, 308
Zooplankton, 307, 308, 315
Zooplankton clearance rates, 308

ERRATA TOZ-C VOL. 4A

To our regret, a number of data concerning the addresses of one of our editors as well as of some of the authors have not been properly represented in the List of Contributors of the previous volume, **4A**, and also in a few other places. Readers are thus requested to correct as follows:

– With regard to the data on DR. M. CHARMANTIER-DAURES, the following changes should be made:
  - on p. 474, in the *List of Contributors* her *former affiliation* should read:
    Equipe Adaptation Ecophysiologique et Ontogenèse
    UMR 5119 Ecosym
    Université Montpellier 2
    Cc 092, Place Eugène Bataillon
    F-34095 MONTPELLIER Cedex 05
    France
    (The additional e-mail address, of the University of Montpellier, should be deleted.)

  - on p. iv, the *Copyright page*, the pertinent phrases should be amended to:
    Mrs. M. Charmantier-Daures (former affiliation: Equipe Adaptation Ecophysiologique et Ontogenèse, UMR 5119 Ecosym, Université Montpellier 2, Cc 092, Place Eugène Bataillon, F-34095 Montpellier Cedex 05, France);

  - on the back cover, under **About the editors**, the phrases characterizing her work should change to:
    M. Charmantier-Daures is a former associate professor at the University of Montpellier and well known for her work on crustacean ecophysiology. She also is an editor of *Crustaceana – International Journal of Crustacean Research*.

– On p. 474, under DR. M.-L. CARIOU, her e-mail address should be mentioned as:
  marie-louise.cariou@legs.cnrs-gif.fr

– On p. 475, under PROF. DR. TH. M. ILIFFE, his e-mail address should be represented as:
  iliffet@tamug.edu

We regret these mistakes and we apologize for any inconvenience caused.

Also on behalf of the Publisher,
THE EDITORS